Life

The Science of Biology

Sixth Edition

Life

Sixth Edition

The Science of Biology

William K. Purves
Emeritus, Harvey Mudd College
Claremont, California

David Sadava
The Claremont Colleges
Claremont, California

Gordon H. Orians
Emeritus, The University of Washington
Seattle, Washington

H. Craig Heller
Stanford University
Stanford, California

 Sinauer Associates, Inc.

 W. H. Freeman and Company

The Cover

Giraffes (*Giraffa camelopardalis*) near Samburu, Kenya.
Photograph © BIOS/Peter Arnold, Inc.

The Opening Page

Soap yucca (*Yucca elata*), White Sands National Monument, New Mexico.
Photograph © David Woodfall/DRK PHOTO.

The Title Page

The endangered Florida panther (*Felis concolor coryi*).
Photograph © Thomas Kitchin/Tom Stack & Associates.

Life: The Science of Biology, Sixth Edition

Copyright © 2001 by Sinauer Associates, Inc. All rights reserved. This
book may not be reproduced in whole or in part without permission.

Address editorial correspondence to:
Sinauer Associates, Inc., 23 Plumtree Road, Sunderland, Massachusetts 01375 U.S.A.
www.sinauer.com

Email: publish@sinauer.com

Address orders to:
VHPS/W. H. Freeman & Co. Order Department, 16365 James Madison Highway,
U.S. Route 15, Gordonsville, VA 22942 U.S.A.
www.whfreeman.com

Examination copy information: 1-800-446-8923
Orders: 1-888-330-8477

Library of Congress Cataloging-in-Publication Data

Life, the science of biology / William K. Purves…[et al.].--6th ed.
      p. cm.
     Includes index.
     ISBN 0-7167-3873-2 (hardcover) – ISBN 0-7167-4348-5 (Volume 1) –
     ISBN 0-7167-4349-3 (Volume 2) – ISBN 0-7167-4350-7 (Volume 3)
     1. Biology I. Purves, William K. (William Kirkwood), 1934–

QH308.2 .L565 2000
570--dc21   00-048235

Printed in U.S.A.

First Printing 2001 Courier Companies Inc.

This book is dedicated to the memory of Angeline Douvas

About the Authors

Gordon Orians   Craig Heller   Bill Purves   David Sadava

William K. Purves is Professor Emeritus of Biology as well as founder and former chair of the Department of Biology at Harvey Mudd College in Claremont, California. He received his Ph.D. from Yale University in 1959 under Arthur Galston. A fellow of the American Association for the Advancement of Science, Professor Purves has served as head of the Life Sciences Group at the University of Connecticut and as chair of the Department of Biological Sciences, University of California, Santa Barbara, where he won the Harold J. Plous Award for teaching excellence. His research interests focused on the chemical and physical regulation of plant growth and flowering. Professor Purves elected early retirement in 1995, after teaching introductory biology for 34 consecutive years, in order to turn his skills to writing and producing multimedia for introductory biology students. That year, he was awarded the Henry T. Mudd Prize as an outstanding member of the Harvey Mudd faculty or administration.

David Sadava is now responsible for *Life*'s chapters on the cell (2–8), in addition to the chapters on genetics and heredity that he assumed in the previous edition. He is the Pritzker Family Foundation Professor of Biology at Claremont McKenna, Pitzer, and Scripps, three of the Claremont Colleges. Professor Sadava received his Ph.D. from the University of California, San Diego in 1972, and has been at Claremont ever since. The author of textbooks on cell biology and on plants, genes, and agriculture, Professor Sadava has done research in many areas of cell biology and biochemistry, ranging from developmental biology, to human diseases, to pharmacology. His current research concerns human lung cancer and its resistance to chemotherapy. Virtually all of the research articles he has published have undergraduates as coauthors. Professor Sadava has taught a variety of courses to both majors and nonmajors, including introductory biology, cell biology, genetics, molecular biology, and biochemistry, and he recently developed a new course on the biology of cancer. For the last 15 years, Professor Sadava has been a visiting professor in the Department of Molecular, Cellular, and Developmental Biology at the University of Colorado, Boulder, and is currently a visiting scientist at the City of Hope Medical Center.

Gordon H. Orians is Professor Emeritus of Zoology at the University of Washington. He received his Ph.D. from the University of California, Berkeley in 1960 under Frank Pitelka. Professor Orians has been elected to the National Academy of Sciences and the American Academy of Arts and Sciences, and is a Foreign Fellow of the Royal Netherlands Academy of Arts and Sciences. He was President of the Organization for Tropical Studies, 1988–1994, and President of the Ecological Society of America, 1995–1996. He is chair of The Board on Environmental Studies and Toxicology of the National Research Council and a member of the board of directors of World Wildlife Fund–US. He is a recipient of the Distinguished Service Award of the American Institute of Biological Sciences. Professor Orians is a leading authority in ecology, conservation biology, and evolution, with research experience in behavioral ecology, plant–herbivore interactions, community structure, the biology of rare species, and environmental policy. He elected early retirement to be able to devote more time to writing and environmental policy activities.

H. Craig Heller is the Lorry Lokey/Business Wire Professor of Biological Sciences and Human Biology at Stanford University. He has served as Director of the popular interdisciplinary undergraduate program in Human Biology and is now Chairman of Biological Sciences. Professor Heller received his Ph.D. from Yale University in 1970 and did postdoctoral work at Scripps Institute of Oceanography on how the brain regulates body temperature of mammals. His current research focuses on the neurobiology of sleep and circadian rhythms. Professor Heller has done research on a great variety of animals ranging from hibernating squirrels to exercising athletes. He teaches courses on animal and human physiology and neurobiology.

Preface

Biologists' understanding of the living world is growing explosively. This isn't the world that the four authors of this book were born into. We never dreamed, as we began our research careers as freshly minted Ph.D.'s, that our science could move so rapidly. Biology has now entered the postgenomic era, allowing biologists and biomedical scientists to tackle once-unapproachable challenges. We are also at the threshold of some experiments that raise ethical concerns so great that we must stand back and participate with others in determining what is right to do and what is not.

The enormous growth and changes in biology create a special challenge for textbook authors. How can a biology textbook provide the basics, keep up with the exciting new discoveries, and not become overwhelming. The increasing bulk of textbooks is of great concern to authors as well as to instructors and their students, who blanch at the prospect of too many pages, too many term papers, and too little sleep. Some reconsideration of what is essential and how that is best presented needs to be made if the proliferation of facts is not to obscure the fundamental principles.

Our major goals were brevity, emphasis on experiments, and better ways to help students learn

In writing the Sixth Edition of *Life*, we committed ourselves to reversing the pattern of ever increasing page lengths in new editions. We wanted a shorter book that brings the subject into sharper focus. We tried to achieve this by judicious reduction of detail, by more concise writing, and by more use of figures as primary teaching sources. It worked! Our efforts were successful. This edition is 200 pages shorter than its predecessor, yet it covers much exciting new material.

While working to tighten and shorten the text, we were also determined to retain and even increase our emphasis on *how* we know things, rather than just *what* we know. To that end, the Sixth Edition inaugurates 72 specially formatted figures that show how experiments, field observations, and comparative methods help biologists formulate and test hypotheses (the figure at right is an example). Another 26 figures highlight some of the many field and laboratory methods created to do this research. These Experiment and Research Methods illustrations are listed on the endpapers at the back of the book.

In the Fifth Edition, we introduced "balloon captions" that guide the reader through the illustrations (rather than having to wade through lengthy captions). This feature was widely applauded and we have worked to refine the balloons' effectiveness. In response to suggestions from users

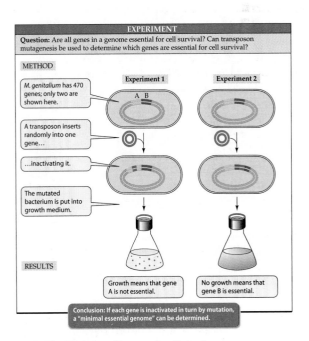

13.22 Using Transposon Mutagenesis to Determine the Minimal Genome
By inactivating genes one by one, scientists can determine which ones are essential for the cell's survival.

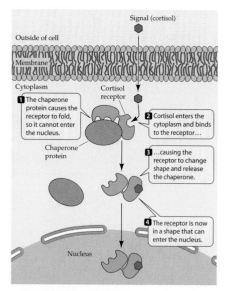

15.9 A Cytoplasmic Receptor
The receptor for cortisol is bound to a chaperone protein. Binding of the signal (which diffuses directly through the membrane) releases the chaperone and allows the receptor protein to enter the cell's nucleus, where it functions as a transcription factor.

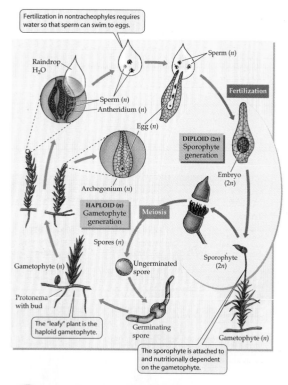

28.3 A Nontracheophyte Life Cycle
The life cycle of nontracheophytes, illustrated here by a moss, is dependent on an external source of liquid water. The visible green structure of nontracheophytes is the gametophyte; in nontracheophyte plants, the "leafy" structures are sporophytes.

of the Fifth Edition, in the Sixth Edition we now number many of the balloons, emphasizing the flow of the figure and making the sequence easier to follow (Figure 15.9 at left is an example).

This edition is accompanied by a comprehensive website, www.thelifewire.com (and an optional CD-ROM that contains the same material) that reinforces the content of every chapter. A key component of the website is a combination of animated tutorials and activities for each chapter, all of which include self-quizzes. Within each book chapter, this 🖐 icon refers students to a tutorial or an activity. An index of the icons begins in the front endpapers of the book. Figure 28.3 (left, below) shows a typical web icon placement.

As part of the ongoing challenge of keeping the writing and illustrations as clear as possible, we frequently employ bulleted lists. We think these lists will help students sort through what is, even after pruning, a daunting amount of material. And we have continued to provide plenty of interim summaries and bridges that link passages of text.

In all the introductory textbooks, the chapters end with summaries. In ours, we have organized the material within the chapter's main headings. In most cases, we tie key concepts to the figure (or figures) that illustrate it. For visual learners, this provides an efficient mode of reviewing the chapter.

From our many decades in the classroom, we know how important it is to motivate students. Each chapter begins with a brief description of some event, phenomenon, or idea that we hope will engage the reader while conveying a sense of the significance and purpose of the chapter's subject.

Evolution Continues to be the Dominant Theme

Evolution continues to be the most important of the themes that link our chapters and provide continuity. As we have written the various editions of the book, however, the emergence of *genomics* as a new paradigm in the late twentieth century has developed, revolutionizing most areas of biology. In this new century, understanding the workings of the genome is of paramount importance in almost any biological discussion.

In this edition, we have moved further toward updating the evolutionary theme to encompass the postgenomic era. Just two examples are the addition of a section on genomic evolution to our coverage of molecular evolution, and a section on "evo/devo" in the chapter on molecular biology of development. In addition, the chapters on the diversity of life reflect the vast changes in our understanding of systematics and phylogenetic relationships thanks to the genomic perspective.

In fact, each chapter of the book has undergone important changes.

The Seven Parts:
Content, Changes, and Themes

In Part One, The Cell, the emphasis in the discussions of biological molecules and thermodynamics has shifted more decisively toward biological aspects and away from pure chemistry. We have made our discussions of enzymes, cell respiration, and photosynthesis less detailed and more focused on the biological applications.

A major addition to Part Two, Information and Heredity, is a new chapter (Chapter 15) on cell signaling and communication, introduced at a place where the students have the necessary grounding in cell biology and molecular genetics. That chapter leads logically into an updated chapter (Chapter 16) on the molecular biology of development, which includes a new section on the intersection of evolutionary and developmental biology—"evo-devo" in the modern jargon. Several chapters incorporate the exciting new work in genomics of prokaryotes, humans, and other eukaryotes.

We have updated all the chapters in Part Three, Evolutionary Processes. In particular, Chapter 24 ("Molecular and Genomic Evolution") reflects the rapid advances in this exciting field. The section on genomic evolution (on pages 446–447) is brand new and includes Figure 24.9 (shown at right).

Part Four, The Evolution of Diversity, now reflects some exciting changes. The chapter on the protists—which can no longer be treated as a single "kingdom"—reflects the continuing uncertainty over the origin and early diversification of eukaryotes. The equally great uncertainty over prokaryote phylogeny, as we deal with the implications of extensive lateral transfer of genes, is evident in the chapter on prokaryote phyla.

We have extended the coverage of the evolution and diversity of plants to two chapters, and that of the animals to three. Recent findings stemming largely from molecular research have led to modifications of the phylogenies of angiosperms and of the animal kingdom. These changes are reflected in the many simplified "trees" that give a broad overview of systematic relationships. Key evolutionary events that separate and unite the different groups are highlighted with red "hot spots" (see Figure 33.1 at right).

We have rearranged Part Five, "The Biology of Flowering Plants," to allow Chapter 39 ("Plant Responses to Environmental Challenges") to serve as a capstone to the whole part, drawing together some of the major threads. We have added sections on hormones and photoreceptors discovered in recent years, and on their signal transduction pathways. The opening chapter (Chapter 34) on "The Flowering Plant Body" has an increased emphasis on meristems.

Part Six, The Biology of Animals, continues to be a broad, comparative treatment of animal physiology with an emphasis on mechanisms of control and regulation. Much new material has been added, including a major revision of Animal

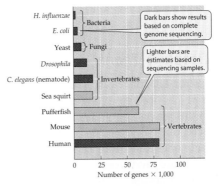

24.9 Complex Organisms Have More Genes than Simpler Organisms
Genome sizes have been measured or estimated in a variety of organisms, ranging from single-celled prokaryotes to vertebrates.

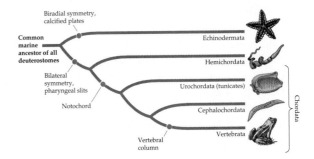

33.1 A Probable Deuterostomate Phylogeny
There are fewer major lineages and many fewer species of deuterostomes than of protostomes.

Development (Chapter 43) to complement and extend the earlier Chapter 16 (Development: Differential Gene Expression). Some other new topics are the role of melatonin in photoperiodism, the role of leptin in the control of food intake, and the discovery in fruit flies of a gene that controls male mating behavior. The extensive coverage of the fast moving field of neurobiology has been substantially updated.

Throughout Part Seven, Ecology and Biogeography, we have added examples of experimental approaches to understanding the dynamics of ecological systems. Some of the examples illustrate the use of experimental and comparative methods. As before, we conclude the book with a chapter on conservation biology (Chapter 58), emphasizing the use of scientific principles to help preserve Earth's vast biological diversity.

There Are Many People to Thank

The reviewing process for *Life*, once a single pass at the stage of draft manuscript, has become an ongoing phenomenon. When the Fifth Edition was still young, we received critiques that influenced our work on this Sixth Edition. The two most penetrating ones came from Zach Gertz, then an undergraduate at Harvard, and Joseph Vanable, a veteran introductory biology professor at Purdue.

Next, still during the Fifth Edition run, 18 instructors recorded their suggestions for improvements in *Life* while teaching from the book. We call these reviews Diary Reviews. The third stage was the Manuscript Reviews. Seventy-three dedicated teachers and researchers read the first-draft chapters and gave us significant and cogent advice. Still another stage has been added to the process and it turned out to be invaluable. We are indebted to 16 Accuracy Reviewers, colleagues who carefully reviewed the almost final page proofs of each chapter to spot lingering errors or imprecisions in the text and art that inevitably escape our weary eyes. Finally, we appreciate the advice given by several experts who reviewed the animations and activities that our publishers developed for the student Web Site/CD-ROM that accompanies this edition of *Life*. We thank all these reviewers and hope this new edition measures up to their expectations. They are listed after this Preface.

J/B Woolsey Associates has again worked closely with each of us to improve an already excellent art program. They helped to refine the very successful "balloon captions" that were introduced in the Fifth Edition. With their creative input we introduced the Experiment and Research Method illustrations found throughout the text.

James Funston joined us again as the developmental editor for the Sixth Edition. As always, James enforced a rigorous standard for clear writing and illustrating. And he contributed significantly to the process of shortening the book. Norma Roche also suggested cuts, and provided incisive copy editing from beginning to end. Her many astute queries often led to rewrites that enhanced the clarity of the presentation. From first draft to final pages, Susan McGlew was tireless in arranging for expert academic reviews of all of the chapters. Since the First Edition, we have profited immeasurably from the work of Carol Wigg, who again coordinated the pre-production process, including illustration editing and copy editing. She wrote many figure captions, suggested several of the chapter-opening stories, orchestrated the flow of the text and art, kept us mostly on schedule, enforced—sometimes with her red pen—the mandate to be concise, and what's more, did it all with good humor, even under pressure. David McIntyre, photo researcher, found many wonderful new photographs to enhance the learning experience and enliven the appearance of the book as a whole.

We again wish to thank the dedicated professionals in W. H. Freeman's marketing and sales group. Their enthusiasm has helped bring *Life* to a wider audience with each edition. We appreciate their continuing support and valuable input on ways to improve the book. A large share of *Life's* success is due to their efforts in this publishing partnership.

We have always respected Sinauer Associates for their outstanding list of biology books at all levels and we have enjoyed having them lead and assist us through yet another edition. Andy Sinauer has been the guiding spirit behind the development of *Life* since two of us first began to write the First Edition. Andy never ceases helping his authors to achieve our goals, while remaining gentle but firm about his agendas. It has been a very satisfying experience for us to work with him yet again, and we look forward to a continuing association.

Bill Purves   David Sadava   Gordon Orians   Craig Heller

November, 2000

Reviewers for the Sixth Edition

Diary Reviewers

Carla Barnwell, University of Illinois

Greg Beaulieu, University of Victoria

Gordon Fain, University of California, Los Angeles

Ruth Finkelstein, University of California, Santa Barbara

Steve Fisher, University of California, Santa Barbara

Alice Jacklet, SUNY, Albany

Clare Hasenkampf, University of Toronto, Scarborough

Werner Heim, Colorado College

David Hershey, Hyattsville, MD

Hans-Willi Honegger, Vanderbilt University

Durrell Kapan, University of Texas, Austin

Cheryl Kerfeld, University of California, Los Angeles

Michael Martin, University of Michigan, Ann Arbor

Murray Nabors, Colorado State University

Ronald Poole, McGill University

Nancy Sanders, Truman State University

Susan Smith, Massasoit Community College

Raymond White, City College of San Francisco

Manuscript Reviewers

John Alcock, Arizona State University

Allen V. Barker, University of Massachusetts, Amherst

Andrew R. Blaustein, Oregon State University

Richard Brusca, University of Arizona

Matthew Buechner, University of Kansas

Warren Burggren, University of North Texas

Jung Choi, Georgia Institute of Technology

Andrew Clark, Pennsylvania State University

Carla D'Antonio, University of California, Berkeley

Alan de Queiroz, University of Colorado

Michael Denbow, Virginia Technological University

Susan Dunford, University of Cincinnati

William Eickmeier, Vanderbilt University

John Endler, University of California, Santa Barbara

Gordon L. Fain, University of California, Los Angeles

Stu Feinstein, University of California, Santa Barbara

Danilo Fernando, SUNY, Syracuse

Steve Fisher, University of California, Santa Barbara

Doug Futuyma, SUNY, Stony Brook

Scott Gilbert, Swarthmore College

Janice Glime, Michigan Technological University

Elizabeth Godrick, Boston University

Robert Goodman, University of Wisconsin, Madison

Nancy Guild, University of Colorado

Jessica Gurevitch, SUNY, Stony Brook

Jeff Hardin, University of Wisconsin, Madison

Joseph Heilig, University of Colorado

David Hershey, Hyattsville, MD

Mark Johnston, Dalhousie University

Walter Judd, University of Florida

Thomas Kane, University of Cincinnati

Laura Katz, Smith College

Elizabeth Kellogg, University of Missouri, St. Louis

Peter Krell, University of Guelph

Thomas Kursar, University of Utah

Wayne Maddison, University of Arizona

William Manning, University of Massachusetts, Amherst

Michael Marcotrigiano, Smith College

Lloyd Matsumoto, Rhode Island College

Stu Matz, The Evergreen State College

D. Jeffrey Meldrum, Idaho State University

Mike Millay, Ohio University (Southern Campus)

David Mindell, University of Michigan, Ann Arbor

Deborah Mowshowitz, Columbia University

Laura Olsen, University of Michigan, Ann Arbor

Guillermo Orti, University of Nebraska

Constance Parks, University of Massachusetts, Amherst

Jane Phillips, University of Minnesota

Ronald Poole, McGill University

Warren Porter, University of Wisconsin, Madison

Thomas Poulson, University of Illinois, Chicago

Loren Rieseberg, Indiana University

Ian Ross, University of California, Santa Barbara

Nancy Sanders, Truman State University

Paul Schroeder, Washington State University

Jim Shinkle, Trinity University

Mitchell Sogin, Marine Biological Laboratory, Woods Hole

Wayne Sousa, University of California, Berkeley

Charles Staben, University of Kentucky

James Staley, University of Washington

Steve Stanley, The Johns Hopkins University

Barbara Stebbins-Boaz, Willamette University

Antony Stretton, University of Wisconsin, Madison

Steven Swoap, Williams College

Gerald Thrush, California State University, San Bernardino

Richard Tolman, Brigham Young University

Mary Tyler, University of Maine

Michael Wade, Indiana University

Bruce Walsh, University of Arizona

Steven Wasserman, University of California, San Diego

Alex Weir, SUNY, Syracuse

Mary Williams, Harvey Mudd College

Jonathan Wright, Pomona College

Accuracy Reviewers

Andrew Clark, Pennsylvania State University

Joanne Ellzey, University of Texas, El Paso

Tejendra Gill, University of Houston, University Park

Paul Goldstein, University of Texas, El Paso

Laura Katz, Smith College

Hans Landel, North Seattle Community College

Sandy Ligon, University of New Mexico

Peter Lortz, North Seattle Community College

Roger Lumb, Western Carolina University

Coleman McCleneghan, Appalachian State University

Janie Milner, Santa Fe Community College

Zack Murrell, Appalachian State University

Ben Normark, University of Massachusetts, Amherst

Mike Silva, El Paso Community College

Phillip Snider, University of Houston, University Park

Steven Wasserman, University of California, San Diego

Media Reviewers

Karen Bernd, Davidson College

Mark Browning, Purdue University

William Eldred, Boston University

Joanne Ellzey, University of Texas, El Paso

Randall Johnson, University of California, San Diego

Coleman McCleneghan, Appalachian State University

Melissa Michael, University of Illinois

Tom Pitzer, Florida International University

Kenneth Robinson, Purdue University

To the Student

Welcome to the study of life! In our student days—and ever since—we have enjoyed studying the fascinating and fast-changing field of biology, and we hope that you will, too.

Getting the Most Out of the Book

There are a few things you can do to help you get the most from this book and from your course. For openers, read the book actively—don't just read passively, but do things that force you to think as you read. If we pose questions, stop and think about them. Ask questions of the text as you go. Do you understand what is being said? Does it relate to something you already know? Is it supported by experimental or other evidence? Does that evidence convince you? How does this passage fit into the chapter as a whole? Annotate the book—write down comments in the margins about things you don't understand, or about how one part relates to another, or even when you find an idea particularly interesting. People remember things they think about much better than they remember things they have read passively. Highlighting is passive; copying is drudge work; questioning and commenting are active and well worthwhile.

"Read" the illustrations actively too. You will find the balloon captions in the illustrations especially useful—they are there to guide you through the complexities of some topics and to highlight the major points.

The chapter summaries will help you quickly review the high points of what you have read. A summary identifies particular illustrations that you should study to help organize the material in your mind. Add concepts and details to the framework by reviewing the text. A way to review the material in slightly more detail after reading the chapter is to go back and look at the boldfaced terms. You can use the boldfaced terms to pose questions—and see if you can answer those questions. The boldfacing will probably be more useful on a second reading than on the first.

Use the "For Discussion" questions at the end of each chapter. These questions are usually open-ended and are intended to cause you to reflect on the material.

The glossary and the index can help you a great deal. When you are uncertain of the meaning of a term, check the glossary first—there are more than 1,500 definitions in it. If you don't find a term in the glossary, or if you want a more thorough discussion of the term, use the index to find where it's discussed.

The Web Site

Use the student Web Site/CD-ROM to help you understand some of the more detailed material and to help you sort out the information we have laid before you. An illustrated guide to the learning resources found on the Web Site/CD-ROM is in the front of this book. Pay particular attention to the activities and animated tutorials on key concepts, and to the self-quizzes. The self-quizzes provide extensive feedback for each correct and incorrect answer, and include hot-linked references to text pages. If you'd like to pursue some topics in greater detail, you'll find a chapter-by-chapter annotated list of suggested readings. We have tried to choose readings from books and magazines, especially *Scientific American*, that should be available in your college library.

What If the Going Gets Tough?

Most students occasionally have difficulty in courses, including biology courses. If you find that you are slipping behind in the course, or if a particular topic is giving you an unreasonable amount of trouble, here are some useful steps you might take. First, the basics: attend class, take careful lecture notes, and read the textbook assignments. Second, note that one of the most important roles of studying is to discover what you *don't* know, so that you can do something about it. Use the index, the glossary, the chapter summaries, and the text itself to try to answer any questions you have and to help you organize the material. Make a habit of looking over your lecture notes within 24 hours of when you take them—find out right away what points are unclear, and get them straightened out in your mind. The web site can help by providing a different perspective.

If none of these self-help remedies does the trick, get help! Other students are often a good source of help, because they are dealing with the material at the same level as you are. Study groups can be very useful, as long as the participants are all committed to learning the material. Tutors are almost always helpful, as are faculty members. The main thing is to *get help when you need it*. It is not a good idea to be strong and silent and drift into a low grade.

But don't make the grade the point of this or any other course. You are in college to learn, to pursue interesting subjects, and to enjoy the subjects you are pursuing. We hope you'll enjoy the pursuit of biology.

Bill Purves   David Sadava   Gordon Orians   Craig Heller

Life's Supplements

For the Student

Web Site/CD-ROM

Student Web Site at www.thelifewire.com
Life 6.0 CD-ROM (optionally bundled with the text)

The Web Site and CD-ROM each support the entire text, offering:

- Over 65 **Animated Tutorials** clarifying key topics from the text
- **Activities**, including flashcards for key terms and concepts, and drag-and-drop exercises
- **Self-quizzes** with extensive feedback, references to the Study Guide, and hot-linked references to *Life: The Science of Biology*, Sixth Edition
- **Glossary** of key terms and concepts
- **End-of-chapter Online Quizzes** (see "Online Quizzing" under "For the Instructor")
- **Lifelines**

   Study Skills (Jerry Waldvogel, *Clemson University*) provides class-tested practical advice on time management, test-taking, note-taking, and how to read the textbook

   Math for Life (Dany Adams, *Smith College*) helps students learn or reacquire basic quantitative skills

- **Suggested Readings** for further study
   Order ISBN 0-7167-3874-0, *Life 6.0* CD-ROM, or
   ISBN 0-7167-3875-9, Text/CD-ROM bundle

Study Guide

Christine Minor, *Clemson University*, Lindsay Goodloe, *Cornell University*, Edward M. Dzialowski and Warren W. Burggren, *University of North Texas*, and Nancy Guild, *University of Colorado at Boulder*.

For each chapter of the text, the study guide offers clearly defined learning objectives, summaries of key concepts, references to *Life* and to the student *Web/CD-ROM*, and review and exam-style self-test questions with answers and explanations.
Order ISBN 0-7167-3951-8

Lecture Notebook

This new tool presents black and white reproductions of all the Sixth Edition's line art and tables (more than 1000 images, with labels). The *Notebook* provides ample ruled spaces for note-taking.
Order ISBN 0-7167-4449-X

For the Instructor

Instructor's Teaching Kit

This **new** comprehensive teaching tool (in a three-ring binder) combines:

1. Instructor's Manual

   Erica Bergquist, *Holyoke Community College*

   The Manual includes:

   - Chapter overviews
   - Chapter outlines
   - A "What's New" guide to the Sixth Edition
   - All the bold-faced key terms from the text
   - Key concepts and facts for each chapter
   - Overviews of the animated tutorials from the Student Web Site/CD-ROM
   - Custom lab ordering information (see "Custom Labs")

2. Enriched Lecture Notes, with diagrams
   Charles Herr, *Eastern Washington University*

3. A PowerPoint® Thumbnail Guide to the PowerPoint® presentations on the Instructor's CD-ROM

Test Bank

Charles Herr, *Eastern Washington University*

The test bank, available in both computerized and printed formats, offers more than 4000 multiple-choice and sentence-completion questions.

The easy-to-use computerized test bank on CD-ROM includes Windows and Macintosh versions in a format that lets instructors add, edit, and resequence questions to suit their needs. From this same CD-ROM, instructors can access *Diploma* Online Testing from the Brownstone Research Group. *Diploma* allows instructors to easily create and administer secure exams over a network and over the Internet, with questions that incorporate multimedia and interactive exercises. More information about *Diploma* is available at http://www.brownstone.net

Online Quizzing

The online quizzing function is accessed via the Student Web Site at www.thelifewire.com. Using Question Mark's *Perception*, instructors can easily and securely quiz students online using multiple-choice questions for each text chapter and its media resources.

Instructor's Resource CD-ROM

The Instructor's Resource CD-ROM employs **Presentation Manager** and includes:

- ▶ All four-color line art and tables from the text (more than 1000 images), resized and reformatted to maximize large-hall projection
- ▶ More than 1500 photographic images, including electron micrographs, from the Biological Photo Service collection—all keyed to *Life* chapters
- ▶ More than 60 animations from the Student Web Site/CD-ROM
- ▶ Exceptional video microscopy from Jeremy Pickett-Heaps and others
- ▶ Chapter outlines and lecture notes from the Instructor's Teaching Kit in editable Microsoft® Word documents

PowerPoint® Presentations

The PowerPoint® slide set for *Life* follows the chapter summaries provided in the Instructor's Teaching Kit and can be used directly or customized. Each slide incorporates a figure from *Life*.

PowerPoint® Tutorials
QuickTime™ movies demonstrate how to use PowerPoint®.

Classroom Management

As a service for adopters using WebCT, we will provide a fully-loaded WebCourselet, including the instructor and student resources for this text. The files can then be customized to fit your specific course needs, or can be used as is. Course outlines, pre-built quizzes, activities, and a whole array of materials are included, eliminating hours of work for instructors interested in creating WebCT courses. For more information and a demo of the WebCourselet for this text, please visit our Web Site (http://bfwpub.com/mediaroom/Index.html) and click "WebCT".

Overhead Transparencies

The transparency set includes all four-color line art and tables from the text (more than 1000 images) in a convenient three-ring binder. Balloon captions (and some labels) are deleted to enhance projection and allow for classroom quizzing. Labels and images have been resized for maximum readability.

Slide Set

The slide set includes selected four-color figures from the text. Labels and images have been resized for maximum readability.

Laboratory Manuals

Biology in the Laboratory, Third Edition

Doris Helms, Robert Kosinski, and John Cummings, *all of Clemson University*

The revised edition of this popular lab manual, which includes a CD-ROM, is available to accompany the Sixth Edition of *Life*.
Order ISBN 0-7167-3146-0

Laboratory Outlines in Biology VI

Peter Abramoff and Robert G. Thomson, *Marquette University*
Order ISBN 0-7167-2633-5

The following manuals are available in a bound volume or as separates:
Anatomy and Dissection of the Rat, Third Edition
Warren F. Walker, Jr., *Oberlin College*, and Dominique Homberger, *Louisiana State University*
Order ISBN 0-7167-2635-1
Anatomy and Dissection of the Fetal Pig, Fifth Edition
Warren F. Walker, Jr., *Oberlin College*, and Dominique Homberger, *Louisiana State University*
Order ISBN 0-7167-2637-8
Anatomy and Dissection of the Frog, Second Edition
Warren F. Walker, Jr., *Oberlin College*
Order ISBN 0-7167-2636-X

Custom Labs

Custom Publishing for Laboratory Manuals at www.custompub.whfreeman.com

With this custom publishing option, instructors can build and order customized lab manuals in just minutes, choosing material from Freeman's acclaimed biology laboratory manuals—lab-tested experiments that have been used successfully by hundreds of thousands of students. Instructors determine the manual's content (with the option to incorporate their own material or blank pages), table of contents or index styles, and cover design, and submit the order. A streamlined production process provides a quick turnaround to meet crucial deadlines.

Contents in Brief

Contents

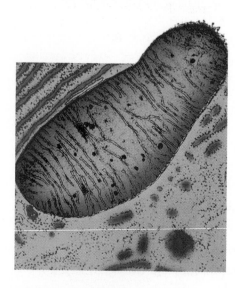

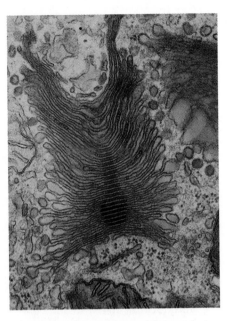

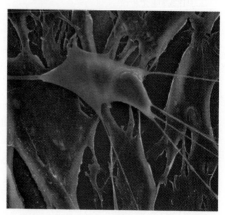

Part Two
INFORMATION AND HEREDITY

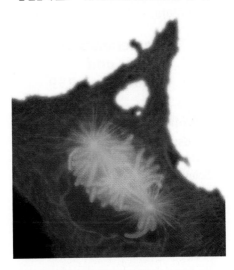

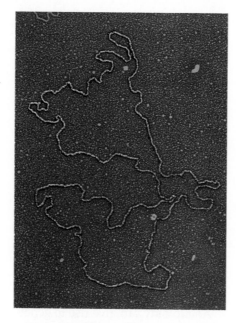

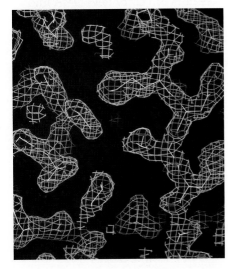

Part Three

EVOLUTIONARY PROCESSES

20 The History of Life on Earth 379

21 The Mechanisms of Evolution 395

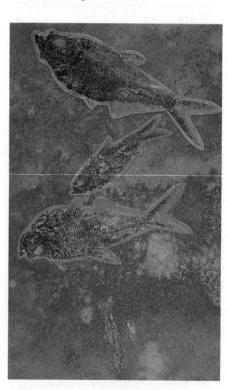

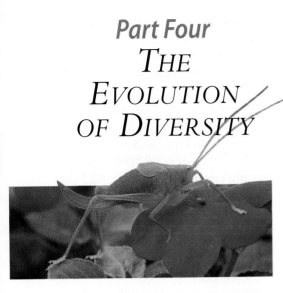

Part Four
THE EVOLUTION OF DIVERSITY

Part Five
THE BIOLOGY OF FLOWERING PLANTS

Part Six
THE BIOLOGY OF ANIMALS

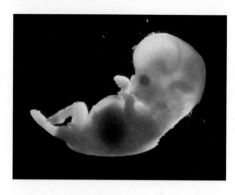

Part Seven

ECOLOGY AND BIOGEOGRAPHY

1 An Evolutionary Framework for Biology

AT MIDNIGHT ON DECEMBER 31, 1999, massive displays of fireworks exploded in many places on Earth as people celebrated a new millennium—the passage from one thousand-year time frame into the next—and the advent of the year 2000. One such millennial display took place above the Egyptian pyramids.

We are impressed with the size of the pyramids, how difficult it must have been to build them, and how ancient they are. The oldest of these awe-inspiring monuments to human achievement was built more than 4,000 years ago; in the human experience, this makes the Egyptian pyramids very, very old. Yet from the perspective of the age of Earth and the time over which life has been evolving, the pyramids are extremely young. Indeed, if the history of Earth is visualized as a 30-day month, recorded human history—the dawn of which coincides roughly with the construction of the earliest pyramids—is confined to the last *30 seconds* of the final day of the month (Figure 1.1).

The development of modern biology depended on the recognition that an immense length of time was available for life to arise and evolve its current richness. But for most of human history, people had no reason to suspect that Earth was so old. Until the discovery of radioactive decay at the beginning of the twentieth century, no methods existed to date prehistoric events. By the middle of the nineteenth century, however, studies of rocks and the fossils they contained had convinced geologists that Earth was much older than had generally been believed. Darwin could not have conceived his theory of evolution by natural selection had he not understood that Earth was very ancient.

In this chapter we review the events leading to the acceptance of the fact that life on Earth has evolved over several billion years. We then summarize how evolutionary mechanisms adapt organisms to their environments, and we review the major milestones in the evolution of life on Earth. Finally, we briefly describe how scientists generate new knowledge, how they develop and test hypotheses, and how that knowledge can be used to inform public policy.

A Celebration of Time
One millennial fireworks display celebrating the year 2000 took place over the ancient pyramids of Egypt, structures that represent more than 4,000 years of human history but an infinitesimal portion of Earth's geologic history.

Organisms Have Changed over Billions of Years

Long before the mechanisms of biological evolution were understood, some people realized that organisms had changed over time and that living organisms had evolved from organisms no longer alive on Earth. In the 1760s, the French naturalist Count George-Louis Leclerc de Buffon (1707–1788) wrote his *Natural History of Animals*, which contained a clear statement of the possibility of evolution. Buffon originally believed that each species had been divinely created for a particular way of life, but as he studied animal anatomy, doubts arose. He observed that the limb bones of all mammals, no matter what their way of life, were re-

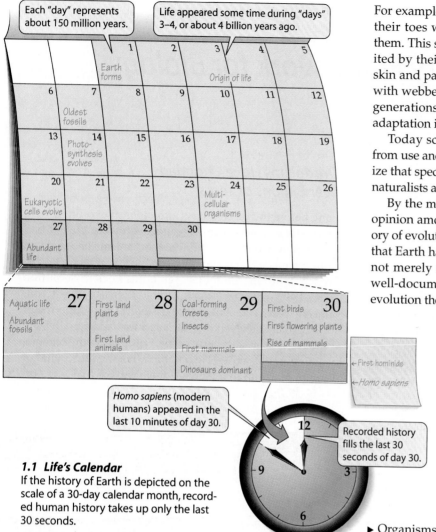

Each "day" represents about 150 million years.

Life appeared some time during "days" 3–4, or about 4 billion years ago.

1.1 Life's Calendar
If the history of Earth is depicted on the scale of a 30-day calendar month, recorded human history takes up only the last 30 seconds.

Homo sapiens (modern humans) appeared in the last 10 minutes of day 30.

Recorded history fills the last 30 seconds of day 30.

markably similar in many details (Figure 1.2). Buffon also noticed that the legs of certain mammals, such as pigs, have toes that never touch the ground and appear to be of no use. He found it difficult to explain the presence of these seemingly useless small toes by special creation.

Both of these troubling facts could be explained if mammals had not been specially created in their present forms, but had been modified over time from an ancestor that was common to all mammals. Buffon suggested that the limb bones of mammals might all have been inherited, and that pigs might have functionless toes because they inherited them from ancestors with fully formed and functional toes. Buffon's idea was an early statement of evolution (descent with modification), although he did not attempt to explain how such changes took place.

Buffon's student Jean Baptiste de Lamarck (1744–1829) was the first person to propose a mechanism of evolutionary change. Lamarck suggested that lineages of organisms may change gradually over many generations as offspring inherit structures that have become larger and more highly developed as a result of continued use or, conversely, have become smaller and less developed as a result of disuse.

For example, Lamarck suggested that aquatic birds extend their toes while swimming, stretching the skin between them. This stretched condition, he thought, could be inherited by their offspring, which would in turn stretch their skin and pass this condition along to their offspring; birds with webbed feet would thereby evolve over a number of generations. Lamarck explained many other examples of adaptation in a similar way.

Today scientists do not believe that changes resulting from use and disuse can be inherited. But Lamarck did realize that species change with time. And after Lamarck, other naturalists and scientists speculated along similar lines.

By the middle of the nineteenth century, the climate of opinion among many scholars was receptive to a new theory of evolutionary processes. By then geologists had shown that Earth had existed and changed over millions of years, not merely a few thousand years. The presentation of a well-documented and thoroughly scientific argument for evolution then triggered a transformation of biology.

The theory of evolution by natural selection was proposed independently by Charles Darwin and Alfred Russel Wallace in 1858. We will discuss evolutionary theory in detail in Chapter 21, but its essential features are easy to understand. The theory rests on two facts and one inference drawn from them. The two facts are:

▶ The reproductive rates of all organisms, even slowly reproducing ones, are sufficiently high that populations would quickly become enormous if mortality rates did not balance reproductive rates.

▶ Organisms of all types are variable, and offspring are similar to their parents because they inherit their features from them.

The inference is:

▶ The differences among individuals influence how well those individuals survive and reproduce. Traits that increase the probability that their bearers will survive and reproduce are more likely to be passed on to their offspring and to their offspring's offspring.

Darwin called the differential survival and reproductive success of individuals **natural selection**. The remarkable features of all organisms have evolved under the influence of natural selection. Indeed, *the ability to evolve by means of natural selection clearly separates life from nonlife.*

Biology began a major conceptual shift a little more than a century ago with the general acceptance of long-term evolutionary change and the recognition that differential survival and reproductive success is the primary process that adapts organisms to their environments. The shift has taken a long time because it required abandoning many components of an earlier worldview. The pre-Darwinian view held that the world was young, and that organisms had been created in their current forms. In the Darwinian view,

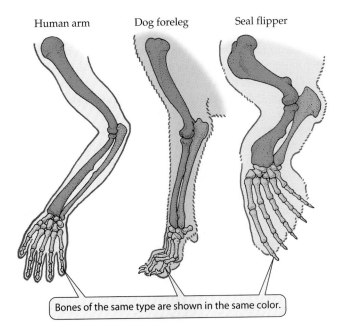

Human arm   Dog foreleg   Seal flipper

Bones of the same type are shown in the same color.

1.2 Mammals Have Similar Limbs
Mammalian forelimbs have different purposes, but the number and types of their bones are similar, indicating that they have been modified over time from a common ancestor.

the world is ancient, and both Earth and its inhabitants have been continually changing. In the Darwinian view of the world, organisms evolved their particular features because individuals with those features survived and reproduced better than individuals with different features.

Adopting this new view of the world means accepting not only the processes of evolution, but also the view that the living world is constantly evolving, and that evolutionary change occurs without any "goals." The idea that evolution is not directed toward a final goal or state has been more difficult for many people to accept than the process of evolution itself. But even though evolution has no goals, evolutionary processes have resulted in a series of profound changes—milestones—over the nearly 4 billion years life has existed on Earth.

Evolutionary Milestones

The following overview of the major milestones in the evolution of life provides both a framework for presenting the characteristics of life that will be described in this book and an overview of how those characteristics evolved during the history of life on Earth.

Life arises from nonlife

All matter, living and nonliving, is made up of chemicals. The smallest chemical units are atoms, which bond together into molecules; the properties of those molecules are the subject of Chapter 2. The processes leading to life began nearly 4 billion years ago with interactions among small molecules that stored useful information.

The information stored in these simple molecules eventually resulted in the synthesis of larger molecules with complex but relatively stable shapes. Because they were both complex and stable, these units could participate in increasing numbers and kinds of chemical reactions. Some of these large molecules—carbohydrates, lipids, proteins, and nucleic acids—are found in all living systems and perform similar functions. The properties of these complex molecules are the subject of Chapter 3.

Cells form from molecules

About 3.8 billion years ago, interacting systems of molecules came to be enclosed in compartments surrounded by membranes. Within these membrane-enclosed units, or **cells**, control was exerted over the entrance, retention, and exit of molecules, as well as the chemical reactions taking place within the cell. Cells and membranes are the subjects of Chapters 4 and 5.

Cells are so effective at capturing energy and replicating themselves—two fundamental characteristics of life—that since the time they evolved, they have been the unit on which all life has been built. Experiments by the French chemist and microbiologist Louis Pasteur and others during the nineteenth century convinced most scientists that, under present conditions on Earth, cells do not arise from noncellular material, but must come from other cells.

For 2 billion years, cells were tiny packages of molecules each enclosed in a single membrane. These **prokaryotic cells** lived autonomous lives, each separate from the other. They were confined to the oceans, where they were shielded from lethal ultraviolet sunlight. Some prokaryotes living today may be similar to these early cells (Figure 1.3).

1.3 Early Life May Have Resembled These Cells
"Rock-eating" bacteria, appearing red in this artificially colored micrograph, were discovered in pools of water trapped between layers of rock more than 1,000 meters below Earth's surface. Deriving chemical nutrients from the rocks and living in an environment devoid of oxygen, they may resemble some of the earliest prokaryotic cells.

To maintain themselves, to grow, and to reproduce, these early prokaryotes, like all cells that have subsequently evolved, obtained raw materials and energy from their environment, using these as building blocks to synthesize larger, carbon-containing molecules. The energy contained in these large molecules powered the chemical reactions necessary for the life of the cell. These conversions of matter and energy are called **metabolism**.

All organisms can be viewed as devices to capture, process, and convert matter and energy from one form to another; these conversions are the subjects of Chapters 6 and 7. *A major theme in the evolution of life is the development of increasingly diverse ways of capturing external energy and using it to drive biologically useful reactions.*

Photosynthesis changes Earth's environment

About 2.5 billion years ago, some organisms evolved the ability to use the energy of sunlight to power their metabolism. Although they still took raw materials from the environment, the energy they used to metabolize these materials came directly from the sun. Early photosynthetic cells were probably similar to present-day prokaryotes called cyanobacteria (Figure 1.4). The energy-capturing process they used—**photosynthesis**—is the basis of nearly all life on Earth today; it is explained in detail in Chapter 8. It used new metabolic reactions that exploited an abundant source of energy (sunlight), and generated a new waste product (oxygen) that radically changed Earth's atmosphere.

The ability to perform photosynthetic reactions probably accumulated gradually during the first billion years or so of evolution, but once this ability had evolved, its effects were dramatic. Photosynthetic prokaryotes became so abundant that they released vast quantities of oxygen gas (O_2) into the atmosphere. The presence of oxygen opened up new avenues of evolution. Metabolic reactions that use O_2, called **aerobic metabolism**, came to be used by most organisms on Earth. The oxygen in the air we breathe today would not exist without photosynthesis.

Over a much longer time, the vast quantities of oxygen liberated by photosynthesis had another effect. Formed from O_2, ozone (O_3) began to accumulate in the upper atmosphere. The ozone slowly formed a dense layer that acted as a shield, intercepting much of the sun's deadly ultraviolet radiation. Eventually (although only within the last 800 million years of evolution), the presence of this shield allowed organisms to leave the protection of the oceans and establish new lifestyles on Earth's land surfaces.

Sex enhances adaptation

The earliest unicellular organisms reproduced by doubling their hereditary (genetic) material and then dividing it into two new cells, a process known as mitosis. The resulting progeny cells were identical to each other and to the parent. That is, they were clones. But **sexual reproduction**—the combining of genes from two cells in one cell—appeared early during the evolution of life. Sexual reproduction is advantageous because an organism that combines its genetic information with information from another individual produces offspring that are more variable. *Reproduction with variation is a major characteristic of life.*

Variation allows organisms to adapt to a changing environment. **Adaptation** to environmental change is one of life's most distinctive features. An organism is adapted to a given environment when it possesses inherited features that enhance its survival and ability to reproduce in that environment. Because environments are constantly changing, organisms that produce variable offspring have an advantage over those that produce genetically identical "clones," because they are more likely to produce some offspring better adapted to the environment in which they find themselves.

Eukaryotes are "cells within cells"

As the ages passed, some prokaryotic cells became large enough to attack, engulf, and digest smaller cells, becoming the first predators. Usually the smaller cells were destroyed within the predators' cells. But some of these smaller cells survived and became permanently integrated into the operation of their hosts' cells. In this manner, cells with complex internal compartments arose. We call these cells **eukaryotic cells**. Their appearance slightly more than 1.5 billion years ago opened more new evolutionary opportunities.

Prokaryotic cells—the Bacteria and Archaea—have no membrane-enclosed compartments. Eukaryotic cells, on the

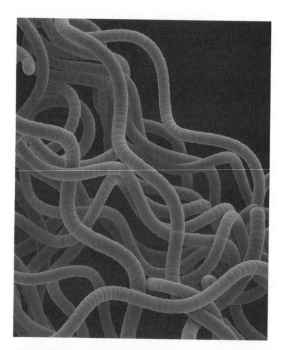

1.4 Oxygen Produced by Prokaryotes Changed Earth's Atmosphere
These modern cyanobacteria are probably very similar to early photosynthetic prokaryotes.

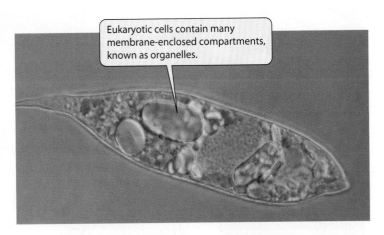

Eukaryotic cells contain many membrane-enclosed compartments, known as organelles.

1.5 Multiple Compartments Characterize Eukaryotic Cells
The nucleus and other specialized organelles probably evolved from small prokaryotes that were ingested by a larger prokaryotic cell. This is a photograph of a single-celled eukaryotic organism known as a protist.

other hand, are filled with membrane-enclosed compartments. In eukaryotic cells, genetic material—genes and chromosomes—became contained within a discrete nucleus and became increasingly complex. Other compartments became specialized for other purposes, such as photosynthesis. We refer to these specialized compartments as **organelles** (Figure 1.5).

Multicellularity permits specialization of cells

Until slightly more than 1 billion years ago, only single-celled organisms existed. Two key developments made the evolution of multicellular organisms—organisms consisting of more than one cell—possible. One was the ability of a cell to change its structure and functioning to meet the challenges of a changing environment. This was accomplished when prokaryotes evolved the ability to change from rapidly growing cells into resting cells called **spores** that could survive harsh environmental conditions. The second development allowed cells to stick together in a "clump" after they divided, forming a multicellular organism.

Once organisms could be composed of many cells, it became possible for the cells to specialize. Certain cells, for example, could be specialized to perform photosynthesis. Other cells might become specialized to transport chemical materials such as oxygen from one part of an organism to another. Very early in the evolution of multicellular life, certain cells began to be specialized for sex—the passage of new genetic information from one generation to the next.

With the presence of specialized sex cells, genetic transmission became more complicated. Simple nuclear division—mitosis—was and is sufficient for the needs of most cells. But among the sex cells, or gametes, a whole new method of nuclear division—meiosis—evolved. Meiosis allows gametes to combine and rearrange the genetic infor-

mation from two distinct parent organisms into a genetic package that contains elements of both parent cells but is different from either. The recombinational possibilities generated by meiosis had great impact on variability and adaptation and on the speed at which evolution could occur.

Mitosis and meiosis are covered in detail in Chapter 9.

Controlling internal environments becomes more complicated

The pace of evolution, quickened by the emergence of sex and multicellular life, was also heightened by changes in Earth's atmosphere that allowed life to move out of the oceans and exploit environments on land. Photosynthetic green plants colonized the land, providing a rich source of energy for a vast array of organisms that consumed them. But whether it is made up of one cell or many, an organism must respond appropriately to its external environment. Life on land presented a new set of environmental challenges.

In any environment, external conditions can change rapidly and unpredictably in ways that are beyond an organism's control. An organism can remain healthy only if its internal environment remains within a given range of physical and chemical conditions. Organisms maintain relatively constant internal environments by making metabolic adjustments to changes in external and internal conditions such as temperature, the presence or absence of sunlight, the presence or absence of specific chemicals, the need for nutrients (food) and water, or the presence of foreign agents inside their bodies. Maintenance of a relatively stable internal condition—such as a constant human body temperature despite variation in the temperature of the surrounding environment—is called **homeostasis**. *A major theme in the evolution of life is the development of increasingly complicated systems for maintaining homeostasis.*

Multicellular organisms undergo regulated growth

Multicellular organisms cannot achieve their adult shapes or function effectively unless their growth is carefully regulated. Uncontrolled growth—one example of which is cancer—ultimately destroys life. *A vital characteristic of living organisms is regulated growth.* Achieving a functional multicellular organism requires a sequence of events leading from a single cell to a multicellular adult. This process is called **development**.

The adjustments that organisms make to maintain constant internal conditions are usually minor; they are not obvious, because nothing appears to change. However, at some time during their lives, many organisms respond to changing conditions not by maintaining their status, but by undergoing major cellular and molecular reorganization. An early form of such developmental reorganization was the prokaryotic spores that were generated in response to environmental stresses. A striking example that evolved much later is **metamorphosis**, seen in many modern in-

1.6 Organisms May Change Dramatically During Their Lives
The caterpillar, pupa, and adult are all stages in the life cycle of a monarch butterfly. The transition from one stage to another is triggered by internal signals.

sects, such as butterflies. In response to internal chemical signals, a caterpillar changes into a pupa and then into an adult butterfly (Figure 1.6).

The activation of gene-based information within cells and the exchange of signal information among cells produce the well-timed events that are required for the transition to the adult form. Genes control the metabolic processes necessary for life. The nature of the genetic material that controls these lifelong events has been understood only within the twentieth century; it is the story to which much of Part Two of this book is devoted.

Altering the timing of development can produce striking changes. Just a few genes can control processes that result in dramatically different adult organisms. Chimpanzees and humans share more than 98 percent of their genes, but the differences between the two in form and in behavioral abilities—most notably speech—are dramatic (Figure 1.7). When we realize how little information it sometimes takes to create major transformations, the still mysterious process of **speciation** becomes a little less of a mystery.

Speciation produces the diversity of life

All organisms on Earth today are the descendants of a kind of unicellular organism that lived almost 4 billion years ago. The preceding pages described the major evolutionary events that have led to more complex living organisms. The course of this evolution has been accompanied by the storage of larger and larger quantities of information and increasingly complex mechanisms for using it. But if that were the entire story, only one kind of organism might exist

on Earth today. Instead, Earth is populated by many millions of kinds of organisms that do not interbreed with one another. We call these genetically independent groups of organisms **species**.

As long as individuals within a population mate at random and reproduce, structural and functional changes may occur, but only one species will exist. However, if a population becomes divided into two or more groups, and individuals can mate only with individuals in their own group, differences may accumulate with time, and the groups may evolve into different species.

The splitting of groups of organisms into separate species has resulted in the great variety of life found on Earth today, as described in Chapter 20. How species form is explained in Chapter 22. From a single ancestor, many species may arise as a result of the repeated splitting of populations. How biologists determine which species have descended from a particular ancestor is discussed in Chapter 23.

1.7 Genetically Similar Yet Very Different
By looking at the two, you might be surprised to learn that chimpanzees and humans share more than 98 percent of their genes.

(a)

(b)

(c)

(d)

1.8 Adaptations to the Environment

(*a*) The long, pointed wings of the peregrine falcon allow it to accelerate rapidly as it dives on its prey. (*b*) The action of a hummingbird's wings allows it to hover in front of a flower while it extracts nectar. (*c*) In a water-limited environment, this saguaro cactus stores water in its fleshy trunk. Its roots spread broadly to extract water immediately after it rains. (*d*) The aboveground root system of mangroves is an adaptation that allows these plants to thrive while inundated by salt water—an environment that would kill most terrestrial plants.

Sometimes humans refer to species as "primitive" or "advanced." These and similar terms, such as "lower" and "higher," are best avoided because they imply that some organisms function better than others. In this book, we use the terms "ancestral" and "derived" to distinguish characteristics that appeared earlier from those that appeared later in the evolution of life.

It is important to recognize that *all* living organisms are successfully adapted to their environments. The wings that allow a bird to fly and the structures that allow green plants to survive in environments where water is either scarce or overabundant are examples of the rich array of adaptations found among organisms (Figure 1.8).

The Hierarchy of Life

Biologists study life in two complementary ways:

▶ They study structures and processes ranging from the simple to the complex and from the small to the large.

▶ They study the patterns of life's evolution over billions of years to determine how evolutionary processes have resulted in lineages of organisms that can be traced back to recent and distant ancestors.

These two themes of biological investigation help us synthesize the hierarchical relationships among organisms and the role of these relationships in space and time. We first describe the hierarchy of interactions among the units of biology from the smallest to the largest—from cells to the biosphere. Then we turn to the hierarchy of evolutionary relationships among organisms.

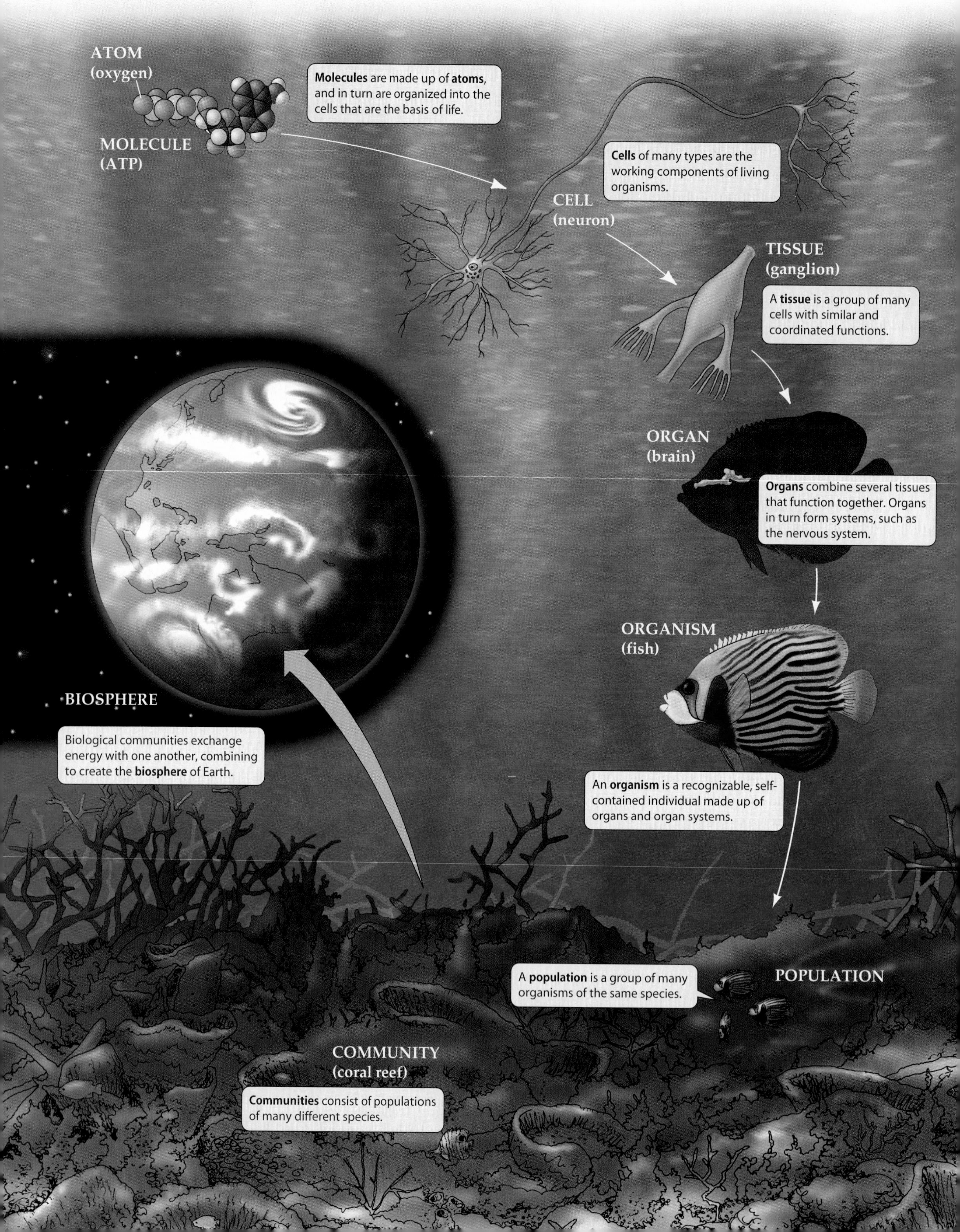

ATOM
(oxygen)

MOLECULE
(ATP)

Molecules are made up of atoms, and in turn are organized into the cells that are the basis of life.

Cells of many types are the working components of living organisms.

CELL
(neuron)

TISSUE
(ganglion)

A tissue is a group of many cells with similar and coordinated functions.

ORGAN
(brain)

Organs combine several tissues that function together. Organs in turn form systems, such as the nervous system.

ORGANISM
(fish)

BIOSPHERE

Biological communities exchange energy with one another, combining to create the biosphere of Earth.

An organism is a recognizable, self-contained individual made up of organs and organ systems.

POPULATION

A population is a group of many organisms of the same species.

COMMUNITY
(coral reef)

Communities consist of populations of many different species.

1.9 The Hierarchy of Life
The individual organism is the central unit of study in biology, but understanding it requires a knowledge of many levels of biological organization both above and below it. At each higher level, additional and more complex properties and functions emerge.

Biologists study life at different levels

Biology can be visualized as a hierarchy in which the units, from the smallest to the largest, include atoms, molecules, cells, tissues, organs, organisms, populations, and communities (Figure 1.9).

The organism is the central unit of study in biology. Parts Five and Six of this book discuss organismal biology in detail. But to understand organisms, biologists must study life at all its levels of organization. Biologists study molecules, chemical reactions, and cells to understand the operations of tissues and organs. They study organs and organ systems to determine how organisms function and maintain internal homeostasis. At higher levels in the hierarchy, biologists study how organisms interact with one another to form social systems, populations, ecological communities, and biomes, which are the subjects of Part Seven of this book.

Each level of biological organization has properties, called **emergent properties**, that are not found at lower levels. For example, cells and multicellular organisms have characteristics and carry out processes that are not found in the molecules of which they are composed.

Emergent properties arise in two ways. First, many *emergent properties of systems result from interactions among their parts*. For example, at the organismal level, developmental interactions of cells result in a multicellular organism whose adult features are vastly richer than those of the single cell from which it grew. Other examples of properties that emerge through complex interactions are memory and emotions. In the human brain, these properties result from interactions among the brain's 10^{12} (trillion) cells with their 10^{15} (quadrillion) connections. No single cell, or even small group of cells, possesses them.

Second, *emergent properties arise because aggregations have collective properties* that their individual units lack. For example, individuals are born and they die; they have a life span. An individual does not have a birth rate or a death rate, but a population (composed of many individuals) does. Birth and death rates are emergent properties of a population. Evolution is an emergent property of populations that depends on variation in birth and death rates, which emerges from the different life spans and reproductive success of individuals in the various populations.

Emergent properties do not violate the principles that operate at lower levels of organization. However, emergent properties usually cannot be detected, predicted, or even suspected by studying lower levels. Biologists could never discover the existence of human emotions by studying single nerve cells, even though they may eventually be able to explain it in terms of interactions among many nerve cells.

Biological diversity is organized hierarchically

As many as 30 million species of organisms inhabit Earth today. Many times that number lived in the past but are now extinct. If we go back four billion years, to the origin of life, all organisms are believed to be descended from a single *common ancestor*. The concept of a common ancestor is crucial to modern methods of classifying organisms. *Organisms are grouped in ways that attempt to define their evolutionary relationships, or how recently the different members of the group shared a common ancestor.*

To determine evolutionary relationships, biologists assemble facts from a variety of sources. Fossils tell us where and when ancestral organisms lived and what they looked like. The physical structures different organisms share—toes among mammals, for example—can be an indication of how closely related they are. But a modern "revolution" in classification has emerged because technologies developed in the past 30 years now allow us to compare the genomes of organisms: We can actually determine how many genes different species share. The more genes species have in common, the more recently they probably shared a common ancestor.

Because no fossil evidence for the earliest forms of life remains, the decision to divide all living organisms into three major **domains**—the deepest divisions in the evolutionary history of life—is based primarily on molecular evidence (Figure 1.10). Although new evidence is constantly being brought to light, it seems clear that organisms belonging to a particular domain have been evolving separately from organisms in the other two domains for more than a billion years.

Organisms in the domains **Archaea** and **Bacteria** are prokaryotes—single cells that lack a nucleus and the other internal compartments found in the Eukarya. Archaea and Bacteria differ so fundamentally from each other in the chemical reactions by which they function and in the products they produce that they are believed to have separated into distinct evolutionary lineages very early during the evolution of life. These domains are covered in Chapter 26.

Members of the third domain have eukaryotic cells containing nuclei and complex cellular compartments called organelles. The **Eukarya** are divided into four groups—the protists and the classical kingdoms Plantae, Fungi, and Animalia (see Figure 1.10). Protists, the subject of Chapter 27, are mostly single-celled organisms. The remaining three kingdoms, whose members are all multicellular, are believed to have arisen from ancestral protists.

Some bacteria, some protists, and most members of the kingdom Plantae (plants) convert light energy to chemical energy by photosynthesis. The biological molecules that they produce are the primary food for nearly all other living organisms. The Plantae are covered in Chapters 28 and 29.

The Fungi, the subject of Chapter 30, include molds, mushrooms, yeasts, and other similar organisms, all of

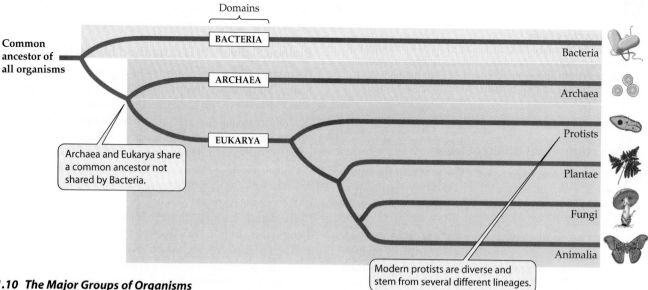

Domains

BACTERIA

**Common
ancestor of
all organisms**

ARCHAEA

EUKARYA

Archaea and Eukarya share
a common ancestor not
shared by Bacteria.

Bacteria

Archaea

Protists

Plantae

Fungi

Animalia

Modern protists are diverse and
stem from several different lineages.

1.10 The Major Groups of Organisms
The classification system used in this book divides Earth's organisms into three domains. The domain Eukarya contains numerous groups of unicellular and multicellular organisms. This "tree" diagram gives information on evolutionary relationships among the groups, as described in Chapter 23.

which are **heterotrophs**: They require a food source of energy-rich molecules synthesized by other organisms. Fungi absorb food substances from their surroundings and break them down (digest them) within their cells. They are important as decomposers of the dead bodies of other organisms.

Members of the kingdom Animalia (animals) are also heterotrophs. These organisms ingest their food source, digest the food outside their cells, and then absorb the products. Animals get their raw materials and energy by eating other forms of life. Perhaps because we are animals ourselves, we are often drawn to study members of this kingdom, which is covered in Chapters 31, 32, and 33.

The biological classification system used today has many hierarchical levels in addition to the ones shown in Figure 1.10. We will discuss the principal levels in Chapter 23. But to understand some of the terms we will use in the intervening chapters, you need to know that each species of organism is identified by two names. The first identifies the **genus**—a group of species that share a recent common ancestor—of which the species is a member. The second name is the species name. To avoid confusion, a particular combination of two names is assigned to only a single species. For example, the scientific name of the modern human species is *Homo sapiens*.

Asking and Answering "How?" and "Why?"

Because biology is an evolutionary science, biological processes and products can be viewed from two different but complementary perspectives. Biologists ask, and try to answer, functional questions: How does it work? They also ask, and try to answer, adaptive questions: Why has it evolved to work that way?

Suppose, for example, that some marine biologists walking on mudflats in the Bay of Fundy, Nova Scotia, Canada, observe many amphipods (tiny relatives of shrimps and lobsters) crawling on the surface of the mud (Figure 1.11). Two obvious questions they might ask are

▶ *How* do these animals crawl?
▶ *Why* do they crawl?

To answer the "how" question, the scientists would investigate the molecular mechanisms underlying muscular contraction, nerve and muscle interactions, and the receipt of stimuli by the amphipods' brains. To answer the "why" question, they would attempt to determine why crawling on the mud is adaptive—that is, why it improves the survival and reproductive success of amphipods.

Is either of these two types of questions more basic or important than the other? Is any one of the answers more fundamental or more important than the other? Not really. The richness of possible answers to apparently simple questions makes biology a complex field, but also an exciting one. Whether we're talking about molecules bonding, cells dividing, blood flowing, amphipods crawling, or forests growing, we are constantly posing both how and why questions. To answer these questions, scientists generate hypotheses that can be tested.

Hypothesis testing guides scientific research

The most important motivator of most biologists is curiosity. People are fascinated by the richness and diversity of life, and they want to learn more about organisms and how they function and interact with one another. Curiosity is probably an adaptive trait. Humans who were motivated to learn about their surroundings are likely to have survived and reproduced better, on average, than their less curious relatives. We hope this book will help you share in the ex-

1.11 An Amphipod from the Mud Flats
Scientists studied this tiny crustacean (whose actual size of approximately 1 centimeter is shown by the scale bar) in an attempt to see whether its behavior changes when it is infected by a parasitic worm. The female of this amphipod species is at the top; the lower specimen is a male.

citement biologists feel as they develop and test hypotheses. There are vast numbers of how and why questions for which we do not have answers, and new discoveries usually engender questions no one thought to ask before. Perhaps *your* curiosity will lead to an important new idea.

Underlying all scientific research is the **hypothetico-deductive (H-D) approach** by which scientists ask questions and test answers. The H-D approach allows scientists to modify and correct their beliefs as new observations and information become available. The method has five stages:

- ▶ Making observations.
- ▶ Asking questions.
- ▶ Forming **hypotheses,** or tentative answers to the questions.
- ▶ Making predictions based on the hypotheses.
- ▶ Testing the predictions by making additional observations or conducting experiments.

The data gained may support or contradict the predictions being tested. If the data support the hypothesis, it is subjected to still more predictions and tests. If they continue to support it, confidence in its correctness increases, and the hypothesis comes to be considered a **theory**. If the data do not support the hypothesis, it is abandoned or modified in accordance with the new information. Then new predictions are made, and more tests are conducted.

Applying the hypothetico-deductive method

The way in which marine biologists answered the question "Why do amphipods crawl on the surface of the mud rather than staying hidden within?" illustrates the H-D approach. As we saw above, the biologists observed something occurring in nature and formulated a question about it. To begin answering the question, they assembled available information on amphipods and the species that eat them.

They learned that during July and August of each year, thousands of sandpipers assemble for four to six weeks on the mudflats of the Bay of Fundy, during their southward migration from their Arctic breeding grounds to their wintering areas in South America (Figure 1.12). On these mud-

1.12 Sandpipers Feed on Amphipods
Migrating sandpipers crowd the exposed tidal flats in search of food. By consuming infected amphipods, the sandpipers also become infected, serving as hosts and allowing the parasitic worm to complete its life cycle.

flats, which are exposed twice daily by the tides, they feed vigorously, putting on fat to fuel their next long flight. Amphipods living in the mud form about 85 percent of the diet of the sandpipers. Each bird may consume as many as 20,000 amphipods per day!

Previous observations had shown that a nematode (roundworm) parasitizes both the amphipods and the sandpipers. To complete its life cycle, the nematode must develop within both a sandpiper and an amphipod. The nematodes mature within the sandpipers' digestive tracts, mate, and release their eggs into the environment in the birds' feces. Small larvae hatch from the eggs and search for, find, and enter amphipods, where they grow through several larval stages. Sandpipers are reinfected when they eat parasitized amphipods.

GENERATING A HYPOTHESIS AND PREDICTIONS. Based on the available information, biologists generated the following hypothesis: *Nematodes alter the behavior of their amphipod hosts in a way that increases the chance that the worms will be*

1.13 Collecting Field Data
Amphipods are collected from the mud to be tested for infection by parasites. Some of these crustaceans will be used in laboratory experiments.

passed on to sandpiper hosts. From this general hypothesis they generated two specific predictions.

► First, they predicted that amphipods infected by nematodes would increase their activity on the surface of the mud during daylight hours, when the sandpipers hunted by sight, but not at night, when the sandpipers fed less and captured prey by probing into the mud.

► Second, they predicted that only amphipods with late-stage nematode larvae—the only stage that can infect sandpipers—would have their behavior manipulated by the nematodes.

For each hypothesis proposing an effect, there is a corresponding **null hypothesis**, which asserts that the proposed effect is absent. For the hypothesis we have just stated, the null hypothesis is that nematodes have no influence on the behavior of their amphipod hosts. The alternative predictions that would support the null hypothesis are (1) that infected amphipods show no increase their activity either during the day or at night and (2) that all larval stages affect their hosts in the same manner. It is important in hypothesis testing to generate and test as many alternate hypotheses and predictions as possible.

TESTING PREDICTIONS. Investigators collected amphipods in the field, taking them from the surface and from within the mud, during the day and at night (Figure 1.13). They found that during the day, amphipods crawling on the surface were much more likely to be infected with nematodes than were amphipods collected from within the mud. At night, however, there was no difference between the proportion of infected amphipods on the surface and those burrowing within the mud. This evidence supported the first prediction.

The field collections also showed that a higher proportion of the amphipods collected on the surface than of those collected from within the mud were parasitized by late-stage nematode larvae. However, amphipods crawling on the surface were no more likely to be infected by early-stage nematode larvae than were amphipods collected from the mud. These findings supported the *second* prediction.

To test the prediction that nematode larvae are more likely to affect amphipod behavior once they become infective, biologists performed laboratory experiments. They artificially infected amphipods with nematode eggs they obtained from sandpipers collected in the field. The infected amphipods established themselves in mud in laboratory containers.

By examining infected amphipods, investigators determined that it took about 13 days for the nematode larvae to reach the late, infective stage. By monitoring the behavior of the amphipods in the test tubes, the researchers determined that the amphipods were more likely to expose themselves on the surface of the mud once the parasites had reached the infective stage (Figure 1.14). This finding supported the second prediction.

Thus a combination of field and laboratory experiments, observation, and prior knowledge all supported the hypothesis that nematodes manipulate the behavior of their amphipod hosts in a way that decreases the survival of the amphipods, but increases the survival of the nematodes.

As is common practice in all the sciences, the researchers gathered all their data and collected them in a report, which they submitted to a scientific journal. Once such a report is published,* other scientists can evaluate the data, make their own observations, and formulate new ideas and experiments.

Experiments are powerful tools

The key feature of **experimentation** is the control of most factors so that the influence of a single factor can be seen clearly. In the laboratory experiments with amphipods, all individuals were raised under the same conditions. As a result, the nematodes reached the infective stage at about the same time in all of the infected amphipods.

Both laboratory and field experiments have their strengths and weaknesses. The advantage of working in a laboratory is that control of environmental factors is more

---

*In the case illustrated here, the data on amphipod behavior were published in the journal *Behavioral Ecology*, Volume 10, Number 4 (1998). D. McCurdy et al., "Evidence that the parasitic nematode *Skrjabinoclava* manipulates host *Corephium* behavior to increase transmission to the sandpiper, *Calidris pusilla*."

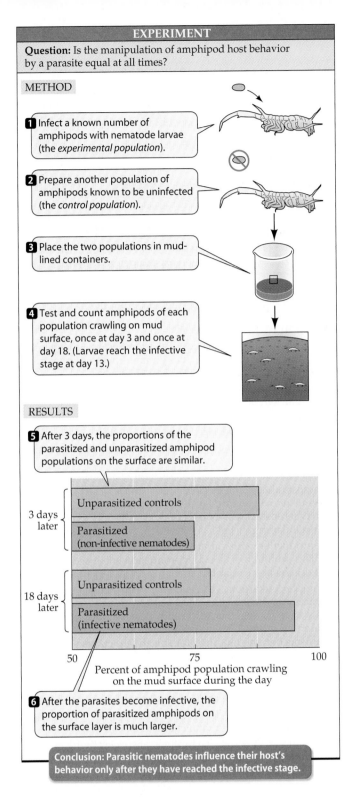

EXPERIMENT

Question: Is the manipulation of amphipod host behavior by a parasite equal at all times?

METHOD

1 Infect a known number of amphipods with nematode larvae (the *experimental population*).

2 Prepare another population of amphipods known to be uninfected (the *control population*).

3 Place the two populations in mud-lined containers.

4 Test and count amphipods of each population crawling on mud surface, once at day 3 and once at day 18. (Larvae reach the infective stage at day 13.)

RESULTS

5 After 3 days, the proportions of the parasitized and unparasitized amphipod populations on the surface are similar.

3 days later
- Unparasitized controls
- Parasitized (non-infective nematodes)

18 days later
- Unparasitized controls
- Parasitized (infective nematodes)

50   75   100

Percent of amphipod population crawling on the mud surface during the day

6 After the parasites become infective, the proportion of parasitized amphipods on the surface layer is much larger.

Conclusion: Parasitic nematodes influence their host's behavior only after they have reached the infective stage.

1.14 An Experiment Demonstrates that Parasites Influence Amphipod Behavior

Amphipods are more likely to crawl on the surface of the mud, exposing themselves to being captured by sandpipers, when their parasitic nematodes have reached the stage at which they can infect a sandpiper.

and field experiments are needed to test most hypotheses about what organisms do.

A single piece of supporting evidence rarely leads to widespread acceptance of a hypothesis. Similarly, a single contrary result rarely leads to abandonment of a hypothesis. Results that do not support the hypothesis being tested can be obtained for many reasons, only one of which is that the hypothesis is wrong. Incorrect predictions may have been made from a correct hypothesis. A negative finding can also result from poor experimental design, or because an inappropriate organism was chosen for the test. For example, a species of sandpiper that fed only by probing in the mud for its prey would have been an unsuitable subject for testing the hypothesis that nematodes alter their hosts in a way to make them more visible to predators.

Accepted scientific theories are based on many kinds of evidence

A general textbook like this one presents hypotheses and theories that have been extensively tested, using a variety of methods, and are generally accepted. When possible, we illustrate hypotheses and theories with observations and experiments that support them, but we cannot, because of space constraints, detail all the evidence. Remember as you read that statements of biological "fact" are mixtures of observations, predictions, and interpretations.

No amount of observation could possibly substitute for experimentation. However, this does not mean that scientists are insensitive to the welfare of the organisms with which they work. Most scientists who work with animals are continually alert to finding ways of getting answers that use the smallest number of experimental subjects and that cause the subjects the least pain and suffering.

Not all forms of inquiry are scientific

If you understand the methods of science, you can distinguish science from non-science. Recently some people have claimed that "creation science," sometimes called "scientific creationism," is a legitimate science that deserves to be taught in schools together with the evolutionary view of the world presented in this book. In spite of these claims, creation science is not science.

Science begins with observations and the formulation of hypotheses that can be tested and that will be rejected if significant contrary evidence is found. Creation science begins with the assertions, derived from religious texts, that Earth is only a few thousand years old and that all species of organisms were created in approximately their present forms. These assertions are not presented as a hypothesis

complete. Field experiments are more difficult because it is usually impossible to control more than a small number of environmental factors. But field experiments have one important advantage: Their results are more readily applicable to what happens where the organisms actually live and evolve. Just because an organism does something in the laboratory does not mean that it behaves the same way in nature. Because biologists usually wish to explain nature, not processes in the laboratory, combinations of laboratory from which testable predictions can be derived. Advocates of creation science assume their assertions to be true and that no tests are needed, nor are they willing to accept any evidence that refutes them.

In this chapter we have outlined the hypotheses that Earth is about 4 billion years old, that today's living organisms evolved from single-celled ancestors, and that many organisms dramatically different from those we see today lived on Earth in the remote past. The rest of this book will provide evidence supporting this scenario. To reject this view of Earth's history, a person must reject not only evolutionary biology, but also modern geology, astronomy, chemistry, and physics. All of this extensive scientific evidence is rejected or misinterpreted by proponents of "creation science" in favor of their particular religious beliefs.

Evidence gathered by scientific procedures does not diminish the value of religious accounts of creation. Religious beliefs are based on faith—not on falsifiable hypotheses, as science is. They serve different purposes, giving meaning and spiritual guidance to human lives. They form the basis for establishing values—something science cannot do. The legitimacy and value of both religion and science is undermined when a religious belief is presented as scientific evidence.

Biology and Public Policy

During the Second World War and immediately thereafter, the physical sciences were highly influential in shaping public policy in the industrialized world. Since then, the biological sciences have assumed increasing importance. One reason is the discovery of the genetic code and the ability to manipulate the genetic constitution of organisms. These developments have opened vast new possibilities for improvements in the control of human diseases and agricultural productivity. At the same time, these capabilities have raised important ethical and policy issues. How much, and in what ways, should we tinker with the genetics of people and other species? Does it matter whether organisms are changed by traditional breeding experiments or by gene transfers? How safe are genetically modified organisms in the environment and in human foods?

Another reason for the importance of the biological sciences is the vastly increased human population. Our use of renewable and nonrenewable natural resources is stressing the ability of the environment to produce the goods and services upon which society depends. Human activities are causing the extinction of a large number of species and are resulting in the spread of new human diseases and the resurgence of old ones. Biological knowledge is vital for determining the causes of these changes and for devising wise policies to deal with them.

Therefore, biologists are increasingly called upon to advise governmental agencies concerning the laws, rules, and regulations by which society deals with the increasing number of problems and challenges that have at least a partial biological basis. We will discuss these issues in many chapters of this book. You will see how the use of biological information can contribute to the establishment and implementation of wise public policies.

Chapter Summary

▶ If the history of Earth were a month with 30 days, recorded human history would occupy only the last 30 seconds. **Review Figure 1.1**

Organisms Have Changed over Billions of Years

▶ Evolution is the theme that unites all of biology. The idea of, and evidence for, evolution existed before Darwin. **Review Figure 1.2**

▶ The theory of evolution by natural selection rests on two simple observations and one inference from them.

Evolutionary Milestones

▶ Life arose from nonlife about 3.8 billion years ago when interacting systems of molecules became enclosed in membranes to form cells.

▶ All living organisms contain the same types of large molecules—carbohydrates, lipids, proteins, and nucleic acids.

▶ All organisms consist of cells, and all cells come from pre-existing cells. Life no longer arises from nonlife.

▶ A major theme in the evolution of life is the development of increasingly diverse ways of capturing external energy and using it to drive biologically useful reactions.

▶ Photosynthetic single-celled organisms released large amounts of oxygen into Earth's atmosphere, making possible the oxygen-based metabolism of large cells and, eventually, multicellular organisms.

▶ Reproduction with variation is a major characteristic of life. The evolution of sexual reproduction enhanced the ability of organisms to adapt to changing environments.

▶ Complex eukaryotic cells evolved when some large prokaryotes engulfed smaller ones. Eukaryotic cells evolved the ability to "stick together" after they divided, forming multicellular organisms. The individual cells of multicellular organisms became modified for specific functions within the organism.

▶ A major theme in the evolution of life is the development of increasingly complicated systems for responding to changes in the internal and external environments and for maintaining homeostasis.

▶ Regulated growth is a vital characteristic of life.

▶ Speciation resulted in the millions of species living on Earth today.

▶ Adaptation to environmental change is one of life's most distinctive features and is the result of evolution by natural selection.

The Hierarchy of Life

▶ Biology is organized into a hierarchy of levels from molecules to the biosphere. Each level has emergent properties that are not found at lower levels. **Review Figure 1.9**

▶ Species are classified into three domains: Archaea, Bacteria, and Eukarya. The domains Archaea and Bacteria consist of prokaryotic cells. The domain Eukarya contains the protists and the kingdoms Plantae, Fungi, and Animalia, all of which have eukaryotic cells. **Review Figure 1.10**

Asking and Answering "How?" and "Why?"

▶ Biologists ask two kinds of questions. "How" questions ask how organisms work. "Why" questions ask why they evolved to work that way.

▶ Both how and why questions are usually answered using a hypothetico-deductive (H-D) approach. Hypotheses are tentative answers to questions. Predictions are made on the basis of a hypothesis. The predictions are tested by observations and experiments, the results of which may support or refute the hypothesis. **Review Figure 1.13**

▶ Science is based on the formulation of testable hypotheses that can be rejected in light of contrary evidence. The acceptance on faith of already refuted, untested, or untestable assumptions is not science.

Biology and Public Policy

▶ Biologists are often called upon to advise governmental agencies on the solution of important problems that have a biological component.

For Discussion

1. According to the theory of evolution by natural selection, a species evolves certain features because they improve the chances that its members will survive and reproduce. There is no evidence, however, that evolutionary mechanisms have foresight or that organisms can anticipate future conditions. What, then, do biologists mean when they say, for example, that wings are "for flying"?

2. Why is it so important in science that we design and perform tests capable of rejecting a hypothesis?

3. One hypothesis about the manipulation of a host's behavior by a parasite was discussed in this chapter, and some tests of that hypothesis were described. Suggest some other hypotheses about the ways in which parasites might change the behavior and physiology of their hosts. Develop some critical tests for one of these alternatives. What are the appropriate associated null hypotheses?

4. Some philosophers and scientists believe that it is impossible to prove any scientific hypothesis—that we can only fail to find a reason to reject it. Evaluate this view. Can you think of reasons why we can be more certain about rejecting a hypothesis than about accepting it?

5. Discuss one current environmental problem whose solution requires the use of biological knowledge. How well is biology being used? What factors prevent scientific data from playing a more important role in finding a solution to the problem?

Part One

THE CELL

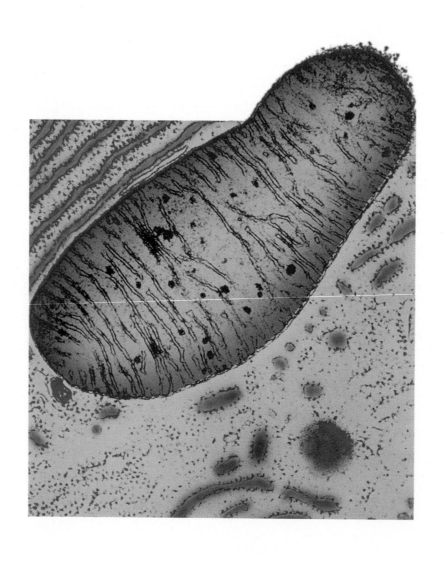

2 Small Molecules: Structure and Behavior

IN RECENT YEARS, SOME STARTLING DISCOVeries have shown that life can exist in places we never dreamed possible. There are organisms living in hot springs at temperatures above the boiling point of water, beneath the frozen Antarctic ice, 2 miles below Earth's surface, 3 miles below the surface of the sea, in extremely acid environments, in extremely salty conditions, and even inside nuclear reactors. Such findings have rekindled interest in astrobiology, the science of and search for life outside Earth.

The one absolute requirement for life is water. Without water to act as a solvent for biochemicals, to receive wastes, to absorb heat, and to participate directly in chemical reactions, life would not exist as we know it. With strong recent evidence that there was once flowing water on Mars, and that Europa (one of Jupiter's moons) may have a thin crust of ice with liquid water below it, there is great excitement about the possibility of life on nearby extraterrestrial bodies.

But what form would this life take? A major discovery of biology is that living things are composed of the same types of chemical elements as the vast nonliving portion of the universe. This *mechanistic* view—that life is chemically based and obeys universal physicochemical laws—is a relatively recent one in human history. The concept of a "vital force" responsible for life, different from the forces found in physics and chemistry, was common in Western culture until the nineteenth century, and many people still assume such a force exists. However, most scientists adhere to a mechanistic view of life.

Before describing how chemical elements are arranged in living creatures, we examine some fundamental chemical concepts. The first part of this chapter will address the constituents of matter: atoms. We examine their variety, their properties, and their capacity to combine with other atoms. Then we consider how matter changes. In addition to changes in state (solid to liquid to gas), substances undergo changes that transform both their composition and their characteristic properties. Then we return to a consideration of the structure and properties of water and its relationship to acids and bases. We close with a consideration of characteristic groups of atoms that contribute specific properties to larger molecules of which they are part, and which will be the subject of Chapter 3.

Atoms: The Constituents of Matter

More than a trillion (10^{12}) atoms could fit over the period at the end of this sentence. Each atom consists of a dense, positively charged nucleus, around which one or more negatively charged electrons move. The nucleus contains one or more protons and may contain one or more neutrons. Atoms and their component particles have **mass**, a property of all matter. Mass measures the quantity of matter present; the greater the mass, the greater the quantity of matter.

The mass of a proton serves as a standard unit of measure: the *atomic mass unit* (amu), or *dalton* (named after the English chemist John Dalton). A single proton or neutron has a mass of about 1 dalton, which is 1.7×10^{-24} grams (0.0000000000000000000000017 g). The mass of an electron is 9×10^{-28} g (0.0005 dalton). Because the mass of an electron is so much less than the mass of a proton or a neutron, the contribution of electrons to the mass of an atom can usually be ignored.

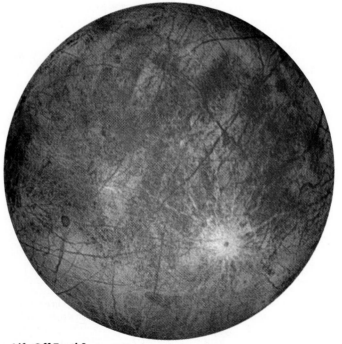

Life Off Earth?
Orbiting 400,000 miles above the giant planet Jupiter, Europa has a surface of water ice, possibly covering a slushy ocean. Where there is water, there could be, or could have been, life.

Even the smallest atoms, such as helium, have measurable mass:

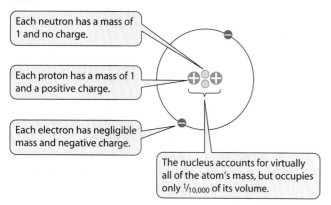

Each neutron has a mass of 1 and no charge.

Each proton has a mass of 1 and a positive charge.

Each electron has negligible mass and negative charge.

The nucleus accounts for virtually all of the atom's mass, but occupies only $1/10,000$ of its volume.

The positive electric charge of a proton is defined as a unit of charge. An electron has a negative charge equal and opposite to that of a proton. Thus the charge of a proton is +1 unit, and that of an electron is –1 unit. Unlike charges (+/–) attract each other; like charges (+/+ or –/–) repel each other. The neutron, as its name suggests, is electrically neutral, so its charge is 0 unit. When the number of protons in an atom equals the number of electrons, the atom is electrically neutral. An atom with more or fewer electrons than protons has an electric charge and is called an ion; we will discuss ions in detail later in the chapter.

An element is made up of only one kind of atom

An **element** is a pure substance that contains only one type of atom. The element hydrogen consists only of hydrogen atoms; the element iron consists only of iron atoms. The atoms of each element have certain characteristics or properties that distinguish them from the atoms of other elements. The more than 100 elements found in the universe are arranged in the periodic table (Figure 2.1). These elements are not found in equal amounts. Earth's crust is half oxygen; 28% silicon; 8% aluminum; 3–5% each of sodium, magnesium, potassium, calcium, and iron; and much smaller amounts of the other elements.

About 98% of the mass of every living organism (bacterium, turnip, or human) is composed of just six elements: carbon, hydrogen, nitrogen, oxygen, phosphorus, and sulfur. Other elements are present in small amounts. The chemistry of the six major elements will be our primary concern here, but the others are not unimportant. Sodium and potassium, for example, are essential for nerves to function; calcium can act as a biological signal; iodine is a component of a vital hormone; and plants need molybdenum in order to incorporate nitrogen into biologically useful substances.

2.1 The Periodic Table
The periodic table groups the elements according to their physical and chemical properties.

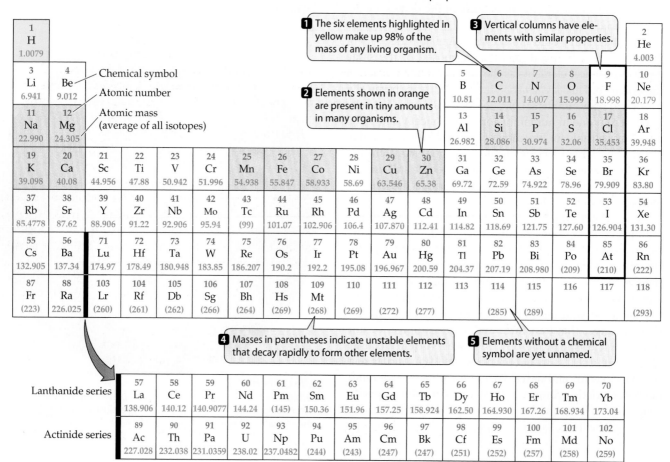

1 The six elements highlighted in yellow make up 98% of the mass of any living organism.

2 Elements shown in orange are present in tiny amounts in many organisms.

3 Vertical columns have elements with similar properties.

4 Masses in parentheses indicate unstable elements that decay rapidly to form other elements.

5 Elements without a chemical symbol are yet unnamed.

Chemical symbol
Atomic number
Atomic mass (average of all isotopes)

The number of protons identifies the element

An element is distinguished from other elements by the number of protons in each of its atoms. This number, which does not change, is called the **atomic number**. An atom of hydrogen contains 1 proton, a helium atom has 2 protons, carbon has 6 protons, and plutonium has 94 protons. The atomic numbers of these elements are thus 1, 2, 6, and 94, respectively.

Every element except hydrogen has one or more neutrons in its nucleus. The **mass number** of an atom equals the total number of protons and neutrons in its nucleus. Because the mass of an electron is infinitesimal compared with that of a neutron or proton, electrons are ignored in calculating the mass number. The nucleus of a helium atom contains 2 protons and 2 neutrons; oxygen has 8 protons and 8 neutrons. Helium, therefore, has a mass number of 4 and oxygen a mass number of 16. The mass number may be thought of as the mass of the atom in daltons.

Each element has its own one- or two-letter chemical symbol. For example, H stands for hydrogen, He for helium, and O for oxygen. Some symbols come from other languages: Fe (from the Latin *ferrum*) stands for iron, Na (Latin *natrium*) for sodium, and W (German *Wolfram*) for tungsten. The periodic table (see Figure 2.1) gives the symbols for the 92 natural elements, as well as showing 26 elements (elements 93–118) that have been synthesized in laboratories but have not been found in nature.

In text, the atomic number and mass number of an element are written to the left of the element's symbol:

Mass number 12 C Symbol of element
Atomic number 6

Thus, hydrogen, carbon, and oxygen are written as $^{1}_{1}H$, $^{12}_{6}C$, and $^{16}_{8}O$, respectively.

Isotopes differ in number of neutrons

We have been speaking of elements as if each had only one atomic form, but this is not true. **Isotopes** of the same element all have the same number of protons, but differ in the number of neutrons in the atomic nucleus (Figure 2.2).

In nature, many elements exist as several isotopes. For example, the natural isotopes of carbon are ^{12}C, ^{13}C, and ^{14}C. Unlike the isotopes of hydrogen, which have special names (see Figure 2.2), the isotopess of most elements do not have distinct names. Rather, they are written in the form shown above and are referred to as carbon-12, carbon-13, and carbon-14, respectively. Most carbon atoms are ^{12}C, about 1.1 percent are ^{13}C, and a tiny fraction are ^{14}C. An element's **atomic mass**, or **atomic weight**,* is the average of the mass numbers of a representative sample of atoms of the element, with all isotopes in their normally occurring

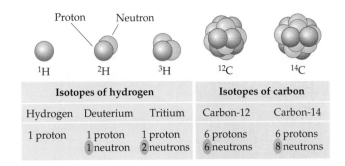

Isotopes of hydrogen			Isotopes of carbon	
Hydrogen	Deuterium	Tritium	Carbon-12	Carbon-14
1 proton	1 proton 1 neutron	1 proton 2 neutrons	6 protons 6 neutrons	6 protons 8 neutrons

2.2 Isotopes Have Different Numbers of Neutrons
Deuterium and tritium are rare isotopes of hydrogen. Unlike these two isotopes, isotopes of other elements do not have have distinct names. Carbon-12 is the most common isotope of carbon; carbon-14 is a rare form.

proportions. The atomic weight of carbon is thus calculated to be 12.011.

Some isotopes, called **radioisotopes**, are unstable and spontaneously give off energy as α (alpha), β (beta), or γ (gamma) radiation from the atomic nucleus. Such radioactive decay transforms the original atom into another atom, usually of another element. For example, carbon-14 loses a beta particle (actually an electron) to form nitrogen-14. Biologists and physicians can incorporate radioisotopes into molecules and use the emitted radiation as a tag to locate those molecules or to identify changes that the molecules undergo inside the body (Figure 2.3). Three radioisotopes commonly used in this way are ^{3}H (tritium), ^{14}C (carbon-14), and ^{32}P (phosphorus-32). In addition to these applications, radioisotopes can be used to date fossils (see Chapter 20).

Although radioisotopes are useful for experiments and in medicine, even low doses of their radiation have the potential to damage molecules and cells. Gamma radiation from cobalt-60 (^{60}Co) is used medically to damage or kill rapidly dividing cancer cells.

Electron behavior determines chemical bonding

When considering atoms, biologists are concerned primarily with electrons because the behavior of electrons explains how chemical changes occur in living cells. These changes, called **chemical reactions** or just **reactions**, are changes in the atomic composition of substances. The characteristic number of electrons in each atom of an element determines how its atoms react with other atoms. All chemical reactions involve changes in the relationships of electrons with one another.

The location of a given electron in an atom at any given time is impossible to determine. We can only describe a volume of space within the atom where the electron is likely to be. The region of space where the electron is found at least

---

*The concepts of "weight" and "mass" are not identical. Weight is the measure of the Earth's gravitational attraction for mass; on another planet, the same quantity of mass would have a different weight. On Earth, however, the term "weight" is often used as a measure of mass, and in biology one encounters the terms "weight" and "atomic weight" more frequently than "mass" and "atomic mass." Therefore, we will use "weight" for the remainder of this book.

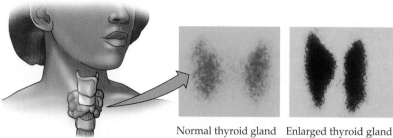

Normal thyroid gland   Enlarged thyroid gland

2.3 A Radioisotope Used in Medicine
The thyroid gland takes up iodine and uses it in the synthesis of thyroid hormone. A patient suspected of having thyroid disease is injected with radioactive iodine, which allows the thyroid gland to be visualized by a scanning device.

90 percent of the time is the electron's **orbital** (Figure 2.4). In an atom, a given orbital can be occupied by at most two electrons. Thus any atom larger than helium (atomic number 2) must have electrons in two or more orbitals. As Figure 2.4 shows, the different orbitals have characteristic forms and orientations in space.

The orbitals in turn constitute a series of **electron shells**, or energy levels, around the nucleus (Figure 2.5). The first, or innermost, electron shell consists of only one orbital, called an *s* orbital. Hydrogen ($_1$H) has one electron in its first shell; helium ($_2$He) has two. All other elements have two first-shell electrons, as well as electrons in other shells.

The second shell is made up of four orbitals (an *s* orbital and three *p* orbitals) and hence can hold up to eight electrons. The *s* orbitals fill with electrons first, and their electrons have the lowest energy. Subsequent shells have different numbers of orbitals, but the outermost shells usually hold only eight electrons.

In any atom, the outermost electron shell determines how the atom combines with other atoms; that is, how an atom behaves chemically. When an outermost shell consisting of four orbitals contains eight electrons, there are no unpaired electrons (see Figure 2.5). Such an atom is stable and will not react with other atoms. Examples of chemically inert elements are helium, neon, and argon.

The atoms of chemically reactive elements seek to attain the stable condition of having no unpaired electrons in their outer shells. They attain this stability by sharing electrons with other atoms, or by gaining or losing one or more electrons from their outermost shells. When they share electrons, atoms are bonded together. Such bonds create stable associations of atoms called molecules.

A **molecule** can be defined as two or more atoms linked by chemical bonds. The tendency of atoms in stable molecules to have eight electrons in their outermost shell is known as the *octet rule*. Many atoms in biologically important molecules—for example, carbon (C) and nitrogen (N)—follow the octet rule. However, some biologically important atoms are exceptions to the rule. Hydrogen (H) is an obvious exception, attaining stability when only two electrons occupy its single shell.

Chemical Bonds: Linking Atoms Together

A **chemical bond** is an attractive force that links two atoms to form a molecule. There are several kinds of chemical bonds (Table 2.1). In this section, we first discuss covalent bonds, the strong bonds that result from the sharing of electrons. Then we examine other kinds of interactions, including hydrogen bonds, that are weaker than covalent bonds but enormously important to biology. Finally, we consider ionic bonding, which results as a consequence of the loss or gain of electrons by atoms.

Covalent bonds consist of shared pairs of electrons

When two atoms attain stable electron numbers in their outer shells by sharing one or more pairs of electrons, a **covalent bond** forms. Consider two hydrogen atoms in close proximity, each with a single unpaired electron in the outer shell. Each positively charged nucleus exerts some attraction on the other atom's un-

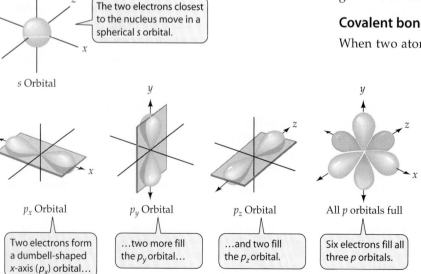

s Orbital

The two electrons closest to the nucleus move in a spherical *s* orbital.

p_x Orbital

Two electrons form a dumbell-shaped *x*-axis (p_x) orbital...

p_y Orbital

...two more fill the p_y orbital...

p_z Orbital

...and two fill the p_z orbital.

All *p* orbitals full

Six electrons fill all three *p* orbitals.

2.4 Electron Orbitals
Each orbital holds a maximum of two electrons. The *s* orbitals have a lower energy level and fill with electrons before the *p* orbitals do.

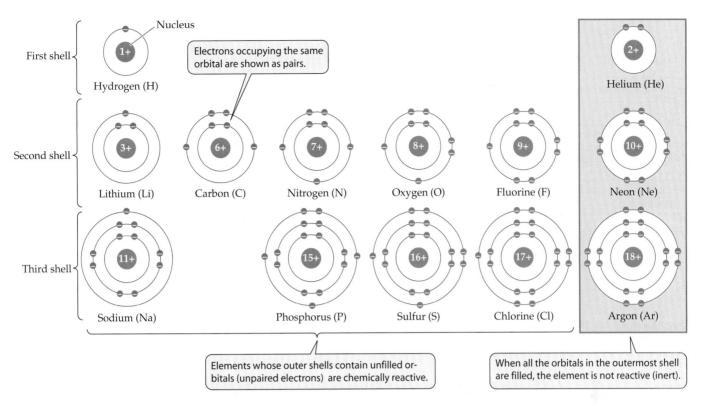

Elements whose outer shells contain unfilled orbitals (unpaired electrons) are chemically reactive.

When all the orbitals in the outermost shell are filled, the element is not reactive (inert).

2.5 Electron Shells Determine the Reactivity of Atoms

Each orbital holds a maximum of two electrons, and each shell can hold a specific maximum number of electrons. Each shell must be filled before electrons move into the next shell. The energy level of electrons is higher in shells farther from the nucleus. An atom with unpaired electrons in its outermost shell may react (bond) with other atoms.

paired electron, but this attraction is balanced by each electron's attraction to its own nucleus. So the two unpaired electrons become shared by both atoms, filling the outer shells of both of them (Figure 2.6).

A carbon atom has a total of six electrons; two electrons fill its inner shell and four are in its outer shell. Because the outer shell can hold up to eight electrons, this atom can share electrons with up to four other atoms. Thus it can

2.1 Chemical Bonds and Interactions

NAME	BASIS OF INTERACTION	STRUCTURE	BOND ENERGY[a] (KCAL/MOL)
Covalent bond	Sharing of electron pairs		50–110
Hydrogen bond	Sharing of H atom		3–7
Ionic interaction	Attraction of opposite charges		3–7
van der Waals interaction	Interaction of electron clouds		1
Hydrophobic interaction	Interaction of nonpolar substances		1–2

[a]*Bond energy* is the amount of energy needed to separate two bonded or interacting atoms under physiological conditions.

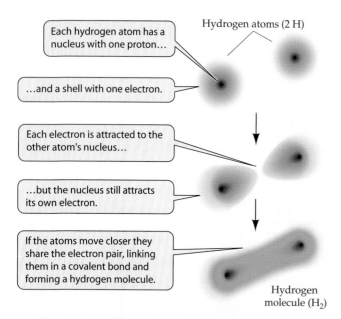

2.6 Electrons Are Shared in Covalent Bonds
Two hydrogen atoms combine to form a hydrogen molecule. Each electron is attracted to both protons. A covalent bond forms when the electron orbitals of the two atoms overlap.

form four covalent bonds. When an atom of carbon reacts with four hydrogen atoms, a substance called methane (CH_4) forms (Figure 2.7a and b). Thanks to electron sharing, the outer shell of methane's carbon atom is filled with eight electrons, and the outer shell of each hydrogen atom is also filled. Thus four covalent bonds—each consisting of a shared pair of electrons—hold methane together. Table 2.2 shows the covalent bonding capacities of some biologically significant elements.

ORIENTATION OF COVALENT BONDS. Covalent bonds are very strong. The thermal energy that biological molecules ordinarily have at body temperature is less than 1 percent of that needed to break covalent bonds. So biological molecules, most of which are put together with covalent bonds, are quite stable. A second property of covalent bonds is that, for a given pair of atoms, they are the same in length, angle, and direction, regardless of the larger molecule of which the particular bond is a part. The four filled orbitals around the carbon nucleus of methane, for example, distribute themselves in space so that the bonded hydrogens are directed to the corners of a regular tetrahedron with carbon in the center (Figure 2.7c). This three-dimensional structure of carbon and hydrogen is the same in complicat-

2.7 Covalent Bonding with Carbon
Different representations of covalent bond formation in methane (CH_4). (a) Diagram illustrating the filling and stabilizing of the outer electron shells in carbon and hydrogen atoms. (b) Two common ways of representing bonds. (c) The spatial orientation of methane's bonds, represented in two ways.

	USUAL NUMBER OF
ELEMENT	**COVALENT BONDS**
Hydrogen (H)	1
Oxygen (O)	2
Sulfur (S)	2
Nitrogen (N)	3
Carbon (C)	4
Phosphorus (P)	5

2.2 Covalent Bonding Capabilities of Some Biologically Important Elements

ed proteins as it is in the simple methane molecule. It makes the prediction of biological structure possible.

Although the orientation of orbitals and the shapes of molecules differ depending on the kinds of atoms involved and how they are linked together, it is essential to remember that all molecules occupy space and have three-dimensional shapes. The shapes of molecules contribute to their biological functions, as we will see in Chapter 3.

MULTIPLE COVALENT BONDS. A covalent bond is represented by a line between the chemical symbols for the atoms. A bond in which a single pair of electrons is shared is called a

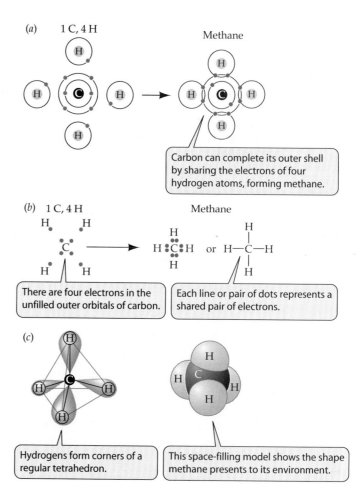

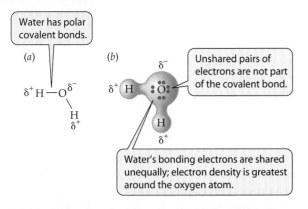

Water has polar covalent bonds.

Unshared pairs of electrons are not part of the covalent bond.

Water's bonding electrons are shared unequally; electron density is greatest around the oxygen atom.

2.8 The Polar Covalent Bond in the Water Molecule
(*a*) A covalent bond between atoms with different electronegativities is a polar covalent bond, and has partial (δ) charges at the ends. (*b*) In water, the electrons are displaced toward the oxygen atom and away from the hydrogen atoms.

single bond (for example, H—H, C—H). When four electrons (two pairs) are shared, the link is called a double bond (C=C). In the gas ethylene ($H_2C=CH_2$), two carbon atoms share two pairs of electrons. Triple bonds (six shared electrons) are rare, but there is one in nitrogen gas (N≡N), the chief component of the air we breathe. In the covalent bonds in these five examples, the electrons are shared more or less equally between the nuclei; consequently, all regions of the bonds are identical.

UNEQUAL SHARING OF ELECTRONS. If two atoms of the same element are covalently bonded, there is an equal sharing of the pair(s) of electrons in the outer shell. However, when the two atoms are of different elements, the sharing is not necessarily equal. One nucleus may exert a greater attractive force on the electron pair than the other nucleus, and so the pair tends to be closer to that atom.

The attractive force that an atom exerts on electrons is its **electronegativity**. It depends on how many positive charges a nucleus has (nuclei with more protons are more positive and thus more attractive to electrons) and how far away the electrons are from the nucleus (closer means more electronegativity). *The closer two atoms are in electronegativity, the more equal their sharing of electrons will be.*

Table 2.3 shows the electronegativities of some elements important in biological systems. Looking at the table, it is

obvious that two oxygen atoms, both with electronegativity of 3.5, will share electrons equally in a covalent bond. So will two hydrogen atoms (both with 2.1). But when hydrogen bonds with oxygen to form water, the electrons involved are *unequally* shared: They tend to be nearer to the oxygen nucleus because it is the more electronegative of the two. The result is called a *polar covalent bond* (Figure 2.8).

Because of this unequal sharing of electrons, the oxygen end of the hydrogen–oxygen bond has a slightly negative charge (symbolized δ⁻ and spoken as "delta negative," meaning a partial unit of charge), and the hydrogen end is slightly positive (δ⁺). The bond is polar because these opposite charges are separated at the two ends of the bond. The partial charges that result from polar covalent bonds produce **polar molecules** or polar regions of large molecules. Polar bonds greatly influence the interactions between molecules that contain them.

Hydrogen bonds may form between molecules

In liquid water, the negatively charged oxygen (δ⁻) atom of one water molecule is attracted to the positively charged hydrogen (δ⁺) atoms of another water molecule. (Remember, negative charges attract positive charges.) The bond resulting from this attraction is called a **hydrogen bond**.

Hydrogen bonds are not restricted to water molecules. They may form between an electronegative atom and a hydrogen covalently bonded to a different electronegative atom (Figure 2.9).

A hydrogen bond is a weak bond; it has about one-tenth (10%) of the strength of a covalent bond between a hydrogen atom and an oxygen atom (see Table 2.1). However, where many hydrogen bonds form, they have considerable strength and greatly influence the structure and properties of substances. Later in this chapter we'll see how hydrogen bonding in water contributes to many of the properties that make water significant for living systems. Hydrogen bonds also play important roles in determining and maintaining the three-dimensional shapes of giant molecules such as DNA and proteins (see Chapter 3).

2.3	Some Electronegativities	
ELEMENT		**ELECTRONEGATIVITY**
Oxygen (O)		3.5
Chlorine (Cl)		3.1
Nitrogen (N)		3.0
Carbon (C)		2.5
Phosphorus (P)		2.1
Hydrogen (H)		2.1
Sodium (Na)		0.9
Potassium (K)		0.8

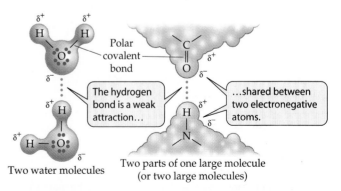

Polar covalent bond

The hydrogen bond is a weak attraction…

…shared between two electronegative atoms.

Two water molecules

Two parts of one large molecule (or two large molecules)

2.9 Hydrogen Bonds Can Form Between or within Molecules
Hydrogen bonds can form between two molecules or, if a molecule is large, between two different parts of the same molecule. Covalent and polar covalent bonds, on the other hand, are always found within molecules.

Ions form bonds by electrical attraction

When one interacting atom is much more electronegative than the other, a complete transfer of one or more electrons may take place. Consider sodium (electronegativity 0.9) and chlorine (3.1). A sodium atom has only one electron in its outermost shell; this condition is unstable. A chlorine atom has seven electrons in its outer shell—another unstable condition. Since the electronegativities of these elements are so different, any electrons involved in bonding will tend to be much nearer to the chlorine nucleus—so near, in fact, that there is a complete transfer of the electron from one element to the other (Figure 2.10). This reaction between sodium and chlorine makes both atoms more stable. The result is two ions. **Ions** are electrically charged particles that form when atoms gain or lose one or more electrons.

▶ The sodium ion (Na^+) has a +1 unit charge because it has one less electron than it has protons. The outermost electron shell of the sodium ion is full, with eight electrons, so the ion is stable. Positively charged ions are called **cations**.

▶ The chloride ion (Cl^-) has a −1 unit charge because it has one more electron than it has protons. This additional electron gives Cl^- an outer shell with a stable load of eight electrons. Negatively charged ions are called **anions**.

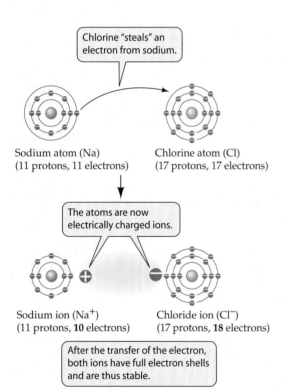

Chlorine "steals" an electron from sodium.

Sodium atom (Na)
(11 protons, 11 electrons)

Chlorine atom (Cl)
(17 protons, 17 electrons)

The atoms are now electrically charged ions.

Sodium ion (Na^+)
(11 protons, **10** electrons)

Chloride ion (Cl^-)
(17 protons, **18** electrons)

After the transfer of the electron, both ions have full electron shells and are thus stable.

2.10 Formation of Sodium and Chloride Ions
When a sodium atom reacts with a chlorine atom, the more electronegative chlorine acquires a more stable, filled outer shell by obtaining an electron from the sodium. In so doing, the chlorine atom becomes a negatively charged chloride ion (Cl^-). The sodium atom, upon losing the electron, becomes a positively charged sodium ion (Na^+).

Some elements form ions with multiple charges by losing or gaining more than one electron. Examples are Ca^{2+} (calcium ion, created from a calcium atom that has lost two electrons) and Mg^{2+} (magnesium ion). Two biologically important elements each yield more than one stable ion: Iron yields Fe^{2+} (ferrous ion) and Fe^{3+} (ferric ion), and copper yields Cu^+ (cuprous ion) and Cu^{2+} (cupric ion). Groups of covalently bonded atoms that carry an electric charge are called **complex ions**; examples include NH_4^+ (ammonium ion), SO_4^{2-} (sulfate ion), and PO_4^{3-} (phosphate ion).

The charge from an ion radiates from it in all directions. Once they form, ions are usually stable, and no more electrons are lost or gained. Ions can form stable bonds, resulting in stable solid compounds such as sodium chloride (NaCl) and potassium phosphate (K_3PO_4).

Ionic bonds are bonds formed by electrical attractions between ions bearing opposite charges. In sodium chloride—familiar to us as table salt—cations and anions are held together by ionic bonds. In solids, the ionic bonds are strong because the ions are close together. However, when ions are dispersed in water, the distance between them can be large; the strength of their attraction is thus greatly reduced. Under the conditions that exist in the cell, an ionic attraction is less than one-tenth as strong as a covalent bond that shares electrons equally (see Table 2.1).

Not surprisingly, ions with one or more units of charge can interact with polar molecules as well as with other ions. Such interaction results when table salt, or any other ionic solid, dissolves in water: "Shells" of water molecules surround the individual ions, separating them (Figure 2.11). The hydrogen bond that we described earlier is a type of ionic bond, because it is formed by electrical attractions. However, it is weaker than most ionic bonds because the hydrogen bond is formed by partial charges (δ^+ and δ^-) rather than by whole-unit charges (+1 unit, −1 unit).

Polar and nonpolar substances interact best among themselves

"Like attracts like" is an old saying, and nowhere is it more true than in polar and nonpolar molecules, which tend to interact with their own kind. Just as water molecules interact with one another through their polarity-induced hydrogen bonds, any molecule that is itself polar will interact with other polar molecules by weak (δ^+ to δ^-) attractions in hydrogen bonds. If a polar molecule interacts with water in this way, it is called *hydrophilic* ("water-loving").

What about nonpolar molecules? For example, carbon (electronegativity 2.5) forms nonpolar bonds with hydrogen (electronegativity 2.1). The resulting *hydrocarbon* molecule—ethane—is nonpolar (Figure 2.12), and in water it will tend to aggregate with other nonpolar molecules rather than with polar water. Such molecules are called *hydrophobic* ("water-hating"), and the interactions between them are hydrophobic interactions. It is important to realize that hydrophobic substances do not really "hate" water; they can form weak interactions with it (recall that the electronegativities of carbon and hydrogen are not exactly the same).

But these interactions are far weaker than the hydrogen bonds between the water molecules, and so the nonpolar substances keep to themselves.

These weak interactions between nonpolar substances are enhanced by **van der Waals forces**, which occur when two atoms are in close proximity. These forces result from random variations in the electron distribution in one mole-

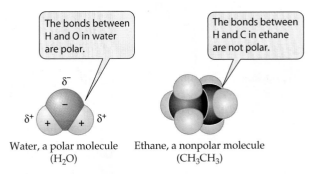

Water, a polar molecule (H_2O) Ethane, a nonpolar molecule (CH_3CH_3)

2.12 Polar and Nonpolar Molecules
Because the hydrocarbon ethane is nonpolar, it does not interact with water, but tends to interact with other nonpolar substances.

cule, which create an opposite charge distribution in the adjacent molecule. The result is a brief, weak attraction. Although each such interaction is brief and weak at any one site, the summation of many such interactions over the entire span of a large nonpolar molecule can produce substantial attraction. van der Waals forces are important in maintaining the structures of many biologically important substances.

Chemical Reactions: Atoms Change Partners

A **chemical reaction** occurs when atoms combine or change bonding partners. Consider the combustion reaction that takes place in the flame of a propane stove. When propane (C_3H_8) reacts with oxygen gas (O_2), the carbon atoms become bonded to oxygen atoms instead of to hydrogen atoms, and the hydrogen atoms become bonded to oxygen instead of carbon (Figure 2.13). As the covalently bonded atoms change partners, the composition of the matter changes, and propane and oxygen gas become carbon dioxide and water. This chemical reaction can be represented by the balanced equation

$$C_3H_8 + 5\,O_2 \rightarrow 3\,CO_2 + 4\,H_2O$$

In this equation, the propane and oxygen are the **reactants**, and the carbon dioxide and water are the **products**. In this case, the reaction is complete: All the propane and oxygen are used up in forming the two products. The arrow symbolizes the chemical reaction. The numbers preceding the molecular formulas balance the equation and indicate how many molecules are used or are produced.

In this and all other chemical reactions, matter is neither created nor destroyed. The total number of carbons on the left equals the total number on the right. However, there is another product of this reaction: energy. The heat of the stove's flame and its blue light reveal that the reaction of propane and oxygen releases a great deal of energy. **Energy** is defined as the capacity to do work, but on a more intuitive level, it can be thought of as the capacity for change. Chemical reactions do not create or destroy energy, but *changes* in energy usually accompany chemical reactions.

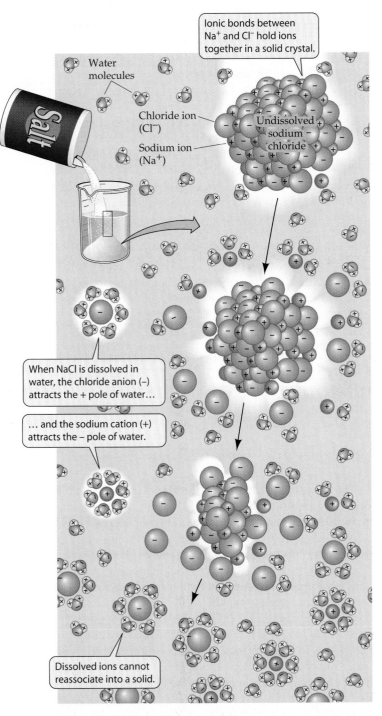

2.11 Water Molecules Surround Ions
When an ionic solid dissolves in water, polar water molecules cluster around cations or anions, blocking their reassociation into a solid and forming a solution.

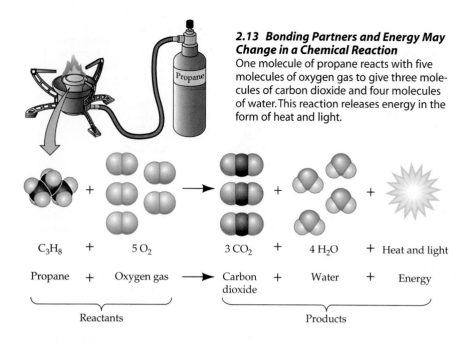

2.13 Bonding Partners and Energy May Change in a Chemical Reaction
One molecule of propane reacts with five molecules of oxygen gas to give three molecules of carbon dioxide and four molecules of water. This reaction releases energy in the form of heat and light.

C_3H_8 + $5 O_2$ → $3 CO_2$ + $4 H_2O$ + Heat and light

Propane + Oxygen gas → Carbon dioxide + Water + Energy

Reactants ———— Products

In the reaction between propane and oxygen, the energy that was released as heat and light was already present in the reactants in another form, called *potential energy*. In some chemical reactions, energy must be supplied from the environment (for example, some substances will react only after being heated), and some of this supplied energy becomes stored as potential chemical energy in the bonds formed in the products.

We can measure the energy associated with chemical reactions using a unit called a **calorie (cal)**. A calorie* is the amount of heat energy needed to raise the temperature of 1 gram of pure water from 14.5°C to 15.5°C. Another unit of energy that is increasingly used is the **joule (J)**. When you compare data on energy, always compare joules to joules and calories to calories. The two units can be interconverted: 1 J = 0.239 cal, and 1 cal = 4.184 J. Thus, for example, 486 cal = 2,033 J, or 2.033 kJ. Although defined in terms of heat, the calorie and the joule are measures of any form of energy—mechanical, electric, or chemical.

Within living cells, chemical reactions called *oxidation–reduction reactions* take place. These biological reactions have much in common with the combustion of propane. The fuel is different (the sugar glucose, rather than propane), and the reactions proceed by many intermediate steps that permit the energy released from the glucose to be harvested and put to use by the cell. But the products are the same: carbon dioxide and water.

We will present and discuss energy changes, oxidation–reduction reactions, and several other types of chemical reactions that are prevalent in living systems in the chapters that follow.

*The nutritionist's or dieter's Calorie, with a capital C, is what biologists call a kilocalorie (kcal) and is equal to 1,000 heat-energy calories.

Water: Structure and Properties

Water, like all other matter, can exist in three states: solid (ice), liquid, and gas (vapor) (Figure 2.14). Liquid water is the medium in which life originated on Earth more than 3.8 billion years ago, and it is in water that life evolved for its first billion years. Today, water covers three-fourths of Earth's surface, and the bodies of all active organisms contain between 45 and 95 percent water.

No organism can remain biologically active without water. Within cells, water participates directly in many chemical reactions, and it is the medium (or solvent) in which most biological reactions take place. In this section we will consider the structure and interactions of water molecules, exploring how these generate properties essential to life.

Water has a unique structure and special properties

Each water molecule is composed of one oxygen atom bonded to two hydrogen atoms (H_2O). In the molecule, the four pairs of electrons in the outer shell of oxygen repel each other, producing a tetrahedral shape:

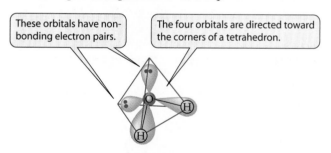

These orbitals have non-bonding electron pairs.

The four orbitals are directed toward the corners of a tetrahedron.

The shape of the water molecule, its polar nature, and its capacity to form hydrogen bonds give water its unusual properties. For example, ice floats, and compared with other liquids, water is an excellent solvent, making it an ideal medium for biochemical reactions. Water is both cohesive (sticking to itself) and adhesive (sticking to other things). And the energy changes that accompany its transitions from solid to liquid to gas are significant in living systems.

ICE FLOATS. In its solid state (ice), water is held by its hydrogen bonds in a rigid, crystalline structure in which each water molecule is hydrogen-bonded to four others (Figure 2.15a). Although these molecules are held firmly in place, they are not as tightly packed as they are in liquid water (Figure 2.15b). In other words, *solid water is less dense than liquid water*, which is why ice floats in water.

If ice sank in water, as almost all other solids do in their corresponding liquids, ponds and lakes would freeze from

2.14 Water: Solid and Liquid
Solid water from a glacier floats in its liquid form. The clouds are also water, but not in its gaseous phase: They are composed of fine drops of liquid water.

the bottom up, becoming solid blocks of ice in winter and killing most of the organisms living in them. Once the whole pond had frozen, its temperature could drop well below the freezing point of water. However, because ice floats, it forms a protective insulating layer on the top of the pond, reducing heat flow to the cold air above. Thus fish, plants, and other organisms in the pond are not subjected to temperatures lower than 0°C, the freezing point of pure water.

MELTING AND FREEZING. Compared with other nonmetallic substances of the same size, molecular ice requires a great

deal of heat energy to melt. Melting 1 mole (a standard quantity—6.02×10^{23}; see page 28) of water molecules requires the addition of 5.9 kJ of energy. This value is high because more than a mole of hydrogen bonds must be broken for 1 mole of water to change from solid to liquid. In the opposite process, freezing, a great deal of energy must be lost for water to transform from liquid to solid. These properties help make water a moderator of temperature changes.

HEAT AND COOLING. Another property of water that moderates temperature is the high heat capacity of liquid water. The **specific heat** of a substance is the amount of heat energy required to raise the temperature of 1 gram of that substance by 1°C. Raising the temperature of liquid water takes a relatively large amount of heat because much of the heat energy is used to break the hydrogen bonds that hold the liquid together. Compared with other small molecules that are liquids, water has a high specific heat. This phenomenon contributes to the surprising constancy of the temperature of the oceans and other large bodies of water through the seasons of the year. The temperature changes of coastal land masses are also moderated by large bodies of water. Indeed, water helps minimize variations in atmospheric temperature throughout the planet.

EVAPORATION AND COOLING. Water also has a high **heat of vaporization**, which means that a lot of heat is required to change water from its liquid state to its gaseous state (the process of evaporation). This heat is absorbed from the environment in contact with the water. Once again, much of the heat energy is used to break hydrogen bonds. Evaporation thus has a cooling effect on the environment—whether a leaf, a forest, or an entire land mass. This effect

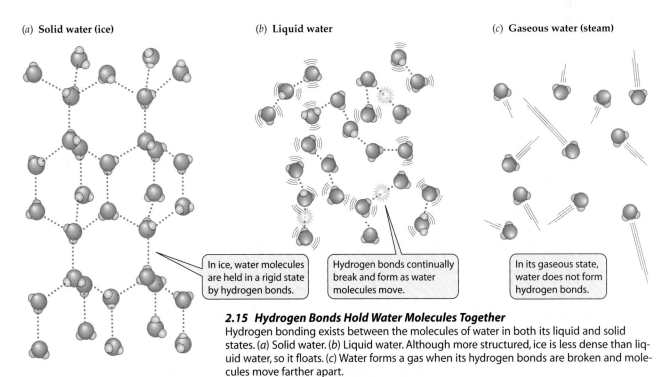

(a) **Solid water (ice)**   (b) **Liquid water**   (c) **Gaseous water (steam)**

In ice, water molecules are held in a rigid state by hydrogen bonds.

Hydrogen bonds continually break and form as water molecules move.

In its gaseous state, water does not form hydrogen bonds.

2.15 Hydrogen Bonds Hold Water Molecules Together
Hydrogen bonding exists between the molecules of water in both its liquid and solid states. (a) Solid water. (b) Liquid water. Although more structured, ice is less dense than liquid water, so it floats. (c) Water forms a gas when its hydrogen bonds are broken and molecules move farther apart.

explains why sweating cools the human body: As sweat evaporates off the skin, it uses up some of the adjacent body heat.

COHESION AND SURFACE TENSION.   In liquid water, the molecules are free to move about. The hydrogen bonds between the water molecules continually form and break. In other words, liquid water has a dynamic structure. On average, every water molecule forms 3.4 hydrogen bonds with other water molecules. This number represents fewer bonds than exist in ice, but it is still a high number.

These hydrogen bonds explain the cohesive strength of liquid water. The **cohesive strength** of water is what permits narrow columns of water to stretch from the roots to the leaves of trees more than 100 meters high. When water evaporates from leaves, the entire column moves upward in response to the pull of the molecules at the top.

Water also has a high **surface tension**, which means that the surface of liquid water exposed to the air is difficult to puncture. The water molecules in this surface layer are hydrogen-bonded to other water molecules below. The surface tension of water permits a container to be filled slightly above its rim without overflowing, and it permits small animals to walk on the surface of water (Figure 2.16).

Most biological substances are dissolved in water

A **solution** is produced when a substance is dissolved in water (an aqueous solution) or another liquid. Many of the important molecules in biological systems are polar, and therefore are soluble in water. Much of biochemistry takes place in an aqueous solution.

One branch of the study of solutions is *qualitative analysis*, which deals with substances dissolved in a solvent (in this case, water) and the chemical reactions that occur there. Qualitative analysis is the subject of much of the next few chapters.

Solutions can also be studied by *quantitative analysis*, in which concentrations—the amount of substance in a given amount of solution—are measured. What follows is a brief introduction to some of the quantitative chemical terms you will see in this text.

▶ A *molecular formula* uses chemical symbols to identify the different atoms in a compound, and subscript numbers to show how many of each type of atoms are present. Thus, the formula for sucrose—table sugar—is $C_{12}H_{22}O_{11}$.

▶ Each compound has a **molecular weight (molecular mass)** that is the sum of the atomic weights of all atoms in the molecule. Looking at the periodic table in Figure 2.1, you can calculate the molecular weight of table sugar to be approximately 342. Molecular weights are usually related to the molecule's size (Figure 2.17).

▶ A **mole** is the amount of an ion or compound in grams whose weight is numerically equal to its molecular weight. So one mole of sugar weighs 342 grams.

One aim of quantitative analysis is to study the behaviors of precise numbers of molecules in solution. But it is

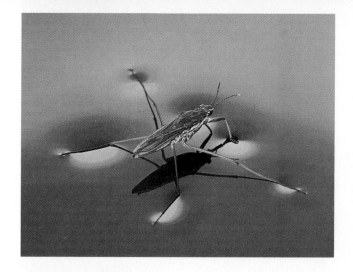

2.16 Surface Tension
Water striders "skate" along, supported by the surface tension of the water that is their home.

not possible to count molecules directly. Instead, chemists use a constant that relates the *weight* of any substance to the *number* of molecules of that substance. This constant is called *Avogadro's number*, which is 6.02×10^{23} *molecules per mole*. It allows chemists to work with moles of substances (which can be weighed out in the laboratory) instead of actual molecules. The mole concept is analogous to the concept of a dozen: We buy a dozen eggs or a dozen doughnuts, knowing that we will get 12 of whichever we buy.

In the same way, chemists can dissolve a mole of sugar in water to make 1 liter, knowing that the mole contains 6.02×10^{23} individual sugar molecules. This solution—1 mole of a substance dissolved in water to make 1 liter—is called a 1 molar (1 *M*) solution.

The many molecules that dissolve in water in living tissues are not present at anything close to a 1 molar concentration. Most are in the micromolar (millionths of a mole; μM) to millimolar (thousandths of a mole; m*M*) range. Some, such as hormones, are far less concentrated than this.

While these abbreviations seem to indicate very low concentrations, remember that even a 1 μM solution has 6.02×10^{17} molecules of the solute per liter.

Acids, Bases, and the pH Scale

Some substances dissolve in water and release hydrogen ions (H^+), which are actually single, positively charged protons. These tiny bits of charged matter can attach to other molecules, and in doing so, change their properties. In this section, we examine the properties of substances that release H^+ (called acids) and attach to H^+ (called bases). We will distinguish strong and weak acids and bases, and provide a quantitative means for stating the concentration of H^+ in solutions: the pH scale.

Acids donate H⁺, bases accept H⁺

If hydrochloric acid (HCl) is added to water, it dissolves and ionizes, releasing the ions H^+ and Cl^-:

$$HCl \rightarrow H^+ + Cl^-$$

2.17 Weights and Sizes of Atoms and Molecules
The color conventions used here are standard for the atoms.
(Yellow is used for sulfur and phosphorus atoms, which are
not depicted.)

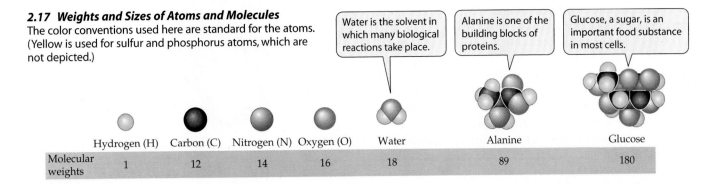

Water is the solvent in which many biological reactions take place.

Alanine is one of the building blocks of proteins.

Glucose, a sugar, is an important food substance in most cells.

	Hydrogen (H)	Carbon (C)	Nitrogen (N)	Oxygen (O)	Water	Alanine	Glucose
Molecular weights	1	12	14	16	18	89	180

Because its H^+ concentration has increased, such a solution is *acidic*. Just like the burning reaction of propane and oxygen (see Figure 2.13), the dissolution of HCl to form its ions is a complete reaction. HCl is therefore called a *strong acid*.

An **acid** *releases* H^+ ions in solution. HCl is an acid, as is H_2SO_4 (sulfuric acid). One molecule of sulfuric acid may ionize to yield two H^+ and one SO_4^{2-}. Biological compounds that contain —COOH (the carboxyl group; see Figure 2.20) are also acids (such as acetic acid and pyruvic acid), because

$$—COOH \rightarrow —COO^- + H^+$$

Not all acids dissolve fully in water. For example, if acetic acid is added to water, at the end of the reaction, there are not just the two ions, but some of the original acid as well. Because the reaction is not complete, acetic acid is a weak acid.

Bases *accept* H^+. Like acids, there are strong and weak bases. If NaOH (sodium hydroxide) is added to water, the NaOH dissolves and ionizes, releasing OH^- and Na^+ ions:

$$NaOH \rightarrow Na^+ + OH^-$$

Because the concentration of OH^- increases, such a solution is *basic*, and because this reaction is complete, NaOH is a strong base.

Weak bases include the bicarbonate ion (HCO_3^-), which can accept a H^+ ion and become carbonic acid (H_2CO_3), and ammonia (NH_3), which can accept a H^+ and become an ammonium ion (NH_4^+). Amino groups in biological molecules can also accept protons, acting as bases:

$$—NH_2 + H^+ \rightarrow —NH_3^+$$

When acetic acid is dissolved in water, two reactions happen. First, acetic acid forms its ions:

$$CH_3COOH \rightarrow CH_3COO^- + H^+$$

Then, once ions are formed, they re-form acetic acid:

$$CH_3COO^- + H^+ \rightarrow CH_3COOH$$

This pair of reactions is reversible. The formula for a reversible reaction can be written with two arrows:

$$CH_3COOH \rightleftharpoons CH_3COO^- + H^+$$

A **reversible reaction** can proceed in either direction—left to right or right to left—depending on the relative starting concentrations of the reactants and products.

In principle, *all* chemical reactions are reversible. In terms of acids and bases, there are two types of reactions, depending on the extent of reversibility:

▶ Ionization of strong acids and bases is virtually irreversible.
▶ Ionization of weak acids and bases is somewhat reversible.

Many of the acid and base groups on large molecules in biological systems are weak.

Water is a weak acid

The water molecule has a slight but significant tendency to ionize into a hydroxide ion (OH^-) and a hydrogen ion (H^+). Actually, *two* water molecules participate in this ionization. One of the two molecules "captures" a hydrogen ion from the other, forming a hydroxide ion and a hydronium ion:

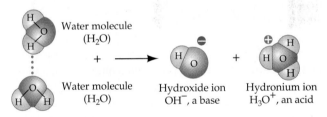

Water molecule (H_2O)

Water molecule (H_2O)

Hydroxide ion OH^-, a base

Hydronium ion H_3O^+, an acid

The hydronium ion is in effect a hydrogen ion bound to a water molecule. For simplicity, biochemists tend to use a modified representation of the ionization of water:

$$H_2O \rightarrow H^+ + OH^-$$

The ionization of water is very important for all living creatures. This fact may seem surprising, since only about one water molecule in 500 million is ionized at any given time. But we are less surprised if we focus on the abundance of water in living systems and the reactive nature of the H^+ produced by ionization.

pH is the measure of hydrogen ion concentration

The terms "acid*ic*" and "bas*ic*" refer only to *solutions*. How acidic or basic a solution is depends on the relative concentrations of H^+ and OH^- ions in it. "Acid" and "base" refer to *compounds* and *ions*. A compound or ion that is an acid can donate H^+; one that is a base can accept H^+.

How do we specify how acidic or basic a solution is? First, let's look at the H^+ concentrations of a few contrasting solutions. In 1 liter of pure water, the H^+ concentration is 10^{-7} M. In 1 M hydrochloric acid, the H^+ concentration is 1 M; and in 1 M sodium hydroxide, the H^+ concentration is 10^{-14} M. Because its values range so widely, the H^+ concentration itself is an inconvenient quantity to measure. It is easier to work with the logarithm of the concentration, because logarithms compress this range.

We indicate how acidic or basic a solution is by its pH (*"potential of Hydrogen"*). The **pH value** is defined as the negative logarithm of the hydrogen ion concentration in moles per liter (molar concentration). In chemical notation, molar concentration is often indicated by putting square brackets around the symbol for a substance; thus $[H^+]$ stands for the molar concentration of H^+. The equation for pH is

$$pH = -\log_{10}[H^+]$$

Since the H^+ concentration of pure water is 10^{-7} M, its pH is $-\log(10^{-7}) = -(-7)$, or 7. A smaller negative logarithm means a larger number. In practical terms, a lower pH means a higher H^+ concentration, or greater acidity. In 1 M HCl, the H^+ concentration is 1 M, so the pH is the negative logarithm of 1 ($-\log 10^0$), or 0. The pH of 1 M NaOH is the negative logarithm of 10^{-14}, or 14.

A solution with a pH of less than 7 is acidic—it contains more H^+ ions than OH^- ions. A solution with a pH of 7 is neutral, and a solution with a pH value greater than 7 is basic. Figure 2.18 shows the pH values of some common substances.

Buffers minimize pH change

An organism must control the pH of the separate compartments within its cells. Animals must also control the pH of their blood. The normal pH of human blood is 7.4, and deviations of even a few tenths of a pH unit can be fatal. The control of pH is made possible in part by buffers—chemical systems that maintain a relatively constant pH even when substantial amounts of acid or base are added.

A **buffer** is a mixture of a weak acid and its corresponding base—for example, carbonic acid (H_2CO_3) and bicarbonate ions (HCO_3^-). If acid is added to this buffer, not all the H^+ ions from that acid stay in solution. Instead, many of them combine with the bicarbonate ions to produce more carbonic acid. This reaction uses up some of the H^+ ions in the solution and decreases the acidifying effect of the added acid:

$$HCO_3^- + H^+ \rightleftharpoons H_2CO_3$$

If a base is added, the reaction essentially reverses. Some of the carbonic acid ionizes to produce bicarbonate ions and more H^+, which counteracts some of the added base. In this way, the buffer minimizes the effects of an added acid or base on pH. A given amount of acid or base causes a

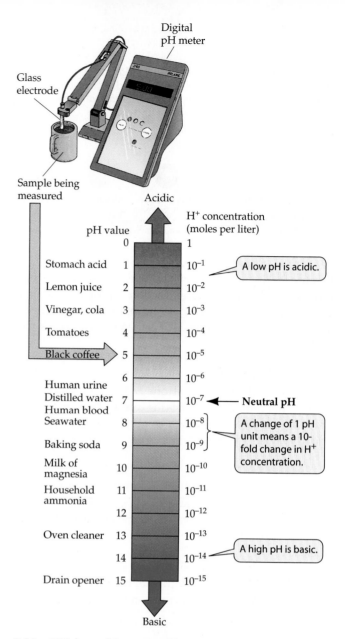

2.18 pH Values of Some Familiar Substances
An electronic instrument similar to the one drawn at the top of the figure is used to measure the pH of a solution.

smaller change in pH in a buffered solution than in an unbuffered one (Figure 2.19).

Buffers illustrate an important chemical principle in reversible reactions called the *law of mass action*. Addition of a component on one side of a reversible system drives the reaction in the direction that uses up that compound. In this case, addition of an acid drives the reaction in one direction; addition of a base drives it in the other.

The Properties of Molecules

Some molecules are small, such as H_2 and CH_4. Others are larger, such as a molecule of table sugar (sucrose), which has 45 atoms. Still other molecules, such as proteins, are gigantic, sometimes containing tens of thousands of atoms bonded together in specific ways.

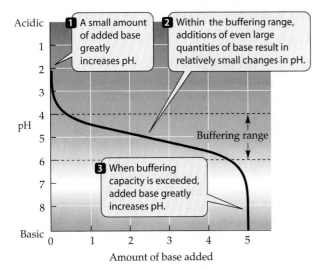

2.19 Buffers Minimize Changes in pH
With increasing amounts of added base, the overall slope of a graph of pH is downward. In the buffering range, however, the slope is shallow. At high and low values of pH, where the buffer is ineffective, the slopes are much steeper.

Whether large, medium, or small, most of the molecules in living systems contain carbon atoms and are thus referred to as **organic molecules**. Most organic molecules include hydrogen and oxygen atoms as well as carbon, and many also include nitrogen and phosphorus.

All molecules have a specific three-dimensional shape. For example, the orientation of the bonding orbitals around the carbon atom gives the methane molecule (CH_4) the shape of a regular tetrahedron (see Figure 2.7c). In carbon dioxide (CO_2), the three atoms are in line. Larger molecules have complex shapes that result from the numbers and kinds of atoms present and the ways in which they are linked together. Some large molecules have compact, ball-like shapes. Others are long, thin, ropelike structures. Their shapes relate to the roles these molecules play in living cells.

In addition to size and shape, molecules have certain properties that characterize them and determine their biological roles. Chemists use the characteristics of composition, structure (three-dimensional shape), reactivity, and solubility to distinguish a sample of one pure molecule from another. That certain groups of atoms are found together in a variety of different molecules simplifies our understanding of the reactions that molecules undergo in living cells.

Functional groups give specific properties to molecules

Functional groups are groups of atoms that make up part of a larger molecule and have particular chemical properties (shape, polarity, reactivity, solubility). The same functional group may be part of very different molecules. You will encounter several functional groups in your study of biology (Figure 2.20).

2.20 Some Functional Groups Important to Living Systems
These functional groups (highlighted in white boxes) are the most common ones found in biologically important molecules. R represents the "remainder" of the molecule, which may be any of a large number of carbon skeletons or other chemical group.

An important kind of biological molecule containing functional groups is the *amino acids*, which have both a carboxyl group and an amino group attached to the same carbon atom, the α (alpha) carbon. Also attached to the α carbon atom are a hydrogen atom and a side chain, designated by the letter R:

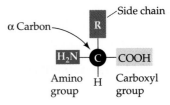

Different side chains have different chemical compositions, structures, and properties. Each of the 20 amino acids found in proteins has a different side chain that gives it its distinctive chemical properties, as we'll see in Chapter 3. Because they possess both carboxyl and amino groups, amino acids are simultaneously acids and bases. At the pH values commonly found in cells, both the carboxyl and the amino groups are ionized: The carboxyl group has lost a proton, and the amino group has gained one.

Isomers have different arrangements of the same atoms

Isomers are molecules that have the same chemical formula but different arrangements of the atoms. (The prefix "iso-" means "same" and is encountered in many biological terms.) Of the different kinds of isomers, we will consider two: structural isomers and optical isomers.

Structural isomers differ in how their atoms are joined together. Consider two simple molecules, each composed of 4 carbon and 10 hydrogen atoms bonded covalently, with the formula C_4H_{10}. These atoms can be linked together in two different ways, resulting in two forms of the molecule:

$$H_3C-\underset{\underset{H}{|}}{\overset{\overset{H}{|}}{C}}-\underset{\underset{H}{|}}{\overset{\overset{H}{|}}{C}}-CH_3 \qquad H_3C-\underset{\underset{H}{|}}{\overset{\overset{CH_3}{|}}{C}}-CH_3$$

$$\text{Butane} \qquad\qquad\qquad \text{Isobutane}$$

The different bonding relationships of butane and isobutane are distinguished in structural formulas, and the compounds have different chemical properties.

Many molecules of biological importance, particularly the sugars and amino acids, have **optical isomers**. Optical isomers occur whenever a carbon atom has four *different* atoms or groups attached to it. This pattern allows two different ways of making the attachments, each the mirror image of the other (Figure 2.21). Such a carbon atom is an *asymmetric carbon*, and the pair of compounds are optical isomers of each other. Your right and left hands are optical isomers. Just as a glove is specific for a particular hand, some biochemical molecules can interact with one optical isomer of a compound, but are unable to "fit" the other.

The α carbon in an amino acid is an asymmetric carbon because it is bonded to four different functional groups. Therefore, amino acids exist in two isomeric forms, called D-amino acids and L-amino acids. "D" and "L" are abbreviations for the Latin terms for right (*dextro*) and left (*levo*), respectively. Only L-amino acids are commonly found in most organisms.

Between the small molecules we have discussed in this chapter and the world of the living cell stands another level, that of the macromolecules. These huge molecules—the proteins, lipids, carbohydrates, and nucleic acids—are the subject of the next chapter.

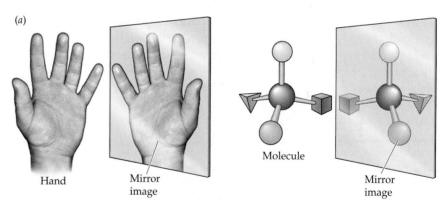

(a)

Hand   Mirror image   Molecule   Mirror image

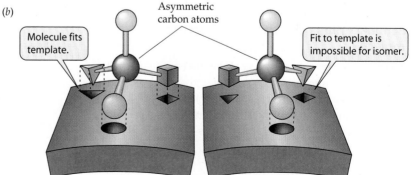

(b)

Asymmetric carbon atoms

Molecule fits template.   Fit to template is impossible for isomer.

2.21 Optical Isomers
(a) Optical isomers are mirror images of each other. (b) Molecular optical isomers result when four different groups are attached to a single carbon atom. (c) If a template is laid out to match the groups on one carbon atom, the groups on the mirror-image isomer cannot be rotated to fit the same template.

Chapter Summary

Atoms: The Constituents of Matter

▶ Matter is composed of atoms. Each atom consists of a positively charged nucleus of protons and neutrons, surrounded by electrons bearing negative charges.

▶ There are many elements in nature, but only a few of them make up the bulk of living systems. **Review Figure 2.1**

▶ Isotopes of an element differ in their numbers of neutrons. Some isotopes are radioactive, emitting radiation as they decay. **Review Figure 2.2**

▶ Electrons are distributed in shells consisting of orbitals. Each orbital contains a maximum of two electrons. **Review Figures 2.4, 2.5**

▶ In losing, gaining, or sharing electrons to become more stable, an atom can combine with other atoms to form molecules. **Review Table 2.1**

Chemical Bonds: Linking Atoms Together

▶ Covalent bonds are strong bonds formed when two atomic nuclei share one or more pairs of electrons. Covalent bonds have spatial orientations that give molecules three-dimensional shapes. **Review Figures 2.6, 2.7, Table 2.2**

▶ Nonpolar covalent bonds are formed when the electronegativities of two atoms are approximately equal. When atoms with strong electronegativity (such as oxygen) bond to atoms with weaker electronegativity (such as hydrogen), a polar covalent bond is formed, in which one end is δ^+ and the other is δ^-. **Review Figure 2.8, Table 2.3**

▶ Hydrogen bonds are weak electrical attractions that form between a δ^+ hydrogen atom in one molecule and a δ^- nitrogen or oxygen atom in another molecule or in another part of a large molecule. Hydrogen bonds are abundant in water. **Review Figure 2.9**

▶ Ions are electrically charged bodies that form when an atom gains or loses one or more electrons. Ionic bonds are electrical attractions between oppositely charged ions. Ionic bonds are strong in solids, but weaker when the ions are separated from one another in solution. **Review Figures 2.10, 2.11**

▶ Nonpolar molecules do not interact directly with polar substances, including water. Nonpolar molecules are attracted to each other by very weak bonds called van der Waals forces. **Review Figure 2.12**

Chemical Reactions: Atoms Change Partners

▶ In chemical reactions, substances change their atomic compositions and properties. Energy is released in some reactions, whereas in others energy must be provided. Neither matter nor energy is created or destroyed in a chemical reaction, but both change form.

▶ Combustion reactions are oxidation–reduction reactions in which a fuel is converted to carbon dioxide and water, while energy is released as heat and light. In living cells, combustion reactions take place in multiple steps so that the released energy can be harvested for cellular activities. **Review Figure 2.13**

Water: Structure and Properties

▶ Water's molecular structure and its capacity to form hydrogen bonds give it unusual properties that are significant for life. Water is an excellent solvent; solid water floats in liquid water; and water gains or loses a great deal of heat when it changes its state, a property that moderates environmental temperature changes. **Review Figure 2.15**

▶ The cohesion of water molecules permits liquid water to rise to great heights in narrow columns and produces a high surface tension. Water's high heat of vaporization assures effective cooling when water evaporates.

▶ Solutions are produced when substances dissolve in water. The concentration of a solution is the amount of a given substance in a given amount of solution. Most biological substances are dissolved in water at very low concentrations.

Acids, Bases, and the pH Scale

▶ Acids are substances that donate hydrogen ions (H^+). Bases are substances that accept hydrogen ions.

▶ The pH of a solution is the negative logarithm of the hydrogen ion concentration. Values lower than pH 7 indicate an acidic solution; values above pH 7 indicate a basic solution. **Review Figure 2.18**

▶ Buffers are systems of weak acids and bases that limit the change in pH when hydrogen ions are added or removed. **Review Figure 2.19**

The Properties of Molecules

▶ Molecules vary in size, shape, reactivity, solubility, and other chemical properties.

▶ Functional groups make up part of a larger molecule and have particular chemical properties. The consistent chemical behavior of functional groups helps us understand the properties of the molecules that contain them. **Review Figure 2.20**

▶ Structural and optical isomers have the same kinds and numbers of atoms, but differ in their structures and properties. **Review Figure 2.21**

For Discussion

1. Would you expect the elemental composition of Earth's crust to be the same as that of the human body? How could you find out?

2. Lithium (Li) is the element with atomic number 3. Draw the electronic structures of the Li atom and of the Li^+ ion.

3. Draw the structure of a pair of water molecules held together by a hydrogen bond. Your drawing should indicate the covalent bonds.

4. The molecular weight of sodium chloride (NaCl) is 58.45. How many grams of NaCl are there in 1 liter of a 0.1 M NaCl solution? How many in 0.5 liter of a 0.25 M NaCl solution?

5. The side chain of the amino acid glycine is simply a hydrogen atom (—H). Are there two optical isomers of glycine? Explain.

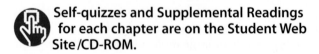 **Self-quizzes and Supplemental Readings for each chapter are on the Student Web Site/CD-ROM.**

3 Macromolecules: Their Chemistry and Biology

A SPIDER WEB IS AN AMAZING STRUCTURE. NOT only is it beautiful to look at, but it is an architectural wonder that serves as the spider's home, its mating ground, and its means of hunting and capturing food. Consider a fly that happens to intersect with a spider web. The fibers of the web must slow down the fly, but they cannot break, so they must stretch in order to dissipate the energy of the fly's movement. On the other hand, the fibers holding the web together cannot stretch too much, because they must be strong enough to hold the entire structure in place and not let the web wobble out of control. Web fibers are far thinner than human hair, yet they are stronger than steel and, in some cases, more elastic than nylon. In fact, spider silk may be as strong as Kevlar, a synthetic substance used to make bulletproof vests and the cords attached to parachutes.

Spider silk is composed of slight variations on a single type of huge molecule—a macromolecule—called a protein. The many types of proteins in biological systems are composed of different amounts of the 20 molecules known as amino acids, and spider silks have their own unique selections of these molecules. The silk protein that stretches contains amino acids that allow it to curl into a spiral, and when these spirals associate into silk fibers, they can slip along each other to change the fiber's length. The strong fibers, in contrast, are made up of amino acids that fold the individual proteins into flat sheets, with ratchets that fit parallel sheets together (much like Lego blocks), so that the fibers are hard to pull apart. The relationship between chemical structure and biological function is the theme of this chapter and many of the succeeding ones in this section.

The four major types of biological macromolecules—proteins, carbohydrates, lipids, and nucleic acids—are composed of building blocks called monomers. In the case of proteins like spider silk, the monomers are amino acids; carbohydrate monomers are sugars, and nucleic acid monomers are nucleotides. Some lipids are composed of a small molecule, glycerol, covalently bonded to larger fatty acids. Lipids interact to form huge macromolecular aggregates, such as the membranes that surround cells.

The four kinds of large molecules are made the same way in all living things, and are present in roughly the same proportions in all organisms (Figure 3.1). Although an apple tree is obviously different from a person, their basic chemistry is the same, demonstrating the unity of life. A protein that has a certain role in the apple probably has a similar role in the human. One important advantage of biochemical unity is that organisms can eat one another. When you eat an apple, the molecules you take in include carbohydrates, lipids, and proteins that can be re-fashioned into the special varieties of those molecules used by humans.

Macromolecules: Giant Polymers

Macromolecules are giant **polymers** (poly-, "many"; -mer, "unit") constructed by the covalent linking of smaller molecules called **monomers** (Table 3.1). These monomers may or may not be identical, but they always have similar chemical structures. Molecules with molecular weights exceeding 1,000 are usually considered macromolecules, and the proteins, polysaccharides (large carbohydrates), and nucleic acids of living systems certainly fall into this category.

Each type of macromolecule performs some combination of a diversity of functions: energy storage, structural

A Complex Macromolecule
Spider silk (purple) being spun into web material from a gland by the shiny black spider, *Castercantha*.

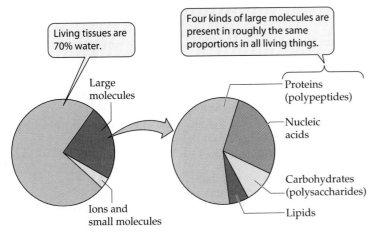

3.1 Substances Found in Living Tissues
The substances shown here make up the nonmineral components of living tissue (bone would be an example of a "mineral tissue"). Most tissues are at least 70 percent water.

support, protection, catalysis, transport, defense, regulation, movement, and heredity. These roles are not necessarily exclusive. For example, both carbohydrates and proteins can play structural roles, supporting and protecting tissues and organisms. However, only nucleic acids specialize in information storage and function as hereditary material, carrying both species and individual traits from generation to generation.

The functions of macromolecules are directly related to their shapes and to the chemical properties of their monomers. Some macromolecules, such as catalytic and defensive proteins, fold into compact spherical forms with surface features that make them water-soluble and capable of intimate interaction with other molecules. Other proteins and carbohydrates form long, fibrous systems that provide strength and rigidity to cells and organisms. Still other long, thin assemblies of proteins can contract and cause movement.

Because macromolecules are so large, they contain many different functional groups (see Figure 2.20). For example, a large protein may contain hydrophobic, polar, and charged functional groups that give specific properties to local sites on a macromolecule. As we will see, this diversity of properties determines the shapes of macromolecules and their interactions with both other macromolecules and smaller molecules.

3.1 The Building Blocks of Organisms

MONOMER	SIMPLE POLYMER	COMPLEX POLYMER (MACROMOLECULE)
Amino acid	Peptide or oligopeptide	Polypeptide (protein)
Nucleotide	Oligonucleotide	Nucleic acid
Monosaccharide (sugar)	Oligosaccharide	Polysaccharide (carbohydrate)

Condensation Reactions

The polymers of living things are constructed from monomers by a series of reactions called **condensation reactions** or *dehydration reactions* (both words refer to the loss of water). Condensation reactions result in covalently bonded monomers (Figure 3.2*a*). The condensation reactions that produce the different kinds of macromolecules differ in detail, but in all cases, polymers will form only if energy is added to the system. In living systems, specific energy-rich molecules supply this energy.

The reverse of a condensation reaction is a **hydrolysis reaction** (hydro-, "water"; -lysis, "break"). These reactions digest polymers and produce monomers. Water reacts with the bonds that link the polymer together, and the products are free monomers. The elements (H and O) of H_2O become part of the products (Figure 3.2*b*). Like condensation reactions, hydrolysis requires the addition of energy.

We begin our study of biological macromolecules with a very diverse group of polymers, the proteins.

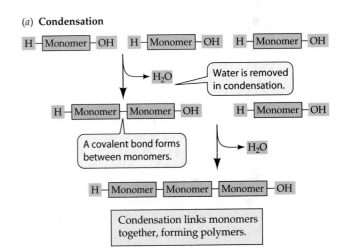

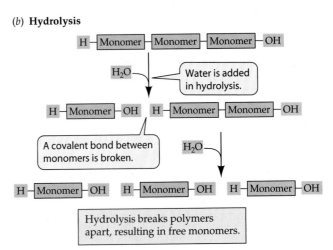

3.2 Condensation and Hydrolysis of Polymers
(*a*) A condensation reaction links monomers into polymers.
(*b*) A hydrolysis reaction digests polymers into individual monomers. In living tissues, these reactions do not occur spontaneously, but require added energy.

Proteins: Polymers of Amino Acids

Proteins are involved in structural support, protection, catalysis, transport, defense, regulation, and movement. Among the functions of macromolecules listed earlier, only energy storage and heredity are not usually performed by proteins.

Of particular importance are proteins called **enzymes** that increase the rates of chemical reactions in cells, a function known as *catalysis*. In general, each chemical reaction requires a different enzyme, because proteins show great specificity for the smaller molecules with which they interact.

Proteins range in size from small ones such as the RNA-digesting enzyme ribonuclease A, which has a molecular weight of 5,733 and 51 amino acid residues, to huge molecules such as the cholesterol transport protein apolipoprotein B, which has a molecular weight of 513,000 and 4,636 amino acid residues. (The word "residue" refers to a monomer when it is part of a polymer.) Each of these proteins consists of a single chain of amino acids (a *polypeptide chain*) folded into a specific three-dimensional shape that is required for protein function.

Some proteins have more than one polypeptide chain. For example, the oxygen-carrying protein hemoglobin has four chains that are folded separately and associate together to make the functional protein. As we will see later in this book, there are many such "multi-protein machines," composed of dozens of interacting polypeptides.

Each of these proteins has a characteristic amino acid composition. But not every protein contains all kinds of amino acids, nor an equal number of different ones. The diversity in amino acid content and sequence is the source of the diversity in protein structures and functions. In some cases, additional chemical structures called **prosthetic groups** may be attached covalently to the protein. These groups include carbohydrates, lipids, phosphate groups, the iron-containing heme group that binds to hemoglobin, and metal ions such as copper and zinc. Prosthetic groups are discussed further in Chapter 6.

The next several chapters will describe the many functions of proteins. To understand them, we must first explore protein structure. First, we will examine the properties of the amino acids and how they link to form proteins. Then we will systematically examine protein structure and look at how a linear chain of amino acids is consistently folded into a compact three-dimensional shape. Finally, we will see how this structure provides a specific physical and chemical environment for other molecules that can interact with the protein.

Proteins are composed of amino acids

The 20 amino acids commonly found in proteins have a wide variety of properties. In Chapter 2, we looked at the structure of amino acids and identified four different groups attached to a central (α) carbon atom: a hydrogen atom, an amino group (NH_3^+), a carboxyl group (COO^-), and a **side chain**, or **R group**. The R groups (R stands for

"remainder" or "residue") of amino acids are important in determining the three-dimensional structure and function of the macromolecule. They are highlighted in white in Table 3.2.

As Table 3.2 shows, amino acids are grouped and distinguished by their side chains. Some side chains are electrically charged (+1, −1), while others are polar (δ+, δ−) or nonpolar and hydrophobic.

▶ The five amino acids that have *electrically charged* side chains attract water and oppositely charged ions of all sorts.

▶ The five amino acids that have *polar* side chains tend to form weak hydrogen bonds with water and with other polar or charged substances.

▶ Seven amino acids have side chains that are *nonpolar* hydrocarbons or very slightly modified hydrocarbons. In the watery environment of the cell, the hydrophobic side chains may cluster together.

▶ Three amino acids—cysteine, glycine, and proline—are special cases, although their R groups are generally hydrophobic.

The cysteine side chain, which has a terminal —SH group, can react with another cysteine side chain to form a covalent bond called a **disulfide bridge** (—S—S—) (Figure 3.3). Disulfide bridges help determine how a protein chain folds. When cysteine is not part of a disulfide bridge, its side chain is hydrophobic.

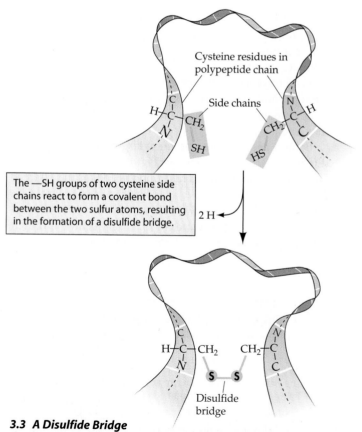

Cysteine residues in polypeptide chain

Side chains

The —SH groups of two cysteine side chains react to form a covalent bond between the two sulfur atoms, resulting in the formation of a disulfide bridge.

2 H

Disulfide bridge

3.3 A Disulfide Bridge
Disulfide bridges (—S—S—) are important in maintaining the proper three-dimensional shapes of some protein molecules.

3.2 **Twenty Amino Acids Found in Proteins**

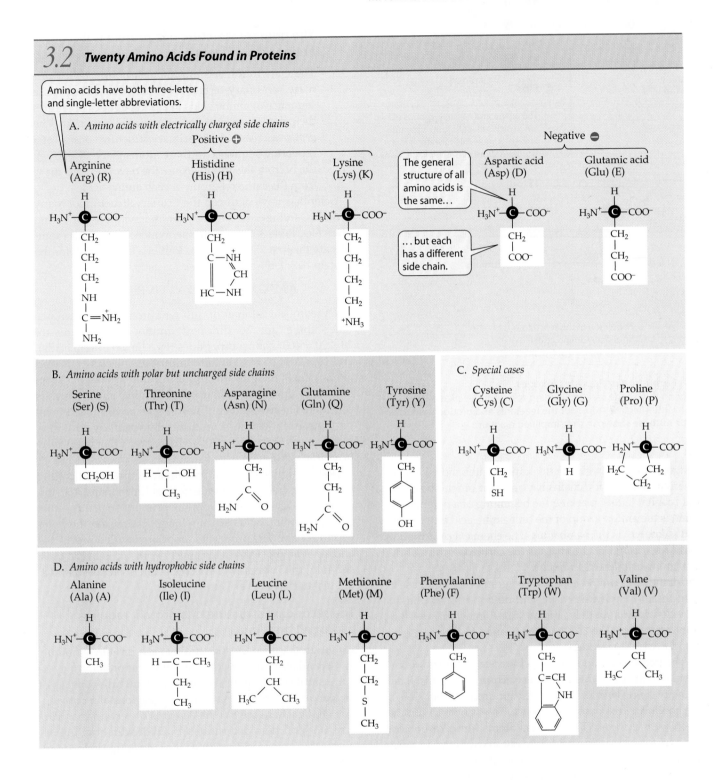

Amino acids have both three-letter and single-letter abbreviations.

A. *Amino acids with electrically charged side chains*

Positive ⊕

The general structure of all amino acids is the same…

…but each has a different side chain.

Negative ⊖

Arginine (Arg) (R)   Histidine (His) (H)   Lysine (Lys) (K)   Aspartic acid (Asp) (D)   Glutamic acid (Glu) (E)

B. *Amino acids with polar but uncharged side chains*

Serine (Ser) (S)   Threonine (Thr) (T)   Asparagine (Asn) (N)   Glutamine (Gln) (Q)   Tyrosine (Tyr) (Y)

C. *Special cases*

Cysteine (Cys) (C)   Glycine (Gly) (G)   Proline (Pro) (P)

D. *Amino acids with hydrophobic side chains*

Alanine (Ala) (A)   Isoleucine (Ile) (I)   Leucine (Leu) (L)   Methionine (Met) (M)   Phenylalanine (Phe) (F)   Tryptophan (Trp) (W)   Valine (Val) (V)

The glycine side chain consists of a single hydrogen atom and is small enough to fit into tight corners in the interior of a protein molecule, where a larger side chain could not fit.

Proline differs from other amino acids because it possesses a modified amino group lacking a hydrogen on its nitrogen, which limits its hydrogen bonding ability. Also, the ring system of proline limits rotation about its α carbon, so proline is often found at bends or loops in a protein.

Peptide linkages covalently bond amino acids together

When amino acids polymerize, the carboxyl group of one amino acid reacts with the amino group of another, undergoing a condensation reaction that forms a **peptide linkage**. Figure 3.4 gives a simplified description of this reaction. (In reality, other molecules must activate the amino acids in order for this reaction to proceed, and there are intermediate steps in the process. We will examine these in Chapter 12).

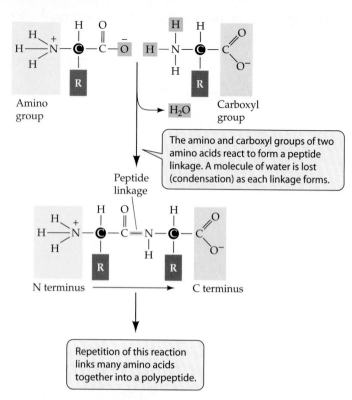

3.4 Formation of Peptide Linkages
In living things, the reaction leading to a peptide linkage has many intermediate steps, but the reactants and products are the same as those shown in this simplified diagram.

Just as a sentence begins with a capital letter and ends with a period, polypeptide chains have a linear order. The chemical "capital letter" marking the beginning of a polypeptide chain is the amino group of the first amino acid in the chain and is known as the *N terminus*. The chemical punctuation mark for the end of the chain is the carboxyl group of the last amino acid (the *C terminus*).

All the other amino and carboxyl groups (except those in side chains) are involved in peptide bond formation, so they do not exist in the chain as "free," intact groups. Biochemists refer to the "N → C," or "amino-to-carboxyl" orientation of polypeptides.

The peptide linkage has two characteristics that are important in the three-dimensional structure of proteins. First, in many single covalent bonds, the groups on either side of the bonds are free to rotate in space. This is not so with the C—N peptide bond. The adjacent atoms (the α carbons of the two adjacent amino acids) are not free to rotate because of the partial double-bond character of the peptide bond. Chemists will realize that this is due to the resonance between the strong electronegativity of the oxygen bound to the carbon and the weak electronegativity of the hydrogen bound to the nitrogen. This characteristic limits the folding of the polypeptide.

Second, the oxygen bound to the carbon carries a slight negative charge ($\delta-$), whereas the hydrogen bound to the nitrogen is slightly positive ($\delta+$). This asymmetry of charge favors hydrogen bonding within the protein molecule itself and with other molecules, contributing to both the structure and the function of many proteins.

The primary structure of a protein is its amino acid sequence

There are four levels of protein structure, called primary, secondary, tertiary, and quaternary. The precise sequence of amino acids in a polypeptide constitutes the **primary structure** of a protein (Figure 3.5*a*). The *peptide backbone* of this primary structure consists of a repeating sequence of three atoms (—N—C—C—): the N from the amino group, the α carbon, and the C from the carboxyl group of each amino acid.

Scientists have deduced the primary structure of many proteins, and use the single-letter abbreviations for amino acids (see Table 3.2) to record the sequence. Here, for example, are the first 25 amino acids (out of a total of 457) for the protein hexokinase, from baker's yeast:

AASXDXSLVEVHXXVFIVPPXILQA

The theoretical number of different proteins is enormous. Since there are 20 different amino acids, there are $20 \times 20 = 400$ distinct dipeptides (two linked amino acids), and $20 \times 20 \times 20 = 8,000$ different tripeptides (three linked amino acids). Imagine this process of multiplying by 20 extended to a protein made up of 100 amino acids (which is considered a small protein). There could be 20^{100} such small proteins, each with its own distinctive primary structure. How large is the number 20^{100}? There aren't that many electrons in the entire universe!

At the higher levels of protein structure, local coiling and folding give the molecule its final functional shape, but all of these levels derive from the primary structure—that is, which amino acids are at which locations on the polypeptide chain. The properties associated with a precise sequence of amino acids determine how the protein can twist and fold, thus adopting a specific stable structure that distinguishes it from every other protein.

The secondary structure of a protein requires hydrogen bonding

Although the primary structure of each protein is unique, the secondary structures of many different proteins may be quite similar. A protein's **secondary structure** consists of regular, repeated patterns in different regions of a polypeptide chain. There are two basic types of secondary structure, both of them determined by hydrogen bonding between the amino acid residues that make up the primary structure.

THE α HELIX. The α (alpha) **helix** is a right-handed coil that is "threaded" in the same direction as a standard wood screw (Figure 3.5*b*). The R groups extend outward from the peptide backbone of the helix. The coiling results from hydrogen bonds between the slightly positive hydrogen of the N—H of one amino acid residue and the slightly negative oxygen of the C=O of another. When this pattern of hydrogen bonding is established repeatedly over a segment of the protein, it stabilizes the coil, resulting in an α helix. Amino acids with large R groups that distort the coil or

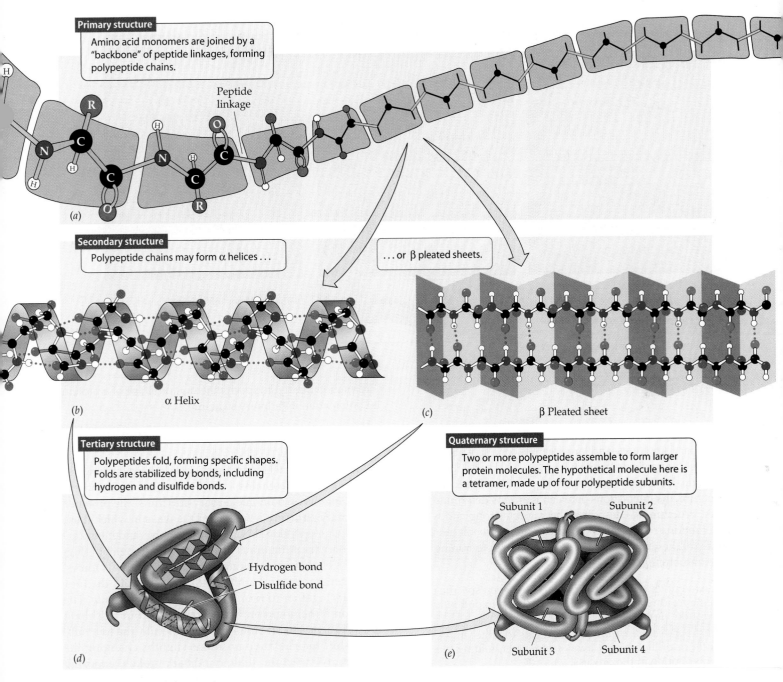

Primary structure
Amino acid monomers are joined by a "backbone" of peptide linkages, forming polypeptide chains.

Peptide linkage

(a)

Secondary structure
Polypeptide chains may form α helices ...

... or β pleated sheets.

α Helix

(b)

β Pleated sheet

(c)

Tertiary structure
Polypeptides fold, forming specific shapes. Folds are stabilized by bonds, including hydrogen and disulfide bonds.

Hydrogen bond
Disulfide bond

(d)

Quaternary structure
Two or more polypeptides assemble to form larger protein molecules. The hypothetical molecule here is a tetramer, made up of four polypeptide subunits.

Subunit 1   Subunit 2

Subunit 3   Subunit 4

(e)

3.5 The Four Levels of Protein Structure
Secondary, tertiary, and quaternary structure all arise from the primary structure of the protein.

otherwise prevent the formation of the necessary hydrogen bonds will keep the α helix from forming.

Alpha-helical secondary structure is particularly evident in the insoluble fibrous structural proteins called keratins, which make up hair, hooves, and feathers. Hair can be stretched because stretching requires that only the hydrogen bonds of the α helix, not the covalent bonds, be broken; when the tension on the hair is released, both the hydrogen bonds and the helix re-form.

THE β PLEATED SHEET.   The β (beta) pleated sheet is formed from two or more polypeptide chains that are almost com-

pletely extended and lying next to one another. The sheet is stabilized by hydrogen bonds between the N—H groups on one chain and the C=O groups on the other (Figure 3.5c). A β pleated sheet may form between separate polypeptide chains, as in spider silk, or between different regions of the same polypeptide that is bent back on itself. Many proteins contain regions of both α helix and β pleated sheet in the same polypeptide chain.

The tertiary structure of a protein is formed by bending and folding

In many proteins, the polypeptide chain is bent at specific sites and folded back and forth, resulting in the **tertiary structure** of the protein (Figure 3.5d). Although the α helices and β pleated sheets contribute to the tertiary struc-

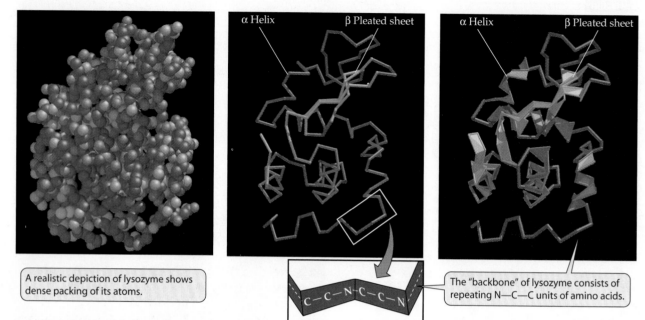

A realistic depiction of lysozyme shows dense packing of its atoms.

α Helix   β Pleated sheet

α Helix   β Pleated sheet

C—C—N—C—C—N

The "backbone" of lysozyme consists of repeating N—C—C units of amino acids.

3.6 Three Representations of Lysozyme
Different molecular representations of a protein emphasize different aspects of its tertiary structure. These three representations of lysozyme are similarly oriented.

ture, only parts of the macromolecule usually have these secondary structures, and large regions consist of structures unique to a particular protein.

While hydrogen bonding is responsible for secondary structure, the interactions between R groups determine tertiary structure. We described the various strong and weak interactions between atoms in Chapter 2 (see Table 2.1). Many of these interactions are involved in determining tertiary structure:

▶ Covalent *disulfide bridges* can form between specific cysteine residues (see Figure 3.3), holding a folded polypeptide in place.
▶ *Hydrophobic side chains* can aggregate together in the interior of the protein, away from water, folding the polypeptide in the process.
▶ *Van der Waals forces* can stabilize the close interactions between the hydrophobic residues.
▶ *Ionic interactions* can occur between positively and negatively charged side chains buried deep within a protein, away from water, forming a *salt bridge*.

A complete description of a protein's tertiary structure specifies the location of every atom in the molecule in three-dimensional space, in relation to all the other atoms. The tertiary structure of the protein lysozyme is represented in Figure 3.6.

Bear in mind that both tertiary structure and secondary structure derive from the protein's primary structure. If lysozyme is heated slowly, the heat energy will disrupt only the weak interactions and cause only the tertiary structure to break down. But the protein will return to its normal tertiary structure when it cools, demonstrating that all the information needed to specify the unique shape of a protein is contained in its primary structure.

The quaternary structure of a protein consists of subunits

As mentioned earlier, many functional proteins have two or more polypeptide chains, called **subunits**, each of them folded into its own unique tertiary structure. The protein's **quaternary structure** results from the ways in which these multiple polypeptide subunits bind together and interact.

Quaternary structure is illustrated by hemoglobin (Figure 3.7). Hydrophobic interactions, van der Waals forces, hydrogen bonds, and ionic bonds all help hold the four subunits together to form the hemoglobin molecule. The function of hemoglobin is to carry oxygen in red blood cells. As hemoglobin binds one O_2 molecule, the four subunits shift their relative positions slightly, changing the quaternary structure. Ionic bonds are broken, exposing buried side chains that enhance the binding of additional O_2 molecules. The structure changes again when hemoglobin releases its oxygen molecules to the cells of the body.

The surfaces of proteins have specific shapes

Small molecules in a solution are in constant motion. They vibrate, rotate, and move from place to place like corn in a popper. If two of them collide in the right circumstances, a chemical reaction can occur. The specific shapes of proteins allow them to bind noncovalently with other molecules, which in turn allows other important biological events to occur. For example:

▶ Two adjacent cells can stick together because proteins protruding from each of the cells interact with each other (see Chapter 5).
▶ A substance can enter a cell by binding to a carrier protein in the cell surface membrane (see Chapter 5).

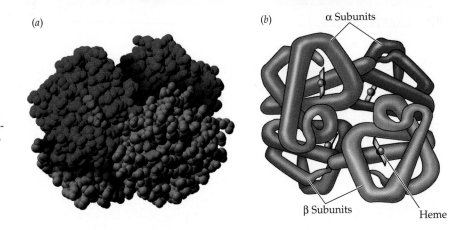

3.7 Quaternary Structure of a Protein
Hemoglobin consists of four folded polypeptide subunits that assemble themselves into the quaternary structure shown here. In these two graphic representations, each type of subunit is a different color. The heme groups contain iron and are the oxygen-carrying sites.

(a) *(b)* α Subunits β Subunits Heme

- A chemical reaction can be speeded up when an enzyme protein binds to one of the reactants (see Chapter 6).
- A multi-protein "machine," DNA polymerase, can catalyze the replication of DNA (see Chapter 11).
- Another multi-protein "machine," the ribosome, can synthesize proteins (see Chapter 12).
- Proteins on a cell's outer surface can bind to chemical signals such as hormones (see Chapter 15).
- Defensive proteins called antibodies can recognize the shape of a virus coat and bind to it (see Chapter 19).

When a small molecule collides with and binds to a much larger protein, it is like a baseball being caught by a catcher: The catcher's mitt has a shape that binds to the ball and fits around it. A hockey puck or a ping-pong ball would not fit the baseball mitt. Thus, the binding of a small molecule to a protein involves a general interaction between two three-dimensional objects that becomes more specific after initial binding. When two large polypeptide chains bind to each other, the interactions are more complicated because extensive surfaces of each macromolecule must come into contact, but the principle is the same.

Biological specificity depends not just on the shape of a protein, but also on the surface chemical groups that it presents to a substance attempting to bind to it (Figure 3.8). The groups on the surface are the R groups of the exposed amino acids, and are therefore a property of the protein's primary structure.

Look again at the structures of the 20 amino acids in Table 3.2, noting the properties of the R groups. Exposed hydrophobic groups will bind to similarly nonpolar groups in the substance with which the protein interacts (often called the *ligand*). Charged R groups will bind to oppositely charged groups on the ligand. Polar R groups containing a hydroxyl (—OH) group can form a hydrogen bond with an incoming ligand. These three types of interactions—hydrophobic, ionic, and hydrogen bonding—are weak by themselves, but strong when all of them act together. So the exposure of appropriate amino acid R groups on the protein surface allows specific binding of a ligand to occur.

Protein shapes are sensitive to the environment

The three-dimensional structure of a protein determines what it binds and, therefore, its function. The primary structure of a protein constrains its secondary, tertiary, and (if subunits exist) quaternary structures. A major effort in

biochemistry is to try to predict three-dimensional protein structure from amino acid sequence. For some short sequences, this is relatively straightforward: For example, certain amino acid sequences will fold into a β pleated sheet. But for large polypeptide chains, the multitude of potential interactions make structure prediction a problem, approachable only by computer. Indeed, a whole new field of computational biochemistry has emerged to tackle the challenge of structure prediction.

Knowing the exact shape of a protein and what can bind to it is important not only in understanding basic biology, but in applied fields such as medicine as well. For example, the three-dimensional structure of a protease, a protein es-

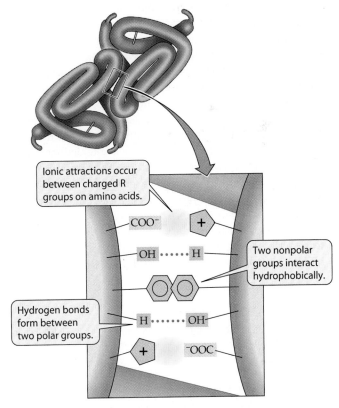

Ionic attractions occur between charged R groups on amino acids.

COO⁻ +

OH ····· H

Two nonpolar groups interact hydrophobically.

Hydrogen bonds form between two polar groups.

H ······ OH

+ ⁻OOC

3.8 Noncovalent Interactions between Polypeptides and Other Molecules
Noncovalent interactions allow a protein to bind tightly to another molecule with specific properties, or allow regions within a protein to interact with one another.

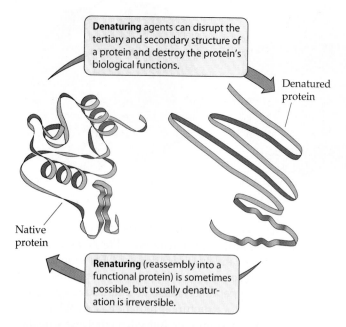

Denaturing agents can disrupt the tertiary and secondary structure of a protein and destroy the protein's biological functions.

Denatured protein

Native protein

Renaturing (reassembly into a functional protein) is sometimes possible, but usually denaturation is irreversible.

3.9 Denaturation Is the Loss of Tertiary Protein Structure and Function
Agents that can cause denaturation include high temperatures and certain chemicals.

sential for the replication of HIV—the virus that causes AIDS—was first determined in this way. Then specific inhibitors were designed to interact with its surface. These protease inhibitors have prolonged the lives of countless people living with HIV.

Because it is determined by weak forces, protein shape is sensitive to environmental conditions that would not break covalent bonds but do upset weaker noncovalent interactions. Elevated temperatures, pH changes, or altered salt concentrations can cause a protein to adopt a different, biologically inactive tertiary structure. Increases in temperature cause more rapid molecular movements and thus can break hydrogen bonds and hydrophobic interactions. Alterations in pH can change the pattern of ionization of carboxyl and amino groups in the R groups of amino acids, thus disrupting the pattern of ionic attractions and repulsions that contributes to normal tertiary structure.

The loss of normal tertiary structure is called **denaturation**, and it is always accompanied by a loss of the normal biological function of the protein (Figure 3.9). Denaturation

can be caused by heat or by high concentrations of polar substances such as urea, which disrupt the hydrogen bonding that is crucial to protein structure. Nonpolar solvents may also disrupt normal structure.

Usually denaturation is irreversible, because amino acids that were buried may now be exposed and vice versa, causing a new structure to form or different molecules to bind to the protein. Boiling an egg denatures its proteins and is, as you know, not reversible. However, as we saw earlier, denaturation is often reversible in the laboratory, especially if it was caused originally by disruption of weak forces. If the denaturing chemicals are removed, the protein returns to its "native" shape and normal function.

Chaperonins help shape proteins

There are two occasions when a polypeptide chain is in danger of binding the wrong ligand. First, following denaturation, hydrophobic R groups, previously on the inside of the protein away from water, become exposed on the surface. Since these groups can interact with similar groups on other molecules, the denatured proteins may aggregate and become insoluble, losing their function. Second, when a protein has just been synthesized and has not yet folded completely, it could present a surface that binds the wrong molecule.

Living systems limit inappropriate protein interactions by making a class of proteins called, appropriately, **chaperonins** (recall the chaperones—usually teachers—at school dances who try to prevent "inappropriate interactions" among the students). Chaperonins were first identified in fruit flies as "heat shock" proteins, which prevented denaturing proteins from clumping together when the flies' temperature was raised.

Some chaperonins work by trapping proteins in danger of inappropriate binding inside a molecular "cage" (Figure 3.10). This cage is composed of identical subunits, and is itself a good example of quaternary protein structure. Inside the cage, the targeted protein folds into the right shape, and then is released at the appropriate time and place.

3.10 Chaperonins Protect Proteins from Inappropriate Binding
Chaperonins surround new or denatured proteins and prevent them from binding to the wrong ligand.

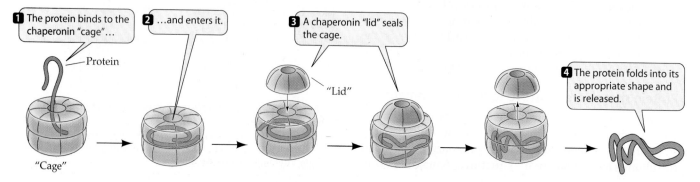

1 The protein binds to the chaperonin "cage"...

Protein

"Cage"

2 ...and enters it.

3 A chaperonin "lid" seals the cage.

"Lid"

4 The protein folds into its appropriate shape and is released.

Carbohydrates: Sugars and Sugar Polymers

Carbohydrates are a diverse group of compounds containing primarily carbon atoms flanked by hydrogen and hydroxyl groups (H—C—OH). They have two major biochemical roles:

▶ They act as a source of energy that can be released in a form usable by body tissues. This energy is stored in strong C—C and C=O covalent bonds.

▶ They serve as carbon skeletons that can be rearranged to form other molecules important for biological structures and functions.

Some carbohydrates are relatively small, with molecular weights less than 100. Others are true macromolecules, with molecular weights in the hundreds of thousands.

There are four categories of biologically important carbohydrates, which we will discuss in turn:

▶ *Monosaccharides* (mono-, "one"; saccharide, "sugar"), such as glucose, ribose, or fructose, are simple sugars and are the monomers out of which the larger forms are constructed.

▶ *Disaccharides* (di-, "two") consist of two monosaccharides.

▶ *Oligosaccharides* (oligo-, "several") have several monosaccharides (3 to 20).

▶ *Polysaccharides* (poly-, "many"), such as starch, glycogen, and cellulose, are large polymers composed of hundreds of thousands of monosaccharide units.

The relative proportions of carbon, hydrogen, and oxygen indicated by the general formula for carbohydrates, CH_2O (i.e., the proportions of these atoms are 1:2:1), apply to monosaccharides. In disaccharides, oligosaccharides, and polysaccharides, these proportions differ slightly from the general formula because two hydrogens and an oxygen are lost during the condensation reactions that form them.

Monosaccharides are simple sugars

Green plants produce monosaccharides through photosynthesis, and animals acquire them directly or indirectly from plants. All living cells contain the monosaccharide glucose. Cells use glucose as an energy source, breaking it down through a series of hydrolysis reactions that release stored energy and produce water and carbon dioxide.

Glucose exists in two forms, the straight chain and the ring; the ring structure predominates in more than 99 percent of circumstances. There are also two forms of the ring structure (α-glucose and β-glucose), which differ only in the placement of the —H and —OH attached to carbon 1 (Figure 3.11). The α and β forms interconvert and exist in equilibrium when dissolved in water.

Different monosaccharides contain different numbers of carbons. (The standard convention for numbering carbons shown in Figure 3.11 is used throughout this book.) Most of the monosaccharides found in living systems belong to the D series of optical isomers (see Chapter 2). But some monosaccharides are *structural* isomers, which have the same kinds and numbers of atoms, but arranged differently by bonding. For example, the *hexoses* (hex-, "six"), a group of structural isomers, all have the formula $C_6H_{12}O_6$. Included among the hexoses are glucose, fructose (so named because it was first found in fruits), mannose, and galactose (Figure 3.12).

Pentoses (pent-, "five") are five-carbon sugars. Some pentoses are found primarily in the cell walls of plants. Two pentoses are of particular biological importance: Ribose and deoxyribose form part of the backbones of the nucleic acids RNA and DNA, respectively. These two pentoses are not isomers; rather, one oxygen atom is missing from carbon 2 in deoxyribose (de-, "absent") (see Figure 3.12). As we will see in Chapter 12, the absence of this oxygen atom has important consequences for the functional distinction of RNA and DNA.

3.11 Glucose: From One Form to the Other

All glucose molecules have the formula $C_6H_{12}O_6$, but their structures vary. When dissolved in water, the α and β "ring" forms of glucose interconvert. The dark line at the bottom of each ring indicates that that edge of the molecule extends toward you; the upper, lighter edge extends back into the page.

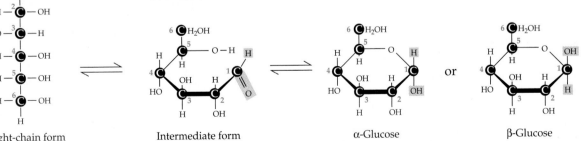

Straight-chain form   Intermediate form   α-Glucose   β-Glucose

The straight-chain form of glucose has an aldehyde group at carbon 1.

A reaction between this aldehyde group and the hydroxyl group at carbon 5 gives rise to a ring form.

Depending on the orientation of the aldehyde group when the ring closes, either of two rapidly and spontaneously interconverting molecules—α-glucose and β-glucose—forms.

Three-carbon sugar

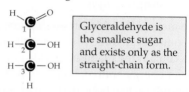

Glyceraldehyde is the smallest sugar and exists only as the straight-chain form.

Five-carbon sugars

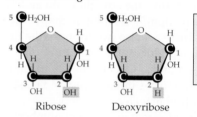

Ribose and deoxyribose each have five carbons, but very different chemical properties and biological roles.

Ribose   Deoxyribose

Six-carbon sugars

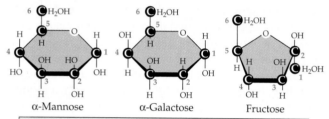

α-Mannose   α-Galactose   Fructose

These hexoses are isomers. All have the formula $C_6H_{12}O_6$, but each has distinct chemical properties and biological roles.

3.12 Monosaccharides Are Simple Sugars
Monosaccharides are made up of varying numbers of carbons. Some are structural isomers, which have the same number of carbons, but arranged differently. Fructose, for example, is a hexose but forms a five-sided ring like the pentoses.

Glycosidic linkages bond monosaccharides together

Monosaccharides are covalently bonded together by condensation reactions that form *glycosidic linkages*. Such a linkage between two monosaccharides forms a disaccharide. For example, a molecule of sucrose (table sugar) is formed from a glucose and a fructose molecule, while lactose (milk sugar) contains glucose and galactose.

The disaccharide maltose contains two glucose molecules, but it is not the only disaccharide that can be made from two glucoses. When glucose molecules form glycosidic linkages, the disaccharide product will be one of two types: α-linked or β-linked, depending on whether the molecule that bonds by its carbon 1 is α-glucose or β-glucose (see Figure 3.11). An α linkage with carbon 4 of a second glucose molecule gives maltose, whereas a β linkage gives cellobiose (Figure 3.13).

Maltose and cellobiose are disaccharide isomers, both having the formula $C_{12}H_{22}O_{11}$. However, they are different compounds with different properties. They undergo different chemical reactions and are recognized by different enzymes. For example, maltose can be hydrolyzed to its monosaccharides in the human body, whereas cellobiose cannot. Certain microorganisms have the chemistry to break down cellobiose.

Oligosaccharides contain several monosaccharides linked by glycosidic linkages at various sites. Many oligosaccharides have additional functional groups, which give them special properties. Oligosaccharides are often covalently bonded to proteins and lipids on the outer cell surface, where they serve as cell recognition signals. The human blood groups (such as ABO) get their specificity from oligosaccharide chains.

Maltose is produced when an α-1,4 glycosidic linkage forms between two glucose molecules. The hydroxyl group on carbon 1 of one glucose in the α (down) position reacts with the hydroxyl group on carbon 4 of the other glucose.

In **cellobiose**, two glucoses are linked by a β-1,4 glycosidic linkage.

3.13 Disaccharides Are Formed by Glycosidic Linkages
Glycosidic linkages between two monosaccharides create many different disaccharides. Which disaccharide is formed depends on which monosaccharides are linked, and on the site (which carbon atom is linked) and form (α or β) of the linkage.

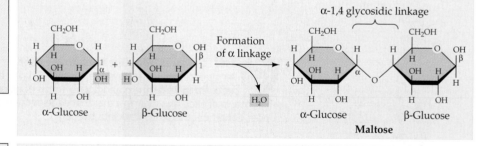

Polysaccharides serve as energy stores or structural materials

Polysaccharides are giant chains of monosaccharides connected by glycosidic linkages. *Starch* is a polysaccharide of glucose with glycosidic linkages in the α-orientation. *Cellulose*, too, is a giant polysaccharide made up solely of glucose, but its individual monosaccharides are connected by β linkages (Figure 3.14*a*). Cellulose is the predominant component of plant cell walls, and is by far the most abundant organic compound on Earth. Both starch and cellulose are composed of nothing but glucose, but their very different chemical and physical properties give them distinct biological functions.

Starch can be more or less easily degraded by the actions of chemicals or enzymes. Cellulose, however, is chemically more stable because of its β-glycosidic linkages. Thus starch

(a) Molecular structure

Cellulose

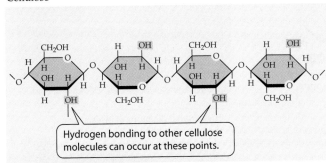

Hydrogen bonding to other cellulose molecules can occur at these points.

Cellulose is an unbranched polymer of glucose with β-1,4 glycosidic linkages that are chemically very stable.

Starch and glycogen

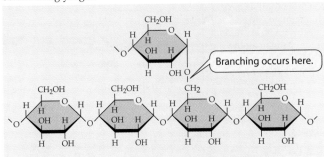

Branching occurs here.

Glycogen and starch are polymers of glucose with α-1,4 glycosidic linkages. α-1,6 glycosidic linkages produce branching at carbon 6.

(b) Macromolecular structure

Linear (cellulose)

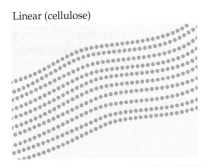

Parallel cellulose molecules hydrogen-bond to form long thin fibrils.

Branched (starch)

Branching limits the number of hydrogen bonds that can form in starch molecules, making starch less compact than cellulose.

Highly branched (glycogen)

The high amount of branching in glycogen makes its solid deposits less compact than starch.

(c) Polysaccharides in cells

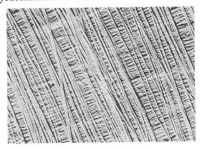

Layers of cellulose fibrils, as seen in this scanning electron micrograph, give plant cell walls great strength.

Dyed red in this micrograph, starch deposits have a large granular shape within cells.

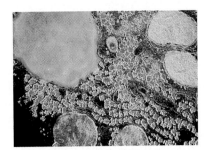

Colored pink in this electron micrograph of human liver cells, glycogen deposits have a small granular shape.

3.14 Representative Polysaccharides
Cellulose, starch, and glycogen demonstrate different levels of branching and compaction in polysaccharides.

(a) **Sugar phosphate**

Fructose 1,6 bisphosphate is involved in the reactions that liberate energy from glucose. (The numbers in its name refer to the carbon sites of phosphate bonding; *bis-* indicates that two phosphates are present.)

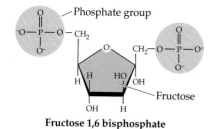

Phosphate group

Fructose

Fructose 1,6 bisphosphate

(b) **Amino sugars**

The monosaccharides glucosamine and galactosamine are amino sugars with an amino group in place of a hydroxyl group.

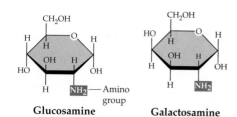

Amino group

Glucosamine **Galactosamine**

(c) **Chitin**

Chitin is a polymer of *N*-acetylglucosamine; *N*-acetyl groups provide additional sites for hydrogen bonding between the polymers.

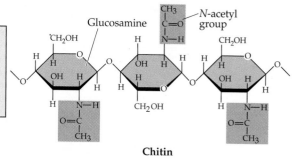

Glucosamine

N-acetyl group

Chitin

3.15 Chemically Modified Carbohydrates
Added functional groups modify the form and properties of a carbohydrate.

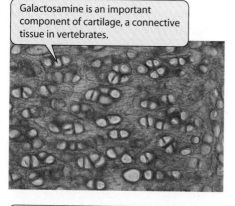

Galactosamine is an important component of cartilage, a connective tissue in vertebrates.

The external skeletons of insects are made up of chitin.

---

is a good storage medium that can be easily broken down to supply glucose for energy-producing reactions, while cellulose is an excellent structural material that can withstand harsh environmental conditions without changing.

Starch actually comprises a large family of giant molecules of broadly similar structure. While all starches are large polymers of glucose with α-linkages, the different starches can be distinguished by the amount of branching that occurs at carbons 1 and 6 (Figure 3.14*b*). Some starches are highly branched; others are not. Plant starches, called *amylose*, are not highly branched. The polysaccharide *glycogen*, which stores glucose in animal livers and muscles, is highly branched.

Starch and glycogen serve as energy storage compounds for plants and animals, respectively. These polysaccharides are readily hydrolyzed to glucose monomers, which in turn can be further degraded to liberate and convert their stored energy to forms that can be used for cellular activities. If it is glucose that is actually needed for fuel, why must it be stored as a polymer? The reason is that 1,000 glucose molecules would exert 1,000 times the osmotic pressure (causing water to enter the cells; see Chapter 5) of a single glycogen molecule. If it were not for polysaccharides, many organisms would expend a lot of time and energy expelling excess water.

Chemically modified carbohydrates contain other groups

Some carbohydrates are chemically modified by adding functional groups such as phosphate and amino groups (Figure 3.15). For example, carbon 6 in glucose may be oxidized from —CH_2OH to a carboxyl group (—COOH), producing glucuronic acid. Or a phosphate group may be added to one or more of the —OH sites. Some of these *sugar phosphates*, such as fructose 1,6-bisphosphate, are important intermediates in cellular energy reactions.

When an amino group is substituted for an —OH group, *amino sugars*, such as glucosamine and galactosamine, are produced. These compounds are important in the extracellular matrix, where they form parts of proteins involved in keeping tissues together. Galactosamine is a major component of cartilage, the material that forms caps on the ends of bones and stiffens the protruding parts of the ears and nose. A derivative of glucosamine produces the polymer *chitin*, which is the principal structural polysaccharide in the skeletons of insects, crabs, and lobsters, as well as in the cell walls of fungi. Fungi and insects (and their relatives) constitute more than 80 percent of the species ever described, and chitin is one of the most abundant substances on Earth.

Nucleic Acids: Informational Macromolecules

The **nucleic acids** are linear polymers specialized for the storage, transmission, and use of information. There are two types of nucleic acids: DNA (deoxyribonucleic acid) and RNA (ribonucleic acid). DNA molecules are giant polymers that encode hereditary information and pass it from generation to generation. Through an RNA intermediate, the information encoded in DNA is also used to make specific proteins. RNA molecules of various types copy the information in segments of DNA to specify the sequence of amino acids in proteins. Information flows from DNA to DNA in reproduction, but in the nonreproductive activities of the cell, information flows from DNA to RNA to proteins, which ultimately carry out these functions. What compositions, structures, and properties of nucleic acids permit them to play these fundamental roles in living systems?

The nucleic acids have characteristic properties

Nucleic acids are composed of monomers called **nucleotides**, each of which consists of a pentose sugar, a phosphate group, and a nitrogen-containing base—either a pyrimidine or a purine (Figure 3.16). (Molecules consisting

The base may be either

a pyrimidine:

or a purine:

3.16 Nucleotides Have Three Components
A nucleotide consists of a phosphate group, a pentose sugar, and a nitrogen-containing base—all linked together by covalent bonds. The nitrogenous bases fall into two categories: Purines have two fused rings, and the smaller pyrimidines have a single ring.

Base + ⬠ = ⬠ + Ⓟ = Ⓟ⬠
Ribose or   **Nucleoside**   Phosphate   **Nucleotide**
deoxyribose

of a pentose sugar and a nitrogenous base, but no phosphate group, are called *nucleosides*.) In DNA, the pentose sugar is deoxyribose, which differs from the ribose found in RNA by one oxygen atom (see Figure 3.12).

In both RNA and DNA, the backbone of the molecule consists of alternating pentose sugars and phosphates (sugar—phosphate—sugar—phosphate—). The bases are attached to the sugars and project from the chain (Figure 3.17). The nucleotides are joined by covalent bonds in what are

3.17 Distinguishing Characteristics of DNA and RNA
RNA is usually a single strand. DNA usually consists of two strands running in opposite directions.

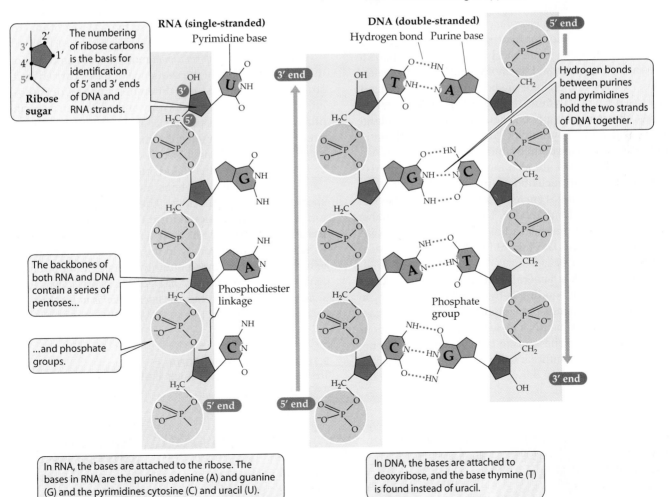

In RNA, the bases are attached to the ribose. The bases in RNA are the purines adenine (A) and guanine (G) and the pyrimidines cytosine (C) and uracil (U).

In DNA, the bases are attached to deoxyribose, and the base thymine (T) is found instead of uracil.

called **phosphodiester linkages** between the sugar of one nucleotide and the phosphate of the next ("-diester" refers to the two bonds formed by —OH groups reacting with acidic phosphate groups). The phosphate groups link carbon 3 in one pentose sugar to carbon 5 in the adjacent sugar.

Most RNA molecules consist of only one polynucleotide chain. DNA, however, is usually double-stranded; it has two polynucleotide chains held together by hydrogen bonding between their nitrogenous bases. The two strands of DNA run in opposite directions. You can see what this means by drawing an arrow through the phosphate group from carbon 5 to carbon 3 in the next ribose. If you do this for both strands, the arrows point in opposite directions. This antiparallel orientation is necessary for the strands to fit together in three-dimensional space.

The uniqueness of a nucleic acid resides in its base sequence

Only four nitrogenous bases—and thus only four nucleotides—are found in DNA. The DNA bases and their abbreviations are adenine (A), cytosine (C), guanine (G), and thymine (T).

A key to understanding the structures and functions of nucleic acids is the principle of **complementary base pairing** through hydrogen bond formation. In double-stranded DNA, *adenine and thymine always pair (AT)*, and *cytosine and guanine always pair (CG)*.

Base pairing is complementary because of three factors: the corresponding sites for hydrogen bonding, the geometry of the sugar–phosphate backbone that brings opposite bases near each other, and the molecular sizes of the paired bases. Adenine and guanine are both purines, consisting of two fused rings. Thymine and cytosine are both pyrimidines, consisting of only one ring. The pairing of a large purine with a small pyrimidine ensures a stable and consistent dimension to the double-stranded molecule of DNA.

Ribonucleic acids are also made up of four different monomers, but the nucleotides differ from those of DNA. In RNA the nucleotides are termed *ribonucleotides* (the ones in DNA are *deoxyribonucleotides*). They contain ribose rather than deoxyribose, and instead of the base thymine, RNA uses the base uracil (U) (Table 3.3). The other three bases are the same as in DNA.

Although RNA is generally single-stranded, complementary hydrogen bonding between ribonucleotides can take place. These bonds play important roles in determining the shapes of some RNA molecules and in associations between RNA molecules during protein synthesis. During the DNA-directed synthesis of RNA, complementary base pairing also takes place between ribonucleotides and the

3.3 **Distinguishing RNA from DNA**		
NUCLEIC ACID	SUGAR	BASES
RNA	Ribose	Adenine Cytosine Guanine **Uracil**
DNA	Deoxyribose	Adenine Cytosine Guanine **Thymine**

bases of DNA. In RNA, guanine and cytosine pair (GC) as in DNA, but adenine pairs with uracil (AU). Adenine in an RNA strand can pair either with uracil (in another RNA strand) or with thymine (in a DNA strand).

The three-dimensional appearance of DNA is strikingly uniform. The segment shown in Figure 3.18 could be from any DNA molecule. Through hydrogen bonding, the two complementary polynucleotide strands pair and twist to form a double helix. When compared with the complex and varied tertiary structures of different proteins, this uniformity is surprising. But this structural contrast makes sense in terms of the functions of these two classes of macromolecules.

DNA is a purely *informational* molecule. The information in DNA is encoded in the sequence of bases carried in its strands. Its variations—the different sequences of bases—are "internal." It can be read easily and reliably, in a specific order. A uniformly shaped molecule like DNA can be interpreted by standard molecular machinery, and any cell's machinery can read any molecule of DNA.

Proteins, on the other hand, have good reason to be so varied. In particular, proteins must recognize their own specific "target" molecules. They do this by having a unique three-dimensional form that can match at least a portion of the surface of their target molecules. In other words, structural diversity in the molecules to which proteins bind requires corresponding diversity in the structure of the proteins themselves. *In DNA, the information is in the sequence of bases; in proteins, the information is in the shape.*

3.18 The Double Helix of DNA
The backbones of the two strands in a DNA molecule are coiled in a double helix.

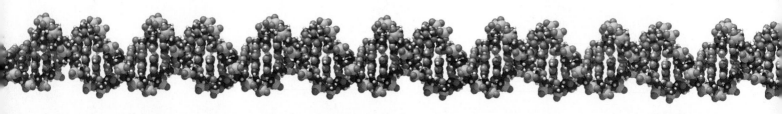

DNA is a guide to evolutionary relationships

Because DNA carries hereditary information between generations, a theoretical series of DNA molecules with changes in base sequences stretches back through evolutionary time. Of course, we cannot study all of these DNA molecules, because many of their organisms have become extinct. However, we can study the DNA of living organisms, which are judged to have changed little through millions of years. Comparisons and contrasts of these DNA molecules can be added to evidence from fossils and other sources to reveal the evolutionary record, as we will see in Chapter 24.

Closely related living species should have more similar base sequences than species judged by other criteria to be more distantly related. Indeed, this is the case. The examination of base sequences confirms many of the evolutionary relationships that have been inferred from the more traditional study of body structures or studies of biochemistry and physiology. For example, the closest living relative of humans (*Homo sapiens*) is the chimpanzee (genus *Pan*), which shares more than 98 percent of its DNA base sequence with human DNA.

This confirmation of well-established evolutionary relationships gives credibility to the use of DNA to elucidate relationships when studies of structure are not possible or are not conclusive. For example, DNA studies revealed a close evolutionary relationship between starlings and mockingbirds that was not expected on the basis of anatomy or behavior.

DNA studies support the division of the prokaryotes into two domains, Bacteria and Archaea. Each of these two groups of prokaryotes is as distinct from the other as either is from the Eukarya, the third domain into which living things are classified (see Chapter 1). In addition, DNA comparisons support the hypothesis that certain subcellular compartments of eukaryotes (the organelles called mitochondria and chloroplasts) evolved from early bacteria that established a stable and mutually beneficial way of life inside larger cells.

Nucleotides have other important roles in the cell

Nucleotides are more than just the building blocks of nucleic acids. As we will describe in later chapters, there are several nucleotides with other functions:

▸ ATP (adenosine triphosphate) acts as an energy transducer in many biochemical reactions (see Chapter 6).

▸ GTP (guanosine triphosphate) serves as an energy source, especially in protein synthesis. It also has a role in the transfer of information from the environment to the body tissues (see Chapter 12 and 15).

▸ cAMP (cyclic AMP), a special nucleotide in which a bond forms between the sugar and phosphate groups within adenosine monophosphate, is essential in many processes, including the actions of hormones and the transmission of information by the nervous system (see Chapter 15).

Lipids: Water-Insoluble Molecules

Lipids are a chemically diverse group of hydrocarbons. The property they all share is an insolubility in water, which is due to the presence of many nonpolar covalent bonds. As we saw in Chapter 2, nonpolar hydrocarbon molecules preferentially aggregate among themselves, away from water, which is polar. When the nonpolar molecules are sufficiently close together, weak but additive van der Waals forces hold them together. These huge macromolecular aggregations are not polymers in a strict chemical sense, since their units (lipid molecules) are not held together by covalent bonds, as are, for example, amino acids in proteins. But they can be considered polymers of individual lipid units.

In this section, we will describe the different types of lipids. Lipids have a number of roles in living organisms:

▸ Fats and oils store energy.
▸ Phospholipids play important structural roles in cell membranes.
▸ The carotenoids help plants capture light energy.
▸ Steroids and modified fatty acids play regulatory roles as hormones and vitamins.
▸ The fat in animal bodies serves as thermal insulation.
▸ A lipid coating around nerves acts as electrical insulation.
▸ Oil or wax on surfaces of skin, fur, and feathers repel water.

Fats and oils store energy

Chemically, fats and oils are **triglycerides**, also known as *simple lipids*. Triglycerides that are solid at room temperature (20°C) are called *fats*; those that are liquid at room temperature are called *oils*. Triglycerides are composed of two types of building blocks: fatty acids and glycerol. Glycerol is a small molecule with three hydroxyl (—OH) groups. Fatty acids are made up of a long nonpolar hydrocarbon

Allow your eyes to follow the yellow phosphorus atoms and their attached red oxygen atoms in the two helical backbones.

The paired bases are stacked in the center of the coil; concentrate on the lighter blue nitrogen and darker blue carbon atoms.

The small white atoms are hydrogens.

Triglyceride

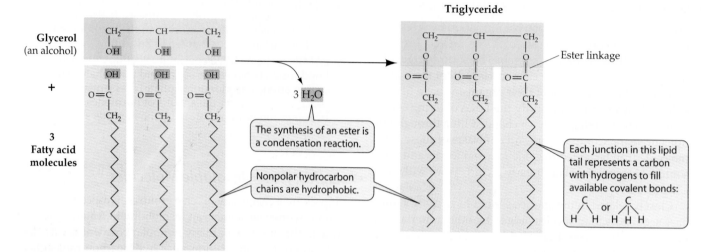

3.19 Synthesis of a Triglyceride
In living things, the reaction that forms triglycerides is more complex, but the end result is as shown here.

chain and a polar carboxyl functional group (—COOH). A triglyceride contains three fatty acid molecules and one molecule of glycerol (Figure 3.19).

The carboxyl group of a fatty acid can react with the hydroxyl group of glycerol to form an **ester** (the reaction product of an acid and an alcohol) and water. The three fatty acids in a triglyceride molecule need not all have the same hydrocarbon chain length or structure.

In *saturated* fatty acids, all the bonds between the carbon atoms in the hydrocarbon chain are single bonds—there are no double bonds. That is, all the bonds are saturated with hydrogen atoms (Figure 3.20a). These fatty acid molecules are relatively rigid and straight, and they pack together tightly, like pencils in a box.

In *unsaturated* fatty acids, the hydrocarbon chain contains one or more double bonds. Oleic acid, for example, is a *monounsaturated* fatty acid that has one double bond near the middle of the hydrocarbon chain, which causes a kink in the molecule (Figure 3.20b). Some fatty acids have more than one double bond—are *polyunsaturated*—and have multiple kinks. These kinks prevent the molecules from packing together tightly.

The kinks are important in determining the fluidity and melting point of a lipid. Animal fats are usually solids at room temperature, and their triglycerides tend to have many long-chain saturated fatty acids, packed well together. The triglycerides of plants, such as corn oil, tend to have short or unsaturated fatty acids. Because of their kinks, these fatty acids pack poorly together, and these triglycerides are usually liquids at room temperature.

3.20 Saturated and Unsaturated Fatty Acids
(a) In saturated fatty acids, the straight chain allows the molecule to pack tightly among other similar molecules. (b) In unsaturated fatty acids, kinks in the chain prevent close packing.

(a)

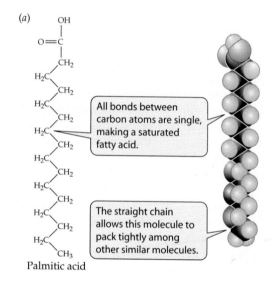

Palmitic acid

(b)

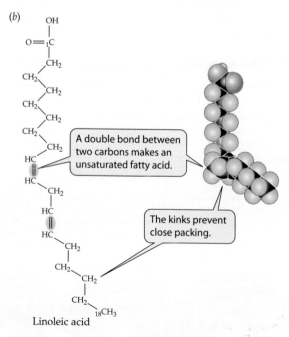

Linoleic acid

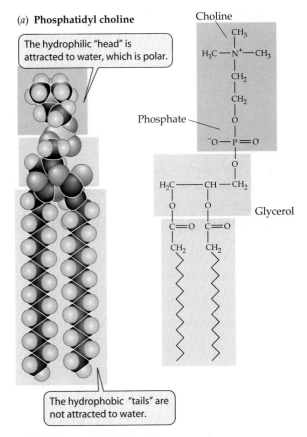

(a) **Phosphatidyl choline**

> The hydrophilic "head" is attracted to water, which is polar.

Choline

Phosphate

Glycerol

> The hydrophobic "tails" are not attracted to water.

(b) **Membrane phospholipid**

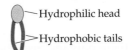

Hydrophilic head

Hydrophobic tails

> Throughout this book, phospholipids in membranes are shown with this symbol.

3.21 Phospholipid Structure

(a) Phosphatidyl choline (lecithin) demonstrates the structure of a phospholipid molecule. In other phospholipids, the amino acid serine, the sugar alcohol inositol, or other compounds replace choline. *(b)* This generalized symbol is used in this book to represent a membrane phospholipid (see Figure 3.22).

Fats and oils are marvelous storehouses for energy. By taking in excess food, many animal species deposit fat droplets in their cells as a means of storing energy. Some plant species, such as olives, avocados, sesame, castor beans, and all nuts, have substantial amounts of lipids in their seeds or fruits that serve as energy reserves for the next generation. This energy can be tapped by people who eat these plant oils or use them for fuel. Indeed, the famous German engineer Rudolf Diesel used peanut oil to power one of his early automobile engines in 1900. As petroleum stores become depleted, there is some interest in using plant oils commercially as fuel.

Phospholipids form the core of biological membranes

Because lipids and water do not interact, a mixture of water and lipids forms two distinct phases. Many biologically im-

portant substances—such as ions, sugars, and free amino acids—that are soluble in water are insoluble in lipids.

Like triglycerides, **phospholipids** contain fatty acids bound to glycerol by ester linkages. In phospholipids, however, any one of several phosphate-containing compounds replaces one of the fatty acids (Figure 3.21). The phosphate functional group has a negative electric charge, so this portion of the molecule is hydrophilic, attracting polar water molecules. But the two fatty acids are hydrophobic, so they aggregate away from water.

In an aqueous environment, phospholipids line up in such a way that the nonpolar, hydrophobic "tails" pack tightly together, and the phosphate-containing "heads" face outward, where they interact with water. The phospholipids thus form a *bilayer*, a sheet two molecules thick, from which water is excluded (Figure 3.22). Biological membranes have this kind of lipid bilayer structure, and we will devote all of Chapter 5 to their biological functions.

Because the word "lipid" defines compounds in terms of their solubility rather than their structural similarity, a great variety of different chemical structures are included as lipids.

Carotenoids and steroids

The next two lipid classes we'll discuss—the carotenoids and the steroids—have chemical structures very different from those of triglycerides and phospholipids and from each other. Both carotenoids and steroids are synthesized by covalent linking and chemical modification of isoprene to form a series of isoprene units:

$$CH_2 = C - CH = CH_2$$
$$\overset{|}{CH_3}$$

CAROTENOIDS TRAP LIGHT ENERGY. The **carotenoids** are a family of light-absorbing pigments found in plants and animals. Beta-carotene (β-carotene) is one of the pigments that traps light energy in leaves during photosynthesis. In humans, a molecule of β-carotene can be broken down into

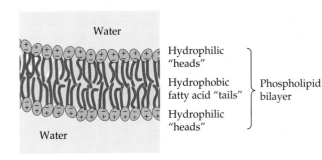

Water

Hydrophilic "heads"

Hydrophobic fatty acid "tails"

Hydrophilic "heads"

Phospholipid bilayer

Water

3.22 Phospholipids Form a Bilayer

In an aqueous environment, hydrophobic interactions bring the "tails" of phospholipids together in the interior of a phospholipid bilayer. The hydrophilic "heads" face outward on both sides of the bilayer, where they interact with the surrounding water molecules.

3.23 β-Carotene is the Source of Vitamin A
The carotenoid β-carotene is symmetrical around its central double bond; when split, β-carotene becomes two vitamin A molecules. The simplified structural formula used here is standard chemical shorthand for large organic molecules with many carbon atoms. Structural formulas are simplified by omitting the C's (for carbon atoms) at the intersections of the lines representing covalent bonds. H's to fill all the available bonding sites on each C are assumed.

two vitamin A molecules (Figure 3.23), from which we make the pigment rhodopsin, which is required for vision. Carotenoids are responsible for the colors of carrots, tomatoes, pumpkins, egg yolks, and butter.

STEROIDS ARE SIGNAL MOLECULES. The **steroids** are a family of organic compounds whose multiple rings share carbons (Figure 3.24). The steroid cholesterol is an important constituent of membranes. Other steroids function as hormones, chemical signals that carry messages from one part of the body to another. Testosterone and the estrogens are steroid hormones that regulate sexual development in vertebrates. Cortisol and related hormones play many regulatory roles in the digestion of carbohydrates and proteins, in the maintenance of salt balance and water balance, and in sexual development.

Cholesterol is synthesized in the liver and is the starting material for making testosterone and other steroid hormones, as well as the bile salts that help break down dietary fats so that they can be digested. Cholesterol is absorbed from foods such as milk, butter, and animal fats. An

excess of cholesterol in the blood can lead to its deposition (along with other substances) in the arteries, a condition that may lead to arteriosclerosis and heart attack.

Some lipids are vitamins

Vitamins are small organic molecules that are not synthesized in the body and must be acquired from dietary sources, and whose deficiencies lead to defined diseases.

Vitamin A is formed from the β-carotene found in green and yellow vegetables (see Figure 3.23). In humans, a deficiency of vitamin A leads to dry skin, eyes, and internal body surfaces; retarded growth and development; and night blindness, which is a diagnostic symptom for the deficiency. Vitamin D regulates the absorption of calcium from the intestines. It is necessary for the proper deposition of calcium in bones; a deficiency of vitamin D can lead to rickets, a bone-softening disease.

Vitamin E seems to protect cells from damaging effects of oxidation–reduction reactions. For example, it has an important role in preventing unhealthy changes in the double bonds in the unsaturated fatty acids of membrane phospholipids. Commercially, vitamin E is added to some foods to slow spoilage. Vitamin K is found in green leafy plants and is also synthesized by bacteria normally present in the human intestine. This vitamin is essential to the formation of blood clots. Predictably, a deficiency of vitamin K leads to slower clot formation and potentially fatal bleeding from a wound.

Wax coatings repel water

The sheen on human hair is not there only for cosmetic purposes. Glands in the skin secrete a waxy coating that repels water and keeps the hair pliable. Birds that live near water have a similar waxy coating on their feathers. The shiny leaves of holly plants, familiar during winter holidays, also have a waxy coating. Finally, bees make their honeycombs out of wax. All waxes have the same basic structure: They are formed by an ester linkage between a saturated, long-chain fatty acid and a saturated, long-chain

3.24 All Steroids Have the Same Ring Structure
The steroids shown, all important in vertebrates, are composed of carbon and hydrogen and are highly hydrophobic. However, small chemical variations, such as the presence or absence of a methyl or hydroxyl group, can produce enormous functional differences.

Cholesterol is a constituent of membranes and is the source of steroid hormones.

Vitamin D₂ can be produced in the skin by the action of light on a cholesterol derivative.

Cortisol is a hormone secreted by the adrenal glands.

Testosterone is a male sex hormone.

alcohol. The result is a very long molecule, with 40–60 CH_2 groups. For example, here is the structure of beeswax:

$$CH_3 — (CH_2)_{14} — \overset{\overset{\displaystyle O}{\|}}{C} — O—CH_2—(CH_2)_{28}—CH_3$$

$\underbrace{\phantom{CH_3 — (CH_2)_{14} — C}}_{\textbf{Fatty acid}}$ $\underbrace{\phantom{O—CH_2—(CH_2)_{28}—CH_3}}_{\textbf{Alcohol}}$

This highly nonpolar structure accounts for the impermeability of wax to water.

The Interactions of Macromolecules

We have treated the classes of macromolecules as if each were separate from the others. In cells, however, certain macromolecules of different classes may be covalently bonded to one another. Proteins with attached oligosaccharides are called *glycoproteins* (glyco-, "sugar"). The specific oligosaccharide chain attached can determine where within the cell a newly synthesized protein will reside. Other carbohydrate chains covalently bond to lipids, resulting in *glycolipids*, which reside in the cell surface membrane, with the carbohydrate chain extending out into the cell's environment. The carbohydrates that determine a person's blood type (A, B, AB, or O) are attached to either proteins or lipids sticking out from the surfaces of red blood cells.

We have already mentioned the fact that proteins can bind noncovalently to other proteins in quaternary structures. But proteins can bind noncovalently to the other types of macromolecules as well. For example, there are hundreds of different proteins that recognize and bind to DNA, regulating its function. Other proteins, in combination with cholesterol and other lipids, form lipoproteins. Some lipoproteins serve as carrier proteins, which make it possible to move very hydrophobic lipids such as cholesterol through water-rich environments such as the blood.

Summary

Macromolecules: Giant Polymers

▶ Macromolecules are constructed by the formation of covalent bonds between smaller molecules called monomers. Macromolecules include polysaccharides, proteins, and nucleic acids. **Review Figure 3.1 and Table 3.1**

▶ Macromolecules have specific, characteristic three-dimensional shapes that depend on the structures, properties, and sequence of their monomers. Different functional groups give local sites on macromolecules specific properties that are important for their biological functioning and their interactions with other macromolecules.

Condensation Reactions

▶ Monomers are joined by condensation reactions, which release a molecule of water for each bond formed. Hydrolysis reactions use water to break polymers into monomers. **Review Figure 3.2**

Proteins: Polymers of Amino Acids

▶ The functions of proteins include support, protection, catalysis, transport, defense, regulation, and movement. Protein function sometimes requires an attached prosthetic group.

▶ There are 20 amino acids found in proteins. Each amino acid consists of an amino group, a carboxyl group, a hydrogen, and a side chain bonded to the α carbon atom. **Review Table 3.2**

▶ The side chains of amino acids may be charged, polar, or hydrophobic; there are also "special cases," such as the —SH groups, which can form disulfide bridges. The side chains give different properties to each of the amino acids. **Review Table 3.2 and Figure 3.3**

▶ Amino acids are covalently bonded together by peptide linkages, which form by condensation reactions between the carboxyl and amino groups. **Review Figure 3.4**

▶ The polypeptide chains of proteins are folded into specific three-dimensional shapes. Four levels of structure are possible: primary, secondary, tertiary, and quaternary.

▶ The primary structure of a protein is the sequence of amino acids bonded by peptide linkages. This primary structure determines both the higher levels of structure and protein function. **Review Figure 3.5a**

▶ Secondary structures of proteins, such as α helices and β pleated sheets, are maintained by hydrogen bonds between atoms of the amino acid residues. **Review Figure 3.5b,c**

▶ The tertiary structure of a protein is generated by bending and folding of the polypeptide chain. **Review Figures 3.5d, 3.6**

▶ The quaternary structure of a protein is the arrangement of polypeptides in a single functional unit consisting of more than one polypeptide subunit. **Review Figures 3.5e, 3.7**

▶ Weak chemical interactions are important in the binding of proteins to other molecules. **Review Figure 3.8**

▶ Proteins denatured by heat, acid, or certain chemicals lose their tertiary and secondary structure as well as their biological function. Renaturation is not always possible. **Review Figure 3.9**

▶ Chaperonins assist protein folding by preventing binding to inappropriate ligands. **Review Figure 3.10**

Carbohydrates: Sugars and Sugar Polymers

▶ All carbohydrates contain carbon bonded to H and OH groups.

▶ Hexoses are monosaccharides that contain six carbon atoms. Examples of hexoses include glucose, galactose, and fructose, which can exist as chains or rings. **Review Figures 3.11, 3.12**

▶ The pentoses are five-carbon monosaccharides. Two pentoses, ribose and deoxyribose, are components of the nucleic acids RNA and DNA, respectively. **Review Figure 3.12**

▶ Glycosidic linkages may have either α or β orientation in space. They covalently link monosaccharides into larger units such as disaccharides (for example, cellobiose), oligosaccharides, and polysaccharides. **Review Figures 3.13, 3.14**

▶ Cellulose, a very stable glucose polymer, is the principal component of the cell walls of plants. It is formed by glucose units linked together by β-glycosidic linkages between carbons 1 and 4. **Review Figure 3.14**

▶ Starches, less dense and less stable than cellulose, store energy in plants. Starches are formed by α-glycosidic linkages between carbons 1 and 4 and are distinguished by the amount of branching that occurs through glycosidic bond formation at carbon 6. **Review Figure 3.14**

▶ Glycogen contains α-1,4 glycosidic linkages and is highly branched. Glycogen stores energy in animal livers and muscles. **Review Figure 3.14**

▶ Chemically modified monosaccharides include the sugar phosphates and amino sugars. A derivative of the amino sugar glucosamine polymerizes to form the polysaccharide chitin, which is found in the cell walls of fungi and the exoskeletons of insects. **Review Figure 3.15**

Nucleic Acids: Informational Macromolecules

▶ In cells, DNA is the hereditary material. Both DNA and RNA play roles in the formation of proteins. Information flows from DNA to RNA to protein.

▶ Nucleic acids are polymers made up of nucleotides. Nucleotides consist of a phosphate group, a sugar (ribose in RNA and deoxyribose in DNA), and a nitrogen-containing base. In DNA the bases are adenine, guanine, cytosine, and thymine, but in RNA uracil substitutes for thymine. **Review Figure 3.16 and Table 3.3**

▶ In the nucleic acids, the bases extend from a sugar–phosphate backbone. The information content of DNA and RNA resides in their base sequences.

▶ RNA is single-stranded. DNA is a double-stranded helix in which there is complementary, hydrogen-bonded base pairing between adenine and thymine (AT) and guanine and cytosine (GC). The two strands of the DNA double helix run in opposite directions. **Review Figures 3.17, 3.18**

▶ Comparing the DNA base sequences of different living species provides information on their evolutionary relatedness.

Lipids: Water-Insoluble Molecules

▶ Although lipids can form gigantic structures, such as lipid droplets and membranes, these aggregations are not chemically macromolecules because the individual units are not linked by covalent bonds.

▶ Fats and oils are composed of three fatty acids covalently bonded to a glycerol molecule by ester linkages. **Review Figure 3.19**

▶ Saturated fatty acids have a hydrocarbon chain with no double bonds. The hydrocarbon chains of unsaturated fatty acids have one or more double bonds that bend the chain, making close packing less possible. **Review Figure 3.20**

▶ Phospholipids have a hydrophobic hydrocarbon "tail" and a hydrophilic phosphate "head." **Review Figure 3.21**

▶ In water, the interactions of the hydrophobic tails and hydrophilic heads generate a phospholipid bilayer that is two molecules thick. The head groups are directed outward, where they interact with the surrounding water. The tails are packed together in the interior of the bilayer. **Review Figure 3.22**

▶ Carotenoids trap light energy in green plants. β-Carotene can be split to form vitamin A, a lipid vitamin. **Review Figure 3.23**

▶ Some steroids, such as testosterone, function as hormones. Cholesterol is synthesized by the liver and has a role in some cell membranes, as well as in the digestion of other fats. Too much cholesterol in the diet can lead to arteriosclerosis. **Review Figure 3.24**

▶ Vitamins are substances that are required for normal functioning but that must be acquired from the diet.

The Interactions of Macromolecules

▶ Both covalent and noncovalent linkages are found between the various classes of macromolecules.

▶ Glycoproteins contain an oligosaccharide "label" that directs the protein to the proper cell destination. The carbohydrate groups of glycolipids are displayed on the cell's outer surface, where they serve as recognition signals.

▶ Hydrophobic interactions bind cholesterol to the protein that transports it in the blood.

For Discussion

1. Phospholipids make up a major part of every biological membrane; cellulose is the major constituent of the cell walls of plants. How do the chemical structures and physical properties of phospholipids and cellulose relate to their functions in cells?

2. Suppose that, in a given protein, one lysine is replaced by aspartic acid (see Table 3.2). Does this change occur in the primary structure or in the secondary structure? How might it result in a change in tertiary structure? In quaternary structure?

3. If there are 20 different amino acids commonly found in proteins, how many different dipeptides are there? How many different tripeptides? How many different trinucleotides? How many different single-stranded RNAs composed of 200 nucleotides?

4. Contrast the following three structures: hemoglobin, a DNA molecule, and a protein that spans a biological membrane.

 Self-quizzes and Supplemental Readings for each chapter are on the Student Web Site /CD-ROM.

This scale is logarithmic. Each unit is ten times bigger than the previous unit.

| 0.1 nm | 1 nm | 10 nm | 100 nm | 1 μm | 10 μm | 100 μm | 1 mm | 1 cm | 0.1 m | 1 m | 10 m | 100 m | 1 km |

Unaided eye

Light microscope

Electron microscope

Atoms

Lipids

Small molecules

Protein

T2 phage

Chloroplast

Most bacteria

Plant and animal cells

Fish egg

Hummingbird

Human

Blue whale

Giant redwood tree

Most cell diameters are in the range of 1–100 μm.

4.1 The Scale of Life
The scale shows the relative sizes of molecules, cells, and multicellular organisms. The measurements are defined in the table on the inside front cover.

As a cell increases in volume, its surface area also increases, but not to the same extent (Figure 4.2). This phenomenon has great biological significance because the volume of a cell determines the amount of chemical activity it carries out per unit of time, but the surface area determines the amount of substances the cell can take in from the outside environment and the amount of waste products it can release to the environment. As the living cell grows, its rate of waste production and its need for resources increase faster than the surface area This explains why large organisms must consist of many small cells: *Cells are small in volume in order to maintain a large surface area-to-volume ratio.*

In a multicellular organism, the large surface area represented by the multitude of small cells that make up the whole organism enables the multitude of functions required for survival. Special structures transport food, oxygen, and waste materials to and from the small cells that are distant from the external surface of the organism.

Microscopes are needed to visualize cells

Cell are usually invisible to the human eye. The smallest object a person can typically discern is about 0.2 mm (200 μm) in size. We refer to this measure as *resolution*, the distance apart two objects must be in order for them to be distinguished as separate; if they are closer together, they appear as a single blur. Many cells are much smaller than 200 μm. *Microscopes* are used to improve resolution so that cells and their internal structures can be seen.

There are two basic types of microscopes: light microscopes and electron microscopes. The **light microscope** (LM) uses glass lenses and visible light to form a magnified image of an object. In its contemporary form, the light microscope has a resolving power of about 0.2 μm, which is 1,000 times that of the human eye. This allows visualization of cell sizes and shapes and some internal cell structures. The latter are hard to see under ordinary light, so cells are often killed and stained with dyes to make the structures stand out.

An **electron microscope** (EM) uses powerful magnets to focus an electron beam, much as the light microscope employs glass lenses to focus a beam of light. Since we cannot see electrons, the electron microscope directs them at a fluorescent screen or a photographic film to create a visible image. The resolving power of electron microscopes is about 0.5 nm, which is 250,000 times finer than that of the human eye. This permits the details of many subcellular structures to be distinguished.

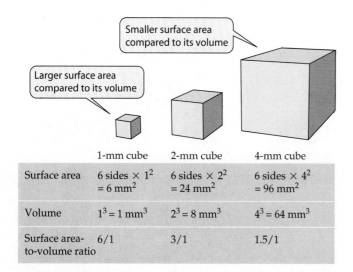

Smaller surface area compared to its volume

Larger surface area compared to its volume

	1-mm cube	2-mm cube	4-mm cube
Surface area	6 sides × 1² = 6 mm²	6 sides × 2² = 24 mm²	6 sides × 4² = 96 mm²
Volume	1³ = 1 mm³	2³ = 8 mm³	4³ = 64 mm³
Surface area-to-volume ratio	6/1	3/1	1.5/1

4.2 Why Cells Are Small
As an object grows, its volume increases more rapidly than its surface area. Cells must maintain a large surface area-to-volume ratio in order to function, which explains why large organisms must be composed of many small cells rather than a few huge ones.

4

The Organization of Cells

JANE, A RESIDENT OF A NURSING HOME, ENJOYED VISits from her grandchildren. They often brought her gifts, but on this winter day, an unprotected sneeze from one of them led to her catching a cold. Over the next week, she stayed in bed while her body fought off this infection. But while her body was weakened by the cold, a more dangerous event occurred.

Tiny bacteria called *Streptococcus pneumoniae* had been living in Jane's throat for several months. Now they began to multiply, invading healthy tissues in her lungs. The lung tissues reacted by swelling—the hallmark of pneumonia. Jane's doctor arrived to find her with shaking chills, chest pain, and a cough that produced a greenish ooze. After capturing some of the green material in a test tube, the physician sent it to a laboratory where the bacteria were clearly seen and a diagnosis of bacterial pneumonia was confirmed. The doctor prescribed antibiotics, which killed the bacteria, and Jane recovered within a week.

Both the bacteria that caused Jane's pneumonia and the lung tissues they invaded are made up of cells, the units of biological structure and function. Bacteria and lung cells are very different in size and structure. The fundamental differences between them allowed the antibiotic to kill the bacteria while sparing the lung cells. Nevertheless, both bacteria and lung cells perform similar biological functions, taking in substances from their environment and refashioning them for their own uses.

The next four chapters are about how cells perform their basic functions. We begin by describing the structural features of simple cells, such as bacteria, and more complex ones, like lung cells. We will discuss cell sizes and shapes, and we will describe the surface structures and internal compartments that permit cells to transform energy, move, change shape, communicate, and maintain internal conditions that are different from their immediate surroundings.

The Cell: The Basic Unit of Life

All organisms are composed of cells. All cells come from preexisting cells. These two statements constitute the **cell theory**. Just as atoms are the units of chemistry, cells are the building blocks of life. They are composed of water molecules and the small and large molecules we examined in the pre-

vious two chapters. Each cell contains at least 10,000 different types of molecules, most of them present in many copies. Cells use these molecules to transform matter and energy, to respond to their environment, and to reproduce themselves.

The cell theory has two important implications. First, it means that studying cell biology is in some sense the same as studying life. The principles that underlie the functions of the single cell in a bacterium are similar to those governing the 60 trillion cells your body. Second, it means that life is continuous. All those cells in your body came from a single cell, the fertilized egg, which came from the fusion of two cells, a sperm and an egg from your parents, whose cells came from *their* fertilized eggs, and so on.

Cell size is limited by the surface area-to-volume ratio

Most cells are tiny. The volume of cells ranges from 1 to 1,000 µm^3 (Figure 4.1). The eggs of some birds are enormous exceptions, to be sure, and individual cells of several types of algae and bacteria are large enough to be viewed with the unaided eye. And although neurons (nerve cells) have a volume that is within the "normal" cell range, they often have fine projections that may extend for meters, carrying signals from one part of a large animal to another. But by and large, cells are minuscule. The reason for this relates to the change in the *surface area-to-volume ratio (SA/V)* of any object as it increases in size.

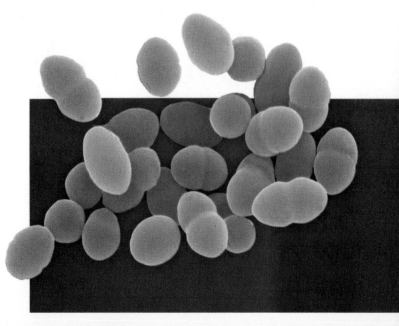

Units of Life, Agents of Disease
Streptococcus pneumoniae are bacterial cells—units of biological structure and function and, in the human lung, a cause of disease.

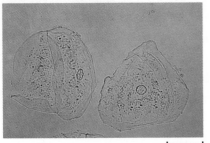

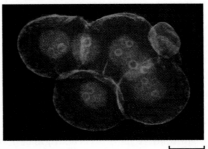

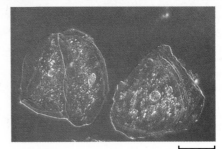

25 μm

25 μm

25 μm

In **bright-field microscopy**, light passes directly through the cells. Unless natural pigments are present, there is little contrast and details are not distinguished.

In **phase-contrast microscopy**, contrast in the image is increased by emphasizing differences in refractive index (the capacity to bend light), thereby enhancing light and dark regions in the cell.

Differential interference-contrast microscopy (Nomarski optics) uses two beams of polarized light. The combined images look as if the cell is casting a shadow on one side.

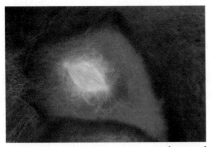

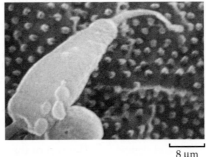

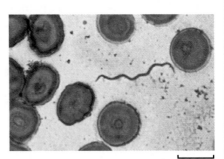

40 μm

40 μm

75 μm

In **fluorescence microscopy**, a natural substance in the cell or a fluorescent dye that binds to a specific cell material is stimulated by a beam of light, and the longer-wavelength fluorescent light is observed coming directly from the dye.

Confocal microscopy uses fluorescent materials but adds a system of focusing both the stimulating and emitted light so that a single plane through the cell is seen. The result is a sharper two-dimensional image than with standard fluorescent microscopy.

In **stained bright-field microscopy**, a stain added to preserved cells enhances contrast and reveals details not otherwise visible. Stains differ greatly in their chemistry and their capacity to bind to cell materials, so many choices are available.

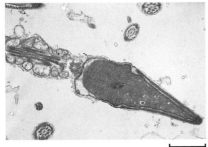

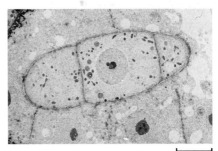

8.5 μm

8 μm

5 μm

In **transmission electron microscopy** (TEM), a beam of electrons is focused on the object by magnets. Objects appear darker if they absorb the electrons. If the electrons pass through, they are detected on a fluorescent screen.

Scanning electron microscopy (SEM) directs electrons to the surface of the sample, where they cause other electrons to be emitted. These electrons are viewed on a screen. The three-dimensional surface of the object can be visualized.

Cryo electron microscopy uses quickly frozen samples to reduce aberrations that are seen when samples are treated chemically. Computer analyses of thick sections can reconstruct a sample in three dimensions.

4.3 *Looking at Cells*

The top six panels represent some of the techniques used in light microscopy. The attributes of light are manipulated in various ways to enhance the images, and many types of stains, dyes, and reagents are used to visualize different cellular components in color. The lower three images were created using electron microscopes, which produce images in black-and-white. Artificial coloration is sometimes added to such micrographs, as in Figure 4.9b.

Many techniques have been developed to enhance the view of cells under both the light and the electron microscope (Figure 4.3). For example, specific dyes will form a colored complex with specific types of molecules (e.g., proteins or DNA), allowing the general composition of various cell structures to be estimated. There are even reagents that will bind to a specific molecule, such as a particular protein, allowing its distribution in the cell to be ascertained.

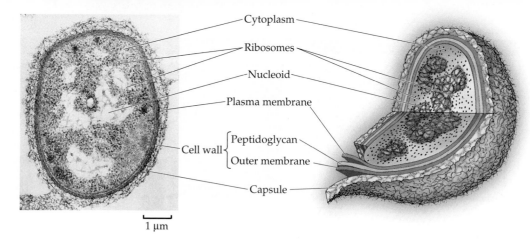

4.4 A Prokaryotic Cell
The bacterium *Pseudomonas aeruginosa* illustrates typical prokaryotic cell structures. The electron micrograph on the left is magnified about 80,000 times. Note the existence of several protective structures external to the plasma membrane.

Cytoplasm
Ribosomes
Nucleoid
Plasma membrane
Cell wall { Peptidoglycan / Outer membrane
Capsule

1 μm

All cells are surrounded by a plasma membrane

A **plasma membrane** separates each cell from its environment, creating a segregated (but not isolated) compartment. The plasma membrane is composed of a phospholipid bilayer, with the hydrophilic ends of the lipids facing the cell's aqueous interior on one side and the extracellular environment on the other (see Figure 3.22). Proteins are embedded in the lipids. In many cases, the proteins protrude into the cytoplasm and into the extracellular environment. We will devote most of the next chapter to the structure and functions of the plasma membrane, but summarize its roles here:

▶ The plasma membrane acts as a *selectively permeable barrier*, preventing some substances from crossing while permiting other substances to enter and leave the cell.

▶ As the cell's boundary with the outside environment, the plasma membrane is important in *communicating with adjacent cells and receiving extracellular signals*. We will describe this function in Chapter 15.

▶ The plasma membrane allows the cell to maintain a more or less *constant internal environment*. A self-maintaining, constant internal environment is a key characteristic of life and will be discussed in detail in Chapter 40.

Cells show two organizational patterns

Once the microscope was applied to biological samples, it soon became apparent that there are two types of cell structures in the living world.

Prokaryotic cell organization is characteristic of the domain Bacteria and Archaea. Organisms in these domains are called *prokaryotes*. Their cells do not have membrane-enclosed internal compartments.

Eukaryotic cell organization is found in the domain Eukarya, which includes the protists, plants, fungi, and animals. The genetic material (DNA) of eukaryotic cells is contained in a special membrane-enclosed compartment called the nucleus. Eukaryotic cells also contain other membrane-enclosed compartments in which specific chemical reactions take place. Organisms with this type of cell are known as *eukaryotes*.

Both prokaryotes and eukaryotes have prospered for many hundreds of millions of years of evolution, and both are great success stories. Let's look first at prokaryotic cells.

Prokaryotic Cells

Prokaryotes can live off more different and diverse energy sources than any other living creatures, and they inhabit greater environmental extremes, such as very hot springs and very salty water. The vast diversity within the prokaryotic domains is the subject of Chapter 26.

Prokaryotic cells are generally smaller than eukaryotic cells, ranging from 0.25×1.2 μm to 1.5×4 μm. So they are generally visible by light microscopy, although their substructures are visible only by electron microscopy. Each prokaryote is a single cell, but many types of prokaryotes are usually seen in chains, small clusters, or even clusters containing hundreds of individuals.

In this section, we will first consider the features that cells in the domains Bacteria and Archaea have in common. Then we will examine structural features that are found in some, but not all, prokaryotes.

All prokaryotic cells share certain features

All prokaryotic cells have the same basic structure (Figure 4.4):

▶ The plasma membrane encloses the cell, regulating the traffic of materials into and out of the cell and separating it from its environment.

▶ A region called the **nucleoid** contains the hereditary material (DNA) of the cell.

The rest of the material enclosed in the plasma membrane is called the **cytoplasm**. Cytoplasm is composed of two parts: the liquid cytosol, and insoluble suspended particles, including ribosomes.

▶ The **cytosol** consists mostly of water that contains dissolved ions, small molecules, and soluble macromolecules such as proteins.

▶ **Ribosomes** are granules about 25 nm in diameter that are sites of protein synthesis.

Although structurally less complicated than eukaryotic cells, prokaryotic cells are functionally complex, carrying out thousands of biochemical transformations.

Some prokaryotic cells have specialized features

Many prokaryotic cells have at least a few structural complexities. For example, most prokaryotes have a **cell wall** lo-

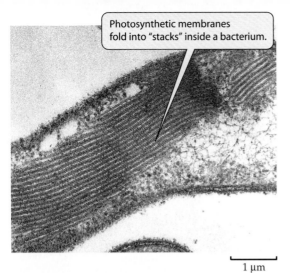

Photosynthetic membranes fold into "stacks" inside a bacterium.

1 μm

4.5 Some Prokaryotes Have Internal Membrane Systems
The presence of internal membranes contradicts the notion that prokaryotes are nothing more than tiny bags of molecules. These photosynthetic membranes contain compounds needed for photosynthesis.

cated outside the plasma membrane (see Figure 4.4). The rigidity of the cell wall supports the cell and determines its shape. The cell walls of most bacteria, but not archaea, contain *peptidoglycan*, a polymer of amino sugars, cross-linked by covalent bonds to form a single giant molecule around the entire cell. In some bacteria, another layer—the *outer membrane* (a polysaccharide-rich phospholipid membrane)—encloses the cell wall. Unlike the plasma membrane, this outer membrane is not a major permeability barrier, and some of its polysaccharides are disease-causing toxins.

Enclosing the cell wall and outer membrane in some bacteria is a layer of slime, composed mostly of polysaccharides and referred to as a **capsule**. The capsules of some bacteria may protect them from attack by white blood cells in the animals they infect. The capsule helps keep the cell from drying out, and sometimes it traps other cells for the bacterium to attack. Many prokaryotes produce no capsule, and those that do have capsules can survive even if they lose them, so the capsule is not essential to cell life.

Some groups of bacteria—the cyanobacteria and some others—carry on photosynthesis. In *photosynthesis*, the energy of sunlight is converted to chemical energy that can be used for a variety of energy-requiring reactions, such as the synthesis of cellular proteins and DNA. In these photosynthetic bacteria, the plasma membrane folds into the cytoplasm to form an internal membrane system that contains bacterial chlorophyll and other compounds needed for photosynthesis (Figure 4.5).

Other groups of prokaryotes possess different types of membranous structures called **mesosomes**, which may function in cell division or in various energy-releasing reactions. Like the photosynthetic membrane systems, mesosomes are formed by infolding of the plasma membrane. They remain attached to the plasma membrane and never form the free-floating, separate membranous organelles that are characteristic of eukaryotic cells.

Some prokaryotes swim by using appendages called **flagella** (Figure 4.6*a*). A single flagellum, made of a protein called flagellin, looks at times like a tiny corkscrew. It spins on its axis like a propeller, driving the cell along. Ring structures anchor the flagellum to the plasma membrane and, in some bacteria, to the outer membrane of the cell wall (Figure 4.6*b*). We know that the flagella cause the motion of the cell because if they are removed, the cell cannot move.

Pili project from the surface of some groups of bacteria (Figure 4.6*c*). Shorter than flagella, these threadlike structures help bacteria adhere to one another during mating, as well as to animal cells for protection and food.

Eukaryotic Cells

Animals, plants, fungi, and protists have cells that are usually larger and structurally more complex than those of the prokaryotes (Figure 4.7). To get a sense of the most promi-

4.6 Prokaryotic Projections
Surface projections such as these bacterial flagella (*a, b*) and pili (*c*) contribute to movement, to adhesion, and to the complexity of prokaryotic cells.

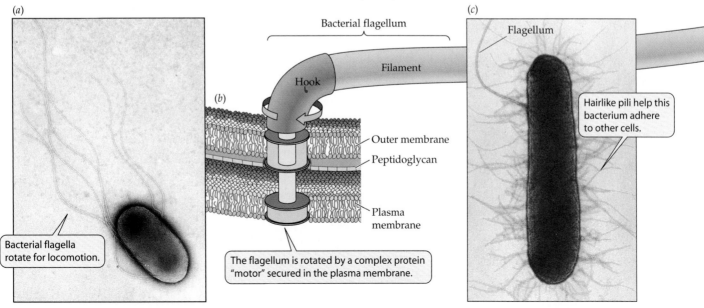

(*a*)

Bacterial flagella rotate for locomotion.

(*b*)

Bacterial flagellum

Filament

Hook

Outer membrane

Peptidoglycan

Plasma membrane

The flagellum is rotated by a complex protein "motor" secured in the plasma membrane.

(*c*)

Flagellum

Hairlike pili help this bacterium adhere to other cells.

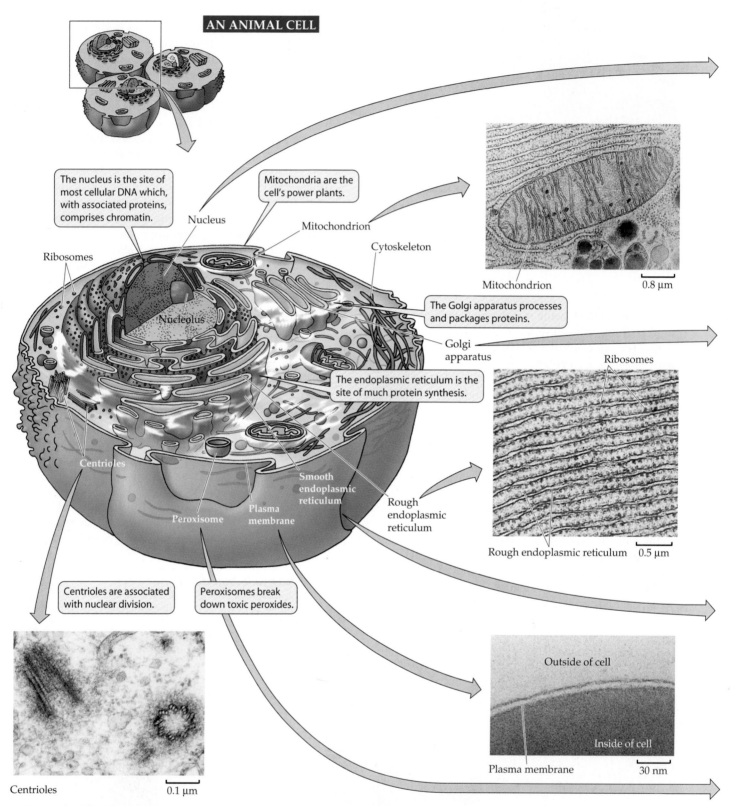

AN ANIMAL CELL

The nucleus is the site of most cellular DNA which, with associated proteins, comprises chromatin.

Mitochondria are the cell's power plants.

Nucleus

Mitochondrion

Cytoskeleton

Ribosomes

Nucleolus

Mitochondrion

0.8 μm

The Golgi apparatus processes and packages proteins.

Golgi apparatus

Ribosomes

The endoplasmic reticulum is the site of much protein synthesis.

Centrioles

Smooth endoplasmic reticulum

Rough endoplasmic reticulum

Rough endoplasmic reticulum

0.5 μm

Peroxisome

Plasma membrane

Centrioles are associated with nuclear division.

Peroxisomes break down toxic peroxides.

Outside of cell

Inside of cell

Centrioles

0.1 μm

Plasma membrane

30 nm

4.7 Eukaryotic Cells

In electron micrographs, many plant cell organelles are nearly identical in form to those observed in animal cells. Cellular structures unique to plant cells include the cell wall and the chloroplasts. Animal cells contain centrioles, which are not found in plant cells.

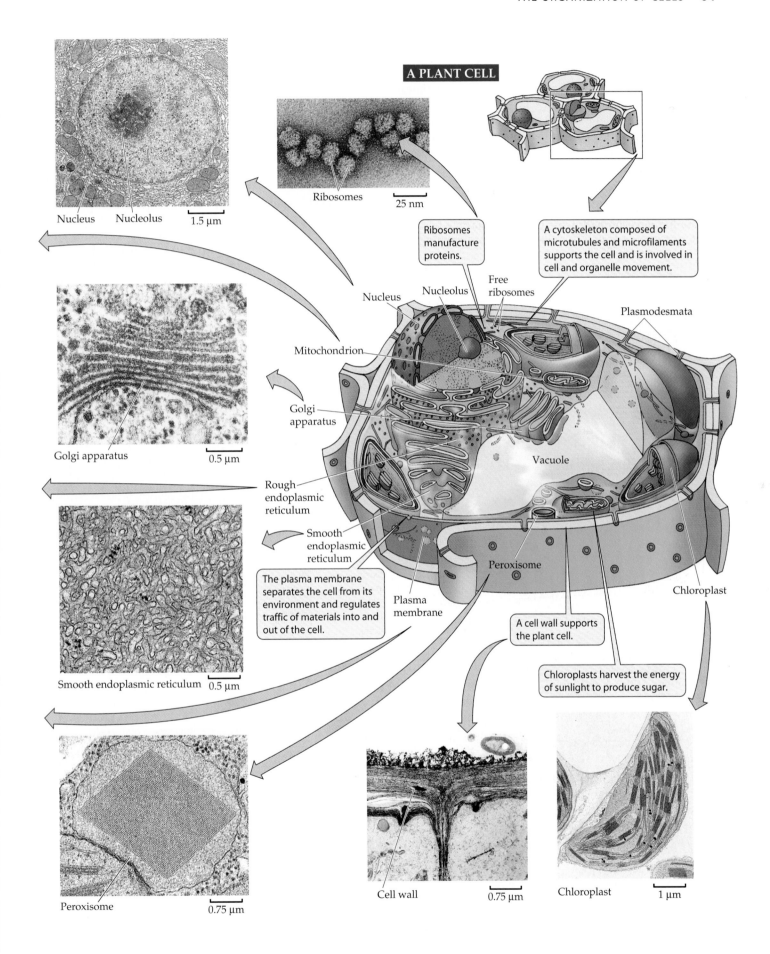

Nucleus   Nucleolus   1.5 μm

Ribosomes   25 nm

A PLANT CELL

Golgi apparatus   0.5 μm

Smooth endoplasmic reticulum   0.5 μm

Peroxisome   0.75 μm

Cell wall   0.75 μm

Chloroplast   1 μm

Ribosomes manufacture proteins.

A cytoskeleton composed of microtubules and microfilaments supports the cell and is involved in cell and organelle movement.

The plasma membrane separates the cell from its environment and regulates traffic of materials into and out of the cell.

A cell wall supports the plant cell.

Chloroplasts harvest the energy of sunlight to produce sugar.

Nucleus   Nucleolus   Free ribosomes   Plasmodesmata

Mitochondrion

Golgi apparatus

Rough endoplasmic reticulum

Smooth endoplasmic reticulum

Plasma membrane

Vacuole

Peroxisome

Chloroplast

nent differences, compare the eukaryotic plant and animal cells on the preceding two pages with the prokaryotic cell in Figure 4.4.

Eukaryotic cells generally have dimensions ten times greater than those of prokaryotes; for example, the spherical yeast cell has a diameter of 8 µm. Like prokaryotic cells, eukaryotic cells have a plasma membrane, cytoplasm, and ribosomes. But added on to this basic organization are two elements not found in prokaryotes:

▶ An internal **cytoskeleton** that maintains cell shape and moves materials
▶ **Membranous compartments** in the cytoplasm whose interiors are separated from the cytosol by a membrane

Compartmentalization is the key to eukaryotic cell function

Recall that prokaryotic cells are surrounded by a plasma membrane that regulates molecular traffic into and out of the cell. In addition, eukaryotic cells have "cells within cells"—interior compartments surrounded by membranes that regulate what enters or leaves that compartment. The membranes ensure that conditions inside the compartment are different from those in the surrounding cytoplasm.

Some of the compartments are like little factories that make specific products. Others are like power plants that take in energy in one form and convert it to a more useful form. These membranous compartments, as well as other structures (such as ribosomes) that lack membranes but possess distinctive shapes and functions, are called **organelles** (see Figure 4.7). Each of these organelles has specific roles in its particular cell. These roles are defined by chemical reactions.

4.8 The Nucleus Is Enclosed by a Double Membrane
The electron micrograph shows the nucleus of a nondividing animal cell. The double-membraned nuclear envelope, nucleolus, nuclear lamina, and nuclear pores are common features of all cell nuclei.

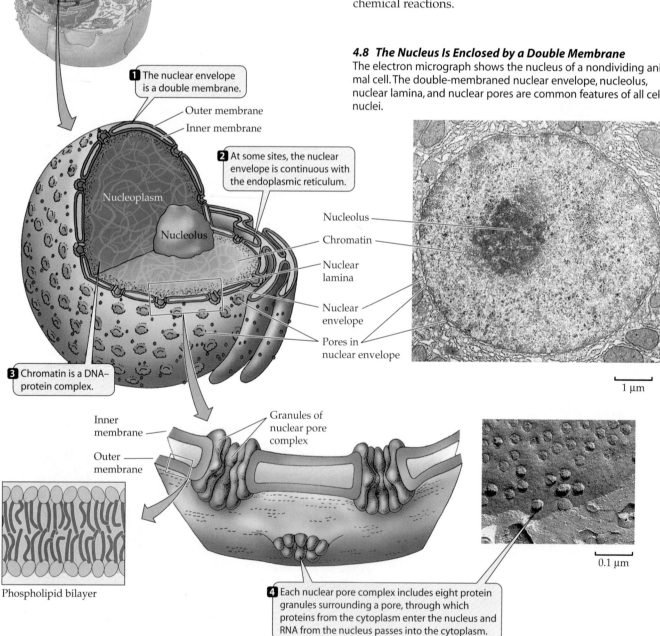

1 The nuclear envelope is a double membrane.

Outer membrane
Inner membrane

2 At some sites, the nuclear envelope is continuous with the endoplasmic reticulum.

Nucleoplasm
Nucleolus

3 Chromatin is a DNA–protein complex.

Nucleolus
Chromatin
Nuclear lamina
Nuclear envelope
Pores in nuclear envelope

1 µm

Inner membrane
Outer membrane

Granules of nuclear pore complex

0.1 µm

Phospholipid bilayer

4 Each nuclear pore complex includes eight protein granules surrounding a pore, through which proteins from the cytoplasm enter the nucleus and RNA from the nucleus passes into the cytoplasm.

- The **nucleus** contains most of the cell's genetic material (DNA). It determines the expression of this material as cell functions and its duplication when the cell reproduces.
- The **mitochondrion** is a power plant and industrial park, where energy stored in the bonds of carbohydrates is converted to a form more useful to the cell and certain essential biochemical conversions of amino acids and fatty acids occur.
- The **endoplasmic reticulum** and **Golgi apparatus** make up a compartment where proteins are packaged and sent to appropriate locations in the cell.
- The **lysosome** and **vacuole** are cellular digestive systems, where large molecules are hydrolyzed into usable monomers.
- The **chloroplast** performs photosynthesis.

All of these organelles have unique chemical compositions and functions. The membrane surrounding each does two essential things: First, it keeps the organelle's molecules away from other molecules in the cell with which they might react inappropriately. Second, it acts as a traffic regulator, letting important raw materials into the organelle and releasing its products to the cytoplasm.

Organelles that Process Information

Living things depend on accurate, appropriate information—internal signals, environmental cues, and stored instructions—to respond appropriately to changing conditions and maintain a constant internal environment. In the cell, information is *stored* as the sequence of nucleotides in DNA molecules. Most DNA in eukaryotic cells resides in the nucleus. Information is *translated* from the language of DNA into the language of proteins at the ribosomes. This process is described in detail in Chapter 12.

The nucleus stores most of the cell's DNA

The single nucleus is usually the largest organelle in a cell (Figure 4.8; see also Figure 4.7). The nucleus of most animal cells is approximately 5 μm in diameter—substantially larger than most entire prokaryotic cells. The nucleus has several roles in the cell:

- The nucleus is the site of DNA duplication to support cell reproduction.
- The nucleus is the site of DNA control of cellular activities.
- A region within the nucleus, called the **nucleolus**, begins the assembly of ribosomes from specific proteins and RNA.

The nucleus is surrounded by *two* membranes, which together form the **nuclear envelope**. The two membranes of the nuclear envelope are separated by only a few tens of nanometers and are perforated by **nuclear pores** approximately 9 nm in diameter, which connect the interior of the nucleus with the cytoplasm. At these pores, the outer membrane of the nuclear envelope is continuous with the inner membrane. Each pore is surrounded by a *pore complex*: eight large protein granules arranged in an octagon where the inner and outer membranes merge. RNA and proteins pass through these pores to enter or leave the nucleus.

At certain sites, the outer membrane of the nuclear envelope folds outward into the cytoplasm and is continuous with the membrane of another organelle, the endoplasmic reticulum (discussed later in the chapter).

Inside the nucleus, DNA combines with proteins to form a fibrous complex called **chromatin**. These are exceedingly long, thin, entangled threads that, prior to cell division, condense to form readily visible objects called **chromosomes** (Figure 4.9).

Surrounding the chromatin are water and dissolved substances collectively referred to as the **nucleoplasm**. Within the nucleoplasm, a network of apparently structural proteins called the *nuclear matrix* organizes the chromatin. At the periphery of the nucleus, the chromatin attaches to a protein meshwork, called the **nuclear lamina**, which is formed by the polymerization of proteins called *lamins* into filaments (Figure 4.10). The nuclear lamina maintains the

4.9 Chromatin and Chromosomes
(*a*) When a cell is not dividing, the nuclear DNA and proteins are aggregated as chromatin, which is dispersed throughout the nucleus. (*b*) The chromatin in a dividing cell is packed into dense bodies called chromosomes.

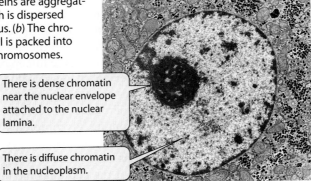

(*a*)

There is dense chromatin near the nuclear envelope attached to the nuclear lamina.

There is diffuse chromatin in the nucleoplasm.

1 μm

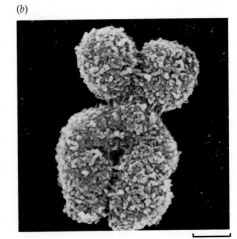

(*b*)

0.5 μm

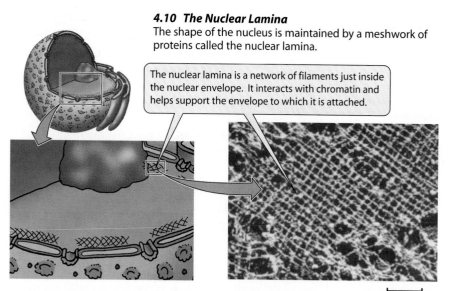

4.10 The Nuclear Lamina
The shape of the nucleus is maintained by a meshwork of proteins called the nuclear lamina.

The nuclear lamina is a network of filaments just inside the nuclear envelope. It interacts with chromatin and helps support the envelope to which it is attached.

0.25 μm

shape of the nucleus by its attachment to both chromatin and the nuclear envelope.

During most of the life cycle of the cell, the nuclear envelope is a stable structure. When the cell divides, however, the nuclear envelope fragments into pieces of membrane with attached pore complexes. The envelope re-forms when distribution of the duplicated DNA to the daughter cells is completed.

Ribosomes are the sites of protein synthesis

In both eukaryotic and prokaryotic cells, proteins are synthesized on thousands of ribosomes. Ribosomes are tiny granules found in three places in almost all eukaryotic cells: free in the cytoplasm, attached to the surface of endoplasmic reticulum (as will be described later in this chapter), and inside the mitochondria, where energy is processed. Ribosomes are also found in chloroplasts, the photosynthetic organelles of plant cells. In each of these locations, the ribosomes provide the sites where proteins are synthesized under the direction of nucleic acids. Although they seem small in comparison to the cell in which they are contained, ribosomes are huge machines composed of several dozen kinds of molecules.

The ribosomes of prokaryotes and of eukaryotes are similar in that both consist of two different-sized subunits. Eukaryotic ribosomes are somewhat larger, but the structure of prokaryotic ribosomes is better understood. Chemically, ribosomes consist of a special type of RNA, called *ribosomal RNA*, to which more than 50 different protein molecules are noncovalently bound.

The Endomembrane System

Much of the volume of some eukaryotic cells is taken up by an extensive **endomembrane system**. This system includes two main components, the endoplasmic reticulum and the Golgi apparatus. Continuities between the nuclear envelope and the endomembrane system are visible by electron microscopy. Tiny vesicles appear to shuttle between the various components of the endomembrane system. This system has various structures, but all of them are essentially compartments, closed off by their membranes from the soluble cytoplasm.

In this section, we will examine the functional significance of these compartments, and show that materials synthesized in the endoplasmic reticulum can be transferred to another organelle, the Golgi apparatus, for further processing, storage, or transport. We will also describe the role of the lysosome in cell digestion.

The endoplasmic reticulum is a complex factory

Electron micrographs reveal a network of interconnected membranes branching throughout the cytoplasm, forming tubes and flattened sacs. These membranes are collectively called the **endoplasmic reticulum**, or **ER**. The interior compartment of the ER, referred to as the *lumen*, is separate and distinct from the surrounding cytoplasm (Figure 4.11). The surface area of the ER can occupy up to 15 percent of the entire interior volume of the cell, and its foldings result in a surface area many times greater than that of the plasma membrane. At certain sites, the ER is continuous with the outer membrane of the nuclear envelope.

Parts of the ER are liberally sprinkled with ribosomes, which are temporarily attached to the outer faces of the flattened sacs. Because of their appearance in the electron microscope, these regions are called **rough ER**, or **RER**. RER has two roles:

▶ As a compartment, it segregates certain newly synthesized proteins away from the cytoplasm and transports them to other locations in the cell.

▶ While inside the RER, proteins can be chemically modified so as to alter their function and intracellular destination.

The attached ribosomes are sites for the synthesis of proteins that function outside the cytosol—that is, proteins that are to be exported from the cell, incorporated into membranes, or moved into organelles of the endomembrane system. These proteins enter the lumen of the ER as they are synthesized. Once in the lumen of the ER, these proteins undergo several changes, including the formation of disulfide bridges and folding into their tertiary structures (see Figure 3.5). Proteins gain carbohydrate groups in the RER, thus becoming glycoproteins. The carbohydrate groups are part of an "addressing" system that ensures that the right proteins are directed to the right parts of the cell.

Some parts of the endoplasmic reticulum, called the **smooth ER** or **SER**, are more tubular (less like flattened sacs) and lack ribosomes (see Figure 4.11). Within the lumen of the SER, proteins that have been synthesized on the RER are chemically modified. In addition, the SER has two other important roles:

▶ It is responsible for chemically modifying small molecules taken in by the cell. This is especially true for drugs and pesticides.
▶ It is the site for the hydrolysis of glycogen and the synthesis of steroids.

Cells that synthesize a lot of protein for export are usually packed with ER. Examples include glandular cells that secrete digestive enzymes and plasma cells that secrete antibodies. In contrast, cells that carry out less protein synthesis (such as storage cells) contain less ER.

The Golgi apparatus stores, modifies, and packages proteins

In 1898, the Italian microscopist Camillo Golgi discovered a delicate structure in nerve cells, which came to be known as the **Golgi apparatus**. Because of the resolution limits of light microscopy, and because the staining techniques of the time often failed to reveal the structure, many biologists regarded it as a product of Golgi's imagination. In the late 1950s, however, the electron microscope showed clearly that the Golgi apparatus does exist—and not just in nerve cells, but in most eukaryotic cells.

The exact appearance of the Golgi apparatus varies from species to species, but it always consists of flattened membranous sacs called *cisternae* and small membrane-enclosed *vesicles*. The cisternae appear to be lying together like a stack of saucers (Figure 4.12*a*). The entire apparatus is about 1 μm long.

The Golgi apparatus has several roles:

▶ It receives proteins from the ER and chemically modifies them.
▶ Proteins within the Golgi apparatus are concentrated, packaged, and sorted before being sent to their cellular or extracellular destinations.
▶ The Golgi apparatus is where some polysaccharides for the plant cell wall are synthesized.

In the cells of plants, protists, fungi, and many invertebrate animals, the stacks of cisternae are individual units scattered throughout the cytoplasm. In vertebrate cells, a few such stacks usually form a larger, single, more complex Golgi apparatus.

The Golgi appears to have three functionally distinct parts: a bottom, a middle, and a top. The bottom cisternae, constituting the *cis* region of the Golgi apparatus, lie nearest to the nucleus or a patch of RER (see Figure 4.12). The top cisternae, constituting the *trans* region, lie closest to the surface of the cell. The cisternae in the middle make up the *medial* region of the complex. These three parts of the Golgi apparatus contain different enzymes and perform different functions.

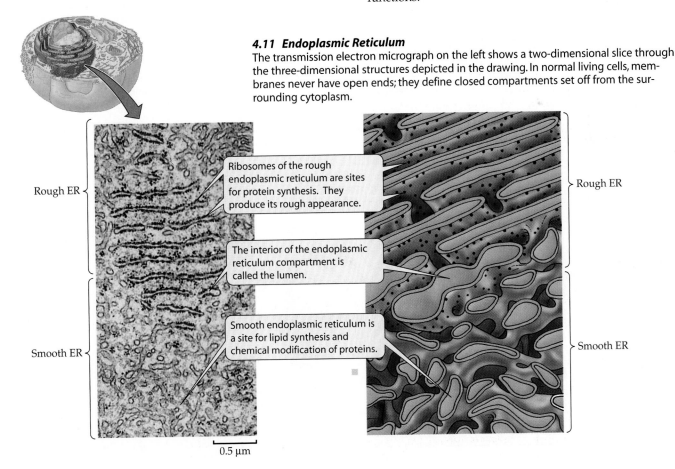

4.11 Endoplasmic Reticulum
The transmission electron micrograph on the left shows a two-dimensional slice through the three-dimensional structures depicted in the drawing. In normal living cells, membranes never have open ends; they define closed compartments set off from the surrounding cytoplasm.

Rough ER

Smooth ER

Ribosomes of the rough endoplasmic reticulum are sites for protein synthesis. They produce its rough appearance.

The interior of the endoplasmic reticulum compartment is called the lumen.

Smooth endoplasmic reticulum is a site for lipid synthesis and chemical modification of proteins.

Rough ER

Smooth ER

0.5 μm

The Golgi apparatus receives proteins from the ER, packages them, and sends them on their way. The chemical modifications made to proteins within the Golgi apparatus generally "tag" them to their proper destinations—a process we will describe further in Chapter 12. So in some sense the Golgi apparatus is a "post office" for the cell.

Since there is often no direct membrane continuity between ER and Golgi apparatus, how does a protein get from one organelle to the other? The protein could simply leave the ER, travel across the cytoplasm, and enter the Golgi apparatus. But this would expose the protein to interactions with other molecules in the cytoplasm. On the other hand, segregation from the cytoplasm could be maintained if a piece of the ER could "bud off," forming a vesicle that contains the target protein—and this is in fact exactly what happens. The protein makes the passage from ER to Golgi apparatus safely enclosed in the vesicle. Once it arrives, the vesicle fuses with the membrane of the Golgi apparatus, releasing its cargo .

Vesicles form from the rough ER, move through the cytoplasm, and fuse with the *cis* region of the Golgi appara-

tus, where their contents are released into the lumen of the Golgi. Other small vesicles may move between the cisternae, transporting proteins. Associated with the cisternae, particularly those toward the *trans* region, are tiny vesicles that pinch off and move to other cisternae or away from the Golgi (see Figure 4.12*b*).

The membranes of two vesicles can sometimes make contact with each other and fuse, resulting in a larger vesicle and a mixing of the contents. Vesicles may also fuse with other organelles, or with the plasma membrane, where they release their contents to the outside of the cell. The formation, transport, and fusing behavior of vesicles is essential to the function of the Golgi apparatus. Structurally, *vesicles are the transport vehicles into and out of the Golgi apparatus and to the ultimate destinations of the proteins.*

Lysosomes contain digestive enzymes

Originating in part from the Golgi apparatus are organelles called **lysosomes**. They contain digestive enzymes, and they are the sites of hydrolysis of macromolecules—proteins, polysaccharides, nucleic acids, and lipids—to their monomers (see Figure 3.2). Lysosomes are about 1 μm in diameter, are surrounded by a single membrane, and have a densely staining, featureless interior (Figure 4.13*a*). There may be dozens of lysosomes in a cell, depending on its needs.

Lysosomes are sites for the breakdown of food and foreign objects taken up by the cell. How do these materials get into the cell in the first place? In a process called *phagocytosis* (phago-, "eating"; cytosis, "cellular"), a pocket forms in the plasma membrane and eventually deepens and en-

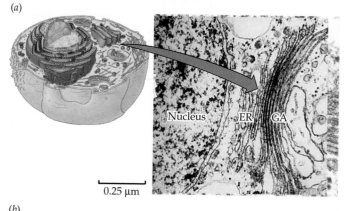

(a)

Nucleus   ER   GA

0.25 μm

4.12 The Golgi Apparatus
(*a*) The Golgi apparatus appears as stacked disks in an electron micrograph. (*b*) The Golgi apparatus modifies proteins from the ER and "targets" them to the correct addresses.

(b)

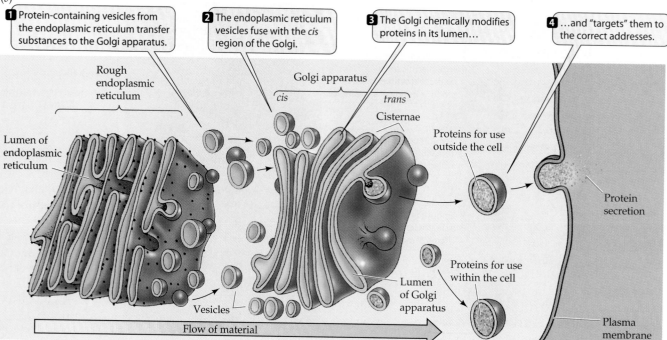

1 Protein-containing vesicles from the endoplasmic reticulum transfer substances to the Golgi apparatus.

2 The endoplasmic reticulum vesicles fuse with the *cis* region of the Golgi.

3 The Golgi chemically modifies proteins in its lumen…

4 …and "targets" them to the correct addresses.

Rough endoplasmic reticulum

Golgi apparatus

cis   *trans*

Cisternae

Proteins for use outside the cell

Lumen of endoplasmic reticulum

Protein secretion

Vesicles

Lumen of Golgi apparatus

Proteins for use within the cell

Plasma membrane

Flow of material

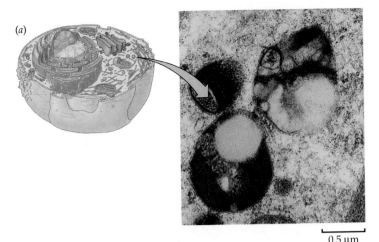

(a)

0.5 µm

4.13 Lysosomes Isolate Digestive Enzymes from the Cytoplasm
(a) In this electron micrograph of a rat cell, the darkly stained organelles are secondary lysosomes in which digestion is taking place. (b) The origin and action of lysosomes and lysosomal digestion.

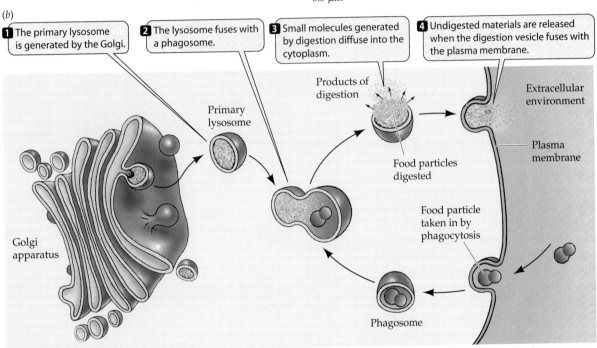

(b)

1 The primary lysosome is generated by the Golgi.

2 The lysosome fuses with a phagosome.

3 Small molecules generated by digestion diffuse into the cytoplasm.

4 Undigested materials are released when the digestion vesicle fuses with the plasma membrane.

Products of digestion

Primary lysosome

Food particles digested

Extracellular environment

Plasma membrane

Golgi apparatus

Food particle taken in by phagocytosis

Phagosome

closes material from outside the cell. This pocket becomes a small vesicle and breaks free of the plasma membrane to move into the cytoplasm as a *phagosome* containing food or other material (Figure 4.13b). The phagosome fuses with a *primary lysosome* to form a *secondary lysosome*, where digestion occurs.

The effect of this fusion is rather like releasing hungry foxes into a chicken coop. The enzymes in the secondary lysosome quickly hydrolyze the food particles. These reactions are enhanced by the mild acidity of the lysosome's interior, where the pH is lower than in the surrounding cytoplasm. The products of digestion exit through the membrane of the lysosome, providing fuel molecules and raw materials for other cell processes. The "used" secondary lysosome containing undigested particles then moves to the plasma membrane, fuses with it, and releases the undigested contents to the environment.

Lysosomes are also where the cell digests its own material in a process called *autophagy*. Autophagy is an ongoing process, in which macromolecules such as proteins are en-

gulfed by lysosomes and hydrolyzed to amino acids, which pass out of the lysosome through its membrane into the cytoplasm for reuse.

Plant cells do not appear to contain lysosomes, but the central vacuole of a plant cell may function in an equivalent capacity because it, like lysosomes, contains many digestive enzymes.

Organelles that Process Energy

A cell uses energy to transform raw materials into cell-specific materials that it can use for activities such as growth, reproduction, and movement. Energy is transformed from one form to another in mitochondria (found in all eukaryotic cells) and in chloroplasts (found in eukaryotic cells that harvest energy from sunlight). In contrast, energy transformations in prokaryotic cells are associated with enzymes attached to the inner surface of the plasma membrane or extensions of the plasma membrane that protrude into the cytoplasm.

Mitochondria are energy transformers

In eukaryotic cells, the utilization of food molecules such as glucose begins in the cytosol. The fuel molecules that result from partial degradation of this food enter the **mitochondria** (singular mitochondrion), whose primary function is to *convert the potential chemical energy of fuel molecules into a form that the cell can use:* the energy-rich molecule called *ATP*, or *adenosine triphosphate*. ATP is not a long-term energy storage form, but rather a kind of energy currency. Its role in the cell is analogous to the role of paper money in an economy. Chemically, ATP can participate in a great number of different cellular reactions and processes that require energy. In the mitochondria, the production of ATP using fuel molecules and O_2 is called *cellular respiration.*

Typical mitochondria are small—somewhat less than 1.5 μm in diameter and 2–8 μm in length—about the size of many bacteria. Mitochondria are visible with a light microscope, but almost nothing was known of their precise structure until they were examined with the electron microscope. Electron micrographs revealed that mitochondria have two membranes. The *outer membrane* is smooth and protective, and it offers little resistance to the movement of substances into and out of the mitochondrion. Immediately inside the outer mitochondrial membrane is an *inner membrane*, which folds inward in many places, giving it a much greater surface area than that of the outer membrane (Figure 4.14). These folds tend to be quite regular, giving rise to shelflike structures called **cristae**.

The inner mitochondrial membrane contains many large protein molecules that participate in cellular respiration and the production of ATP. The inner membrane exerts much more control over what enters and leaves the mitochondrion than does the outer membrane. The region enclosed by the inner membrane is referred to as the **mitochondrial matrix**. In addition to many proteins, the matrix contains some ribosomes and DNA that are used to make some of the proteins needed for cellular respiration.

The number of mitochondria per cell ranges from one contorted giant in some unicellular protists to a few hundred thousand in large egg cells. An average human liver cell contains more than a thousand mitochondria. Cells that require the most chemical energy tend to have the most mitochondria per unit of volume. In Chapter 7 we will see how the different parts of the mitochondrion work together in cellular respiration.

Plastids photosynthesize or store materials

One class of organelles—the **plastids**—is produced only in plants and certain protists. There are several types of plastids, with different functions.

CHLOROPLASTS. The most familiar of the plastids is the **chloroplast**, which contains the green pigment chlorophyll and is the site of photosynthesis (Figure 4.15). In photosynthesis, light energy is converted into the chemical energy of bonds between atoms. The molecules formed in photosynthesis provide food for plants themselves and for other

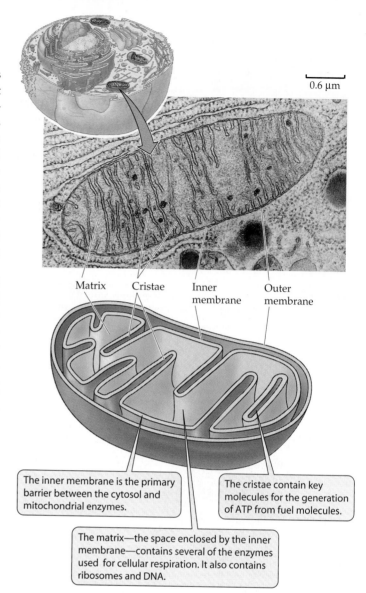

0.6 μm

Matrix | Cristae | Inner membrane | Outer membrane

The inner membrane is the primary barrier between the cytosol and mitochondrial enzymes.

The cristae contain key molecules for the generation of ATP from fuel molecules.

The matrix—the space enclosed by the inner membrane—contains several of the enzymes used for cellular respiration. It also contains ribosomes and DNA.

4.14 A Mitochondrion Converts Energy from Fuel Molecules into ATP
The electron micrograph is a two-dimensional slice through a three-dimensional reality. As the drawing emphasizes, the cristae are extensions of the inner mitochondrial membrane.

organisms that eat plants. Directly or indirectly, photosynthesis is the energy source for most of the living world.

Chloroplasts are quite variable in size and shape (Figure 4.16a,b). Like the mitochondrion, the chloroplast is surrounded by two membranes. Arising from the inner membrane is a series of discrete internal membranes whose structure and arrangement vary from one group of photosynthetic organisms to another. Here we concentrate on the chloroplasts of the flowering plants. Even these show some variation, but the pattern shown in Figure 4.15 is typical.

As seen in electron micrographs, the internal membranes of chloroplasts look like stacks of flat, hollow pita bread. These stacks, called **grana** (singular granum), consist of a series of flat, closely packed, circular compartments called **thylakoids**. In addition to phospholipids and proteins, the membranes of the thylakoids contain molecules

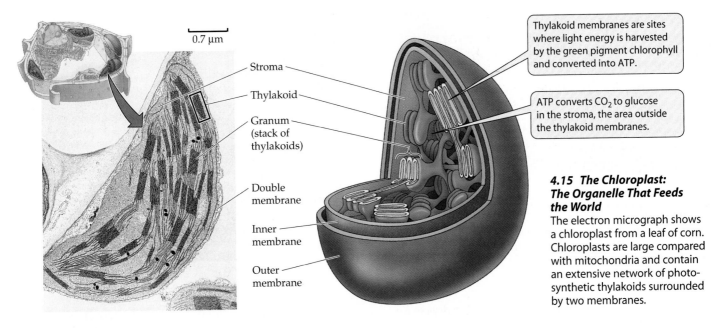

Thylakoid membranes are sites where light energy is harvested by the green pigment chlorophyll and converted into ATP.

ATP converts CO_2 to glucose in the stroma, the area outside the thylakoid membranes.

Stroma

Thylakoid

Granum (stack of thylakoids)

Double membrane

Inner membrane

Outer membrane

4.15 The Chloroplast: The Organelle That Feeds the World

The electron micrograph shows a chloroplast from a leaf of corn. Chloroplasts are large compared with mitochondria and contain an extensive network of photosynthetic thylakoids surrounded by two membranes.

of the green pigment chlorophyll and the yellow-orange carotenoids. These two pigment families harvest light for photosynthesis. Thylakoids of one granum may be connected to those of other grana, making the interior of the chloroplast a highly developed network of membranes, much like the ER.

The fluid in which the grana are suspended is referred to as **stroma**. Like the mitochondrial matrix, the chloroplast stroma contains ribosomes and DNA, and these are used to synthesize some, but not all, of the proteins that make up the chloroplast.

Animal cells do not *produce* chloroplasts, but some do *contain* functional chloroplasts. These are either taken up as free chloroplasts derived from the partial digestion of green plants, or contained within unicellular algae that live within the animal's tissues. The green color of some corals and sea anemones results from chloroplasts in algae that live within those animals (Figure 4.16c). The animals derive some of their nutrition from the photosynthesis that their chloroplast-containing "guests" carry out. Such an intimate relationship between two different organisms is called *symbiosis*.

OTHER TYPES OF PLASTIDS.   The red color of a flower or a ripe tomato results from the presence of legions of plastids called **chromoplasts** (Figure 4.17a). Just as chloroplasts derive their color from the pigment chlorophyll, chromoplasts are red, orange, or yellow depending on the kinds of carotenoid pigments present. The chromoplasts have no known chemical function in the cell, but the colors they give to some petals and fruits probably help attract animals that assist in pollination or seed dispersal. (On the other hand, carrot roots gain no apparent advantage from being orange.)

4.16 Being Green

(a) In green plants, chloroplasts are concentrated in the leaf cells.
(b) Green algae are photosynthetic and filled with chloroplasts.
(c) No animal species produces its own chloroplasts, but in this symbiotic arrangement, unicellular green algae nourish a giant sea anemone.

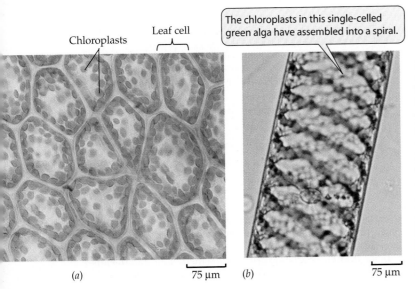

Chloroplasts

Leaf cell

The chloroplasts in this single-celled green alga have assembled into a spiral.

(a)   75 µm   (b)   75 µm

Chloroplast-filled green algae live in the tissues of this sea anemone.

(c)

4.17 Chromoplasts and Leucoplasts
(a) Colorful pigments stored in the chromoplasts of flowers like this begonia may help attract pollinating insects. (b) Leucoplasts in the cells of a potato are filled with white starch grains.

5 µm

Leucoplast

Starch grains

1 µm

Other plastids, called **leucoplasts**, are storage depots for starch and fats (Figure 4.17*b*).

Mitochondria and chloroplasts may have an endosymbiotic origin

Chloroplasts and mitochondria are about the size of prokaryotic cells. They contain DNA and have ribosomes that are similar to prokaryotic ribosomes, and they reproduce and divide within the cell to produce additional mitochondria and chloroplasts.

But, these organelles, even though although these have the genetic material and protein synthesis machinery needed to make some of their own components, they are not independent of control by the nucleus. The vast majority of their proteins are encoded by nuclear DNA, made in the cytoplasm, and imported into the organelle. These observations have led to speculation on the origin of these organelles. One proposal for this origin is the **endosymbiosis theory** of the origin of mitochondria and chloroplasts, which envisions the following scenario.

About 2 billion years ago, only prokaryotes inhabited Earth. Some of them absorbed their food directly from the environment. Others were photosynthetic. Still others fed on smaller prokaryotes by engulfing them (Figure 4.18).

Suppose that a small, photosynthetic prokaryote was *ingested* by a larger one, but was not *digested*. Instead, it survived trapped within a vesicle in the cytoplasm of the larger cell. The smaller, ingested prokaryote divided at about the same rate as the larger one, so successive generations of the larger cell also contained the offspring of the smaller one. We call this phenomenon *endosymbiosis* (endo-, "within"; symbiosis, "living together"); it is comparable to the algae that live within sea anemones (see Figure 4.16*c*).

According to this scenario, endosymbiosis provided benefits for both organisms. The larger cell obtained the photosynthetic products from the smaller cell, and the smaller cell was protected by the larger one. The smaller cell gradually lost much of its DNA to the nucleus, resulting in the modern chloroplast.

Much circumstantial evidence favors the endosymbiosis theory. Chloroplast DNA sequences are more like certain prokaryotic sequences than like any plant DNA. Moreover, on an evolutionary time scale of millions of years, there is evidence for DNA moving between organelles in a cell. Finally, there are many biochemical similarities between chloroplasts and modern bacteria.

Similar evidence and arguments also support the proposition that mitochondria are the descendants of respiring prokaryotes engulfed by larger prokaryotes. The benefits of this endosymbiotic relationship might have been due to the capacity of the engulfed prokaryote to detoxify molecular oxygen (O_2), which was increasing in Earth's atmosphere because of photosynthesis.

However, mitochondria and chloroplasts are not enough to turn a prokaryote into a eukaryote. The endosymbiosis theory is still incomplete. For example, the origins of the nuclear envelope and other important structures—including those responsible for nuclear division—still need to be understood. We discuss further aspects of the origin of the eukaryotic cell in Chapter 27.

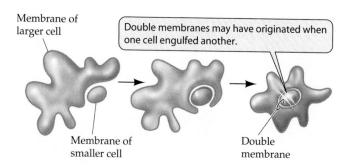

Membrane of larger cell

Double membranes may have originated when one cell engulfed another.

Membrane of smaller cell

Double membrane

4.18 The Endosymbiosis Theory
The double membrane that encloses mitochondria and chloroplasts may have arisen from two different sources: the outer membrane from the engulfing cell's plasma membrane and the inner membrane from the engulfed cell's plasma membrane.

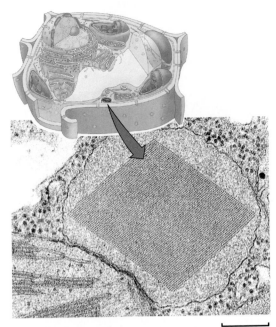

4.19 A Peroxisome   0.25 µm
A diamond-shaped crystal, composed of an enzyme, almost
entirely fills this rounded peroxisome in a leaf cell. The enzyme
catalyzes one of the reactions fulfilling the special function of the
peroxisome.

Other Organelles

In addition to the information-processing organelles (nucleus and ribosomes), the energy-processing organelles (mitochondria and chloroplasts), and the organelles of the endomembrane system (endoplasmic reticulum, Golgi apparatus, and lysosomes), there are two other kinds of membrane-enclosed organelles: peroxisomes and vacuoles. Both are surrounded by a single membrane.

Peroxisomes house specialized chemical reactions

Peroxisomes are small organelles—0.2 to 1.7 µm in diameter. They have a single membrane and a granular interior (Figure 4.19). Peroxisomes are found at one time or another in at least some of the cells of almost every eukaryotic species. Peroxisomes are organelles within which toxic peroxides (such as hydrogen peroxide, H_2O_2) are formed as unavoidable side products of chemical reactions. Subsequently, the peroxides are safely broken down within the peroxisomes without mixing with other parts of the cell.

A structurally similar organelle, the **glyoxysome**, is found only in plants. Glyoxysomes, which are most prominent in young plants, are the sites where stored lipids are converted into carbohydrates for transport to growing cells.

Vacuoles are filled with water and soluble substances

Many eukaryotic cells, but particularly those of plants and protists, contain membrane-enclosed organelles that look empty under the electron microscope. These organelles are called **vacuoles** (Figure 4.20). They are not actually empty; rather, they are filled with aqueous solutions that contain many dissolved substances.

Despite their structural simplicity, vacuoles have a variety of functions. For example, like animals and other organisms, plant cells produce a number of toxic by-products and waste materials. Animals have specialized excretory mechanisms for getting rid of such wastes, but plants do not. Although plants can secrete some wastes to their environment, many are simply stored within vacuoles. And since they are poisonous or distasteful, these stored materials deter some animals from eating the plants. Thus stored wastes may contribute to plant survival.

In many plant cells, enormous vacuoles take up more than 90 percent of the cell volume and grow as the cell grows. But vacuoles are by no means a waste of space, for the dissolved substances in the vacuole, working together with the vacuolar membrane, provide the *turgor*, or stiffness, of the cell, which in turn provides support for the structure of nonwoody plants. The presence of the dissolved substances causes water to enter the vacuole, making it tend to swell like a balloon. Plant cells have a rigid cell wall, which acts like a box, resisting the swelling of the vacuole but providing strength in the process.

Vacuoles even play a role in the sex life of plants. Some pigments (especially blue and pink ones) in petals and fruits are contained in vacuoles. These pigments—the anthocyanins—are visual cues that encourage animals to visit flowers and thus aid in pollination, or to eat fruits and thus aid in seed dispersal.

Food vacuoles are found in some simple and evolutionarily ancient groups of organisms: single-celled protists and simple multicellular organisms such as sponges. In these organisms, the cells engulf food particles by phagocytosis, generating a food vacuole. Fusion of this vacuole with a

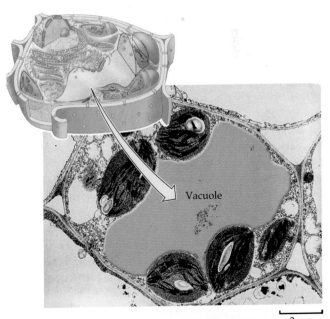

2 µm

4.20 Vacuoles in Plant Cells Are Usually Large
The large central vacuole in this cell is typical of mature plant cells. Smaller vacuoles are visible toward each end of the cell.

4.21 The Cytoskeleton

Three highly visible and important structural components of the cytoskeleton are shown in detail. These structures maintain and reinforce cell shape, and contribute to cell movement.

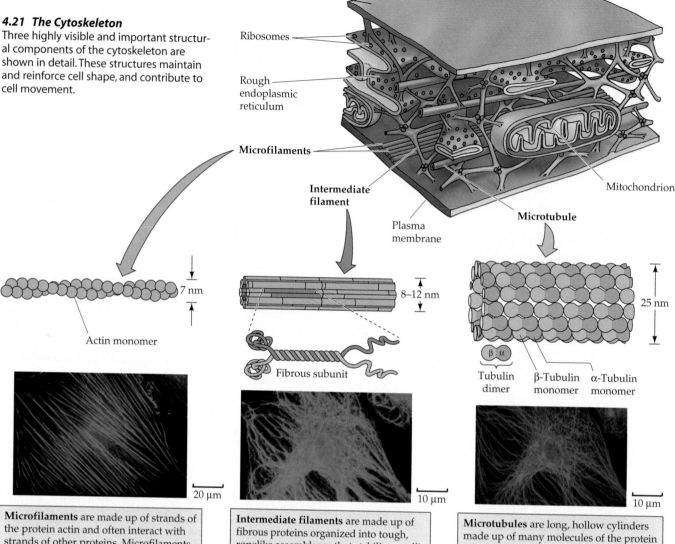

Microfilaments are made up of strands of the protein actin and often interact with strands of other proteins. Microfilaments may occur . They change cell shape and drive cellular motion, including contraction, cytoplasmic streaming, and the "pinched" shape changes that occur during cell division. Microfilaments and myosin strands together drive muscle action.

Intermediate filaments are made up of fibrous proteins organized into tough, ropelike assemblages that stabilize a cell's structure and help maintain its shape. Some intermediate filaments hold neighboring cells together. Others make up the nuclear lamina.

Microtubules are long, hollow cylinders made up of many molecules of the protein tubulin. Tubulin consists of two subunits, α-tubulin and β-tubulin. Microtubules lengthen or shorten by adding or subtracting tubulin dimers. Microtubule shortening moves chromosomes. Interactions between microtubules drive the movement of cells. Microtubules serve as "tracks" for the movement of vesicles.

lysosome results in digestion, and small molecules leave the vacuole and enter the cytoplasm for use or distribution to other organelles.

Many freshwater protists have a highly specialized *contractile vacuole*. Its function is to rid the cell of the excess water that rushes in because of the imbalance in salt concentration between the relatively salty interior of the cell and its freshwater environment. The contractile vacuole enlarges as water enters, then abruptly contracts, forcing the water out of the cell through a special pore structure.

The Cytoskeleton

In addition to the many membrane-enclosed organelles, the eukaryotic cytoplasm has a set of long, thin fibers called the **cytoskeleton**, which fills at least three important roles:

▶ It maintains cell shape and support.
▶ It provides for various types of cell movement.
▶ Some of its fibers act as tracks or supports for "motor proteins," which help the cell move or move things within the cell.

In the discussion that follows, we'll look at three components of the cytoskeleton: microfilaments, intermediate filaments, and microtubules (Figure 4.21).

Microfilaments function in support and movement

Microfilaments can exist as single filaments, in bundles, or in networks. They are about 7 nm in diameter and several μm long. They are assembled from **actin**, a protein that exists in several forms and has many functions among members of the animal phyla. The actin found in microfilaments (which are also known as actin filaments) is extensively folded and has distinct "head" and "tail" sites. These sites interact with similar actin molecules to assemble into a long chain (see Figure 4.21). Two of these chains interact to form the double helix of a microfilament. The polymerization of actin into microfilaments is reversible, and they can disappear from cells, breaking down into units of free actin.

Microfilaments have two major roles:

▸ They help the entire cell or parts of the cell to contract.
▸ They stabilize cell shape.

In muscle cells, actin fibers are associated with another protein called *myosin*, and their interactions account for the contraction of muscles. In nonmuscle cells, actin fibers are associated with localized changes of shape in cells. For example, microfilaments are involved in a flowing movement of the cytoplasm called *cytoplasmic streaming*, in movements of specific organelles and particles within cells, and in the "pinching" contractions that divide an animal cell into two daughter cells. Microfilaments are also involved in the formation of cellular extensions, called *pseudopodia* (pseudo-, "false;" podia, "feet"), that enable cells to move (Figure 4.22).

In some cell types, microfilaments form a meshwork just inside the plasma membrane. Actin-binding proteins then cross-link the microtubules to form a rigid structure that supports the

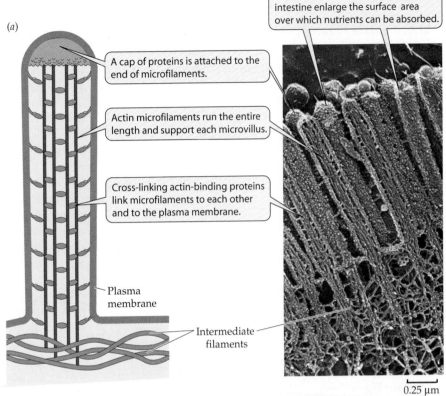

The microvilli of the cells lining the intestine enlarge the surface area over which nutrients can be absorbed.

A cap of proteins is attached to the end of microfilaments.

Actin microfilaments run the entire length and support each microvillus.

Cross-linking actin-binding proteins link microfilaments to each other and to the plasma membrane.

Plasma membrane

Intermediate filaments

0.25 µm

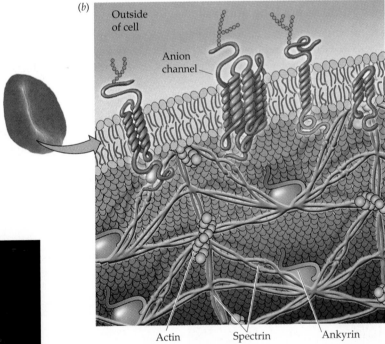

Outside of cell

Anion channel

Actin   Spectrin   Ankyrin

4.23 Microfilaments for Support
(a) Microfilaments form the backbone of the microvilli that increase the surface area of some cells, such as intestinal cells that absorb nutrients. (b) Actin microfilaments, along with ankyrin and spectrin proteins, support the "doughnut" shape of red blood cells.

20 µm

4.22 Microfilaments for Motion
The green-stained microfilaments in these cells provide a way for the cell to move.

cell. For example, microfilaments support the tiny *microvilli* that line the intestine, giving it a larger surface area through which to absorb nutrients. Such a "submembrane skeleton" also helps keep the red blood cell in its familiar doughnut shape (Figure 4.23).

Intermediate filaments are tough supporting elements

Intermediate filaments (see Figure 4.21) are found only in multicellular organisms. Although there are at least five distinct types of intermediate filaments, all share the same general structure and are composed of fibrous proteins of the keratin family, similar to the protein that makes up hair and fingernails. In cells, these proteins are organized into tough, ropelike assemblages 8 to 12 nm in diameter.

Intermediate filaments have two major structural functions:

▶ They stabilize cell structure.
▶ They resist tension.

In some cells, intermediate filaments radiate from the nuclear envelope and may maintain the positions of the nucleus and other organelles in the cell. The lamins of the nuclear lamina are intermediate filaments. Other kinds of intermediate filaments help hold a complex apparatus of microfilaments in place in muscle cells. Still other kinds stabilize and help maintain rigidity in surface tissues by connecting "spot welds" called *desmosomes* between adjacent cells (see Figure 5.6*b*).

Microtubules are long and hollow

Microtubules are long, hollow, unbranched cylinders about 25 nm in diameter and up to several micrometers long. Assembled from molecules of the protein **tubulin**, microtubules have two roles:

▶ They form a rigid internal skeleton for some cells, especially at cell extensions.
▶ They act as a framework on which motor proteins can move structures in the cell.

Tubulin is a dimer made up of two polypeptide monomers, called α-tubulin and β-tubulin. Thirteen rows, of tubulin dimers surround the central cavity of the microtubule (see Figure 4.21). The two ends of a microtubule are different. One end is designated the + end, the other the – end. Tubulin dimers can be added or subtracted mainly at the + end, lengthening or shortening the microtubule. This capacity to change length rapidly makes microtubules dynamic structures.

This dynamic property is seen in animal cells, where microtubules are often found in parts of the cell that are changing shape. Many microtubules radiate from a region of the cell called the *microtubule organizing center*. Tubule polymerization results in a rigid cell, and tubule depolymerization leads to a collapse of this rigid structure. In plants, microtubules help control the arrangement of the cellulose fibers of the cell wall. Electron micrographs of plants frequently show microtubules lying just inside the plasma membrane of cells that are forming or extending their cell walls. Experimental alteration of the orientation of these microtubules leads to a similar change in the cell wall, and a new shape for the cell. In many cells, microtubules serve as tracks for *motor proteins*, specialized molecules that use energy to change their shape and move. Motor proteins bond to and move along the microtubules, carrying materials from one part of the cell to another. Microtubules are also essential in distributing chromosomes to daughter cells during cell division. And they are intimately associated with movable cell appendages: the flagella and cilia.

Microtubules power cilia and flagella

Many eukaryotic cells possess flagella and/or cilia. These whiplike organelles push or pull the cell through its aque-

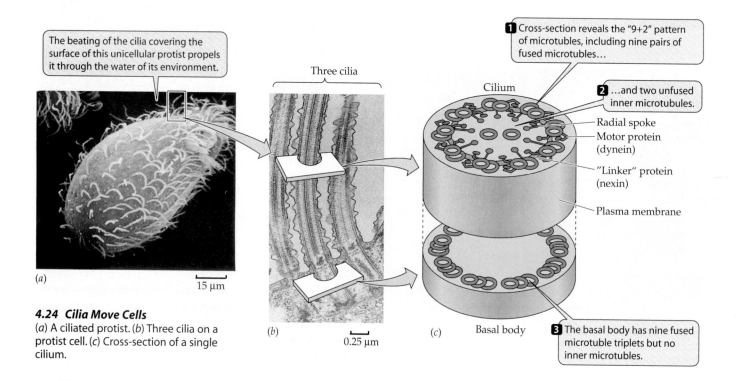

The beating of the cilia covering the surface of this unicellular protist propels it through the water of its environment.

Three cilia

1 Cross-section reveals the "9+2" pattern of microtubules, including nine pairs of fused microtubules…

Cilium

2 …and two unfused inner microtubules.

Radial spoke
Motor protein (dynein)
"Linker" protein (nexin)
Plasma membrane

Basal body

3 The basal body has nine fused microtuble triplets but no inner microtubules.

(a) 15 μm

4.24 Cilia Move Cells
(*a*) A ciliated protist. (*b*) Three cilia on a protist cell. (*c*) Cross-section of a single cilium.

(b) 0.25 μm *(c)*

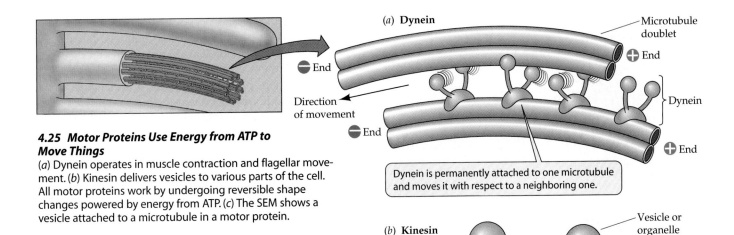

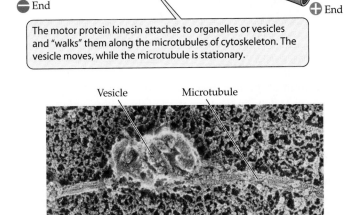

(a) Dynein

Microtubule doublet

− End   + End

Dynein

+ End

− End

Direction of movement

Dynein is permanently attached to one microtubule and moves it with respect to a neighboring one.

4.25 Motor Proteins Use Energy from ATP to Move Things
(a) Dynein operates in muscle contraction and flagellar movement. (b) Kinesin delivers vesicles to various parts of the cell. All motor proteins work by undergoing reversible shape changes powered by energy from ATP. (c) The SEM shows a vesicle attached to a microtubule in a motor protein.

(b) Kinesin

Vesicle or organelle

Direction of movement   Kinesin

Microtubule of cytoskeleton

− End   + End

The motor protein kinesin attaches to organelles or vesicles and "walks" them along the microtubules of cytoskeleton. The vesicle moves, while the microtubule is stationary.

Vesicle   Microtubule

25 nm

ous environment, or they may move surrounding liquid over the surface of the cell (Figure 4.24a). Cilia and eukaryotic (but not prokaryotic*) flagella are both assembled from specialized microtubules and have identical internal structures, but they differ in their relative lengths and their patterns of beating:

▶ **Flagella** are longer than cilia and are usually found singly or in pairs. Waves of bending propagate from one end of a flagellum to the other in snakelike undulation.

▶ **Cilia** are shorter than flagella and are usually present in great numbers. They beat stiffly in one direction and recover flexibly in the other direction (like a swimmer's arm), so that the recovery stroke does not undo the work of the power stroke.

Observed by electron microscopy in cross section, a typical cilium or eukaryotic flagellum is surrounded by the plasma membrane and contains a "9 + 2" array of microtubules. As Figure 4.24b shows, nine fused pairs of microtubules—called **doublets**—form an outer cylinder, and one pair of unfused microtubules runs up the center. A spoke radiates from one microtuble of each pair and connects the doublet to the center of the structure.

In the cytoplasm at the base of every eukaryotic flagellum or cilium is an organelle called a **basal body**. The nine microtubule doublets extend into the basal body. In the basal body, each doublet is accompanied by another microtubule, making nine sets of *three* microtubules. The central, unfused microtubules do not extend into the basal body.

The microtuble doublets of cilia and flagella are linked by proteins. The motion of cilia and flagella results from the sliding of the microtubules past each other, driven by a motor protein called **dynein**, which can undergo changes

*Some prokaryotes have flagella, as we saw earlier, but prokaryotic flagella lack microtubules and dynein. The flagella of prokaryotes are neither structurally nor evolutionarily related to those of eukaryotes. The prokaryotic flagellum is assembled from a protein called *flagellin*, and it has a much simpler structure and a smaller diameter than a single eukaryotic microtubule. And whereas eukaryotic flagella beat in a wavelike motion, prokaryotic flagella rotate (see Figure 4.7).

in its shape driven by energy from ATP. Dynein molecules attached to one microtubule bind to a neighboring microtubule. As the dynein molecules change shape, they move the microtubule past its neighbor (Figure 4.25a). Blocking the motor action of dynein is the idea behind a new class of spermicides used for contraception: Because these spermicides inhibit dynein, the sperm cannot swim toward the egg, and fertilization cannot occur.

Dynein and another motor protein, **kinesin**, are responsible for carrying protein-laden vesicles from one part of the cell to another. Recall that microtubules have a + end and a − end. Dynein binds to a microtubule and moves attached vesicles and other organelles toward the − end, while kinesin moves them toward the + end (Figure 4.25b).

Centrioles are almost identical to basal bodies. Centrioles are found in all eukaryotes except the flowering plants, pine trees and their relatives, and some protists. Under the light microscope, a centriole looks like a small, featureless particle, but the electron microscope reveals that it is made up of a precise bundle of microtubules, arranged as nine sets of three fused microtubules each (Figure 4.26). Centri-

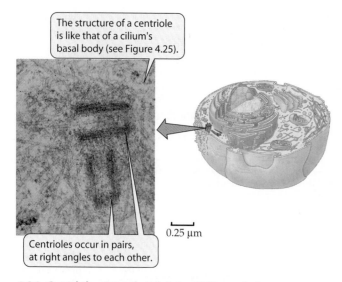

The structure of a centriole is like that of a cilium's basal body (see Figure 4.25).

0.25 μm

Centrioles occur in pairs, at right angles to each other.

4.26 Centrioles Contain Triplets of Microtubules
Centrioles are found in the microtubule organizing center, a region near the nucleus. The electron micrograph shows a pair of centrioles at right angles to each other.

oles lie in the microtubule organizing center in cells that are about to undergo division. As you will see in Chapter 9, they are involved in the formation of the mitotic spindle, to which the chromosomes attach.

Extracellular Structures

Although the plasma membrane is the functional barrier between the inside and outside of a cell, many structures outside the plasma membrane are produced by cells, secreted to the outside, and play essential roles in protecting, supporting, or attaching cells. These structures are said to be **extracellular** because they are outside the plasma membrane. The peptidoglycan cell wall of bacteria is such an extracellular structure. In eukaryotes, other extracellular structures play the same roles: in plants, the cellulose cell wall, and in multicellular animals, the extracellular matrix found between cells. Both of these structures are made up of a prominent fibrous macromolecule embedded in a jelly-like medium.

The plant cell wall consists largely of cellulose

The **cell wall** of plant cells is a semirigid structure outside the plasma membrane (Figure 4.27). It consists of cellulose fibers embedded in other complex polysaccharides and proteins. The cell wall has two major roles in plants:

▶ It provides support for the cell and limits its volume by remaining rigid.
▶ It acts as a barrier to infections by fungi and other organisms that can cause plant diseases.

Because of their thick cell walls, plant cells viewed under a light microscope appear to be entirely isolated from each other. But electron microscopy reveals that this is not the

case. The cytoplasm of adjacent plant cells is connected by numerous plasma membrane-lined channels, called *plasmodesmata*, that are about 20 to 40 nm in diameter and extend through the walls of adjoining cells (see Figure 4.27). These connections permit the diffusion of water, ions, small molecules, and RNA and proteins between connected cells. Such diffusion ensures that the cells of a plant have uniform concentrations of these substances.

Animal cells have elaborate extracellular matrices

The cells of multicellular animals lack the semirigid cell wall that is characteristic of plant cells, but many animal cells are surrounded by, or are in contact with, an **extracellular matrix**. This matrix is composed of fibrous proteins such as collagen (the most abundant protein in mammals) and glycoproteins (Figure 4.28). These proteins, as well as other substances particular to certain body tissues, are secreted by cells that are present in or near the matrix. In the human body, some tissues, such as those in the brain, have very little extracellular matrix; other tissues, such as bone and cartilage, have large amounts of extracellular matrix. The functions of the extracellular matrix are many:

▶ It holds cells together in tissues.
▶ It contributes to the physical properties of cartilage, skin, and other tissues.
▶ It helps filter materials passing between different tissues.
▶ It helps orient cell movements during embryonic development and during tissue repair.
▶ It plays a role in chemical signaling from one cell to another.

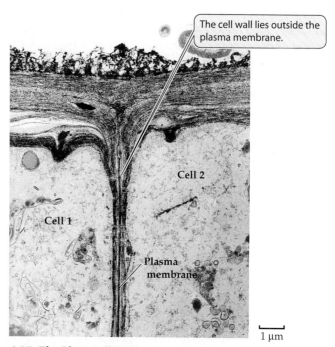

The cell wall lies outside the plasma membrane.

Cell 2

Cell 1

Plasma membrane

1 μm

4.27 The Plant Cell Wall
The semirigid cell wall provides support for plant cells.

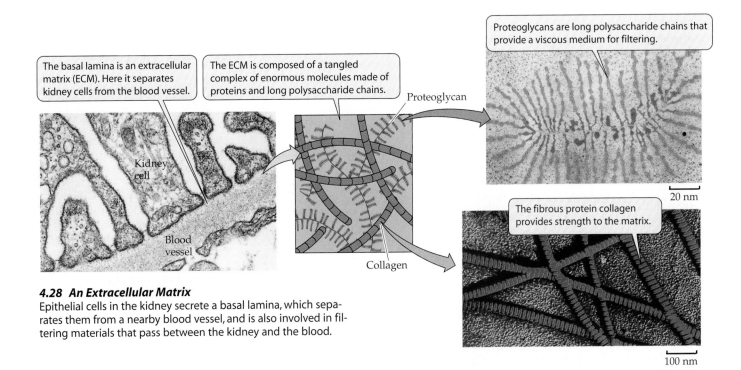

The basal lamina is an extracellular matrix (ECM). Here it separates kidney cells from the blood vessel.

The ECM is composed of a tangled complex of enormous molecules made of proteins and long polysaccharide chains.

Proteoglycans are long polysaccharide chains that provide a viscous medium for filtering.

Proteoglycan

Kidney cell

Blood vessel

Collagen

20 nm

The fibrous protein collagen provides strength to the matrix.

100 nm

4.28 An Extracellular Matrix
Epithelial cells in the kidney secrete a basal lamina, which separates them from a nearby blood vessel, and is also involved in filtering materials that pass between the kidney and the blood.

The cells embedded in bone and cartilage, for example, secrete and maintain the extracellular material that makes up these structures. Bone cells are embedded in an extracellular matrix that consists primarily of collagen and calcium phosphate. This matrix gives bone its familiar rigidity. Epithelial cells, which line body cavities, lie together as a sheet spread over a **basal lamina**, or basement membrane, a form of extracellular matrix (see Figure 4.28).

Some extracellular matrices are made up, in part, of an enormous **proteoglycan**. A single molecule of this proteoglycan consists of many hundreds of polysaccharides covalently attached to about a hundred proteins, all of which are attached to one enormous polysaccharide. The molecular weight of this proteoglycan can exceed 100 million; the molecule takes up as much space as an entire prokaryotic cell.

Chapter Summary

The Cell: The Basic Unit of Life

▶ All cells come from preexisting cells and have certain processes, types of molecules, and structures in common.

▶ To maintain adequate exchanges with its environment, a cell's surface area must be large compared with its volume. **Review Figure 4.2**

▶ Microscopes are needed to visualize cells. Because of their greater resolving power, electron microscopes allow observation of greater detail than can be seen with light microscopes. **Review Figure 4.3**

▶ Prokaryotic cell organization is characteristic of the kingdoms Eubacteria and Archaebacteria. Prokaryotic cells lack internal compartments. **Review Figure 4.4**

▶ Eukaryotic cell organization is characteristic of cells in the other four kingdoms. Eukaryotic cells have many membrane-enclosed compartments, including a nucleus that contains DNA. **Review Figure 4.7**

Prokaryotic Cells

▶ All prokaryotic cells have a plasma membrane, a nucleoid region with DNA, and a cytoplasm that contains ribosomes, dissolved enzymes, water, and small molecules. Some prokaryotes have additional protective structures: cell wall, outer membrane, and capsule. Some prokaryotes contain photosynthetic membranes, and some have mesosomes. **Review Figures 4.4, 4.5**

▶ Projecting from the surface of some prokaryotes are rotating flagella, which move the cells from place to place. Pili are projections by which prokaryotic cells attach to one another or to environmental surfaces. **Review Figure 4.6**

Eukaryotic Cells

▶ Like prokaryotic cells, eukaryotic cells have a plasma membrane, cytoplasm, and ribosomes. However, eukaryotic cells are larger and contain many membrane-enclosed organelles. **Review Figure 4.7**

▶ The membranes that envelop organelles in the eukaryotic cell are partial barriers, ensuring that the chemical composition of the interior of the organelle differs from that of the surrounding cytoplasm.

Organelles that Process Information

▶ The nucleus is usually the largest organelle in a cell. It is surrounded by a double membrane, the nuclear envelope, which disassembles during cell division. Within the nucleus, the nucleolus is the source of the ribosomes found in the cytoplasm. **Review Figure 4.8**

▶ Nuclear pores have complex structures that govern what enters and leaves the nucleus. **Review Figure 4.8**

▶ The nucleus contains most of the cell's DNA, which associates with protein to form chromatin. Chromatin is diffuse throughout the nucleus until just before cell division, when it condenses to form chromosomes. **Review Figure 4.9**

The Endomembrane System

▶ The endomembrane system is made up of a series of interrelated membranes and compartments.

▶ The rough endoplasmic reticulum has attached ribosomes that synthesize proteins. The smooth endoplasmic reticulum lacks ribosomes and is associated with the synthesis of lipids. **Review Figures 4.7, 4.11**

▶ The Golgi apparatus adds signal molecules to proteins, directing them to their proper destinations. It receives materials from the rough ER by means of vesicles that fuse with the *cis* region of the Golgi. **Review Figures 4.7, 4.12, 4.13**

▶ Vesicles originating from the *trans* region of the Golgi contain proteins for different cellular locations. Some of these vesicles fuse with the plasma membrane and release their contents outside the cell. **Review Figure 4.12**

▶ Lysosomes contain many digestive enzymes. Lysosomes fuse with the phagosomes produced by phagocytosis to form secondary lysosomes, in which engulfed materials are digested. Undigested materials are secreted from the cell when the secondary lysosome fuses with the plasma membrane. **Review Figure 4.13**

Organelles that Process Energy

▶ Mitochondria are enclosed by an outer membrane and an inner membrane that folds inward to form cristae. Mitochondria contain the proteins needed for cellular respiration and the generation of ATP. **Review Figure 4.14**

▶ All eukaryotic cells contain mitochondria space. Green plant cells also contain chloroplasts These organelles are enclosed by double membranes and contain an internal system of thylakoids organized as grana. **Review Figures 4.7, 4.16**

▶ Thylakoids within chloroplasts contain the chlorophyll and proteins that harvest light energy for photosynthesis. **Review Figure 4.16**

▶ Both mitochondria and chloroplasts contain their own DNA and ribosomes and are capable of making some of their own proteins.

▶ The endosymbiosis theory of the evolutionary origin of mitochondria and chloroplasts states that these organelles originated when larger prokaryotes engulfed, but did not digest, smaller prokaryotes. Mutual benefits permitted this symbiotic relationship to be maintained and to evolve into the eukaryotic organelles observed today. **Review Figure 4.18**

Other Organelles Enclosed by Membranes

▶ Peroxisomes and glyoxysomes contain special enzymes and carry out specialized chemical reactions inside the cell.

▶ Vacuoles are prominent in many plant cells and consist of a membrane-enclosed compartment full of water and dissolved substances. By taking in water, vacuoles enlarge and provide the pressure needed to stretch the cell wall and provide structural support for the plant.

The Cytoskeleton

▶ The cytoskeleton within the cytoplasm of eukaryotic cells provides shape, strength, and movement. It consists of three interacting types of protein fibers. **Review Figure 4.21**

▶ Microfilaments consist of two chains of actin units that together form a double helix. Microfilaments strengthen cellular structures and provide the movement in animal cell

division, cytoplasmic streaming, and pseudopod extension. Microfilaments may be found as individual fibers, bundles of fibers, or networks of fibers joined by linking proteins. **Review Figures 4.21, 4.23**

▶ Intermediate filaments are formed of keratins and are organized into tough, ropelike structures that add strength to cell attachments in multicellular organisms. **Review Figure 4.21**

▶ Microtubules are composed of dimers of the protein tubulin. They can lengthen and shorten by adding and losing tubulin dimers. They are involved in the structure and function of cilia and flagella, both of which have a characteristic 9 + 2 pattern of microtubules. **Review Figures 4.21, 4.24**

▶ The movements of cilia and flagella are due to the binding of the motor protein dynein to the microtubules. Microtubules also bind motor proteins, including kinesin and dynein, that move organelles through the cell. **Review Figure 4.25**

▶ Centrioles, made up of triplets of microtubules, are involved in the distribution of chromosomes during nuclear division. **Review Figure 4.26**

Extracellular Structures

▶ Materials external to the plasma membrane provide protection, support, and attachment for cells in multicellular systems.

▶ The cell wall of plants consists principally of cellulose. It is pierced by plasmodesmata that join the cytoplasm of adjacent cells. **Review Figure 4.27**

▶ In multicellular animals, the extracellular matrix consists of different kinds of proteins, including proteoglycan. In bone and cartilage, the protein collagen predominates. **Review Figure 4.28**

For Discussion

1. Which organelles and other structures are found in both plant and animal cells? Which are found in plant but not animal cells? In animal but not plant cells? Discuss these differences in relation to the activities of plants and animals.

2. Through how many membranes would a molecule have to pass in going from the interior of a chloroplast to the interior of a mitochondrion? From the interior of a lysosome to the outside of a cell? From one ribosome to another?

3. How does the possession of double membranes by chloroplasts and mitochondria relate to the endosymbiosis theory of the origins of these organelles? What other evidence supports the theory?

4. What kinds of cells and subcellular structures would you choose to examine by transmission electron microscopy? By scanning electron microscopy? By light microscopy? What are the advantages and disadvantages of each of these modes of microscopy?

5 Cellular Membranes

"NO SWEAT" MAY DESCRIBE YOUR REACTION TO a course with a light workload. But it certainly does not apply to a professional athlete—or to anyone else—who is engaging in vigorous activity. The harder we work physically, the hotter we get, and soon we start to sweat. Sweating is a way to reduce body heat by using the excess heat to evaporate water. At peak activity, we lose as much as 2 liters of water in an hour.

The sweat glands lie just below the surface of the skin. They are essentially tubes bathed in extracellular fluid. When stimulated by physical activity or other signals, these tubes fill with water and dissolved solutes. To get from the extracellular fluid into the tubes, water must pass into and through the cells that line the tube.

A hallmark of living cells is their ability to regulate what enters and leaves their cytoplasm. This is a function of the plasma membrane, which is composed of a hydrophobic lipid bilayer with associated proteins. When a person engages in normal activities, the membranes of the cells lining their sweat glands do not allow much water to enter or leave. But when the same person exercises, special pore proteins in the membrane, called aquaporins, open and allow water from the extracellular fluid to pass through the cells into a tube that leads to the surface of the skin.

Membranes are dynamic structures whose components move, change, and perform vital physiological roles as they allow cells to interact with other cells and molecules in the environment. We describe the structural aspects of these interactions here. Membranes also regulate ionic and molecular traffic into and out of the cell. This selective permeability, which we describe in this chapter, is an important characteristic of life. Later, we will see it in action in such diverse situations as the transduction of light energy into chemical energy in the chloroplast and the retention of water and ions in the mammalian kidney.

Membrane Composition and Structure

The chemical makeup, physical organization, and functioning of a biological membrane depend on three classes of biochemical compounds: lipids, proteins, and carbohy-drates (Figure 5.1). The lipids establish the physical integrity of the membrane and create an effective barrier to the rapid passage of hydrophilic materials such as water and ions. In addition, the phospholipid bilayer serves as a lipid "lake" in which a variety of proteins "float." This general design is known as the *fluid mosaic model* of the membrane. Membrane proteins embedded in the phospholipid bilayer have a number of functions, including moving materials through the membrane and receiving chemical signals from the cell's external environment.

Like some proteins, carbohydrates—the third class of compounds important in membranes—are crucial in recognizing specific molecules. The carbohydrates attach either to lipid or to protein molecules on the outside of the plasma membrane, where they protrude into the environment, away from the cell.

Lipids constitute the bulk of a membrane

Nearly all of the lipids in biological membranes are phospholipids. Recall from Chapter 2 that some compounds are hydrophilic ("water-loving") and others are hydrophobic

Sweating: A Regulated Membrane Activity
Tennis star Venus Williams, shown here winning a gold medal at the 2000 Olympic Games, can lose up to 2 liters of water in an hour by sweating. The excess body heat generated by her physical activity is used to evaporate the sweat, helping to keep her body temperature at normal levels.

("water-hating"). Phospholipids are both: They have both hydrophilic regions and hydrophobic regions. The long, nonpolar fatty acid "tails" of a phospholipid molecule are hydrophobic and associate easily with other nonpolar materials, but they do not dissolve in water or associate with hydrophilic substances. The phosphorus-containing "head" of the phospholipid is electrically charged and hence very hydrophilic.

As a consequence of these properties, one way in which phospholipids can coexist with water is to form a double layer, with the fatty acids of the two layers interacting with each other and the polar regions facing the outside aqueous environment (Figure 5.2). It is easy to make artificial membranes with the same bilayered arrangement in the laboratory. Both artificial and natural membranes form continuous sheets. Because of the tendency of the nonpolar fatty acids to associate with one another and exclude water, small holes or rips in a membrane seal themselves spontaneously. This property helps membranes fuse during vesicle fusion, phagocytosis, and related processes.

The phospholipid bilayer stabilizes the entire membrane structure. At the same time, the fatty acids of the phospholipids make the hydrophobic interior of the membrane somewhat fluid—about as fluid as lightweight machine oil. This fluidity permits some molecules to move laterally within the plane of the membrane. A given phospholipid molecule in the plasma membrane may travel from one end of the cell to the other in a little more than a second. On the other hand, seldom does a phospholipid molecule in one half of the bilayer flip over to the other side and trade places with another phospholipid molecule. For such a swap to happen, the polar part of each molecule would have to move through the hydrophobic interior of the membrane. Since phospholipid flip-flops are rare, the inner and outer halves of the bilayer may be quite different in the kinds of phospholipids present.

All biological membranes have a similar structure, but membranes from different cells or organelles may differ greatly in their lipid composition. For example, 25% of the lipid in some membranes is cholesterol (see Chapter 3), but

5.1 The Fluid Mosaic Model

The general molecular structure of biological membranes is a continuous phospholipid bilayer in which proteins are embedded.

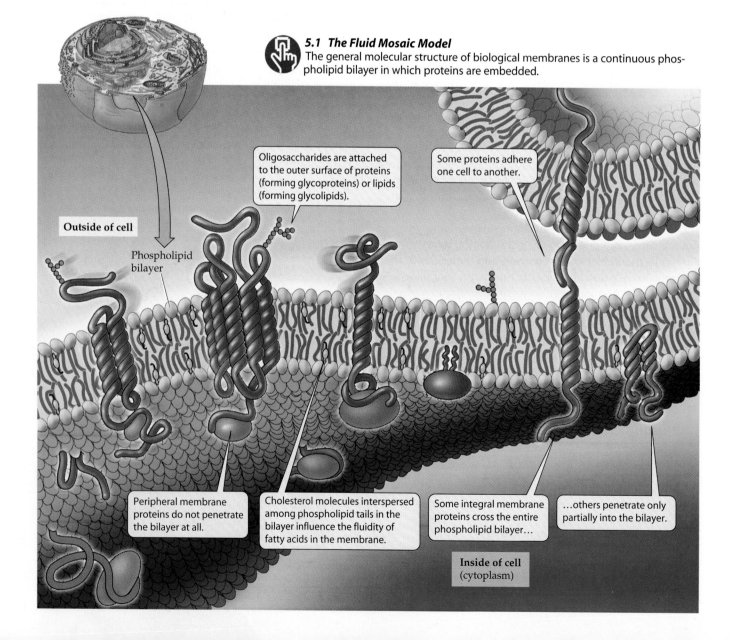

Oligosaccharides are attached to the outer surface of proteins (forming glycoproteins) or lipids (forming glycolipids).

Some proteins adhere one cell to another.

Outside of cell

Phospholipid bilayer

Peripheral membrane proteins do not penetrate the bilayer at all.

Cholesterol molecules interspersed among phospholipid tails in the bilayer influence the fluidity of fatty acids in the membrane.

Some integral membrane proteins cross the entire phospholipid bilayer…

…others penetrate only partially into the bilayer.

Inside of cell
(cytoplasm)

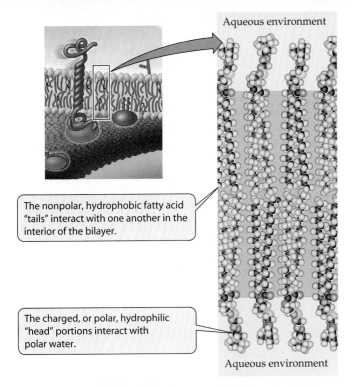

The nonpolar, hydrophobic fatty acid "tails" interact with one another in the interior of the bilayer.

The charged, or polar, hydrophilic "head" portions interact with polar water.

Aqueous environment

Aqueous environment

5.2 A Phospholipid Bilayer Separates Two Aqueous Regions
The eight phospholipid molecules shown here represent a small cross section of a membrane bilayer.

other membranes have no cholesterol at all. When present, cholesterol is important to membrane integrity; most cholesterol in membranes is not hazardous to your health. A molecule of cholesterol is commonly situated next to an unsaturated fatty acid, and the polar hydroxyl end of the cholesterol extends into the surrounding aqueous layer (see Figure 5.1).

Cholesterol may either increase or decrease membrane fluidity, depending on other factors, such as fatty acid composition. Shorter fatty acid chains make for a more fluid membrane, as do unsaturated fatty acids. Adequate membrane fluidity is essential for many membrane functions. Since molecules move more slowly and fluidity decreases at reduced temperatures, membrane functions may decline in organisms that cannot keep their bodies warm. To address this problem, some organisms simply change the lipid compositions of their membranes, replacing saturated with unsaturated fatty acids and using fatty acids with shorter tails. Such changes play a part in the survival of plants and hibernating animals and bacteria during the winter.

Membrane proteins are asymmetrically distributed

All biological membranes contain proteins. Typically, plasma membranes have 1 protein molecule for every 25 phospholipid molecules. This ratio varies, depending on membrane function. In the inner membrane of the mitochondrion, which is specialized for energy processing, there

5.3 Membrane Proteins Revealed by the Freeze-Fracture Technique
This membrane from a spinach chloroplast was first frozen and then separated so that the membrane bilayer was split open.

is 1 protein for every 15 lipids; myelin, which encloses nerve cells and uses the properties of lipids to act as an electrical insulator, has only 1 protein per 70 lipids.

Many membrane proteins are embedded in, and/or extend across, the lipid bilayer. Like phospholipids, these proteins have polar and nonpolar regions. In the polar regions, amino acids with hydrophilic R groups (side chains) predominate, while the nonpolar regions have amino acids with hydrophobic R groups (see Table 3.2). Like the phospholipids, these proteins are positioned in the membrane so that the polar ends stick out into the aqueous environment and the nonpolar regions aggregate with one another (and with the nonpolar fatty acid tails of the lipids) away from water.

A special preparation method for electron microscopy, freeze-fracturing, reveals membrane proteins embedded in the lipid bilayer (Figure 5.3). The bumps that can be seen protruding from the interior of a membrane are not observed in pure lipid bilayers.

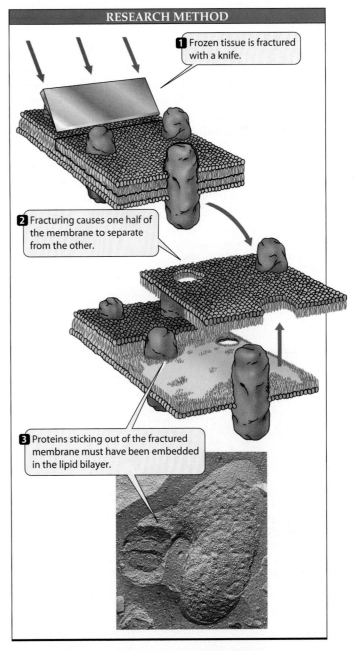

RESEARCH METHOD

1 Frozen tissue is fractured with a knife.

2 Fracturing causes one half of the membrane to separate from the other.

3 Proteins sticking out of the fractured membrane must have been embedded in the lipid bilayer.

According to the fluid mosaic model, the proteins and lipids in a membrane are independent of each other, and interact only noncovalently. The polar ends of proteins can interact with polar ends of lipids, and the nonpolar regions of both molecules interact hydrophobically (see Figure 5.1).

There are two general types of membrane proteins:

▶ **Integral** membrane proteins have hydrophobic regions, and penetrate the phospholipid bilayer. Many of these proteins have long hydrophobic α-helical regions that span the hydrophobic core of the bilayer. Their hydrophilic ends protrude into the aqueous environments on either side of the membrane (Figure 5.4).

▶ **Peripheral** membrane proteins lack hydrophobic regions, and are not embedded in the bilayer. Instead, they have polar or charged regions that interact with similar regions on exposed parts of the integral membrane proteins or phospholipid molecules.

Some membrane proteins are covalently attached to fatty acids or other lipid groups. These proteins can be classified as a special type of integral protein, as their hydrophobic lipid component allows them to insert themselves into the lipid bilayer.

Like the lipids, many membrane proteins move relatively freely within the phospholipid bilayer. Experiments using the technique of cell fusion illustrate this migration dramatically. When two cells are fused, a single continuous membrane forms and surrounds both cells, and some proteins from each cell distribute themselves uniformly around this membrane.

Although many proteins are mobile in the membrane, some are not free to migrate but appear to be "anchored" by components of the cytoskeleton. This anchoring can re-

sult in a segregation of these proteins, resulting in functional specialization to different regions on the cell surface. For example, in certain muscle cells, the plasma membrane protein that serves as a receptor for the chemical signal from nerve cells is normally found only at the site where a nerve cell meets the muscle cell. None of this protein is found elsewhere on the surface of the muscle cell.

Proteins are asymmetrically distributed on the inner and outer surfaces of a membrane. Transmembrane proteins show different "faces" on the two membrane surfaces. Such proteins have certain specific domains (or regions) of their primary structure on the outer side of the membrane, other domains within the membrane, and still other domains on the inner side of the membrane. Peripheral membrane proteins are localized on one side of the membrane or the other, but not both. This arrangement gives the two surfaces of the membrane different properties. As we will see when we discuss active transport, these differences have great functional significance.

Membrane carbohydrates are recognition sites

In addition to lipids and proteins, all plasma membranes and some internal cytoplasmic membranes contain significant amounts of carbohydrates. The carbohydrates are located on the outer surface of the membrane and serve as recognition sites for other cells and molecules (see Figure 5.1).

Membrane-associated carbohydrates may be covalently bound to lipids or to proteins. A carbohydrate bound to a lipid forms a *glycolipid*. The carbohydrate units of glycolipids often extend to the outside of the membrane, where they serve as recognition signals for interactions between cells. For example, the carbohydrate of some glycolipids changes when a cell becomes cancerous. This change may allow white blood cells to target the cancer cells for destruction.

Most of the carbohydrate in membranes is covalently bound to proteins, forming *glycoproteins*. The bound carbohydrates are oligosaccharide chains, usually not exceeding 15 monosaccharide units in length. The chains are added to membrane proteins inside the endoplasmic reticulum and are modified in the Golgi apparatus.

Glycoproteins enable a cell to be recognized by other cells and proteins. An "alphabet" of monosaccharides can be used to generate a diversity of messages. Recall from Chapter 3 that sugar molecules can be formed from 3–7 carbons attached at different sites to one another, forming linear or branched oligosaccharides with many different three-dimensional shapes. An oligosaccharide of a specific shape from one cell can bind to a mirror-image shape on an adjacent cell. This binding forms the basis of cell-to-cell adhesion.

Cell Adhesion

A living sponge is a multicellular marine animal with a simple body plan (see Chapter 31). The cells of the sponge are stuck together, but they can be disaggregated mechanically by passing the animal several times through a fine wire screen. What was an animal is now hundreds of individual of cells, suspended in seawater. Remarkably, if the

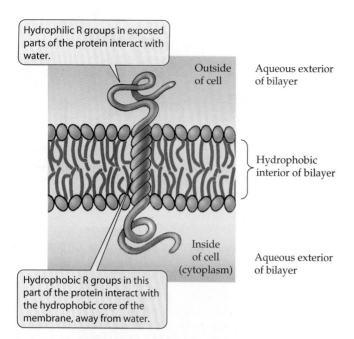

5.4 Interactions of Integral Membrane Proteins
An integral membrane protein is held in the membrane by the distribution of the hydrophilic and hydrophobic side chains of its amino acids.

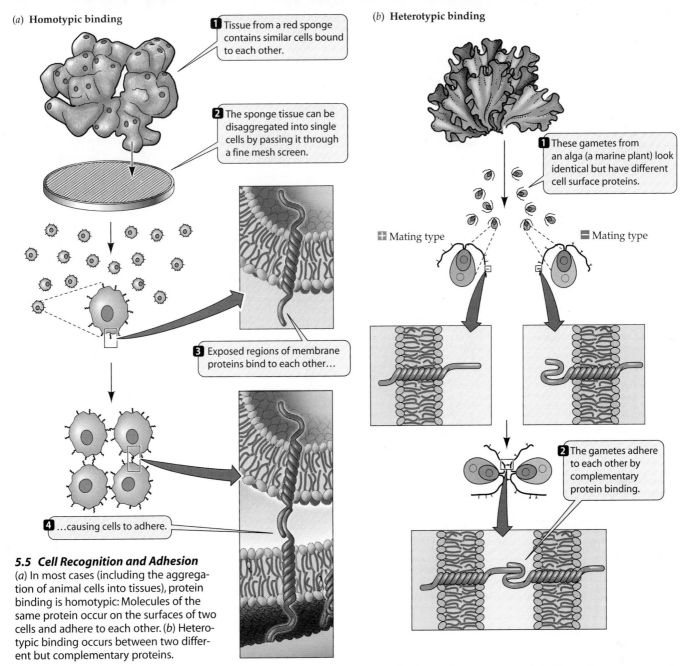

(a) Homotypic binding

1 Tissue from a red sponge contains similar cells bound to each other.

2 The sponge tissue can be disaggregated into single cells by passing it through a fine mesh screen.

3 Exposed regions of membrane proteins bind to each other…

4 …causing cells to adhere.

(b) Heterotypic binding

1 These gametes from an alga (a marine plant) look identical but have different cell surface proteins.

➕ Mating type   ➖ Mating type

2 The gametes adhere to each other by complementary protein binding.

5.5 Cell Recognition and Adhesion
(a) In most cases (including the aggregation of animal cells into tissues), protein binding is homotypic: Molecules of the same protein occur on the surfaces of two cells and adhere to each other. (b) Heterotypic binding occurs between two different but complementary proteins.

cell suspension is shaken for a few hours, the cells reaggregate into a sponge!

If two different species of sponge are disaggregated together, the cells of each species reaggregate into that species only. Such tissue-specific and species-specific *cell adhesions* are essential in the formation and maintenance of tissues and organisms. Think of the skin in your arm: What keeps its cells together and separates the skin from the underlying bones? You will see many examples of specific cell adhesion throughout this book; here, we describe its general principles.

Cell adhesion involves recognition proteins

The factor responsible for cell–cell recognition and adhesion in sponges was the first such molecule to be identified and purified. It is a huge membrane glycoprotein (80% sugar), partly embedded in the plasma membrane, with the recognition part sticking out and exposed to the environ-

ment (and to other sponge cells). Since then, many such recognition proteins have been purified.

As we saw in Chapter 3, a macromolecule such as a protein has not only a specific shape, but also specific chemical groups. Both features allow binding to other specific molecules (Figure 5.5). In most cases, the binding of cells in a tissue is *homotypic*; that is, the same molecule sticks out of both cells, and the exposed surfaces bind to each other. But *heterotypic* binding (of cells with different proteins) also can occur. For example, when the mammalian sperm meets the egg, different proteins on the two types of cells have complementary binding surfaces.

Plant cells also have recognition proteins. Some single-celled plants form similar-appearing reproductive cells (analogous to male and female) that have flagella to propel them toward each other. Male and female cells can recognize each other by heterotypic proteins on their flagella. In the majority of plant cells, the plasma membrane is covered with a thick cell wall, and this, too, has adhesion proteins that allow cells to bind to one another.

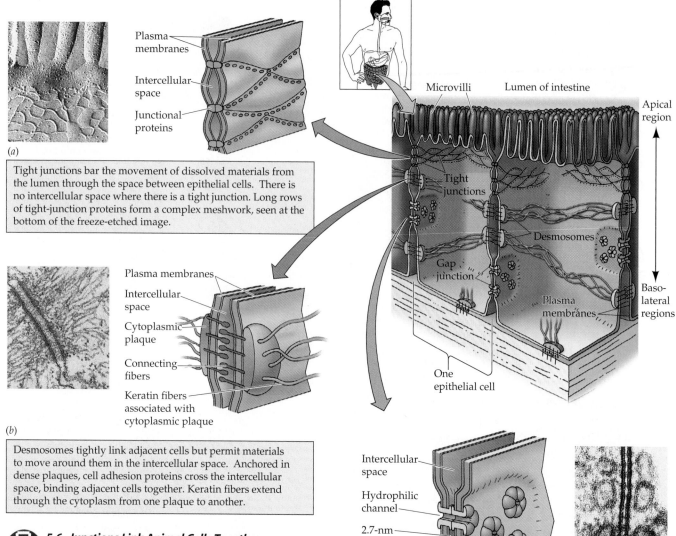

(a)

Tight junctions bar the movement of dissolved materials from the lumen through the space between epithelial cells. There is no intercellular space where there is a tight junction. Long rows of tight-junction proteins form a complex meshwork, seen at the bottom of the freeze-etched image.

(b)

Desmosomes tightly link adjacent cells but permit materials to move around them in the intercellular space. Anchored in dense plaques, cell adhesion proteins cross the intercellular space, binding adjacent cells together. Keratin fibers extend through the cytoplasm from one plaque to another.

5.6 Junctions Link Animal Cells Together
(*a,b*) Tight junctions and desmosomes are abundant in epithelial tissues. (*c*) Gap junctions are also found in muscle and nerve tissues.

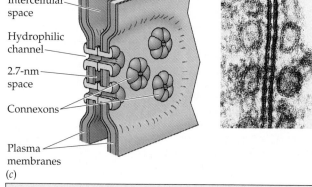

(c)

Gap junctions let adjacent cells communicate. Dissolved molecules and electric signals may pass from one cell to the other through the channels formed by two connexons extending from adjacent cells.

Cell adhesion proteins from many multicellular organisms have been characterized. Some of them do not just bind the two cells together, but initiate the formation of specialized cell junctions.

Specialized Cell Junctions

In a complex multicellular organism, cell–cell recognition proteins allow specific kinds of cells to adhere to each other. Often, both cells contribute material to additional membrane structures that "cement" their relationship. These specialized structures, called *cell junctions*, are most evident in electron micrographs of epithelial tissues, which are layers of cells that line body cavities or cover body surfaces. We will examine three types of cell surface junctions that enable cells to make direct physical contact and link with one another: tight junctions, desmosomes, and gap junctions

Tight junctions seal tissues and prevent leaks

Tight junctions are specialized structures at the plasma membrane that link adjacent epithelial cells. They result

from the mutual binding of strands of specific membrane proteins that form a series of joints encircling the epithelial cells. They are found in the region surrounding the lumen (cavity) of organs such as the intestine (Figure 5.6a).

Tight junctions have two functions:

▶ They prevent substances from moving through the intercellular space. Thus, any substance entering the body from the lumen must pass through the epithelial cells.

▶ They restrict the migration of membrane proteins and phospholipids from one region of the cell to another. Thus, the proteins and phospholipids in the region facing the lumen (apical) can be different from those in the regions facing the sides and bottom of the cell (basolateral).

By forcing materials to enter some cells, and by allowing different ends of cells to have different membrane proteins with different functions, tight junctions help ensure the directional movement of materials into the body.

Desmosomes hold cells together

Desmosomes are specialized structures associated with the plasma membrane at certain sites in epithelial tissues. They hold adjacent cells firmly together, acting like spot welds or rivets (Figure 5.6*b*). Each desmosome has a dense plaque on the cytoplasmic surface of the plasma membrane. This plaque is attached to fibers in the cytoplasm and special cell adhesion proteins in the plasma membrane. These proteins stretch from the plaque through the plasma membrane of one cell, across the intercellular space, and through the plasma membrane of the adjacent cell, where they bind to the plaque proteins in that cell.

The cytoplasmic fibers of a desmosome, which are intermediate filaments of the cytoskeleton (see Figure 4.21), are made of a protein called *keratin*. They stretch from one cytoplasmic plaque across the cell to connect with another plaque on the other side of the cell. Anchored thus on both sides of the cell, these extremely strong keratin fibers provide great mechanical stability to epithelial tissues, which often receive rough wear in protecting the organism's body surface integrity.

Gap junctions are a means of communication

Whereas tight junctions and desmosomes have mechanical roles, **gap junctions** facilitate communication between cells. Each gap junction is made up of specialized protein channels, called connexons, that span the plasma membranes of two adjacent cells and the intercellular space between them (Figure 5.6*c*). We will describe their role in more detail, as well as that of plasmodesmata, which perform a similar role in plants, when we discuss cell communication in Chapter 15.

Passive Processes of Membrane Transport

We have examined membrane structure and how it is used to perform one membrane function: the binding of one cell to another. Now we turn to the second major membrane function, selective permeability: the ability to allow some substances, but not others, to pass through the plasma membrane and enter or leave the cell.

There are two fundamentally different processes by which substances cross biological membranes to enter and leave cells or organelles: passive processes and active processes. Passive processes include the different types of diffusion: simple diffusion through the phospholipid bilayer, and facilitated diffusion through channel proteins or by means of carrier molecules. Active processes, on the other hand, require the input of energy. We'll discuss the active processes later in this chapter, after first focusing on the passive processes. Before considering diffusion as it works across a membrane, however, we must understand the basic principles of diffusion.

The physical nature of diffusion

Nothing in this world is ever absolutely at rest. Everything is in motion, though the motions may be very small. As the temperature of a solution rises, its molecules and ions move faster—they vibrate, rotate, and move from place to place more quickly. An important consequence of this random jiggling is that all the components of a solution tend eventually to become evenly distributed throughout the system. For example, if a drop of ink is allowed to fall into a container of water, the pigment molecules of the ink are initially very concentrated. Without human intervention such as stirring, the pigment molecules of the ink move about at random, spreading slowly through the water until eventually the concentration of pigment—and thus the intensity of color—is exactly the same in every drop of liquid in the container. A solution in which the particles are uniformly distributed is said to be at *equilibrium*, because there will be no future net change in concentration.

Diffusion is the process of random movement toward a state of equilibrium. Although the motion of each individual particle is absolutely random, in diffusion the *net movement* of particles is directional until equilibrium is reached. Diffusion is thus net movement from regions of *greater concentration* to regions of *lesser concentration* (Figure 5.7).

In a complex solution (one with many different solutes), the diffusion of each substance is independent of that of the others. How fast a substance diffuses depends on four factors: (1) the diameter of the molecules or ions; (2) the temperature of the solution; (3) the electric charge, if any, of the diffusing material; and (4) the **concentration gradient** in the system. The concentration gradient is the change in concentration with distance in a given direction. The greater the concentration gradient, the more rapidly a substance diffuses.

DIFFUSION WITHIN CELLS AND TISSUES. Within cells, or wherever distances are very short, solutes distribute themselves rapidly by diffusion. Small molecules and ions may move from one end of an organelle to another in a millisecond (10^{-3} s). On the other hand, the usefulness of diffusion as a transport mechanism declines drastically as distances become greater. In the absence of mechanical stirring, diffusion across more than a centimeter may take an hour or more, and diffusion across meters may take years! Diffusion would not be adequate to distribute materials over the length of the human body, but within our cells or across layers of one or two cells, diffusion is rapid enough to distribute small molecules and ions almost instantaneously.

DIFFUSION ACROSS MEMBRANES. In a solution without barriers, all the solutes diffuse at rates determined by temperature, their physical properties, and the concentration gradient of each solute. If a biological membrane is introduced as a barrier, the movement of the different solutes can be affected by the properties of the membrane. The membrane is said to be *permeable* to solutes that can cross it more or less easily, but *impermeable* to substances that cannot move across it. Molecules to which the membrane is permeable

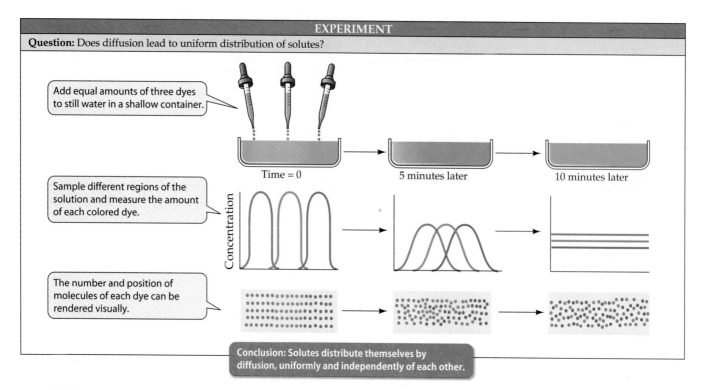

EXPERIMENT

Question: Does diffusion lead to uniform distribution of solutes?

Add equal amounts of three dyes to still water in a shallow container.

Time = 0   5 minutes later   10 minutes later

Sample different regions of the solution and measure the amount of each colored dye.

Concentration

The number and position of molecules of each dye can be rendered visually.

Conclusion: Solutes distribute themselves by diffusion, uniformly and independently of each other.

5.7 Diffusion Leads to Uniform Distribution of Solutes

Diffusion is the net movement of a solute from regions of greater concentration to regions of lesser concentration. The speed of diffusion varies with the substances involved, but the process continues until the solution reaches equilibrium.

diffuse from one compartment to the other until their concentrations are equal on both sides of the membrane. Molecules to which the membrane is impermeable remain in separate compartments, and their concentrations remain different on the two sides of the membrane. Equilibrium is reached when the concentrations of the diffusing substance are identical on both sides of the permeable membrane. Individual molecules are still passing through the membrane when equilibrium is established, but equal numbers of molecules are moving in each direction, so there is no net change in concentration.

Simple diffusion takes place through the membrane bilayer

In **simple diffusion**, small molecules pass through the lipid bilayer of the membrane. The more lipid-soluble the molecule, the more rapidly it diffuses through the bilayer. This statement holds true over a wide range of molecular weights. Only water and the smallest of molecules seem to deviate from this rule, passing through bilayers much more rapidly than their lipid solubilities would predict.

Charged and/or polar molecules such as amino acids, sugars, and ions do not pass readily through a membrane, for two reasons. First, cells are made up of, and exist in, water, and polar or charged substances form many hydrogen bonds with water, preventing their "escape" to the membrane. Second, the interior of the membrane is hy-

drophobic, and hydrophilic substances tend to be excluded from it. On the other hand, a molecule that is itself hydrophobic, and hence soluble in lipids, enters the membrane readily and is thus able to pass through it.

Osmosis is the diffusion of water across membranes

Water molecules are abundant enough and small enough that they move through membranes by a diffusion process called **osmosis**. This completely passive process uses no metabolic energy and can be understood in terms of the concentrations of solutions. Osmosis depends on the *number* of solute particles present—not the kind of particles. We will describe osmosis using red blood cells and plant cells as examples.

Red blood cells are normally suspended in a fluid called *plasma*, which contains salts, proteins, and other solutes. If a drop of blood is examined under the light microscope, the red cells are seen to have their characteristic donut shape. If pure water is added to the drop of blood, the cells quickly swell and burst (Figure 5.8a). Similarly, if slightly wilted lettuce is put in pure water, it soon becomes crisp; by weighing it before and after, we can show that it has taken up water (Figure 5.8b).

If, on the other hand, red blood cells or crisp lettuce leaves are placed in a relatively concentrated solution of salt or sugar, the leaves become limp (wilt) and the red blood cells pucker and shrink. From analyses of such observations, we know that the difference in solute concentrations is the principal factor that determines whether water will move from the surrounding environment into cells, or out of cells into the environment.

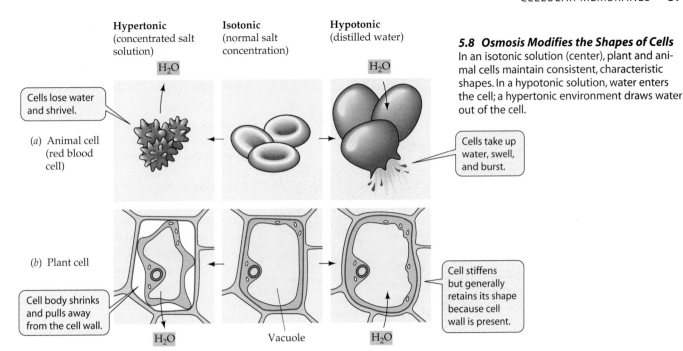

Hypertonic (concentrated salt solution)

Isotonic (normal salt concentration)

Hypotonic (distilled water)

5.8 Osmosis Modifies the Shapes of Cells
In an isotonic solution (center), plant and animal cells maintain consistent, characteristic shapes. In a hypotonic solution, water enters the cell; a hypertonic environment draws water out of the cell.

Cells lose water and shrivel.

(a) Animal cell (red blood cell)

Cells take up water, swell, and burst.

(b) Plant cell

Cell body shrinks and pulls away from the cell wall.

Cell stiffens but generally retains its shape because cell wall is present.

Vacuole

Other things being equal, if two different solutions are separated by a membrane that allows water, but not solutes, to pass through, water molecules will move across the membrane toward the solution with a higher solute concentration. In other words, water will diffuse from a region of *its higher concentration* (lower concentration of solutes) to a region of *its lower concentration* (higher concentration of solutes).

Three terms are used to compare the solute concentrations of two solutions separated by a membrane:

▶ *Isotonic solutions* have equal total solute concentrations.
▶ A *hypertonic solution* has a higher total solute concentration than the other solution with which it is being compared.
▶ A *hypotonic* solution has a lower total solute concentration than the other solution with which it is being compared.

Water moves from a hypotonic solution across a membrane to a hypertonic solution.

When we say that "water moves," bear in mind that we are referring to the *net* movement of water. Since it is so abundant, water is constantly moving across the plasma membrane into and out of cells. Whether the overall movement is greater in one direction or the other is what concerns us here.

The concentration of solutes in the environment determines the direction of osmosis in all animal cells. A red blood cell takes up water from a solution that is hypotonic to the cell's contents. The cell bursts because its plasma membrane cannot withstand the swelling of the cell (see Figure 5.8a). The integrity of red blood cells (and other blood cells) is absolutely dependent on the maintenance of a constant solute concentration in the plasma in which they are suspended: The plasma must be isotonic with the cells if the cells are not to burst or shrink.

In contrast to animal cells, the cells of plants, archaea, bacteria, fungi, and some protists have cell walls that limit the volume of the cells and keep them from bursting. Cells with sturdy cell walls take up a limited amount of water and, in so doing, build up internal pressure against the cell wall that prevents further water from entering. This pressure within the cell, called *turgor pressure*, is the driving force for the enlargement of plant cells—it is a normal and essential component of plant development.

Diffusion may be aided by channel proteins

As we saw earlier, polar substances such as amino acids and sugars and charged substances such as ions do not diffuse across membranes. Instead, they cross the hydrophobic lipid barrier through protein-lined channels in a process called **facilitated diffusion**. Integral membrane proteins form these channels (Figure 5.9), which are lined with polar amino acids and water on the inside (to bind to the polar or charged substance and allow it to pass) and nonpolar amino acids on the outside (to allow the protein channel to insert itself into the lipid bilayer).

The best-studied protein channels are the *ion channels*. As you will see, the movement of ions into and out of cells is important in many biological processes, ranging from the electrical activity of the nervous system to the opening of pores in leaves that allow gas exchange with the environment. Hundreds of these channels have been identified, and all show the basic structure of a water-lined pore that just fits the ion that moves through it.

Ion channels are *gated*: they can be closed to ion passage, or open. A gated channel opens when something happens to change the shape of the protein. Depending on the channel, this stimulus can range from the binding of a chemical signal to an electrical charge caused by an imbalance of ions. Once the channel opens, millions of ions can rush through it

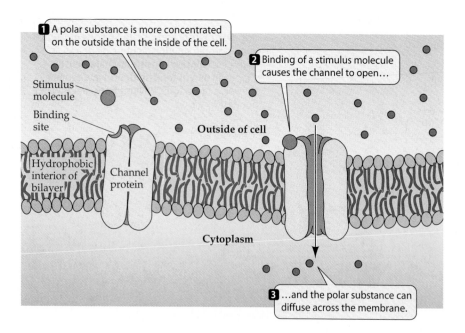

5.9 A Gated Channel Protein Opens in Response to a Stimulus
The membrane protein changes its three-dimensional shape when the stimulus binds.

favoring glucose entry, with a higher concentration outside the cell (in blood capillaries or the intestine) than inside.

Transport by carrier proteins is different from simple diffusion. In both processes, the *rate* of movement depends on the concentration gradient across the membrane. However, in facilitated diffusion, a point is reached at which further increases in the concentration gradient are not accompanied by an increased rate of diffusion. At this point, the facilitated diffusion system is said to be *saturated*. Because there are only a limited number of carrier protein molecules per unit of membrane area, the rate of movement reaches a maximum when all the carrier molecules are fully loaded with solute molecules. In other words, when the differences in solute concentration across the membrane are sufficiently high, not enough carrier molecules are free at a given moment to handle all the solute molecules.

Active Transport

In many biological situations, an ion or molecule must be moved across a membrane from a region of lower concentration to a region of higher concentration. In these cases,

per second. How fast this happens, and in which direction (into or out of the cell), depends on the concentration gradient of the ion between the cytoplasm and the exterior environment of the cell. For example, if the concentration of potassium ion is much higher outside of the cell than inside, potassium will enter the cell through a potassium channel by diffusion; if it is higher inside the cell, potassium ion will diffuse out of the cell.

As we mentioned, water crosses the plasma membrane at a rate far in excess of expectations, given its polarity. One way that water can do this is by hydrating ions as they pass through ion channels. Up to 12 water molecules may coat an ion as it traverses a channel. Another way that water enters cells rapidly is through water channels called *aquaporins*. Membrane proteins that allow water to pass through them have been characterized in many cells, from the plant vacuole, where they are important in maintaining turgor, to the mammalian kidney, where they act in retaining water that would otherwise be lost through urine.

Carrier proteins aid diffusion by binding substances

Another kind of facilitated diffusion involves not just the opening of a channel, but the actual binding of the transported substance to a membrane protein. These proteins are called *carriers*, and, like channel proteins, they allow diffusion both into and out of the cell. They are used to transport polar molecules such as sugars and amino acids.

Glucose, for example, is the major energy source for most mammalian cells, and those cells have a carrier protein called the glucose transporter that facilitates the uptake of glucose (Figure 5.10). Since glucose is rapidly altered as soon as it gets into a cell, there is almost always a strong concentration gradient

5.10 A Carrier Protein Facilitates Diffusion
The carrier protein allows glucose to enter the cell at a faster rate than would be possible by simple diffusion across the membrane barrier.

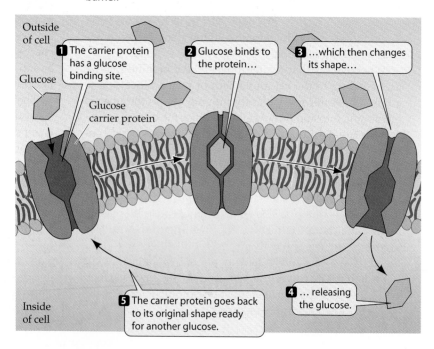

5.1 Membrane Transport Mechanisms	SIMPLE DIFFUSION	FACILITATED DIFFUSION	ACTIVE TRANSPORT
Direction	With concentration gradient	With concentration gradient	Against concentration gradient
Energy source	Concentration gradient	Concentration gradient	ATP hydrolysis (primary)
Membrane protein required?	No	Yes	Yes
Specificity	Not specific	Specific	Specific

the substance cannot not rush into or out of cells by diffusion. The movement of a substance across a biological membrane *against* a concentration gradient—called **active transport**—requires the expenditure of energy. The differences between diffusion and active transport are summarized in Table 5.1.

Active transport is directional

Three types of proteins are involved in active transport (Figure 5.11):

▶ **Uniport** transporters move a single solute in one direction. For example, a Ca^{2+}-binding protein found in the plasma membrane and endoplasmic reticulum membranes of many cells actively transports this ion to regions of higher concentration either outside the cell or inside the ER.

▶ **Symport** transporters move two solutes in the same direction. For example, the uptake of amino acids from the intestine into the cells that line it requires the simultaneous binding of Na^+ and amino acid to the same carrier protein.

▶ **Antiport** transporters move two solutes in opposite directions, one into the cell and the other out of the cell. For example, many cells have an "Na^+–K^+ pump" that moves Na^+ out of the cell and K^+ into it.

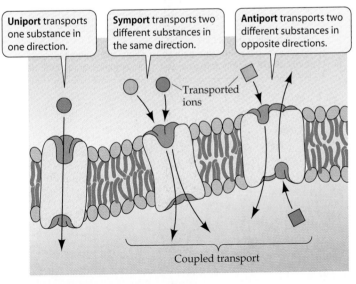

Uniport transports one substance in one direction.

Symport transports two different substances in the same direction.

Antiport transports two different substances in opposite directions.

Transported ions

Coupled transport

5.11 Proteins for Active Transport
Note that in each of the three cases, transport is directional.

Primary and secondary active transport rely on different energy sources

There are two basic types of active transport processes. The first, **primary active transport**, requires the direct participation of ATP. Energy released by the hydrolysis of ATP drives the movement of specific ions against a concentration gradient. For example, if we compare the concentrations of potassium ions (K^+) and sodium ions (Na^+) inside a nerve cell and in the fluid bathing the nerve (Table 5.2), we can see that the K^+ concentration is much higher inside the cell, whereas the Na^+ concentration is much higher outside. Nevertheless, a protein in the nerve cells continues to pump Na^+ out and K^+ in, against these concentration gradients, ensuring that the gradients are maintained. This *sodium–potassium pump* is found in all animal cells and is an integral membrane glycoprotein. It breaks down a molecule of ATP to ADP and phosphate (P_i), and uses the energy released to bring two K^+ ions into the cell and export three Na^+ ions (Figure 5.12). The Na^+–K^+ pump is thus an antiport transport system.

Only cations are transported directly by pumps in primary active transport. Other solutes are transported by **secondary active transport**. This form of active transport does not use ATP directly; rather, the transport of the solute is tightly coupled to an ion concentration gradient established by primary active transport. The movement of the solute *against* its concentration gradient is accomplished using energy "regained" by letting ions move across the membrane *with* their concentration gradient.

For example, energy from ATP is used in primary active transport to establish concentration gradients of potassium and sodium ions; then the passive diffusion of some sodium ions in the opposite direction provides energy for the secondary active transport of the sugar glucose (Figure 5.13). Other secondary active transporters aid in the uptake

5.2 Concentration of Major Ions Inside and Outside the Nerve Cell of a Squid	CONCENTRATION (MOLAR)	
ION	INSIDE	OUTSIDE
K^+	0.400	0.020
Na^+	0.050	0.440
Cl^-	0.120	0.560

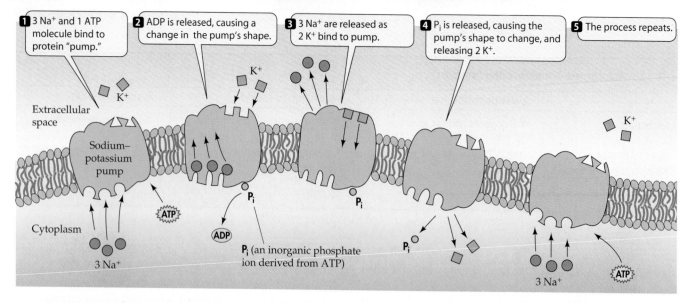

1. 3 Na⁺ and 1 ATP molecule bind to protein "pump."

2. ADP is released, causing a change in the pump's shape.

3. 3 Na⁺ are released as 2 K⁺ bind to pump.

4. P$_i$ is released, causing the pump's shape to change, and releasing 2 K⁺.

5. The process repeats.

Extracellular space

Sodium–potassium pump

K⁺

Cytoplasm

3 Na⁺

ATP

ADP

P$_i$

P$_i$ (an inorganic phosphate ion derived from ATP)

P$_i$

3 Na⁺

ATP

K⁺

5.12 Primary Active Transport: The Na⁺–K⁺ Pump

In active transport, energy is used to move a solute against its concentration gradient. Even though the Na⁺ concentration is higher outside the cell and the K⁺ concentration is higher inside the cell, for each molecule of ATP used, two K⁺ are pumped *into* the cell and three Na⁺ are pumped *out of* the cell.

of amino acids and other sugars, which are essential raw materials for cell maintenance and growth. Both types of coupled transport proteins—symports and antiports—are used for secondary active transport.

Endocytosis and Exocytosis

Macromolecules such as proteins, polysaccharides, and nucleic acids are simply too large and too charged or polar to pass through membranes. This is a fortunate property. Think of the consequences if these molecules could diffuse out of cells: A red blood cell would not retain its hemoglobin! On the other hand, cells must sometimes take up or secrete intact large molecules. As we saw in Chapter 4, this occurs by means of vesicles that either pinch off from the plasma mem-

brane and enter the cell (endocytosis) or fuse with the plasma membrane and release their contents (exocytosis).

Macromolecules and particles enter the cell by endocytosis

Endocytosis is a general term for a group of processes that bring macromolecules, large particles, small molecules, and even small cells into the eukaryotic cell (Figure 5.14*a*). There are three types of endocytosis: phagocytosis, pinocytosis, and receptor-mediated endocytosis. In all three, the plasma membrane invaginates (folds inward) around materials from the environment, forming a small pocket. The pocket deepens, forming a vesicle. This vesicle separates from the surface of the cell and migrates with its contents to the cell's interior.

5.13 Secondary Active Transport

The sodium ion concentration gradient established by primary active transport (right) powers the secondary active transport of glucose (left). The movement of glucose across the membrane against its concentration gradient is coupled by a symport protein to the movement of Na⁺ into the cell.

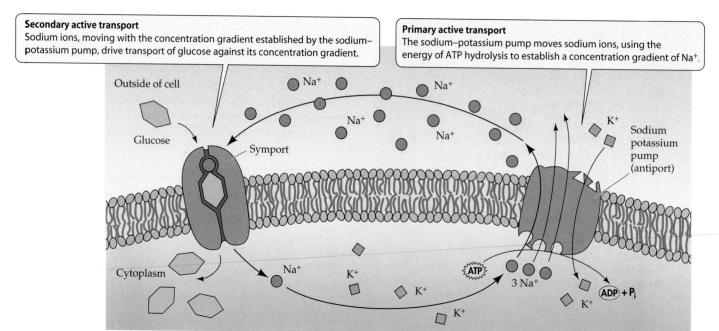

Secondary active transport
Sodium ions, moving with the concentration gradient established by the sodium–potassium pump, drive transport of glucose against its concentration gradient.

Primary active transport
The sodium–potassium pump moves sodium ions, using the energy of ATP hydrolysis to establish a concentration gradient of Na⁺.

Outside of cell

Glucose

Symport

Na⁺

Na⁺

Na⁺

Na⁺

K⁺

Sodium potassium pump (antiport)

Cytoplasm

Na⁺

K⁺

K⁺

ATP

3 Na⁺

ADP + P$_i$

K⁺

K⁺

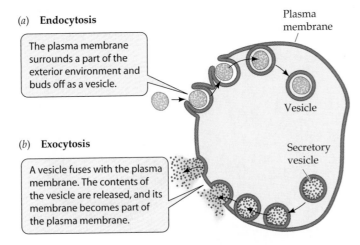

(a) Endocytosis

The plasma membrane surrounds a part of the exterior environment and buds off as a vesicle.

Plasma membrane

Vesicle

(b) Exocytosis

A vesicle fuses with the plasma membrane. The contents of the vesicle are released, and its membrane becomes part of the plasma membrane.

Secretory vesicle

5.14 Endocytosis and Exocytosis
Endocytosis and exocytosis are used by all eukaryotic cells to take up substances from and release substances to the outside environment.

In **phagocytosis**, part of the plasma membrane engulfs fairly large particles or even entire cells. Phagocytosis is used as a cellular feeding process by unicellular protists and by some white blood cells that defend the body against foreign cells and substances. The food vacuole or phagosome formed usually fuses with a lysosome, and its contents are digested (see Figure 4.13b).

In **pinocytosis** ("cellular drinking"), vesicles also form. However, these vesicles are smaller, and the process operates to bring in small dissolved substances or fluids. It is relatively nonspecific as to what it brings into the cell. For example, pinocytosis goes on constantly in the endothelium, the single layer of cells that separates a tiny blood capillary from its surrounding tissue (Figure 5.15), and is a way for the cells to rapidly acquire the fluids of the blood.

In **receptor-mediated endocytosis**, specific reactions at the cell surface trigger the uptake of specific materials. Let's take a closer look at this process.

Receptor-mediated endocytosis is highly specific

Receptor-mediated endocytosis is used by animal cells to capture specific macromolecules from the cell's environment. The uptake process is similar to nonspecific endocytosis, as already described. However, in receptor-mediated endocytosis, receptor proteins at particular sites on the outer surface of the plasma membrane bind to specific substances in the environment outside the cell. These sites are called **coated pits** because they form a slight depression in the plasma membrane whose cytoplasmic surface is coated by fibrous proteins, such as *clathrin*.

When a receptor protein binds to its specific macromolecule outside the cell, its coated pit invaginates and forms a **coated vesicle** around the bound macromolecule. Strengthened and stabilized by clathrin molecules, this vesicle carries the macromolecule into the cell (Figure 5.16). Once inside, the vesicle loses its clathrin coat and may fuse with a lysosome, where the engulfed material is processed and released into the cytoplasm. Because of its specificity for particular macromolecules, receptor-mediated endocytosis is a rapid and efficient method of taking up what may be minor constituents of the cell's environment.

Receptor-mediated endocytosis is the method by which cholesterol is taken up by most mammalian cells. Water-insoluble cholesterol is synthesized in the liver and transported in the blood attached to a protein, forming a lipoprotein called *low-density lipoprotein*, or *LDL*. The uptake of cholesterol begins with the binding of LDL to specific receptor proteins in coated pits. After being engulfed by endocytosis, the LDL particle is freed from the receptors. The receptors segregate to a region of the vesicle that buds off to form a new vesicle, which is recycled to the plasma membrane. The freed LDL particle remains in the original vesicle, which fuses with a lysosome in which the LDL is digested and the cholesterol made available for cell use. Persons with the inherited disease *hypercholesterolemia* (-*emia*, "blood") have dangerously high levels of cholesterol in their blood because of a deficient receptor for LDL.

Exocytosis moves materials out of the cell

Exocytosis is the process by which materials packaged in vesicles are secreted from a cell when the vesicle membrane fuses with the plasma membrane (see Figure 5.14b). The initial event in this process is the binding of a membrane protein protruding from the cytoplasmic side of the vesicle with a membrane protein on the cytoplasmic side of the target site on the plasma membrane. The phospholipid regions of the two membranes merge, and an opening to the outside of the cell develops. The contents of the vesicle are released to the environment, and the vesicle membrane is smoothly incorporated into the plasma membrane.

In Chapter 4, we encountered exocytosis as the last step in the processing of material engulfed by phagocytosis: the secretion of indigestible materials to the environment. Exocytosis is also important in the secretion of many different sub-

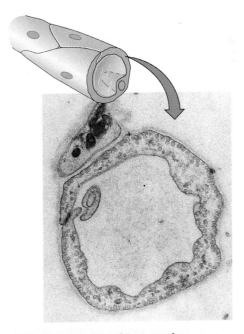

5.15 Pinocytosis and Exocytosis
The single endothelial cell that surrounds a blood capillary uses pinocytosis and exocytosis to transport substances between the blood and the surrounding tissue.

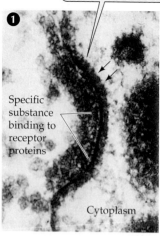
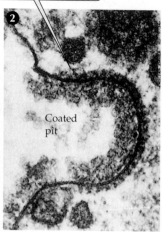
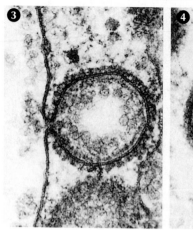
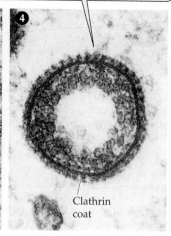

The protein clathrin coats the cytoplasmic side of the plasma membrane at a coated pit.

The endocytosed contents are surrounded by a clathrin-coated vesicle.

❶ Specific substance binding to receptor proteins

Cytoplasm

❷ Coated pit

❸

❹ Clathrin coat

0.1 μm

5.16 Formation of a Coated Vesicle
In receptor-mediated endocytosis, the receptor proteins in a coated pit bind specific macromolecules, which are then carried into the cell by the coated vesicle.

stances, including digestive enzymes from the pancreas, neurotransmitters from nerve cells, and materials for the construction of the plant cell wall.

Membranes Are Not Simply Barriers

We have discussed several functions of membranes—the compartmentalization of cells, the regulation of traffic between compartments, and the movement of materials into and out of cells—but there are more. In Chapter 4, we described how the membrane of the rough endoplasmic reticulum serves as a site for ribosome attachment. Newly formed proteins are passed from the ribosomes through the membrane and into the interior of the ER for modification and delivery to other parts of the cell. On the other hand, the membranes of nerve cells, muscle cells, some eggs, and other cells are electrically excitable. In nerve cells, the plasma membrane is the conductor of the nerve impulse from one end of the cell to the other.

Numerous other biological activities and properties discussed in the chapters to follow are associated with membranes. We review three of these here.

INFORMATION PROCESSING. As we have seen, the plasma membranes at cell surfaces and the membranes within cells may have protruding integral membrane proteins or attached carbohydrates that can bind to specific substances in the environment. The binding of a specific substance can serve as a signal to initiate, modify, or turn off a cell function (Figure 5.17a).

In this type of information processing, specificity in binding is essential. We have already seen the role of a specific receptor protein in the endocytosis of LDL and its cargo of cholesterol (see Figure 5.17). Another example is the binding of a hormone, such as insulin, to specific receptors on a target cell, such as a liver cell, to elicit a response—in this case, the uptake of glucose. There are many other examples, which we will discuss in Chapter 15.

ENERGY TRANSFORMATION. In a variety of cells, the membranes of organelles are specialized for processing energy (Figure 5.17b). For example, the inner mitochondrial membrane helps convert the energy of fuel molecules to the energy in ATP, and the thylakoid membranes of chloroplasts participate in the conversion of light energy to the energy of chemical bonds. The two characteristics of membranes that enable them to participate in these processes are their structural organization and their separation of electric charges.

ORGANIZING CHEMICAL REACTIONS. Many processes in cells depend on a series of enzyme-catalyzed reactions in which the products of one reaction serve as the reactants for the next. For such a reaction to occur, all the necessary molecules must come together. In a solution, the reactants and enzymes are all randomly distributed, and collisions among them are random. For this reason, a complete series of chemical reactions in solution may occur very slowly. However, if the different enzymes are bound to a membrane in sequential order, the product of one reaction can be released close to the enzyme for the next reaction. With such an "assembly line," reactions proceed more rapidly and efficiently (Figure 5.17c).

Membranes Are Dynamic

As we have seen in this chapter, membranes participate in numerous physiological and biochemical processes. Membranes are dynamic in another sense as well: They are constantly forming, transforming from one type to another, fusing with one another, and breaking down.

In eukaryotes, phospholipids are synthesized on the surface of the smooth endoplasmic reticulum and rapidly distributed to membranes throughout the cell as vesicles form from the ER, move away, and fuse with other organelles. Membrane proteins are inserted into the rough endoplasmic reticulum as they form on ribosomes. Functioning membranes also move about within eukaryotic cells. For example, portions of the rough ER bud away from the ER and join the *cis* faces of the Golgi apparatus (see Chapter 4). Rapidly—often in less than an hour—these segments of membrane

(a) Information processing

1 Signal binding induces a change in the receptor protein…

Signal molecule

Signal binding site

Outside of cell

Cytoplasm

2 …causing some effect inside the cell.

(b) Energy transformation

1 A membrane protein or pigment absorbs energy.

Outside energy source (such as light)

Energy-rich protein

P_i + ADP

ATP

2 The membrane protein transfers the energy to ADP to form ATP, where it is in a form for use by the cell.

5.17 More Membrane Functions

(a) Membrane proteins conduct signals from outside the cell that trigger changes inside the cell. (b) The membranes of organelles such as mitochondria and chloroplasts are specialized for the transformation of energy. (c) When a series of biochemical reactions must take place in sequence, the membrane can sometimes arrange the enzymes in an "assembly line" to ensure that the reactions occur in proximity to each other.

(c) Organizing chemical reactions

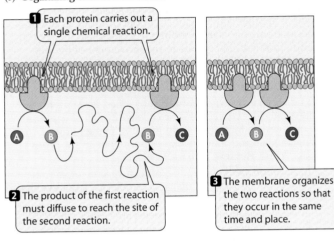

1 Each protein carries out a single chemical reaction.

A B B C

2 The product of the first reaction must diffuse to reach the site of the second reaction.

A B C

3 The membrane organizes the two reactions so that they occur in the same time and place.

find themselves in the *trans* regions of the Golgi, from which they bud away to join the plasma membrane (Figure 5.18)

During this journey, changes in the membrane's proteins and phospholipids occur. Membrane from vesicles is constantly merging with the plasma membrane by exocytosis, but this process is largely balanced by the removal of membrane in endocytosis, affording a recovery path by which internal membranes are replenished. In sum, there is a steady flux of membranes and membrane components in cells.

Because all membranes appear similar under the electron microscope, and because they interconvert readily, we might expect all subcellular membranes to be chemically identical. However, that is not the case, for there are major chemical differences among the membranes of even a single cell. Membranes are changed chemically when they form parts of certain organelles. In the Golgi apparatus, for example, the membranes of the *cis* face closely resemble those of the endoplasmic reticulum in chemical composition, but the *trans*-face membranes are more similar to the plasma membrane. As a vesicle is formed, the mix of pro-

5.18 Dynamic Continuity of Cellular Membranes

Membranes continually form, move, and fuse in cells.

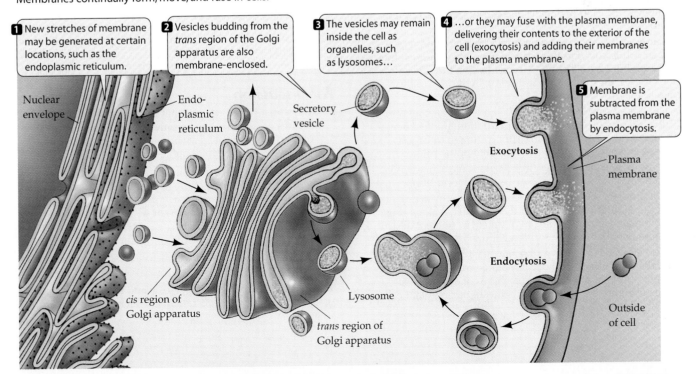

1 New stretches of membrane may be generated at certain locations, such as the endoplasmic reticulum.

2 Vesicles budding from the *trans* region of the Golgi apparatus are also membrane-enclosed.

3 The vesicles may remain inside the cell as organelles, such as lysosomes…

4 …or they may fuse with the plasma membrane, delivering their contents to the exterior of the cell (exocytosis) and adding their membranes to the plasma membrane.

5 Membrane is subtracted from the plasma membrane by endocytosis.

Nuclear envelope

Endoplasmic reticulum

Secretory vesicle

Exocytosis

Plasma membrane

Endocytosis

cis region of Golgi apparatus

Lysosome

trans region of Golgi apparatus

Outside of cell

teins and lipids in its membrane is selected, just as its internal contents are selected, to correspond with the vesicle's target membrane.

Ceaselessly moving, functioning, changing their composition and roles, biological membranes are central to life.

Chapter Summary

Membrane Composition and Structure

▶ Biological membranes consist of lipids, proteins, and carbohydrates. The fluid mosaic model of membrane structure describes a phospholipid bilayer in which membrane proteins can move about laterally within the membrane. **Review Figures 5.1, 5.2**

▶ Integral membrane proteins are at least partially inserted into the phospholipid bilayer. Peripheral proteins attach to the surface of the bilayer by ionic bonds. **Review Figure 5.1**

▶ The two surfaces of a membrane may have different properties because of their different phospholipid composition, exposed domains of integral membrane proteins, and their peripheral membrane proteins. **Review Figures 5.1, 5.2**

▶ Carbohydrates attached to proteins or phospholipids project from the external surface of the plasma membrane and function as recognition signals for interactions between cells. **Review Figure 5.1**

Cell Adhesion

▶ In an organism or tissue, cells recognize and bind to each other by means of membrane proteins that protrude from the cell surface. **Review Figure 5.5**

▶ Tight junctions prevent the passage of molecules through the space around cells, and they define functional regions of the plasma membrane by restricting the migration of membrane proteins uniformly over the cell surface. Desmosomes allow cells to adhere strongly to one another. Gap junctions provide channels for chemical and electrical communication between adjacent cells. **Review Figure 5.6**

Passive Processes of Membrane Transport

▶ Substances can diffuse passively across a membrane by three processes: unaided diffusion through the phospholipid bilayer, facilitated diffusion through protein channels, or facilitated diffusion by means of a carrier protein. **Review Table 5.1**

▶ Solutes diffuse across a membrane from a region with a greater solute concentration to a region with a lesser solute concentration. Equilibrium is reached when the concentrations of a given solute are identical on both sides of the membrane. **Review Figure 5.7**

▶ The rate of simple diffusion of a solute across a membrane is directly proportional to the concentration gradient across the membrane. An important factor in simple diffusion across a membrane is the lipid solubility of the solute.

▶ In osmosis, water diffuses from regions of higher water concentration to regions of lower water concentration.

▶ In hypotonic solutions, cells tend to take up water, while in hypertonic solutions, cells tend to lose water. Animal cells must remain isotonic to the environment to prevent destructive loss or gain of water. **Review Figure 5.8a**

▶ The cell walls of plants and some other organisms prevent the cells from bursting under hypotonic conditions. The turgor pressure that develops under these conditions keeps plants upright and stretches the cell wall during plant cell growth. **Review Figure 5.8b**

▶ Channel proteins and carrier proteins function in facilitated diffusion. **Review Figures 5.9, 5.10**

▶ The rate of carrier-mediated facilitated diffusion reaches a maximum when a solute concentration is reached that saturates the carrier proteins so that no increase in rate is observed with further increases in solute concentration.

Active Transport

▶ Active transport requires the use of energy to move substances across a membrane against a concentration gradient. **Review Table 5.1**

▶ Active transport proteins may be uniports, symports, or antiports. **Review Figure 5.11**

▶ In primary active transport, energy from the hydrolysis of ATP is used to move ions into or out of cells against their concentration gradients. **Review Figure 5.12**

▶ Secondary active transport couples the passive movement of one solute with its concentration gradient to the movement of another solute against its concentration gradient. Energy from ATP is used indirectly to establish the concentration gradient that results in the movement of the first solute. **Review Figure 5.13**

Endocytosis and Exocytosis

▶ Endocytosis transports macromolecules, large particles, and small cells into eukaryotic cells by means of engulfment by and vesicle formation from the plasma membrane. Phagocytosis and pinocytosis are both nonspecific types of endocytosis. **Review Figures 5.14, 5.15**

▶ In receptor-mediated endocytosis, a specific membrane receptor binds to a particular macromolecule. **Review Figure 5.16**

▶ In exocytosis, materials in vesicles are secreted from the cell when the vesicles fuse with the plasma membrane. **Review Figure 5.14**

Membranes Are Not Simply Barriers

▶ Membranes function as sites for recognition and initial processing of extracellular signals, for energy transformations, and for organizing chemical reactions. **Review Figure 5.17**

Membranes Are Dynamic

▶ Although not all cellular membranes are identical, ordered modifications in membrane composition accompany the conversions of one type of membrane into another type. **Review Figure 5.18**

For Discussion

1. In Chapter 47, we will see that the functioning of muscles requires calcium ions to be pumped into a subcellular compartment against a calcium concentration gradient. What types of molecules are required for this to happen?

2. Some algae have complex glassy structures in their cell walls. These structures form within the Golgi apparatus. How do these structures reach the cell wall without having to pass through a membrane?

3. Organisms that live in fresh water are almost always hypertonic to their environment. In what way is this a serious problem? How do some organisms cope with this problem?

4. Contrast nonspecific endocytosis and receptor-mediated endocytosis with respect to mechanism and to performance.

6 Energy, Enzymes, and Metabolism

THE 5-YEAR-OLD BOY PRESENTED BOSTON SURgeon Joseph Upton with a problem. Dr. Upton had sewn the boy's ear back on after a dog had bitten it off, but after 4 days, blood flow to and from the ear was blocked. If blood flow was not restored quickly, the reattachment would fail. To open up the blood vessels, Dr. Upton tried an old technique: He applied 24 leeches to the wound. The leeches used their sucking mouthparts to attach to the boy's ear and drank his blood. In the process, they released a molecule called hirudin (after the scientific name of the leech, *Hirudo medicinalis*) into the boy's blood. A potent anti-clotting agent, the hirudin acted slowly but consistently over 24 hours to clear the obstructed blood vessels, and the boy's ear was saved.

The medical use of leeches to prevent blood clotting goes back thousands of years. One of the most powerful anti-clotting agents known, hirudin is a small protein of 65 amino acids that folds into a specific shape that allows it to bind tightly to thrombin, a protein present in human blood. In the absence of hirudin, thrombin has a three-dimensional structure that allows it to bind to fibrinogen (yet another blood plasma protein). When this binding occurs, a peptide bond between two of the amino acids in fibrinogen is broken, forming fibrin—the protein that forms blood clots. If hirudin binds to thrombin, thrombin cannot act on fibrinogen, and blood clots do not form.

In chemical terms, thrombin is an enzyme, or biological catalyst, for fibrin formation. The hydrolysis of peptide bonds to form fibrin would happen whether thrombin was there or not; it's just that it would happen much more slowly, certainly too slowly to have any benefit in the lifetime of the organism! Thousands of such reactions go on all the time in every organism, each reaction catalyzed by a specific protein with a particular three-dimensional structure. Taken together, these reactions make up metabolism, which is the sum total of all of the chemical conversions in a cell.

Many metabolic reactions can be classified as either (1) building up complexity in the cell, using energy to do so; or (2) breaking down complex substances into simpler ones, releasing energy in the process.

Biomedical Medicine from a Natural Source
Leeches are the source of hirudin, a molecule that prevents blood coagulation by inhibiting the action of an enzyme, thrombin, in mammalian blood.

This chapter is concerned with energy and enzymes. Without them, neither we nor any other organism would be able to function. Indeed, when an enzyme is inactivated, either by the binding of an inhibitor such as hirudin that keeps the enzyme from binding to its target, or by some error leading to an alteration in its three-dimensional structure, its function is destroyed. This can have dire consequences: What if we had hirudin in our blood all the time?

Before considering how enzymes perform their molecular wizardry, let us consider the general principles of energy in biological systems.

Energy and Energy Conversions

Physicists define *energy* as the capacity to do work, which occurs when a force operates on an object over a distance. In biochemistry, energy represents the capacity for change. All living things must obtain energy from the environment—no cell manufactures energy. Indeed, one of the fundamental physical laws is that energy can neither be created nor destroyed. However, energy can be transformed from one kind into another. Energy transformations are linked to the chemical transformations that occur in cells. *Metabolism* is the total chemical activity of a living organism; at any instant, metabolism consists of thousands of individual chemical reactions.

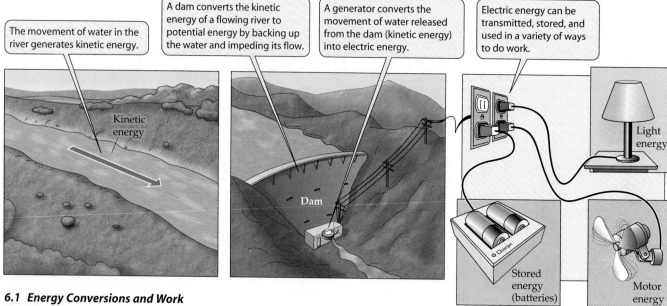

The movement of water in the river generates kinetic energy.

A dam converts the kinetic energy of a flowing river to potential energy by backing up the water and impeding its flow.

A generator converts the movement of water released from the dam (kinetic energy) into electric energy.

Electric energy can be transmitted, stored, and used in a variety of ways to do work.

Kinetic energy

Dam

Light energy

Stored energy (batteries)

Motor energy

6.1 Energy Conversions and Work
The kinetic energy of a flowing river can be converted to potential energy by a dam. Release of water from the dam converts the potential energy back into kinetic energy, which a generator can convert into electric energy.

Energy changes are related to changes in matter

Energy comes in many forms, such as chemical energy, light energy, and mechanical energy. But all forms of energy can be considered as one of two basic types:

▶ **Kinetic energy** is the energy of movement. This type of energy does work that alters the state or motion of matter. It can exist in the form of heat, light, electric, and mechanical energy, among others.

▶ **Potential energy** is the energy of state or position—that is, stored energy. It can be stored in chemical bonds, as a concentration gradient, and as electric potential, among other ways.

Water stored behind a dam has potential energy. When the water is released from the dam, some of this potential energy is converted into kinetic energy (Figure 6.1). Likewise, fatty acids, with their many C—C and C—H bonds, store chemical energy, which can be released to do biochemical work. In Chapter 5, we saw an example of the potential energy of a concentration gradient in secondary active transport, in which the gradient of one substance (Na^+) across a plasma membrane powers the transport of another (glucose) (see Figure 5.13).

In all cells of all organisms, two types of metabolic reactions occur:

▶ **Anabolic reactions** link together simple molecules to form more complex molecules. The synthesis of a protein from amino acids is anabolic. Anabolic reactions store energy in the chemical bonds that are formed.

▶ **Catabolic reactions** (catabolism) break down complex molecules into simpler ones and release stored energy.

Catabolic and anabolic reactions are often linked. The energy released in catabolic reactions is used to do biological work and drive anabolic reactions (Figure 6.2). The energy needed during anabolism to form the peptide bonds that link amino acids together into proteins comes from catabolism.

Cellular activities such as growth, motion, and active transport of ions across a membrane all require energy, and

6.2 Biological Energy Transformations
Cavorting lionesses convert chemical energy, obtained from the prey they have eaten, into a burst of kinetic energy of motion. Their prey obtained chemical energy by consuming plants. The plants trapped light energy and produced the prey's food by photosynthesis.

(a)

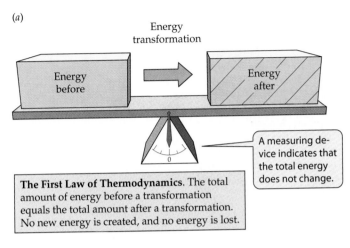

Energy transformation

A measuring device indicates that the total energy does not change.

The First Law of Thermodynamics. The total amount of energy before a transformation equals the total amount after a transformation. No new energy is created, and no energy is lost.

(b)

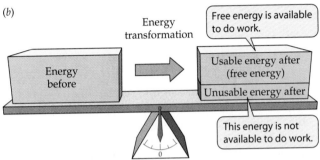

Energy transformation

Free energy is available to do work.

Usable energy after (free energy)

Unusable energy after

This energy is not available to do work.

The Second Law of Thermodynamics. Although a transformation does not change the total amount of energy within a closed system, after any transformation the amount of free energy available to do work is always less than the original amount of energy.

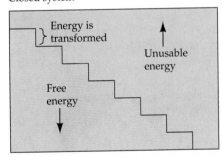

Closed system

Energy is transformed

Unusable energy

Free energy

Another statement of the Second Law is that in a closed system, with repeated energy transformations, free energy decreases and unusable energy increases—a phenomenon known as **entropy**.

6.3 The Laws of Thermodynamics
(a) The first law is that energy cannot be created or destroyed.
(b) The second law is that during energy transformations, free energy is lost.

none of them would proceed without a source of energy. In the discussion that follows, you will discover the physical laws that govern all energy transformations, identify the energy available to do work, and consider the direction of energy flow.

The first law: Energy is neither created nor destroyed

Energy can be converted from one form to another. For example, by striking a match, you convert potential chemical energy to light and heat. In any conversion of energy from one form to another (chemical to light, mechanical to electric), energy is neither created nor destroyed. This is the **first law of thermodynamics** (Figure 6.3a).

The first law applies to the universe as a whole or to any closed system within the universe. By "system" we mean any part of the universe containing specified matter and energy. A *closed system* is one that is not exchanging energy with its surroundings. For example, a thermos bottle does not gain or lose heat, and so the material inside it is a closed system (Figure 6.4a).

Open systems, such as living cells, exchange matter and energy with their surroundings (Figure 6.4b). Does this mean that cells disobey the first law, or that the first law does not apply to living organisms? Not at all. It means that an open system is merely one part of a larger closed system and receives energy from other parts of that larger system.

The first law tells us that in any interconversion of the forms of energy, the total energy before and after the conversion is the same. As you will see in the next two chapters, potential energy in the chemical bonds of carbohydrates and lipids can be converted to potential energy in ATP. This energy can then be used to produce potential energy in the concentration gradients established by active transport, which can be converted to kinetic energy and used to do mechanical work, such as muscle contraction.

The second law: Not all energy can be used, and disorder tends to increase

The **second law of thermodynamics** states that, although energy cannot be created or destroyed, when energy is converted from one form to another, some of the energy becomes unavailable to do work (Figure 6.3b). In other words, no physical process or chemical reaction is 100 percent efficient, and not all the energy released can be converted to work. Some energy is lost to a form associated with disorder. The second law applies to all energy transformations, but we will focus here on chemical reactions in living systems.

NOT ALL ENERGY CAN BE USED.   In any system, the total energy includes the *usable* energy that can do work and the *unusable* energy that is lost to disorder:

$$\text{total energy} = \text{usable energy} + \text{unusable energy}$$

In biological systems, the total energy is called **enthalpy** (*H*). The usable energy that can do work is called **free energy** (*G*). Free energy is what cells require for all the chemical reactions of cell growth, cell division, and the maintenance of cell health. The unusable energy is represented by **entropy** (*S*), which is the disorder of the system, multiplied

(a) **A closed system**

(b) **An open system**

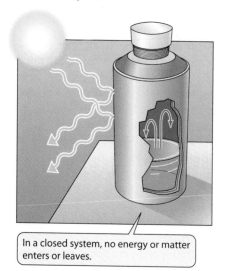

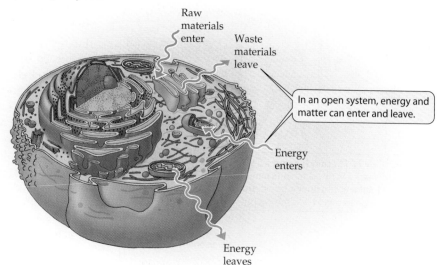

Raw materials enter

Waste materials leave

In an open system, energy and matter can enter and leave.

Energy enters

In a closed system, no energy or matter enters or leaves.

Energy leaves

6.4 Closed Systems and Open Systems

(a) A thermos bottle, sealed and insulated from its surroundings, is an example of a closed system. (b) A living cell, like a living individual, is an open system that must obtain energy and raw materials from its surroundings.

by the absolute temperature (T). Thus we can rewrite the word equation above more precisely as

$$H = G + TS$$

Because we are interested in usable energy, we rearrange this expression:

$$G = H - TS$$

Although we cannot measure G, H, or S absolutely, we can determine the *change* of each at a constant temperature. These energy changes are measured as calories (cal) or joules (J) (see Chapter 2). A change in a value is represented by the Greek letter delta (Δ), and it can be negative or positive. Therefore, the change in free energy (ΔG) of any reaction at constant temperature is defined in terms of the change in total energy (ΔH) and the change in entropy (ΔS):

$$\Delta G = \Delta H - T\Delta S$$

This equation tells us whether free energy is released or consumed by a chemical reaction. If ΔG is negative $\Delta G<0$, free energy is released. If ΔG is positive ($\Delta G>0$), free energy is required (consumed). If the needed free energy is not available, the reaction does not occur.

The sign and magnitude of ΔG depend on the two factors on the right of the equation: ΔH and $T\Delta S$. In a chemical reaction, ΔH is the total amount of energy added to the system ($\Delta H>0$) or released ($\Delta H<0$). In determining the free energy released, the sign of ΔH is obviously important. But the change in entropy must also be considered. Depending on the magnitude and sign of ΔS, the entire term, $T\Delta S$, may be negative or positive, large or small. In other words, in biological systems at constant temperature (no change

in T), the magnitude and sign of ΔG can depend a lot on changes in entropy.

If a chemical reaction increases entropy, its products are more disordered or random. If there are more products than reactants, as in the hydrolysis of a protein to its amino acids, the products have considerable freedom to move around. The disorder and the entropy in a solution of amino acids will be large compared with that in the protein, in which peptide bonds and other forces prevent free movement. So in hydrolysis, the change in entropy (ΔS) will be positive. If there are fewer products, and they are more restrained in their movements than the reactants, ΔS will be negative. For example, a large protein linked by peptide bonds is less free in its movements than a solution of the thousands of amino acids from which it was synthesized.

DISORDER TENDS TO INCREASE. The second law of thermodynamics also predicts that, as a result of energy conversions, disorder tends to increase in the universe or a closed system. Chemical changes, physical changes, and biological processes all tend to increase entropy and therefore tend toward disorder, or randomness (Figure 6.3b). This tendency for disorder to increase gives a directionality to physical processes and chemical reactions. It explains why some reactions proceed in one direction rather than another. The constant thermonuclear reactions in the sun will eventually result in its "running down" several billion years from now.

The second law does not say that ordered systems cannot be formed inside a large, complex closed system. In a closed system such as our solar system,* free energy can be used to create order in one part of the system. The entire

*Some energy does enter and leave the solar system—we do see the light of stars, after all—but the solar system is very nearly closed. The universe itself is a perfectly closed system (perhaps the *only* perfectly closed system).

Earth, or a single cell on Earth, is an open system that receives energy from its environment. In the case of Earth, the sun provides most of that energy. In a cell, the breakdown of complex molecules in food provides the energy to create order. For example, an input of free energy results in the formation of a protein from amino acids in a solution, and an input of free energy maintains the order in your cells and in your room.

Chemical reactions release or take up energy

How are the laws of thermodynamics relevant to our understanding of the chemical reactions that occur in living things? Chemical reactions in cells are accompanied by changes in energy and changes in order. Anabolic reactions may make a single product, such as a protein (a highly ordered substance), out of many smaller amino acids (less ordered), and such reactions require or consume energy. Catabolic reactions may reduce an organized substance, such as a glucose molecule, to smaller, more randomly distributed substances, such as carbon dioxide and water, and in the process give off energy. In other words, some reactions release free energy, and others take it up.

The amount of energy released ($-\Delta G$) or taken up ($+\Delta G$) is related directly to the tendency of a reaction to run to completion (for all the reactants to form products). When a reaction goes more than halfway to completion without an input of energy, we say that it is a **spontaneous reaction.** **Nonspontaneous reactions** proceed only with the addition of free energy from the environment.

▶ Spontaneous reactions release free energy. A reaction that releases free energy is said to be **exergonic** and has a negative ΔG (Figure 6.5a).

▶ Nonspontaneous reactions require free energy from the environment. Such reactions are said to be **endergonic** and have a positive ΔG (Figure 6.5b).

If a reaction runs spontaneously in one direction (from reactant A to product B, for example), then the reverse reaction (from B to A) requires a steady supply of energy to drive it:

If A → B is spontaneous and exergonic ($\Delta G<0$),

then

B → A is nonspontaneous and endergonic ($\Delta G>0$).

So protein hydrolysis to amino acids is spontaneous and exergonic, while protein synthesis is nonspontaneous and endergonic.

In principle, chemical reactions can run both forward and backward. For example, if compound A can be converted into compound B (A → B), then B, in principle, can be converted into A (B → A), although at given concentrations of A and B, only one of these directions will be favored. Think of the overall reaction as resulting from competition between forward and reverse reactions (A ⇌ B). Increasing the concentration of the reactants (A) speeds up the forward reaction, and increasing the concentration of the products (B) favors the reverse reaction. At some concentration of A and B, the forward and reverse reactions take place at the same rate. At this point, no further net change in the system is observable, although individual

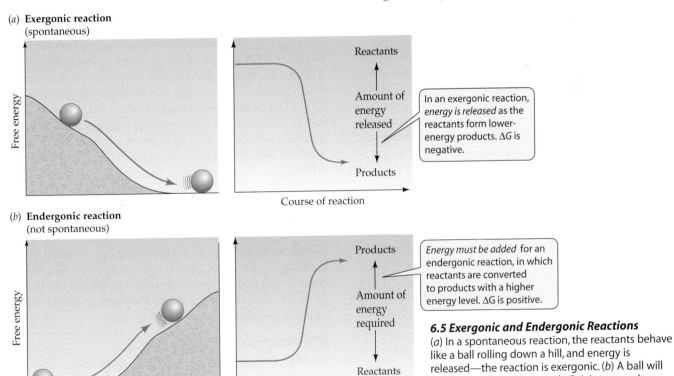

(a) **Exergonic reaction** (spontaneous)

In an exergonic reaction, *energy is released* as the reactants form lower-energy products. ΔG is negative.

(b) **Endergonic reaction** (not spontaneous)

Energy must be added for an endergonic reaction, in which reactants are converted to products with a higher energy level. ΔG is positive.

6.5 Exergonic and Endergonic Reactions
(a) In a spontaneous reaction, the reactants behave like a ball rolling down a hill, and energy is released—the reaction is exergonic. (b) A ball will not roll uphill spontaneously. Driving an endergonic reaction, like moving a ball uphill, requires adding free energy.

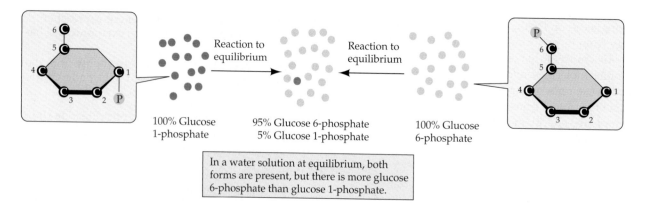

100% Glucose 1-phosphate

95% Glucose 6-phosphate 5% Glucose 1-phosphate

100% Glucose 6-phosphate

In a water solution at equilibrium, both forms are present, but there is more glucose 6-phosphate than glucose 1-phosphate.

6.6 Concentration at Equilibrium

No matter what quantities of glucose 1-phosphate and glucose 6-phosphate are dissolved in water, when equilibrium is attained, there will always be 95 percent glucose 6-phosphate and 5 percent glucose 1-phosphate.

molecules are still forming and breaking apart. This balance between forward and reverse reactions is known as *chemical equilibrium*.

Chemical equilibrium and free energy are related

Every chemical reaction proceeds to a certain extent, but not necessarily to completion. In other words, all the reactants present are not necessarily converted to products. Each reaction has a specific equilibrium point, and that equilibrium point is related to the free energy released by the reaction under specified conditions. To understand the principle of equilibrium, consider the following example.

Every living cell contains glucose 1-phosphate, which is converted in the cell to glucose 6-phosphate. Imagine that we start out with an aqueous solution of glucose 1-phosphate that has a concentration of 0.02 M. (M stands for molar concentration; see Chapter 2.) The solution is maintained under constant environmental conditions (25°C and pH 7). As the reaction proceeds to equilibrium, the concentration of the product, glucose 6-phosphate, rises from 0 to 0.019 M, while the glucose 1-phosphate concentration falls to 0.001 M. The reaction proceeds until equilibrium is reached at these concentrations (Figure 6.6). From then on, the reverse reaction, from glucose 6-phosphate to glucose 1-phosphate, progresses at the same rate as the forward reaction.

At equilibrium, then, this reaction has a product-to-reactant ratio of 19:1 (0.019/0.001), so the forward reaction has gone 95 percent of the way to completion ("to the right," as written). Therefore, the forward reaction is a spontaneous reaction. This result is obtained every time the experiment is run under the same conditions. The reaction is described by the equation

$$\text{glucose 1-phosphate} \rightleftharpoons \text{glucose 6-phosphate}$$

The change in free energy (ΔG) for any reaction is related directly to its point of equilibrium. The further toward completion the point of equilibrium lies, the more

free energy is given off. In an exergonic reaction, such as the conversion of glucose 1-phosphate to glucose 6-phosphate, ΔG is a negative number (in this example, $\Delta G = -1.7$ kcal/mol, or -7.1 kJ/mol).

A large, positive ΔG for a reaction means that it proceeds hardly at all to the right (A → B). But if the product is present, such a reaction runs backward, or "to the left" (A ← B), to near completion (nearly all B is converted to A). A ΔG value near zero is characteristic of a readily reversible reaction: Reactants and products have almost the same free energies.

The principles of thermodynamics we have been discussing apply to all energy exchanges in the universe—no exceptions have ever been found. Thus these principles are very powerful and useful. Next, we'll apply them to reactions in cells that involve the biological energy currency, ATP.

ATP: Transferring Energy in Cells

The previous chapters have mentioned **adenosine triphosphate**, or **ATP**, and its role in cells. All living cells rely on ATP for the capture, transfer, and storage of the free energy needed to do chemical work and maintain the cells (Figure 6.7). ATP operates as a kind of energy currency. That is, just as your professor teaches you, receives money for it, and then spends the money on food, some of the free energy released by certain exergonic reactions is captured in ATP, which can then release free energy to drive endergonic reactions.

ATP is produced by cells in a number of ways, which we will describe in the next two chapters, and it is used in many ways. ATP is not an unusual molecule. In fact, it has another important use in the cell: It can be converted into a building block for DNA and RNA. But there are two things about it that make it especially useful to cells: It can be hydrolyzed, and it can donate a phosphate group to many different molecules.

ATP hydrolysis releases energy

An ATP molecule consists of the nitrogenous base adenine bonded to ribose (a sugar), which is attached to a sequence of three phosphate groups (Figure 6.8). The hydrolysis of ATP yields **ADP (adenosine diphosphate)** and an inor-

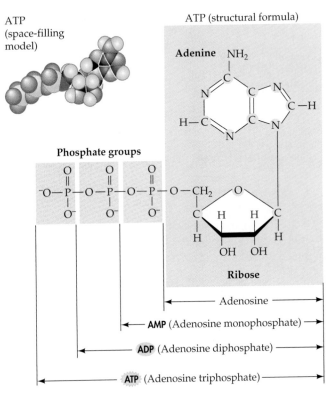

6.7 Using ATP to Make Light
Fireflies convert the energy of chemical bonds in ATP into light energy, emitting rhythmic flashes that signal the insect's readiness to mate. Very little of the energy in this conversion is lost as heat.

6.8 Structures of ATP, ADP, and AMP
ATP is richer in energy than its relatives ADP and AMP. The hydrolysis of ATP releases this energy.

ganic phosphate ion (abbreviated P_i, short for HPO_4^{2-}), as well as free energy:

$$ATP + H_2O \rightarrow ADP + P_i + \text{free energy}$$

Two important properties of this reaction are:

▸ It is exergonic, releasing free energy. The change in free energy (ΔG) is about –12 kcal/mol (–50 kJ/mol) at the temperature, pH, and substrate concentrations typical of living cells.*

▸ The equilibrium for this reaction is far to the right—that is, toward ADP production. At equilibrium in the cell, there is 10 million times as much ADP as ATP.

What characteristics of ATP account for the free energy released by the hydrolysis of its phosphates? Consider how phosphates are added to adenosine monophosphate—AMP—to make ADP. The phosphate on AMP is negatively charged; so is the free phosphate to be added. It takes a lot of free energy to overcome the tendencies of these phosphates to repel each other. The same thing happens with the addition of the third phosphate to ADP (adenosine diphosphate) to make ATP. Once again, the two negatively charged molecules repel each other unless a lot of energy is added. An analogy is the springs that suspend a car: To compress the springs, you need to lean on the car (input of energy), and the car bounces up (the energy is released) when the springs extend. Likewise, the energy required to make ATP is released when it is hydrolyzed.

ATP couples exergonic and endergonic reactions

As we have just seen, the hydrolysis of ATP is exergonic and yields ADP, P_i, and free energy. The reverse reaction, the formation of ATP from ADP and P_i, is endergonic and

*The "standard" ΔG for ATP hydrolysis is –7.3 kcal/mol or –30 kJ/mol, but that value is valid only at pH 7 and with ATP, ADP, and phosphate present at concentrations of 1 M—concentrations that differ greatly from those found in cells.

consumes as much free energy as is released by the breakdown of ATP:

$$ADP + P_i + \text{free energy} \rightarrow ATP + H_2O$$

Many different enzyme-catalyzed exergonic reactions in the cell can provide the energy to convert ADP to ATP. In eukaryotes, the most important of these reactions is cellular respiration, in which energy is released from fuel molecules and trapped in ATP. The formation and hydrolysis of ATP constitute what might be called an "energy-coupling cycle," in which ATP shuttles energy from exergonic reactions to endergonic reactions.

How does this ATP cycle trap and release energy? An exergonic reaction is coupled to the endergonic reaction that forms ATP from ADP and P_i (Figure 6.9). Coupling of exergonic and endergonic reactions is very common in biochemistry. When it forms, ATP captures free energy. ATP then diffuses to another site in the cell, where its hydrolysis releases free energy to drive an endergonic reaction.

A specific example of this energy-coupling cycle is shown in Figure 6.10. The formation of the amino acid glutamine has a positive ΔG (is endergonic, nonspontaneous) and will not proceed without the input of free energy from ATP hydrolysis, which has a negative ΔG (exergonic, spontaneous). The total ΔG for the *coupled reactions* is negative (the two ΔG's are added together). Hence the reactions proceed spontaneously when they are coupled, and glutamine is synthesized.

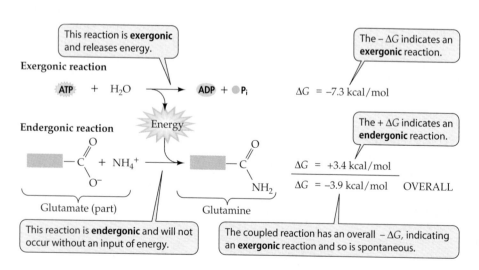

6.9 Formation and Use of ATP
Exergonic cellular processes release the energy needed to create ATP from ADP. The energy released from the conversion of ATP back to ADP can be used to fuel endergonic processes.

Actually, the overall reaction proceeds in two steps and involves a phosphorylated intermediate that is common to both reactions:

▶ In the first reaction, ATP transfers a phosphate group to glutamate (glutamic acid), producing the higher-energy product glutamyl phosphate (the phosphorylated intermediate).

▶ In the second reaction, the hydrolysis of glutamyl phosphate provides sufficient free energy to drive the reaction with an ammonium ion (NH_4^+) to form glutamine.

An active cell requires millions of molecules of ATP per second to drive its biochemical machinery. An ATP molecule is consumed within a minute following its formation, on average. At rest, an average person hydrolyzes and produces about 40 kg of ATP per day—as much as some people weigh! This means that each ATP molecule undergoes about 10,000 cycles of synthesis and hydrolysis every day.

Enzymes: Biological Catalysts

When we know the change in free energy (ΔG) of a reaction, we know where the equilibrium point of the reaction lies: The more negative ΔG is, the further the reaction proceeds toward completion. However, ΔG tells us nothing about the *rate* of a reaction—the speed at which it moves toward equilibrium. As we will see, some exergonic reac-

tions are very rapid; others are slower. Living cells cope with this variability by using biological catalysts to increase the rates of almost all chemical reactions.

A **catalyst** is any substance that speeds up a chemical reaction without itself being used up. A catalyst does not cause a reaction to take place that would not take place eventually without it, but merely speeds up the rates of both forward and backward reactions, allowing equilibrium to be approached faster. Most biological catalysts are proteins called **enzymes**. Although we will focus here on proteins, you should know that certain RNA molecules, called *ribozymes*, are also catalytic. Indeed, in the evolution of life, catalytic RNA may have preceded catalytic proteins (see Chapter 25).

In the discussion that follows, we will identify the energy barrier that controls the rate of reactions. Then we'll focus on the role of enzymes: how they interact with reactants, how they lower the activation energy barrier, and how they permit reactions to proceed faster. After exploring the nature and significance of enzyme specificity, we'll look at how enzymes contribute to the coupling of reactions.

For a reaction to proceed, an energy barrier must be overcome

An exergonic reaction may release a great deal of free energy, but the reaction may take place very slowly. Some reactions are slow because there is an energy barrier between reactants and products. Think about a butane lighter. The burning of the butane gas (butane + $O_2 \rightarrow CO_2 + H_2O$) is obviously exergonic—heat and light are released. Once started, the reaction goes to completion: All of the butane reacts with oxygen to form carbon dioxide and water vapor.

6.10 Coupling ATP Hydrolysis to an Endergonic Reaction
The synthesis of the amino acid glutamine from glutamate and an ammonium ion is endergonic and must be coupled with the exergonic hydrolysis of ATP.

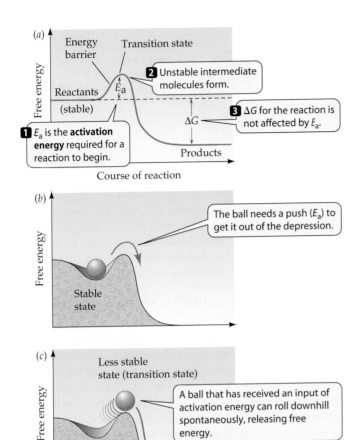

6.11 Activation Energy Initiates Reactions
(a) In any chemical reaction, an initial stable state must become less stable before change is possible. (b,c) A ball on a hillside provides a physical analogy to the biochemical principle graphed in (a).

In a chemical reaction, the activation energy is the energy needed to change the reactants into unstable molecular forms called *transition-state species*. Transition-state species have higher free energies than either the reactants or the products. Their bonds may be stretched and hence unstable. Although the amount of activation energy needed for different reactions varies, it is often small compared with the change in free energy of the reaction. The activation energy that starts a reaction is recovered during the ensuing "downhill" phase of the reaction, so it is not a part of the net free energy change, ΔG (see Figure 6.11a).

Where does the activation energy come from? In any collection of reactants at room or body temperature, molecules are moving around and could use their kinetic energy of motion to overcome the energy barrier, enter the transition state, and react (Figure 6.12). However, at normal temperatures, only a few molecules have enough energy to do this; most have insufficient kinetic energy for activation. If the system were heated, all the reactant molecules would move faster and have more kinetic energy. Since more of them would have energy exceeding the required activation energy, the reaction would speed up.

However, adding enough heat to increase the average kinetic energy of the molecules is not an effective option for living systems. Such a general, nonspecific approach would accelerate all the reactions, including destructive ones, such as the denaturation of proteins (see Chapter 3). Another, biologically more effective way to speed up a reaction is to lower the activation energy barrier. In living cells, enzymes accomplish this task.

Enzymes bind specific reactant molecules

All types of catalysts speed chemical reactions. Most non-biological catalysts are nonspecific and work on a variety

Because burning butane liberates so much energy, you might expect this reaction to proceed rapidly whenever butane is exposed to oxygen. But this does not happen. Simply mixing butane with air produces no reaction. Butane will start burning only if a spark—an input of energy—is provided. (In the butane lighter, this is supplied by friction.) The need for this spark to start the reaction shows that there is an *energy barrier between the reactants and the products*.

In general, exergonic reactions proceed only after they are pushed over the energy barrier by a small amount of added energy. The energy barrier thus represents the amount of energy needed to start the reaction. This amount is called the *activation energy*, and is symbolized E_a (Figure 6.11). Recall the ball rolling down the hill in Figure 6.5. The ball has a lot of potential energy at the top of the hill. However, if the ball is stuck in a small depression, it won't roll down the hill, even though that action is exergonic (Figure 6.11b). To start the ball rolling, a small amount of energy (activation energy) is needed to get the ball out of the depression (Figure 6.11c).

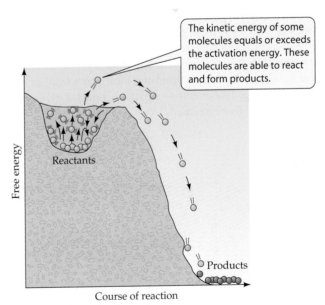

6.12 Over the Energy Barrier
Some molecules have enough kinetic energy to surmount the energy barrier and react, forming products. At the temperatures of most organisms, this is a small proportion of the molecules.

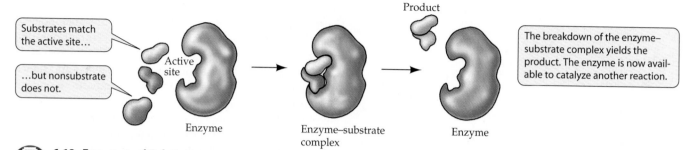

Substrates match the active site...

...but nonsubstrate does not.

Active site

Enzyme

Enzyme–substrate complex

Product

The breakdown of the enzyme–substrate complex yields the product. The enzyme is now available to catalyze another reaction.

Enzyme

6.13 Enzyme and Substrate
An enzyme is a protein catalyst with an active site capable of binding one or more substrate molecules. The enzyme–substrate complex yields product and free enzyme.

of reactants. For example, powdered platinum catalyzes virtually any reaction in which molecular hydrogen (H_2) is a reactant. In contrast, most biological catalysts are proteins called enzymes, and they are *highly specific*. An enzyme usually recognizes and binds to only one or a few closely related reactants, and it catalyzes only a single chemical reaction.

In an enzyme-catalyzed reaction, the reactants are called **substrates**. Substrate molecules bind to a particular site on the enzyme surface, called the **active site**, where catalysis takes place (Figure 6.13). The specificity of an enzyme results from the exact three-dimensional shape and structure of its active site, into which only a narrow range of substrates fit. Other molecules—with different shapes, different functional groups, and different properties—cannot properly fit and bind to the active site.

The names of enzymes reflect the specificity of their functions. For example, the enzyme RNA polymerase will catalyze the formation of RNA but not DNA, and the enzyme hexokinase accelerates the phosphorylation of hexose sugars. but not pentose sugars. Most, but not all, names of enzymes end in the suffix "-ase."

The binding of a substrate to the active site produces an *enzyme–substrate complex* held together by one or more means, such as hydrogen bonding, ionic attraction, or covalent bonding. The enzyme–substrate complex gives rise to product and free enzyme:

$$E + S \rightarrow ES \rightarrow E + P$$

where E is the enzyme, S is the substrate, P is the product, and ES is the enzyme–substrate complex. The free enzyme (E) is in the same chemical form at the end of the reaction as at the beginning. While bound to the substrate, it may change chemically, but by the end of the reaction it has been restored to its initial form.

Enzymes lower the activation energy barrier but do not affect equilibrium

The formation of an enzyme–substrate complex results in a lower activation energy than the transition-state species of the corresponding uncatalyzed reaction (Figure 6.14). Thus the enzyme lowers the energy barrier for the reaction—it offers the reaction an easier path. When an enzyme lowers the activation energy barrier, both the forward and the reverse reactions speed up, so the enzyme-catalyzed overall reaction proceeds toward equilibrium more rapidly than the uncatalyzed reaction. The final equilibrium (and ΔG) is the same with or without the enzyme.

The enzyme lactate dehydrogenase, for example, catalyzes the highly reversible reaction

$$\text{pyruvate} \rightleftharpoons \text{lactate}$$

We will study this reaction in the next chapter. But for now, what is the substrate for this reaction? The answer is, *either* pyruvate *or* lactate. When we exercise vigorously, the reaction proceeds to the right because there is a lot of pyruvate around. Lactate builds up in our muscles (and we get cramps). Lactate moves into the blood, and the blood system takes it from the muscles to the liver, where lactate dehydrogenase catalyzes its conversion back to pyruvate. In the meantime, the lactate concentration in our muscles is now lower, so lactate dehydrogenase in the muscles catalyzes the conversion of pyruvate to lactate. This is an excellent example of how concentrations of substrates affect the net direction of a reversible reaction.

Adding an enzyme to a reaction does not change the difference in free energy (ΔG) be-

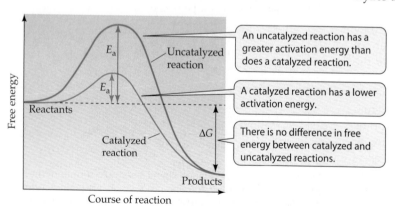

E_a

Uncatalyzed reaction

E_a

Reactants

Catalyzed reaction

ΔG

Products

Course of reaction

Free energy

An uncatalyzed reaction has a greater activation energy than does a catalyzed reaction.

A catalyzed reaction has a lower activation energy.

There is no difference in free energy between catalyzed and uncatalyzed reactions.

6.14 Enzymes Lower the Activation Energy Barrier
Although the activation energy is lower in an enzyme-catalyzed reaction than in an uncatalyzed reaction, the energy released is the same with or without catalysis. In other words, E_a is lower, but ΔG is unchanged.

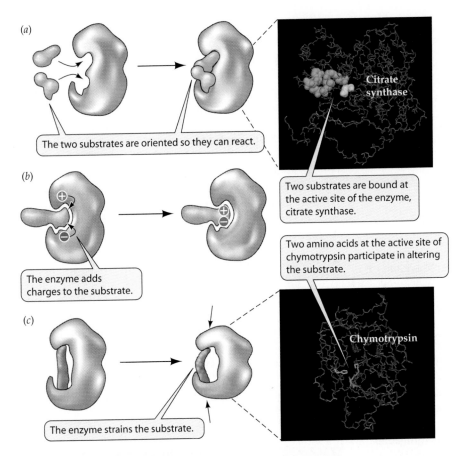

(a) The two substrates are oriented so they can react.

(b) The enzyme adds charges to the substrate.

(c) The enzyme strains the substrate.

Two substrates are bound at the active site of the enzyme, citrate synthase.

Two amino acids at the active site of chymotrypsin participate in altering the substrate.

Citrate synthase

Chymotrypsin

6.15 Life at the Active Site
Enzymes have several ways of causing their substrates to enter the transition state: (a) Orientation, (b) Chemical charge, (c) Physical strain.

tween the reactants and the products (see Figure 6.14). It does change the activation energy and, consequently, the rate of reaction. If 600 molecules of a protein with arginine as its terminal amino acid just sit in solution, thermodynamically the proteins tend toward disorder and the terminal peptide bonds break, releasing the arginines (ΔS increases). After 7 years, about half (300) of the proteins will have undergone this spontaneous reaction. With the enzyme carboxypeptidase A catalyzing the reaction, the 300 arginines are released in half a second!

What are the chemical events at active sites of enzymes?

How does an enzyme speed up the rate of a reaction? Like any substance that binds to a protein, a substrate interacts with the active site of an enzyme by shape and by chemical interactions. The chemical interactions contribute directly to the breaking of old bonds and the formation of new ones (Figure 6.15). In catalyzing a reaction, an enzyme may use one or more of the following mechanisms.

ENZYMES ORIENT SUBSTRATES. While free in solution, substrates are rotating and tumbling around and may not have the proper orientation to interact when they collide. Part of

the activation energy is used to make the substrates collide with the right atoms for bond formation next to each other.

When proteins are synthesized, for example, a peptide bond is formed between the carboxyl group of one amino acid and the amino group of the next (see Figure 3.4). For two amino acids to collide and form a peptide bond, these two chemical groups must be the sites of collision. When the active site of an enzyme binds to one amino acid, it is held in the right orientation to react with a second amino acid substrate when it binds to the enzyme.

ENZYMES ADD CHARGES TO SUBSTRATES. The side chains (R groups) of an enzyme's amino acids may be direct participants in making its substrates more chemically reactive. For example, in acid-base catalysis, the acidic or basic side chains of the amino acids forming the active site may transfer H^+ to or from the substrate, destabilizing a covalent bond in the substrate and permitting it to break. In covalent catalysis, a functional group in a side chain forms a temporary covalent bond with a portion of the substrate. In metal ion catalysis, metal ions such as copper, zinc, iron, and manganese, which are firmly bound to side chains of the protein, can lose or gain electrons without altering the bonds that hold them to the protein. This ability makes them important participants in oxidation–reduction reactions, which involve loss or gain of electrons.

ENZYMES INDUCE STRAIN IN THE SUBSTRATE. Once a substrate has bound to the active site, the enzyme can cause bonds in the substrate to stretch, putting it in an unstable transition state. For example, the carbohydrate substrate for the enzyme lysozyme (see Figure 6.17) enters the active site in a flat-ringed "chair" shape, but the active site quickly causes it to flatten out into a "sofa." The resulting stretching of its bonds causes them to be less stable and more reactive to the other substrate, water.

Substrate concentration affects reaction rate

For a reaction of the type A → B, the rate of the uncatalyzed reaction is directly proportional to the concentration of A (Figure 6.16). The higher the concentration of substrate, the more collisions, and the more reactions per unit of time (higher rate). Addition of the appropriate enzyme speeds up the reaction, of course, but it also changes the shape of the plot of rate versus substrate concentration. At first, the rate of the enzyme-catalyzed reaction increases as the substrate concentration increases, but then it levels off.

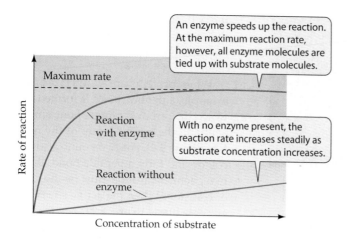

An enzyme speeds up the reaction. At the maximum reaction rate, however, all enzyme molecules are tied up with substrate molecules.

Maximum rate

Reaction with enzyme

With no enzyme present, the reaction rate increases steadily as substrate concentration increases.

Reaction without enzyme

Rate of reaction

Concentration of substrate

6.16 Enzymes Speed Up Reaction Rates
Because there is usually less enzyme than substrate present, the reaction rate levels off when the enzyme becomes saturated.

When further increases in the substrate concentration do not significantly increase the reaction rate, the maximum rate is attained.

Since the concentration of an enzyme is usually much lower than that of its substrate, what we are seeing is a saturation phenomenon like the one that occurs in facilitated diffusion (see Chapter 5). When all the enzyme molecules are bound to substrate molecules, the enzyme is working as fast as it can—at its maximum rate. Nothing is gained by adding more substrate, because no free enzyme molecules are left to act as catalysts.

The maximum rate of an enzyme reaction can be used to measure how efficient the enzyme can be—that is, how many molecules of substrate are converted to product per unit of time when there is an excess of substrate present. This *turnover number* ranges from 1 molecule every 2 seconds for lysozyme (see Figure 6.17) to an amazing 40 *million* molecules per second for the liver enzyme catalase.

Molecular Structure Determines Enzyme Function

Most enzymes are much larger than their substrates. An enzyme is typically a protein with hundreds of amino acids, and its substrate is generally not a macromolecule at all, but a small-molecule metabolite. The active site of the enzyme is usually quite small, not more than 6–12 amino acids. Two questions arise from this observation:

▸ What is the nature of the active site that allows it to recognize and bind the substrate?
▸ What is the role of the rest of the huge protein?

The active site is specific to the substrate

The remarkable ability of an enzyme to select exactly the right substrate depends on a precise interlocking of molecular shapes and interactions of chemical groups at the binding site. The binding of the substrate to the active site

depends on the same kinds of forces that maintain the tertiary structure of the enzyme: hydrogen bonds, the attraction and repulsion of electrically charged groups, and hydrophobic interactions (see Chapter 3).

In 1894, the German chemist Emil Fischer compared the fit between an enzyme and its substrate to that of a lock and key. Fischer's model persisted for more than half a century with only indirect evidence to support it. The first direct evidence came in 1965, when David Phillips and his colleagues at the Royal Institution in London succeeded in crystallizing the enzyme lysozyme and determined its tertiary structure using the techniques of X-ray crystallography (described in Chapter 11). They observed a pocket in lysozyme that neatly fits its substrate (Figure 6.17).

An enzyme changes shape when it binds a substrate

Fischer's idea has turned out to be largely correct, with one modification. Studies on *enzyme inhibitors* were done to see if molecules similar to a substrate could fit into an active site on an enzyme and prevent the real substrate from binding. The first such experiments were successful in that the mimic "substrates" did bind to the enzyme, but they did not react. Likewise, a false key can fit into a lock, but the lock will not open.

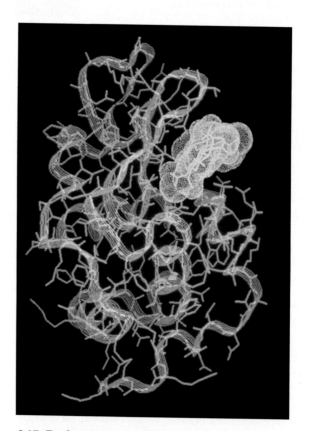

6.17 Tertiary Structure of Lysozyme
Lysozyme is an enzyme that protects the animals that produce it by destroying invading bacteria. To destroy the bacteria, it cleaves certain polysaccharide chains in their cell walls. Lysozyme is found in tears and other bodily secretions, and it is particularly abundant in the whites of bird eggs. The active site of lysozyme appears as an indentation filled with the substrate (shown in yellow).

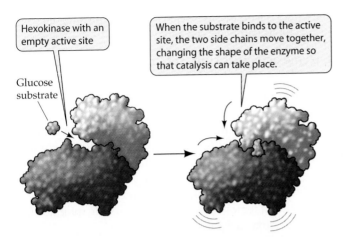

Hexokinase with an empty active site

Glucose substrate

When the substrate binds to the active site, the two side chains move together, changing the shape of the enzyme so that catalysis can take place.

6.18 Some Enzymes Change Shape When Substrate Binds to Them
Shape changes result in an induced fit between enzyme and substrate, improving the catalytic ability of the enzyme.

Then, remarkably, some enzyme inhibitors were found that were considerably larger than the real substrate, yet *were* effective. How could a key twice as large as the real one fit into the lock? It is here that the Fischer lock and key analogy breaks down.

The answer is that enzymes are flexible, and their active sites can change (expand) to fit substrates. When the substrate binds, the enzyme changes shape, exposing the parts of itself that react with the substrate. This change in enzyme shape caused by substrate binding is called *induced fit.*

Induced fit can be observed in the enzyme hexokinase (Figure 6.18) when it is studied with and without one of its substrates, glucose (its other substrate is ATP). It catalyzes the reaction

Glucose + ATP → glucose 6-phosphate + ADP

Induced fit brings reactive side chains from the enzyme's active site into alignment with the substrates, facilitating the catalytic mechanisms described earlier (see Figure 6.15).

Equally important, the folding of hexokinase to fit around the glucose substrate excludes water from the active site. This is essential, because the two molecules binding to the active site are glucose and ATP. If water were present, ATP would be rapidly hydrolyzed to ADP and phosphate. But since water is absent, the transfer of a phosphate from ATP to glucose is favored.

Induced fit at least partly explains why enzymes are so large. The rest of the molecule may have two roles:

▶ It provides a framework so that the amino acids of the active site are properly positioned in relation to the substrate.

▶ It participates in the small but significant changes in protein shape and structure that result in induced fit.

To operate, some enzymes require added molecules

Whether they consist of a single folded polypeptide chain or several subunits, many enzymes require other, nonprotein molecules in order to function (Table 6.1).

▶ **Cofactors** are inorganic ions such as copper, zinc, and iron that bind temporarily to certain enzymes and are essential to their function.

▶ **Coenzymes** are carbon-containing molecules that are required for the action of one or more enzymes. Coenzymes are usually relatively small compared with the enzyme to which they temporarily bind (Figure 6.19).

6.1 A Few Examples of Nonprotein Molecular "Partners" of Enzymes

TYPE OF MOLECULE	ROLE IN CATALYZED REACTIONS
Cofactors	
Iron (Fe^{2+} or Fe^{3+})	Oxidation/reduction
Copper (Cu^+ or Cu^{2+})	Oxidation/reduction
Zinc (Zn^{2+})	Helps bind NAD
Coenzymes	
Biotin	Carries —COO^-
Coenzyme A	Carries —CH_2—CH_3
NAD	Carries electrons
FAD	Carries electrons
Prosthetic groups	
Heme	Binds ions, O_2, and electrons; contains iron cofactor
Flavin	Binds electrons
Retinal	Converts light energy

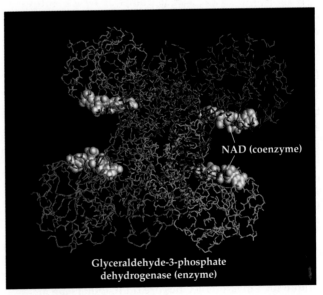

NAD (coenzyme)

Glyceraldehyde-3-phosphate dehydrogenase (enzyme)

6.19 An Enzyme with a Coenzyme
Some enzymes require coenzymes in order to function. This illustration shows the relative sizes of the four subunits (red, orange, green, and purple) of the enzyme glyceraldehyde-3-phosphate dehydrogenase and its coenzyme, NAD (white).

▶ **Prosthetic groups** are permanently bound to their enzymes. They include the heme groups that are attached to the oxygen-carrying protein hemoglobin (see Figure 3.7).

Because coenzymes are not permanently bound to the enzyme, they must react with it as a substrate does. For the catalyzed reaction to proceed, coenzymes must collide with the enzyme and bind to its active site just as the substrate must. A coenzyme can be considered a substrate because it changes chemically during the reaction and then separates from the enzyme to participate in other reactions. Coenzymes move from enzyme molecule to enzyme molecule, adding or removing chemical groups from the substrate.

ATP and ADP can be considered coenzymes, because they are necessary for some reactions, are changed by reactions, and bind to and detach from the enzyme. In the next chapter, we will encounter coenzymes that function in energy processing by accepting or donating electrons or hydrogen atoms. In animals, some coenzymes are produced from vitamins that must be obtained from food—they cannot be synthesized by the body.

Metabolism and the Regulation of Enzymes

All organisms need to maintain stable internal conditions, or *homeostasis*. Thus we and all other organisms must regulate our metabolisms. The regulation of the rates at which our thousands of different enzymes operate contributes to metabolic homeostasis.

In the remainder of this chapter, we will investigate the role of enzymes in organizing and regulating metabolism. In living cells, the activity of enzymes can be inhibited in various ways, so the presence of an enzyme does not necessarily ensure that it is functioning. There are mechanisms to alter the rate at which some enzymes catalyze reactions, making enzymes the target points at which entire sequences of chemical reactions can be regulated. Finally, we examine how the environment—namely, temperature and pH—affects enzyme activity.

Metabolism is organized into pathways

An organism's metabolism is the totality of the biochemical reactions that take place within it. Metabolism transforms raw materials and stored potential energy into forms that can be used by living cells. Metabolism consists of sequences of enzyme-catalyzed chemical reactions called *pathways*. In these sequences, the product of one reaction is the substrate for the next:

$$A \xrightarrow{\text{enzyme}} B \xrightarrow{\text{enzyme}} C \xrightarrow{\text{enzyme}} D$$

Some metabolic pathways are anabolic, and synthesize the important chemical building blocks from which macromolecules are built. Others are catabolic, breaking down molecules for usable free energy. The balance among these anabolic and catabolic pathways may change depending on the cell's (and the organism's) needs. So a cell must regulate all its metabolic pathways constantly.

Enzyme activity is subject to regulation

Various *inhibitors* can bind to enzymes, slowing down the rates of enzyme-catalyzed reactions. Some inhibitors occur naturally in cells; others are artificial. Naturally occurring inhibitors regulate metabolism; artificial ones can be used to treat disease, to kill pests, or in the laboratory to study how enzymes work. Some inhibitors irreversibly inhibit the enzyme by permanently binding to it. Others have reversible effects; that is, they can become unbound from the enzyme. The removal of a natural reversible inhibitor increases an enzyme's rate of catalysis.

IRREVERSIBLE INHIBITION. Some inhibitors irreversibly covalently bond to certain side chains at active sites of enzymes, thereby inactivating the enzymes by destroying their capacity to interact with the normal substrate. Such inhibitors are generally not natural products. A compound called DIPF (diisopropylphosphorofluoridate), for example, reacts with a hydroxyl group of the amino acid serine at an enzyme's active site, preventing the use of this side chain in catalytic reactions (Figure 6.20). DIPF is an *irreversible inhibitor* for the protein-digesting enzyme trypsin and for many other enzymes whose active sites contain serine. Another DIPF-inhibited enzyme is acetylcholinesterase, which is essential for the orderly propagation of impulses from one nerve cell to another. Because of their effect on acetylcholinesterase, DIPF and other similar compounds are classified as nerve gases. One of them, Sarin, was used in an attack on the Tokyo subway in 1995, resulting in a dozen deaths and hundreds hospitalized. The widely used insecticide malathion is a derivative of DIPF that inhibits only insect acetylcholinesterase and not the mammalian enzyme.

REVERSIBLE INHIBITION. Not all inhibition is irreversible. Some inhibitors are similar enough to a particular enzyme's natural substrate to bind to the active site, yet dif-

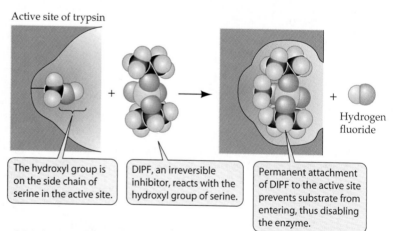

Active site of trypsin

Hydrogen fluoride

The hydroxyl group is on the side chain of serine in the active site.

DIPF, an irreversible inhibitor, reacts with the hydroxyl group of serine.

Permanent attachment of DIPF to the active site prevents substrate from entering, thus disabling the enzyme.

6.20 Irreversible Inhibition
DIPF forms a stable covalent bond with the side chain of the amino acid serine at the active site of the enzyme trypsin.

ferent enough that the enzyme catalyzes no chemical reaction. While such a molecule is bound to the enzyme, the natural substrate cannot enter the active site; thus the inhibitor effectively wastes the enzyme's time, preventing its catalytic action. Such molecules are called *competitive inhibitors* because they compete with the natural substrate for the active site (Figure 6.21a). In these cases, the inhibition is reversible. When the concentration of the competitive inhibitor is reduced, it detaches from the active site, and the enzyme is again active.

The enzyme succinate dehydrogenase is subject to competitive inhibition. This enzyme, found in all mitochondria, catalyzes the conversion of the compound succinate to another compound, fumarate. The compound oxaloacetate is similar to succinate and can act as a competitive inhibitor of succinate dehydrogenase by binding to its active

site. However, having bound to oxaloacetate, the enzyme can do nothing more with it—no reaction occurs. An enzyme molecule cannot bind a succinate molecule until the oxaloacetate molecule has moved out of the active site.

Some inhibitors that do not react with the active site are called *noncompetitive inhibitors*. Noncompetitive inhibitors bind to the enzyme at a site distinct from the active site. Their binding can cause a conformational change in the enzyme that alters the active site (Figure 6.21b). In this case, the active site may still bind substrate molecules, but the rate of product formation may be reduced. Noncompetitive inhibitors, like competitive inhibitors, can become unbound, so their effects are reversible.

Allosteric enzymes have interacting subunits

Many important enzymes have a quaternary structure consisting of two or more polypeptide subunits, each with a molecular weight in the tens of thousands (see Chapter 3). These subunits are bound together by various weak bonds that permit changes in the shape of one subunit to influence the shape and properties of the others. Multisubunit enzymes that undergo such changes in shape and function

6.21 Reversible Inhibition
(a) In competitive inhibition, an inhibitor binds temporarily to the active site. Succinate dehydrogenase, for example, is subject to competitive inhibition by oxaloacetate. (b) A noncompetitive inhibitor binds temporarily to the enzyme at a site away from the active site, but still prevents the enzyme from functioning.

(a) Competitive inhibition

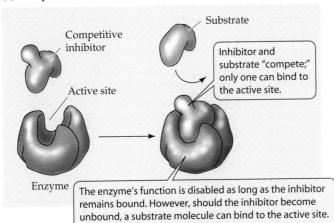

Competitive inhibitor

Active site

Substrate

Inhibitor and substrate "compete;" only one can bind to the active site.

Enzyme

The enzyme's function is disabled as long as the inhibitor remains bound. However, should the inhibitor become unbound, a substrate molecule can bind to the active site.

(b) Noncompetitive inhibition

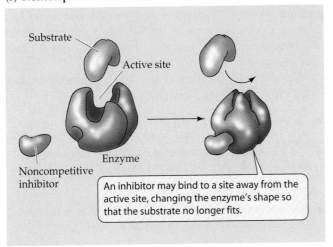

Substrate

Active site

Noncompetitive inhibitor

Enzyme

An inhibitor may bind to a site away from the active site, changing the enzyme's shape so that the substrate no longer fits.

Competitive inhibition of succinate dehydrogenase

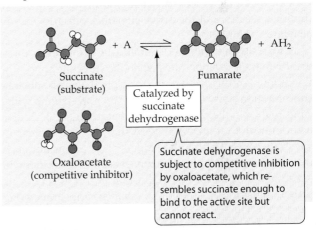

Succinate (substrate)   + A ⇌   Fumarate   + AH$_2$

Catalyzed by succinate dehydrogenase

Oxaloacetate (competitive inhibitor)

Succinate dehydrogenase is subject to competitive inhibition by oxaloacetate, which resembles succinate enough to bind to the active site but cannot react.

Noncompetitive inhibition of threonine dehydratase

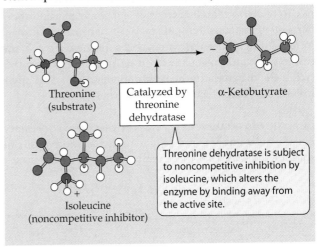

Threonine (substrate)

Catalyzed by threonine dehydratase

α-Ketobutyrate

Isoleucine (noncompetitive inhibitor)

Threonine dehydratase is subject to noncompetitive inhibition by isoleucine, which alters the enzyme by binding away from the active site.

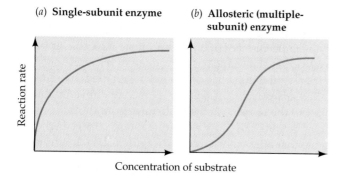

(a) **Single-subunit enzyme**   (b) **Allosteric (multiple-subunit) enzyme**

Reaction rate

Concentration of substrate

6.22 Allostery and Reaction Rate
How the rate of an enzyme-catalyzed reaction changes with increasing substrate concentration depends on whether the enzyme consists of one or more than one polypeptide subunit.

are called **allosteric enzymes** (allo-, "different"; -steric, "shape"). (Note that not all allosteric enzymes have multiple subunits; there are some single-chain enzymes that are allosterically regulated.)

The activity of allosteric enzymes is controlled by molecules called *effectors*, which may have no structural similarity either to the reactants or to the products of the reaction being catalyzed. Effectors bind to an *allosteric site* that is separate from the active site, changing the structure of the enzyme and thus its activity. Their binding may enhance or diminish reactions at the active site; thus, effectors may be activators or inhibitors.

Allosteric enzymes and nonallosteric enzymes differ greatly in their reaction rates when the substrate concentration is low. Graphs of reaction rate plotted against substrate concentration show this relationship. For an enzyme with a single subunit, the plot looks like that in Figure 6.22a. The reaction rate first increases very sharply with increasing substrate concentration, then tapers off to a constant maximum rate as the supply of enzyme becomes saturated with substrate. The plot for many allosteric enzymes is radically different, having a sigmoidal (S-shaped) appearance (Figure 6.22b). The increase in rate with increasing substrate concentration is slight at low substrate concentrations, but within a certain range the reaction rate is extremely sensitive to relatively small changes in substrate concentration. Because of this sensitivity, allosteric enzymes are important in regulating entire pathways and activities of a cell. We can understand this behavior in terms of interactions between the different kinds of subunits that make up an allosteric enzyme, which we'll examine next.

CATALYTIC AND REGULATORY SUBUNITS INTERACT AND COOPERATE. An allosteric enzyme not only usually has more than one subunit, it usually has more than one *type* of subunit:

▶ A *catalytic subunit* has an active site that binds the enzyme's substrate.

▶ A *regulatory subunit* has one or more allosteric sites that bind specific effector molecules.

Binding of either a substrate or an effector affects the structure of the enzyme as a whole.

An allosteric enzyme can exist in two forms. The active form has catalytic activity, whereas the inactive form lacks activity. When the enzyme is in its active form, the active sites on the catalytic subunits can accept substrate. When the enzyme is in its inactive form, the allosteric sites on the regulatory subunits can accept inhibitor.

An allosteric enzyme usually consists of two or more catalytic subunits and one or more regulatory subunits. The existence of two or more linked catalytic subunits allows for cooperativity, in which one subunit's activity influences the activity of its neighbors.

When a molecule of substrate binds to the active site of one catalytic subunit of an allosteric enzyme, it causes a change in the other catalytic subunits, making it *easier* for substrate to bind to them (Figure 6.23). Conversely, when an allosteric inhibitor binds to the allosteric site of a regulatory subunit, it causes a different change in the catalytic subunits, making it *harder* for substrate to bind to them. This cooperativity between subunits makes an enzyme exquisitely sensitive to its molecular environment. The binding of just one substrate molecule makes it easier for further substrate molecules to react.

ALLOSTERIC EFFECTS REGULATE METABOLISM. Metabolic pathways typically involve a starting material, various intermediates, and a product, which is used for some purpose by the cell. In each pathway, there are a number of reactions, each forming an intermediate, and each catalyzed by a different enzyme. The first step in a pathway is called the *commitment step*, meaning that once this enzyme-catalyzed conversion occurs, the "ball is rolling," and the other conversions happen in sequence, leading to the final product. But what if the cell has no need for that product—for example, if it takes it up from its environment in adequate amounts? It would be energetically wasteful for the cell to continue making something it does not need.

One way that cells solve this problem is to shut down the metabolic pathway by having the final product allosterically inhibit the enzyme that catalyzes the commitment step (Figure 6.24). This mechanism is known as *end-product inhibition*. When the end product is present in a high concentration, some of it binds to an allosteric site on the commitment step enzyme, thereby causing it to become inactive.

We will describe other examples of allosteric interactions in later chapters. These include:

▶ The binding of oxygen to hemoglobin, which shows a sigmoid relationship and cooperativity (see Chapter 49).

▶ The binding of a hormone to its cellular receptor protein, which causes the protein to change shape and provides the signal to initiate reactions within the cell (see Chapter 15).

▶ The binding of an inducer to a protein that regulates DNA expression (see Chapter 12).

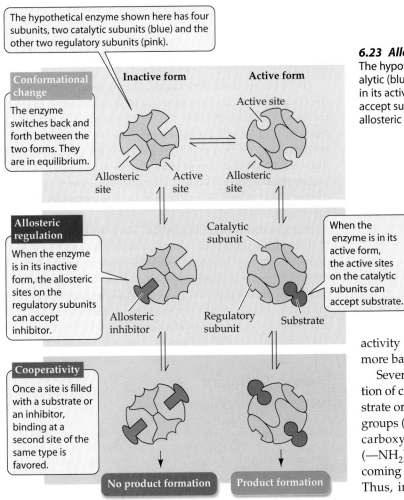

The hypothetical enzyme shown here has four subunits, two catalytic subunits (blue) and the other two regulatory subunits (pink).

Conformational change

The enzyme switches back and forth between the two forms. They are in equilibrium.

Inactive form   **Active form**

Active site

Allosteric site   Active site   Allosteric site

Allosteric regulation

When the enzyme is in its inactive form, the allosteric sites on the regulatory subunits can accept inhibitor.

Catalytic subunit

Allosteric inhibitor   Regulatory subunit   Substrate

When the enzyme is in its active form, the active sites on the catalytic subunits can accept substrate.

Cooperativity

Once a site is filled with a substrate or an inhibitor, binding at a second site of the same type is favored.

No product formation   Product formation

6.23 Allosteric Regulation of Enzymes
The hypothetical enzyme shown here has four subunits: two catalytic (blue), the other two regulatory (pink). When the enzyme is in its active form, the active sites on the catalytic subunits can accept substrate. When the enzyme is in its inactive form, the allosteric sites on the regulatory subunits can accept inhibitor.

Enzymes and their environment

Enzymes enable cells to perform chemical reactions and carry out complex processes without using the extremes of temperature and pH employed by chemists in the laboratory. However, because of their three-dimensional structures and the chemistry of the side chains in their active sites, enzymes are highly sensitive to temperature and pH.

We described the general effects of these factors on proteins in Chapter 3. Here, we will examine their effects on enzyme function, which, of course, depends on enzyme structure and chemistry.

PH AFFECTS ENZYME ACTIVITY.   The rates of most enzyme-catalyzed reactions depend on the pH of the medium in which they occur. Each enzyme is most active at a particular pH; its activity decreases as the solution is made more acidic or more basic than its "ideal" pH (Figure 6.25).

Several factors contribute to this effect. One is the ionization of carboxyl, amino, and other groups on either the substrate or the enzyme. In neutral or basic solutions, carboxyl groups ($-COOH$) release H^+ to become negatively charged carboxylate groups ($-COO^-$). Similarly, amino groups ($-NH_2$) accept H^+ ions in neutral or acidic solutions, becoming positively charged $-NH_3^+$ groups (see Chapter 2). Thus, in a neutral solution, a molecule with an amino group will be attracted electrically to another molecule that has a carboxyl group, because both groups are ionized and they have opposite charges.

If the pH changes, however, the ionization of these groups may change. For example, at a low pH (high H^+ concentration), the excess H^+ may react with the $-COO^-$ to form COOH. If this happens, the group is no longer charged and cannot interact with other charged groups in the protein, so the folding of the protein is altered. If this occurs at the active site of an enzyme, the enzyme may no longer have the correct shape to bind to its substrate.

The process of evolution matches enzymes and their work environments. For example, the protein-digesting enzyme pepsin, found only in the stomach, works best at the very low pH values that prevail in the stomach after a meal. In contrast, salivary amylase works best at neutral pH, which is characteristic of the mouth.

1 The first reaction is the commitment step.

2 Each of these reactions is catalyzed by a different enzyme, and each forms a different intermediate product.

NH$_3^+$
|
H—C—COO$^-$
|
H—C—OH
|
CH$_3$

Threonine
(starting material)

O
||
C—COO$^-$
|
CH$_2$
|
CH$_3$

α-Ketobutyrate
(intermediate product)

NH$_3^+$
|
H—C—COO$^-$
|
H—C—CH$_3$
|
CH$_2$
|
CH$_3$

Isoleucine
(end product)

3 Buildup of the end product allosterically inhibits the enzyme catalyzing the commitment step, thus shutting down its own production.

6.24 Inhibition of Metabolic Pathways
The commitment step is catalyzed by an allosteric enzyme that can be inhibited by the end product of the pathway. The specific pathway shown here is the synthesis of isoleucine, an amino acid, from threonine. This particular reaction series is performed by bacteria, but it is typical of many enzyme-catalyzed biological reactions.

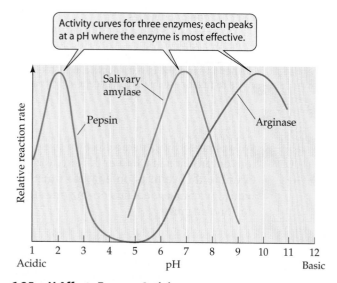

Activity curves for three enzymes; each peaks at a pH where the enzyme is most effective.

Pepsin

Salivary amylase

Arginase

1 Acidic   2   3   4   5   6   7   pH   8   9   10   11   12 Basic

Relative reaction rate

6.25 pH Affects Enzyme Activity
Each enzyme catalyzes at a maximum rate at a particular pH.

TEMPERATURE AFFECTS ENZYME ACTIVITY. In general, warming increases the rate of an enzyme-catalyzed reaction, because at higher temperatures a greater fraction of the reactant molecules have enough energy to provide the activation energy for the reaction (Figure 6.26). Temperatures that are too high, however, inactivate enzymes, because at high temperatures enzyme molecules vibrate and twist so rapidly that some of their noncovalent bonds break. When heat destroys their tertiary structure, enzymes become inactivated, or denatured (see Chapter 3). Some enzymes denature at temperatures only slightly above that of the human body, but a few are stable even at the boiling or freezing points of water. All enzymes, however, show an optimal temperature for activity.

Individual organisms adapt to changes in the environment in many ways, one of which is based on groups of enzymes, called *isozymes*, that catalyze the same reaction but have different chemical compositions and physical

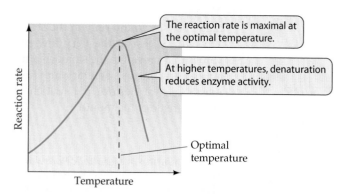

The reaction rate is maximal at the optimal temperature.

At higher temperatures, denaturation reduces enzyme activity.

Optimal temperature

Reaction rate

Temperature

6.26 Temperature Affects Enzyme Activity
Each enzyme is most active at a particular optimal temperature.

properties. Different isozymes within a given group may have different optimal temperatures.

The rainbow trout, for example, has several isozymes of the enzyme acetylcholinesterase, whose operation is essential to normal transmission of nerve impulses. If a rainbow trout is transferred from warm water to near-freezing water (2°C), the fish produces an isozyme of acetylcholinesterase that is different from the one it produces at the higher temperature. The new isozyme has a lower optimal temperature, which helps the fish to perform normally in the colder water.

In general, enzymes adapted to warm temperatures fail to denature because their tertiary structures are held together largely by covalent bonds, such as disulfide links, instead of the more heat-sensitive weak chemical forces. Most enzymes in humans are more stable at high temperatures than those of the bacteria that infect us, so that a moderate fever tends to denature bacterial enzymes, but not our own.

Chapter Summary

Energy and Energy Conversions

▶ Energy is the capacity to do work. Potential energy is the energy of state or position; it includes the energy stored in chemical bonds. Kinetic energy is the energy of motion (and related forms such as electric energy, light, and heat).

▶ Potential energy can be converted to kinetic energy, which does work. **Review Figure 6.1**

▶ The first law of thermodynamics tells us that energy cannot be created or destroyed. The second law of thermodynamics tells us that, in any closed system, the quantity of energy available to do work (free energy) decreases and unusable energy (associated with entropy) increases. **Review Figure 6.3**

▶ Living things, like everything else, obey the laws of thermodynamics. Organisms are open systems that are part of a larger closed system. **Review Figure 6.4**

▶ Changes in free energy, total energy, temperature, and entropy are related by the equation $\Delta G = \Delta H - T\Delta S$.

▶ Spontaneous, exergonic reactions release free energy and have a negative ΔG. Nonspontaneous, endergonic reactions take up free energy and have a positive ΔG. Endergonic reactions proceed only if free energy is provided. **Review Figure 6.5**

▶ The change in free energy of a reaction determines its point of chemical equilibrium, at which the forward and reverse reactions proceed at the same rate. For spontaneous, exergonic reactions, the equilibrium point lies toward completion (the conversion of all reactants into products). **Review Figure 6.6**

ATP: Transferring Energy in Cells

▶ ATP (adenosine triphosphate) serves as an energy currency in cells. Hydrolysis of ATP releases a relatively large amount of free energy. **Review Figure 6.8**

▶ The ATP cycle couples exergonic and endergonic reactions, transferring free energy from the exergonic to the endergonic reaction. **Review Figures 6.9, 6.10**

Enzymes: Biological Catalysts

▶ The rate of a chemical reaction is independent of ΔG but is determined by the size of the activation energy barrier. Catalysts speed reactions by lowering the activation energy barrier. **Review Figures 6.11, 6.12**

▶ Enzymes are biological catalysts, proteins that are highly specific for their substrates. Substrates bind to the active site, where catalysis takes place, forming an enzyme–substrate complex. **Review Figure 6.13**

▶ At the active site, a substrate can be oriented correctly, chemically modified, or strained. As a result, the substrate readily forms its transition state, and the reaction proceeds. **Review Figures 6.14, 6.15**

▶ Substrate concentration affects the rate of an enzyme-catalyzed reaction. **Review Figure 6.16**

Molecular Structure Determines Enzyme Function

▶ The active site where substrate binds determines the specificity of an enzyme. Upon binding to substrate, some enzymes change shape, facilitating catalysis. **Review Figures 6.13, 6.18**

▶ Some enzymes require cofactors to carry out catalysis. Prosthetic groups are permanently bound to the enzyme. Coenzymes are not usually bound to the enzyme. They enter into the reaction as a "cosubstrate," as they are changed by the reaction and then released from the enzyme. **Review Table 6.1 and Figure 6.19**

Metabolism and the Regulation of Enzymes

▶ Metabolism is organized into pathways, in which the product of one reaction is a reactant for the next reaction. Each reaction is catalyzed by an enzyme.

▶ Enzyme activity is subject to regulation. Some compounds react irreversibly with enzymes and reduce their catalytic activity. Others react reversibly, inhibiting enzyme action only temporarily. A compound closely similar in structure to an enzyme's normal substrate may competitively inhibit the action of the enzyme. **Review Figures 6.20, 6.21**

▶ For allosteric enzymes, plots of reaction rate versus substrate concentration are sigmoidal, in contrast to plots of the same variables for non-allosteric enzymes. **Review Figure 6.22**

▶ Allosteric inhibitors bind to a site different from the active site and stabilize the inactive form of the enzyme. The multiple catalytic subunits of many allosteric enzymes interact cooperatively. **Review Figure 6.23**

▶ The end product of a metabolic pathway may inhibit the allosteric enzyme that catalyzes the commitment step of the pathway. **Review Figure 6.24**

▶ Enzymes are sensitive to their environment. Both pH and temperature affect enzyme activity. **Review Figures 6.25, 6.26**

For Discussion

1. How can endergonic reactions proceed in organisms?

2. Consider two proteins: One is an enzyme dissolved in the cytosol; the other is an ion channel in a membrane. Contrast the structures of the two proteins, indicating at least two important differences.

3. Plot free energy versus the course of an endergonic reaction and that of an exergonic reaction. Include the activation energy in both plots. Label E_a and ΔG on both graphs.

4. Consider an enzyme that is subject to allosteric regulation. If a competitive inhibitor (not an allosteric inhibitor) is added to a solution of such an enzyme, the ratio of enzyme molecules in the active form to those in the inactive form increases. Explain this observation.

7 Cellular Pathways That Harvest Chemical Energy

THE USE OF CRUSHED PLANT MATERIALS, SUCH as grapes and barley, to make alcoholic beverages, such as wine and beer, is as ancient as recorded history. But *how* these transformations come about has only recently been deciphered. The arts of the winemaker and brewer came under the scrutiny of science in the nineteenth century. Things came to a head (so to speak) when the German chemist Justus von Liebig claimed that these transformations were simply chemical reactions, and not some special property of the once-living plant material. Biologists, on the other hand, armed with microscopes and their cell theory, said that grape and barley extracts were converted to wine and beer by cells. Initially, these two ideas seemed to conflict.

Louis Pasteur, the great French scientist, tackled the issue in 1860. A group of distillers wanted to use sugar beets, which grow abundantly in regions of France, to produce alcohol. With careful observations, Pasteur noted three things: (1) Nothing happened to the sugar beet mash unless tiny, living yeast cells were present; (2) in the presence of fresh air, yeast cells grew vigorously on the mash and bubbles of carbon dioxide were formed; and (3) without fresh air, the yeasts grew slowly, less carbon dioxide was produced, and alcohol was formed. Pasteur had shown that the production of alcohol by ground-up, sugary extracts was a property of living cells, thereby introducing the concept of biochemistry.

Biochemists eventually identified the intermediate substances in the pathway between sugar and carbon dioxide and showed that each intermediate step is catalyzed by a specific enzyme. This examples sums up the concept of metabolism: the sum total of all of the chemical transformations in living systems as they break down simple sugars and other molecules in order to liberate energy and build up complex molecules.

In this chapter, we will describe some aspects of metabolism, especially as they relate to the breakdown of sugars. The metabolism of sugars is important not only in making alcoholic beverages, but in providing the energy

that organisms store in ATP—the energy you use all the time to fuel both conscious actions such as turning the pages of this book, and automatic ones such as the beating of your heart.

Several principles govern metabolic pathways in the cell:

▶ Complex chemical transformations in the cell do not occur in a single reaction, but in a number of small steps that are connected in a pathway.
▶ Each reaction is catalyzed by a specific enzyme.
▶ Metabolic pathways are similar in all organisms.
▶ Many metabolic pathways are compartmentalized, with certain steps occurring inside an organelle.
▶ Metabolic pathways in organisms are regulated by the activities of a few enzymes.

Obtaining Energy and Electrons from Glucose

The most common fuel for living cells is the sugar **glucose** ($C_6H_{12}O_6$). Many other compounds serve as foods, but almost all of them are converted to glucose, or to intermediate compounds in the step-by-step metabolism of glucose. As you will see in this section, cells obtain energy from glucose by the chemical process of oxidation.

Cells trap free energy while metabolizing glucose

If glucose is burned in a flame, it readily forms carbon dioxide, water, and a lot of energy—but only if oxygen gas (O_2)

Grape Harvest
The conversion of grape sugars into alcohol is mediated by enzymes in yeast cells.

is present. The balanced equation for this combustion reaction is

$$C_6H_{12}O_6 + 6 O_2 \rightarrow 6 CO_2 + 6 H_2O + \text{energy}$$
$$\text{(heat and light)}$$

The same equation applies to the metabolism of glucose in cells, except that metabolism is a multi-step, controlled series of reactions, ending up with almost half of the energy captured in ATP.

The change in free energy (ΔG) for the *complete* conversion of glucose and oxygen to carbon dioxide and water, whether by combustion or by metabolism, is –686 kcal/mol (–2,870 kJ/mol). Thus the overall reaction is highly exergonic and can drive the endergonic formation of a great deal of ATP from ADP and phosphate.

Some kinds of cells, unable to obtain or use oxygen gas, metabolize glucose *incompletely*, thus obtaining less ATP per glucose molecule. Not all of the carbon atoms of glucose are converted to carbon dioxide in this incomplete breakdown, which is called *fermentation*.

Three metabolic processes play roles in the utilization of glucose for energy: glycolysis, cellular respiration, and fermentation (Figure 7.1). All of these processes consist of metabolic pathways made up of many distinct, but coupled, chemical reactions.

▶ *Glycolysis* is a series of reactions that begins the metabolism of glucose in all cells and produces the three-carbon product **pyruvate**. A small amount of the energy stored in glucose is released in usable form.

▶ *Cellular respiration* occurs when the environment is **aerobic** (contains oxygen gas, O_2), and essentially converts pyruvate to carbon dioxide (CO_2). In the process, a great deal of the energy stored in the covalent bonds of pyruvate is released and trapped in ATP.

▶ *Fermentation* occurs when the environment is **anaerobic** (lacking oxygen gas). Instead of energy-poor CO_2, relatively energy-rich molecules such as lactic acid or ethanol are produced, so the energy extracted from glucose is far less than under aerobic conditions.

Redox reactions transfer electrons and energy

In Chapter 6, we described the addition of phosphate groups to ADP to make ATP as an endergonic reaction that can extract and store energy from exergonic reactions. Another way of transferring energy is to transfer electrons. A reaction in which one substance transfers one or more electrons to another substance is called an oxidation–reduction reaction, or **redox reaction**.

The *gain* of one or more electrons by an atom, ion, or molecule is called **reduction**. The *loss* of one or more electrons is called **oxidation**. Although oxidation and reduction are always defined in terms of traffic in *electrons*, we may also think in these terms when hydrogen atoms (*not* hydrogen ions) are gained or lost, because transfers of hydrogen atoms involve transfers of electrons ($H = H^+ + e^-$). Thus, when a molecule loses hydrogen atoms, it becomes oxidized:

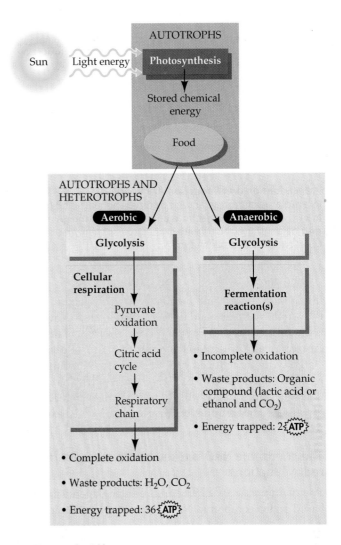

7.1 Energy for Life
Both heterotrophic ("other feeding") and autotrophic ("self-feeding") organisms obtain energy from the food compounds that autotrophs produce by photosynthesis. They convert these compounds to glucose, then metabolize glucose by glycolysis, fermentation, and cellular respiration.

$$\overset{\text{oxidation}}{\overbrace{AH_2 + B \rightarrow \underset{\text{reduction}}{\underbrace{BH_2 + A}}}}$$

Oxidation and reduction *always* occur together: As one material is oxidized, the electrons it loses are transferred to another material, reducing that material. In a redox reaction, we call the reactant that becomes reduced an *oxidizing agent* and the one that becomes oxidized a *reducing agent* (Figure 7.2).

▶ An oxidizing agent accepts electrons; in the process of oxidizing the reducing agent, the oxidizing agent itself becomes reduced.

▶ Conversely, the reducing agent donates electrons; it becomes oxidized as it reduces the oxidizing agent.

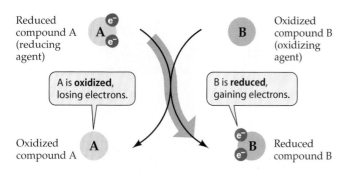

7.2 Oxidation and Reduction Are Coupled
In a redox reaction, reactant A is oxidized and reactant B reduced. In the process, A loses electrons and B gains electrons. A proton may be transferred along with an electron, so that what is actually transferred is a hydrogen atom.

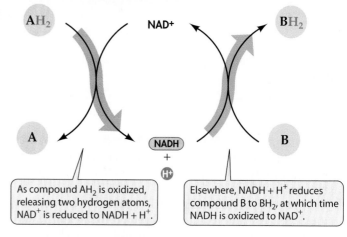

7.4 NAD Is an Energy Carrier
Thanks to its ability to carry free energy and electrons, NAD is a major and universal energy intermediary in cells.

In both the burning and the metabolism of glucose, glucose is the reducing agent and oxygen gas is the oxidizing agent.

In a redox reaction, energy is transferred. Some of the energy originally present in the reducing agent becomes associated with the reduced product. (The rest remains in the reducing agent or is lost.) The overall ΔG of a redox reaction is negative. As we will see, some of the key reactions of glycolysis and cellular respiration are highly exergonic redox reactions.

The coenzyme NAD is a key electron carrier in redox reactions

The main pair of oxidizing and reducing agents in cells is based on the compound **NAD** (nicotinamide adenine dinu-

cleotide). NAD exists in two chemically distinct forms, one oxidized (**NAD⁺**) and the other reduced (**NADH + H⁺**) (Figure 7.3). NAD⁺ and NADH + H⁺ participate in biological redox reactions. The reduction reaction

$$NAD^+ + 2\,H \rightarrow NADH + H^+$$

is formally equivalent to the transfer of two hydrogen atoms ($2\,H^+ + 2\,e^-$). However, what is actually transferred is a hydride ion (H^-, a proton and two electrons), leaving a free proton (H^+).

Oxygen gas (O_2) is highly electronegative (see Table 2.3) and readily accepts electrons from NADH. The oxidation of NADH + H⁺ by O_2 is highly exergonic:

$$NADH + H^+ + \tfrac{1}{2}\,O_2 \rightarrow NAD^+ + H_2O$$
$$\Delta G = -52.4 \text{ kcal/mol } (-219 \text{ kJ/mol})$$

(Note that the oxidizing agent appears here as "½ O₂" instead of "O." This notation emphasizes that it is oxygen gas, O_2, that acts as the oxidizing agent.)

In the same way that ATP can be thought of as packaging free energy in bundles of about 12 kcal/mol (50 kJ/mol), NAD can be thought of as packaging free energy in bundles of approximately 50 kcal/mol (200 kJ/mol) (Figure 7.4).

NAD is not the only electron carrier in cells. As you will see, another carrier, FAD (flavin adenine dinucleotide), is also involved in transferring electrons during the metabolism of glucose.

1 Two hydrogen atoms ($2\,e^- + 2\,H^+$) are released by another molecule.

2 The ring structure of NAD acquires $2\,e^-$ and $1\,H^+$...

Oxidized form (**NAD⁺**)   Reduced form (**NADH** + **H⁺**)

$+ 2\,H$ Reduction / Oxidation

3 ...leaving 1 H⁺ free.

7.3 Oxidized and Reduced Forms of NAD
NAD⁺ is the oxidized form and NADH the reduced form of NAD. The unshaded portion of the molecule remains unchanged by the redox reaction.

An Overview: Releasing Energy from Glucose

The three energy-extracting processes of cells may be divided into distinct pathways:

▶ When O_2 is available as the final electron acceptor, four pathways operate. Glycolysis takes place first, and is

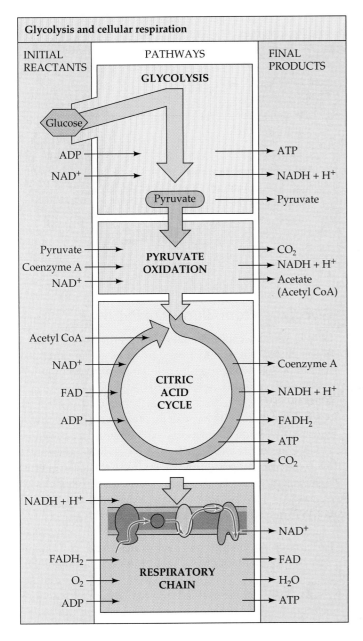

Glycolysis and cellular respiration

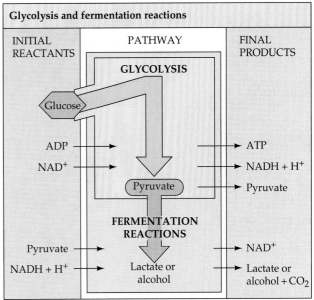

Glycolysis and fermentation reactions

7.5 An Overview of the Cellular Energy Pathways
Energy-producing reactions can be grouped into five pathways: glycolysis, pyruvate oxidation, the citric acid cycle, the respiratory chain, and fermentation. The three middle pathways occur only in the presence of oxygen and are collectively referred to as cellular respiration.

In eukaryotes, glycolysis and fermentation take place in the cytoplasm outside of the mitochondria. The enzymes for these pathways were once believed to be soluble in the cytosol, but more recent discoveries suggest that at least some of them may be bound to components of the cytoskeleton. The other reactions are associated with the mitochondria. Pyruvate oxidation and the respiratory chain are both associated with the inner membrane of mitochondria, where their enzymes are bound. The enzymes and reactions of the citric acid cycle are found in the mitochondrial matrix.

Glycolysis begins the breakdown of glucose. It is a sequence of ten separate chemical reactions in which glucose is incompletely oxidized to pyruvate. It contains an oxidative step in which the electron carrier NAD^+ becomes reduced, acquiring electrons. The major products of glycolysis are ATP, pyruvate, and the electrons acquired by NAD. Both the pyruvate and the electrons must be processed further.

Cellular respiration operates when O_2 is available, yielding CO_2 and H_2O as products. It is made up of three pathways: pyruvate oxidation, the citric acid cycle, and the respiratory chain.

In **pyruvate oxidation**, the end product of glycolysis (pyruvate) is oxidized to acetate, which is activated by the addition of a coenzyme and further metabolized by the citric acid cycle.

The **citric acid cycle** is a cyclic series of reactions in which the acetate becomes *completely* oxidized, forming CO_2 and transferring electrons (along with their hydrogen

followed by the three pathways of cellular respiration: pyruvate oxidation, the citric acid cycle, and the respiratory chain.

▶ When O_2 is unavailable, pyruvate oxidation, the citric acid cycle, and the respiratory chain do not function, and fermentation is added to the glycolytic pathway.

These five chemical pathways, which we will consider one at a time, have different locations in the cell (Table 7.1). Figure 7.5 summarizes the starting reactants and products of these pathways.

In prokaryotes, the enzymes used in glycolysis, fermentation, and the citric acid cycle are soluble in the cytosol. The enzymes involved in pyruvate oxidation and the respiratory chain are associated with the inner surface of the plasma membrane or inward elaborations of that membrane (see Chapter 4).

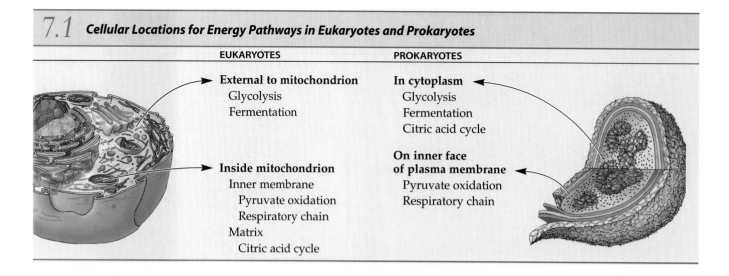

7.1 Cellular Locations for Energy Pathways in Eukaryotes and Prokaryotes

EUKARYOTES

External to mitochondrion
Glycolysis
Fermentation

Inside mitochondrion
Inner membrane
    Pyruvate oxidation
    Respiratory chain
Matrix
    Citric acid cycle

PROKARYOTES

In cytoplasm
Glycolysis
Fermentation
Citric acid cycle

**On inner face
of plasma membrane**
Pyruvate oxidation
Respiratory chain

nuclei) to carrier molecules. The citric acid cycle produces many more electrons than are produced in glycolysis. And, as we are about to see, harvesting more electrons means a greater ultimate harvest of ATP.

In glycolysis, pyruvate oxidation, and the citric acid cycle, the electron carriers NAD$^+$ and FAD become reduced and acquire hydrogen atoms. Through these pathways, energy originally present in the covalent bonds of glucose becomes associated with reduced forms of these carriers (NADH + H$^+$ and FADH$_2$).

The fourth energy-extracting pathway for aerobic cells is the **respiratory chain**, which releases energy from the reduced NADH + H$^+$ in such a way that it can be used to form ATP. This pathway consists of a series of redox reactions in which electrons derived from hydrogen atoms are passed from one type of carrier to another and finally are allowed to react with O$_2$ to produce water. Hydrogen is an outstanding fuel. When it reacts with O$_2$, a great deal of free energy is released; better still, the "waste" product of this reaction—water—is not toxic to the environment or to any organism that produces it.

The transfer of electrons along the respiratory chain drives the active transport of hydrogen ions (protons) from the mitochondrial matrix into the space between the inner and outer mitochondrial membrane. This active transport sets up an imbalance of both concentration and charge across the membrane. As we saw in Chapter 6, such imbalances represent potential energy. This energy is recaptured by the subsequent diffusion of protons back into the matrix, which is coupled to the synthesis of ATP from ADP and P$_i$.

Overall, the inputs to the respiratory chain are hydrogen atoms and O$_2$, and the outputs are water and energy captured as ATP.

In the discussion that follows, we will examine in more detail the four pathways of aerobic energy metabolism (glycolysis, pyruvate oxidation, the citric acid cycle, and the respiratory chain), and fermentation.

Glycolysis: From Glucose to Pyruvate

Glycolysis (also called the glycolytic pathway) takes place in the cytoplasm. It may be regarded as a common pathway to be followed either by cellular respiration or, under anaerobic conditions, fermentation.

In glycolysis, glucose is only partly oxidized. A molecule of glucose taken in by a cell enters the glycolytic pathway, which consists of 10 reactions that convert the six-carbon glucose molecule, step by step, into two molecules of the three-carbon compound pyruvate (pyruvic acid)* (Figure 7.7, which appears on pages 120–121.). These reactions are accompanied by the *net* formation of two molecules of ATP and by the reduction of two molecules of NAD$^+$ to two molecules of NADH + H$^+$. At the end of the glycolytic pathway, then, energy has been transferred to ATP, and four hydrogen atoms have been transferred to NADH + H$^+$.

Glycolysis can be divided into two groups of reactions: *energy-investing* reactions that use ATP, and *energy-harvesting* reactions that produce ATP.

The energy-investing reactions of glycolysis require ATP

Using Figure 7.7, let us work our way through the glycolytic pathway. The first five reactions are endergonic; that is, the cell is *investing* free energy rather than gaining it during the early reactions of glycolysis. In separate reactions, two molecules of ATP are invested in attaching two phosphate groups to the sugar molecule (reactions 1 and 3), thereby raising its free energy by about 15 kcal/mol (62.7 kJ/mol) (Figure 7.6). Later, these phosphate groups will be transferred to ADP to make new molecules of ATP.

*We tend to use words like "pyruvate" and "pyruvic acid" interchangeably. However, at the pH values commonly found in cells, the ionized form—pyruvate—is present rather than the acid—pyruvic acid. Similarly, all carboxylic acids are present as ions (the "-ate" forms) at these pH values.

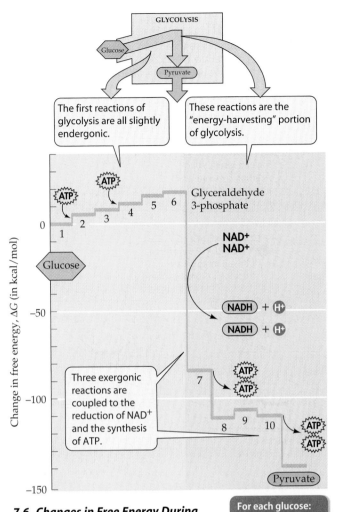

7.6 Changes in Free Energy During Glycolysis
Each reaction of glycolysis changes the free energy available, as shown by the different energy levels of the series of reactants and products from glucose to pyruvate.

For each glucose:
2 Pyruvate
2 NADH + 2 H⁺
2 ATP are produced.

Although both of these first steps of glycolysis use ATP as one of their substrates, each is catalyzed by a different, specific enzyme. The enzyme hexokinase catalyzes reaction 1, in which a phosphate group from ATP is attached to the six-carbon glucose molecule, forming glucose 6-phosphate. (A *kinase* is any enzyme that catalyzes the transfer of a phosphate group from ATP to another substrate.) In reaction 2, the six-membered glucose ring is rearranged into a five-membered fructose ring. Then, in reaction 3, the enzyme phosphofructokinase adds a second phosphate (taken from another ATP) to the fructose ring, forming a six-carbon sugar bisphosphate.*

*The root *bis-* means "two." A sugar bisphosphate has two phosphate groups attached to two different carbons, as opposed to the prefix *di-*, which implies the serial attachment of two phosphate groups to one carbon, as in ADP (adenosine *di*phosphate).

The fourth reaction opens up and cleaves the six-carbon sugar ring to give two different three-carbon sugar phosphates. In reaction 5, one of them, dihydroxyacetone phosphate, is converted into a second molecule of the other one, glyceraldehyde 3-phosphate (G3P).

By this time—the halfway point of the glycolytic pathway—the following things have happened: Two molecules of ATP have been invested, and the six-carbon glucose molecule has been converted into two molecules of a three-carbon sugar phosphate, G3P. No ATP has been gained, and nothing has been oxidized.

The energy-harvesting reactions of glycolysis yield ATP and NADH + H⁺

In the discussion that follows, remember that each reaction occurs twice for each glucose molecule going through glycolysis, because in the first five reactions, each glucose molecule has been split into two molecules of G3P.

Reaction 6 is a three-step reaction catalyzed by the enzyme triose phosphate dehydrogenase. Its end product is a phosphate ester, 1,3-bisphosphoglycerate (BPG). Reaction 6 is accompanied by an enormous drop in free energy—more than 100 kcal of energy per mole of glucose is released in this extremely exergonic reaction. What has happened here? Why the big energy change?

In the first step of reaction 6, a phosphate ion is snatched from the surroundings (but not, this time, from ATP) and tacked onto G3P.

In the second step, G3P is converted to an acid:

$$R - \overset{\overset{\displaystyle O}{\|}}{C} - H + (O) \rightarrow R - \overset{\overset{\displaystyle O}{\|}}{C} - OH$$

Since this is an oxidation reaction, it is very exergonic.

The third step, the formation of a phosphate ester (BPG) from an acid

$$R - \overset{\overset{\displaystyle O}{\|}}{C} - OH + HPO_4^{2-} \rightarrow R - \overset{\overset{\displaystyle O}{\|}}{C} - O - \overset{\overset{\displaystyle O}{\underset{\underset{\displaystyle O^-}{|}}{\|}}}{P} - O^- + H_2O$$

is slightly endergonic, but not nearly enough to offset the drop in free energy from the oxidation.

If this big energy drop were simply a loss of heat, glycolysis would not provide useful energy to the cell. However, rather than being lost, this energy is stored by reducing two molecules of NAD⁺ to make two molecules of NADH + H⁺. This stored energy is regained later—either in the respiratory chain, by the formation of ATP, or in the last step of fermentation, in which pyruvate or its product is reduced by the two molecules of NADH + H⁺, forming NAD⁺. Because NAD⁺ is present in small amounts in the cell, it must be recycled to keep glycolysis going; if none of the NADH is oxidized back to NAD⁺, glycolysis comes to a halt.

The remaining steps of glycolysis are simpler. The two phosphate groups of BPG are transferred, one at a time, to

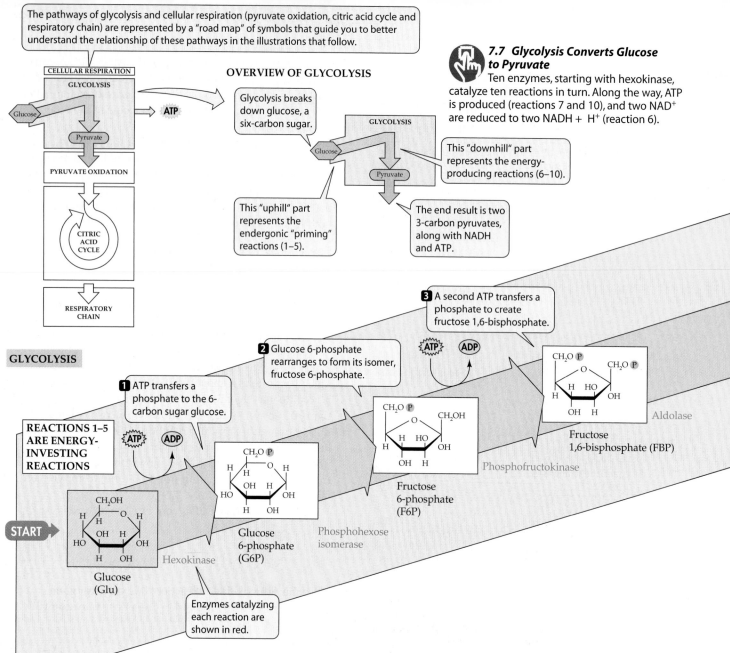

7.7 Glycolysis Converts Glucose to Pyruvate

Ten enzymes, starting with hexokinase, catalyze ten reactions in turn. Along the way, ATP is produced (reactions 7 and 10), and two NAD^+ are reduced to two $NADH + H^+$ (reaction 6).

minal phosphate-to-phosphate bond of ATP. A second enzyme, phosphoglycerate kinase, catalyzes the transfer of this phosphate group from BPG to ADP in reaction 7, forming ATP. Both reactions are exergonic, even though a substantial amount of energy is consumed in the formation of ATP.

GLYCOLYSIS MAY BE FOLLOWED BY FERMENTATION. A review of the glycolytic pathway shows that at the beginning of glycolysis, two molecules of ATP are used per molecule of glucose, but that ultimately four molecules of ATP are produced (two for each of the two BPG molecules)—a net gain of two ATP molecules and two $NADH + H^+$.

When fermentation follows glycolysis, the total usable energy yield is just these two ATP molecules per glucose molecule. Under anaerobic conditions, the $NADH + H^+$ is rapidly recycled to NAD^+ by the reduction of pyruvate. The NAD^+ is then available for the glycolytic reaction catalyzed by the enzyme triose phosphate dehydrogenase (reaction 6 in Figure 7.7).

molecules of ADP, with a rearrangement in between. More than 20 kcal (83.6 kJ/mol) of free energy is stored in ATP for every mole of BPG broken down. Finally, we are left with two moles of pyruvate for every mole of glucose that entered glycolysis.

SUBSTRATE-LEVEL PHOSPHORYLATION. The enzyme-catalyzed transfer of phosphate groups from donor molecules to ADP molecules (reaction 7) is called **substrate-level phosphorylation**. This process is driven by energy obtained from oxidation. For example, when G3P reacts with a phosphate group (P_i) and NAD^+, becoming BPG, an aldehyde is oxidized to a carboxylic acid, with NAD^+ acting as the oxidizing agent. The oxidation provides so much energy that the newly added phosphate group is linked to the rest of the molecule by a bond that has even more energy than the ter-

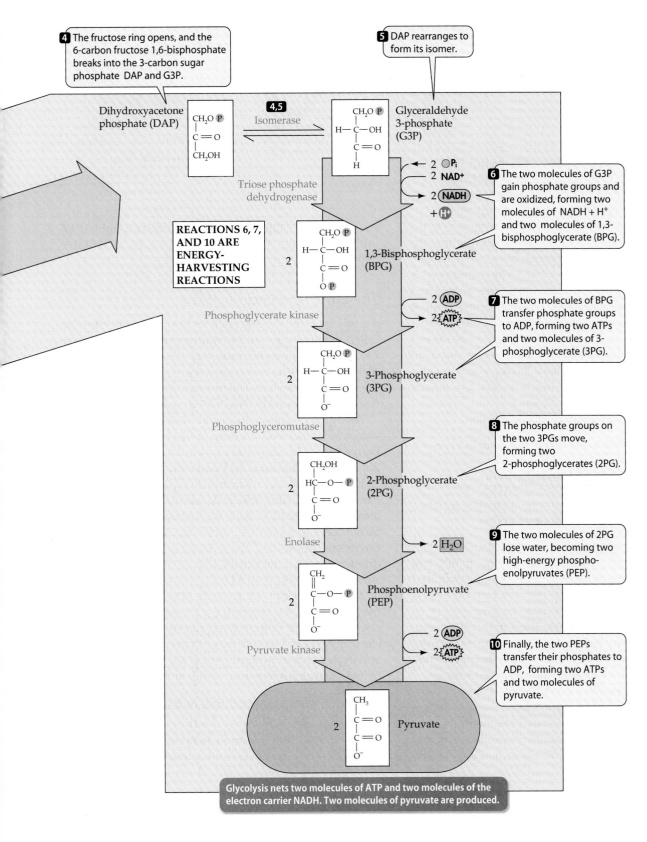

4 The fructose ring opens, and the 6-carbon fructose 1,6-bisphosphate breaks into the 3-carbon sugar phosphate DAP and G3P.

5 DAP rearranges to form its isomer.

Dihydroxyacetone phosphate (DAP)

4,5 Isomerase

Glyceraldehyde 3-phosphate (G3P)

Triose phosphate dehydrogenase

2 P_i
2 NAD+
2 NADH
+ H+

REACTIONS 6, 7, AND 10 ARE ENERGY-HARVESTING REACTIONS

6 The two molecules of G3P gain phosphate groups and are oxidized, forming two molecules of NADH + H+ and two molecules of 1,3-bisphosphoglycerate (BPG).

2   1,3-Bisphosphoglycerate (BPG)

Phosphoglycerate kinase

2 ADP
2 ATP

7 The two molecules of BPG transfer phosphate groups to ADP, forming two ATPs and two molecules of 3-phosphoglycerate (3PG).

2   3-Phosphoglycerate (3PG)

Phosphoglyceromutase

8 The phosphate groups on the two 3PGs move, forming two 2-phosphoglycerates (2PG).

2   2-Phosphoglycerate (2PG)

Enolase

2 H_2O

9 The two molecules of 2PG lose water, becoming two high-energy phospho-enolpyruvates (PEP).

2   Phosphoenolpyruvate (PEP)

Pyruvate kinase

2 ADP
2 ATP

10 Finally, the two PEPs transfer their phosphates to ADP, forming two ATPs and two molecules of pyruvate.

2   Pyruvate

Glycolysis nets two molecules of ATP and two molecules of the electron carrier NADH. Two molecules of pyruvate are produced.

On the other hand, in the presence of oxygen, eukaryotes and some bacteria reap far more energy by completely oxidizing pyruvate and by oxidizing NADH + H+ through the respiratory chain, as we will see in the sections that follow. In eukaryotes, these reactions take place in the mitochondria.

7.8 The Pyruvate Dehydrogenase Complex Catalyzes Pyruvate Oxidation

A massive multiprotein complex, pyruvate dehydrogenase converts pyruvate to acetyl CoA by transferring electrons, removing a carboxyl group, and adding a coenzyme (CoA). Energy is stored temporarily in acetyl CoA.

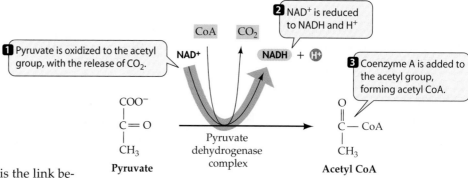

1 Pyruvate is oxidized to the acetyl group, with the release of CO_2.

2 NAD^+ is reduced to NADH and H^+

3 Coenzyme A is added to the acetyl group, forming acetyl CoA.

Pyruvate Oxidation

The oxidation of pyruvate to acetate is the link between glycolysis and cellular respiration. Pyruvate oxidation is a multistep reaction catalyzed by an enormous enzyme complex that is attached to the inner mitochondrial membrane. Pyruvate diffuses into the mitochondrion, where it is oxidized. In this reaction, the pyruvate, a three-carbon compound, loses two hydrogen atoms and a carboxyl group ($-COO^-$). The reaction yields the two-carbon acetyl group, as well as free energy and CO_2. The free energy is captured when the acetyl group is linked to a coenzyme, called **coenzyme A (CoA)**, producing **acetyl coenzyme A (acetyl CoA)** (Figure 7.8). Acetyl CoA has 7.5 kcal/mol (31.4 kJ/mol) more energy than simple acetate. (Acetyl CoA can donate the acetyl group to acceptors such as oxaloacetate, much as ATP can donate phosphate groups to various acceptors.)

There are three steps in this oxidation reaction:

▶ Pyruvate is oxidized to the acetyl group, and CO_2 is released.

▶ Part of the energy from the oxidation in the first step is saved by the reduction of NAD^+ to NADH + H^+.

▶ Some of the remaining energy is stored temporarily by the combining of the acetyl group with CoA.

An analogous three-step reaction occurs in reaction 6 of the glycolytic pathway, when G3P is converted to BPG (see Figure 7.6). In that reaction, an aldehyde group is oxidized to an acid, some of the energy released by oxidation is stored in NADH + H^+, and some of the remaining energy is preserved in a second phosphate bond in the BPG molecule. As the similarity between these two three-step reactions shows, a good metabolic idea is likely to appear more than once; we will see it again in the citric acid cycle.

As you might suspect, a complex set of steps such as those in pyruvate oxidation requires more than one type of catalytic protein. This reaction is catalyzed by the *pyruvate dehydrogenase complex*, a huge multi-protein machine that consists of 72 polypeptide chains—24 each of three differ-

ent protein molecules, for a total molecular weight of 4.6 million. The three component enzymes use a total of five different coenzymes.

The Citric Acid Cycle

Acetyl CoA is the starting point for the citric acid cycle (also called the *Krebs cycle* or the *tricarboxylic acid cycle*) (Figure 7.9). This pathway, which consists of eight reactions, completely oxidizes the two-carbon acetyl group to two molecules of carbon dioxide. The free energy released from these reactions is captured by NAD, FAD, and ADP.

As Figure 7.7 shows, the metabolism of glucose to pyruvate is accompanied by a drop in free energy of about 140 kcal/mol (585 kJ/mol). About a third of this energy is captured in the formation of ATP and reduced NAD (NADH + H^+). Oxidizing pyruvate to acetate yields much additional free energy. The citric acid cycle takes the acetyl group and breaks it down to CO_2, using the hydrogen atoms to reduce carrier molecules and passing chemical free energy to those carriers in the process. The reduced carriers are later oxidized in the respiratory chain, which transfers an enormous amount of free energy to ATP.

The principal inputs to the citric acid cycle are acetate (in the form of acetyl CoA), water, and oxidized electron carriers. The principal outputs are carbon dioxide and reduced electron carriers. Overall, for each acetyl group, the citric acid cycle removes two carbons as CO_2 and uses four pairs of hydrogen atoms to reduce carrier molecules.

The citric acid cycle produces two CO_2 molecules and reduced carriers

At the beginning of the citric acid cycle, acetyl CoA, which has two carbon atoms in its acetyl group, reacts with a four-

7.9 The Citric Acid Cycle

Pyruvate diffuses into the mitochondrion and is oxidized to acetyl CoA, which enters the citric acid cycle. The two carbons from acetyl CoA are shown in black circles. Reactions 3, 4, 6, and 8 accomplish the major overall effects of the cycle—the trapping of energy—by passing electrons to NAD or FAD. Reaction 5 traps energy directly in ATP. ▶

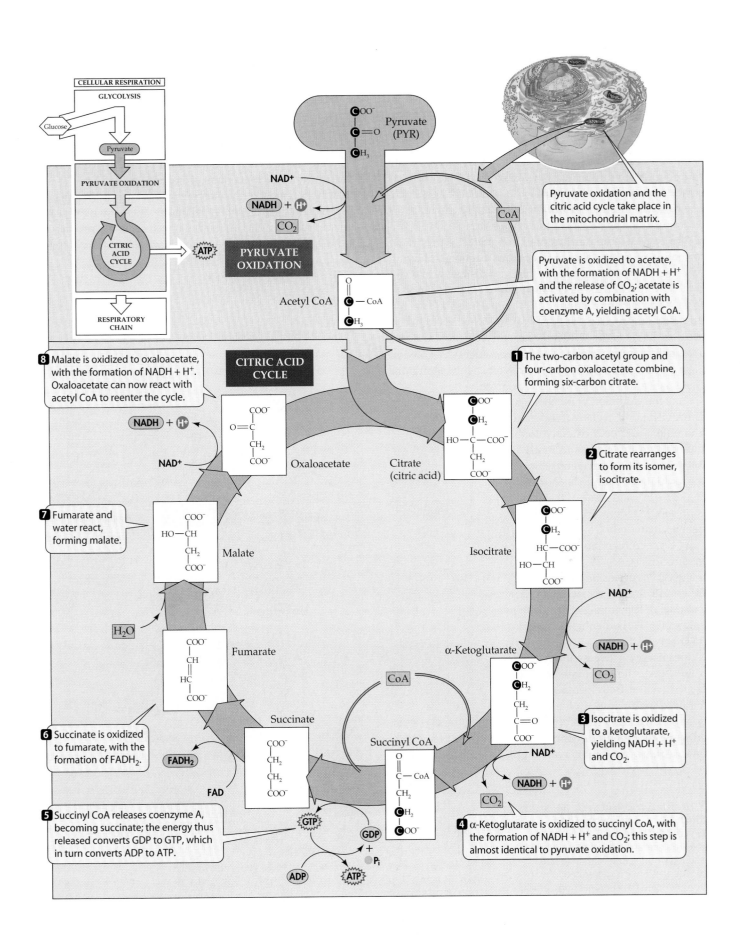

CELLULAR RESPIRATION

GLYCOLYSIS

Glucose

Pyruvate

PYRUVATE OXIDATION

CITRIC ACID CYCLE

ATP

RESPIRATORY CHAIN

Pyruvate (PYR)

NAD$^+$

NADH + H$^+$

CO$_2$

PYRUVATE OXIDATION

Pyruvate oxidation and the citric acid cycle take place in the mitochondrial matrix.

Acetyl CoA

Pyruvate is oxidized to acetate, with the formation of NADH + H$^+$ and the release of CO$_2$; acetate is activated by combination with coenzyme A, yielding acetyl CoA.

CITRIC ACID CYCLE

8 Malate is oxidized to oxaloacetate, with the formation of NADH + H$^+$. Oxaloacetate can now react with acetyl CoA to reenter the cycle.

NADH + H$^+$

NAD$^+$

Oxaloacetate

Citrate (citric acid)

1 The two-carbon acetyl group and four-carbon oxaloacetate combine, forming six-carbon citrate.

2 Citrate rearranges to form its isomer, isocitrate.

7 Fumarate and water react, forming malate.

Malate

Isocitrate

H$_2$O

Fumarate

α-Ketoglutarate

NAD$^+$

NADH + H$^+$

CO$_2$

CoA

3 Isocitrate is oxidized to α ketoglutarate, yielding NADH + H$^+$ and CO$_2$.

6 Succinate is oxidized to fumarate, with the formation of FADH$_2$.

FADH$_2$

FAD

Succinate

Succinyl CoA

NAD$^+$

NADH + H$^+$

CO$_2$

5 Succinyl CoA releases coenzyme A, becoming succinate; the energy thus released converts GDP to GTP, which in turn converts ADP to ATP.

GTP

GDP

+ P$_i$

ADP

ATP

4 α-Ketoglutarate is oxidized to succinyl CoA, with the formation of NADH + H$^+$ and CO$_2$; this step is almost identical to pyruvate oxidation.

carbon acid, oxaloacetate, to form the six-carbon compound citrate (citric acid). The remainder of the cycle consists of a series of enzyme-catalyzed reactions in which citrate is degraded to a new four-carbon molecule of oxaloacetate. This new oxaloacetate can react with a second acetyl CoA, producing a second molecule of citrate and thus enabling the cycle to continue. Acetyl CoA enters the cycle from pyruvate oxidation, and CO_2 exits.

The citric acid cycle is maintained in a steady state—that is, although materials enter and leave and intermediate compounds are formed as they are metabolized, the concentrations of molecules in the cycle do not change much. Pay close attention to the numbered reactions in Figure 7.9 as you read the next several paragraphs.

The energy temporarily stored in acetyl CoA drives the formation of citrate from oxaloacetate (reaction 1). During this reaction, the coenzyme A molecule falls away, to be recycled. In reaction 2, the citrate molecule is rearranged to form isocitrate. In reaction 3, a CO_2 molecule and two hydrogen atoms are removed, converting isocitrate to α-ketoglutarate. As Figure 7.10 indicates, this reaction produces a large drop in free energy. The released energy is stored in NADH + H$^+$ and can be recovered later in the respiratory chain, when the NADH + H$^+$ is reoxidized.

Like the oxidation of pyruvate to acetyl CoA, reaction 4 of the citric acid cycle is complex. The five-carbon α-ketoglutarate molecule is oxidized to the four-carbon molecule succinate. In the process, CO_2 is given off, some of the oxidation energy is stored in NADH + H$^+$, and some of the energy is preserved temporarily by combining succinate with CoA to form succinyl CoA. In reaction 5, the energy in succinyl CoA is harvested to make GTP (guanosine triphosphate) from GDP and P$_i$, which is another example of substrate-level phosphorylation. GTP is then used to make ATP from ADP.

Free energy is released in reaction 6, in which the succinate released from succinyl CoA in reaction 5 is oxidized to fumarate. In the process, two hydrogens are transferred to an enzyme that contains the carrier FAD. After a molecular rearrange-

ment (reaction 7), one more NAD$^+$ reduction occurs, producing oxaloacetate from malate (reaction 8). These two reactions illustrate a common biochemical mechanism: Water (H_2O) is added in reaction 7 to form an —OH group, and then the H from that —OH group is removed in reaction 8 to reduce NAD$^+$ to NADH + H$^+$. Essentially, these two reactions provide energy by using a very abundant substance (H_2O). The final product, oxaloacetate, is ready to combine with another acetyl group from acetyl CoA and go around the cycle again. The citric acid cycle operates twice for each glucose molecule that enters glycolysis.

Although most of the enzymes of the citric acid cycle are dissolved in the mitochondrial matrix, there are two exceptions: succinate dehydrogenase, which catalyzes reaction 6, and α-ketoglutarate dehydrogenase, which catalyzes reac-

7.10 The Citric Acid Cycle Releases Much More Free Energy Than Glycolysis Does
Electron carriers (NAD in glycolysis; NAD and FAD in the citric acid cycle) are reduced and ATP is generated in reactions coupled to other reactions, producing major drops in free energy as metabolism proceeds.

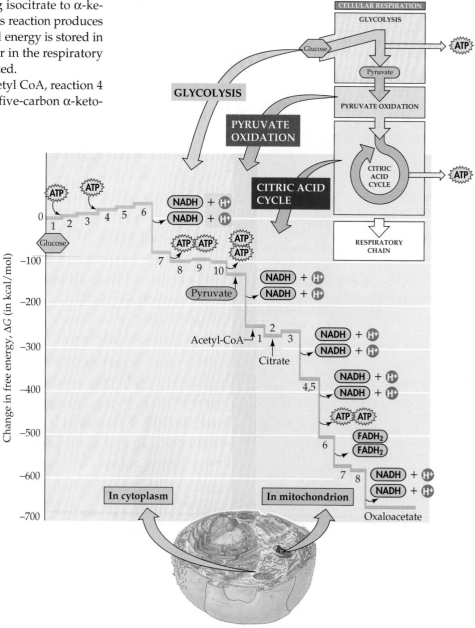

tion 4. These enzymes are integral membrane proteins of the inner mitochondrial membrane.

Generations of students have asked the question, "Why did this complicated system evolve to achieve the simple goal of oxidizing two carbon acetyl units to CO_2?" There are two reasons. First, the cycle includes molecules that have other roles in the cell. As we will see later in this chapter, the intermediates of the citric acid cycle are themselves catabolic (breakdown) products or anabolic (synthesis) sources of other molecules, such as amino acids and nucleotides. Second, the citric acid cycle is far more efficient at tapping off energy than any single reaction could be.

The Respiratory Chain: Electrons, Proton Pumping, and ATP

Without NAD$^+$ and FAD, the oxidative steps of glycolysis, pyruvate oxidation, and the citric acid cycle could not occur. Once reduced, these carriers must have some place to donate their hydrogens (H$^+$ + e$^-$). The fate of these protons and electrons is the rest of the story of cellular respiration.

The story has three parts:

▶ First, the electrons pass through a series of membrane-associated electron carriers called the respiratory chain.
▶ Second, the flow of electrons along the chain causes the active transport of protons across the inner mitochondrial membrane, out of the matrix, creating a concentration gradient.

▶ Third, the protons diffuse back into the mitochondrial matrix through a proton channel, which couples this diffusion to the synthesis of ATP.

The overall process of ATP synthesis resulting from electron transport through the respiratory chain is called **oxidative phosphorylation**.

The respiratory chain transports electrons and releases energy

The respiratory chain contains three components:

▶ Three large protein complexes, containing carrier molecules and their associated enzymes
▶ A small protein called **cytochrome c**
▶ A nonprotein component called **ubiquinone** (abbreviated **Q**)

The large protein complexes are bound to the folds of the inner mitochondrial membranes, the cristae (see Figure 4.16), in eukaryotes, or to the plasma membrane of aerobic prokaryotes. Cytochrome c is a peripheral membrane protein that lies in the space between the inner and outer mitochondrial membranes, loosely attached to the inner membrane. Ubiquinone (Q) is a small, nonpolar molecule that moves freely within the hydrophobic interior of the phospholipid bilayer of the inner membrane (Figure 7.11).

NADH + H$^+$ passes hydrogens to Q by way of the first large protein complex, **NADH-Q reductase**, which contains 26 polypeptides and attached prosthetic groups.

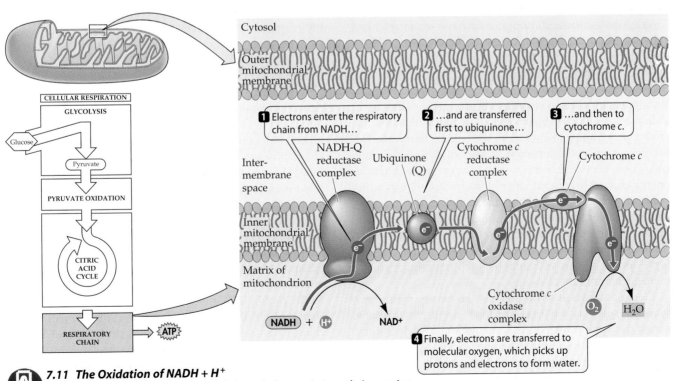

7.11 The Oxidation of NADH + H$^+$
Electrons from NADH + H$^+$ are passed through the respiratory chain, a series of carrier molecules in the inner mitochondrial membrane. The carriers gain free energy when they become reduced and release free energy when they are oxidized.

NADH-Q reductase passes the hydrogens to Q, forming QH_2. **Cytochrome c reductase**, with 10 subunits, receives hydrogens from QH_2 and passes them to cytochrome c. **Cytochrome c oxidase**, with 8 subunits, receives electrons from cytochrome c and passes them to oxygen. Reduced oxygen $(\frac{1}{2}O_2^-)$ picks up two hydrogen ions (H^+) to form H_2O. Different subunits within each of these large protein complexes bear different electron carriers, so electrons are transported *within* each complex, as well as from complex to complex.

The electron carriers of the respiratory chain (including those contained in the three protein complexes) differ as to how they change when they become reduced. NAD^+, for example, accepts H^- (a hydride ion—one proton and two electrons), leaving the proton from the other hydrogen atom to float free: $NADH + H^+$. Other carriers, including Q, bind both protons and both electrons, becoming, for example, QH_2. The remainder of the chain, however, is only an electron transport process. Electrons, but not protons, are passed from Q to cytochrome c. An electron from QH_2 reduces a cytochrome's Fe^{3+} to Fe^{2+}.

Electrons pour into the pool of Q molecules from the $NADH + H^+$ pathway, and some come from another source: the succinate-to-fumarate reaction of the citric acid cycle (reaction 6 in Figure 7.9). Another protein complex, *succinate-Q reductase*, links the oxidation of succinate to the reduction of Q (Figure 7.12). The enzyme that constitutes the first part of succinate-Q reductase has attached to it an FAD carrier molecule, which is reduced by succinate to $FADH_2$. Later, hydrogen atoms are transferred to the Q molecules. No protons are pumped, and hence no ATP is generated in the succinate-to-Q branch of the respiratory chain.

Why should the respiratory chain have so many links? Why, for example, don't cells just use the following single step?

$$NADH + H^+ + \frac{1}{2}O_2 \rightarrow NAD^+ + H_2O$$

Well, to begin with, no enzyme will catalyze the direct oxidation of $NADH + H^+$ by oxygen. More fundamentally, this would be an untamable reaction. It would be terrifically exergonic—rather like setting off a stick of dynamite in the cell. There is no biochemical way to harvest that burst of energy efficiently and put it to physiological use (that is, no metabolic reaction is so endergonic as to consume a significant fraction of that energy in a single step). To control the release of energy during oxidation of glucose in a cell, evolution has produced the lengthy respiratory chain we observe today: a *series* of reactions, each releasing a small, manageable amount of energy.

Electron transport within each of the three protein complexes results, as we'll see, in the pumping of protons across the inner mitochondrial membrane, and the return of the protons across the membrane leads to the formation of ATP. Thus the energy originally contained in glucose and other foods is finally tucked into the cellular energy currency, ATP. For each pair of electrons passed along the res-

7.12 The Complete Respiratory Chain
Electrons enter the chain from two sources, but they follow the same pathway from Q onward.

piratory chain from $NADH + H^+$ to oxygen, three molecules of ATP are formed.

If only electrons are carried through the final reactions of the respiratory chain, what happens to H^+? And how is electron transport coupled to ATP production?

Active proton transport is followed by diffusion coupled to ATP synthesis

As we have seen, all the carriers and enzymes of the respiratory chain (except cytochrome c) are embedded in the inner mitochondrial membrane (see Figure 7.11). The operation of the respiratory chain results in the active transport of protons (H^+) *against* their concentration gradient, across the inner membrane of the mitochondrion from inside to outside ("outside" being the space between the inner and outer mitochondrial membranes). This occurs because the carriers are arranged on the three large complexes such that protons are produced on one side (the intermembrane space) and transported along with electrons to the other side (facing the mitochondrial matrix) (Figure 7.13). Because of the charge on the proton (H^+), this transport causes not only a difference in proton concentration, but also a difference in electric charge across the membrane, with the inside of the organelle (the matrix) more negative than the outside.

Together, the proton concentration gradient and the charge difference constitute a source of potential energy called the *proton-motive force*. This force tends to drive the

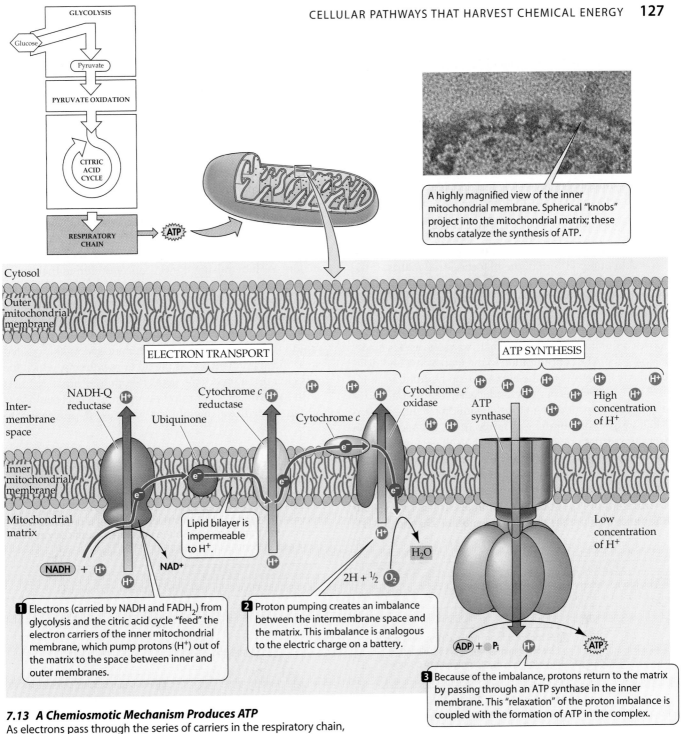

A highly magnified view of the inner mitochondrial membrane. Spherical "knobs" project into the mitochondrial matrix; these knobs catalyze the synthesis of ATP.

1 Electrons (carried by NADH and FADH$_2$) from glycolysis and the citric acid cycle "feed" the electron carriers of the inner mitochondrial membrane, which pump protons (H$^+$) out of the matrix to the space between inner and outer membranes.

2 Proton pumping creates an imbalance between the intermembrane space and the matrix. This imbalance is analogous to the electric charge on a battery.

3 Because of the imbalance, protons return to the matrix by passing through an ATP synthase in the inner membrane. This "relaxation" of the proton imbalance is coupled with the formation of ATP in the complex.

7.13 A Chemiosmotic Mechanism Produces ATP
As electrons pass through the series of carriers in the respiratory chain, protons are pumped from the mitochondrial matrix into the intermembrane space. As the protons return to the matrix through an ATP synthase, ATP forms.

protons back across the membrane, just as the charge on a battery drives the flow of electrons, discharging the battery.

The conversion of the proton-motive force into kinetic energy is prevented by the fact that the lipid bilayer of the inner membrane is impermeable to protons. However, they *can* diffuse across the membrane by passing through a specific channel protein, called **ATP synthase**, that couples proton movement to the synthesis of ATP. This coupling of

proton-motive force and ATP synthesis is called the **chemiosmotic mechanism**.

THE CHEMIOSMOTIC MECHANISM COUPLES ELECTRON TRANSPORT TO ATP SYNTHESIS. The chemiosmotic mechanism has three parts:

▶ The flow of electrons from NADH (or FADH$_2$) from one electron carrier to another in the respiratory chain is a series of exergonic reactions that occurs in the inner mitochondrial membrane.

▶ These exergonic reactions drive the endergonic pumping of H^+ out of the mitochondrial matrix and across the inner membrane into the intermembrane space. This pumping forms a H^+ gradient.

▶ The potential energy of the H^+ gradient, or proton-motive force, is harnessed by ATP synthase. This protein has two roles: It acts as a channel allowing the H^+ to diffuse back into the matrix, and it uses the energy of that diffusion to make ATP from ADP and P_i.

ATP synthesis is a reversible reaction, and ATP synthase can also act as an ATPase, hydrolyzing ATP to ADP and P_i:

$$ATP \rightleftharpoons ADP + P_i + \text{free energy}$$

If the reaction goes to the right, free energy is released, and is used to pump H^+ out of the mitochondrial matrix. If the reaction goes to the left, it uses free energy from H^+ diffusion into the matrix to make ATP. What makes it prefer ATP synthesis?

There are two answers to this question. First, ATP is removed from the mitochondrial matrix as soon as it is made, keeping the ATP concentration in the matrix low and driving the reaction toward the left. A person hydrolyzes about 10^{25} ATP molecules per day, and clearly the vast majority are recycled. Second, the H^+ gradient is constantly replenished by electron pumping. (The electrons, you recall, come from the oxidation of NADH, which itself gets reduced by the oxidations of glycolysis and the citric acid cycle. So, one reason for eating food is to replenish the H^+ gradient!)

ATP synthase is a large multi-protein machine, having 16 different polypeptides in mammals. It has two visible and functional parts. One is the membrane channel for H^+.

The other, which sticks out into the mitochondrial matrix like a lollipop (see Figure 7.13), is the actual ATP synthesis (or ATP hydrolysis) active site. The actual mechanism of energy transduction involves the physical rotation of the core of the enzyme, with this rotational energy transferred to ATP.

TWO EXPERIMENTS DEMONSTRATE THE CHEMIOSMOTIC MECHANISM. Two key experiments have shown that (1) a proton (H^+) gradient and proton-motive force across a membrane can drive ATP synthesis; and (2) the enzyme ATP synthase is the catalyst for it.

Experiment 1 in Figure 7.14 "fooled" mitochondria into making ATP by raising the H^+ concentration in their environment. A sample of isolated mitochondria with a low H^+ concentration was suddenly put in an acidic medium with a high concentration of H^+. The outer mitochondrial membrane, unlike the inner one, is freely permeable to H^+, so H^+ rapidly diffused into the intermembrane space. This created an artificial gradient across the inner membrane, which the mitochondrion used to make ATP from ADP and P_i.

In Experiment 2, a proton pump from a bacterium was added to artificial membrane vesicles. It proceeded to

7.14 Two Experiments Demonstrate the Chemiosmotic Mechanism

These two experiments show that an H^+ gradient across a membrane is all that is needed to drive the synthesis of ATP by the enzyme ATP synthase. Whether the H^+ gradient is produced by the electron transport chain found in nature, or artificially as in these experiments, does not matter.

EXPERIMENT 1
Question: Can an H^+ gradient drive ATP synthesis by isolated mitochondria?

1 Mitochondria are isolated from cells and placed in a medium at pH 8. This results in a low H^+ concentration both outside and inside the organelles.

pH 8

Mitochondrion

pH 8

2 The mitochondria are moved to an acidic medium (pH 4; high H^+ concentration).

3 H^+ movement into mitochondria drives the synthesis of ATP in the absence of continuous electron transport.

pH 4

pH 4

ADP + P_i ATP

pH 8

pH 8

Conclusion: In the absence of electron transport, an artificial H^+ gradient is sufficient for ATP synthesis by mitochondria.

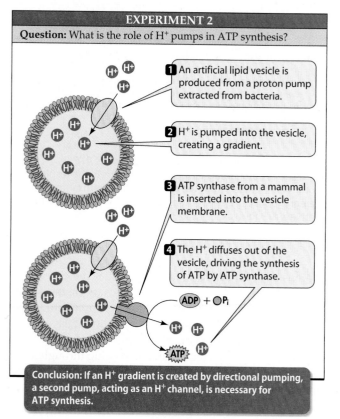

EXPERIMENT 2
Question: What is the role of H^+ pumps in ATP synthesis?

1 An artificial lipid vesicle is produced from a proton pump extracted from bacteria.

2 H^+ is pumped into the vesicle, creating a gradient.

3 ATP synthase from a mammal is inserted into the vesicle membrane.

4 The H^+ diffuses out of the vesicle, driving the synthesis of ATP by ATP synthase.

ADP + P_i

ATP

Conclusion: If an H^+ gradient is created by directional pumping, a second pump, acting as an H^+ channel, is necessary for ATP synthesis.

pump H⁺ into the vesicles, creating a gradient. If mammalian ATP synthase was then put into the membranes of these vesicles, it made ATP even in the absence of the usual electron carriers.

These experiments show that *the key to ATP synthesis is the H⁺ gradient*; it does not matter whether this gradient is produced naturally by the electron transport chain, or artificially by an experimenter.

PROTON DIFFUSION CAN BE UNCOUPLED FROM ATP PRODUCTION. For the chemiosmotic mechanism to work, the diffusion of H⁺ and the formation of ATP must be tightly *coupled*; that is, the protons must pass through the ATP synthase channel in order to move inward. If a simple H⁺ diffusion channel (*not* ATP synthase) is inserted into the membrane, the energy of the H⁺ gradient is released as heat, rather than being coupled to the synthesis of ATP. Such uncoupling molecules are deliberately used by some organisms to generate heat. For example, uncoupling the protein *thermogenin* plays an important role in regulating the temperature of some mammals, especially newborn human infants, who lack the hair to keep warm, and of hibernating animals. We will describe this process in more detail in Chapter 40.

Fermentation: ATP from Glucose, without O₂

Suppose the supply of oxygen to a respiring cell is cut off (an anaerobic condition), perhaps by drowning or by extreme exertion. As we can deduce from Figure 7.13, the first consequence of an insufficient supply of O₂ is that the cell cannot reoxidize cytochrome *c*, so all of that compound is soon in the reduced form. When this happens, QH₂ cannot be oxidized back to Q, and soon all the Q is in the reduced form. So it goes, until the entire respiratory chain is reduced. Under these circumstances, no NAD⁺ and no FAD are generated from their reduced forms. Therefore, the oxidative steps in glycolysis, pyruvate oxidation, and the citric acid cycle also stop. If the cell has no other way to obtain energy from its food, it will die.

Under anaerobic conditions, many (but not all) cells can continue to carry out glycolysis and produce a limited amount of ATP by **fermentation**. This process occurs in the cytoplasm with glycolysis.

Fermentation has two defining characteristics. First, it uses NADH + H⁺ formed by glycolysis to reduce pyruvate or one of its metabolites, and consequently NAD⁺ is regenerated. NAD⁺ is required for reaction 6 of glycolysis (see Figure 7.6), so once the cell has replenished its supply in this way, it can carry more glucose through glycolysis.

Second, fermentation enables glycolysis to produce a small but sustained amount of ATP. Only as much ATP is produced as can be obtained from substrate-level phosphorylation—not the much greater yield of ATP obtained by cellular respiration.

When cells capable of fermentation become anaerobic, the rate of glycolysis speeds up tenfold or even more. Thus a substantial rate of ATP production is maintained, although

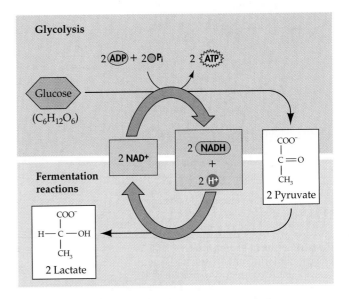

7.15 Lactic Acid Fermentation
Glycolysis produces pyruvate, as well as ATP and NADH + H⁺, from glucose. Lactic acid fermentation, using NADH + H⁺ as the reducing agent, then reduces pyruvate to lactic acid (lactate).

efficiency in terms of ATP molecules per glucose molecule is greatly reduced as compared with aerobic respiration.

Some organisms are confined to totally anaerobic environments and use only fermentation. Other organisms carry on fermentation even in the presence of oxygen. And several bacteria carry on *cellular respiration*—not fermentation—*without using oxygen gas as an electron acceptor*. Instead, to oxidize their cytochromes, these bacteria reduce nitrate ions (NO₃⁻) to nitrite ions (NO₂⁻). We'll put that observation into a broader context in Chapter 26.

Some fermenting cells produce lactic acid and others produce alcohol

Many different types of fermentation are carried out by different bacteria and eukaryotic body cells. These different fermentations are distinguished by the final product produced. For example, in **lactic acid fermentation**, pyruvate is reduced to lactate (Figure 7.15). Lactic acid fermentation takes place in many microorganisms as well as in our muscle cells. Unlike muscle cells, however, nerve cells (neurons) are incapable of fermentation because they lack the enzyme that reduces pyruvate to lactate. For this reason, without adequate oxygen, the human nervous system (including the brain) is rapidly destroyed; it is the first part of the body to die.

Certain yeasts and some plant cells carry on a process called **alcoholic fermentation** under anaerobic conditions (Figure 7.16). This process requires two enzymes to metabolize pyruvate. First, carbon dioxide is removed from pyruvate, leaving the compound *acetaldehyde*. Second, the acetaldehyde is reduced by NADH + H⁺, producing NAD⁺ and *ethyl alcohol* (ethanol). The brewing industry relies on alcoholic fermentation to produce wine and beer.

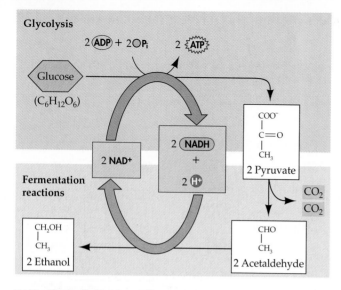

7.16 Alcoholic Fermentation

In alcoholic fermentation (the basis for the brewing industry), pyruvate from glycolysis is converted to acetaldehyde and CO_2 is released. The NADH + H⁺ from glycolysis acts as a reducing agent, reducing acetaldehyde to ethanol.

Contrasting Energy Yields

The total net energy yield from fermentation is two molecules of ATP per molecule of glucose oxidized. In contrast, the maximum yield that can be obtained from a molecule of glucose through glycolysis followed by complete aerobic respiration is much greater—about 36 molecules of ATP (Figure 7.17). (You can study Figures 7.6, 7.9, and 7.13 to review where these ATP molecules come from.)

Why is so much more ATP produced by aerobic respiration? Fermentation is an incomplete oxidation of glucose. Much more energy remains in the end products of fermentation, such as lactic acid and ethanol, than in CO_2. In cellular respiration, carriers (mostly NAD⁺) are reduced in pyruvate oxidation and the citric acid cycle, then oxidized by the respiratory chain, with the accompanying production of ATP (three for each NADH + H⁺ and two for each FADH₂) by the chemiosmotic mechanism. In an aerobic environment, an organism capable of this type of metabolism will be at an advantage (in terms of energy availability per glucose molecule) over one limited to fermentation.

The total gross yield of ATP from one molecule of glucose taken through glycolysis and cellular respiration is 38. However, we may subtract two from that gross—for a net yield of 36 ATP—because in some animal cells the inner mitochondrial mem-

brane is impermeable to NADH, and a "toll" of one ATP must be paid for each NADH produced in glycolysis that is shuttled into the mitochondrial matrix.

Metabolic Pathways

Glycolysis and the respiratory pathways do not operate in isolation from the rest of metabolism. Rather, there is an in-

7.17 Cellular Respiration Yields More Energy Than Glycolysis Does

Carriers are reduced in pyruvate oxidation and the citric acid cycle, then oxidized by the respiratory chain. These reactions produce ATP via the chemiosmotic mechanism.

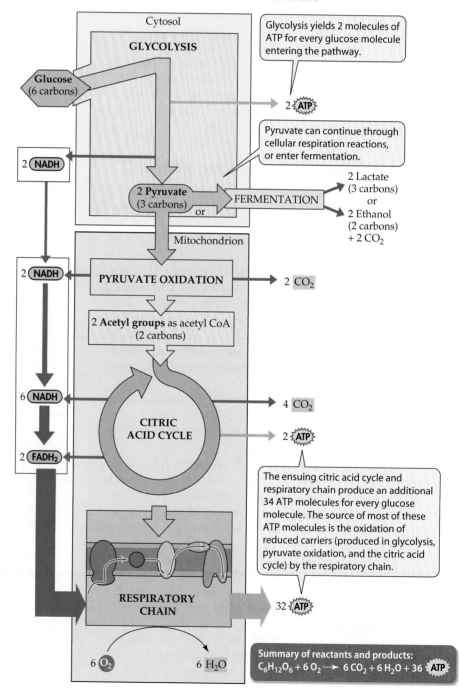

terchange, with biochemical traffic flowing both into these pathways and out of them, to and from the synthesis and breakdown of amino acids, nucleotides, fatty acids, and so forth. Indeed, the energy-harvesting pathways can be thought of, in railway terms, as the "central switching yard," where carbon skeletons enter from other molecules that are broken down to release their energy (catabolism), and carbon skeletons leave to form the major macromolecular constituents of the cell (anabolism). These relationships are summarized in Figure 7.18.

Catabolism and anabolism involve interconversions using carbon skeletons

A typical hamburger or vegiburger contains three major sources of carbon skeletons for the person who eats it: carbohydrates, mostly as starch (a polysaccharide); lipids,

mostly as triglycerides (three fatty acids attached to glycerol); and proteins (polymers of amino acids). Looking at Figure 7.18, you can see how each of these three types of macromolecules can be used in catabolism or anabolism.

CATABOLIC INTERCONVERSIONS. Polysaccharides, lipids, and proteins can all be broken down to provide energy.

▸ *Polysaccharides* are hydrolyzed to glucose phosphate, an intermediate in glycolysis. This molecule then passes through the rest of glycolysis and the citric acid cycle, where its energy is extracted in NADH and ATP.

▸ *Lipids* are converted to their substituents, glycerol and fatty acids. Glycerol is converted to dihydroxyacetone phosphate, an intermediate in glycolysis, and fatty acids to acetate and then acetyl CoA in the mitochondria. In both cases, further oxidation to CO_2 and release of energy then occur.

▸ *Proteins* are hydrolyzed to their amino acid building blocks. The 20 amino acids feed into glycolysis or the citric acid cycle at different points. A specific example is shown in Figure 7.19, in which an amino acid is converted to an intermediate in the citric acid cycle.

ANABOLIC INTERCONVERSIONS. As you can see in Figure 7.19, many of the pathways for catabolism can operate in reverse. That is, glycolytic and citric acid cycle intermediates, instead of being *oxidized* to form CO_2, can be *reduced* and used to form glucose in a process called *gluconeogenesis* (which means "new formation of glucose"). Likewise, acetyl CoA can form fatty acids. The most common fatty acids have an even number of carbons: 14, 16, 18, and so forth. These molecules are formed by adding two-carbon acetyl CoA "units" one at a time until the appropriate chain length is reached. Amino acids can be formed by reversing reactions such as the one shown in Figure 7.19, and these amino acids can then be polymerized into proteins.

Some intermediates of the citric acid cycle are used in the synthesis of various important cellular constituents. α-Ketoglutarate is a starting point for purines and oxaloacetate for pyrimidines, both constituents of the nucleic acids DNA and RNA. Succinyl CoA is a starting point for chlorophyll synthesis. Acetyl

7.18 Relationships Among the Major Metabolic Pathways of the Cell
Note the central place of glycolysis and the citric acid cycle in this network of metabolic pathways.

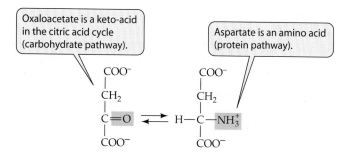

7.19 Coupling Metabolic Pathways
This reaction, in which oxaloacetate (a keto-acid) and aspartate (an amino acid) interconvert, is called a transamination.

CoA is a building block for various pigments, plant growth substances, rubber, and the steroid hormones of animals, among other functions.

Catabolism and anabolism are integrated

A carbon atom from a protein in your burger can end up in DNA or fat or CO_2, among other fates. How does the cell "decide" which metabolic pathway to follow? With all of the possible interconversions, you might expect that the cellular concentrations of various biochemical molecules would vary widely. For example, the level of oxaloacetate in your cells might depend on what you eat (some food molecules form oxaloacetate) and whether oxaloacetate is used up (in the citric acid cycle or in forming the amino acid aspartate). Remarkably, the levels of these substances in what is called the "metabolic pool" are quite constant. The cell regulates the enzymes of catabolism and anabolism so as to maintain a balance. This metabolic homeostasis gets upset only in unusual circumstances. Let's look at one of them: undernutrition.

Glucose is an excellent source of energy. From Figure 7.18, you can see that fats and proteins are also energy sources. Any one, or all three, can be used to provide the energy you need. In reality, things are not so simple. Proteins, for example, have essential roles in your body as enzymes and structural elements, and using them for energy might deprive you of a catalyst for a vital reaction.

Polysaccharides and fats have no such catalytic roles. But polysaccharides, because they are somewhat polar, can bind a lot of water. Because they are nonpolar, fats do not bind as much water as polysaccharides. So, in water, fats weigh less than polysaccharides. Also, fats are more reduced than carbohydrates (more C—H bonds as opposed to C—OH) and have more energy stored in their bonds. For these two reasons, fats are a better way for an organism to store energy than polysaccharides. It is not surprising that a typical person has about one day's worth of food energy stored as glycogen, a week's food energy as usable proteins in blood, and over a month's food energy stored as fats.

What happens if a person does not eat enough food to produce sufficient ATP and NADH for anabolism and biological activities? This situation can be the result of a deliberate decision to lose weight, but for too many people, it is forced upon them because not enough food is available. In either case, the first energy stores in the body to be used are the glycogen stores in muscle and liver cells. This doesn't last long, and next come fats.

The level of acetyl CoA rises as fatty acids are broken down. However, a problem remains: Because fatty acids cannot get from the blood to the brain, the brain can use *only* glucose as its energy source. With glucose stores already depleted, the body must convert something else to make glucose for the brain. This gluconeogenesis uses mostly amino acids, largely from the breakdown of proteins. So, without sufficient food intake, both proteins (for glucose) and fats (for energy) are used up. After several weeks of starvation, fat stores become depleted, and the only energy source left is proteins, some of which have already been degraded to supply the brain with glucose. At this point, essential proteins, such as antibodies used to fight off infections, get broken down, both for energy and for gluconeogenesis. The loss of these proteins can lead to severe illnesses.

Regulating Energy Pathways

We have described the relationships between metabolic pathways and noted that they work together to provide homeostasis in the cell and organism. But how does the cell regulate these interconversions to maintain constant metabolic pools?

Consider what happens to the starch in your burger bun. In the digestive system, starch is hydrolyzed to glucose, which enters the blood for distribution to the rest of the body. Before this happens, however, a "decision" must be made: Is there already enough glucose in the blood to supply the body's needs? If there is, the excess glucose is con-

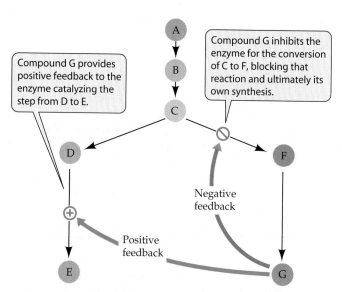

7.20 Regulation by Negative and Positive Feedback
Allosteric regulation plays an important role in metabolic pathways. Excess accumulation of some products can shut down their synthesis or stimulate the synthesis of other products.

7.21 Feedback Regulation of Glycolysis and the Citric Acid Cycle
Feedback controls glycolysis and the citric acid cycle at crucial early steps in the pathways, increasing their efficiency and preventing the excessive buildup of intermediates.

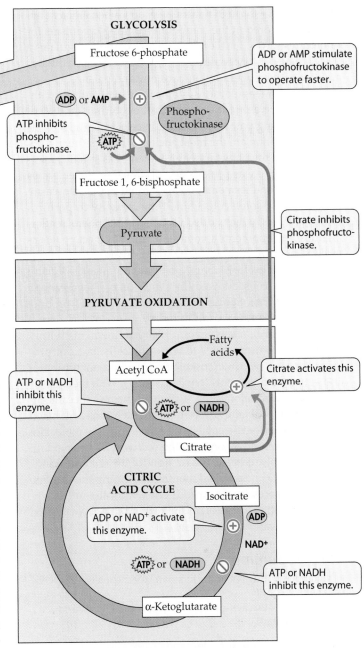

verted to stored glycogen in the liver. If not enough glucose is supplied by food, liver glycogen is broken down to supply it, or other molecules are used to make glucose by gluconeogenesis.

The end result is that the level of glucose in the blood is remarkably constant. We will describe the details of how this happens in Part Six of this book. For now, it is important to realize that the interconversions of glucose involve many steps, each catalyzed by an enzyme, and it is here that the controls often reside.

Allostery regulates metabolism

Glycolysis, the citric acid cycle, and the respiratory chain are regulated by allosteric control of the enzymes involved. In metabolic pathways, as we saw in Chapter 6, a high concentration of the products of a later reaction can suppress the action of enzymes that catalyze an earlier reaction. On the other hand, an excess of the product of one branch of a synthetic chain can speed up reactions in another branch, diverting raw materials away from synthesis of the first product (Figure 7.20). These negative and positive feedback control mechanisms are used at many points in the energy-harvesting processes, which are summarized in Figure 7.21.

The main control point in glycolysis is the enzyme *phosphofructokinase* (reaction 3 in Figure 7.6). This enzyme is allosterically inhibited by ATP and activated by ADP or AMP. As long as fermentation proceeds, yielding a relatively small amount of ATP, phosphofructokinase operates at full efficiency. But when aerobic respiration begins producing ATP 18 times faster than fermentation does, the abundant ATP allosterically inhibits the enzyme, and the conversion of fructose 6-phosphate to fructose 1,6-bisphosphate declines, as does the rate of glucose utilization.

The main control point in the citric acid cycle is the enzyme *isocitrate dehydrogenase*, which converts isocitrate to α-ketoglutarate (reaction 3 in Figure 7.9). NADH + H$^+$ and ATP are feedback inhibitors of this reaction; ADP and NAD$^+$ are activators (Figure 7.21). If too much ATP is accumulating, or if NADH + H$^+$ is being produced faster than it can be used by the respiratory chain, the conversion of isocitrate is slowed, and the citric acid cycle is essentially shut down. A shutdown of the citric acid cycle would cause large amounts of isocitrate and citrate to accumulate, except that the conversion of acetyl CoA to citrate is also slowed by abundant ATP and NADH + H$^+$.

However, a certain excess of citrate does accumulate, and this excess acts as an additional negative feedback inhibitor to slow the fructose 6-phosphate reaction early in glycolysis. Consequently, if the citric acid cycle has been slowed down

because of abundant ATP (and not because of a lack of oxygen), glycolysis is shut down as well. Both processes resume when the ATP level falls and they are needed again. Allosteric control keeps these processes in balance.

Another control point involves a method for storing excess acetyl CoA. If too much ATP is being made and the citric acid cycle shuts down, the accumulation of citrate switches acetyl CoA to the synthesis of fatty acids for storage. This is one reason why people who eat too much accumulate fat. These fatty acids may be metabolized later to produce more acetyl CoA.

Evolution has led to metabolic efficiency

Allosteric control of the sort illustrated in Figure 7.21 is one of the most impressive examples of the tight organization that can evolve through natural selection when efficient operation is favored in the competition among organisms for limited resources. Each of the feedback controls regulates a part or various parts of the energy-harvesting pathways and keeps them operating in harmony and balance.

Not just the regulatory systems have evolved; the pathways themselves are the products of evolution. Of the energy-harvesting pathways present in cells today, glycolysis and fermentation are the most ancient. These pathways appeared when the planetary environment was strictly anaerobic and all life was prokaryotic. To this day, the enzymes of glycolysis and fermentation are located in the cytoplasm.

Eventually some cells gained the capacity to perform photosynthesis, which added O_2 to the atmosphere, rendering most environments aerobic. Evolution in the aerobic environment led to the appearance of pyruvate oxidation and the citric acid cycle. Elaboration of membranes, especially in eukaryotic cells, allowed the evolution of chemiosmotic mechanisms for coupling electron transport to ATP production, as in oxidative phosphorylation.

There have also been evolutionary refinements within the pathways themselves. Eukaryotic cells, but not prokaryotic ones, have a cytoskeleton based on microtubules and actin microfilaments (see Figure 4.21). With the appearance of a cytoskeleton, some glycolytic enzymes became attached to cytoskeletal components, which thus organized the enzymes into efficient associations that allow molecules to move from one enzyme in the pathway to the next. In eukaryotes, hexokinase, the first glycolytic enzyme, binds to the outer mitochondrial membrane, giving the enzyme immediate access to ATP produced within the mitochondria. In metabolism, as in all the rest of biology, evolution leads to adaptation.

Chapter Summary

▶ Metabolic pathways occur in small steps, each catalyzed by a specific enzyme.

▶ Metabolic pathways are often compartmentalized and are highly regulated.

Obtaining Energy and Electrons from Glucose

▶ When glucose burns, energy is released as heat and light:

$$C_6H_{12}O_6 + 6 O_2 \rightarrow 6 CO_2 + 6 H_2O + energy$$

The same equation applies to the metabolism of glucose by cells, but the reaction is accomplished in many separate steps so that the energy can be captured as ATP. **Review Figure 7.1**

▶ As a material is oxidized, the electrons it loses are transferred to another material, which is thereby reduced. Such redox reactions transfer large amounts of energy. Much of the energy liberated by the oxidation of the reducing agent is captured in the reduction of the oxidizing agent. **Review Figure 7.2**

▶ The coenzyme NAD is a key electron carrier in biological redox reactions. It exists in two forms, one oxidized (NAD^+) and the other reduced ($NADH + H^+$). **Review Figures 7.3, 7.4**

An Overview: Releasing Energy from Glucose

▶ Glycolysis operates in the presence or absence of O_2. Under aerobic conditions, cellular respiration continues the breakdown process. **Review Figure 7.5**

▶ Pyruvate oxidation and the citric acid cycle produce CO_2 and hydrogen atoms carried by NADH and $FADH_2$. The respiratory chain combines these hydrogens with O_2, releasing enough energy for the synthesis of ATP. **Review Figure 7.5**

▶ In some cells under anaerobic conditions, pyruvate can be reduced by NADH to form lactate and regenerate the NAD^+ needed to sustain glycolysis. **Review Figure 7.5**

▶ In eukaryotes, glycolysis and fermentation take place in the cytoplasm outside of the mitochondria; pyruvate oxidation, the citric acid cycle, and the respiratory chain operate in association with mitochondria. In prokaryotes, glycolysis, fermentation, and the citric acid cycle take place in the cytoplasm; and pyruvate oxidation and the respiratory chain operate in association with the plasma membrane. **Review Table 7.1**

Glycolysis: From Glucose to Pyruvate

▶ Glycolysis is a pathway of ten enzyme-catalyzed reactions located in the cytoplasm. Glycolysis provides starting materials for both cellular respiration and fermentation. **Review Figure 7.7**

▶ The energy-investing reactions of glycolysis use two ATPs per glucose molecule and eventually yield two glyceraldehyde 3-phosphate molecules. In the energy-harvesting reactions, two NADH molecules are produced, and four ATP molecules are generated by substrate-level phosphorylation. Two pyruvates are produced for each glucose molecule. **Review Figures 7.6, 7.7**

Pyruvate Oxidation

▶ The pyruvate dehydrogenase complex catalyzes three reactions: (1) Pyruvate is oxidized to the acetyl group, releasing one CO_2 molecule and considerable energy; (2) some of this energy is captured when NAD^+ is reduced to $NADH + H^+$; and (3) the remaining energy is captured when the acetyl group is combined with coenzyme A, yielding acetyl CoA. **Review Figure 7.8**

The Citric Acid Cycle

▶ The energy in acetyl CoA drives the reaction of acetate with oxaloacetate to produce citrate. The citric acid cycle is a series of reactions in which citrate is oxidized and oxaloacetate regenerated (hence a "cycle"). It produces two CO_2, one $FADH_2$, three NADH, and one ATP for each acetyl CoA. **Review Figures 7.9, 7.10**

The Respiratory Chain: Electrons, Proton Pumping, and ATP

▶ $NADH + H^+$ and $FADH_2$ from glycolysis, pyruvate oxidation, and the citric acid cycle are oxidized by the respiratory chain, regenerating NAD^+ and FAD. Most of the enzymes and other electron carriers of the chain are part of the inner mitochondrial membrane. Oxygen (O_2) is the final acceptor of electrons and protons, forming water (H_2O). **Review Figures 7.11, 7.12**

▶ The chemiosmotic mechanism couples proton transport to oxidative phosphorylation. As the electrons move along the respiratory chain, they lose energy, which is captured by proton pumps that actively transport H^+ out of the mitochondrial matrix, establishing a gradient of both proton concentration and electric charge—the proton-motive force. **Review Figure 7.13**

▶ The proton-motive force causes protons to diffuse back into the mitochondrial interior through the membrane channel protein ATP synthase, which couples that diffusion to the production of ATP. Several key experiments demonstrate that chemiosmosis produces ATP. **Review Figure 7.14**

Fermentation: ATP from Glucose, without O_2

▶ Many organisms and some cells live without O_2, deriving all their energy from glycolysis and fermentation. Together, these pathways partly oxidize glucose and generate energy-containing products such as lactic acid or ethanol. Fermentation reactions anaerobically oxidize the NADH + H^+ produced in glycolysis. **Review Figures 7.15, 7.16**

Contrasting Energy Yields

▶ For each molecule of glucose used, fermentation yields 2 molecules of ATP. In contrast, glycolysis operating with pyruvate oxidation, the citric acid cycle, and the respiratory chain yields up to 36 molecules of ATP per molecule of glucose. **Review Figure 7.17**

Metabolic Pathways

▶ Catabolic pathways feed into the respiratory pathways. Polysaccharides are broken down into glucose, which enters glycolysis. Glycerol from fats also enters glycolysis, and acetyl CoA from fatty acid degradation enters the citric acid cycle. Proteins enter glycolysis and the citric acid cycle via amino acids. **Review Figures 7.18, 7.19**

▶ Anabolic pathways use intermediate components of respiratory metabolism to synthesize fats, amino acids, and other essential building blocks for cellular structure and function. **Review Figures 7.18, 7.19**

Regulating Energy Pathways

▶ The rates of glycolysis and the citric acid cycle are increased or decreased by the actions of ATP, ADP, NAD^+, or NADH + H^+ on allosteric enzymes.

▶ Inhibition of the glycolytic enzyme phosphofructokinase by abundant ATP from oxidative phosphorylation slows down glycolysis. ADP activates this enzyme, speeding up glycolysis. The citric acid cycle enzyme isocitrate dehydrogenase is inhibited by ATP and NADH and activated by ADP and NAD^+. **Review Figures 7.20, 7.21**

For Discussion

1. Trace the sequence of chemical changes that occurs in mammalian brain tissue when the oxygen supply is cut off. (The first change is that the cytochrome *c* oxidase system becomes totally reduced, because electrons can still flow from cytochrome *c* but there is no oxygen to accept electrons from cytochrome *c* oxidase. What are the remaining steps?)

2. Trace the sequence of chemical changes that occurs in mammalian *muscle* tissue when the oxygen supply is cut off. (The first change is exactly the same as that in Question 1.)

3. Some cells that use the citric acid cycle and the respiratory chain can also thrive by using fermentation under anaerobic conditions. Given the lower yield of ATP (per molecule of glucose) in fermentation, why can these cells function so efficiently under anaerobic conditions?

4. Describe the mechanisms by which the rates of glycolysis and of aerobic respiration are kept in balance with one another.

8 _Photosynthesis: Energy from the Sun_

FOR SEVERAL DECADES, CORN GROWERS IN THE United States competed to see who could coax the highest yield of grain from their acreage. After rising rapidly in the first half of the twentieth century, yields continued to increase, albeit somewhat more slowly. But the trend was clearly up—until the last decade of the century. From 1990 on, crop yields per acre have leveled off for corn, rice, and wheat—three grains which together supply over half the human race with food.

Although overall food production continues to rise as more land is put into production and the environment of the crops is more intensively manipulated with fertilizers and pesticides, the increase of the human population is wiping out any per capita gains. Per person food production has not improved much since 1960–1980, the peak of the so-called "Green Revolution" in agriculture. This was a period when new genetic strains and more intensive environmental management combined to more than double crop yields.

To coax crop plants to grow more and produce more on the available land, scientists are now focusing on photosynthesis, the biochemical process by which plants turn sunlight into carbohydrates, sugars, and starch. Photosynthesis is the very basis of life on Earth.

The basic transformation of photosynthesis—the conversion of solar energy into chemical energy—is a familiar example of the laws of thermodynamics. As the first law tells us, when the form of energy changes from sunlight to plant, no energy is lost. However, as the second law states, the conversion is relatively inefficient, with only about 4 percent of the incident solar energy ending up in chemical bonds. Moreover, the use of solar energy initially captured as ATP and reduced electron carriers to reduce carbon dioxide to sugars is also inefficient.

How can these efficiencies be improved? An important first step is a thorough understanding of photosynthesis. The process of photosynthesis can be neatly broken down into two steps. The first step is the conversion of energy from light to chemical bonds in reduced electron carriers and ATP. In the second step, these two sources of chemical energy are used to drive the synthesis of carbohydrates from carbon dioxide. In this chapter, we will examine these two processes, and show how they are related to each other and to plant growth.

Identifying Photosynthetic Reactants and Products

By the beginning of the nineteenth century, scientists understood the broad outlines of photosynthesis. It was known to use three principal ingredients—water, carbon dioxide (CO_2), and light—and to produce not only carbohydrate but also oxygen gas (O_2). Scientists had learned that:

▶ The water for photosynthesis in land plants comes primarily from the soil and must travel from the roots to the leaves.
▶ Carbon dioxide is taken in, and water and O_2 are released, through tiny openings in leaves, called *stomata* (singular *stoma*) (Figure 8.1)
▶ Light is absolutely necessary for the production of oxygen and carbohydrate.

Primary Producers
Powered by sunlight, corn plants (*Zea mays*) convert atmospheric CO_2 and water into an energy source (food) for humans and animals.

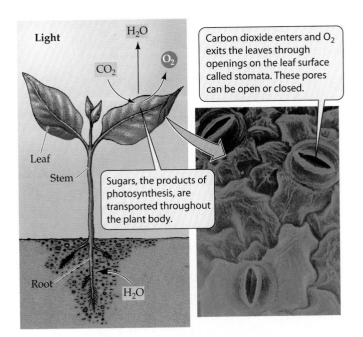

8.1 Ingredients for Photosynthesis
A typical terrestrial plant uses light from the sun, water from the soil, and carbon dioxide from the atmosphere to form organic compounds by photosynthesis.

By 1804, scientists could summarize photosynthesis as follows:

carbon dioxide + water + light energy → sugar + oxygen

which turns into an equation that is the *reverse* of the overall equation for cellular respiration given in Chapter 7:

$$6\ CO_2 + 6\ H_2O \rightarrow C_6H_{12}O_6 + 6\ O_2$$

Although correct, these statements say nothing about the details of the process. What roles does light play? How do the carbons become linked? And does the oxygen gas come from the CO_2 or from the H_2O?

Almost a century and a half passed before the source of the O_2 released during photosynthesis was determined. Its identification was one of the first uses of an isotopic tracer in biological research. In these experiments, two groups of green plants were allowed to carry on photosynthesis. Plants in the first group were supplied with water containing the heavy-oxygen isotope ^{18}O and with CO_2 containing only the common oxygen isotope ^{16}O; plants in the second group were supplied with CO_2 labeled with ^{18}O and water containing only ^{16}O.

When oxygen gas was collected from each group of plants and analyzed, it was found that O_2 containing ^{18}O was produced in abundance by the plants that had been given ^{18}O-labeled water, but not by the plants given labeled CO_2. These results showed that all the oxygen gas produced during photosynthesis comes from water (Figure 8.2). This discovery is reflected in a revised balanced equation:

$$6\ CO_2 + 12\ H_2O \rightarrow C_6H_{12}O_6 + 6\ O_2 + 6\ H_2O$$

Water appears on both sides of the equation because water is both used as a reactant (the twelve molecules on the left) and released as a product (the six new ones on the right). In this revised equation, there are now sufficient water molecules to account for all the oxygen gas produced.

The photosynthetic production of oxygen by green plants is an important source of atmospheric oxygen, which most organisms—including plants themselves—require in order to complete their respiratory chains and obtain the energy for life.

The Two Pathways of Photosynthesis: An Overview

The overall photosynthetic reaction takes place in the chloroplasts of photosynthetic cells, which in most plants are found in the leaves. But photosynthesis does not proceed in a single step. In fact, in all of chemistry, no such complex reaction is accomplished in a single step. Rather, a series of simpler steps is required.

By the middle of the twentieth century, it was clear that photosynthesis consists of many reactions that can be divided into two pathways:

▶ The first pathway, called the *light reactions*, is driven by light energy. It produces ATP and a reduced electron carrier (NADPH + H⁺).

▶ The second pathway, called the *Calvin–Benson cycle*, does not use light directly. It uses ATP, NADPH + H⁺, and CO_2 to produce sugar.

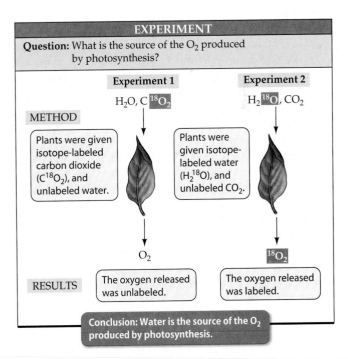

8.2 Water Is the Source of the Oxygen Produced by Photosynthesis
Because only plants given isotope-labeled water released labeled O_2, this experiment showed that water is the source of the oxygen released during photosynthesis.

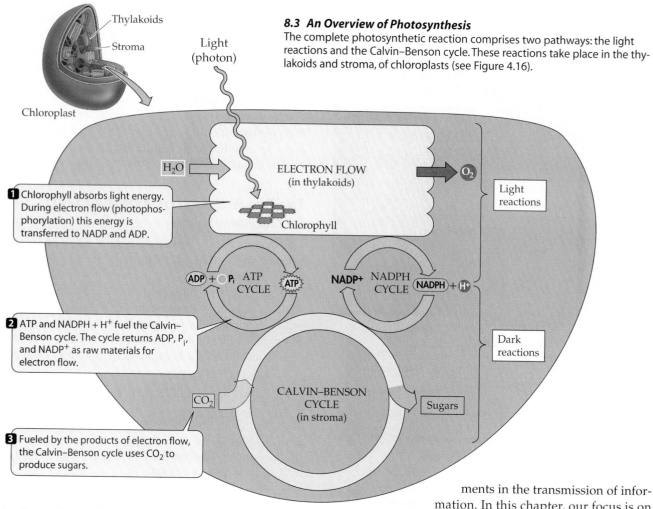

8.3 An Overview of Photosynthesis
The complete photosynthetic reaction comprises two pathways: the light reactions and the Calvin–Benson cycle. These reactions take place in the thylakoids and stroma, of chloroplasts (see Figure 4.16).

1 Chlorophyll absorbs light energy. During electron flow (photophosphorylation) this energy is transferred to NADP and ADP.

2 ATP and NADPH + H$^+$ fuel the Calvin–Benson cycle. The cycle returns ADP, P$_i$, and NADP$^+$ as raw materials for electron flow.

3 Fueled by the products of electron flow, the Calvin–Benson cycle uses CO_2 to produce sugars.

In the first pathway of photosynthesis—the **light reactions**—light energy is captured by pigment molecules and used to produce ATP from ADP and P$_i$. The the light reactions are mediated by molecular assemblies called **photosystems**. These systems pass electrons from one molecule to another, and some of this electron flow is coupled to ATP synthesis. Because light is the ultimate energy source, the synthesis of ATP in this pathway is called *photophosphorylation*.

The NADPH + H$^+$ and ATP produced by the light reactions are used in the second pathway, the **Calvin–Benson cycle**, whose reactions trap CO_2 and reduce the resulting acid to sugar. This pathway is also known as the photosynthetic carbon reduction cycle, or simply the **dark reactions** (because none of its reactions uses light directly).

The reactions of both pathways proceed within the chloroplast, but they reside in different parts of that organelle (Figure 8.3). Both pathways stop in the dark because ATP synthesis and NADP$^+$ reduction require light. The rate of each set of reactions depends on the rate of the other. They are linked by the exchange of ATP and ADP, and of NADP$^+$ and NADPH.

Properties of Light and Pigments

Light is a source of both energy and information. In later chapters, we'll examine the many roles of light and pig-

ments in the transmission of information. In this chapter, our focus is on light as a source of energy. We start by examining the physical nature of light.

Light comes in packets called photons

Light is a form of electromagnetic radiation. It comes in discrete packets called **photons**. Light also behaves as if it were propagated in waves. The **wavelength** of light is the distance from the peak of one wave to the peak of the next (Figure 8.4). Light and other forms of electromagnetic radi-

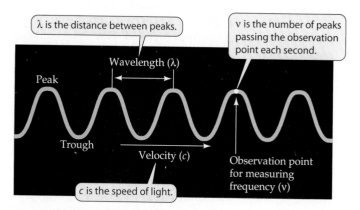

λ is the distance between peaks.

ν is the number of peaks passing the observation point each second.

Wavelength (λ)

Peak

Trough

Velocity (c)

Observation point for measuring frequency (ν)

c is the speed of light.

8.4 Light Has Wavelike Properties
Light can be envisioned as a series of waves whose peaks pass a fixed observation point with uniform frequency.

ation—cosmic rays, gamma rays, X rays, ultraviolet radiation, infrared radiation, microwaves, and radio waves—can be classified according to their wavelengths. Visible light fits into this *electromagnetic spectrum* between ultraviolet and infrared radiation (Figure 8.5).

Humans perceive light as having distinct colors. The colors relate to the wavelengths of the light, as shown in Figure 8.5. Most people can see electromagnetic radiation in the range of wavelengths from 400 to 700 nm. The wavelength at 400 nm marks the violet end of the visible spectrum; the one at 700 nm marks the red end. Wavelengths in the range from about 100 to 400 nm are *ultraviolet radiation*; those immediately above 700 nm are referred to as *infrared*.

The speed of light in a vacuum is one of the universal constants of nature. In a vacuum, light travels at 3×10^{10} centimeters per second (or 186,000 miles per second), a value symbolized as *c*. In air, glass, water, and other media, light travels slightly more slowly.

Let's consider light as a long train of waves moving in a straight line and see what the train would look like to a stationary observer. Successive peaks of the waves pass the observer with a uniform **frequency** determined by the wavelength and the speed of light. The exact relationship is

$$\nu = c/\lambda$$

where ν (the Greek letter nu) is the frequency; *c* is the speed of light; and λ (Greek lambda) is the wavelength. Often ν is expressed in hertz (Hz), *c* in centimeters per second (cm/s), and λ in nanometers (nm) (1 nm = 10^{-9} m or 10^{-7} cm).

The amount of energy, *E*, contained in a single photon is directly proportional to its frequency. The constant of proportionality that describes this relationship, *h*, is named *Planck's constant*, after Max Planck, who first introduced the concept of the photon. With this information we can write the equation

$$E = h\nu$$

where ν is the frequency in Hz.

Substituting c/λ for ν (from the equation above relating λ, ν, and *c*), we see that

$$E = hc/\nu$$

Thus shorter wavelengths mean greater energies; that is, energy is inversely proportional to wavelength. A photon of red light of wavelength 660 nm has less energy than a photon of blue light of 430 nm; an ultraviolet photon of 284 nm is much more energetic than either of these. For a photon to be active in any light-driven biological process—such as photosynthesis—it must have enough energy to perform the work required.

The brightness, or **intensity**, of light at a given point is the amount of energy falling on a defined area—such as 1 cm^2—per second. Light intensity is usually expressed in energy units (such as calories) per square centimeter per second, but the intensity of pure light of a single wavelength may also be expressed in terms of photons per square centimeter per second.

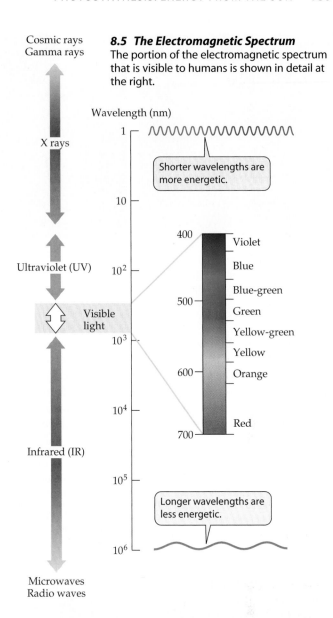

8.5 The Electromagnetic Spectrum
The portion of the electromagnetic spectrum that is visible to humans is shown in detail at the right.

Absorption of a photon puts a pigment in an excited state

When a photon meets a molecule, one of three things happens:

▶ The photon may bounce off the molecule—it may be *reflected*.
▶ The photon may pass through the molecule—it may be *transmitted*.

Neither of these outcomes causes any change in the molecule, and neither has any chemical consequences.

▶ The photon may be *absorbed* by the molecule.

In this case, the photon disappears. Its energy, however, cannot disappear, because energy is neither created nor destroyed.

When a molecule absorbs a photon, it acquires the energy of that photon. It is thereby raised from a *ground state* (lower energy) to an *excited state* (higher energy) (Figure

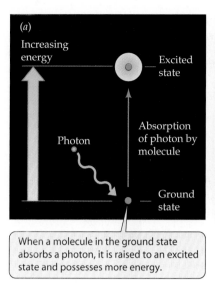

(a)

Increasing energy

Excited state

Photon

Absorption of photon by molecule

Ground state

When a molecule in the ground state absorbs a photon, it is raised to an excited state and possesses more energy.

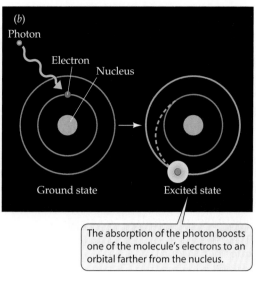

(b)

Photon

Electron

Nucleus

Ground state

Excited state

The absorption of the photon boosts one of the molecule's electrons to an orbital farther from the nucleus.

8.6 Exciting a Molecule
(a) When a molecule absorbs the energy of a photon, it is raised from a ground state to an excited state. (b) In the excited state, one of the molecule's electrons is boosted to a higher orbital, where it is held less firmly by the molecule.

8.6a). The difference in energy between the excited state and the ground state is precisely equal to the energy of the absorbed photon. The increase in energy boosts one of the electrons in the molecule into an orbital farther from the nucleus; this electron is now held less firmly by the molecule (Figure 8.6b), with chemical consequences that we will discuss later in this chapter.

All molecules absorb electromagnetic radiation. The specific wavelengths absorbed by a particular molecule are characteristic of that type of molecule. Molecules that absorb wavelengths in the visible region of the spectrum are called **pigments**.

When a beam of white light (light containing visible light of all wavelengths) falls on a pigment, certain wavelengths of the light are absorbed. The remaining wavelengths, which are reflected or transmitted, make the pigment appear to us to be colored. For example, if a pigment absorbs both blue and red light—as chlorophyll does—what we see is the remaining light—primarily green.

Light absorption and biological activity vary with wavelength

A given type of molecule can absorb radiant energy of only certain wavelengths. If we plot a compound's absorption of light as a function of the wavelengths of the light, the result is an *absorption spectrum* (Figure 8.7). Absorption spectra are good "fingerprints" of compounds; sometimes an absorption spectrum contains enough information to enable us to identify an unknown compound.

Light can also be analyzed for the magnitude of its effect on a particular

— Visible spectrum —

Notice how much of the visible spectrum would go to waste if chlorophyll *a* were the only pigment absorbing light for photosynthesis.

Chlorophyll *b*

Phycoerythrin

Phycocyanin

β-Carotene

Chlorophyll *a*

Absorption

| 250 | 300 | 350 | 400 | 450 | 500 | 550 | 600 | 650 | 700 |
| | UV | | Violet | Blue | Green | | Yellow | Red | IR |

Wavelength (nm)

8.7 Photosynthetic Pigments Have Distinct Absorption Spectra
Photosynthesis uses most of the visible spectrum because the participating pigments absorb photons most strongly at different wavelengths.

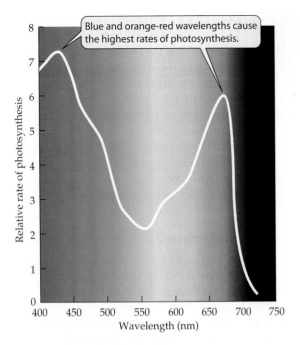

8.8 Action Spectrum of Photosynthesis
An action spectrum plots the biological effectiveness of different wavelengths of radiation. Here the rate of photosynthesis in the freshwater plant *Anacharis* is plotted against wavelengths of visible light. If we compare this action spectrum with the absorption spectra of specific pigments, such as those in Figure 8.7, we can identify which pigments are responsible for the process in *Anacharis*.

activity, such as photosynthesis. A plot of the effectiveness of light as a function of wavelength is called an *action spectrum*. Figure 8.8 shows the action spectrum for photosynthesis by *Anacharis*, a freshwater plant. All wavelengths of visible light are at least somewhat effective in causing photosynthesis, but the blue and orange-red wavelengths are the most effective. Action spectra are helpful in determining what pigment or pigments are being used in a particular photobiological process, such as photosynthesis. We should be able to find which pigment or pigments have absorption spectra that match the action spectrum of the process we are observing.

Photosynthesis uses chlorophylls and accessory pigments

Certain pigments are important in biological processes, and we will discuss them as they appear in this book. Here we discuss the pigments that play roles in photosynthesis. Of these, the most important ones are the **chlorophylls**. Chlorophylls occur universally in the plant kingdom, in photosynthetic protists, and in photosynthetic bacteria. A mutant individual that lacks chlorophyll is unable to perform photosynthesis and will starve to death.

In plants, two chlorophylls predominate: **chlorophyll *a*** and **chlorophyll *b***. These two molecules differ only slightly in their molecular structure. Both have a complex ring

structure similar to the heme group of hemoglobin. In the center of each chlorophyll ring is a magnesium atom, and at a peripheral location on the ring is attached a long hydrocarbon "tail" that can adhere the chlorophyll to the hydrophobic portion of the thylakoid membrane (Figure 8.9).

We saw in Figures 8.7 and 8.8 that the chlorophylls absorb blue and red wavelengths, which are near the two ends of the visible spectrum. Thus, if only chlorophyll pigments were active in photosynthesis, much of the visible spectrum would go unused. However, all photosynthetic organisms possess *accessory pigments*, which absorb photons intermediate in energy between the red and the blue wavelengths, then transfer a portion of that energy to the chlorophylls.

Among these accessory pigments are *carotenoids*, such as β-carotene (see Figure 3.23), which absorb photons in the blue and blue-green wavelengths and appear deep yellow. The *phycobilins* (phycoerythrin and phycocyanin), which are found in red algae and in cyanobacteria (contributing to their respective colors), absorb various yellow-green, yellow, and orange wavelengths. Such accessory pigments, in collaboration with the chlorophylls, constitute an energy-absorbing antenna system covering much of the visible spectrum.

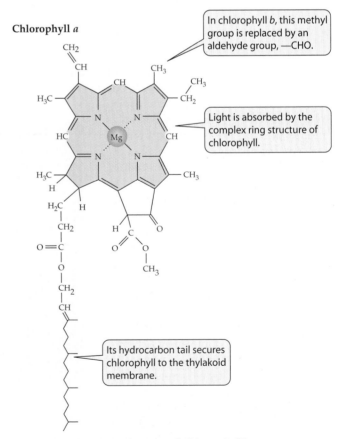

8.9 The Molecular Structure of Chlorophyll
Chlorophyll consists of a complex ring with a magnesium atom (shaded area) at the center, plus a hydrocarbon "tail."

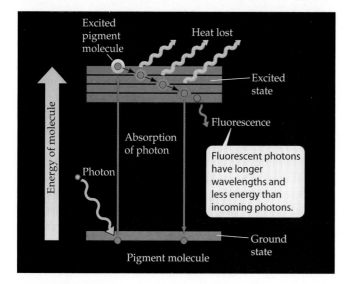

8.10 Fluorescence
An excited pigment molecule may give off some of its absorbed energy as fluorescent light when the molecule returns to its ground state.

Light Reactions: Light Absorption

A pigment molecule enters an excited state when it absorbs a photon (see Figure 8.6). The excited state is an unstable potential energy state, and the molecule usually does not stay in it very long. One of two things happens:

▶ The molecule returns to the ground state, *emitting much of the absorbed energy as fluorescence.*
▶ The molecule passes some of the absorbed energy to another pigment molecule.

In *fluorescence*, the boosted electron falls back from its higher orbital to its original, lower one (Figure 8.10). This process is accompanied by a loss of energy, which is given off as another photon. The energy of this photon, however, is somewhat less than the energy the pigment absorbed (recall the second law of thermodynamics), and so the emitted photon has a longer wavelength than the absorbed one. In any case, there can be no chemical changes or biological consequences—no chemical work is done.

However, rather than emitting the photon's energy as fluorescence, the pigment molecules may *pass the absorbed energy along.* The pigments in photosynthetic organisms are arranged into energy-absorbing *antenna systems.* In these systems, the molecules are held by a complex of proteins in the right orientation for light absorption. Any pigment molecule with a suitable absorption spectrum can absorb an incoming photon and become excited. The excitation passes from one pigment molecule in the antenna to another, moving from pigments that absorb shorter wavelengths (higher energies) to pigments that absorb longer wavelengths (lower energies) of light. Thus the excitation ends up in the one pigment molecule in the antenna that absorbs the longest wavelength; this molecule occupies the *reaction center* of the antenna (Figure 8.11).

The reaction center is the part of the antenna that converts the light absorbed into chemical energy. In plants, the pigment molecule in the reaction center is always a molecule of chlorophyll *a*. There are many other chlorophyll *a* molecules in the antenna, but all of them absorb light at shorter wavelengths than does the molecule in the reaction center.

Excited chlorophyll acts as a reducing agent

The light energy absorbed by the antenna system is transferred from one pigment molecule to another as an electron. When this happens, the second molecule is reduced by the first. Recall that reduction is the addition of electrons and oxidation is their removal. Ultimately, however, photosynthesis conserves energy by using the excited chlorophyll molecule in the reaction center as a reducing agent (Figure 8.11).

Ground-state chlorophyll (symbolized as Chl) is not much of a reducing agent, but excited chlorophyll (Chl*) is a good one. To understand the reducing capability of Chl*, recall that in an excited molecule, one of the electrons is zipping about in an orbital farther away from its nucleus. Less tightly held, this electron can be passed on in a redox reaction to an oxidizing agent. Thus Chl* (but not Chl) can react with an oxidizing agent A in a reaction like this:

$$Chl^* + A \rightarrow Chl^+ + A^-$$

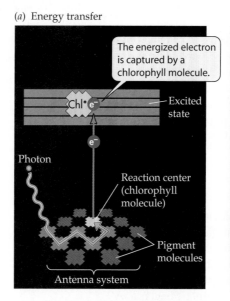
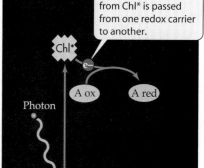

8.11 Energy Transfer and Electron Flow
Rather than being lost as fluorescence, energy may be transferred from one molecule to another, preserving the energy for biochemical work. (*a*) An excited molecule can transfer energy to a chlorophyll molecule in the reaction center. (*b*) In electron flow, the energetic electron from the excited chlorophyll molecule is transferred from one redox carrier to another.

This, then, is the first biochemical consequence of light absorption by chlorophyll: The chlorophyll becomes a reducing agent and participates in a redox reaction that would not have occurred in the dark.

As we are about to see, the further adventures of the electrons from chlorophyll reduce NADP$^+$ and generate a proton-motive force that is eventually used to synthesize ATP.

Electron Flow, Photophosphorylation and Reductions

The high energy stored in the electrons of excited chlorophyll can be transferred to suitably oxidized nonpigment acceptor molecules. When these molecules are reduced, they can in turn reduce other molecules, setting up an **electron flow**. Electrons can flow through a series of carriers where the reduced form of one has more energy than the reduced form of the next; this system is similar to electron transport in the mitochondria (Chapter 7), and releases the energy that is captured by the chemiosmotic synthesis of ATP in a process called **photophosphorylation**. ATP is used

in the dark reactions as a source of energy for the endergonic synthesis of carbohydrate (see Figure 8.3).

A second energy-rich product of the light reactions that is used in the dark reactions is a reduced coenzyme, NADPH + H$^+$. Just as NAD$^+$ couples the pathways of cellular respiration, a similar compound, NADP^+ (nicotinamide adenine dinucleotide *phosphate*) couples the photosynthetic pathways. NADP$^+$ is identical to NAD (see Figure 7.3), except that the latter has another phosphate group attached to the ribose. Whereas NAD participates in catabolism, NADP is used in synthetic reactions (anabolism), such as carbohydrate synthesis from CO_2, that require energy through reducing power.

There are two different systems of electron flow in photosynthesis:

▶ **Noncyclic electron flow** produces NADPH + H$^+$ and ATP.
▶ **Cyclic electron flow** produces only ATP.

We'll consider these noncyclic and cyclic reactions before considering the role of chemiosmosis in phosphorylation—a process that is very similar to that discovered for oxidative phosphorylation in mitochondria.

Noncyclic electron flow produces ATP and NADPH

In noncyclic electron flow, light energy is used to oxidize water, forming O_2, H$^+$, and electrons (Figure 8.12). Electrons from water replenish the electrons that chlorophyll molecules lose when they are excited by light. As the electrons are passed from water to chlorophyll, and ultimately to

8.12 Noncyclic Electron Flow Uses Two Photosystems
Photosystems I and II both make use of the excited chlorophyll molecules of their respective reaction centers.

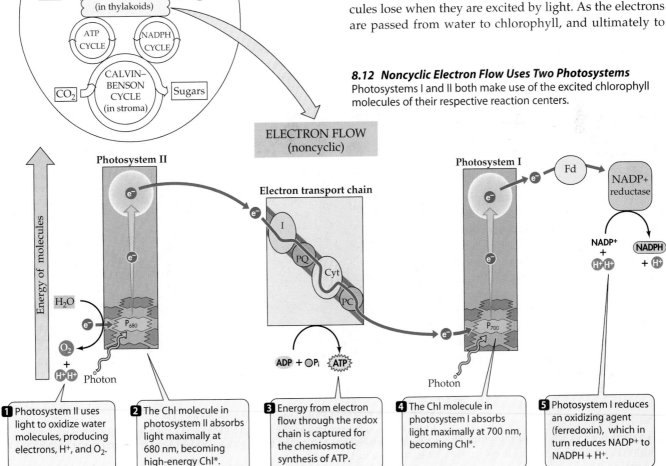

1. Photosystem II uses light to oxidize water molecules, producing electrons, H$^+$, and O_2.

2. The Chl molecule in photosystem II absorbs light maximally at 680 nm, becoming high-energy Chl*.

3. Energy from electron flow through the redox chain is captured for the chemiosmotic synthesis of ATP.

4. The Chl molecule in photosystem I absorbs light maximally at 700 nm, becoming Chl*.

5. Photosystem I reduces an oxidizing agent (ferredoxin), which in turn reduces NADP$^+$ to NADPH + H$^+$.

NADP$^+$, they pass through a series of electron carriers. These redox reactions are exergonic, and some of the free energy released is used ultimately to form ATP by a chemiosmotic mechanism.

Noncyclic electron flow requires the participation of two distinct molecules of chlorophyll. These molecules are associated with two different photosystems, each of which consists of many chlorophyll molecules and accessory pigments in separate energy-absorbing antennas:

▶ **Photosystem I** uses light energy to reduce NADP$^+$ to NADPH + H$^+$.

▶ **Photosystem II** uses light energy to oxidize water molecules, producing electrons, protons (H$^+$), and O$_2$.

The reaction center for photosystem I contains a chlorophyll *a* molecule in a form called P$_{700}$ because it can best absorb light of wavelength 700 nm. The reaction center for photosystem II contains a chlorophyll *a* molecule in a form called P$_{680}$ because it absorbs light maximally at 680 nm. Thus photosystem II requires photons that are somewhat more energetic (i.e., lower wavelengths) than those required by photosystem I. To keep noncyclic electron flow going, *both* photosystems I and II must constantly be absorbing light, thereby boosting electrons to higher orbitals from which they may be captured by specific oxidizing agents.

The reactions of noncyclic electron flow from water to NADP$^+$ are depicted in Figure 8.12. Photosystem II absorbs photons, sending electrons from P$_{680}$ to pheophytin-I—the first carrier in the redox chain—and causing P$_{680}$ to become oxidized to P$_{680}^+$. Electrons from the oxidation of water are passed to P$_{680}^+$, reducing it once again to P$_{680}$, which can absorb more photons. The electron from photosystem II passes through a series of exergonic reactions in the redox chain, which are coupled to proton pumping. This pumping creates a proton gradient that stores energy for ATP synthesis.

In photosystem I, P$_{700}$ absorbs photons, becoming excited to P$_{700}^*$, which then leads to the reduction of an oxidizing agent, *ferredoxin* (Fd), while being oxidized to P$_{700}^+$. Then P$_{700}^+$ returns to the ground state by accepting electrons passed through the redox chain from photosystem II. Now electron flow in photosystem II is accounted for, and we must consider only the electrons from photosystem I. These electrons are used in the last step of noncyclic electron flow, in which two electrons and two protons are used to reduce a molecule of NADP$^+$ to NADPH + H$^+$.

In sum, noncyclic electron flow uses a molecule of water, four photons (two

each absorbed by photosystems I and II), one molecule each of NADP$^+$ and ADP, and one P$_i$. From these ingredients it produces one molecule each of NADPH + H$^+$ and ATP, and half a molecule of oxygen ($\frac{1}{2}$ O$_2$). A substantial fraction of the light energy absorbed in noncyclic electron flow is lost as heat, but another significant fraction is trapped in ATP and NADPH + H$^+$.

Cyclic electron flow produces ATP but no NADPH

Noncyclic electron flow produces equal quantities of ATP and NADPH + H$^+$. However, as we will see, the Calvin–Benson cycle uses more ATP than NADPH + H$^+$. In order to keep things in balance, plants sometimes make use of a supplementary form of electron flow that does not generate NADPH + H$^+$.

Electron flow that produces only ATP is called *cyclic* because an electron passed from an excited chlorophyll molecule at the outset cycles back to the same chlorophyll molecule at the end of the chain of reactions (Figure 8.13). Water, which supplies electrons to restore chlorophyll molecules to the ground state in noncyclic electron flow, does not enter these reactions; thus they produce no O$_2$.

Before cyclic flow begins, P$_{700}$, the reaction center chlorophyll of photosystem I, is in the ground state. It absorbs a photon and becomes P$_{700}^*$. The P$_{700}^*$ then reacts with oxidized ferredoxin (Fd$_{ox}$) to produce reduced ferredoxin (Fd$_{red}$). The reaction is exergonic, releasing free energy.

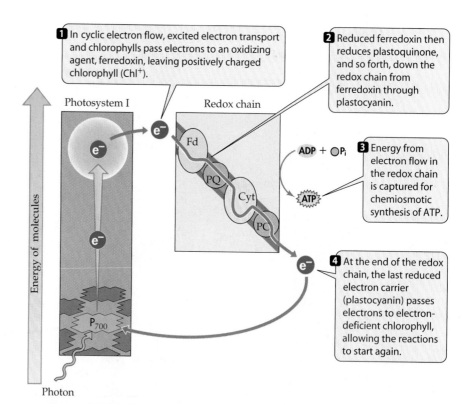

1 In cyclic electron flow, excited electron transport and chlorophylls pass electrons to an oxidizing agent, ferredoxin, leaving positively charged chlorophyll (Chl$^+$).

2 Reduced ferredoxin then reduces plastoquinone, and so forth, down the redox chain from ferredoxin through plastocyanin.

3 Energy from electron flow in the redox chain is captured for chemiosmotic synthesis of ATP.

4 At the end of the redox chain, the last reduced electron carrier (plastocyanin) passes electrons to electron-deficient chlorophyll, allowing the reactions to start again.

Photosystem I

Redox chain

ADP + P$_i$

ATP

Energy of molecules

P$_{700}$

Photon

Fd

PQ

Cyt

PC

8.13 Cyclic Electron Flow Traps Light Energy as ATP
Cyclic electron flow produces ATP but no NADPH + H$^+$. The same chlorophyll molecule passes on the electrons that start the reactions and receives the electrons at the end to start the process over again.

In noncyclic electron flow, Fd_{red} reduces $NADP^+$ to form $NADPH + H^+$. However, Fd_{red} can also pass its added electron to a different oxidizing agent, plastoquinone (PQ, a small organic molecule). This is what happens in cyclic flow, which occurs in some organisms when the ratio of $NADPH + H^+$ to $NADP^+$ in the chloroplast is high.

Thus, Fd_{red} reduces PQ, and PQ_{red} passes the electron to a cytochrome complex (Cyt). The electron continues down the redox chain until it completes its cycle by returning to P_{700}. This cycle is a series of redox reactions, each exergonic, and the released energy is stored in a form that ultimately can be used to produce ATP.

When P_{700}^* passed its electron on to Fd, it became positively charged P_{700}^+. In due course, P_{700}^+ interacts with the last reducing agent in the redox chain, plastocyanin (PC), which donates an electron to P_{700}^+, resulting in a restoration of its uncharged form. By the time the electron from P_{700}^* travels through the redox chain and comes back to reduce P_{700}^+, all the energy from the original photon has been released. In each of the redox reactions, some free energy is used to form ATP and some free energy is lost as heat.

Chemiosmosis is the source of ATP

In Chapter 7 we considered the chemiosmotic mechanism for ATP formation in the mitochondrion. The chemiosmotic mechanism also operates in photophosphorylation (Figure 8.14). In chloroplasts, as in mitochondria, electrons move through a series of redox reactions, releasing energy, which is used to transport protons (H^+) across a membrane. This

8.14 Chloroplasts Form ATP Chemiosmotically Protons (H^+) pumped across the thylakoid membrane from the stroma during photophosphorylation make the interior of the thylakoid more acidic than the stroma. Driven by this pH difference, the protons diffuse back to the stroma through ATP synthase channels, which couple the energy of proton flow to the formation of ATP from $ADP + P_i$.

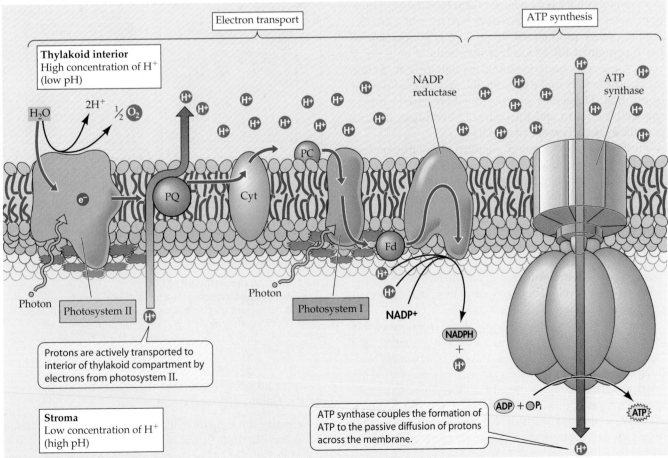

active proton transport results in a proton-motive force—a difference in pH and in electric charge across the membrane.

In the mitochondrion, protons are pumped out of the matrix, across the inner membrane, and into the space between the inner and outer mitochondrial membranes (see Figure 7.13). Similarly, in the chloroplast, the electron carriers in the thylakoid membranes are oriented so that protons move into the interior of the thylakoid, and the inside becomes acidic with respect to the outside. The ratio of H^+ inside versus outside a thylakoid is usually 10,000:1, which is a difference of 4 pH units. This difference in pH leads to the diffusion of H^+ back out of the thylakoid through specific protein channels in the membrane. These channels are enzymes—ATP synthases—that couple the formation of ATP to the diffusion of protons back across the membrane, just as in mitochondria.

Photosynthetic pathways are the products of evolution

The first photosynthetic organisms were probably anaerobic bacteria that used hydrogen sulfide, rather than water, as a source of electrons:

$$CO_2 + 2 H_2S \rightarrow (CH_2O) + 2 S + H_2O$$

Many bacteria still use this system, which releases sulfur rather than oxygen.

Nearly 3 billion years ago, the evolution of new pigments in certain bacteria allowed them to extract electrons from water and use them to reduce $NADP^+$ while producing O_2 as a by-product. At this time, the atmosphere of Earth contained little O_2. Over hundreds of millions of years, these cyanobacteria poured enough oxygen gas into the atmosphere to make possible the evolution of cellular respiration. These new photosynthetic reactions forever changed the Earth and the course of evolution, making possible a great diversification of life.

Furthermore, if a larger cell engulfed one of these cyanobacteria, that cell could perform photosynthesis. This was probably the first event in the evolution of eukaryotic, photosynthetic plant cells.

Making Sugar from CO_2: The Calvin–Benson Cycle

The second main pathway of photosynthesis is the Calvin–Benson cycle. The reactions of this pathway incorporate CO_2 into sugars.

Most of the enzymes that catalyze the reactions of this pathway are dissolved in the chloroplast stroma (the "soup" outside the thylakoids), and this is where the reactions take place. These reactions are sometimes called the "dark reactions" because they do not directly require light energy. However, they use the energy in ATP and NADPH, produced in the thylakoids during the light reactions, to reduce CO_2 to carbohydrate. So these reactions require light indirectly, and *they take place only in the light*.

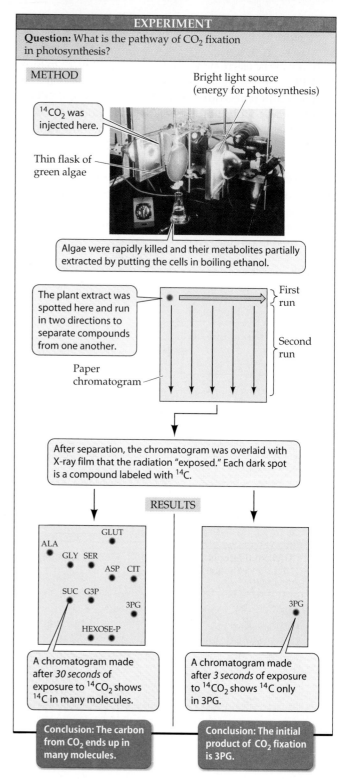

8.15 Tracing the Pathway of CO_2
The historical photograph at the top shows the apparatus Calvin and his colleagues used to follow labeled carbon dioxide molecules ($^{14}CO_2$) as they were transformed by photosynthesis.

Isotope labeling experiments reveal the steps of the Calvin–Benson cycle

To identify the sequence of reactions by which CO_2 ends up in carbohydrate, it was necessary to label CO_2 so that it could be followed after presentation to a plant cell. In the

1950s, Melvin Calvin, Andrew Benson, and their colleagues used radioactively labeled CO_2 in which some of the carbon atoms were not the normal ^{12}C, but its radioisotope ^{14}C. Although ^{14}C is distinguished by its emission of radiation, chemically it behaves virtually identically to nonradioactive ^{12}C. In general, enzymes do not distinguish between isotopes of an element in their substrates, so $^{14}CO_2$ is treated the same way by photosynthesizing cells as $^{12}CO_2$.

Calvin and his colleagues exposed cultures of the unicellular green alga *Chlorella* to $^{14}CO_2$ for 30 seconds. They killed the cells, extracted their carbohydrates, and separated the different compounds from one another by paper chromatography. Many compounds, including monosaccharides and amino acids, contained ^{14}C (Figure 8.15). However, if they stopped the exposure after just 3 seconds, only one compound was labeled—a three-carbon sugar phosphate called 3-phosphoglycerate (3PG):

3-Phosphoglycerate (3PG)

By tracing the steps in this manner, they soon discovered a cycle, similar to the citric acid cycle, that "fixes" CO_2 in a larger molecule, produces carbohydrate, and regenerates the initial CO_2 acceptor. This cycle was appropriately named the Calvin–Benson cycle.

The initial reaction in the Calvin–Benson cycle fixes the one-carbon CO_2 in a five-carbon compound, ribulose 1,5-bisphosphate (**RuBP**). An intermediate six-carbon compound forms, which quickly breaks down, forming two three-carbon molecules of 3PG, which is what Calvin and colleagues had seen (Figure 8.16). The enzyme that catalyzes the fixation reaction, ribulose bisphosphate carboxy-

lase/oxygenase (**rubisco**), is the most abundant protein in the world, comprising about 20 percent of all the protein in every plant leaf.

The Calvin–Benson cycle is composed of three processes

The Calvin–Benson cycle uses the high-energy compounds made in the thylakoids during the light reactions (ATP, NADPH) to reduce CO_2 to carbohydrate. There are three processes that make up the cycle (Figure 8.17):

▶ Fixation of CO_2. As we saw, this reaction is catalyzed by rubisco, and its product is 3PG.

▶ Conversion of fixed CO_2 into carbohydrate (G3P). This series of reactions involves a phosphorylation (using the ATP made in the light reactions) and a reduction (using the NADPH made in the light reactions).

▶ Regeneration of the CO_2 acceptor, RuBP. Most of the 3PG ends up as RuMP (ribulose monophosphate), and ATP is used to convert this to RuBP. So for every "turn" of the cycle, with one CO_2 fixed, the acceptor gets regenerated.

The end product of this cycle is glyceraldehyde 3-phosphate (G3P), which is a three-carbon sugar phosphate, also called triose phosphate:

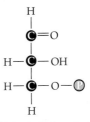

Glyceraldehyde 3-phosphate (G3P)

There are two fates for the G3P that ends up as a product of the Calvin–Benson cycle. In a typical leaf, about a third of it ends up in the polysaccharide starch, which is stored in the chloroplast and serves as a source of glucose. Two-thirds of the G3P product is converted to the disaccharide sucrose,

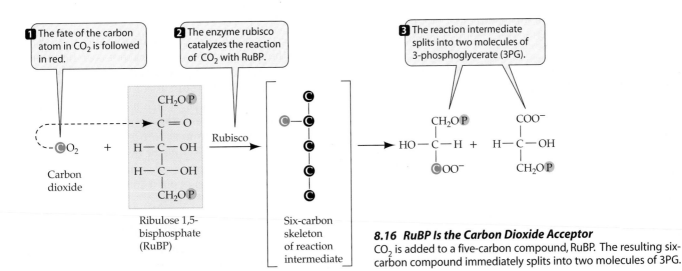

1 The fate of the carbon atom in CO_2 is followed in red.

2 The enzyme rubisco catalyzes the reaction of CO_2 with RuBP.

3 The reaction intermediate splits into two molecules of 3-phosphoglycerate (3PG).

8.16 RuBP Is the Carbon Dioxide Acceptor
CO_2 is added to a five-carbon compound, RuBP. The resulting six-carbon compound immediately splits into two molecules of 3PG.

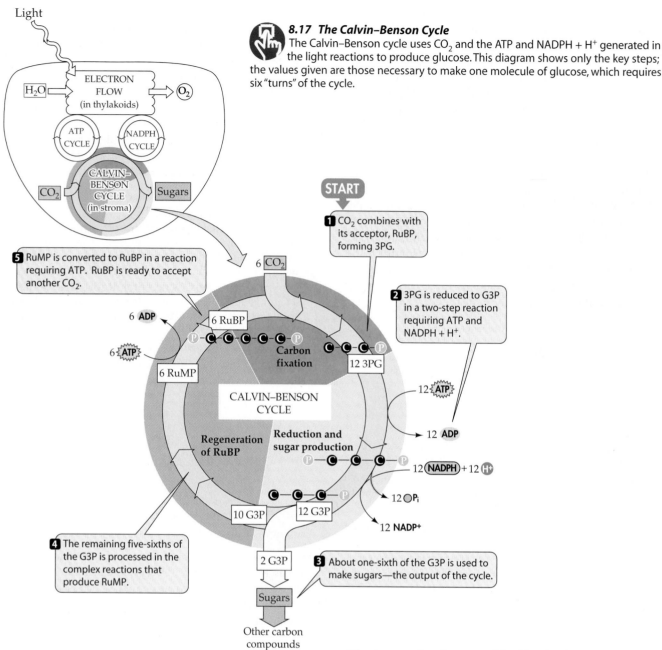

8.17 The Calvin–Benson Cycle
The Calvin–Benson cycle uses CO_2 and the ATP and NADPH + H⁺ generated in the light reactions to produce glucose. This diagram shows only the key steps; the values given are those necessary to make one molecule of glucose, which requires six "turns" of the cycle.

5 RuMP is converted to RuBP in a reaction requiring ATP. RuBP is ready to accept another CO_2.

START

1 CO_2 combines with its acceptor, RuBP, forming 3PG.

2 3PG is reduced to G3P in a two-step reaction requiring ATP and NADPH + H⁺.

4 The remaining five-sixths of the G3P is processed in the complex reactions that produce RuMP.

3 About one-sixth of the G3P is used to make sugars—the output of the cycle.

Carbon fixation

CALVIN–BENSON CYCLE

Regeneration of RuBP

Reduction and sugar production

Other carbon compounds

which is transported out of the leaf to other organs in the plant, where it is hydrolyzed to its constituent monosaccharides: glucose and fructose.

The glucose produced in photosynthesis is subsequently used by the plant to make other compounds besides sugars. The carbon of glucose is incorporated into amino acids, lipids, and the building blocks of the nucleic acids.

The products of the Calvin–Benson cycle are of crucial importance to the entire biosphere, for the covalent bonds of these products represent the total energy yield from the harvesting of light by plants. Most of this stored energy is released by glycolysis and cellular respiration during plant growth, development, and reproduction. However, much plant matter ends up being consumed by animals. Glycolysis and cellular respiration in the animals releases free energy from the plant matter for use in the animal cells.

Photorespiration and Its Evolutionary Consequences

The properties of rubisco are remarkably identical in all photosynthetic organisms, from bacteria to flowering plants. However, some properties of this enzyme severely limit its effectiveness. In the discussion that follows, we will identify and explore some of these limitations and see how evolution has constructed bypasses around them. First we'll look at photorespiration, a process in which rubisco fixes O_2 instead of CO_2, lowering the overall rate of CO_2 fixation and plant growth. Then we'll examine some biochemical pathways and features of plant anatomy that compensate for the limitations of rubisco.

RuBP reacts with O_2 in photorespiration

As its full name indicates, rubisco is a carboxylase (adding CO_2 to an acceptor molecule, RuBP) as well as an oxyge-

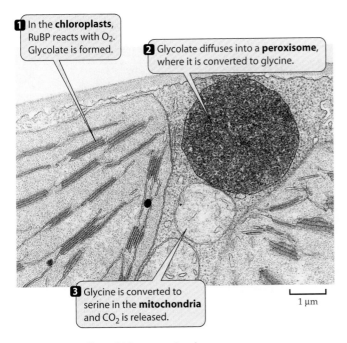

1 In the **chloroplasts**, RuBP reacts with O_2. Glycolate is formed.

2 Glycolate diffuses into a **peroxisome**, where it is converted to glycine.

3 Glycine is converted to serine in the **mitochondria** and CO_2 is released.

1 µm

8.18 Organelles of Photorespiration
The reactions of photorespiration take place in the chloroplasts, peroxisomes, and, finally, in the mitochondria.

nase (adding O_2 to RuBP). These two reactions compete with each other.

When RuBP and O_2 react, one of the products is a two-carbon compound, glycolate:

$$RuBP + O_2 \rightarrow glycolate$$

The glycolate diffuses into membrane-enclosed organelles called peroxisomes (Figure 8.18). There, a series of reactions converts it to the amino acid glycine:

$$glycolate \rightarrow \rightarrow glycine$$

The glycine then diffuses into a mitochondrion, where two glycine molecules are converted to another amino acid, serine:

$$2 \; glycine \rightarrow serine + CO_2$$

This pathway, called **photorespiration**, uses ATP and NADPH produced in the light reactions, just like the Calvin–Benson cycle. But the net effect of photorespiration essentially undoes what the Calvin–Benson cycle accomplishes: CO_2 is *released* instead of being fixed into carbohydrate. In many plants, photorespiraton reduces the amount of carbon fixed into carbohydrate by 25 percent.

How does rubisco "decide" whether to act as an oxygenase or a carboxylase? The prime consideration is the relative concentrations of CO_2 and O_2 in the leaf. If O_2 is relatively abundant, rubisco acts as an oxygenase, and photorespiration ensues. If CO_2 predominates, rubisco fixes it, and the Calvin–Benson cycle occurs.

The level of O_2 in a leaf becomes especially high on a hot, dry day. To prevent water loss, the stomata that allow water to evaporate from the leaf close (see Figure 8.1). But this also prevents gases from entering and leaving the leaf. So, the CO_2 concentration falls because it is being

used up by the light-driven photosynthetic reactions, and the O_2 concentration rises because of these same reactions.

If photorespiration is wasteful of the photosynthetic process, why does it exist? Apparently, it is not essential for life. Many plants get along without it, and even plants with it can grow well if it is inhibited chemically. One explanation is that the active site of rubisco evolved to bind both CO_2 and O_2. This was not a problem originally, as there was little O_2 in the atmosphere, and the CO_2 binding activity was the only one used. When O_2 appeared, so did the photorespiration pathway.

Some plants have evolved systems to bypass photorespiration

A solution to the problem of photorespiration has evolved in a number of plants, most of them related to tropical species such as sugarcane. The objective is to raise the level of CO_2 in relation to O_2 around rubisco, so that carboxylation is favored. This is not an easy task, since the air surrounding a leaf is 21 percent O_2 and only 0.036 percent CO_2.

In the leaves of plants such as roses, wheat, and rice, the mesophyll cells just below the surface are full of chloroplasts that contain abundant rubisco (Figure 8.19a). On a

(a) **Arrangement of cells in a C_3 leaf**

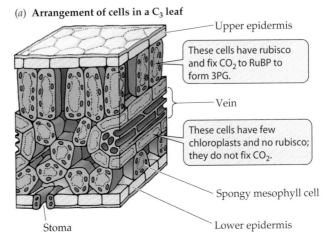

Upper epidermis

These cells have rubisco and fix CO_2 to RuBP to form 3PG.

Vein

These cells have few chloroplasts and no rubisco; they do not fix CO_2.

Spongy mesophyll cell

Stoma

Lower epidermis

(b) **Arrangement of cells in a C_4 leaf**

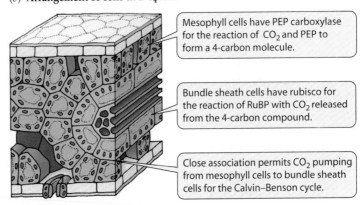

Mesophyll cells have PEP carboxylase for the reaction of CO_2 and PEP to form a 4-carbon molecule.

Bundle sheath cells have rubisco for the reaction of RuBP with CO_2 released from the 4-carbon compound.

Close association permits CO_2 pumping from mesophyll cells to bundle sheath cells for the Calvin–Benson cycle.

 8.19 Leaf Anatomy of C_3 and C_4 Plants
Carbon dioxide fixation occurs in different organelles and cells of the leaves in the two types of plants.

hot day, these leaves close their stomata to conserve water. The level of CO_2 in the air spaces of the leaves falls, and that of O_2 rises, as photosynthesis goes on. Because the first product of CO_2 fixation in these plants is the three-carbon molecule 3PG, they are called **C$_3$ plants**. As we have seen, photorespiration occurs under these conditions.

In corn, sugarcane, and other tropical grasses, the chloroplasts, with their abundant rubisco, are in a cell layer in the interior of the leaf (Figure 8.19*b*). Like C$_3$ plants, these plants close their stomata on a hot day, but their rate of photosynthesis does not fall, nor does photorespiration occur. They have a way to keep the ratio of CO_2 to O_2 around rubisco high, so that rubisco acts as a carboxylase. They do this in part by having a four-carbon compound, oxaloacetate, as the first product of CO_2 fixation, and so are called **C$_4$ plants**.

C$_4$ plants perform the normal Calvin–Benson cycle, but they have an additional early reaction that fixes CO_2 without losing carbon to photorespiration, greatly increasing the overall photosynthetic yield. Because this initial CO_2 fixation step can function even at low levels of CO_2 and high temperatures, C$_4$ plants very effectively optimize photosynthesis under conditions that inhibit the photosynthesis of C$_3$ plants.

C$_4$ plants have two separate enzymes for CO_2 fixation in different chloroplasts in two different locations in the leaf (Figure 8.20). One, present in the mesophyll cells near the surface of the leaf, fixes CO_2 to a three-carbon acceptor compound (phosphoenolpyruvate) to produce the four-carbon fixation product (oxaloacetate). This enzyme, **PEP carboxylase**, has two advantages over rubisco:

▶ It does not have oxygenase activity.
▶ It fixes CO_2 at very low levels.

So even on a hot day when the stomata are closed, CO_2 is low, and O_2 is high, PEP carboxylase just keeps on fixing CO_2.

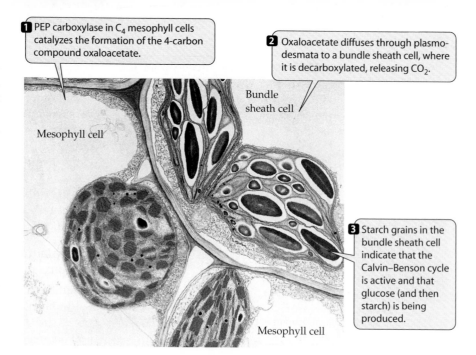

1 PEP carboxylase in C$_4$ mesophyll cells catalyzes the formation of the 4-carbon compound oxaloacetate.

2 Oxaloacetate diffuses through plasmodesmata to a bundle sheath cell, where it is decarboxylated, releasing CO_2.

Bundle sheath cell

Mesophyll cell

3 Starch grains in the bundle sheath cell indicate that the Calvin–Benson cycle is active and that glucose (and then starch) is being produced.

Mesophyll cell

8.20 C$_4$ Photosynthesis
Carbon dioxide is fixed initially in the mesophyll cells, but enters the Calvin–Benson cycle in the bundle sheath cells.

Oxaloacetate diffuses out of the mesophyll cells and into the bundle sheath cells in the interior of the leaf. The chloroplasts in bundle sheath cells contain abundant rubisco. There, the four-carbon oxaloacetate is decarboxylated, losing CO_2 and regenerating the three-carbon acceptor, which diffuses back out to the mesophyll cells. The role of this acceptor is to bind CO_2 from the air in the leaf and carry it to the interior cells, where it is "dropped off" at rubisco. This process essentially pumps up the CO_2 concentration around rubisco, so that it acts as a carboxylase and begins the Calvin–Benson cycle.

Kentucky bluegrass, a C$_3$ plant, thrives on lawns in April and May. But in the heat of summer, it does not do as well, and crabgrass, a C$_4$ plant, takes over the lawn. The same is true on a global scale for crops: C$_3$ plants, such as soybeans, rice, wheat, and barley, have been adapted for human food production in temperate climates, while C$_4$ plants, such as corn and sugarcane, originated and are grown in the tropics. Table 8.1 compares C$_3$ and C$_4$ photosynthesis.

8.1 Comparison of Photosynthesis in C$_3$ and C$_4$ Plants

VARIABLE	C$_3$ PLANTS	C$_4$ PLANTS
Photorespiration	Extensive	Minimal
Perform Calvin–Benson cycle?	Yes	Yes
Primary CO_2 acceptor	RuBP	PEP
CO_2-fixing enzyme	Rubisco (RuBP carboxylase/oxygenase)	PEP carboxylase
First product of CO_2 fixation	3PG (3-carbon compound)	Oxaloacetate (4-carbon compound)
Affinity of carboxylase for CO_2	Moderate	High
Photosynthetic cells of leaf	Mesophyll	Mesophyll + bundle sheath
Classes of chloroplasts	One	Two

C_3 plants are certainly more ancient than C_4 plants. While C_3 photosynthesis appears to have begun about 3.5 billion years ago, C_4 plants appeared about 12 million years ago. A possible factor in the emergence of the C_4 pathway is the decline in atmospheric CO_2. When dinosaurs ruled the Earth 100 million years ago, the concentration of CO_2 was four times what it is now. As CO_2 levels then declined, the more efficient C_4 plants would have had an advantage over their C_3 counterparts.

CAM plants also use PEP carboxylase

Other plants besides the C_4 species use PEP carboxylase to fix and accumulate CO_2 while their stomata are closed. Such plants include some water-storing plants (called *succulents*) of the family Crassulaceae, many cacti, pineapples, and several other kinds of flowering plants. These plants conserve water by keeping their stomata closed during the daylight hours, thus minimizing water loss by evaporation. How, then, can they perform photosynthesis? Their trick is to open their stomata at night and store CO_2 by a different mechanism.

The CO_2 metabolism of these plants is called **crassulacean acid metabolism**, or **CAM**, after the family of succulents in which it was discovered. CAM is much like the metabolism of C_4 plants in that CO_2 is initially fixed into four-carbon compounds. In CAM plants, however, the processes of initial CO_2 fixation and the Calvin–Benson cycle are separated in time rather than in space (Figure 8.21). And CAM plants lack the specialized cell relationships of C_4 plants.

In CAM plants, CO_2 is fixed initially in mesophyll cells to form the four-carbon compound oxaloacetate, which is converted to malic acid. This fixation occurs during the night, when less water is lost through open stomata. When daylight arrives, the accumulated malic acid is shipped to the chloroplasts, where decarboxylation supplies the CO_2 for operation of the Calvin–Benson cycle, and the light reactions supply the necessary ATP and NADPH + H$^+$.

Metabolic Pathways in Plants

Green plants are autotrophs, and can synthesize all the molecules they need from three simple starting materials: CO_2, H_2O, and NH_4^+. The latter is needed for amino acids, and comes either from the conversion of nitrogen-containing molecules taken up in soil water by the roots or from the bacterial conversion of N_2 gas.

Although the light reactions of photosynthesis generate some ATP and NADPH, the products are used for the Calvin–Benson cycle and are far less than plant cells need to fuel their endergonic reactions. Also, not all plant cells are photosynthetic. To satisfy their need for ATP, plants, like all other organisms, carry out respiration. Both aerobic respiration and fermentation can occur in plants, although the former is far more common.

Plant cellular respiration, unlike photosynthesis, takes place *both* in the light and in the dark. Because glycolysis occurs in the cytosol, respiration in the mitochondria, and photosynthesis in the chloroplasts, all these processes can proceed simultaneously.

Photosynthesis and respiration are closely linked through the Calvin–Benson cycle (Figure 8.22). Two linkages are particularly important:

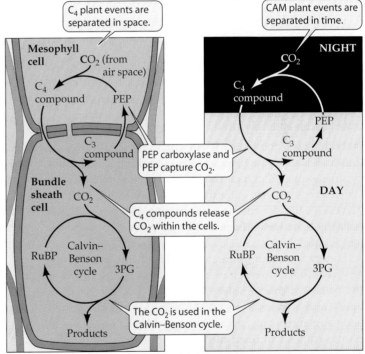

Sempervivum tectorum (CAM)

8.21 C_4 and CAM Plants Separate Two Sets of Reactions Differently
Both plant types use four-carbon compounds whose production is separate from the Calvin–Benson cycle. The separation is spatial in C_4 plants, temporal in CAM plants.

Sorghum (C_4)

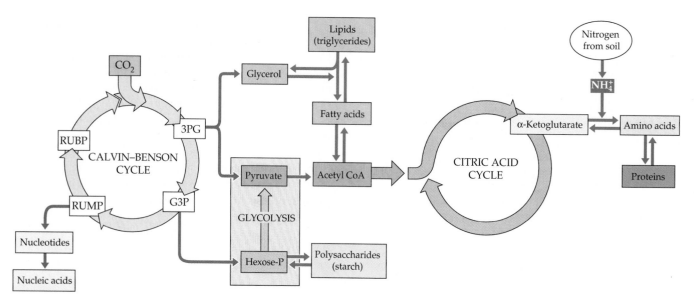

8.22 Metabolic Interactions in a Plant Cell
Note the relationships among the Calvin–Benson cycle, the citric acid cycle, and glycolysis.

▶ 3PG from the Calvin–Benson cycle can be converted to pyruvate, the end product of glycolysis.

▶ G3P from the Calvin–Benson cycle can be converted to hexose phosphates (such as glucose 1-phosphate), which can enter glycolysis.

In both cases, the result is the catabolic breakdown of Calvin–Benson cycle products to CO_2, with the associated synthesis of ATP. Once the carbon skeletons from the Calvin–Benson cycle enter the "central switching yard" of glycolysis and the citric acid cycle, they can be used anabolically to make lipids, proteins, and other carbohydrates (see Figure 7.19).

Energy flows from sunlight to reduced carbon in photosynthesis to ATP in respiration. Energy can also be stored in the bonds of macromolecules such as polysaccharides, lipids, and proteins. For a plant to grow, energy storage (as body structures) must exceed energy release; that is, overall photosynthesis to fixed carbon must exceed respiration. This is the aim of the farmers growing corn whom we described in the opening of this chapter. And it is the basis of the ecological food chain, as we will see in later chapters.

Chapter Summary

▶ Life on Earth depends on the absorption of light energy from the sun.

▶ In plants, photosynthesis takes place in chloroplasts.

Identifying Photosynthetic Reactants and Products

▶ Photosynthesizing plants take in CO_2, water, and light energy, producing O_2 and carbohydrate. The overall reaction is

$$6\ CO_2 + 12\ H_2O + \text{light} \rightarrow C_6H_{12}O_6 + 6\ O_2 + 6\ H_2O$$

The oxygen atoms in O_2 come from water, not from CO_2.
Review Figures 8.1, 8.2

The Two Pathways of Photosynthesis: An Overview

▶ In the light reactions of photosynthesis, electron flow and photophosphorylation produce ATP and reduce $NADP^+$ to $NADPH + H^+$. **Review Figure 8.3**

▶ ATP and $NADPH + H^+$ are needed for the reactions that fix and reduce CO_2 in the Calvin–Benson cycle, forming sugars. **Review Figure 8.3**

Properties of Light and Pigments

▶ Light energy comes in packets called photons, but it also has wavelike properties. **Review Figure 8.4**

▶ Pigments absorb light in the visible spectrum. **Review Figure 8.5**

▶ Absorption of a photon puts a pigment molecule in an excited state that has more energy than its ground state. **Review Figure 8.6**

▶ Each compound has a characteristic absorption spectrum. An action spectrum reveals the biological effectiveness of different wavelengths of light. **Review Figures 8.7, 8.8**

▶ Chlorophylls and accessory pigments form antenna systems for absorption of light energy. **Review Figures 8.7, 8.9, 8.11**

Light Reactions: Light Absorption

▶ An excited pigment molecule may lose its energy by fluorescence, or by transferring it to another pigment molecule. **Review Figures 8.10, 8.11**

Electron Flow, Photophosphorylation, and Reductions

▶ Noncyclic electron flow uses two photosystems (I and II), producing ATP, $NADPH + H^+$, and O_2. Photosystem II uses P_{680} chlorophyll, from which light-excited electrons are passed to a redox chain that drives chemiosmotic ATP production. Light-driven oxidation of water releases O_2 and passes electrons from water to the P_{680} chlorophyll. Photosystem I passes electrons from P_{700} chlorophyll to another redox chain and then to $NADP^+$, forming $NADPH + H^+$. **Review Figure 8.12**

▶ Cyclic electron flow uses P_{700} chlorophyll and produces only ATP. Its operation maintains the proper balance of ATP and $NADPH + H^+$ in the chloroplast. **Review Figure 8.13**

▶ Chemiosmosis is the source of ATP in photophosphorylation. Electron transport pumps protons from the stroma into the thylakoids, establishing a proton-motive force. Diffusion of the protons back to the stroma via ATP synthase channels drives ATP formation from ADP and P_i. **Review Figure 8.14**

▶ Photosynthesis probably originated in anaerobic bacteria that used H_2S as a source of electrons instead of H_2O. Oxygen production by bacteria was an important event in the evolution of eukaryotes.

Making Sugar from CO$_2$: The Calvin–Benson Cycle

▶ The Calvin–Benson cycle makes sugar from CO_2. This pathway was elucidated through the use of radioactive tracers. **Review Figure 8.15**

▶ The Calvin–Benson cycle consists of three phases: fixation of CO_2, reduction and carbohydrate production, and regeneration of RuBP. RuBP is the initial CO_2 acceptor, and 3PG is the first stable product of CO_2 fixation. The enzyme rubisco catalyzes the reaction of CO_2 and RuBP to form 3PG. **Review Figures 8.16, 8.17**

Photorespiration and Its Evolutionary Consequences

▶ The enzyme rubisco can catalyze a reaction between O_2 and RuBP in addition to the reaction between CO_2 and RuBP. This consumption of O_2 is called photorespiration and significantly reduces the efficiency of photosynthesis. The reactions that constitute photorespiration are distributed over three organelles: chloroplasts, peroxisomes, and mitochondria. **Review Figure 8.18**

▶ At high temperatures and low CO_2 concentrations, the oxygenase function of rubisco is favored.

▶ C_4 plants bypass photorespiration with special chemical reactions and specialized leaf anatomy. In C_4 plants, PEP carboxylase in mesophyll chloroplasts initially fixes CO_2 in four-carbon acids, which then diffuse into bundle sheath cells, where their decarboxylation produces locally high concentrations of CO_2. **Review Figures 8.19, 8.20**

▶ CAM plants operate much like C_4 plants, but their initial CO_2 fixation by PEP carboxylase is temporally separated from the Calvin–Benson cycle, rather than spatially separated as in C_4 plants. **Review Figure 8.21**

Metabolic Pathways in Plants

▶ Plants respire both in the light and in the dark, but photosynthesize only in the light. To survive, a plant must photosynthesize more than it respires, giving it a net gain of reduced energy-rich compounds.

▶ Photosynthesis and respiration are linked through the Calvin–Benson cycle, the citric acid cycle, and glycolysis. **Review Figure 8.22**

For Discussion

1. Both electron flow and the Calvin–Benson cycle stop in the dark. Which specific reaction stops first? Which stops next? Continue answering the question "Which stops next?" until you have explained why both pathways have stopped.

2. In what principal ways are the reactions of electron flow in photosynthesis similar to the respiratory chain and oxidative phosphorylation discussed in Chapter 7? Differentiate between cyclic and noncyclic electron flow in terms of (1) the products and (2) the source of electrons for the reduction of oxidized chlorophyll.

3. The development of what two experimental techniques made it possible to elucidate the Calvin–Benson cycle? How were these techniques used in the investigation?

4. If water labeled with ^{18}O is added to a suspension of photosynthesizing chloroplasts, which of the following compounds will first become labeled with ^{18}O: ATP, NADPH, O_2, or 3PG? If water labeled with ^3H is added to a suspension of photosynthesizing chloroplasts, which of the same compounds will first become radioactive? If CO_2 labeled with ^{14}C is added to a suspension of photosynthesizing chloroplasts, which of those compounds will first become radioactive?

Part Two

INFORMATION AND HEREDITY

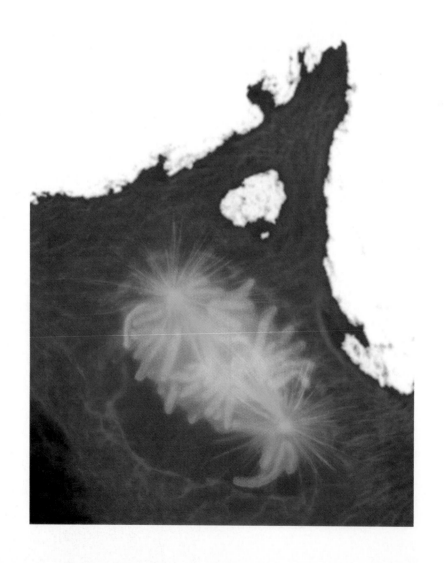

9 Chromosomes, the Cell Cycle, and Cell Division

IN 1951, 31-YEAR-OLD HENRIETTA LACKS ENtered Johns Hopkins Hospital to be treated for a cancerous tumor. Although she died a few months later, her tumor cells are alive today. Scientists found that, given adequate nourishment, cancerous cells from the tumor reproduced themselves indefinitely in a laboratory dish. These "HeLa cells" became a test-tube model for studies of human cell biology and biochemistry. Over the past half-century, tens of thousands of research articles have been published using information obtained from Henrietta's cells. But are these "immortal" cells really a good model for human biology?

In one sense, they are. Most multicellular organisms come from a single cell: the fertilized egg. This cell reproduces itself to make two cells, these in turn divide to become four cells, and so on until all the cells of a new organism have been produced. An organism is not just a ball of many cells, however; the cells must specialize into tissues and organs, each with specific roles to perform. This process of specialization, or differentiation, is a subject we will return to in later chapters of Part Two.

In normal tissues, cell reproduction ("births") is offset by cell loss ("deaths"). We know cell death is important from careful studies of a tiny worm, in which 1,090 cells are produced from the fertilized egg and exactly 131 of them die before the worm is born. If they do not die, the worm's organs are severely malformed. Another example occurs in the mammalian brain. Young mice, for instance, lose hundreds of thousands of brain cells each day; if these cells do not die, the mouse's overcrowded brain simply does not work.

A cell's death is often programmed into its genetic message; normal cells "sacrifice" themselves for the greater good of the organism. Once an organism reaches its adult size, it stays that way through a combination of cell division and programmed cell death. Like most cancerous cells, Henrietta Lacks's tumor cells keep growing because they have a genetic imbalance that heavily favors cell reproduction over cell death.

Unicellular organisms use cell division primarily to reproduce themselves, whereas in multicellular organisms cell division also plays important roles in the growth and repair of tissues (Figure 9.1) In this chapter, we first describe how prokaryotic cells produce two new organisms from the original single-celled organism. Then we describe two types of cell and nuclear division—mitosis and meiosis—and relate these two modes of cell division to asexual and sexual reproduction in eukaryotic organisms. Finally, to balance our discussion of cell "birth" through division, we will describe the important process of programmed cell death, also known as apoptosis.

Systems of Cell Reproduction

In order for any cell to divide, four events must occur:

▶ There must be a *reproductive signal*. This signal, which may come either from inside or outside the cell, initiates the cellular reproductive events.
▶ *Replication* of DNA, the genetic material, and other vital cell components must occur so that each of the two new cells will have complete cell functions.
▶ The cell must *distribute* (segregate) the replicated DNA to each of the two new cells.
▶ The cell membrane (and the cell wall, in organisms that have one) must grow to separate the two new cells in a process called *cytokinesis*.

Prokaryotes divide by fission

In prokaryotes, cell division often means reproduction of the entire single-celled organism. The cell grows in size, replicates its DNA, and then essentially divides into two new cells—a process called *fission.*

HeLa Cells: More Births Than Deaths
These cells have been cultured in a laboratory since 1951. They are the source of much data relating the reproduction of human cells.

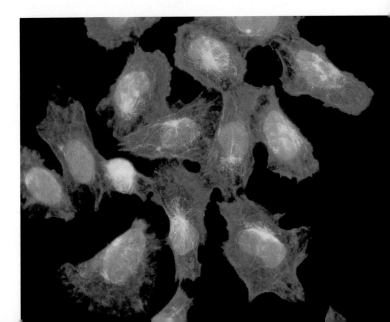

(a) Cell division contributes to the growth of this root tissue.

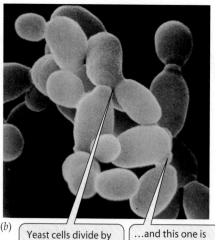

(b) Yeast cells divide by budding. This one has nearly divided…

…and this one is beginning to bud.

9.1 Important Consequences of Cell Division

Cell division is the basis for growth, reproduction, and regeneration.

(c) Cell division contributes to the regeneration of a lizard's tail.

REPRODUCTIVE SIGNALS. The reproductive rates of many prokaryotes respond to conditions in the environment. The bacterium *Escherichia coli*, a species that is commonly used in genetic studies, is a "cell division machine" that essentially divides continuously. Typically, cell division takes 40 minutes at 37°C. But if there are abundant sources of carbohydrates and salts available, the division cycle speeds up so that cells may divide in 20 minutes. Another bacterium, *Bacillus subtilis*, stops dividing under adverse nutritional conditions, then resumes dividing when things improve. These observations suggest that the initiation of cell division in prokaryotes is under the control of metabolic intermediates, such as carbohydrates, in the environment.

REPLICATION OF DNA. A **chromosome**, as we saw in Chapter 4, is a DNA molecule containing genetic information. When a cell divides, its chromosomes must be copied, or *replicated*, and each of the two resulting copies must find its way into one of the two new cells.

Most prokaryotes have only one chromosome, a single long DNA molecule with proteins bound to it. In the bacterium *E. coli*, the DNA is a circular molecule about 1.6 million nm (1.6 mm) in circumference. The bacterium itself is only about 1 μm (1,000 nm) in diameter and about 4 μm long. Thus the long thread of DNA, which could form a circle over 100 times larger if fully expanded, is packed into a very small space. So it is not surprising that the molecule usually appears in electron micrographs as a hopeless tangle of fibers (Figure 9.2). The DNA molecule accomplishes some packing by folding in on itself, and positively charged (basic) proteins bound to negatively charged (acidic) DNA contribute to this packing. Circular chromosomes appear to be characteristic of all prokaryotes, as well as some viruses, and are also found in the chloroplasts and mitochondria of eukaryotic cells.

Functionally, the prokaryotic DNA molecule has two regions that are important for cell reproduction:

▶ *Ori* is the origin of replication, where replication of the circle starts.
▶ *Ter* is the terminus of replication, where it ends.

The process of chromosome replication occurs as the DNA is threaded through a "replication complex" of proteins at the center of the cell.

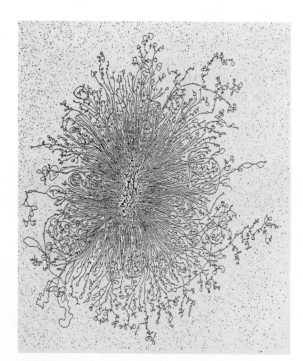

9.2 The Prokaryotic Chromosome Is a Circle

These long, looping fibers of DNA from a cell of the bacterium *Escherichia coli* are all part of one continuous circular chromosome.

DISTRIBUTION OF DNA. DNA replication actively drives the parceling out of the two new DNA molecules to the new cells. The first region to be replicated is *ori*. The two *ori* regions are attached to the plasma membrane, and they separate as the new chromosome forms and new plasma membrane forms between them (Figure 9.3). By the end of replication, there are two chromosomes, one at either end of the bacterial cell.

CYTOKINESIS. Cell partition, or **cytokinesis**, begins 20 minutes after chromosome duplication is finished. The first event of cytokinesis is a pinching in of the plasma membrane to form a ring similar to a purse string. Fibers composed of a protein similar to eukaryotic tubulin (which makes up microtubules) are major components of this ring. As the membrane pinches in, new cell wall materials are synthesized, which finally separate the two cells.

Eukaryotic cells divide by mitosis or meiosis

Cell reproduction in eukaryotes also involves reproductive signals, DNA replication, segregation, and cytokinesis. But, as you might expect, events in eukaryotes are somewhat more complex.

First, unlike prokaryotes, eukaryotic cells do not constantly divide whenever environmental conditions are adequate. In fact, eukaryotic cells that have *differentiated* (become specialized) seldom divide. So the signals for cell division are related not to the physiology of the single cell, but to the needs of the entire organism. Second, instead of a single chromosome, eukaryotes usually have many (humans have 46), so the processes of replication and segregation, while basically the same as in prokaryotes, are more intricate. Third, eukaryotic cells have a distinct nucleus, which has to be replicated and then divided into two new nuclei. Finally, cytokinesis is different in plant cells (which have a cell wall) than in animal cells (which do not).

Mitosis is a nuclear division mechanism that operates in most types of cells. Mitosis sorts the genetic material into two new nuclei and ensures that both contain exactly the same genetic information. A second mechanism of nuclear division, **meiosis**, occurs in the gametes—those cells that will contribute to the reproduction of a new organism. Meiosis generates diversity by shuffling the genetic material, resulting in new gene combinations. It plays a key role in sexual life cycles.

The duplication of a eukaryotic cell typically consists of three steps:

▶ The replication of the genetic material within the nucleus
▶ The packaging and separation of the genetic material into two new nuclei
▶ The division of the cytoplasm

What determines whether a cell will divide? How does mitosis lead to identical cells, and meiosis to diversity? Why do we need both identical copies and diverse cells? Why do most eukaryotic organisms reproduce sexually? In the pages that follow, we will describe the details of mitosis, meiosis, and interphase, as well as their consequences for heredity, development, and evolution.

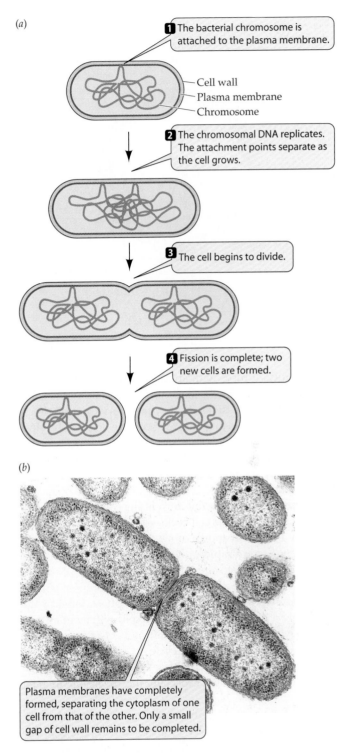

(a)

1 The bacterial chromosome is attached to the plasma membrane.

Cell wall
Plasma membrane
Chromosome

2 The chromosomal DNA replicates. The attachment points separate as the cell grows.

3 The cell begins to divide.

4 Fission is complete; two new cells are formed.

(b)

Plasma membranes have completely formed, separating the cytoplasm of one cell from that of the other. Only a small gap of cell wall remains to be completed.

9.3 Prokaryotic Cell Division
(*a*) The steps of cell division in prokaryotes. (*b*) These two cells of the bacterium *Pseudomonas aeruginosa* have almost completed fission. Each cell contains a complete chromosome, visible as the nucleoid in the center of the cell.

Interphase and the Control of Cell Division

Between divisions of the cytoplasm—that is, for most of its life—a eukaryotic cell is in a condition called **interphase**. A cell lives and functions until it divides or dies—or, if it is a

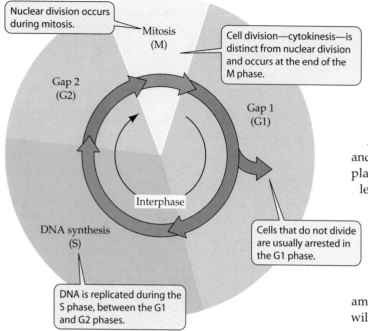

Nuclear division occurs during mitosis.

Cell division—cytokinesis—is distinct from nuclear division and occurs at the end of the M phase.

Cells that do not divide are usually arrested in the G1 phase.

DNA is replicated during the S phase, between the G1 and G2 phases.

9.4 The Eukaryotic Cell Cycle

The cell cycle consists of a mitotic (M) phase, during which first nuclear division (mitosis) and then cell division (cytokinesis) take place. The M phase is followed by a long period of growth known as interphase. Interphase has three subphases (G1, S, and G2) in cells that divide.

Although one key event—DNA replication—dominates and defines the S phase, important cell cycle processes take place in the gap phases as well. G1 is quite variable in length in different cell types. Some rapidly dividing embryonic cells dispense with it entirely, while other cells may remain in G1 for weeks or even years. The biochemical hallmark of a G1 cell is that it is preparing for the S phase. It is at the G1-to-S transition that the commitment to enter another cell cycle is made. During G2, the cell makes preparations for mitosis—for example, synthesizing components of the microtubules that will form the spindle.

Cyclins and other proteins signal events in the cell cycle

How are appropriate decisions to enter the S or M phases made? These transitions—from G1 to S and from G2 to M—depend on the activation of a protein called **cyclin-dependent kinase**, or **Cdk**. A *kinase* is an enzyme that catalyzes the transfer of a phosphate group from ATP to another molecule; this phosphate transfer is called *phosphorylation*. Cdk is a kinase that can catalyze the phosphorylation of certain amino acids in proteins.* Activated Cdk's are important in initiating the steps of the cell cycle.

The discovery that Cdk's induce cell division is a beautiful example of research on different organisms and different cell types converging on a single mechanism. One group of scientists was studying immature sea urchin eggs, trying to find out how they are stimulated to divide and form mature eggs. A protein called *maturation promoting factor* was purified from the maturing eggs, which by itself prodded the immature eggs into division. At the same time, other scientists studying the cell cycle in yeast, a single-celled eukaryote, found a strain that was stalled at the G1–S boundary because it lacked a Cdk. This yeast Cdk was very similar to the sea urchin's maturation promoting factor. Similar Cdk's were soon found to control the G1–S transition in many other organisms, including humans.

But Cdk's are not active by themselves. They must be bound to a second type of protein, called **cyclin**. This binding—an example of allosteric interaction—causes the Cdk to alter its shape and exposes its active site. It is the cyclin-Cdk *complex* that acts as a protein kinase and triggers the transition from G1 to S phase. Then the cyclin breaks down and the Cdk becomes inactive (Figure 9.5).

*Phosphorylation changes the three-dimensional structure of the targeted protein, sometimes simultaneously changing that protein's function. This important biochemical process is discussed further in Chapters 12 and 15.

sex cell, until it fuses with another sex cell. Some types of cells, such as red blood cells, muscle cells, and nerve cells, lose the capacity to divide as they mature. Other cell types, such as cortical cells in plant stems, divide only rarely. Most cells, however, have some probability of dividing, and some are specialized for rapid division. For most types of cells, we may speak of a **cell cycle** that has two phases: mitosis and interphase.

A given cell lives for one turn of the cell cycle and then becomes two cells. The cell cycle, when repeated again and again, is a constant source of new cells. However, even in tissues engaged in rapid growth, cells spend most of their time in interphase. Examination of any collection of dividing cells, such as the tip of a root or a slice of liver, will reveal that most of the cells are in interphase most of the time; only a small percentage of the cells will be in mitosis at any given moment. We can confirm this fact by watching a single cell through its entire cycle.

In this section, we will describe the cell cycle events that occur during interphase, especially the "decision" to enter mitosis.

Interphase consists of three subphases, identified as G1, S, and G2. The cell's DNA replicates during the **S phase** (the S stands for *synthesis*). The period between the end of mitosis and the onset of the S phase is called **G1**, or Gap 1. Another gap—**G2**—separates the end of the S phase and the beginning of mitosis, when nuclear and cytoplasmic division take place and two new cells are formed. Mitosis and cytokinesis are referred to as the **M phase** of the cell cycle (Figure 9.4).

The process of DNA replication is a major topic by itself, and will be discussed in Chapter 11. But its result is, at the end of S phase, where there was formerly one chromosome, there are now two, joined together and awaiting segregation into two new cells by mitosis or meiosis.

1 Cyclin D is made at the beginning of G1.

Cdk4 molecule

2 Cyclin D binds to Cdk4, changing its structure and activating it.

Cyclin D

Cyclin D–Cdk4 complex

5 Cdk4 is present throughout the cell cycle but is active only during G1.

3 Protein kinase activity stimulates progress through G1.

4 Cyclin D is broken down at the end of G1, rendering Cdk4 inactive.

9.5 Cyclin-Dependent Kinase and Cyclin Trigger Decisions in the Cell Cycle
A human cell makes the decision to enter the cell cycle during G1, when cyclin D binds to a cyclin-dependent kinase (Cdk4). There are four such cyclin-Cdk controls during the typical cell cycle in humans.

Several different cyclin-Cdk combinations act at various stages of the mammalian cell cycle:

▶ Cyclin D-Cdk4 acts during the middle of G1. This is the *restriction point*, a key decision point beyond which the rest of the cell cycle is normally inevitable (see Figure 9.5).
▶ Cyclin E-Cdk2 acts at the G1–S boundary, initiating DNA replication.
▶ Cyclin A-Cdk2 acts during S, and also stimulates DNA replication.
▶ Cyclin B-Cdk1 acts at the G2–M boundary, initiating the transition to chromosome condensation and mitosis.

The cyclin-Cdk complexes act as *checkpoints*, points at which cell cycle progress can be monitored to determine if the next step can be taken. For example, if DNA is damaged by radiation during G1, a protein called p21 is made. (The p stands for "protein," and the 21 stands for its molecular weight—about 21,000 daltons.) The p21 protein then binds to the two G1 Cdk's, preventing their activation by cyclins. So the cell cycle stops while repairs are made to DNA. The p21 protein itself is targeted for degradation, so that it breaks down after the DNA is repaired, allowing cyclins to bind to the Cdk's and the cell cycle to proceed.

What molecules do cyclin-Cdk complexes target for phosphorylation? Some important targets are known. For example, the cyclin B-Cdk1 complex catalyzes the phosphorylation of target proteins that then bind to DNA and initiate chromosome condensation. Phosphorylation of other target proteins results in the disaggregation of the nuclear envelope early in mitosis.

Because cancer results from inappropriate cell division, it is not surprising that the cyclin-Cdk controls are disrupted in cancer cells. For example, some fast-growing breast cancers have too much cyclin D, which overstimulates Cdk4 and cell division. As we will describe in Chapter 18, a major protein in normal cells that prevents them from dividing is p53, which leads to inhibition of Cdk's. More than half of all human cancers contain defective p53, resulting in the absence of cell cycle controls.

Growth factors can stimulate cells to divide

Cyclin-Cdk complexes provide an *internal* control for progress through the cell cycle. But there are situations in the body in which cells that are slowly cycling, or not cycling at all, must be stimulated to divide through *external* controls, called **growth factors**. When you cut yourself and bleed, specialized cell fragments called platelets gather at the wound and help initiate blood clotting. The platelets also produce and release a protein, called *platelet-derived growth factor*, that diffuses to the adjacent cells in the skin and stimulates them to divide and heal the wound.

Other growth factors include *interleukins*, which are made by one type of white blood cell and promote cell division in other cells that are essential for the body's immune system defenses. *Erythropoietin*, made by the kidney, stimulates the division of bone marrow cells and the production of red blood cells. In addition, many hormones promote division in specific cell types.

We will describe the physiological roles of these external mitotic inducers in later chapters, but all growth factors act in a similar way. They bind to their target cells via specialized receptor proteins on the target cell surface. This specific binding triggers events within the target cell that initiate a cell division cycle. Cancer cells often cycle inappropriately because they make their own growth factors, or because they no longer require growth factors to start cycling.

Eukaryotic Chromosomes

Most human cells other than eggs and sperm contain two full sets of genetic information, one from the mother and the other from the father. As in prokaryotes, this genetic information consists of molecules of DNA packaged as chromosomes. However, unlike prokaryotes, eukaryotes have more than one chromosome, and during interphase these chromosomes reside within a membrane-enclosed organelle, the nucleus.

The basic unit of the eukaryotic chromosome is a gigantic, linear, double-stranded molecule of DNA complexed

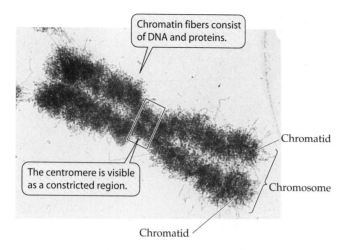

Chromatin fibers consist of DNA and proteins.

The centromere is visible as a constricted region.

Chromatid

Chromosome

Chromatid

9.6 Chromosomes, Chromatids, and Chromatin
A human chromosome, shown as the cell prepares to divide.

with many proteins. During most of the eukaryotic cell cycle, each chromosome contains only one such double-stranded DNA molecule. However, after the DNA molecule replicates during the S phase, the chromosome consists of two joined **chromatids**, each made up of one double-stranded DNA molecule complexed with proteins (Figure 9.6). The two chromatids are joined together at a specific small region called the **centromere**.

Chromatin consists of DNA and proteins

The complex of DNA and proteins that makes up a eukaryotic chromosome is referred to as **chromatin**. The DNA carries the genetic information; the proteins organize the chromosome physically and regulate the activities of the DNA. By mass, the amount of chromosomal protein is equivalent to that of DNA.

Chromatin changes form dramatically during mitosis and meiosis. During interphase, the chromatin is strung out so thinly that the chromosome cannot be seen in the nucleus under the light microscope. But during most of mitosis and meiosis, the chromatin is highly coiled and compacted, so that the chromosome appears as a dense, bulky object (see Figure 9.6).

This alternation of forms relates to the function of chromatin during different phases of the cell cycle. Before each mitosis, the genetic material is replicated. Mitosis separates this replicated genetic material into two new nuclei. This separation is easier to accomplish if the DNA is neatly arranged in compact units rather than being tangled up like a plate of spaghetti. During interphase, however, the DNA must direct the activities of the cell. Such functions require that portions of the DNA be unwound and exposed so that it can interact with enzymes.

Chromatin proteins organize the DNA in chromosomes

The DNA of a typical human cell has a total length of 2 meters. Yet the nucleus is only 5 μm (0.000005 meters) in diam-

eter. So, although the DNA in an interphase nucleus is "unwound," it is still impressively packed! This packing is achieved largely by proteins associated closely with the chromosomal DNA (Figure 9.7).

During interphase, chromosomes contain large quantities of proteins called **histones** (from the Greek word meaning "web"). There are five classes of histones. All of them have a positive charge at cellular pH levels because of their high content of the basic amino acids lysine and arginine. These positive charges electrostatically attract the negative phosphate groups on DNA. These interactions, as well as interactions among the histones themselves, form beadlike units called **nucleosomes**. Each nucleosome contains:

▶ Eight histone molecules, two each of four of the histone classes, united to form a core or spool.

▶ 146 base pairs of DNA, 1.65 turns of it wound around the histone core.

▶ Histone H1 (the remaining histone class) on the outside of the DNA, which may clamp it to the histone core.

Interphase chromatin is made up of a single DNA molecule running around vast numbers of nucleosomes like beads on a string. Between the nucleosomes stretches a variable amount of non-nucleosomal "linker" DNA. Since this DNA is exposed to the nuclear environment, it is accessible to proteins involved in its duplication and the regulation of its expression, as we will see in Chapter 14. There are also proteins that bind to nucleosomal DNA.

The many nucleosomes of a mitotic chromatid may pack together and coil. During both mitosis and meiosis, the chromatin becomes ever more coiled and condensed, with further folding of the chromatin continuing up to the time at which chromosomes begin to move apart.

Mitosis: Distributing Exact Copies of Genetic Information

In mitosis, a single nucleus gives rise to two nuclei that are genetically identical to each other and to the parent nucleus. This process ensures the accurate distribution of the eukaryotic cell's multiple chromosomes to the daughter nuclei. In reality, mitosis is a continuous process in which each event flows smoothly into the next. For discussion, however, it is convenient to look at mitosis—the M phase of the cell cycle—as a series of separate events, or subphases: prophase, prometaphase, metaphase, anaphase, and telophase, as shown on pages 162–163.

The centrosomes determine the plane of cell division

Once the commitment to enter mitosis has been made, the cell enters S phase, and DNA is replicated. At the same time, a pair of **centrosomes** ("central bodies") forms from a single centrosome that lies near the nucleus. This duplication is under the control of cyclin E-Cdk2, whose concentration peaks at the G1-to-S transition. This is the key event in orienting the direction of mitosis.

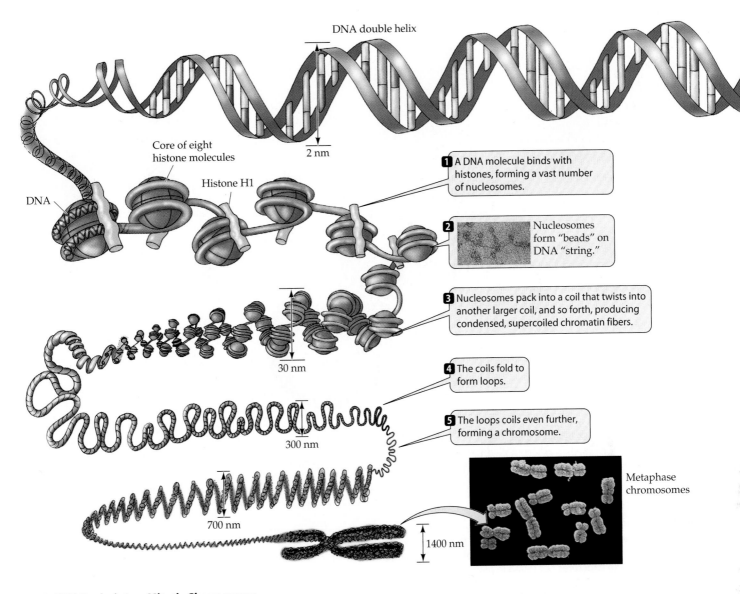

DNA double helix

2 nm

Core of eight
histone molecules

Histone H1

DNA

1 A DNA molecule binds with
histones, forming a vast number
of nucleosomes.

2 Nucleosomes
form "beads" on
DNA "string."

3 Nucleosomes pack into a coil that twists into
another larger coil, and so forth, producing
condensed, supercoiled chromatin fibers.

30 nm

4 The coils fold to
form loops.

300 nm

5 The loops coils even further,
forming a chromosome.

700 nm

Metaphase
chromosomes

1400 nm

9.7 DNA Packs into a Mitotic Chromosome
The nucleosome, formed by DNA and histones, is the essential
building block in this highly packed structure.

At the G2-to-M transition, the two centrosomes separate from each other, moving to opposite ends of the nuclear envelope. The orientation of the centrosomes determines the plane at which the cell will divide, and therefore the spatial relationship of the two new cells to the parent cell. This relationship may be of little consequence to single free-living cells such as yeasts, but it is important for cells that make up part of a body tissue.

In many organisms, each centrosome contains a pair of **centrioles**. Each pair consists of one "parent" centriole and one smaller "daughter" centriole at right angles to the parent centriole (see Figure 4.27). Their role is not clear, although they do appear to be necessary for centrosome function. Centrioles, if present, replicate during interphase: The two paired centrioles first separate, and then each acts as a "parent" for the formation of a new "daughter" centriole at right angles to it.

The centrosomes are the regions of the cell that initiate the formation of microtubules, which will orchestrate chromosomal movement. These regions are not enclosed by membranes and are not visible as discrete objects, but their positions are evident from the arrangement of nearby microtubules. Plant cells lack centrosomes, but distinct microtubule organizing centers at either end of the cell serve the same role.

The spindle forms during prophase

During interphase, only the nuclear envelope, the nucleoli, and a barely discernible tangle of chromatin are visible under the light microscope. The appearance of the nucleus changes as the cell enters **prophase**—the beginning of mitosis (Figure 9.8).

Each of the two centrosomes serves as a **mitotic center** that organizes microtubules. The two mitotic centers can be thought of as two poles toward which the chromosomes will move. **Polar microtubules** that form between the mitotic centers make up the developing **spindle**. The spindle

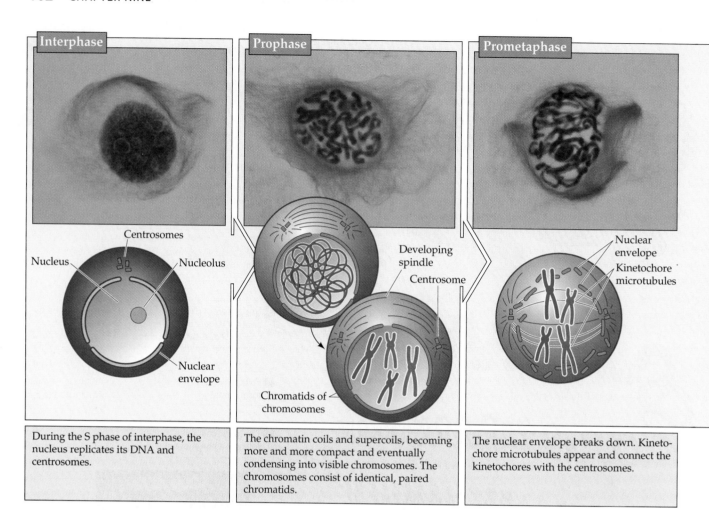

Interphase

Centrosomes

Nucleus

Nucleolus

Nuclear envelope

During the S phase of interphase, the nucleus replicates its DNA and centrosomes.

Prophase

Developing spindle

Centrosome

Chromatids of chromosomes

The chromatin coils and supercoils, becoming more and more compact and eventually condensing into visible chromosomes. The chromosomes consist of identical, paired chromatids.

Prometaphase

Nuclear envelope

Kinetochore microtubules

The nuclear envelope breaks down. Kinetochore microtubules appear and connect the kinetochores with the centrosomes.

9.8 Mitosis

Mitosis results in two new nuclei that are genetically identical to one another and to the nucleus from which they formed. The photomicrographs here are of plant nuclei, which lack centrioles. The diagrams are of corresponding phases in animal cells and introduce the structures not found in plants. In the micrographs, the red dye stains microtubules (and thus the spindle); the blue dye stains the chromosomes. In the diagrams, the chromosomes are stylized to emphasize the fates of the individual chromatids.

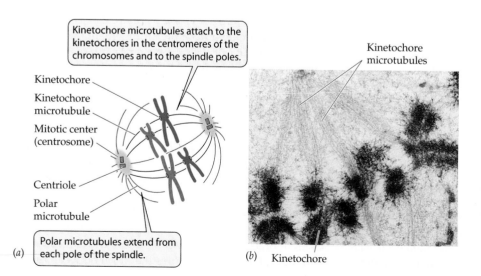

Kinetochore microtubules attach to the kinetochores in the centromeres of the chromosomes and to the spindle poles.

Kinetochore

Kinetochore microtubule

Mitotic center (centrosome)

Centriole

Polar microtubule

Polar microtubules extend from each pole of the spindle.

(a)

Kinetochore microtubules

(b) Kinetochore

9.9 The Mitotic Spindle Consists of Microtubules

(a) Diagram of the spindle apparatus in a cell at metaphase. (b) An electron micrograph of the stage shown in (a).

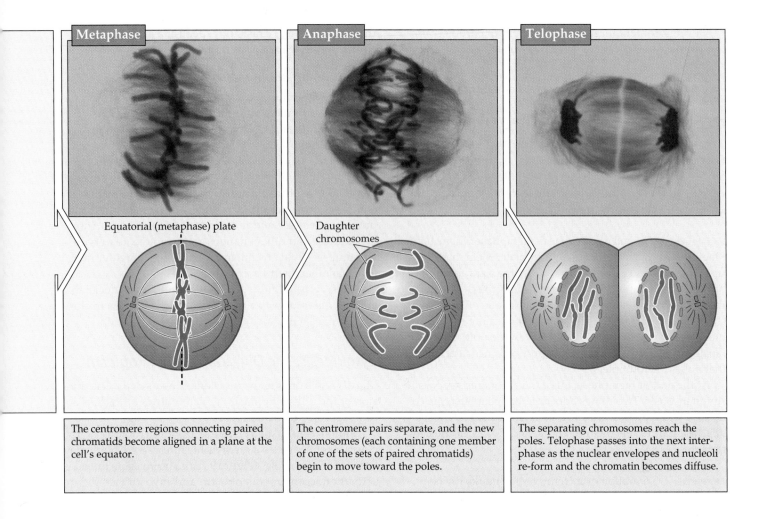

Metaphase

Equatorial (metaphase) plate

The centromere regions connecting paired chromatids become aligned in a plane at the cell's equator.

Anaphase

Daughter chromosomes

The centromere pairs separate, and the new chromosomes (each containing one member of one of the sets of paired chromatids) begin to move toward the poles.

Telophase

The separating chromosomes reach the poles. Telophase passes into the next interphase as the nuclear envelopes and nucleoli re-form and the chromatin becomes diffuse.

serves as a "railroad track" along which the chromosomes will move, as well as a framework keeping the two poles apart.

The spindle is actually two *half spindles*: Each polar microtubule runs from one mitotic center to the middle of the spindle, where it overlaps with polar microtubules of the other half spindle (Figure 9.9). The polar microtubules are initially unstable, constantly forming and falling apart, until they contact polar microtubules from the other half spindle and become more stable.

A prophase chromosome consists of two chromatids

The chromatin also changes during prophase. The extremely long, thin fibers take on a more orderly form as a result of coiling and compacting (see Figures 9.7 and 9.8). Under the light microscope, each prophase chromosome can be seen to consist of two chromatids held tightly together over much of their length. The two chromatids of a single mitotic chromosome are identical in structure, chemistry, and the hereditary information they carry because of the way in which DNA replicates during the S phase.

Within the region of tight binding of the chromatids lies the centromere, which is where chromatids become associated with the microtubules of the spindle. Very late in prophase, specialized three-layered structures called **kinetochores** develop in the centromere region, one on each chromatid (see Figure 9.9). The kinetochores are the sites at which microtubules will attach to the chromatids.

Chromosome movements are highly organized

The next three phases of mitosis—prometaphase, metaphase, and anaphase—are the phases during which chromosomes actually move. During these phases, the centromeres holding the two chromatids together separate, and the former chromatids—now called daughter chromosomes—move away from each other in opposite directions.

PROMETAPHASE. At the beginning of **prometaphase**, the nuclear lamina disintegrates and the nuclear envelope breaks into small vesicles, allowing the developing spindle to "invade" the nuclear region. The polar microtubules begin to attach to chromatids at their kinetochores, at which point they are called **kinetochore microtubules** (see Figure 9.9). The kinetochore of one chromatid is attached to microtubules coming from one pole, while the kinetochore of its sister chromatid is attached to microtubules emanating from the other pole.

During prometaphase, the movement of chromosomes toward the poles is counteracted by two factors:

▶ A repulsive force from the poles pushes the chromosomes toward the middle region, or **equatorial (metaphase) plate**, between the poles.
▶ The two chromatids are held together, apparently by proteins called *cohesins*.

So, during prometaphase, chromosomes appear to move aimlessly back and forth between the poles and the middle of the spindle. Gradually, the kinetochores approach the equatorial plate (see Figure 9.9).

METAPHASE. The cell is said to be in **metaphase** when all the kinetochores arrive at the equatorial plate. Metaphase lasts up to an hour, and is the best time to see the sizes and shapes of chromosomes. Because a microtubule (or a bundle of them) from one of the poles is attached to one of the kinetochores in each chromatid pair, that kinetochore (and chromatid) is oriented toward that pole. By default, the other kinetochore faces the other pole, and becomes attached to that pole's microtubule(s).

At the end of metaphase, all of the chromatids separate simultaneously. Two things appear to happen: First, the cohesins break down, and then, an enzyme called *DNA topoisomerase II* unravels the interconnected DNA's at the centromere.

ANAPHASE. Separation of the chromatids marks the beginning of **anaphase**, the phase of mitosis during which the two sister chromatids of each chromosome—now called *daughter chromosomes*, each containing one double-stranded DNA molecule—move to opposite ends of the spindle.

What propels this highly organized mass migration, which takes about 10 minutes, is not clear. Two things seem to move the chromosomes along. First, at the kinetochores are proteins that act as "molecular motors." These proteins, called *cytoplasmic dynein*, have the ability to hydrolyze ATP to ADP and phosphate, thus releasing energy to move the chromosomes along the microtubules toward the poles. These motor proteins account for about 75 percent of the force of motion. Second, the kinetochore microtubules shorten from the poles, drawing the chromosomes toward them. This accounts for about 25 percent of the motion.

During anaphase the poles of the spindle are pushed farther apart, doubling the distance between them. The distance between poles increases because polar microtubules from opposite ends of the spindle contain motor proteins that cause them to slide past each other, pushing the poles apart in much the same way that microtubules slide in cilia and flagella (see Chapter 4). This polar separation contributes to the separation of one set of daughter chromosomes from the other.

The movements of chromosomes are slow, even in cellular terms. At about 1 μm per minute, it takes about 10–60 minutes for them to complete their journey to the poles. This is like a human taking 7 million years to travel across the United States! This slow speed may ensure that the chromosomes segregate accurately.

Nuclei re-form during telophase

When the chromosomes stop moving at the end of anaphase, the cell enters **telophase**. Two sets of chromosomes containing identical DNA, carrying identical sets of hereditary instructions, are now at the opposite ends of the spindle, which begins to break down. The chromosomes begin to uncoil, continuing until they become the diffuse tangle of chromatin that is characteristic of interphase. The nuclear envelopes and nucleoli, which were disaggregated during prophase, coalesce and re-form their respective structures. When these and other changes are complete, telophase—and mitosis—is at an end, and each of the daughter nuclei enters another interphase.

Mitosis is beautifully precise. Its result is two nuclei that are *identical to each other* and to the parent nucleus in chromosomal makeup, and hence in genetic constitution.

Cytokinesis: The Division of the Cytoplasm

Mitosis refers only to the division of the nucleus. The division of the cell's cytoplasm, which follows mitosis, is accomplished by cytokinesis.

Animal cells usually divide by a furrowing of the plasma membrane, as if an invisible thread were tightening between the two poles (Figure 9.10*a*). The invisible thread is actually microfilaments of actin and myosin (see Figure 4.23*a*) located in a ring just beneath the plasma membrane. These two proteins interact to produce a contraction, just as they do in muscles, thus pinching the cell in two. These microfilaments assemble rapidly from actin monomers that are present in the interphase cytoskeleton. Their assembly appears to be under the control of Ca^{2+} released from storage sites in the center of the cell.

Plant cell cytoplasm divides differently, because plants have cell walls. As the spindle breaks down after mitosis, membranous vesicles derived from the Golgi apparatus appear in the equatorial region roughly midway between the two daughter nuclei. Moving along microtubules, these vesicles fuse to form new plasma membrane and contribute their contents to a *cell plate*, which is the beginning of a new cell wall (Figure 9.10*b*).

Following cytokinesis, both daughter cells contain all the components of a complete cell. A precise distribution of chromosomes is ensured by mitosis. Organelles such as ribosomes, mitochondria, and chloroplasts need not be distributed equally between daughter cells as long as some of each are present in both cells; accordingly, there is no mechanism with a precision comparable to that of mitosis to provide for their equal allocation to daughter cells.

Reproduction: Sexual and Asexual

The mitotic cell cycle repeats itself. By this process, a single cell can give rise to a vast number of others. Meiosis, on the

(*a*)

The division furrow has completely separated the cytoplasm of one daughter cell from another, although their surfaces remain in contact.

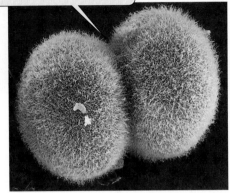

(*b*)

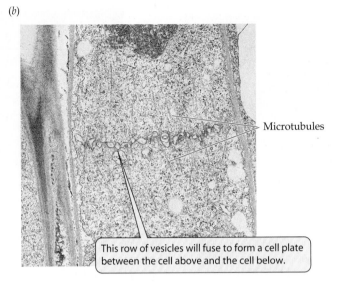

Microtubules

This row of vesicles will fuse to form a cell plate between the cell above and the cell below.

9.10 *Cytokinesis Differs in Animal and Plant Cells*
Plant cells form cell walls and thus must divide differently from animal cells. (*a*) A sea urchin egg that has just completed cytokinesis at the end of the first cell division of its development into an embryo. (*b*) A dividing plant cell in late telophase.

other hand, results in only four progeny, which usually do not undergo further duplications. These two methods of nuclear and cell division have different reproductive roles.

Reproduction by mitosis results in genetic constancy

A single cell that undergoes mitosis may be an entire organism reproducing itself with each cell cycle, or a cell that divides further to produce a multicellular organism. A multicellular organism, in turn, may be able to reproduce itself by releasing cells derived from mitosis and cytokinesis as a spore, or by having a multicellular piece break away and grow on its own (Figure 9.11).

A dividing unicellular organism and a multicellular organism reproducing by releasing cells both provide examples of **asexual reproduction**, sometimes called *vegetative reproduction*. This mode of reproduction is based on mitotic division of the nucleus and, accordingly, produces a **clone** of offspring that are genetically identical to the parent. If

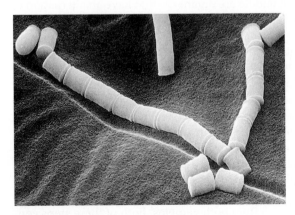

9.11 *Asexual Reproduction*
These spool-shaped cells are asexual spores formed by a fungus. Each spore contains a nucleus produced by a mitotic division. A spore is the same genetically as the parent that fragmented to produce it.

there is any variation among the offspring, it is likely to be due to *mutations*, or changes, in the genetic material. Asexual reproduction is a rapid and effective means of making new individuals, and it is common in nature.

Sexual reproduction, which involves meiosis, is very different. In sexual reproduction two parents, each contributing one cell, produce offspring that differ genetically from each parent as well as from each other. This variety among the offspring means that some of them may be better adapted than others to reproduce in a particular environment.

Reproduction by meiosis results in genetic diversity

Sexual reproduction, which combines genetic information from two different cells, fosters genetic diversity. The hallmarks of all sexual life cycles are:

▶ There are two parents, each of which provides chromosomes to the offspring in the form of a sex cell, or **gamete**.
▶ Each gamete contains a single set of chromosomes.
▶ The two gametes—often identifiable as a female egg and a male sperm—fuse to produce a single cell, the **zygote**, or fertilized egg. This fusion is called **fertilization**. The zygote thus contains two sets of chromosomes.

In each recognizable pair of chromosomes, one chromosome comes from each of two parents. The members of such a **homologous pair** are similar in size and appearance (except for the so-called *sex chromosomes* found in some species; the two members of these pairs are different, and their differences determine the sex of the organism, as we will see in Chapter 10). The two chromosomes (the ho-

Fungus (*Rhizopus oligosporus*)

Fern (*Osmunda cinnamomcea*)

Elephant (*Loxodonta africana*)

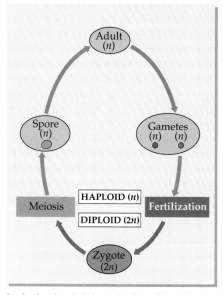

In the **haplontic life cycle**, the adult is haploid and the zygote is the only diploid stage.

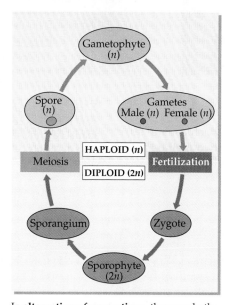

In **alternation of generations**, there are both haploid and diploid adult stages.

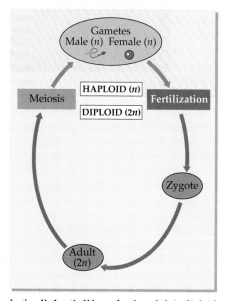

In the **diplontic life cycle**, the adult is diploid and the gametes are the only haploid stage.

9.12 Fertilization and Meiosis Alternate in Sexual Reproduction
In sexual reproduction, haploid (*n*; yellow) cells or organisms alternate with diploid (*2n*; blue) cells or organisms.

mologs) of a homologous pair bear corresponding, though generally not identical, types of genetic information.

Haploid cells contain only one homolog from each pair of chromosomes. The number of chromosomes in such a single set is denoted by *n*. When haploid gametes fuse in fertilization, the resulting zygote has two homologs of each type. It is thus said to be **diploid**, denoted *2n*.

As you can see in Figure 9.12, sexual life cycles exhibit different patterns of development after zygote formation. In *haplontic* organisms, such as protists and many fungi, the mature organism is haploid. The zygote undergoes a reduction division—meiosis—to produce haploid cells, or *spores*. These spores then form the new organism by mitosis of haploid cells, which may be single-celled or multicellular. Gametes are then produced by this organism by mitosis. So in haplontic organisms, the zygote is the only diploid cell in the life cycle.

At the other extreme are *diplontic* organisms, which include animals and some plants. Here, the gametes are the only haploid cells, and the organism itself is diploid. Gametes are formed by meiosis, and the formation of the organism involves mitosis of diploid cells.

In the middle are organisms that have an alternation of haploid and diploid generations. Most plants fall into this category. Here, the zygote divides by mitosis of diploid cells into a diploid organism. Meiosis does not give rise to gametes, but instead to haploid spores that divide by mitosis to form an alternate, haploid life stage. It is this haploid organism that forms gametes by mitosis, and after fertilization the cycle begins anew. We will look at all of these life cycles in greater detail in subsequent chapters.

The essence of sexual reproduction is the random selection of half of a parent's diploid chromosome set to make a haploid gamete, followed by the fusion of two such haploid gametes to produce a diploid cell that contains genetic information from both gametes. Both of these steps contribute to a shuffling of genetic information in the population, so no two individuals have exactly the same genetic constitution. The diversity provided by sexual reproduction opens up enormous opportunities for evolution.

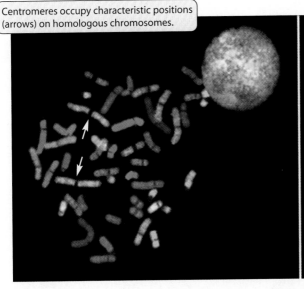

Centromeres occupy characteristic positions (arrows) on homologous chromosomes.

The karyotype shows 23 pairs of chromosomes, including the sex chromosomes. This female's sex chromosomes are X and X; a male would have X and Y chromosomes.

9.13 Human Cells Have 46 Chromosomes
Chromosomes from a human cell are shown in metaphase of mitosis. In this "chromosome painting" technique, each homologous pair shares a distinctive color. The multicolored globe is an interphase nucleus. The karyotype on the right is produced by computerized analysis of the image on the left.

The number, shapes, and sizes of the metaphase chromosomes constitute the karyotype

When nuclei are in metaphase of mitosis, it is often possible to count and characterize the individual chromosomes. This is a relatively simple process in some organisms, thanks to techniques that can capture cells in metaphase and spread out the chromosomes. A photograph of the entire set of chromosomes can then be made, and the images of the individual chromosomes can be placed in an orderly arrangement. Such a rearranged photograph reveals the

number, shapes, and sizes of chromosomes in a cell, which together constitute its **karyotype** (Figure 9.13).

Individual chromosomes can be recognized by their lengths, the positions of their centromeres, and characteristic banding when they are stained and observed at high magnification. When the cell is diploid, the karyotype consists of homologous pairs of chromosomes—23 pairs for a total of 46 chromosomes in humans, and greater or smaller numbers of pairs in other diploid species. There is no simple relationship between the size of an organism and its chromosome number (Table 9.1).

Meiosis: A Pair of Nuclear Divisions

Meiosis consists of two nuclear divisions that reduce the number of chromosomes to the haploid number in preparation for sexual reproduction. Although the *nucleus divides twice* during meiosis, the *DNA is replicated only once*. To understand the process of meiosis and its specific details, it is useful to keep in mind the overall functions of meiosis:

▶ To reduce the chromosome number from diploid to haploid.

▶ To ensure that each of the haploid products has a complete set of chromosomes.

▶ To promote genetic diversity among the products.

Two unique features characterize the first meiotic division, **meiosis I**. The first is that homologous chromosomes pair along their entire lengths. This process, called **synapsis**, lasts from prophase to the end of metaphase. The second is that after this metaphase, the homologous chromosomes separate. The individual chromosomes, each consisting of two joined chromatids, remain intact until the end of the metaphase of **meiosis II**, the second meiotic division. In the discussion that follows, you can refer to Figure 9.14 to help you visualize each step.

9.1	Numbers of Pairs of Chromosomes in Some Plant and Animal Species	
COMMON NAME	**SPECIES**	**NUMBER OF CHROMOSOME PAIRS**
Mosquito	*Culex pipiens*	3
Housefly	*Musca domestica*	6
Toad	*Bufo americanus*	11
Rice	*Oryza sativa*	12
Frog	*Rana pipiens*	13
Alligator	*Alligator mississippiensis*	16
Rhesus monkey	*Macaca mulatta*	21
Wheat	*Triticum aestivum*	21
Human	*Homo sapiens*	23
Potato	*Solanum tuberosum*	24
Donkey	*Equus asinus*	31
Horse	*Equus caballus*	32
Dog	*Canis familiaris*	39
Carp	*Cyprinus carpio*	52

MEIOSIS I

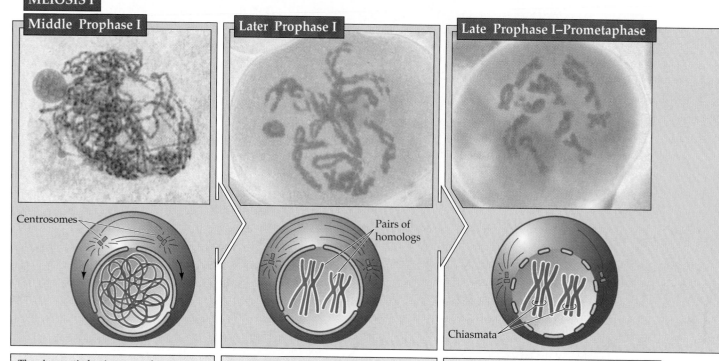

Middle Prophase I

Centrosomes

The chromatin begins to condense following interphase.

Later Prophase I

Pairs of homologs

Synapsis aligns homologs, and chromosomes condense. Homologs are shown in different colors indicating those coming from each parent. In reality, their differences are very small, usually comprising different alleles of some genes.

Late Prophase I–Prometaphase

Chiasmata

The chromosomes continue to coil and shorten. Crossing-over at chiasmata results in an exchange of genetic material. In prometaphase the nuclear envelope breaks down.

MEIOSIS II

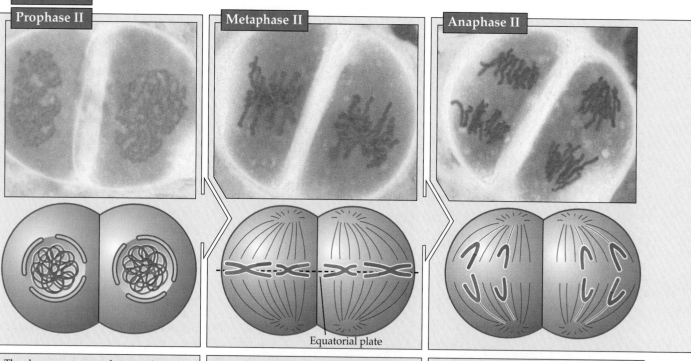

Prophase II

The chromosomes condense again, following a brief interphase (interkinesis) in which DNA does not replicate.

Metaphase II

Equatorial plate

Kinetochores of the paired chromatids line up across the equatorial plates of each cell.

Anaphase II

The chromatids finally separate, becoming chromosomes in their own right, and are pulled to opposite poles. Because of crossing over in prophase I, each new cell will have a different genetic makeup.

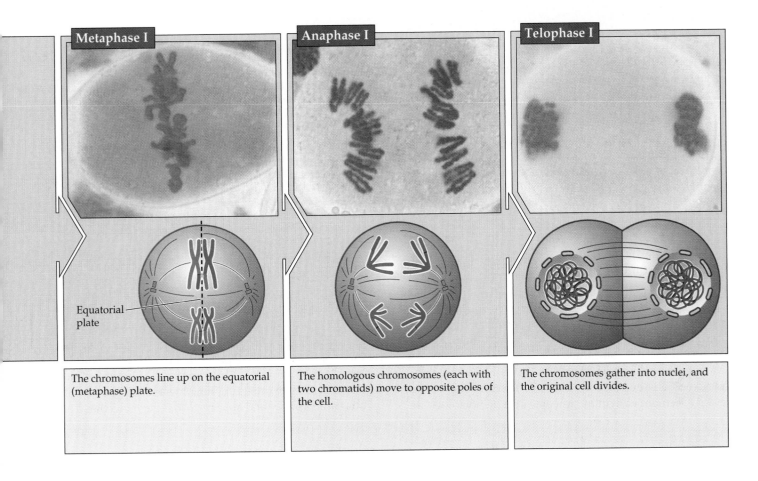

Metaphase I

The chromosomes line up on the equatorial (metaphase) plate.

Equatorial plate

Anaphase I

The homologous chromosomes (each with two chromatids) move to opposite poles of the cell.

Telophase I

The chromosomes gather into nuclei, and the original cell divides.

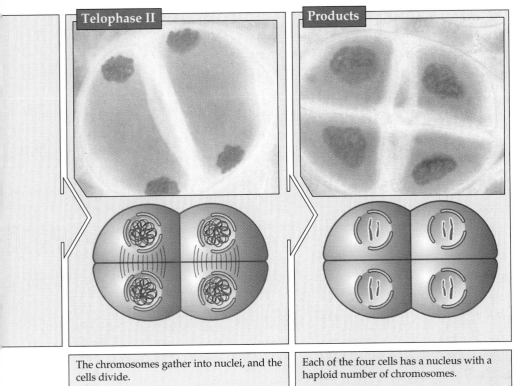

Telophase II

The chromosomes gather into nuclei, and the cells divide.

Products

Each of the four cells has a nucleus with a haploid number of chromosomes.

9.14 Meiosis
In meiosis, two sets of chromosomes are divided among four nuclei, each of which then has half as many chromosomes as the original cell. These four haploid cells are the result of two successive nuclear divisions. The photomicrographs shown here are of meiosis in the male reproductive organ of a lily. As in Figure 9.8, the diagrams show corresponding phases in an animal.

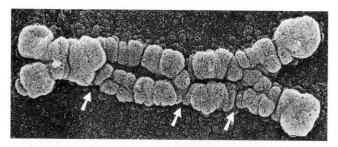

9.15 Chiasmata: Evidence of Exchange between Chromatids
Chiasmata are visible near the middle of some chromatids from a desert locust in this scanning electron micrograph, and near the ends of others. Three chiasmata are indicated with arrows.

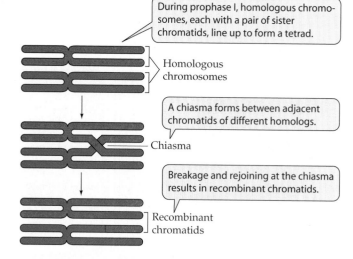

During prophase I, homologous chromosomes, each with a pair of sister chromatids, line up to form a tetrad.

Homologous chromosomes

A chiasma forms between adjacent chromatids of different homologs.

Chiasma

Breakage and rejoining at the chiasma results in recombinant chromatids.

Recombinant chromatids

9.16 Crossing Over Forms Genetically Diverse Chromosomes
The exchange of genetic material by crossing over may result in new combinations of genetic information on the recombinant chromosomes.

The first meiotic division reduces the chromosome number

Like mitosis, meiosis I is preceded by an interphase with an S phase during which each chromosome is replicated. As a result, each chromosome consists of two sister chromatids.

Meiosis I begins with a long prophase I (the first three frames of Figure 9.14), during which the chromosomes change markedly. A key change is that homologous chromosomes join together, or *synapse*. By the time they can be clearly seen under light microscope, the two homologs are already tightly joined. This joining begins at the centromeres and is mediated by a recognition of homologous DNA sequences on homologous chromosomes. In addition, a special group of proteins may form a scaffold called the *synaptonemal complex* that runs lengthwise along the homologous chromosomes and appears to join them together. The

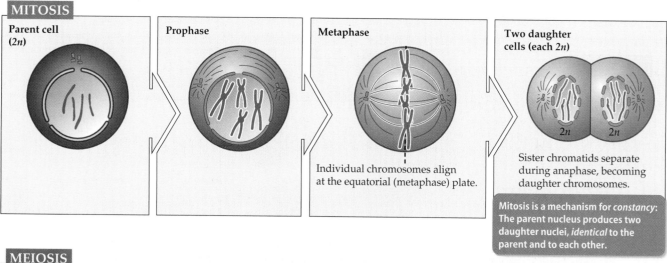

MITOSIS

Parent cell (2n)

Prophase

Metaphase

Individual chromosomes align at the equatorial (metaphase) plate.

Two daughter cells (each 2n)

Sister chromatids separate during anaphase, becoming daughter chromosomes.

Mitosis is a mechanism for *constancy*: The parent nucleus produces two daughter nuclei, *identical* to the parent and to each other.

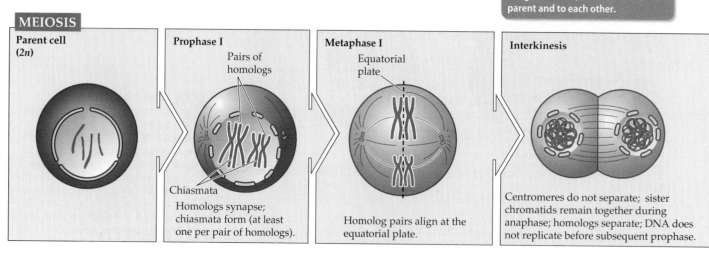

MEIOSIS

Parent cell (2n)

Prophase I

Pairs of homologs

Chiasmata

Homologs synapse; chiasmata form (at least one per pair of homologs).

Metaphase I

Equatorial plate

Homolog pairs align at the equatorial plate.

Interkinesis

Centromeres do not separate; sister chromatids remain together during anaphase; homologs separate; DNA does not replicate before subsequent prophase.

four chromatids of each pair of homologous chromosomes form a *tetrad*, or *bivalent*. To summarize: A tetrad is four chromatids, two each from two homologous chromosomes. For example, there are 46 chromosomes in a human diploid cell, so there are 23 homologous pairs of chromosomes, each with two chromatids, for a total of 92 chromatids during prophase I. In other words, there are 23 tetrads, each containing two homologous chromosomes and four chromatids.

Throughout prophase I and metaphase I, the chromatin continues to coil and compact progressively, so the chromosomes appear ever thicker. At a certain point, the homologous chromosomes seem to *repel* each other, especially near the centromeres, but they are held together by physical attachments. Regions having these attachments take on an X-shaped appearance and are called **chiasmata** (from the Greek word "chiasma," meaning "cross"; Figure 9.15). A chiasma reflects an exchange of material between chromatids on homologous chromosomes—what geneticists call *crossing over* (Figure 9.16). The chromosomes begin exchanging material shortly after synapsis begins, but the chiasmata do not become visible until later, when the homologs are repelling each other.

Crossing over increases the genetic variation among the products of meiosis. We will have a great deal to say about crossing over and its genetic consequences in the coming chapters.

There seems to be plenty of time for the complicated events of prophase I to occur. Whereas mitotic prophase is usually measured in minutes, and all of mitosis seldom takes more than an hour or two, meiosis can take much longer. In human males, the cells in the testis that undergo meiosis take about a week for prophase I and about a month for the entire meiotic cycle. In the cells that will become eggs, prophase I begins long before a woman's birth, during early fetal development, and ends as much as decades later, during the monthly ovarian cycle.

9.17 Mitosis and Meiosis: A Comparison
Meiosis differs from mitosis by synapsis and by the failure of the centromeres to separate at the end of metaphase I.

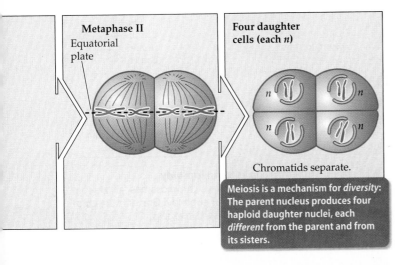

Meiosis is a mechanism for *diversity*: The parent nucleus produces four haploid daughter nuclei, each *different* from the parent and from its sisters.

Prophase I is followed by prometaphase I (not pictured in Figure 9.14), during which the nuclear envelope and the nucleoli disappear. A spindle forms, and microtubules become attached to the kinetochores of the chromosomes. In meiosis I, there is only *one kinetochore per chromosome*, not one per chromatid as in mitosis. Thus the entire chromosome, consisting of two chromatids, will migrate to one pole in the meiotic cell.

By metaphase I, all the chromosomal kinetochores have become connected to polar microtubules, and all the chromosomes have moved to the equatorial plate. Until this point, they have been held together by chiasmata.

The homologous chromosomes separate in anaphase I, when *individual chromosomes*, each still consisting of two chromatids, are pulled to the poles, with one homolog of a pair going to one pole and the other homolog going to the opposite pole. (Note that this process differs from the separation of *chromatids* during mitotic anaphase.) Each of the two daughter nuclei from this division is haploid; that is, it contains only one set of chromosomes, not the two sets that were present in the original diploid nucleus. However, because they consist of two chromatids rather than just one, each of these chromosomes has twice the mass that a chromosome at the end of a mitotic division has.

In some species, but not in others, there is a telophase I, with the reappearance of nuclear envelopes and so forth. When there is a telophase I, it is followed by an interphase, called **interkinesis**, similar to the mitotic interphase. During interkinesis the chromatin is partially uncoiled; however, there is no replication of the genetic material, because each chromosome already consists of two chromatids. Furthermore, the sister chromatids in interkinesis are generally not genetically identical, because crossing over in prophase I has reshuffled genetic material between maternal and paternal chromosomes.

The second meiotic division separates the chromatids

Meiosis II is similar to mitosis in many ways. In each nucleus produced by meiosis I, the chromosomes line up at equatorial plates in metaphase II, the chromatids—each of which has a centromere—separate, and new daughter chromosomes move to the poles in anaphase II.

The three major differences between meiosis II and mitosis are:

► DNA replicates before mitosis, but not before meiosis II.
► In mitosis, the sister chromatids that make up a given chromosome are identical; in meiosis II, they differ over part of their length if they participated in crossing over during prophase of meiosis I.
► The number of chromosomes on the equatorial plate of each of the two nuclei in meiosis II is half the number in the single mitotic nucleus.

Figure 9.17 compares mitosis and meiosis. The result of meiosis is four nuclei; each nucleus is haploid and has a single set of chromosomes that differs from other such sets in its exact genetic composition. The differences, to repeat a very important point, result from crossing over during prophase I and from the segregation of homologous chromosomes during anaphase I.

Meiosis leads to genetic diversity

What are the consequences of the synapsis and separation of homologous chromosomes during meiosis? In mitosis, each chromosome behaves independently of its homolog; its two chromatids are sent to opposite poles at anaphase. If we start a mitotic division with x chromosomes, we end up with x chromosomes in each daughter nucleus, and each chromosome consists of one chromatid. In meiosis, things are very different.

In meiosis, synapsis organizes things so that chromosomes of maternal origin pair with their paternal homologs. Then their separation during meiotic anaphase I ensures that each pole receives one member of each homologous pair. (Remember that each chromosome still consists of two chromatids.) For example, at the end of meiosis I in humans, each daughter nucleus contains 23 of the original 46 chromosomes. In this way, the chromosome number is decreased from diploid to haploid. Furthermore, meiosis I guarantees that each daughter nucleus gets one full set of chromosomes, for it must have one of each homologous pair.

The products of meiosis I are genetically diverse for two reasons. First, synapsis during prophase I allows the maternal chromosome to interact with the paternal one; after crossing over, the recombinant chromatids contain some genetic material from each chromosome. Second, which member of a homologous pair goes to which daughter cell at anaphase I is a matter of pure chance. For example, if there are two pairs of chromosomes in the diploid parent nucleus, a particular daughter nucleus could get paternal chromosome 1 and maternal chromosome 2, or paternal 2 and maternal 1, or both maternals, or both paternals. It all depends on the way in which the homologous pairs line up at metaphase I.

Note that of the four possible chromosome combinations just described, two produce daughter nuclei that are the same as one of the parental types (except for any material exchanged by crossing over). The greater the number of chromosomes, the less probable that the original parental combinations will be reestablished, and the greater the potential for genetic diversity. Most species of diploid organisms do, indeed, have more than two pairs. In humans, with 23 chromosome pairs, 2^{23} different combinations can be produced.

Meiotic Errors

A pair of homologous chromosomes may fail to separate during meiosis I, or sister chromatids may fail to separate during meiosis II or during mitosis. This phenomenon is called **nondisjunction**, and it results in the production of aneuploid cells (Figure 9.18). **Aneuploidy** is a condition in

which one or more chromosomes or pieces of chromosomes are either lacking or present in excess.

Aneuploidy can give rise to genetic abnormalities

One reason for nondisjunction may be a lack of chiasmata. Recall that these structures, formed during prophase I, hold the two homologous chromosomes together into metaphase I. This ensures that one homolog will face one pole and the other homolog the other pole. Without this "glue," the two homologs may line up randomly at metaphase I, just like chromosomes during mitosis, and there is a 50 percent chance that both will go to the same pole.

If, for example, the chromosome 21 pair fails to separate during the formation of a human egg (and thus both mem-

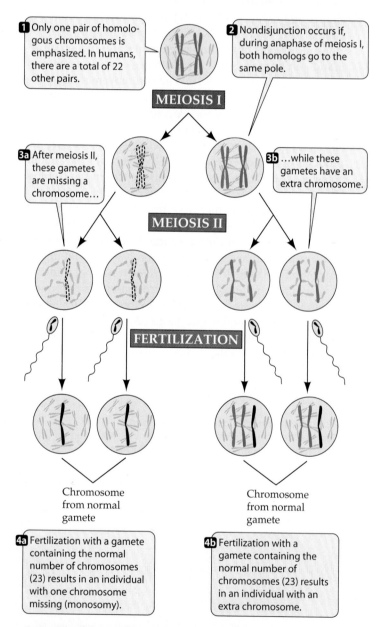

9.18 Nondisjunction Leads to Aneuploidy
Nondisjunction occurs if homologous chromosomes fail to separate during meiosis I. The result is aneuploidy: One or more chromosomes are either lacking or present in excess.

bers of the pair go to one pole during anaphase I), the resulting egg will contain either two of chromosome 21 or none at all. If an egg with two of these chromosomes is fertilized by a normal sperm, the resulting zygote will have three copies of the chromosome: It will be **trisomic** for chromosome 21. A child with an extra chromosome 21 demonstrates the symptoms of *Down syndrome:* impaired intelligence; characteristic abnormalities of the hands, tongue, and eyelids; and an increased susceptibility to cardiac abnormalities and diseases such as leukemia.

Other abnormal events can also lead to aneuploidy. In a process called *translocation,* a piece of a chromosome may break away and become attached to another chromosome. For example, a particular large part of one chromosome 21 may be translocated to another chromosome. Individuals who inherit this translocated piece along with two normal chromosomes 21 will have Down syndrome.

Trisomies (and the corresponding monosomies) are surprisingly common in human zygotes, but most of the embryos that develop from such zygotes do not survive to birth. Trisomies for chromosomes 13, 15, and 18 greatly reduce the probability that an embryo will survive to birth, and virtually all infants who are born with such trisomies die before the age of 1 year. Trisomies and monosomies for other chromosomes are lethal to the embryo. At least one-fifth of all recognized pregnancies spontaneously terminate during the first two months, largely because of such trisomies and monosomies. (The actual proportion of spontaneously terminated pregnancies is certainly higher, because the earliest ones often go unrecognized.)

Polyploids can have difficulty in cell division

Both diploid and haploid nuclei divide by mitosis. Multicellular diploid and multicellular haploid individuals develop from single-celled beginnings by mitotic divisions. Likewise, mitosis may proceed in diploid organisms even when a chromosome from one of the haploid sets is missing or when there is an extra copy of one of the chromosomes (as in Down syndrome).

Under some circumstances, triploid ($3n$), tetraploid ($4n$), and higher-order polyploid nuclei may form. Each of these *ploidy levels* represents an increase in the number of complete sets of chromosomes present. If, by accident, the nucleus has one or more extra full sets of chromosomes—that is, if it is triploid, tetraploid, or of still higher ploidy—this abnormally high ploidy in itself does not prevent mitosis. In mitosis, each chromosome behaves independently of the others.

In meiosis, by contrast, chromosomes synapse to begin division. If even one chromosome has no homolog, anaphase I cannot send representatives of that chromosome to both poles. A diploid nucleus can undergo normal meiosis; a haploid one cannot. A tetraploid nucleus has an even number of each kind of chromosome, so each chromosome can pair with its homolog. But a triploid nucleus cannot undergo normal meiosis, because one-third of the chromosomes would lack partners.

This limitation has important consequences for the fertility of triploid, tetraploid, and other chromosomally unusual organisms that may be produced by plant breeding or by natural accidents. Modern bread wheat plants are hexaploids, the result of the accidental crossing of three different grasses, each having its own diploid set of 14 chromosomes.

Cell Death

As we mentioned at the start of this chapter, an essential role of cell division in complex eukaryotes is to replace cells that die. In humans, billions of cells die each day, mainly in the blood and the epithelia lining organs such as the intestine. Cells die in one of two ways. The first, **necrosis**, occurs when cells either are damaged by poisons or are starved of essential nutrients. These cells usually swell up and burst, releasing their contents into the extracellular environment. This often results in inflammation (see Chapter 19). The scab that forms around a wound is a familiar example of necrotic tissue.

More typically, cell death in an organism is due to **apoptosis** (from the Greek word meaning "falling off"). Apoptosis is a prescribed series of events that constitute genetically programmed cell death. These two ways for cells to die are compared in Table 9.2.

9.2 **Two Different Ways for Cells to Die**		
	NECROSIS	**APOPTOSIS**
Stimuli	Low O_2, toxins, ATP depletion, damage	Specific, genetically programmed physiological signals
ATP required	No	Yes
Cellular pattern	Swelling, organelle disruption, tissue death	Chromatin condensation, membrane blebbing, single-cell death
DNA breakdown	Random fragments	Nucleosome-sized fragments
Plasma membrane	Burst	Blebbed (see Figure 9.19*b*)
Fate of dead cells	Ingested by phagocytes	Ingested by neighboring cells
Reaction in tissue	Inflammation	No inflammation

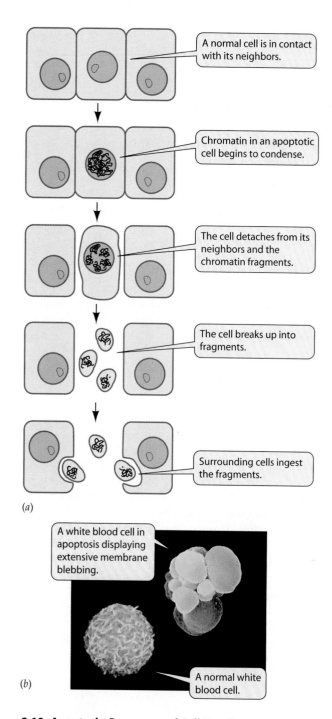

(a)

A normal cell is in contact with its neighbors.

Chromatin in an apoptotic cell begins to condense.

The cell detaches from its neighbors and the chromatin fragments.

The cell breaks up into fragments.

Surrounding cells ingest the fragments.

A white blood cell in apoptosis displaying extensive membrane blebbing.

(b)

A normal white blood cell.

9.19 Apoptosis: Programmed Cell Death
Many cells are genetically programmed to "self-destruct" when they are no longer needed, or when they have lived long enough to accumulate a burden of DNA damage that might harm the organism.

Why would a cell initiate apoptosis, which is essentially "cell suicide"? One reason is that the cell in question is no longer needed by the organism. For example, before birth, a human fetus has weblike hands, with connective tissue between the fingers. As development proceeds, this unneeded tissue disappears as its cells undergo apoptosis (see Figure 16.10).

A second reason for apoptosis is that the longer cells live, the more prone they are to damage that could lead to cancer. This is especially true of cells in the blood and intestine, which are exposed to high levels of toxic substances. In these cases, cells "sacrifice their lives for the good of the organism." Such cells normally die after only days or weeks.

Like the cell division cycle, the cell death cycle has signals controlling its progress. These include the lack of a mitotic signal, such as a growth factor, and recognition of DNA damage. As we will see in Chapter 17, many of the drugs used to treat diseases of cell proliferation such as cancer work via these signals.

The events of apoptosis are very similar in most organisms (Figure 9.19). The cell becomes isolated from its neighbors, chops up its chromatin into nucleosome-sized pieces, and then fragments itself. In a remarkable example of the economy of nature, the surrounding living cells usually ingest the remains of the dead cell. The genetic signals that lead to apoptosis are also common to many organisms.

Chapter Summary

▶ Cell division is necessary for reproduction, growth, and repair of an organism. **Review Figure 9.1**

Systems of Cell Reproduction

▶ Cell division must be initiated by a reproductive signal. Cell division consists of three steps: replication of the genetic material (DNA), partitioning of the two DNA molecules to separate portions of the cell, and division of the cytoplasm.

▶ In prokaryotes, cellular DNA is a single molecule, or chromosome. Prokaryotes reproduce by cell fission. **Review Figure 9.3**

▶ In eukaryotes, nuclei divide by either mitosis or meiosis.

Interphase and the Control of Cell Division

▶ The mitotic cell cycle has two main phases: interphase (during which cells are not dividing) and mitosis (when cells divide).

▶ During most of the cell cycle the cell is in interphase, which is divided into three subphases: S, G1, and G2. DNA is replicated during S phase. **Review Figure 9.4**

▶ Cyclin-Cdk complexes regulate the passage of cells from G1 into S phase and from G2 into M phase. **Review Figure 9.5**

▶ In addition to the internal cyclin-Cdk complexes, controls external to the cell, such as growth factors and hormones, can also stimulate the cell to begin a division cycle.

Eukaryotic Chromosomes

▶ Chromosomes contain DNA and proteins. At mitosis, chromosomes initially appear to be double because two sister chromatids are held together at the centromere. Each sister chromatid consists of one double-stranded DNA molecule complexed with proteins and referred to as chromatin. **Review Figure 9.6**

▶ During interphase, the DNA in chromatin is wound around cores of histones to form nucleosomes. DNA folds over and over again, packing itself within the nucleus. When mitotic chromosomes form, it folds even more. **Review Figure 9.7**

Mitosis: Distributing Exact Copies of Genetic Information

▶ After DNA is replicated during S phase, the first sign of mitosis is the separation of centrosomes, which initiate microtubule formation for the spindle. **Review Figure 9.9**

▶ Mitosis can be divided into several phases, called prophase, prometaphase, metaphase, anaphase, and telophase. **Review Figure 9.8**

▶ During prophase, the chromosomes condense and appear as paired chromatids.

▶ During prometaphase, the chromosomes move toward the middle of the spindle. In metaphase, they gather at the middle of the cell with their centromeres on the equatorial plate. At the end of metaphase, the centromeres holding the chromatid pairs together separate, and during anaphase each member of the pair, now called a daughter chromosome, migrates to its pole along the microtubule track.

▶ During telophase, the chromosomes become less condensed. The nuclear envelopes and nucleoli re-form, thus producing two nuclei whose chromosomes are identical to each other and to those of the cell that began the cycle. **Review Figure 9.8**

Cytokinesis: The Division of the Cytoplasm

▶ Nuclear division is usually followed by cytokinesis. Animal cell cytoplasm usually divides by a furrowing of the plasma membrane, caused by the contraction of cytoplasmic microfilaments. In plant cells, cytokinesis is accomplished by vesicle fusion and the synthesis of new cell wall material. **Review Figure 9.10**

Reproduction: Sexual and Asexual

▶ The cell cycle can repeat itself many times, forming a clone of genetically identical cells.

▶ Asexual reproduction produces a new organism that is genetically identical to the parent. Any genetic variety is the result of mutations.

▶ In sexual reproduction, two haploid gametes—one from each parent—unite in fertilization to form a genetically unique, diploid zygote. **Review Figure 9.12**

▶ In sexually reproducing organisms, certain cells in the adult undergo meiosis, a process by which a diploid cell produces haploid gametes. Each gamete contains a random mix of one of each pair of homologous chromosomes from the parent.

▶ The number, shapes, and sizes of the chromosomes constitute the karyotype of an organism. **Review Figure 9.13**

Meiosis: A Pair of Nuclear Divisions

▶ Meiosis reduces the chromosome number from diploid to haploid and ensures that each haploid cell contains one member of each chromosome pair. It consists of two nuclear divisions. **Review Figure 9.14**

▶ During prophase I of the first meiotic division, homologous chromosomes pair up with each other, and material may be exchanged by crossing over between nonsister chromatids of two adjacent homologs. In metaphase I, the paired homologs gather at the equatorial plate. Each chromosome has only one kinetochore and associates with polar microtubules for one pole. In anaphase I, entire chromosomes, each with two chromatids, migrate to the poles. By the end of meiosis I, there are two nuclei, each with the haploid number of chromosomes with two sister chromatids. **Review Figures 9.14, 9.16**

▶ In meiosis II, the sister chromatids separate. No DNA replication precedes this division, which in other aspects is similar to mitosis. The result of meiosis is four cells, each with a haploid chromosome content. **Review Figures 9.14, 9.17**

▶ Both crossing over during prophase I and the random selection of which homolog of a pair migrates to which pole during anaphase I ensure that the genetic composition of each haploid gamete is different from that of the parent and from that of the other gametes. The more chromosome pairs there are in a diploid cell, the greater the diversity of chromosome combinations generated by meiosis.

Meiotic Errors

▶ In nondisjunction, one member of a homologous pair of chromosomes fails to separate from the other, and both go to the same pole. This event leads to one gamete with an extra chromosome and another other lacking that chromosome. Fertilization with a normal haploid gamete results in aneuploidy and genetic abnormalities that are invariably harmful or lethal to the organism. **Review Figure 9.18**

Cell Death

▶ Cells may die by necrosis or may self-destruct by apoptosis, a genetically programmed series of events that includes the detachment of the cell from its neighbors and the fragmentation of its nuclear DNA. **Review Figure 9.19**

Applying Concepts

1. Compare chromatids and chromosomes. At what stages during mitosis and meiosis are chromatids present?

2. Strains of organisms unable to carry out certain functions in the cell cycle have been invaluable to scientists to determine what happens in the normal cell cycle. Describe the cell cycle in cells
   a lacking the G1 cyclin.
   b. lacking the mitotic spindle.
   c. lacking the microfilaments involved in plasma membrane contraction.

3. Compare the sequence of events in the mitotic cell cycle with the sequence in programmed cell death.

4. The potato plant has 24 pairs of chromosomes. What is the number of
   a. chromatids in a cell at prophase of mitosis?
   b. chromosomes in a cell at anaphase of mitosis?
   c. chromatids in a cell at metaphase I of meiosis?
   d. chromatids in a cell at prophase II of meiosis?

10

Genetics: Mendel and Beyond

In the middle eastern desert 1,800 years ago, a rabbi faced a serious dilemma. A Jewish woman had given birth to a son. As required by laws first set down by God's commandment to Abraham almost 2,000 years previously and reiterated later by Moses, the mother brought her 8-day-old son to the rabbi for ritual penile circumcision. The rabbi knew that the woman's two previous sons had bled to death when their foreskins were cut. Yet the Biblical commandment remained: Unless he was circumcised, the boy could not be counted among those with whom God had made His solemn covenant. After consultation with other rabbis, it was decided to exempt this, the third son.

Almost one thousand years later, in the twelfth century, the physician and biblical commentator Moses Maimonides reviewed this and numerous other cases in the rabbinical literature, and stated that in such instances the third son should not be circumcised. Furthermore, the ban should apply whether the son was "from her first husband or from her second husband." The bleeding disorder, he reasoned, was clearly carried by the mother and passed on to her sons.

Knowing nothing of our modern vision of genetics, these rabbis linked a human disease (which turns out to be hemophilia A) to a pattern of inheritance (which we know as sex linkage). Only in the past several decades have the precise biochemical nature of hemophilia A and its genetic determination been worked out.

How do we account for, and predict, such patterns of inheritance? In this chapter, we will discuss how the units of inheritance, called genes, are transmitted from generation to generation of plants and animals, and show how many of the rules that govern genetics can be explained by the behavior of chromosomes during meiosis. We will also describe the interactions of genes with one another and with the environment, and the consequences of the fact that genes occupy specific positions on chromosomes.

An Ancient Ritual
A male infant undergoes ritual circumcision in accordance with Jewish laws. Sons of Jewish mothers who carry the gene for hemophilia may be exempt from the ritual.

The Foundations of Genetics

Much of the early study of biological inheritance was done with plants and animals of economic importance. Records show that people were deliberately cross-breeding date palm trees and horses as early as 5,000 years ago. By the early 1800s, plant breeding was widespread, especially with ornamental flowers such as tulips. Half a century later, in 1866, Gregor Mendel used the knowledge of plant reproduction to design and conduct experiments on inheritance. Although his published results were neglected by scientists for 40 years, they ultimately became the foundation for the science of genetics.

Plant breeders showed that both parents contribute equally to inheritance

Plants are easily grown in large quantities, many produce large numbers of offspring (in the form of seeds), and many have relatively short generation times. In most plant species, the same individuals have both male and female reproductive organs, permitting each plant to reproduce as a male, as a female, or as both. Best of all, it is often easy to control which individuals mate (Figure 10.1).

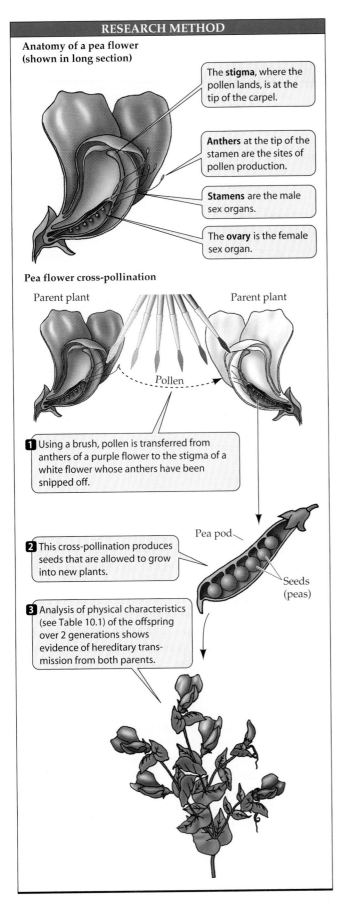

RESEARCH METHOD

Anatomy of a pea flower
(shown in long section)

The **stigma**, where the pollen lands, is at the tip of the carpel.

Anthers at the tip of the stamen are the sites of pollen production.

Stamens are the male sex organs.

The **ovary** is the female sex organ.

Pea flower cross-pollination

Parent plant

Parent plant

Pollen

1 Using a brush, pollen is transferred from anthers of a purple flower to the stigma of a white flower whose anthers have been snipped off.

Pea pod

2 This cross-pollination produces seeds that are allowed to grow into new plants.

Seeds (peas)

3 Analysis of physical characteristics (see Table 10.1) of the offspring over 2 generations shows evidence of hereditary transmission from both parents.

10.1 A Controlled Cross between Two Plants
Plants were widely used in early genetic studies because it is easy to control which individuals mate with which. Mendel used the pea plant, *Pisum sativum*, in many of his experiments.

Some discoveries that Mendel found useful in his studies had been made in the late eighteenth century by a German botanist, Josef Gottlieb Kölreuter. Kölreuter studied the offspring of reciprocal crosses between plants and showed that the two parents contributed equally to the characteristics inherited by their offspring. In a **reciprocal cross**, plants are crossed with (mated with) each other in opposite directions. For example, in one cross, males that have white flowers are mated with females that have red flowers, while in a complementary cross, red-flowered males and white-flowered females are the parents. In Kölreuter's experience, such reciprocal crosses always gave identical results.

Although the concept of equal parental contributions was an important discovery, the nature of what exactly the parents were contributing—the units of inheritance—remained unknown. Laws of inheritance proposed at the time favored the concept of *blending*. If a plant that had one form of a characteristic (say, red flowers) was crossed with one that had a different form of that characteristic (blue flowers), the offspring would be a blended combination of the two parents (purple flowers).

According to the blending concept, it was thought that once heritable elements were combined, they could not be separated again (like combined inks). The red and blue genetic determinants were thought to be forever blended into the new purple one. Then, about a century after Kölreuter completed his work, Mendel began his.

Mendel's discoveries were overlooked for decades

Gregor Mendel was an Austrian monk, not an academic scientist, but he was qualified to undertake scientific investigations. Although in 1850 he had failed an examination for a teaching certificate in natural science, he later undertook intensive studies in physics, chemistry, mathematics, and various aspects of biology at the University of Vienna. His work in physics and mathematics probably led him to apply experimental and quantitative methods to the study of heredity—and these were the key ingredients in his success.

Mendel worked out the basic principles of inheritance in plants over a period of about 9 years. His work culminated in a public lecture in 1865 and a detailed written account published in 1866. Mendel's paper appeared in a journal that was received by 120 libraries, and he sent reprinted copies (of which he had obtained 40) to several distinguished scholars. However, his theory was not accepted. In fact, it was ignored.

The chief difficulty was that the most prominent biologists of Mendel's time were not in the habit of thinking in mathematical terms, even the simple terms used by Mendel. Even Charles Darwin, whose theory of evolution by natural selection depended on genetic variation among individuals, failed to understand the significance of Mendel's findings. In fact, Darwin performed breeding experiments like Mendel's on snapdragons and got data similar to Mendel's, but he missed the point, still relying on the concept of blending. In addition, Mendel had little credibility as a biologist; in-

10.1 Mendel's Results from Monohybrid Crosses

DOMINANT × RECESSIVE			DOMINANT	RECESSIVE	TOTAL	RATIO
	Spherical seeds × Wrinkled seeds		5,474	1,850	7,324	2.96:1
	Yellow seeds × Green seeds		6,022	2,001	8,023	3.01:1
	Purple flowers × White flowers		705	224	929	3.15:1
	Inflated pods × Constricted pods		882	299	1,181	2.95:1
	Green pods × Yellow pods		428	152	580	2.82:1
	Axial flowers × Terminal flowers		651	207	858	3.14:1
	Tall stems × Dwarf stems (1 m) (0.3 m)		787	277	1,064	2.84:1

deed, his lowest grades were in biology! Whatever the reasons, Mendel's pioneering paper had no discernible influence on the scientific world for more than 30 years.

Then, in 1900, Mendel's discoveries burst into prominence as a result of independent experiments by three plant geneticists: the Dutch Hugo de Vries, the German Karl Correns, and the Austrian Erich von Tschermak. Each of these scientists carried out crossing experiments and obtained quantitative data about the progeny; each published his principal findings in 1900; each cited Mendel's 1866 paper. By that time, meiosis had been observed and described. At last the time was ripe for biologists to appreciate the significance of what these four geneticists had discovered.

Mendel's Experiments and the Laws of Inheritance

That Mendel was able to make his discoveries before the discovery of meiosis was due in part to the methods of experimentation he used. Mendel's work is a fine example of preparation, execution, and interpretation. Let's see how he approached each of these steps.

Mendel devised a careful research plan

Mendel chose the garden pea for his studies because of its ease of cultivation, the feasibility of controlled pollination (see Figure 10.1), and the availability of varieties with differing traits. He controlled pollination, and thus fertilization, of his parent plants by manually moving pollen from one plant to another. Thus he knew the parentage of the offspring in his experiments. If untouched, the pea plants Mendel studied naturally *self-pollinate*—that is, the female organ of each flower receives pollen from the male organs of the same flowers—and he made use of this natural phenomenon in some of his experiments.

Mendel began by examining different varieties of peas in a search for heritable characters and traits suitable for study. A **character** is a feature, such as flower color; a **trait** is a particular form of a character, such as white flowers. A **heritable** character trait is one that is passed from parent to offspring. Mendel looked for characters that had well-defined, contrasting alternative traits, such as purple flowers versus white flowers, that were true-breeding.

To be considered **true-breeding**, the observed trait must be the only form present for many generations. In other words, peas with white flowers, when crossed with one another, would have to give rise only to progeny with white flowers for many generations; tall plants bred to tall plants would have to produce only tall progeny.

Mendel isolated each of his true-breeding strains by repeated inbreeding (done by crossing of sibling plants that were seemingly identical, or allowing individuals to self-pollinate) and selection. In most of his work, Mendel concentrated on the seven pairs of contrasting traits shown in Table 10.1. Before performing any given cross, he made sure that each potential parent was from a true-breeding strain—an essential point in his analysis of his experimental results.

10.2 Contrasting Traits
In Experiment 1, Mendel studied the inheritance of seed shape. We know today that the wrinkled seeds possess an abnormal form of starch. Contrast their appearance with that of the spherical seeds below.

Mendel then collected pollen from one parental strain and placed it onto the stigma (female organ) of flowers of the other strain. The plants providing and receiving the pollen were the **parental generation**, designated **P**. In due course, seeds formed and were planted. The resulting new plants constituted the **first filial generation, F_1**. Mendel and his assistants examined each F_1 plant to see which traits it bore and then recorded the number of F_1 plants expressing each trait. In some experiments the F_1 plants were allowed to self-pollinate and produce a **second filial generation**, or F_2. Again, each F_2 plant was characterized and counted.

In sum, Mendel devised a well-organized plan of research, pursued it faithfully and carefully, recorded great amounts of quantitative data, and analyzed the numbers he recorded to explain the relative proportions of the different kinds of progeny. His 1866 paper stands to this day as a model of clarity. His results and the conclusions to which they led are the subject of the next few sections.

Mendel's Experiment 1 examined a monohybrid cross

"Experiment 1" in Mendel's paper involved a **monohybrid cross**—one in which each parent pea plant was true-breeding for a given character, but in this case each displayed a different form of that character (a different trait). He took pollen from plants of a true-breeding strain with wrinkled seeds and placed it on the stigmas of flowers of a true-breeding, spherical-seeded strain (Figure 10.2). He also performed the reciprocal cross, placing pollen from the spherical-seeded strain on the stigmas of flowers of the wrinkled-seeded strain.

In both cases, all the F_1 seeds that were produced were spherical—it was as if the wrinkled trait had disappeared completely. The following spring Mendel grew 253 F_1 plants from these spherical seeds, each of which was allowed to self-pollinate—this was the monohybrid cross—to produce F_2 seeds. In all, there were 7,324 F_2 seeds, of which 5,474 were spherical and 1,850 wrinkled (Figure 10.3).

Mendel concluded that the spherical seed trait was **dominant**: In the F_1 generation, it was always expressed rather

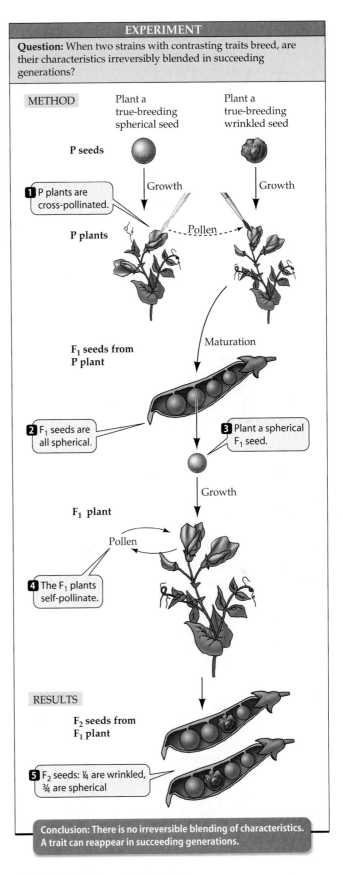

EXPERIMENT

Question: When two strains with contrasting traits breed, are their characteristics irreversibly blended in succeeding generations?

METHOD

Plant a true-breeding spherical seed

Plant a true-breeding wrinkled seed

P seeds

1 P plants are cross-pollinated.

Growth

Growth

P plants

Pollen

F_1 seeds from P plant

Maturation

2 F_1 seeds are all spherical.

3 Plant a spherical F_1 seed.

Growth

F_1 plant

Pollen

4 The F_1 plants self-pollinate.

RESULTS

F_2 seeds from F_1 plant

5 F_2 seeds: ¼ are wrinkled, ¾ are spherical

Conclusion: There is no irreversible blending of characteristics. A trait can reappear in succeeding generations.

10.3 Mendel's Experiment 1
The pattern Mendel observed in the F_2 generation—¼ of the seeds wrinkled, ¾ spherical—was the same no matter which variety contributed the pollen in the parental generation.

than the wrinkled seed trait, which he called **recessive**. In each of the other six pairs of traits Mendel studied, one proved to be dominant over the other. When he crossed plants differing in one of these traits, only one of each pair of traits was evident in the F_1 generation. However, the trait that was not seen in the F_1 reappeared in the F_2.

Of most importance, the ratio of the two traits in the F_2 generation was always the same—approximately 3:1. That is, three-fourths of the F_2 showed the dominant trait and one-fourth showed the recessive trait (see Table 10.1). In Mendel's Experiment 1, the ratio was 5,474:1,850 = 2.96:1. The reciprocal cross in the parental generation gave a similar outcome in the F_2.

By themselves, the results from Experiment 1 disproved the widely held belief that inheritance is always a blending phenomenon. According to the blending theory, Mendel's F_1 seeds should have had an appearance intermediate between those of the two parents—in other words, they should have been slightly wrinkled. Furthermore, the blending theory offered no explanation for the reappearance of the wrinkled trait in the F_2 seeds after its apparent absence in the F_1 seeds.

Mendel proposed that the units responsible for the inheritance of specific traits are present as discrete particles that occur in pairs and segregate (separate) from one another during the formation of gametes. According to this *particulate theory,* the units of inheritance retain their integrity in the presence of other units. The particulate theory is a sharp contrast to the concept of blending, in which the units of inheritance were believed to lose their identities when mixed together.

As he wrestled mathematically with his data, Mendel reached the conclusion that each pea plant has two units of inheritance for each character, one from each parent. During the production of gametes, only one of these paired units for a given character is given to a gamete. Hence each gamete contains one unit, and the resulting zygote contains two, because it was produced by the fusion of two gametes. This conclusion is the core of Mendel's model of inheritance. Mendel's unit of inheritance is now called a **gene**.

Mendel reasoned that in Experiment 1, the spherical-seeded parent had a pair of genes of the same type, which we will call *S,* and the parent with wrinkled seeds had two *s* genes. The *SS* parent produced gametes each containing a single *S,* and the *ss* parent produced gametes each with a single *s.* Each member of the F_1 generation had an *S* from one parent and an *s* from the other; an F_1 could thus be described as *Ss.* We say that *S* is dominant over *s* because the *s* trait is not evident when both forms of the gene are present.

The different forms of a gene (*S* and *s* in this case) are called **alleles**. Individuals that are true-breeding for a trait contain two copies of the same allele. For example, all the individuals in a population of a strain of true-breeding peas with wrinkled seeds must have the allele pair *ss*; if *S* were present, the plants would produce spherical seeds.

We say that the individuals that produce wrinkled seeds are **homozygous** for the allele *s,* meaning that they have

two copies of the same allele (*ss*). Some peas with spherical seeds—the ones with the genotype *SS*—are also homozygous. However, not all plants with spherical seeds have the *SS* genotype. Some spherical-seeded plants, like Mendel's F_1, are **heterozygous**: They have two different alleles of the gene in question (in this case, *Ss*).

To illustrate these terms with a more complex example, one in which there are three gene pairs, an individual with the genotype *AABbcc* is homozygous for the A and C genes—because it has two *A* alleles and two *c* alleles—but heterozygous for the B gene because it contains the *B* and *b* alleles. An individual that is homozygous for a character is sometimes called a *homozygote;* a *heterozygote* is heterozygous for the character in question.

The physical appearance of an organism is its **phenotype**. Mendel correctly supposed the phenotype to be the result of the **genotype**, or genetic constitution, of the organism showing the phenotype. In Experiment 1 we are dealing with two phenotypes (spherical seeds and wrinkled seeds). As we will see in the next section, the F_2 generation contains these two phenotypes and three genotypes. The wrinkled-seed phenotype is produced only by the genotype *ss,* whereas the spherical-seed phenotype may be produced by the genotypes *SS* or *Ss.*

Mendel's first law says that alleles segregate

How does Mendel's model of inheritance explain the composition of the F_2 generation in Experiment 1? Consider first the F_1, which has the spherical-seeded phenotype and the *Ss* genotype. According to Mendel's model, *when any individual produces gametes, the alleles separate, so that each gamete receives only one member of the pair of alleles.* This is Mendel's first law, the **law of segregation**. In Experiment 1, half the gametes produced by the F_1 contained the *S* allele and half the *s* allele.

During self-pollination, the random combination of gametes produces the F_2 generation (Figure 10.4). Three different F_2 genotypes are possible: *SS, Ss* (which is the same thing as *sS*), and *ss.* Our quantitative way of looking at things may lead us to wonder what proportions of these genotypes we might expect to observe in the F_2 progeny. The expected frequencies of these three genotypes may be determined by using the Punnett square, devised in 1905 by the British geneticist Reginald Crundall Punnett.

The **Punnett square** is a device that reminds us to consider all possible combinations of gametes. The square looks like this:

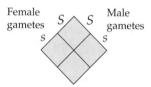

It is a simple grid with all possible sperm genotypes shown across one side and all possible egg genotypes along another side. To complete the grid, we fill in each square with the corresponding sperm genotype and egg genotype, giv-

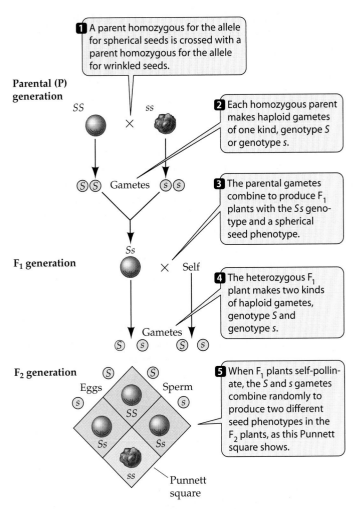

1 A parent homozygous for the allele for spherical seeds is crossed with a parent homozygous for the allele for wrinkled seeds.

Parental (P) generation

2 Each homozygous parent makes haploid gametes of one kind, genotype *S* or genotype *s*.

Gametes

3 The parental gametes combine to produce F₁ plants with the *Ss* genotype and a spherical seed phenotype.

F₁ generation

4 The heterozygous F₁ plant makes two kinds of haploid gametes, genotype *S* and genotype *s*.

Gametes

F₂ generation

Eggs   Sperm

5 When F₁ plants self-pollinate, the *S* and *s* gametes combine randomly to produce two different seed phenotypes in the F₂ plants, as this Punnett square shows.

Punnett square

10.4 Mendel's Explanation of Experiment 1
Mendel concluded that inheritance depends on factors from each parent, and that these factors are discrete units that do not blend in the offspring.

ing the diploid genotype of one member of the F₂ generation. For example, to fill the rightmost square, we put in the *S* from the egg (female gamete) and the *s* from the sperm (male gamete), yielding *Ss*.

Examination of the Punnett square in Figure 10.4 reveals that self-pollination of the F₁ genotype *Ss* will give the three F₂ genotypes in the expected ratio 1 *SS*:2 *Ss*:1 *ss*. Because *S* is dominant and *s* recessive, only two phenotypes result, in the ratio of 3 spherical (*SS* and *Ss*) to 1 wrinkled (*ss*), just as Mendel observed.

Mendel did not live to see his theory placed on a sound physical footing based on chromosomes and DNA. Genes are now known to be regions of the DNA molecules in chromosomes. More specifically, a gene is a portion of the DNA that resides at a particular site, called a **locus** (plural **loci**), within the chromosome and that encodes a particular function. Mendel arrived at his law of segregation with no knowledge of chromosomes or meiosis, but today we can picture the different alleles of a gene segregating as chromosomes separate in meiosis I (Figure 10.5).

Mendel verified his hypothesis by performing a test cross

The **test cross** is a way to test whether a given individual showing a dominant trait is homozygous or heterozygous. In a test cross, the individual in question is crossed with an individual known to be homozygous for the recessive trait—an easy individual to identify, because in order to have the recessive phenotype it must be homozygous for the recessive trait.

For the pea gene that we have been considering, the recessive homozygote used for the test cross is *ss*. The individual being tested may be described initially as *S*– because

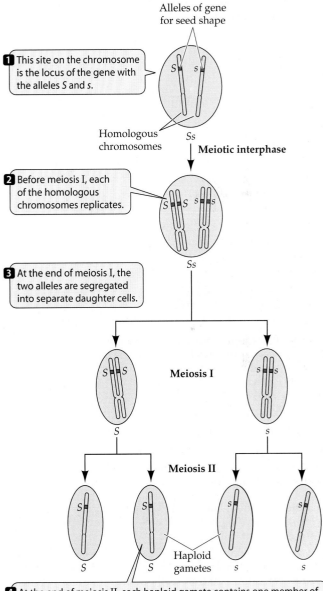

Alleles of gene for seed shape

1 This site on the chromosome is the locus of the gene with the alleles *S* and *s*.

Homologous chromosomes

Meiotic interphase

2 Before meiosis I, each of the homologous chromosomes replicates.

3 At the end of meiosis I, the two alleles are segregated into separate daughter cells.

Meiosis I

Meiosis II

Haploid gametes

4 At the end of meiosis II, each haploid gamete contains one member of each pair of homologous chromosomes, and thus one allele for each pair of genes.

10.5 Meiosis Accounts for the Segregation of Alleles
Although Mendel had no knowledge of chromosomes or meiosis, we now know that a pair of alleles resides on homologous chromosomes, and that meiosis segregates those alleles.

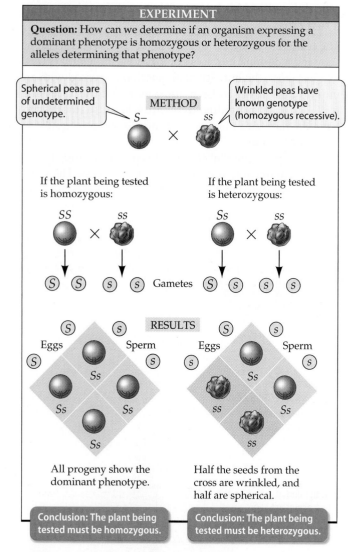

Question: How can we determine if an organism expressing a dominant phenotype is homozygous or heterozygous for the alleles determining that phenotype?

METHOD

Spherical peas are of undetermined genotype.

$S-$ × ss

Wrinkled peas have known genotype (homozygous recessive).

If the plant being tested is homozygous:

SS × ss

If the plant being tested is heterozygous:

Ss × ss

Gametes S S s s S s s s

RESULTS

Eggs Sperm

Ss Ss Ss Ss

All progeny show the dominant phenotype.

Eggs Sperm

Ss Ss ss ss

Half the seeds from the cross are wrinkled, and half are spherical.

Conclusion: The plant being tested must be homozygous.

Conclusion: The plant being tested must be heterozygous.

10.6 Homozygous or Heterozygous?
A plant with a dominant phenotype may be homozygous or heterozygous. Its genotype can be determined by making a test cross, which involves crossing it with a homozygous recessive plant and observing the phenotypes of the progeny produced.

we do not yet know the identity of the second allele. There are two possible results:

▶ If the individual being tested is homozygous dominant (SS), all offspring of the test cross will be Ss and show the dominant trait (spherical seeds).

▶ If the individual being tested is heterozygous (Ss), then approximately half of the offspring of the test cross will show the dominant trait (Ss), but the other half will be homozygous for, and will show, the recessive trait (ss) (Figure 10.6).

These were exactly the results that Mendel obtained; thus Mendel's model accurately predicts the results of such test crosses.

Mendel's second law says that alleles of different genes assort independently

What happens if two parents that differ at two or more loci are crossed? Consider an organism heterozygous for two genes, Ss and Yy, in which S and Y came from its mother and s and y came from its father. When this organism makes gametes, do the alleles of maternal origin (S and Y) go together to one gamete and those of paternal origin (s and y) to another gamete? Or can a single gamete receive one maternal and one paternal allele, S and y (or s and Y)? To answer these questions, Mendel performed a series of **dihybrid crosses**, crosses made between parents that are identical double heterozygotes.

In these experiments, Mendel began with peas that differed for two characters of the seeds: seed shape and seed color. One true-breeding strain produced only spherical, yellow seeds ($SSYY$) and the other strain produced only wrinkled, green ones ($ssyy$). A cross between these two strains produced an F_1 generation in which all the plants were $SsYy$. Because the S and Y alleles are dominant, these F_1 seeds were all yellow and spherical.

Mendel continued this experiment to the next generation—the dihybrid cross. There are two ways in which these doubly heterozygous plants might produce gametes, as Mendel saw it. (Remember that he had never heard of chromosomes or meiosis.)

First, if the alleles maintain the associations they had in the original parents (that is, if they are *linked*), then the F_1 plants should produce two types of gametes (SY and sy), and the F_2 progeny resulting from self-pollination of the F_1 plants should consist of three times as many plants bearing spherical, yellow seeds as ones with wrinkled, green seeds. Were such results to be obtained, there might be no reason to suppose that seed shape and seed color were regulated by two different genes, because spherical seeds would always be yellow, and wrinkled ones always green.

The second possibility is that the segregation of S from s is independent of the segregation of Y from y during the production of gametes (that is, that they are *unlinked*). In this case, four kinds of gametes should be produced, in equal numbers: SY, Sy, sY, and sy. When these gametes combine at random, they should produce an F_2 of nine different genotypes. The progeny could have any of three possible genotypes for shape (SS, Ss, or ss) and any of three possible genotypes for color (YY, Yy, or yy). The combined nine genotypes should produce just four phenotypes (spherical yellow, spherical green, wrinkled yellow, wrinkled green). By using a Punnett square, we can show that these four phenotypes would be expected to occur in a ratio of 9:3:3:1. (Figure 10.7).

Mendel's dihybrid crosses produced the results predicted by the second possibility. Four different phenotypes appeared in the F_2 in a ratio of about 9:3:3:1. The parental traits appeared in new combinations of two of the phenotypic classes (spherical green and wrinkled yellow). Such new combinations are called **recombinant phenotypes**.

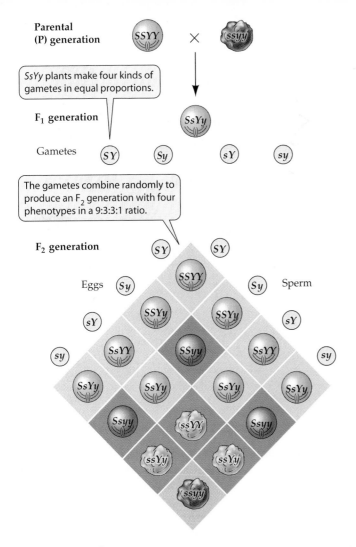

Parental (P) generation

SsYy plants make four kinds of gametes in equal proportions.

F₁ generation

Gametes

The gametes combine randomly to produce an F₂ generation with four phenotypes in a 9:3:3:1 ratio.

F₂ generation

Eggs   Sperm

10.7 Independent Assortment
The 16 possible combinations of gametes result in 9 different genotypes. Because *S* and *Y* are dominant over *s* and *y*, respectively, the 9 genotypes determine 4 phenotypes in the ratio of 9:3:3:1.

These results led Mendel to the formulation of what is now known as Mendel's second law: *Alleles of different genes assort independently of one another during gamete formation.* This **law of independent assortment** is not as universal as the law of segregation, because it applies to genes that lie on separate chromosomes but not necessarily to those that lie on the same chromosome. However, it is correct to say that *chromosomes* segregate independently during the formation of gametes, and so do any two genes on separate chromosome pairs (Figure 10.8).

Punnett squares or probability calculations: A choice of methods

Many people find it easiest to solve genetics problems using probability calculations, perhaps because the principle is familiar. When we flip a coin, for example, we expect that it has an equal probability of landing "heads" or "tails." For a given toss of a fair coin, the probability of heads is independent of what happened in all the previous tosses. A run of ten straight heads implies nothing about

the next toss. No "law of averages" increases the likelihood that the next toss will come up tails, and no "momentum" makes an eleventh occurrence of heads any more likely. On the eleventh toss, the odds are still 50:50.

The basic conventions of probability are simple: If an event is absolutely certain to happen, its probability is 1. If it cannot happen, its probability is 0. Otherwise, its probability lies between 0 and 1. A coin toss results in heads approximately half the time, and the probability of heads is ½—as is the probability of tails.

MULTIPLYING PROBABILITIES.   If two coins (a penny and a dime, say) are tossed, each acts independently of the other. What, then, is the probability of both coins coming up heads? Half the time, the penny comes up heads; of that fraction, half the time the dime also comes up heads. Therefore, the joint probability of two heads is half of one-half, or $\frac{1}{2} \times \frac{1}{2} = \frac{1}{4}$. To find the joint probability of independent events, then, the

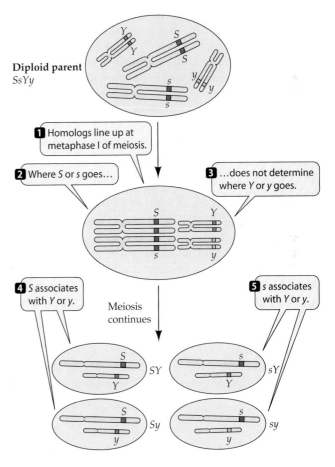

Diploid parent
SsYy

1 Homologs line up at metaphase I of meiosis.

2 Where *S* or *s* goes...

3 ...does not determine where *Y* or *y* goes.

4 *S* associates with *Y* or *y*.

Meiosis continues

5 *s* associates with *Y* or *y*.

Four haploid gametes
SY, Sy, sY, sy

10.8 Meiosis Accounts for Independent Assortment of Alleles
We now know that alleles are segregated independently during metaphase I of meiosis. Thus a parent of genotype *SsYy* can form gametes with four different genotypes; which ones actually form is a matter of chance.

10.9 Joint Probabilities of Independent Events

Like two tosses of a coin, the segregation of each allele into a sperm or an egg is an independent event. The probability of any given combination of alleles from a sperm and an egg is obtained by multiplying the probabilities of each event; this is the probability of producing a homozygote. Since a heterozygote can be formed in two ways, the two probabilities are added together.

general rule is to multiply the probabilities of the individual events (Figure 10.9).

THE MONOHYBRID CROSS. To apply a probabilistic approach to genetics problems, we need only deal with gamete formation and random fertilization instead of coin tosses. A homozygote can produce only one type of gamete, so, for example, an SS individual has a probability equal to 1 of producing gametes with the genotype S. The heterozygote Ss produces S gametes with a probability of $\frac{1}{2}$, and s gametes with a probability of $\frac{1}{2}$.

Consider the F_2 progeny of the cross in Figure 10.4. They are obtained by self-pollination of F_1 plants of genotype Ss. The probability that an F_2 plant will have the genotype SS must be $\frac{1}{2} \times \frac{1}{2} = \frac{1}{4}$ because there is a 50:50 chance that the sperm will have the genotype S, and this chance is independent of the 50:50 chance that the egg will have the genotype S. Similarly, the probability of ss offspring is $\frac{1}{2} \times \frac{1}{2} = \frac{1}{4}$.

ADDING PROBABILITIES. The probability of an F_2 plant getting S from the sperm and s from the egg is also $\frac{1}{4}$, but remember that the same genotype can also result from s in the sperm and S in the egg, with a probability of $\frac{1}{4}$. The probability of an event that can occur in two or more different ways is the sum of the individual probabilities of those ways. Thus the probability that an F_2 plant will be a heterozygote is equal to the sum of the probabilities of each way of forming a heterozygote: $\frac{1}{4} + \frac{1}{4} = \frac{1}{2}$ (see Figure 10.9). The three genotypes are therefore expected in the ratio $\frac{1}{4} SS:\frac{1}{2} Ss:\frac{1}{4} ss$—hence the 1:2:1 ratio of genotypes and the 3:1 ratio of phenotypes seen in Figure 10.4.

THE DIHYBRID CROSS. If F_1 plants heterozygous for two independent characters self-pollinate, the resulting F_2 plants express four different phenotypes. The proportions of these

1 Two coin tosses are individual events.

Probability $(P) = \frac{1}{2}$ 　$P = \frac{1}{2}$

2 Each individual outcome is the result of two independent events, each with a probability of $\frac{1}{2}$; the joint probability is $\frac{1}{2} \times \frac{1}{2} = \frac{1}{4}$ (multiplication rule).

$P = \frac{1}{2}$ 　$P = \frac{1}{2}$

$\frac{1}{2} \times \frac{1}{2} = \frac{1}{4}$ 　$\frac{1}{2} \times \frac{1}{2} = \frac{1}{4}$ 　$\frac{1}{2} \times \frac{1}{2} = \frac{1}{4}$ 　$\frac{1}{2} \times \frac{1}{2} = \frac{1}{4}$

3 There are two ways to arrive at a heterozygote, so we add the probabilities of the two individual outcomes: $\frac{1}{4} + \frac{1}{4} = \frac{1}{2}$ (addition rule).

phenotypes are easily determined by probabilities. Let's see how this works for the experiment shown in Figure 10.7.

The probability that a seed will be spherical is $\frac{3}{4}$, as we have just seen. By the same reasoning, the probability that a seed will be yellow is also $\frac{3}{4}$. The two characters are determined by separate genes and are independent of each other, so the joint probability that a seed will be both spherical and yellow is $\frac{3}{4} \times \frac{3}{4} = \frac{9}{16}$. For the wrinkled, yellow members of the F_2 generation, the probability of being yellow is again $\frac{3}{4}$; the probability of being wrinkled is $\frac{1}{2} \times \frac{1}{2} = \frac{1}{4}$. The joint probability that a seed will be both wrinkled and yellow, then, is $\frac{3}{4} \times \frac{1}{4} = \frac{3}{16}$. The same probability applies, for similar reasons, to the spherical, green F_2 seeds. Finally, the probability that F_2 seeds will be both wrinkled and green must be $\frac{1}{4} \times \frac{1}{4} = \frac{1}{16}$. Looking at all four phenotypes, we see they are expected in the ratio of 9:3:3:1.

10.10 Pedigree Analysis and Dominant Inheritance

A human pedigree showing dominant inheritance. This family carries the allele for Huntington's disease. Everyone who inherits this allele is affected.

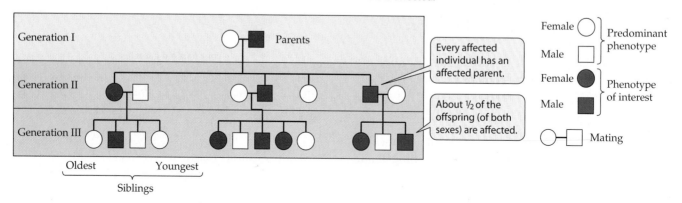

Every affected individual has an affected parent.

About $\frac{1}{2}$ of the offspring (of both sexes) are affected.

Female ◯ ⎫ Predominant
Male ☐ ⎬ phenotype

Female ● ⎫ Phenotype
Male ■ ⎬ of interest

◯—☐ Mating

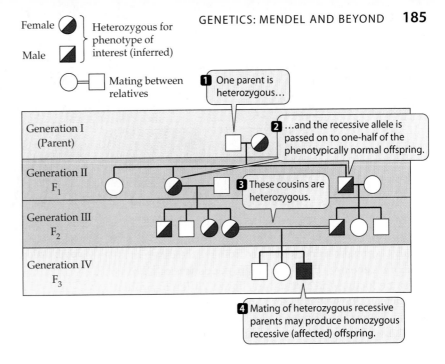

10.11 Recessive Inheritance
This family carries the allele for albinism, a recessive trait. In an affected individual, the trait must be inherited from two heterozygous parents or (rarely) from one homozygous and one heterozygous parent. In this case the heterozygous parents are cousins, but the same result could occur if the parents were unrelated but heterozygous.

Probability calculations and Punnett squares give the same results. Learn to do genetics problems both ways, and then decide which method you prefer.

Mendel's laws can be observed in human pedigrees

A few years after Mendel's work was uncovered by plant breeders, Mendelian inheritance was found in humans. By now, patterns of over 2,500 inherited human characteristics have been described.

Mendel worked out the rules of inheritance by performing many planned crosses and counting many offspring. Neither of these approaches is possible with humans. So human geneticists rely on **pedigrees**, family trees that show the segregation of phenotypes (and alleles) in several generations of related individuals.

Because humans have such small numbers of offspring, human pedigrees do not show the clear proportions of offspring that Mendel saw in his pea plants (see Table 10.1). For example, when two people marry who are both heterozygous for a recessive allele (say, *Aa*), there will be, for each of their children, a 25 percent probability that the child will be homozygous recessive (*aa*). Over many such marriages, one-fourth of all the children will be homozygous recessive (*aa*). But what about a single marriage? In human families, while the odds for each child remain the same, small numbers of children mitigate against getting the exact one-fourth proportion. So, in a family with two children, both could easily be *aa* (or *Aa* or *AA*).

To deal with this ambiguity, human geneticists assume that any allele that causes an abnormal phenotype is rare in the population. This means that in a given family with the rare allele (say, one parent is *Aa*), it is highly unlikely that an outsider marrying into the family will have that same rare allele (the outsider is most likely *AA*).

Human geneticists may wish to know whether a particular rare allele is dominant or recessive. Figure 10.10 depicts a pedigree showing the pattern of inheritance of a rare dominant phenotype. The following are the key features to look for in such a pedigree:

▶ Every affected person has an affected parent.
▶ About half of the offspring of an affected person are also affected.
▶ The phenotype occurs equally in both sexes.

Compare this pattern with Figure 10.11, which shows the pattern of inheritance of a rare recessive phenotype:

▶ Affected people usually have parents who are both not affected.
▶ About one-quarter of the children of unaffected parents can be affected.
▶ The phenotype occurs equally in both sexes.

In pedigrees showing recessive inheritance, it is not uncommon to find a marriage of two relatives. This observation is a result of the rarity of phenotypically abnormal alleles. For two phenotypically normal parents to have an affected child (*aa*), the parents must both be heterozygous (*Aa*). If the *a* allele is rare in the general population, the chance of two people marrying who are both carrying the same rare allele is quite low. On the other hand, if the particular recessive allele is present in a family, two cousins might share it (see Figure 10.10). This is why studies on populations isolated either culturally (by religion, as with the Amish in the United States) or geographically (as on islands) have been so valuable to human geneticists. People in these groups tend either to have large families, or marry among themselves, or both.

Because the major use of pedigree analysis is the clinical evaluation and counseling of patients with inherited abnormalities, a single pair of alleles is usually followed. However, just as pedigree analysis shows the segregation of alleles, it also can show independent assortment if two different allele pairs are considered.

Alleles and Their Interactions

Let's move on to the extensions of Mendelian genetics that have been developed by other researchers, mostly in the early part of the twentieth century. Decades after Mendel's work, others discovered that his hereditary particles—genes—are chemical entities—DNA sequences—that are

Possible genotypes	CC, Cc^{ch}, Cc^h, Cc	c^{ch}, c^{ch}	$c^{ch}c^h, c^{ch}c$	c^hc^h, c^hc	cc
Phenotype	Dark gray	Chinchilla	Light gray	Himalayan	Albino

10.12 Inheritance of Coat Color in Rabbits
There are four alleles of the gene for coat color in rabbits. Different combinations of two alleles give different colors.

usually expressed as proteins. Accordingly, the different alleles of a gene are slightly different sequences of DNA at the same locus, which result in slightly different protein products. In the next chapter we'll see the molecular basis for the distinctions between alleles. In this section we deal with how alleles relate to one another, some of their general properties, and how they arise.

In many cases, alleles do not show simple relationships between dominance and recessiveness. In others, a single allele may have multiple phenotypic effects when it is expressed. Existing alleles can form new alleles by mutation, so there can be many alleles for a single character.

New alleles arise by mutation

Different alleles exist because any gene is subject to *mutation*, which occurs when a gene is changed to a *stable, heritable* new form. In other words, an allele can mutate to become a different allele. Mutation, which will be discussed in detail in Chapter 12, is a random process; different copies of the same gene may be changed in different ways, depending on how and where the DNA sequence changes.

One particular allele of a gene may be defined as the **wild type**, or standard, because it is present in most individuals in nature and gives rise to an expected trait or phenotype. Other alleles of that same gene, often called **mutant** alleles, may produce a different phenotype. The wild-type and mutant alleles reside at the same locus and are inherited according to the rules set forth by Mendel. A genetic locus with a wild-type allele that is present less than 99 percent of the time (the rest of the alleles being mutant) is said to be **polymorphic** (from the Greek *poly*, "many," and *morph*, "form").

Many genes have multiple alleles

Because of random mutations, a group of individuals may have more than two alleles of a given gene. (Any one individual has only two alleles, of course—one from its mother and one from its father.) In fact, there are many examples of such multiple alleles.

Coat color in rabbits is determined by one gene with four alleles. There is a dominance hierarchy in the gene combinations:

$$C > c^{ch} > c^h > c$$

Any rabbit with the C allele (along with any of the four) is gray, and a rabbit that is cc is albino. The intermediate colors result from the different allelic combinations shown in Figure 10.12.

Multiple alleles increase the number of possible phenotypes. In Mendel's monohybrid cross, there was just one pair of alleles (Ss) and two possible phenotypes (resulting from SS or Ss and ss). The four alleles of the rabbit coat color gene produce five phenotypes.

Dominance is usually not complete

In the single-pair alleles studied by Mendel, dominance is complete when an individual is heterozygous. That is, an Ss individual will express the S phenotype. However, many genes have alleles that are not dominant or recessive to one another. Instead, the heterozygotes show an intermediate phenotype—at first glance like that predicted by the old blending theory of inheritance. For example, if a true-breeding red snapdragon is crossed with a true-breeding white one, all the F$_1$ flowers are pink. That this phenomenon can still be explained in terms of Mendelian genetics, rather than blending, is readily demonstrated by a further cross.

According to the blending theory, if one of the pink F$_1$ snapdragons is crossed with a true-breeding white one, all the offspring should be a still lighter pink. In fact, approximately $\frac{1}{2}$ of the offspring are white, and $\frac{1}{2}$ are the same shade of pink as the original F$_1$. When the F$_1$ pink snapdragons are allowed to self-pollinate, the resulting F$_2$ plants are distributed in a ratio of 1 red:2 pink:1 white (Figure 10.13). Clearly the hereditary particles—the genes—have not blended; they are readily sorted out in the F$_2$.

We can understand these results in terms of the Mendelian model. When a heterozygous phenotype is intermediate, as in the snapdragon example, the gene is said to be governed by **incomplete dominance**. All we need to do in cases like this is recognize that the heterozygotes show a phenotype intermediate between those of the two homozygotes.

We can also understand incomplete dominance in molecular terms. Remember that genes code for specific proteins, many of which are enzymes. Different alleles at a locus code for alternative forms of a protein. When the protein is an enzyme, the different forms often have different degrees of catalytic activity. In the snapdragon example, one allele codes for an enzyme that catalyzes a reaction leading to the forma-

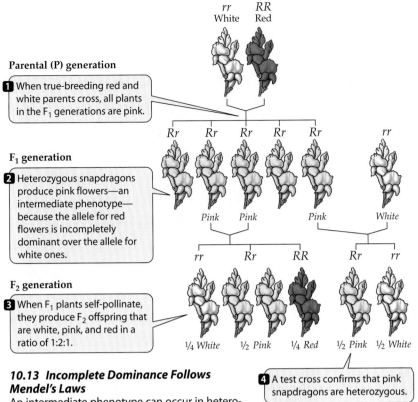

Parental (P) generation

1 When true-breeding red and white parents cross, all plants in the F$_1$ generations are pink.

F$_1$ generation

2 Heterozygous snapdragons produce pink flowers—an intermediate phenotype—because the allele for red flowers is incompletely dominant over the allele for white ones.

F$_2$ generation

3 When F$_1$ plants self-pollinate, they produce F$_2$ offspring that are white, pink, and red in a ratio of 1:2:1.

4 A test cross confirms that pink snapdragons are heterozygous.

10.13 Incomplete Dominance Follows Mendel's Laws
An intermediate phenotype can occur in heterozygotes when neither allele is dominant. The phenotype (here, pink flowers) may give the appearance of a blended trait, but dominant and recessive traits reappear in their original forms in succeeding generations, as predicted by Mendel's laws.

tion of a red pigment in the flowers. The alternative allele codes for an altered enzyme that lacks catalytic activity for pigment production. Plants homozygous for this alternative allele cannot synthesize red pigment, and their flowers are white. Heterozygous plants, with only one allele for the functional enzyme, produce just enough red pigment that their flowers are pink.

There are more examples of incomplete dominance than of complete dominance in nature. Thus an unusual feature of Mendel's report is that all seven of the examples he described (see Table 10.1) are characterized by complete dominance. For dominance to be complete, a single copy of the dominant allele must produce enough of its protein product to give the maximum phenotypic response. For example, just one copy of the dominant allele T at one of the loci studied by Mendel leads to the production of enough of a growth-promoting chemical that Tt heterozygotes are as tall as homozygous dominant plants (TT)—the second copy of T causes no further growth of the stem. Homozygous recessive plants (tt) are much shorter because the allele t does not lead to the production of the growth promoter.

In codominance, both alleles are expressed

Sometimes two alleles at a locus produce two different phenotypes that both appear in heterozygotes. An example of this phenomenon, called **codominance**, is seen in the ABO blood group system in humans.

Early attempts at blood transfusion—made before blood types were understood—frequently killed the patient. Around 1900, however, the Austrian scientist Karl Landsteiner mixed blood cells and serum (blood from which cells have been removed) from different individuals. He found that only certain combinations of blood are compatible. In other combinations, the red blood cells of one individual form clumps because of the presence in the other individual's serum of specific proteins, called *antibodies*, that react with foreign, or "nonself," cells. Proteins on nonself cells, called *antigens*, prompt the synthesis of antibodies. This discovery led to our ability to administer compatible blood transfusions that do not kill the recipient.

Blood compatibility is determined by a set of three alleles (I^A, I^B, and I^O) at one locus, which determines certain proteins (antigens) on the surface of red blood cells. Different combinations of these alleles in different people produce four different blood types, or phenotypes: A, B, AB, and O (Figure 10.14).

Some alleles have multiple phenotypic effects

When a single allele has more than one distinguishable phenotypic effect, we say that the allele is **pleiotropic**. A

Blood type of cells	Genotype	Antibodies made by body	Reaction to added antibodies	
			Anti-A	Anti-B
A	I^AI^A or I^AI^O	Anti-B		
B	I^BI^B or I^BI^O	Anti-A		
AB	I^AI^B	Neither anti-A nor anti-B		
O	I^OI^O	Both anti-A and anti-B		

Red blood cells that do not react with antibody remain evenly dispersed.

Red blood cells that react with antibody clump together (speckled appearance).

10.14 ABO Blood Reactions Are Important in Transfusions
Cells of blood types A, B, AB, and O were mixed with anti-A or anti-B antibodies. As you look down the columns, note that each of the types, when mixed separately with anti-A and with anti-B, gives a unique pair of results; this is the basic method by which blood is typed. A person with type O blood is a good blood donor because O cells do not provoke or react with either anti-A or anti-B antibodies. A person with type AB blood is a good recipient, since neither type of antibody is made.

familiar example of pleiotropy involves the allele responsible for the coloration pattern (light body, darker extremities) of Siamese cats, discussed later in this chapter. The same allele is also responsible for the characteristic crossed eyes of Siamese cats. Although these effects appear to be unrelated, both result from the same protein produced under the influence of the allele.

Gene Interactions

Thus far we have treated the phenotype of an organism, with respect to a given character, as a simple result of its genotype, and we have implied that a single trait results from the alleles of a single gene. In fact, several genes may interact to determine a trait's phenotype. For example, height in people is determined by the actions of many genes, such as those that determine bone growth, hormone concentrations, and other aspects of development. Sometimes several genes act *additively*, so that the phenotype can be predicted by how many of these genes are active. To complicate things further, the physical environment may interact with the genetic constitution of an individual in determining the phenotype. Height in people, for example, is not determined only by their genes. Nutrition is just one environmental factor that undoubtedly has a strong influence on height.

Some genes alter the effects of other genes

Epistasis occurs when the phenotypic expression of one gene is affected by another gene. For example, several genes determine coat color in mice. The wild-type color is agouti, a grayish pattern resulting from bands on the individual hairs. The dominant allele B determines that the hairs will have bands and thus that the color will be agouti, whereas the homozygous recessive genotype bb results in unbanded hairs, and the mouse appears black. On another chromosome, a second locus affects an early step in the formation of hair pigments. The dominant allele A at this locus allows normal color development, but aa blocks all pigment production. Thus, aa mice are all-white albinos, irrespective of their genotype at the B locus (Figure 10.15).

If a mouse with genotype $AABB$ (and thus the agouti phenotype) is crossed with an albino of genotype $aabb$, the F_1 is $AaBb$ and has the agouti phenotype. If the F_1 mice are crossed with each other to produce an F_2 generation, then epistasis will result in an expected phenotypic ratio of 9 agouti:3 black:4 albino. (Can you show why? The underlying ratio is the usual 9:3:3:1 for a dihybrid cross with unlinked genes, but look closely at each genotype and watch out for epistasis.)

In another form of epistasis, two genes are mutually dependent: The expression of each depends on the alleles of the other. The epistatic action of such *complementary genes* may be explained as follows: Suppose gene A codes for enzyme A in the metabolic pathway for purple pigment in flowers, and gene B codes for enzyme B:

colorless precursor → (enzyme A) → colorless intermediate → (enzyme B) → purple pigment

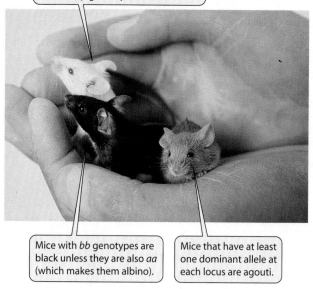

Mice with genotype *aa* are albino regardless of their genotype for the other locus, because the *aa* genotype blocks all pigment production.

Mice with *bb* genotypes are black unless they are also *aa* (which makes them albino).

Mice that have at least one dominant allele at each locus are agouti.

10.15 Genes May Interact Epistatically
Epistasis occurs when one gene alters the phenotypic effect of another gene. In these mice, the presence of the recessive genotype (*aa*) at one locus blocks pigment production, producing an albino mouse no matter what the genotype is at the second locus.

In order for the pigment to be produced, both reactions must take place. The recessive alleles a and b code for nonfunctional enzymes. If a plant is homozygous for either a or b, the corresponding reaction will not occur, no purple pigment will form, and the flowers will be white.

Hybrid vigor results from new gene combinations and interactions

If Mendel's paper was the most important event in genetics in the nineteenth century, perhaps an equally important paper in applied genetics was published early in the twentieth century by G. H. Shull, entitled "The composition of a field of maize". For centuries, it has been known that if one takes two pure, homozygous genetic strains of a plant or animal, and crosses them, the result is offspring that are phenotypically much stronger, larger, and in general more "vigorous" than either of the parents (Figure 10.16).

Conversely, avoidance of *inbreeding* (mating between close relatives) is a time-honored tradition among farmers growing crops and in human societies (where it is called incest). The reason for this is that close relatives tend to have the same recessive alleles, some of which may be harmful, as we saw in our discussion of human pedigrees above.

Shull crossed two of the thousands of existing varieties of corn (maize). Both varieties produced about 20 bushels of corn per acre. But when he crossed them, the yield of their offspring was an astonishing 80 bushels per acre. This phenomenon is known as **heterosis** (short for heterozygosis), or **hybrid vigor**. The cultivation of hybrid corn spread rapidly in the United States and all over the world, quadrupling grain production. The practice of hybridization has spread to many other crops and animals used in agriculture.

10.16 Hybrid Vigor in Corn
The heterozygous F₁ offspring is larger and stronger than either homozygous parent.

Parent   Parent   Hybrid offspring

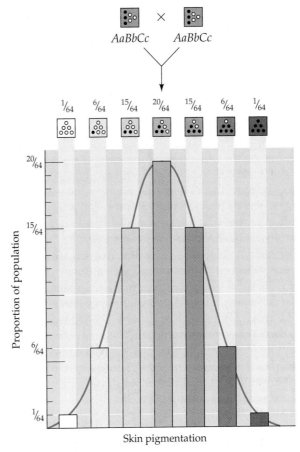

10.17 Polygenes Determine Human Skin Pigmentation
A model of polygenic inheritance based on three genes. Alleles *A*, *B*, and *C* contribute melanin to the skin, but alleles *a*, *b*, and *c* do not. The more *A*, *B*, and *C* alleles an individual possesses, the darker that person's skin will be. If both members of a couple have intermediate pigmentation (in this example, *AaBbCc*), it is unlikely (but not impossible) that their children will have either very light or very dark skin. The actual number of genes involved is much higher.

The actual mechanism by which hybrid vigor works is not known. A widely accepted hypothesis is *overdominance*, a situation in which the heterozygous condition in certain important genes is superior to either homozygote.

Polygenes mediate quantitative inheritance

Individual heritable characters are often found to be controlled by groups of several genes, called **polygenes**, of which each allele intensifies or diminishes the observed phenotype. As a result, variation in such characters is **continuous**, or *quantitative*) rather than, as in the examples we have been considering, **discontinuous** (or *discrete*). Many characters that vary continuously—such as height and other aspects of size, or skin color—are under polygenic control. The polygenes affecting a particular quantitative character are commonly located on many different chromosomes.

Humans differ with respect to the amount of a dark pigment, *melanin*, in their skin (Figure 10.17). There is great variation in the amount of melanin among different people, but much of this variation is determined by alleles at many loci. No alleles at these loci demonstrate dominance. Of course, skin color is not entirely determined by the genotype, since exposure to sunlight in light-skinned people can cause the production of more melanin (that is, a suntan).

The environment affects gene action

The phenotype of an individual does not result from its genotype alone. Genotype and environment interact to determine the phenotype of an organism. Environmental variables such as light, temperature, and nutrition can affect the translation of a genotype into a phenotype. A familiar example involves the Siamese cat. This handsome animal normally has darker fur on its ears, nose, paws, and tail than on the rest of its body. These darkened extremities normally have a lower temperature than the rest of the body.

A few simple experiments show that the Siamese cat has a genotype that results in dark fur, but only at temperatures below the general body temperature. If some dark fur is removed from the tail and the cat is kept at higher than usual

temperatures, the new fur that grows in is light. Conversely, removal of light fur from the back, followed by local chilling of the area, causes the spot to fill in with dark fur.

It is sometimes possible to determine the proportion of individuals in a group with a given genotype that actually show the expected phenotype. This proportion is called the **penetrance** of the genotype. The environment may also affect the **expressivity** of the genotype—that is, the degree to which it is expressed in an individual. For an example of environmental effects on expressivity, consider how Siamese cats kept indoors or outdoors in different climates might look.

Uncertainty over how much of the phenotypic variation we observe is due to the environment and how much to the effects of polygenes complicates the analysis of quantitative inheritance. A useful approach that avoids this difficulty is to study identical twins, which develop from the same fertilized egg. Since such twins are genetically identical, any differences between them can be attributed to environmental effects.

Genes and Chromosomes

The recognition that genes occupy characteristic positions on chromosomes and thus are segregated by meiosis enabled Mendel's successors to provide a physical explanation for his model of inheritance. It soon became apparent that the association of genes with chromosomes has other genetic consequences as well.

In this section we will address the following questions: What is the pattern of inheritance of genes that occupy nearby loci on the same chromosome? How do we determine the order of genes on a chromosome—and the distances between them? Why were all the carriers of hemophilia in Queen Victoria's family women, and why were all of her descendants who had hemophilia men?

The answers to these and many other genetic questions were worked out in studies of the fruit fly *Drosophila melanogaster* (Figure 10.18). Its small size, its ease of cultivation, and its short generation time made this animal an attractive experimental subject. Beginning in 1909, Thomas Hunt Morgan and his students established *Drosophila* as a highly useful laboratory organism in Columbia University's famous "fly room," where they discovered the phenomena described in this section. *Drosophila* remains extremely important in studies of chromosome structure, population genetics, the genetics of development, and the genetics of behavior.

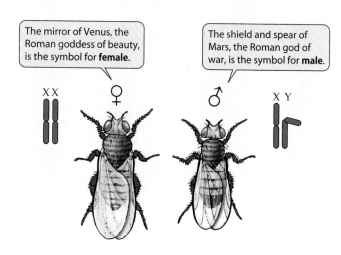

The mirror of Venus, the Roman goddess of beauty, is the symbol for **female**.

The shield and spear of Mars, the Roman god of war, is the symbol for **male**.

10.18 Drosophila melanogaster, the Star of Morgan's Fly Room
The fruit fly (whose Latin name means "vinegar-loving, dark-bodied fly") has a short generation time—a major reason for its widespread use as a laboratory organism in genetics experiments.

Genes on the same chromosome are linked

In the immediate aftermath of the rediscovery of Mendel's laws, the second law—independent assortment—was considered to be generally applicable. However, some investigators, including R. C. Punnett (the inventor of the Punnett square), began to observe strange deviations from the expected 9:3:3:1 ratio in some dihybrid crosses. T. H. Morgan, too, obtained data not in accord with Mendelian ratios, and specifically not in accord with the law of independent assortment.

Morgan crossed *Drosophila* of two known genotypes, *BbVgvg × bbvgvg*, in which *B*, the wild type (gray body), is dominant over *b* (black body), and *Vg* (wild-type wing) is dominant over *vg* (*vestigial*, a very small wing). (Do you recognize this type of cross? It is a test cross for the two gene pairs—see Figure 10.6.) Morgan expected to see four phenotypes in a ratio of 1:1:1:1, but this was not what he observed. The body color gene and the wing size gene were not assorting independently; rather, they were for the most part inherited together (Figure 10.19).

These results became understandable to Morgan when he assumed that the two loci are *on the same chromosome—*

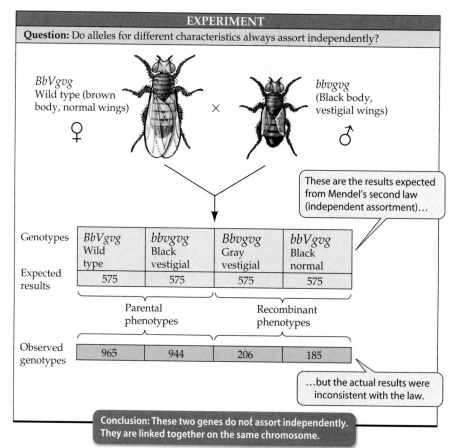

EXPERIMENT

Question: Do alleles for different characteristics always assort independently?

BbVgvg
Wild type (brown body, normal wings) ♀

×

bbvgvg
(Black body, vestigial wings) ♂

These are the results expected from Mendel's second law (independent assortment)...

Genotypes	*BbVgvg* Wild type	*bbvgvg* Black vestigial	*Bbvgvg* Gray vestigial	*bbVgvg* Black normal
Expected results	575	575	575	575

Parental phenotypes — Recombinant phenotypes

| Observed genotypes | 965 | 944 | 206 | 185 |

...but the actual results were inconsistent with the law.

Conclusion: These two genes do not assort independently. They are linked together on the same chromosome.

10.19 Alleles That Do Not Assort Independently
Morgan's studies showed that the genes for body color and wing size in *Drosophila* are linked, so their alleles do not assort independently. Linkage accounts for the departure of the phenotype ratios observed from the results predicted by Mendel's laws.

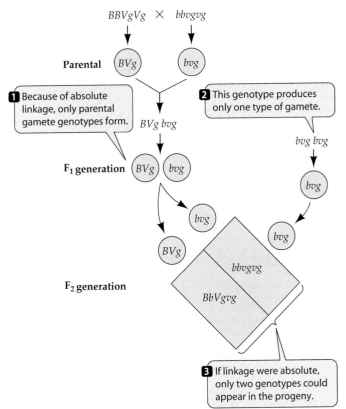

1. **Because of absolute linkage, only parental gamete genotypes form.**

2. **This genotype produces only one type of gamete.**

3. **If linkage were absolute, only two genotypes could appear in the progeny.**

10.20 *If Linkage Were Absolute*
If two genes are absolutely linked on the same chromosome, all the F_2 offspring from a dihybrid test cross would have parental genotypes. If the genes in Morgan's experiment had been absolutely linked, they would have been inherited as if they were a single gene.

Recall that the DNA has been replicated by this stage, and that each chromosome consists of two chromatids. The exchange event involves only two of the four chromatids, one from each member of the chromosome pair. The chiasma can occur at any point along the length of the chromosome. The chromosome sections involved are exchanged reciprocally, so both chromatids involved in crossing over become recombinant (that is, each chromatid ends up with genes from both parents).

When crossing over takes place between two linked genes, not all progeny of a cross will have the parental types. Instead, recombinant offspring appear as well, and

that is, that they are **linked**. After all, since the number of genes in a cell far exceeds the number of chromosomes, each chromosome must contain many genes. The full set of loci on a given chromosome constitutes a *linkage group*. The number of linkage groups in a species equals the number of homologous chromosome pairs.

Suppose, now, that the *Bb* and *Vgvg* loci are indeed located on the same chromosome. If we assume that the linkage is absolute, we expect to see just *two* types of progeny from Morgan's test cross (Figure 10.20). These two would resemble the original (grand)parents. However, this is not always the case.

Genes can be exchanged between chromatids

Absolute linkage is extremely rare. If linkage were absolute, Mendel's second law (independent assortment of alleles at different loci) would apply only to loci on different chromosomes. What actually happens is more complex, and therefore more interesting. The chromosome is not unbreakable, so **recombination** of genes can occur. Genes at different loci on the same chromosome do sometimes separate from one another during meiosis.

Genes may recombine when two homologous chromosomes physically exchange corresponding segments during prophase I of meiosis—that is, by crossing over (see Figure 9.16). In other words, recombination may occur at a chiasma when homologous chromosomes are paired up during meiosis (Figure 10.21).

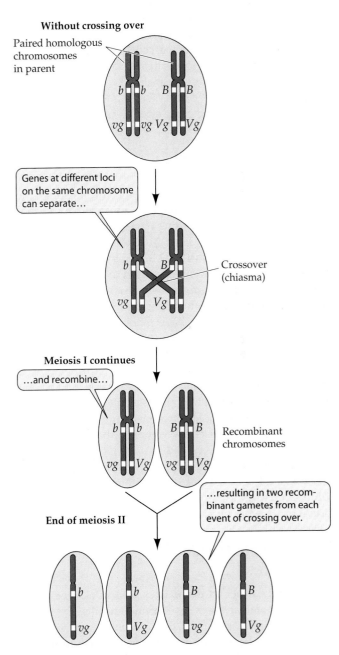

Without crossing over

Paired homologous chromosomes in parent

Genes at different loci on the same chromosome can separate…

Crossover (chiasma)

Meiosis I continues

…and recombine…

Recombinant chromosomes

…resulting in two recombinant gametes from each event of crossing over.

End of meiosis II

10.21 *Crossing Over Results in Genetic Recombination*
Genes at different loci on the same chromosome can separate from one another and recombine by crossing over. Recombination occurs at a chiasma during prophase I of meiosis.

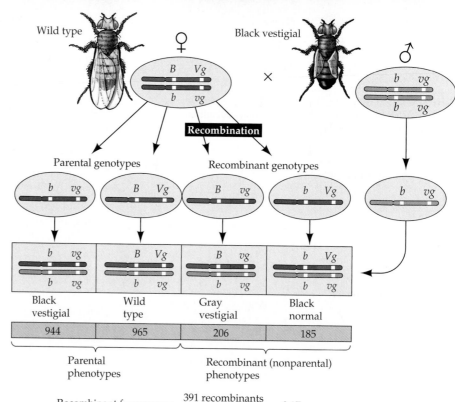

10.22 Recombinant Frequencies
The frequency of recombinant offspring (those with a phenotype different than either parent's) can be calculated. Recombinant frequencies will be larger for loci that are far apart than for those that are close together on the chromosome.

Recombination

Parental genotypes   Recombinant genotypes

Black vestigial	Wild type	Gray vestigial	Black normal
944	965	206	185

Parental phenotypes   Recombinant (nonparental) phenotypes

$$\text{Recombinant frequency} = \frac{391 \text{ recombinants}}{2,300 \text{ total offspring}} = 0.17$$

they appear in repeatable proportions called **recombinant frequencies**, which equal the number of recombinant progeny divided by the total number of progeny (Figure 10.22). Recombinant frequencies will be greater for loci that are far apart on the chromosome than for loci that are closer together, because a chiasma is more likely to cut between genes that are far apart than genes that are close together.

Geneticists make maps of eukaryotic chromosomes

If two loci are very close together on a chromosome, the odds for crossing over between them are small. In contrast, if two loci are far apart, crossing over could occur between them at many points. In 1911, Alfred Sturtevant, then an undergraduate student in T. H. Morgan's fly room, realized how that simple insight could be used to show where different genes lie on the chromosome in relation to one another. He suggested that the farther apart two genes are on a chromosome, the greater the likelihood that they will separate and recombine in meiosis.

The Morgan group had determined recombinant frequencies for many pairs of linked genes. Sturtevant used these recombinant frequencies to create genetic maps that indicated the arrangement of genes along the chromosome (Figure 10.23). Ever since Sturtevant demonstrated this important point, geneticists have mapped the chromosomes of eukaryotes, prokaryotes, and viruses, assigning distances between genes in **map units**. A map unit corresponds to a recombinant frequency of 0.01; it is also referred to as a centimorgan (cM), in honor of the founder of the fly room. You, too, can work out a genetic map (Figure 10.24).

Sex Determination and Sex-Linked Inheritance

In Kölreuter's experience, and later in Mendel's, reciprocal crosses apparently always gave identical results. The reason is that in diploid organisms, chromosomes come in pairs. One member of each chromosome pair derives from each parent; it does not matter, in general, whether a dominant allele was contributed by the mother or by the father. But sometimes the parental origin of a chromosome does matter. To understand the types of inheritance in which parental origin is important, we must consider the ways in which sex is determined in different species.

Sex is determined in different ways in different species

In corn, a plant much studied by geneticists, every diploid adult has both male and female reproductive structures. The two types of tissue are genetically identical, just as roots and leaves are genetically identical. Plants such as maize and Mendel's pea plants, and animals such as earthworms, which produce both male and female gametes in the same organism, are said to be *monoecious* (from the Greek for "one house"). Other plants, such as date palms and oak trees, and most animals are *dioecious* ("two houses"), meaning that some individuals can produce only male gametes and the others can produce only female gametes. In other words, dioecious organisms have two sexes.

In most dioecious organisms, sex is determined by differences in the chromosomes, but such determination operates in different ways in different groups of organisms. For example, the sex of a honeybee depends on whether it develops from a fertilized or an unfertilized egg. A fertilized egg is diploid and gives rise to a female bee—either a worker or a queen, depending on the diet during larval life (again, note how the environment affects the phenotype). An unfertilized egg is haploid and gives rise to a male drone:

Diploid worker   Diploid queen   Haploid drone

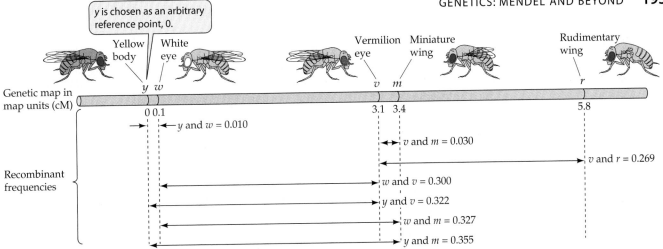

y is chosen as an arbitrary reference point, 0.

10.23 Steps Toward a Genetic Map
Because the chance of a recombinant genotype occurring increases the farther apart two loci fall on a chromosome, Sturtevant was able to derive this partial map of a *Drosophila* chromosome from the Morgan group's data on the recombinant frequencies of five recessive traits. He assigned an arbitrary unit of distance—the map unit, or centimorgan (cM)—equivalent to a recombinant frequency of 0.01.

1 At the outset, we have no idea of the individual distances, and there are several possible sequences (*a-b-c*, *a-c-b*, *b-a-c*).

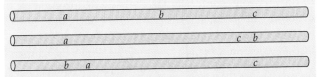

We make a cross *AABB* × *aabb*, and obtain an F_1 generation with a genotype *AaBb*. We test cross these *AaBb* individuals with *aabb*. Here are the genotypes of the first 1,000 progeny:

450 *AaBb*, 450 *aabb*, 50 *Aabb*, and 50 *aaBb*.

2 How far apart are the *a* and *b* genes? Well, what is the recombinant frequency? Which are the recombinant types, and which are the parental types?

Recombinant frequency (*a* to *b*) = (50 + 50)/1,000 = 0.1
So the map distance is
Map distance = 100 × recombinant frequency = 100 × 0.1 = 10 cM

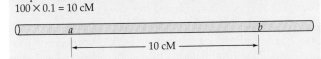

3 Now we make a cross *AACC* × *aacc*, obtain an F_1 generation, and test cross it, obtaining:

460 *AaCc*, 460 *aacc*, 40 *Aacc*, and 40 *aaCc*.

How far apart are the *a* and *c* genes?

Recombinant frequency (*a* to *c*) = (40 + 40)/1,000 = 0.08
Map distance = 100 × recombinant frequency = 100 × 0.08 = 8 cM

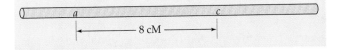

10.24 Map These Genes
We want to determine the order of three loci (*a*, *b*, and *c*) on a chromosome, as well as the map distances (in cM) between them. How do we determine a map distance?

4 How far apart are the *b* and *c* genes?

We make a cross *BBCC* × *bbcc*, obtain an F_1 generation, and test cross it, obtaining:

490 *BbCc*, 490 *bbcc*, 10 *Bbcc*, and 10 *bbCc*.

Determine the map distance between *b* and *c*.

Recombinant frequency (*b* to *c*) = (10 + 10)/1,000 = 0.02
Map distance = 100 × recombinant frequency = 100 × 0.02 = 2 cM

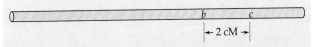

5 Which of the three genes is between the other two?
Because *a* and *b* are the farthest apart, *c* must be between them.

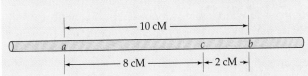

These numbers add up perfectly, but in most real cases they don't add up perfectly because of multiple crossovers.

In many other animals, including humans, sex is determined by a single **sex chromosome** (or by a pair of them). Both males and females have two copies of each of the rest of the chromosomes, which are called **autosomes**.

Female grasshoppers, for example, have two **X chromosomes**, whereas males have only one. Female grasshoppers are described as being XX (ignoring the autosomes) and males as XO (pronounced "ex-oh"):

Females form eggs that contain one copy of each autosome and one X chromosome. Males form approximately equal amounts of two types of sperm: One type contains one copy of each autosome and one X chromosome; the other type contains only autosomes. When an X-bearing sperm fertilizes an egg, the zygote is XX, and develops into a female. When a sperm without an X fertilizes an egg, the zygote is XO, and develops into a male. This chromosomal mechanism ensures that the two sexes are produced in approximately equal numbers.

As in grasshoppers, female mammals have two X chromosomes and males have one. However, male mammals also have a sex chromosome that is not found in females: the **Y chromosome**. Females may be represented as XX and males as XY:

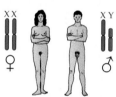

Males produce two kinds of gametes: Each has a complete set of autosomes, but half the gametes carry an X chromosome and the other half carry a Y. When an X-bearing sperm fertilizes an egg, the resulting XX zygote is female; when a Y-bearing sperm fertilizes an egg, the resulting XY zygote is male.

Some subtle but important differences show up clearly in mammals with abnormal sex chromosome constitutions. These conditions, which result from nondisjunctions, as described in Chapter 9, tell us something about the functions of the X and Y chromosomes. In humans, XO individuals sometimes appear. Human XO individuals are females who are physically moderately abnormal but mentally normal; usually they are also sterile. The XO condition in humans is called *Turner syndrome*. It is the only known case in which a human can survive with only one member of a chromosome pair (here, the XY pair), although most XO conceptions terminate spontaneously early in development. XXY individuals also occur; this condition is known as *Klinefelter syndrome*. People with this genotype are sometimes taller than average, always sterile, and always male.

The X and Y chromosomes have different functions

The gene that determines maleness was identified through observations of people with chromosomal abnormalities. For example, some XY individuals who are phenotypically women have been identified and studied; in these people, a small portion of the Y chromosome was missing. In other cases, men who were genetically XX had a small piece of the Y chromosome present, attached to another chromosome. The missing and present Y fragment in these two examples, respectively, contained the maleness-determining gene, which was named *SRY* (sex-determining region on the Y chromosome).

The *SRY* gene codes for a protein involved in *primary sex determination*—that is, the determination of the kinds of gametes that will be produced gametes and the organs that will make them. In the presence of functional SRY protein, the embryo develops sperm-producing testes. If SRY protein is absent, the primary sex determination is female: ovaries and eggs develop. In this case, a gene on the X chromosome called *DAX1* produces an anti-testis factor. So the role of *SRY* in a male is to inhibit the maleness inhibitor made by *DAX1*.

Primary sex determination is not the same as *secondary sex determination*, which results in the outward manifestations of maleness and femaleness (body type, breast development, body hair, and voice). These outward characteristics are not determined directly by the presence or absence of the Y chromosome. Rather, they are determined by the actions of *hormones*, such as testosterone and estrogen.

The Y chromosome functions differently in *Drosophila melanogaster*. Superficially, *Drosophila* follows the same pattern of sex determination as mammals—females are XX and males are XY. However, XO individuals are males (rather than females as in mammals) and almost always are indistinguishable from normal XY males except that they are sterile. XXY *Drosophila* are normal, fertile females:

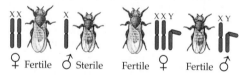

Thus, in *Drosophila*, sex is determined strictly by the ratio of X chromosomes to autosome sets. If there is one X chromosome for each set of autosomes, the individual is a female; if there is only one X chromosome for the two sets of autosomes, the individual is a male. The Y chromosome plays no sex-determining role in *Drosophila*, but it is needed for male fertility.

In birds, moths, and butterflies, males are XX and females are XY. To avoid confusion, these forms are usually expressed as ZZ (male) and ZW (female):

In these organisms, the female produces two types of gametes. Thus the egg determines the sex of the offspring, rather than the sperm, as in humans and fruit flies.

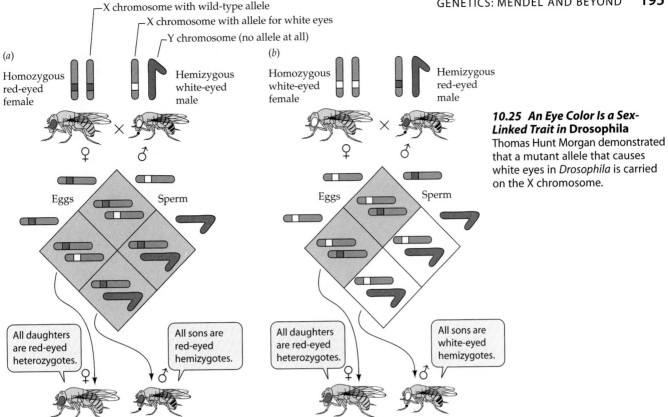

X chromosome with wild-type allele
X chromosome with allele for white eyes
Y chromosome (no allele at all)

(a)

Homozygous red-eyed female ♀ × Hemizygous white-eyed male ♂

Eggs | Sperm

All daughters are red-eyed heterozygotes. ♀

All sons are red-eyed hemizygotes. ♂

(b)

Homozygous white-eyed female ♀ × Hemizygous red-eyed male ♂

Eggs | Sperm

All daughters are red-eyed heterozygotes. ♀

All sons are white-eyed hemizygotes. ♂

10.25 An Eye Color Is a Sex-Linked Trait in Drosophila
Thomas Hunt Morgan demonstrated that a mutant allele that causes white eyes in *Drosophila* is carried on the X chromosome.

Genes on sex chromosomes are inherited in special ways

In *Drosophila* and in humans, the Y chromosome carries few known genes, but a substantial number of genes affecting a great variety of characters are carried on the X chromosome. The result of this arrangement is a deviation from the usual Mendelian ratios for the inheritance of genes located on the X chromosome. Any such gene is present in two copies in females, but in only one copy in males. Therefore, females may be heterozygous for genes that are on the X chromosome, but males will always be **hemizygous** for genes on the X chromosome—they will have only one of each, and it will be expressed.

Kölreuter's historic reciprocal crosses, mentioned at the beginning of this chapter, always gave the same outcome regardless of which parent displayed which trait. However, reciprocal crosses do not give identical results for characters whose genes are carried on the sex chromosomes. This is a sharp deviation from the rules governing the inheritance of alleles on autosomes.

The first and still one of the best examples of **sex-linked inheritance**—inheritance of characters governed by loci on the sex chromosomes—is that of eye color in *Drosophila*. The wild-type eye color of these flies is red. In 1910, Morgon discovered a mutation that causes white eyes. He experimented by crossing flies of the wild-type and mutant phenotypes. His results demonstrated that the eye color locus is on the X chromosome.

When homozygous red-eyed females were crossed with (hemizygous) white-eyed males, all the sons and daughters had red eyes, because red is dominant over white, and all the progeny had inherited a wild-type X chromosome from

their mothers (Figure 10.25*a*). However, in the reciprocal cross, in which a white-eyed female was mated with a red-eyed male, all the sons were white-eyed and all the daughters red-eyed (Figure 10.25*b*).

The sons from the reciprocal cross inherited their only X chromosome from their white-eyed mother; the Y chromosome they inherited from their father does not carry the eye color locus. The daughters, on the other hand, got an X chromosome with the white allele from their mother and an X chromosome bearing the red allele from their father; they were therefore red-eyed heterozygotes.

When Morgan mated heterozygous females with red-eyed males, he observed that half their sons had white eyes, but all their daughters had red eyes. Thus, in this case, eye color was carried on the X chromosome and not on the Y.

Human beings display many sex-linked characters

The human X chromosome carries thousands of genes. The alleles at these loci follow the same pattern of inheritance as those for white eyes in *Drosophila*. One human X chromosome gene, for example, has a mutant recessive allele that leads to red-green color blindness, a hereditary disorder. Red-green color blindness appears in individuals who are homozygous or hemizygous for the mutant allele.

Pedigree analysis of X-linked recessive phenotypes (Figure 10.26) reveals the following patterns:

▶ The phenotype appears much more often in males than in females, because only one copy of the rare allele is needed for its expression in males, while two copies must be present in females.

▶ A male with the mutation can pass it on only to his daughters; all his sons get his Y chromosome.

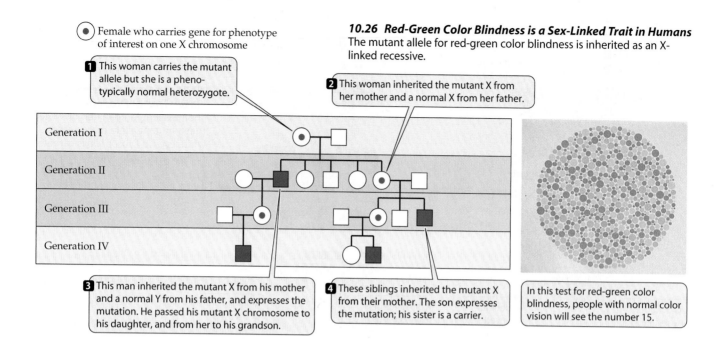

10.26 Red-Green Color Blindness is a Sex-Linked Trait in Humans
The mutant allele for red-green color blindness is inherited as an X-linked recessive.

Female who carries gene for phenotype of interest on one X chromosome

1 This woman carries the mutant allele but she is a phenotypically normal heterozygote.

2 This woman inherited the mutant X from her mother and a normal X from her father.

3 This man inherited the mutant X from his mother and a normal Y from his father, and expresses the mutation. He passed his mutant X chromosome to his daughter, and from her to his grandson.

4 These siblings inherited the mutant X from their mother. The son expresses the mutation; his sister is a carrier.

In this test for red-green color blindness, people with normal color vision will see the number 15.

▶ Daughters who receive one mutant X chromosome are heterozygous *carriers*. They are phenotypically normal, but they can can pass the mutant X to both sons and daughters (only half of the time; half of their X chromosomes carry the normal allele).

▶ The mutant phenotype can skip a generation if the mutation passes from a male to his daughter (phenotypically normal) to her son.

As we will see in later chapters, there are several important human diseases that are inherited as X-linked recessives, including the most common forms of muscular dystrophy and hemophilia. England's Queen Victoria was a heterozygous carrier of hemophilia A, the bleeding disorder mentioned at the beginning of this chapter. She passed it on to some of her male offspring and thereby to several of the royal families of Europe.

Human mutations inherited as X-linked dominants are rarer than recessives, because dominants appear in every generation, and because people carrying the harmful mutations, even as heterozygotes, often fail to survive and/or reproduce. (Look at the four points above and try to determine what would happen if the mutation were dominant.)

The small human Y chromosome carries only about 20 known genes. Among them are the maleness determinants, whose existence was suggested by the phenotypes of the XO and XXY individuals described on page 20. Y-linked alleles are passed from father to son. (You can verify this with a Punnett square.)

Non-Nuclear Inheritance

The nucleus is not the only organelle in a eukaryotic cell that carries genetic material. As we described in Chapter 4, mitochondria and plastids, which may have arisen from bacteria that colonized other cells, contain small numbers of genes. For example, in humans, there are about 60,000 genes in the nucleus and 37 in mitochondria. Plastid genomes are about five times larger than those of mitochondria. In any case, the organelle genes include several that are important for organelle assembly and function, so it is not surprising that mutations of these genes have profound effects on the organism.

The inheritance of organelle genes differs from that of nuclear genes for several reasons. First, mitochondria and plastids are apparently passed on from the mother only. As you will see in later chapters, eggs contain abundant cytoplasm and organelles, but the only part of the sperm that survives to take part in the union of haploid gametes is the nucleus. So you have inherited your mother's mitochondria (with their genes), but not your father's.

Second, there may be hundreds of mitochondria and/or plastids in a cell. So a cell is not diploid for organelle genes; rather, it is highly polyploid. A third factor is that organelle genes tend to mutate at much faster rates than nuclear genes, so there are multiple alleles of organelle genes.

The phenotypes of mutations in the DNA of organelles reflect the organelles' roles. For example, some plastid mutations affect proteins that assemble chlorophyll molecules into the photosystem reaction centers (see Figure 8.11), and result in a phenotype that is essentially a white instead of a green tissue. Mitochondrial mutations that affect one of the complexes in the electron transport chain result in less ATP production. They have especially noticeable effects in tissues with a high energy requirement, such as the nervous system, muscles, and kidneys. In 1995, Greg Lemond, a professional cyclist who had won the famous Tour de France three times, was forced to retire because of muscle weakness suspected to be caused by a mitochondrial mutation.

Chapter Summary
The Foundations of Genetics

▶ Plant breeders can control which plants mate. Although it has long been known that both parent plants contribute equally to the character traits of their offspring, before Mendel's time it was believed that, once they were brought together, the units of inheritance blended and could never be separated. **Review Figure 10.1**

▶ Although Gregor Mendel's work was meticulous and well documented, his discoveries, reported in the 1860s, lay dormant until decades later, when others rediscovered them.

Mendel's Experiments and Laws of Inheritance

▶ Mendel used garden pea plants for his studies because they were easily cultivated and crossed, and because they showed numerous characters (such as seed shape) with clearly different traits (spherical or wrinkled). **Review Table 10.1**

▶ In a monohybrid cross, the offspring showed only one of the two traits. Mendel proposed that the trait observed in the first generation (F_1) was dominant and the other was recessive. **Review Table 10.1**

▶ When the F_1 offspring were self-pollinated, the resulting F_2 generation showed a 3:1 phenotypic ratio, with the recessive phenotype present in one-fourth of the offspring. This reappearance of the recessive phenotype refuted the blending hypothesis. **Review Figure 10.3**

▶ Because some alleles are dominant and some are recessive, the same phenotype can result from different genotypes. Homozygous genotypes have two copies of the same allele; heterozygous genotypes have two different alleles. Heterozygous genotypes yield phenotypes that show the dominant trait.

▶ On the basis of many crosses using different characters, Mendel proposed his first law: that the units of inheritance (now known as genes) are particulate, that there are two copies (alleles) of each gene in every parent, and that during gamete formation the two alleles for a character segregate from each other. **Review Figure 10.4**

▶ Geneticists who followed Mendel showed that genes are carried on chromosomes and that alleles are segregated during meiosis I. **Review Figure 10.5**

▶ Using a test cross, Mendel was able to determine whether a plant showing the dominant phenotype was homozygous or heterozygous. The appearance of the recessive phenotype in half of the offspring of such a cross indicates that the parent is heterozygous. **Review Figure 10.6**

▶ From studies of the simultaneous inheritance of two characters, Mendel concluded that alleles of different genes assort independently. **Review Figures 10.7, 10.8**

▶ We can predict the results of hybrid crosses either by using a Punnett square or by calculating probabilities. To determine the joint probability of independent events, we multiply the individual probabilities. To determine the probability of an event that can occur in two or more different ways, we add the individual probabilities. **Review Figure 10.9**

▶ That humans exhibit Mendelian inheritance can be inferred by the analysis of pedigrees. **Review Figures 10.10, 10.11**

Alleles and Their Interactions

▶ New alleles arise by mutation, and many genes have multiple alleles. **Review Figure 10.12**

▶ Dominance is usually not complete, since both alleles in a heterozygous organism may be expressed in the phenotype. **Review Figures 10.13, 10.14**

Gene Interactions

▶ In epistasis, the products of different genes interact to produce a phenotype. **Review Figure 10.15**

▶ In some cases, the phenotype is the result of the additive effects of several genes (polygenes), and inheritance is quantitative. **Review Figure 10.17**

▶ Environmental variables such as temperature, nutrition, and light affect gene action.

Genes and Chromosomes

▶ Each chromosome carries many genes. Genes located on the same chromosome are said to be linked, and they are often inherited together. **Review Figures 10.19, 10.20**

▶ Linked genes recombine by crossing over in prophase I of meiosis. The result is recombinant gametes, which have new combinations of linked genes because of the exchange. **Review Figures 10.21, 10.22**

▶ The distance between genes on a chromosome is proportional to the frequency of crossing over between them. Genetic maps are based on recombinant frequencies. **Review Figures 10.23, 10.24**

Sex Determination and Sex-Linked Inheritance

▶ Sex chromosomes carry genes that determine whether male or female gametes are produced. The specific functions of X and Y chromosomes differ among species.

▶ In fruit flies and mammals, the X chromosome carries many genes, but its homolog, the Y chromosome, has only a few. Males have only one allele for most X-linked genes, so rare alleles show up phenotypically more often in males than in females. **Review Figures 10.25, 10.26**

Non-Nuclear Inheritance

▶ Cytoplasmic organelles such as plastids and mitochondria contain some heritable genes.

▶ Organelle genes are generally inherited by way of the egg (maternal inheritance), because male gametes contribute only their nucleus to the zygote at fertilization.

Some Genetics Problems

1. Using the Punnett squares below, show that for typical dominant and recessive autosomal traits, it does not matter which parent contributes the dominant allele and which the recessive allele. Cross true-breeding tall plants (*TT*) with true-breeding dwarf plants (*tt*).

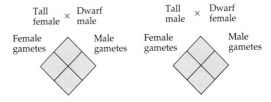

2. The accompanying photograph shows the shells of 15 bay scallops, *Argopecten irradians*. These scallops are hermaphroditic; that is, a single individual can reproduce sexually by self-fertilization, as did the pea plants of the F_1 generation in Mendel's experiments. Three color schemes are evident: yellow, orange, and black and white. The color-determining gene has three alleles. The top row shows a yellow scallop and a representative sample of its offspring, the middle row shows a black-and-white scallop and its offspring, and the bottom row shows an orange scallop and its offspring. Assign a suitable symbol to each of the three alleles participating in color determination;

then determine the genotype of each of the three parent individuals and explain what you can about the genotypes of the different offspring. Explain your results carefully.

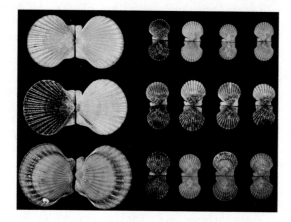

3. Show diagrammatically what occurs when the F_1 offspring of the cross in Question 1 self-pollinate.

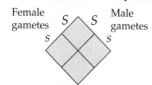

4. A new student of genetics suspects that a particular recessive trait in fruit flies (dumpy wings, which are somewhat smaller and more bell-shaped than the wild type) is sex-linked. A single mating between a fly having dumpy wings (*dp*; female) and a fly with wild-type wings (*Dp*; male) produces three dumpy-winged females and two wild-type males. On the basis of these data, is the trait sex-linked or autosomal? What were the genotypes of the parents? Explain how these conclusions can be reached on the basis of so few offspring.

5. The sex of fishes is determined by the same XY system as in humans. An allele at one locus on the Y chromosome of the fish *Lebistes* causes a pigmented spot to appear on the dorsal fin. A male fish that has a spotted dorsal fin is mated with a female fish that has an unspotted fin. Describe the phenotypes of the F_1 and the F_2 generations from this cross.

6. In *Drosophila melanogaster*, the recessive allele *p*, when homozygous, determines pink eyes. *Pp* or *PP* results in wild-type eye color. Another gene, on another chromosome, has a recessive allele, *sw*, that produces short wings when homozygous. Consider a cross between females of genotype *PPSwSw* and males of genotype *ppswsw*. Describe the phenotypes and genotypes of the F_1 generation and of the F_2 generation produced by allowing the F_1 progeny to mate with one another.

7. On the same chromosome of *Drosophila melanogaster* that carries the *p* (pink eyes) locus, there is another locus that affects the wings. Homozygous recessives, *byby*, have blistery wings, while the dominant allele *By* produces wild-type wings. The *P* and *By* loci are very close together on the chromosome; that is, they are tightly linked. In answering the following questions, assume that no crossing over occurs.

a. For the cross *PPByBy* x *ppbyby*, give the phenotypes and genotypes of the F_1 generation and of the F_2 generation produced by interbreeding of the F_1 progeny.

b. For the cross *PPbyby* x *ppByBy*, give the phenotypes and genotypes of the F_1 and F_2 generations.

c. For the cross in Question 7b, what further phenotype(s) would appear in the F_2 generation if crossing over occurred?

d. Draw a nucleus undergoing meiosis, at the stage in which the crossing over in Question 7c occurred. In which generation (P, F_1, or F_2) did this crossing over take place?

8. Consider the following cross of *Drosophila melanogaster* with alleles as described in Question 6. Males with genotype *Ppswsw* are crossed with females of genotype *ppSwsw*. Describe the phenotypes and genotypes of the F_1 generation.

9. In the Andalusian fowl, a single pair of alleles controls the color of the feathers. Three colors are observed: blue, black, and splashed white. Crosses among these three types yield the following results:

PARENTS	PROGENY
Black × blue	Blue and black (1:1)
Black × splashed white	Blue
Blue × splashed white	Blue and splashed white (1:1)
Black × black	Black
Splashed white × splashed white	Splashed white

a. What progeny would result from the cross blue × blue?

b. If you wanted to sell eggs, all of which would yield blue fowl, how should you proceed?

10. In *Drosophila melanogaster*, white (*w*), eosin (*w^e*), and wild-type red (*w^+*) are multiple alleles of a single locus for eye color. This locus is on the X chromosome. A female that has eosin (pale orange) eyes is crossed with a male that has wild-type eyes. All the female progeny are red-eyed; half the male offspring have eosin eyes, and half have white eyes.

a. What is the order of dominance of these alleles?

b. What are the genotypes of the parents and progeny?

11. Color blindness is a recessive trait. Two people with normal color vision have two sons, one color-blind and one with normal color vision. If the couple also has daughters, what proportion of them will have normal color vision? Explain.

12. A mouse with an agouti coat is mated with an albino mouse of genotype *aabb*. Half of the offspring are albino, one-fourth are black, and one-fourth are agouti. What are the genotypes of the agouti parents and of the various kinds of offspring? (*Hint*: See the section on epistasis.)

13. The disease Leber's optic neuropathy is caused by a mutation in a gene carried on mitochondrial DNA. What would be the result in their first child if a man with this disease married a woman who did not have the disease? What would be the result if the wife had the disease and the husband did not?

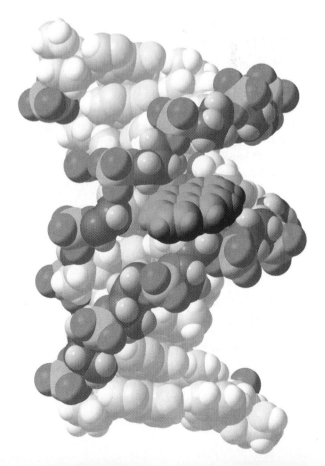

11

DNA and Its Role in Heredity

THE IMAGE OF THE DNA DOUBLE HELIX IS ONE of the great secular icons to emerge in the last half of the twentieth century. Its elegance and simplicity make it instantly recognizable by the general public. The story of how scientists determined that the gene envisioned by Mendel is made of DNA is one of the epics of experimental biology. These studies opened up an entirely new field of natural science: molecular biology, which is concerned with DNA and its expression in cells.

The representation of DNA shown below is the familiar double helix, but with an added chemical shown in green. The green molecule is benzpyrene, one of the toxic chemicals emitted in tobacco smoke. This extra chemical entity has dire consequences. The regular, twisted structure of DNA is just the right shape to allow benzpyrene to wedge into the groove of the helix. A covalent bond forms between the benzpyrene and a DNA monomer, causing a major problem when DNA is expressed and replicated. Ultimately, there are irreversible changes in the DNA, and these changes are passed on to the two daughter cells after a cell division cycle. This damage to the DNA is the key event that begins cancer—in the case of benzpyrene, usually lung cancer.

In this and the next several chapters, we focus on the structure, replication, and function of DNA. As you will see, the structure of DNA determines its functions. This chapter first describes the key experiments that led to the determination that the genetic material is DNA. Then the structure and replication of the molecule are described. Finally, we present two practical applications that have arisen from our knowledge of DNA replication: DNA sequencing, and the polymerase chain reaction.

DNA: The Genetic Material

During the first half of the twentieth century, the hereditary material was generally assumed to be a protein. The impressive chemical diversity of proteins made this assumption seem reasonable. In addition, some proteins—notably enzymes and antibodies—showed great specificity. Nucleic acids, by contrast, were known to have only a few components and seemed too simple to carry the complex information expected in the genetic material.

Circumstantial evidence, however, pointed to DNA. It was in the right place, since it was an important component of the nucleus and chromosomes, which were known to carry genes. And it was present in the right amounts. During the 1920s, a dye was developed that bound specifically to DNA and turned red in direct proportion to the amount of DNA present. When different cells were stained with this dye and their color intensity measured, each species appeared to have its own specific nuclear DNA content. Furthermore, the quantity in somatic cells was twice that in eggs or sperm—as might be expected for diploid and haploid cells, respectively. These two observations were consistent with DNA as the genetic material.

But circumstantial evidence is not a scientific demonstration of cause and effect. After all, proteins are also present in nuclei. The convincing demonstration that DNA is the genetic material came from two lines of experiments, one on bacteria and the other on bacterial viruses.

The Double Helix of DNA
A computer-generated model of DNA. The molecule in green is benzpyrene, a major cancer-causing component of tobacco smoke. The "backbone" of the DNA molecule is visible as a chain of sugars (gray) and phosphate groups (red and orange).

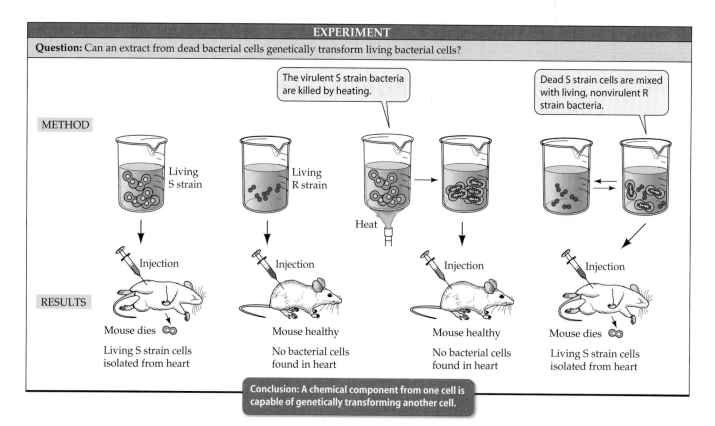

EXPERIMENT

Question: Can an extract from dead bacterial cells genetically transform living bacterial cells?

METHOD

The virulent S strain bacteria are killed by heating.

Dead S strain cells are mixed with living, nonvirulent R strain bacteria.

Living S strain

Living R strain

Heat

RESULTS

Injection

Injection

Injection

Injection

Mouse dies

Living S strain cells isolated from heart

Mouse healthy

No bacterial cells found in heart

Mouse healthy

No bacterial cells found in heart

Mouse dies

Living S strain cells isolated from heart

Conclusion: A chemical component from one cell is capable of genetically transforming another cell.

11.1 Genetic Transformation of Nonvirulent R Pneumococci
Frederick Griffith's experiments demonstrated that something in the virulent S strain could transform nonvirulent R strain bacteria into a lethal form, even when the S strain bacteria had been killed by high temperatures.

DNA from one type of bacterium genetically transforms another type

The history of biology is filled with incidents in which research on one specific topic has—with or without answering the question originally asked—contributed richly to another, apparently unrelated area. Such a case of "serendipity" is the work of Frederick Griffith, an English physician.

In the 1920s, Griffith was studying the disease-causing behavior of the bacterium *Streptococcus pneumoniae*, or pneumococcus, one of the agents that causes pneumonia in humans. He was trying to develop a vaccine against this devastating illness (antibiotics had not yet been discovered). He was working with two strains of pneumococcus. A bacterial *strain* is a population of cells descended from a single parent cell; strains differ in one or more inherited characteristics. Griffith's strains were designated S and R because, when grown in the laboratory, one produces shiny, smooth (S) colonies, and the other produces colonies that look rough (R).

When the S strain was injected into mice, the mice died within a day, and the hearts of the dead mice were found to be teeming with the deadly bacteria. When the R strain was injected, the mice did not become diseased. In other words, the S strain is *virulent* (disease-causing) and the R strain is *nonvirulent*. The virulence of the S strain is caused by a polysaccharide capsule that protects the bacterium from the immune defense mechanisms of the host. The R strain lacks this capsule, so the R strain cells can be inactivated by the defenses of a mouse.

With the hope of developing a vaccine against pneumonia, Griffith inoculated some mice with heat-killed S pneumococci. These heat-killed bacteria did not produce infection. However, when Griffith inoculated other mice with a mixture of living R bacteria and heat-killed S bacteria, to his astonishment, the mice died of pneumonia. When he examined blood from the hearts of these mice, he found it full of living bacteria—many of them with characteristics of the virulent S strain! Griffith concluded that, in the presence of the dead S pneumococci, some of the living R pneumococci had been transformed into virulent S-strain organisms (Figure 11.1).

We now call the phenomenon of the genetic alteration of an organism **transformation**. In terms of Griffith's observations, one could say that transformation is the uptake of information from the environment. As we'll see, today's definition of transformation is more precise. For now, note that living R pneumococci had gained a trait—virulence—from something in their environment.

Did this transformation of the bacteria depend on something the mouse did? No. It was shown that simply incubating living R and heat-killed S bacteria together in a test tube yielded the same transformation. Next it was discovered that a cell-free extract of heat-killed S cells also transformed R cells. (A cell-free extract contains all the contents of ruptured cells, but no intact cells.) This result demonstrated that some substance—called at the time a chemical transforming principle—from the dead S pneumococci could cause a heritable change in the affected R cells. From these observations, some scientists concluded that this transforming principle carried heritable information, and thus was the genetic material.

The transforming principle is DNA

The identification of the transforming principle was a crucial step in the history of biology, accomplished over a period of several years by Oswald T. Avery and his colleagues at what is now Rockefeller University. They treated samples of the transforming principle in a variety of ways to destroy different types of substances—proteins, nucleic acids, carbohydrates, and lipids—and tested the treated samples to see if they had retained transforming activity.

The answer was always the same: If the DNA in the sample was destroyed, transforming activity was lost; everything else was dispensable. As a final step, Avery, with Colin MacLeod and Maclyn McCarty, isolated virtually pure DNA from a sample of pneumococcal transforming principle and showed that it caused bacterial transformation.

The work of Avery, MacLeod, and McCarty, published in 1944, was a milestone in establishing that DNA is the genetic material in cells. However, it had little impact at the time, for two reasons. First, most scientists did not believe that DNA was chemically complex enough to be the hereditary material, especially given the great chemical complexity of proteins. Second, and perhaps more important, it was not yet obvious that bacteria even had genes; bacterial genetics was still to be elucidated (see Chapter 13).

Viral replication experiments confirm that DNA is the genetic material

A report published in 1952 by Alfred D. Hershey and Martha Chase of the Carnegie Laboratory of Genetics had a much greater immediate impact than did Avery's 1944 paper. The Hershey–Chase experiment was carried out with a virus that infects bacteria. This virus, called T2 bacteriophage, consists of little more than a DNA core packed inside a protein coat (Figure 11.2a). The virus is thus made of the two materials that were, at the time, the leading candidates for the genetic material.

When a T2 bacteriophage attacks a bacterium, part (but not all) of the virus enters the bacterial cell. Hershey and Chase set out to determine which part of the virus—protein or DNA—enters the bacterial cell. To trace these two components during the life cycle of the virus (Figure 11.2b), Hershey and Chase labeled each with a specific radioactive tracer.

All proteins contain some sulfur (in the amino acids cysteine and methionine), an element not present in DNA, and sulfur has a radioactive isotope, ^{35}S. The deoxyribose–phosphate "backbone" of DNA, on the other hand, is rich in phosphorus (see Chapter 3), an element that is *not* present in most proteins—and phosphorus also has a radioactive isotope, ^{32}P. Hershey and Chase grew one batch of T2 in a bacterial culture in the presence of ^{32}P, so that all the viral DNA was labeled with ^{32}P. Similarly, all the proteins of another batch of T2 were labeled with ^{35}S.

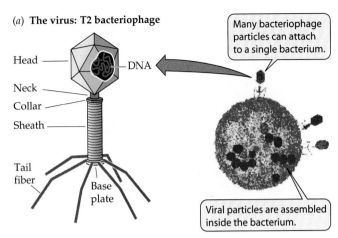

(a) The virus: T2 bacteriophage

Head — DNA

Neck

Collar

Sheath

Tail fiber

Base plate

Many bacteriophage particles can attach to a single bacterium.

Viral particles are assembled inside the bacterium.

(b) Life cycle of the T2 bacteriophage

E. coli

Infection

1 T2 bacteriophage attaches to the surface of *E. coli* and injects its DNA.

2 Viral genes take over the host's synthetic machinery.

(20 minutes)

3 The bacterium breaks open, releasing about 200 viruses.

11.2 *T2 and the Bacteriophage Reproduction Cycle*
(a) The external structures of the bacteriophage T2 consist entirely of protein. This cutaway view shows a strand of DNA within the head. (b) T2 is parasitic on *E. coli*, depending on the bacterium to produce new viruses.

In separate experiments, Hershey and Chase combined radioactive viruses containing either ^{32}P or ^{35}S with bacteria. After a few minutes, they agitated the mixtures vigorously in a kitchen blender, which (without bursting the bacteria) stripped away the parts of the virus that had not penetrated the bacteria. Then, using a centrifuge, Hershey and Chase separated the bacteria from the rest of the mate-

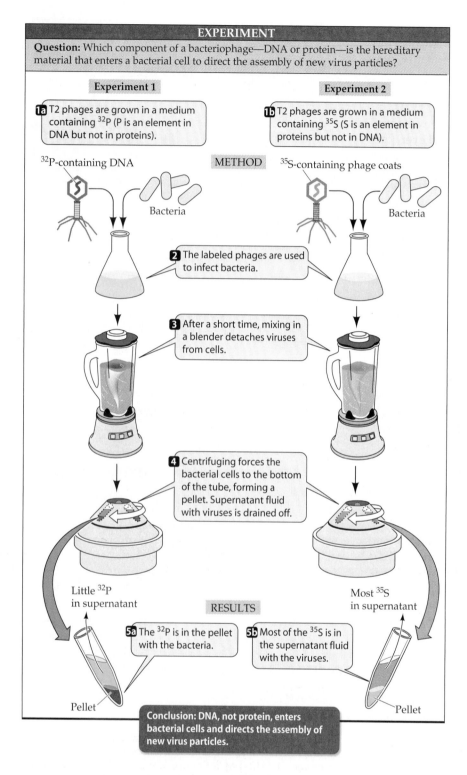

EXPERIMENT

Question: Which component of a bacteriophage—DNA or protein—is the hereditary material that enters a bacterial cell to direct the assembly of new virus particles?

Experiment 1

1a T2 phages are grown in a medium containing ^{32}P (P is an element in DNA but not in proteins).

^{32}P-containing DNA

METHOD

Bacteria

Experiment 2

1b T2 phages are grown in a medium containing ^{35}S (S is an element in proteins but not in DNA).

^{35}S-containing phage coats

Bacteria

2 The labeled phages are used to infect bacteria.

3 After a short time, mixing in a blender detaches viruses from cells.

4 Centrifuging forces the bacterial cells to the bottom of the tube, forming a pellet. Supernatant fluid with viruses is drained off.

Little ^{32}P in supernatant

Most ^{35}S in supernatant

RESULTS

5a The ^{32}P is in the pellet with the bacteria.

5b Most of the ^{35}S is in the supernatant fluid with the viruses.

Pellet

Pellet

Conclusion: DNA, not protein, enters bacterial cells and directs the assembly of new virus particles.

11.3 The Hershey–Chase Experiment
Because only DNA entered the bacterial cell during infection, the experiment demonstrated that DNA, not protein, is the hereditary material.

^{35}S, but about one-third of the original ^{32}P—and thus, presumably, one-third of the original DNA. Because DNA was carried over in the virus from generation to generation but protein was not, a logical conclusion was that the hereditary information of the virus is contained in the DNA.

The Hershey–Chase experiment convinced most scientists that DNA is the carrier of hereditary information. By this time, other researchers had identified mutations—and therefore genes—in viruses and bacteria.

The Structure of DNA

Once scientists agreed that the genetic material is DNA, they wanted to learn its precise chemical structure. In the structure of DNA, they hoped to find the answers to two questions: how DNA is replicated between nuclear divisions, and how it causes the synthesis of specific proteins. Both expectations were fulfilled. X-ray crystallography studies provided the first clues about the dimensions of DNA and hinted that it had a helical form. Dimensionally accurate models built by James Watson and Francis Crick completed the picture.

X-ray crystallography provided clues to DNA structure

The structure of DNA was deciphered only after many types of experimental evidence and theoretical considerations were combined. The most crucial evidence was obtained by X-ray crystallography (Figure 11.4). The positions of atoms in a crystalline substance can be inferred from the pattern of diffraction of X-rays passed through it, but even today this is not an easy task when the substance is of enormous molecular weight.

In the early 1950s, even a highly talented X-ray crystallographer could (and did) look at the best available images from DNA preparations and fail to see what they meant. Nonetheless, the attempt to characterize DNA would have been impossible without the crystallographs prepared by

rial. They found that most of the ^{35}S (and thus the protein) had separated from the bacteria, and that most of the ^{32}P (the DNA) had stayed with the bacteria. These results suggested that the DNA was transferred to the bacteria, whereas the protein remained outside (Figure 11.3).

Hershey and Chase then performed similar but longer experiments, allowing a progeny generation of viruses to be collected. The resulting T2 progeny (the "offspring" of the original viruses) contained almost none of the original

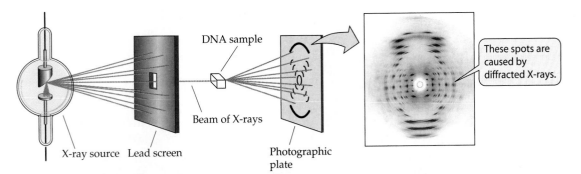

11.4 X-Ray Crystallography Revealed the Basic Helical Nature of DNA Structure
The positions of atoms in DNA can be inferred by the pattern of diffraction of X-rays passed through it, although the task requires tremendous skill.

the English chemist Rosalind Franklin. Franklin's work, in turn, depended on the success of the English biophysicist Maurice Wilkins, who prepared very uniformly oriented DNA fibers, which provided samples for diffraction that were far better than previous ones.

The chemical composition of DNA was known

The chemical composition of DNA also provided important clues about its structure. Biochemists knew that DNA was a polymer of nucleotides. Each nucleotide of DNA consists of a molecule of the sugar deoxyribose, a phosphate group, and a nitrogen-containing base (see Figures 3.16 and 3.17). The only differences among the four nucleotides of DNA are their nitrogenous bases: the purines **adenine** (**A**) and **guanine** (**G**), and the pyrimidines **cytosine** (**C**) and **thymine** (**T**).

In 1950, Erwin Chargaff at Columbia University reported some observations of major importance. He and his colleagues found that DNA from many different species—and from different sources within a single organism—exhibits certain regularities. In almost all DNA the following rule holds: The amount of adenine equals the amount of thymine, and the amount of guanine equals the amount of cytosine (Figure 11.5). As a result, the total abundance of purines equals the total abundance of pyrimidines. The structure of DNA could not have been worked out without this information, yet its significance was overlooked for at least three years. Interestingly, the ratio of A + T to G + C

varies widely among different organisms (Table 11.1). This observation reinforced the importance of Chargaff's rule.

Watson and Crick described the double helix

The solution to the puzzle of the structure of DNA was accelerated by the technique of model building: assembling three-dimensional representations of possible molecular structures using known relative molecular dimensions and known bond angles. This technique, originally exploited in structural studies by the American chemist Linus Pauling, was used by the English physicist Francis Crick and the American geneticist James D. Watson, then both at the Cavendish Laboratory of Cambridge University.

Watson and Crick attempted to combine all that had been learned so far about DNA structure into a single coherent model. The crystallographers' results (see Figure 11.4) convinced Watson and Crick that the DNA molecule is **helical** (cylindrically spiral) and provided the values of certain distances within the helix. The results of density measurements and previous model building suggested that there are two polynucleotide chains in the molecule. The modeling studies had also led to the conclusion that the two chains in DNA run in opposite directions—that is, that they are *antiparallel*. (We'll clarify this point in the next section.)

Crick and Watson built several models. Late in February of 1953, they built the one that established the general structure of DNA. There have been minor amendments to their first published structure, but the principal features remain unchanged.

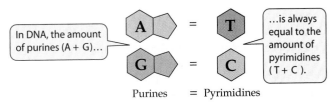

11.5 Chargaff's Rule
The total abundances of purines and pyrimidines are equal in DNA.

	Percentages of Bases in the DNA of Some Well-Studied Species			
	AMOUNT OF BASE (PERCENTAGE OF TOTAL DNA)			
DNA ORIGIN	A	T	G	C
Human (*Homo sapiens*)	31.0	31.5	19.1	18.4
Corn (*Zea mays*)	25.6	25.3	24.5	24.6
Fruit fly (*Drosophila melanogaster*)	27.3	27.6	22.5	22.5
Bacterium (*Escherichia coli*)	26.1	23.9	24.9	25.1

11.1

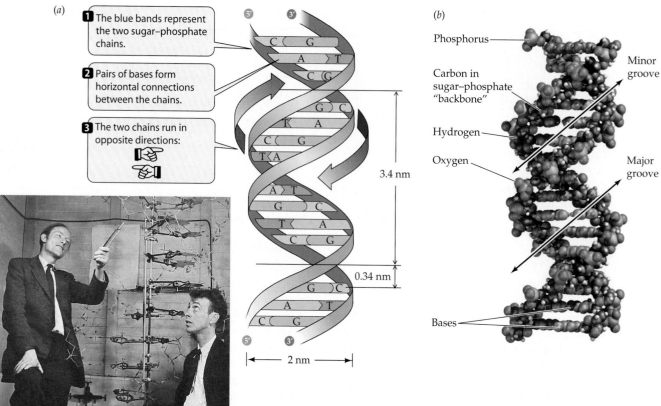

(a)

1 The blue bands represent the two sugar–phosphate chains.

2 Pairs of bases form horizontal connections between the chains.

3 The two chains run in opposite directions:

3.4 nm

0.34 nm

2 nm

(b)

Phosphorus

Minor groove

Carbon in sugar–phosphate "backbone"

Hydrogen

Oxygen

Major groove

Bases

11.6 DNA Is a Double Helix

(a) Watson and Crick proposed that DNA is a double helical molecule. (b) Biochemists can now pinpoint the position of every atom in a DNA macromolecule. To see that the essential features of the original Watson–Crick model have been verified, follow with your eyes the double helical ribbons of sugar–phosphate groups and note the horizontal rungs of the bases (see also Figure 3.18).

Four key features define DNA structure

Four features summarize the molecular architecture of DNA. The DNA molecule is

▶ a double-stranded helix,
▶ of uniform diameter,
▶ right-handed (that is, it twists to the right, as do the threads on most screws), and
▶ antiparallel (the two strands run in opposite directions).

The sugar–phosphate backbones of the polynucleotide chains coil around the outside of the helix, and the nitrogenous bases point toward the center (Figure 11.6).

The two chains are held together by hydrogen bonding between specifically paired bases. Consistent with Chargaff's rule, adenine (A) pairs with thymine (T) by forming two hydrogen bonds, and guanine (G) pairs with cytosine (C) by forming three hydrogen bonds (Figure 11.7). Every base pair consists of one purine (A or G) and one pyrimidine (C or T). This pattern is known as **complementary base pairing**. Because the AT and GC pairs, like rungs of a ladder, are of equal length and fit identically into the double helix, the diameter of the helix is uniform. The base pairs are flat, and their stacking in the center of the molecule is stabilized by hydrophobic interactions (see Chapter 2), contributing to the overall stability of the double helix.

What does it mean to say that the two DNA strands run in opposite directions? The direction of a polynucleotide can be defined by looking at the linkages (called phosphodiester bonds) between adjacent nucleotides. In the sugar–phosphate backbone of DNA, the phosphate groups connect to the 3′ carbon of one deoxyribose molecule and the 5′ carbon of the next, linking successive sugars together (see Figure 11.7). The prime (′) designates the position of a carbon atom in the five-carbon sugar deoxyribose.

Thus the two ends of a polynucleotide differ. Polynucleotides have a free (not connected to another nucleotide) 5′ phosphate group ($-OPO_3^-$) at one end, called the **5′ end**; a free 3′ hydroxyl group ($-OH$) is at the other, the **3′ end**. The 5′ end of one strand in a DNA double helix is paired with the 3′ end of the other strand, and vice versa; in other words, the strands run in opposite directions .

The double helical structure of DNA is essential to its function

The genetic material must perform four important functions, and the DNA molecule modeled by Watson and Crick was elegantly suited to three of them.

▶ *The genetic material should be able to store an organism's genetic information.* With its millions of nucleotides in a sequence that differs in every species and every individual, DNA fits this role nicely.

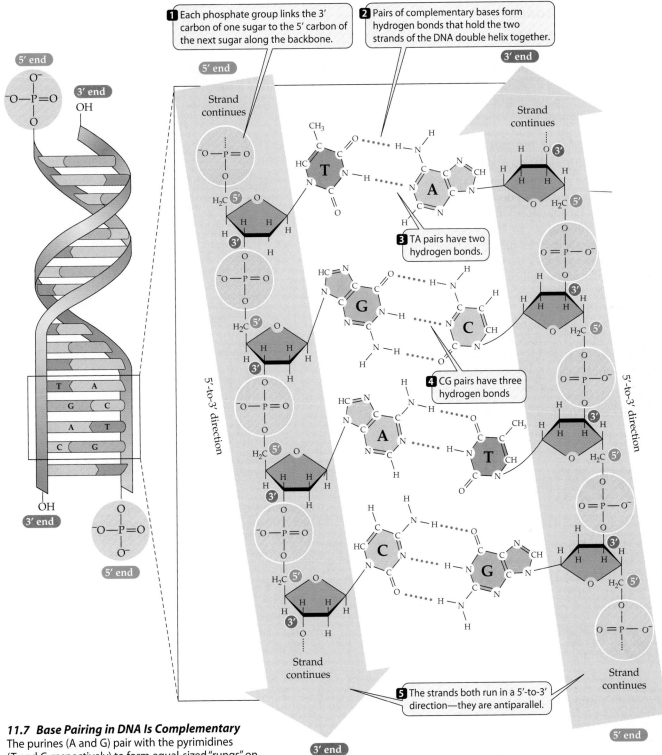

1 Each phosphate group links the 3′ carbon of one sugar to the 5′ carbon of the next sugar along the backbone.

2 Pairs of complementary bases form hydrogen bonds that hold the two strands of the DNA double helix together.

3 TA pairs have two hydrogen bonds.

4 CG pairs have three hydrogen bonds

5 The strands both run in a 5′-to-3′ direction—they are antiparallel.

5′ end

3′ end

Strand continues

5′-to-3′ direction

3′ end

OH

5′ end

3′ end

5′ end

Strand continues

5′-to-3′ direction

Strand continues

5′ end

3′ end

11.7 Base Pairing in DNA Is Complementary
The purines (A and G) pair with the pyrimidines (T and C, respectively) to form equal-sized "rungs" on a "ladder" (the sugar–phosphate backbones). The ladder is twisted into a double helical structure.

▶ *The genetic material must be susceptible to mutation, or permanent changes in its information.* For DNA, mutations might be simple changes in the linear sequence of nucleotide pairs.

▶ *The genetic material must be precisely replicated in the cell division cycle.* Replication could be accomplished by

complementary base pairing, A with T and G with C. In the original publication of their findings in the journal *Nature* in 1953, Watson and Crick coyly pointed out, "It has not escaped our notice that the specific pairing we have postulated immediately suggests a possible copying mechanism for the genetic material."

▶ *The genetic material must be expressed as the phenotype.* This function is not inherent in the structure of DNA; however, as we show in the next chapter, it also turns out to be well served by DNA.

DNA Replication

Watson and Crick's model for DNA replication was soon confirmed. First, experiments showed DNA replicated from template strands in a test tube containing simple substrates and an enzyme. Then an elegant experiment showed that each of the two strands of the double helix serves as a template for a new strand.

Three modes of DNA replication appeared possible

Just three years after Watson and Crick published their paper in *Nature*, their prediction that the DNA molecule contains the information needed for its own replication was demonstrated by the work of Arthur Kornberg, then at Washington University in St. Louis. Kornberg showed that DNA can replicate in a test tube with no cells present. The only requirements are DNA, a specific enzyme (which he obtained from bacteria and called **DNA polymerase**), and a mixture of four precursors: the deoxyribonucleoside triphosphates dATP, dCTP, dGTP, and dTTP. If any one of the four deoxyribonucleoside triphosphates is omitted from the reaction mixture, DNA does not replicate.

Somehow, the DNA itself serves as a template for the reaction—a guide to the exact placement of nucleotides in the

Original DNA   **After one round of replication**

(a)

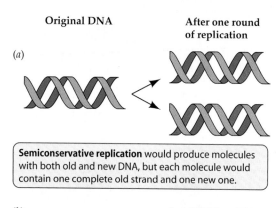

Semiconservative replication would produce molecules with both old and new DNA, but each molecule would contain one complete old strand and one new one.

(b)

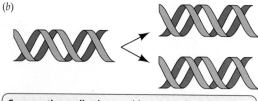

Conservative replication would preserve the original molecule and generate an entirely new molecule.

(c)

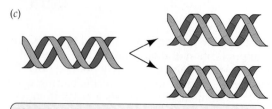

Dispersive replication would produce two molecules with old and new DNA interspersed along each strand.

11.8 Three Models for DNA Replication

In each model, original DNA is shown in blue and newly synthesized DNA in red.

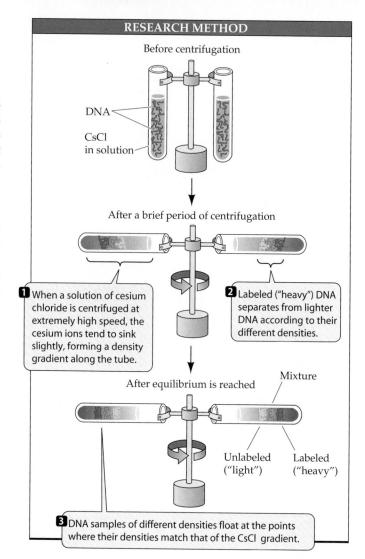

RESEARCH METHOD

Before centrifugation

DNA

CsCl in solution

After a brief period of centrifugation

1 When a solution of cesium chloride is centrifuged at extremely high speed, the cesium ions tend to sink slightly, forming a density gradient along the tube.

2 Labeled ("heavy") DNA separates from lighter DNA according to their different densities.

After equilibrium is reached   Mixture

Unlabeled ("light")   Labeled ("heavy")

3 DNA samples of different densities float at the points where their densities match that of the CsCl gradient.

11.9 Density Gradient Centrifugation

Labeled ("heavy") DNA will separate from lighter DNA in the density gradient formed by a cesium chloride solution.

new strand. Where there is a T in the template, there must be an A in the new strand, and so forth. How does DNA perform the template function; that is, how exactly does the molecule replicate?

There were three possible replication patterns that would result in complementary base pairing:

▶ *Semiconservative replication*, in which each parent strand serves as a template for a new strand and the two new DNA's each have one old and one new strand (Figure 11.8a)

▶ *Conservative replication*, in which the original double helix serves as a template for, but does not contribute to, the new double helix (Figure 11.8b)

▶ *Dispersive replication*, in which fragments of the original DNA molecule serve as templates for assembling two molecules, each containing old and new parts, perhaps at random (Figure 11.8c)

Watson and Crick's original paper suggested that DNA replication was semiconservative, but Kornberg's experiment did not provide a basis for choosing among these three models.

Meselson and Stahl demonstrated that DNA replication is semiconservative

A clever experiment conducted by Matthew Meselson and Franklin Stahl convinced the scientific community that semiconservative replication is the correct model. Working at the California Institute of Technology in 1957, they devised a simple way to distinguish old strands of DNA from new ones: density labeling.

The key to their experiment was the use of a "heavy" isotope of nitrogen. Heavy nitrogen (^{15}N) is a rare, nonradioactive isotope that makes molecules containing it more dense than chemically identical molecules containing the common isotope, ^{14}N. To distinguish DNA of different densities (that is, DNA containing ^{15}N versus DNA containing ^{14}N), Meselson, Stahl, and Jerome Vinograd invented a new centrifugation procedure using a cesium chloride (CsCl) solution.

Spinning solutions or suspensions at high speed in a centrifuge causes the solutes or particles to separate, and they form a gradient according to their density. A concentrated solution of CsCl has a density very close to that of DNA. At high gravitational forces, cesium ions sediment out of the solution to some extent, establishing a gradient from low density at the top of the centrifuge tube to high density at the bottom. When a DNA sample is dissolved in CsCl and centrifuged at about 100,000 times the force of gravity, the DNA gathers in a band at a position in the tube where the density of the CsCl solution equals its own density (Figure 11.9).

After developing this method of distinguishing DNA densities, Meselson and Stahl grew a culture of the bacterium *Escherichia coli* for 17 generations in a medium in which the nitrogen source (ammonium chloride, NH_4Cl) was made with ^{15}N instead of ^{14}N. As a result, all the DNA in the bacteria was "heavy." They grew another culture in a medium with ^{14}N, and extracted DNA from both cultures. When the extracts were combined and centrifuged with CsCl, two separate DNA bands formed, showing that this method could distinguish DNA samples of slightly different densities.

Next, Meselson and Stahl grew another *E. coli* culture on ^{15}N medium, then *transferred* it to normal ^{14}N medium and allowed the bacteria to continue growing (Figure 11.10). Under the conditions they used, *E. coli* replicates its DNA every 20 minutes. Meselson and Stahl collected some of the

11.10 The Meselson–Stahl Experiment
Density gradient centrifugation revealed a pattern that supports the semiconservative model of DNA replication.

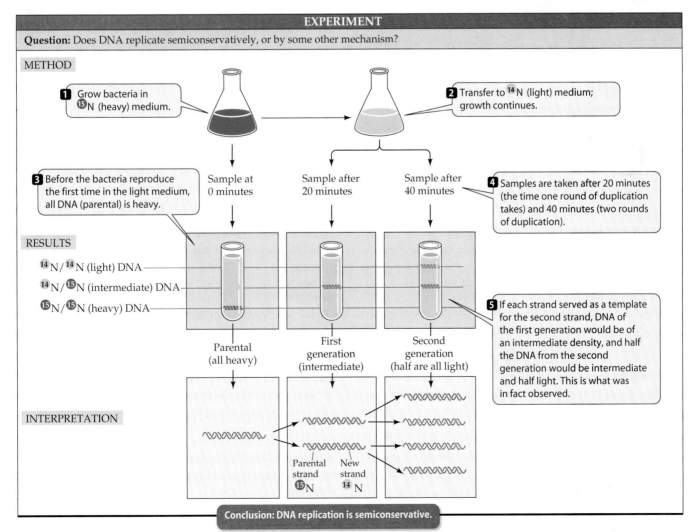

EXPERIMENT

Question: Does DNA replicate semiconservatively, or by some other mechanism?

METHOD

1 Grow bacteria in ^{15}N (heavy) medium.

2 Transfer to ^{14}N (light) medium; growth continues.

3 Before the bacteria reproduce the first time in the light medium, all DNA (parental) is heavy.

Sample at 0 minutes

Sample after 20 minutes

Sample after 40 minutes

4 Samples are taken after 20 minutes (the time one round of duplication takes) and 40 minutes (two rounds of duplication).

RESULTS

$^{14}N/^{14}N$ (light) DNA

$^{14}N/^{15}N$ (intermediate) DNA

$^{15}N/^{15}N$ (heavy) DNA

Parental (all heavy)

First generation (intermediate)

Second generation (half are all light)

5 If each strand served as a template for the second strand, DNA of the first generation would be of an intermediate density, and half the DNA from the second generation would be intermediate and half light. This is what was in fact observed.

INTERPRETATION

Parental strand New strand
^{15}N ^{14}N

Conclusion: DNA replication is semiconservative.

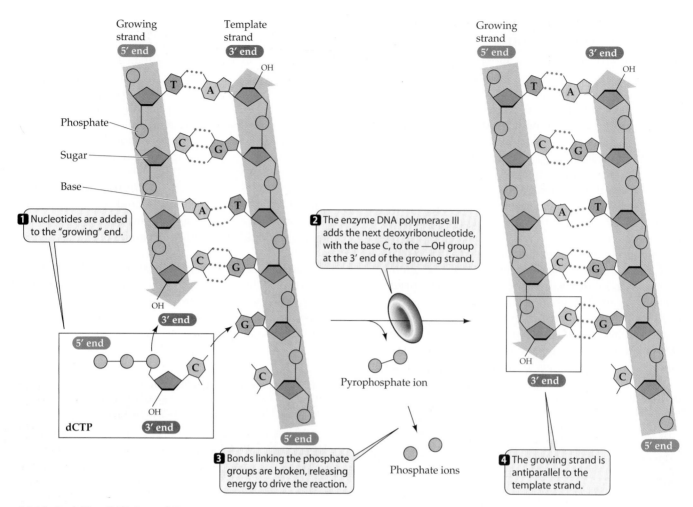

1 Nucleotides are added to the "growing" end.

2 The enzyme DNA polymerase III adds the next deoxyribonucleotide, with the base C, to the —OH group at the 3' end of the growing strand.

3 Bonds linking the phosphate groups are broken, releasing energy to drive the reaction.

4 The growing strand is antiparallel to the template strand.

Pyrophosphate ion

Phosphate ions

dCTP

11.11 Each New DNA Strand Grows from its 5' End to its 3' End
A DNA strand, with its 3' end at the top and its 5' end at the bottom, is the template for the synthesis of the complementary strand at the left.

bacteria after each division and extracted DNA from the samples.

Meselson and Stahl observed that the DNA banding in the density gradient was different in each bacterial generation. At the time of the transfer, the DNA was uniformly labeled with ^{15}N, and hence was relatively dense. After one generation, when the DNA had been duplicated once, all the DNA was of an intermediate density. After two generations, there were two equally large DNA bands: one of low density and one of intermediate density. In samples from subsequent generations, the proportion of low-density DNA increased steadily.

The results of this experiment can be explained by the semiconservative model of DNA replication. The high-density DNA had two ^{15}N strands; the intermediate-density DNA had one ^{15}N and one ^{14}N strand; and the low-density DNA had two ^{14}N strands. In the first round of DNA replication, the strands of the double helix—both heavy with ^{15}N—separated. While separated, each strand

acted as the template for a second strand, which contained only ^{14}N and hence was less dense. Each double helix then consisted of one ^{15}N and one ^{14}N strand and was of intermediate density. In the second replication, the ^{14}N-containing strands directed the synthesis of partners with ^{14}N, creating low-density DNA, and the ^{15}N strands got new ^{14}N partners (see Figure 11.10).

The crucial observation demonstrating the semiconservative model was that intermediate-density DNA (^{15}N–^{14}N) appeared in the first generation and continued to appear in subsequent generations. With the other models, the results would have been quite different. In the conservative model, the first generation would have had both high-density DNA (^{15}N–^{15}N) and low density DNA (^{14}N–^{14}N), but no intermediate DNA. In the dispersive model, the density of the new DNA would have been between low and high, and not exactly intermediate.

Soon after Meselson and Stahl published their work, other scientists showed that semiconservative replication occurred in the DNAs of eukaryotic plant and animal cells. Using labeled DNA, they even demonstrated that chromatids appeared to replicate semiconservatively, providing the first evidence that a chromatid is a single molecule of double-helical DNA.

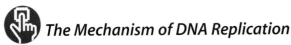

The Mechanism of DNA Replication

But how does DNA get replicated semiconservatively? There are four requirements for this process:

▶ DNA must act as a *template* for complementary base pairing.
▶ The four *deoxyribonucleoside triphosphates*, dATP, dGTP, dCTP, and dTTP, must be present.
▶ A DNA polymerase *enzyme* is needed to bring the substrates to the template and catalyze the reactions.
▶ A source of chemical *energy* is needed to drive this highly endergonic synthesis reaction.

DNA replication takes place in two steps:

▶ The DNA is locally *denatured* (unwound) to separate the two template strands and make them available for base pairing.
▶ The new *nucleotides are linked* by covalent bonding to each growing strand in a sequence determined by complementary base pairing.

A key observation of virtually all DNA replication is that nucleotides are always added to the growing strand at the 3′ end—the end at which the DNA strand has a free hydroxyl (—OH) group on the 3′ carbon of its terminal deoxyribose (Figure 11.11). The three phosphate groups in a deoxyribonucleoside triphosphate are attached to the 5′ position of the sugar (see Figure 11.7). So when a new nucleotide is added to DNA, it can attach only to the 3′ end.

When DNA polymerase brings a new deoxyribonucleoside triphosphate to the 3′ end of a growing chain, the free hydroxyl group on the chain reacts with one of the substrate's phosphate groups. As this happens, the bond linking the terminal two phosphate groups to the rest of the deoxyribonucleoside triphosphate breaks, and thereby releases energy for this reaction. The resulting pyrophosphate ion, consisting of the two terminal phosphate groups, also breaks, forming two phosphate ions and in the process releasing additional free energy. The phosphate group still on the nucleotide becomes part of the sugar–phosphate backbone of the growing DNA molecule.

DNA is threaded through a replication complex

DNA is replicated through the interaction of the template DNA with a huge protein complex that catalyzes the reactions. All chromosomes have at least one sequence of nucleotides, called the **origin of replication**, that is recognized by this replication complex. DNA replicates in both directions from the origin, forming two **replication forks**. Both of the separated strands of the parent molecule act as templates, and the formation of the new strands is guided by complementary base pairing.

Until recently, DNA replication was depicted as a locomotive (the replication complex) moving along a railroad track (the DNA) (Figure 11.12*a*). The current view is that this is not so. Instead, the replication complex is stationary,

attached to nuclear structures, and it is the DNA that moves, essentially threading through the complex as single strands and emerging as double strands (Figure 11.12*b*). During S phase in eukaryotes, there are about 100 replication complexes, and each of them contains as many as 300 individual replication forks.

All replication complexes contain several proteins with different roles. We will describe these proteins as we examine the steps of the process:

▶ *DNA helicase* opens up the double helix.
▶ *Single-strand binding proteins* keep the two strands separated.
▶ *RNA primase* makes the primer strand needed to get replication under way.
▶ *DNA polymerase* adds nucleotides to the primer that are complementary to the template, proofreads the DNA, and repairs it.
▶ *DNA ligase* seals up breaks in the sugar–phosphate backbone.

Small, circular DNA's replicate from a single origin, while large, linear DNA's have many origins

The key event at the origin of replication is the localized unwinding (denaturation) of DNA. There are several forces that hold the two strands together, including hydrogen bonding and the hydrophobic stacking of bases. An enzyme, **DNA helicase**, uses energy from ATP hydrolysis to unwind the DNA, and special proteins bind to the unwound strands to keep them that way, preventing them from reassociating into a double helix. This makes the two template strands available for complementary base pairing.

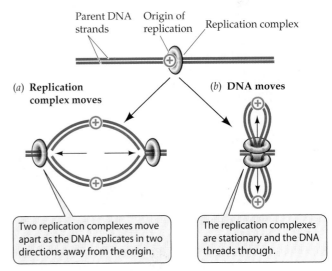

11.12 Two Views of DNA Replication
(*a*) It was once thought that the replication complex moved along DNA. (*b*) Newer evidence suggests that the DNA is threaded through the stationary complex.

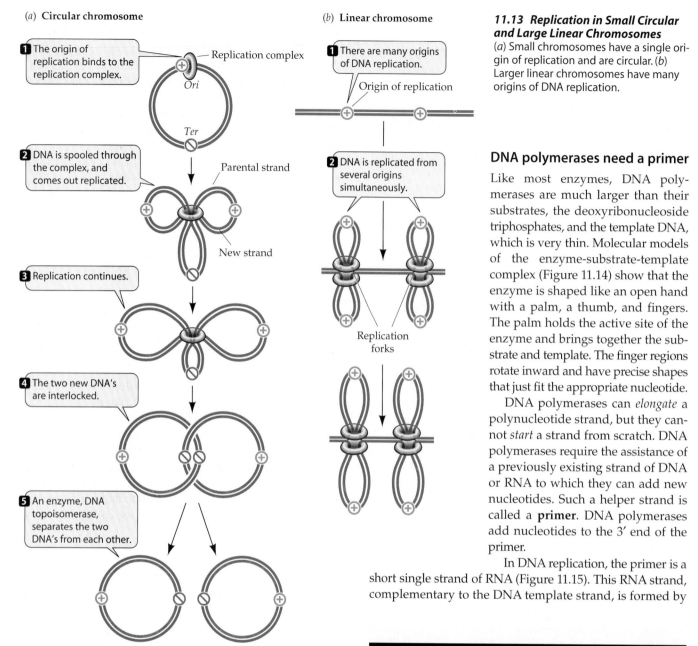

(a) Circular chromosome

1 The origin of replication binds to the replication complex.

Replication complex

Ori

Ter

2 DNA is spooled through the complex, and comes out replicated.

Parental strand

New strand

3 Replication continues.

4 The two new DNA's are interlocked.

5 An enzyme, DNA topoisomerase, separates the two DNA's from each other.

(b) Linear chromosome

1 There are many origins of DNA replication.

Origin of replication

2 DNA is replicated from several origins simultaneously.

Replication forks

11.13 Replication in Small Circular and Large Linear Chromosomes
(a) Small chromosomes have a single origin of replication and are circular. (b) Larger linear chromosomes have many origins of DNA replication.

DNA polymerases need a primer

Like most enzymes, DNA polymerases are much larger than their substrates, the deoxyribonucleoside triphosphates, and the template DNA, which is very thin. Molecular models of the enzyme-substrate-template complex (Figure 11.14) show that the enzyme is shaped like an open hand with a palm, a thumb, and fingers. The palm holds the active site of the enzyme and brings together the substrate and template. The finger regions rotate inward and have precise shapes that just fit the appropriate nucleotide.

DNA polymerases can *elongate* a polynucleotide strand, but they cannot *start* a strand from scratch. DNA polymerases require the assistance of a previously existing strand of DNA or RNA to which they can add new nucleotides. Such a helper strand is called a **primer**. DNA polymerases add nucleotides to the 3' end of the primer.

In DNA replication, the primer is a short single strand of RNA (Figure 11.15). This RNA strand, complementary to the DNA template strand, is formed by

Small chromosomes, such as the 3-million-base-pair DNA of bacteria, have a single origin of replication. As the DNA moves through the replication complex, the replication forks grow around the circle (Figure 11.13a). Finally, two interlocking circular DNA's are formed, and they are separated by an enzyme called *DNA topoisomerase*.

In large chromosomes, such as a human chromosome that has 80 million base pairs, there are hundreds of origins of replication. Because each replicating factory has many adjacent replication complexes, adjacent origins of replication along the linear chromosome can bind at the same time and are replicated simultaneously. So there are many replication forks in eukaryotic DNA (Figure 11.13b).

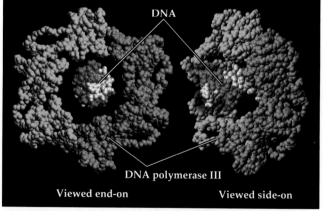

DNA

DNA polymerase III

Viewed end-on   **Viewed side-on**

11.14 DNA Polymerase Binds to Template DNA
The enzyme is much larger than the DNA. It is shaped like a hand, and in side view, "fingers" curl around DNA and can recognize the different shapes of the four bases.

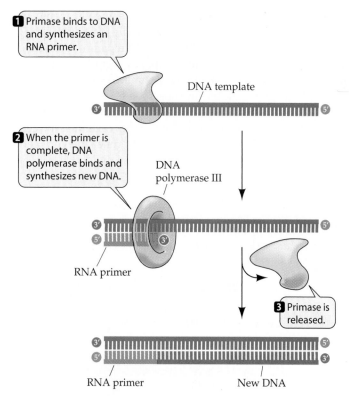

1 Primase binds to DNA and synthesizes an RNA primer.

DNA template

2 When the primer is complete, DNA polymerase binds and synthesizes new DNA.

DNA polymerase III

RNA primer

3 Primase is released.

RNA primer   New DNA

11.15 No DNA Forms without a Primer
In order to synthesize the replicating strand, DNA polymerases require a primer—a previously existing strand of DNA or RNA to which they can add new nucleotides.

an enzyme called a **primase**, which is one of several polypeptides bound together in an aggregate called a **primosome**. DNA polymerase adds nucleotides to the primer until the replication of that section of DNA has been completed. Then the RNA primer is degraded and DNA is added in its place. When DNA replication is complete, each daughter molecule consists only of DNA.

DNA polymerase III extends the new DNA strands

Bacterial cells contain more than one DNA polymerase. **DNA polymerase III** is the enzyme that catalyzes the elongation of new DNA strands at the replication forks in prokaryotes. Besides DNA polymerase III, various other proteins play roles in replacement of the RNA primer and in other replication tasks; some of these are shown in Figure 11.16.

Recall that the two original strands are antiparallel; that is, the 3′ end of one strand is paired with the 5′ end of the other. One template strand is exposed at its 3′ end, which presents no problem—its complementary strand is synthesized continuously as the replication fork proceeds through the replication complex. This daughter strand, called the **leading strand**, elongates from its own 5′ to its own 3′ end, as do all other nucleic acids.

However, the template for the second daughter strand—the **lagging strand**—is exposed from its 5′ end toward its 3′ end. Yet that template must direct the 5′-to-3′ synthesis of the lagging strand. How can it do this, given that new nucleotides are added only at the 3′ end of a polynucleotide chain?

The lagging strand is synthesized from Okazaki fragments

Synthesis of the lagging strand requires working in relatively small, backward-directed, discontinuous stretches (100 to 200 nucleotides at a time in eukaryotes; 1,000 to 2,000 at a time in prokaryotes). Discontinuous stretches are synthesized just as the leading strand is, by adding new nucleotides one at a time to the 3′ end of the daughter strand, but the synthesis of this new daughter DNA moves in the direction opposite to that in which the replication fork is moving (Figure 11.17).

These stretches of new DNA for the lagging strand are called **Okazaki fragments**, after their discoverer, the Japanese biochemist Reiji Okazaki. While the leading strand grows continuously "forward," the lagging strand grows in shorter, "backward" stretches with gaps between them.

A single primer suffices for synthesis of the leading strand, but each Okazaki fragment in the lagging strand requires its own primer. DNA polymerase III synthesizes Okazaki fragments by adding nucleotides to the primer until it reaches the primer of the previous Okazaki fragment. At this point, **DNA polymerase I** (the original DNA polymerase that Kornberg discovered) removes the old RNA primer and replaces it with DNA. Left behind is a tiny nick—the final phosphodiester

1 DNA polymerase III elongates the leading strand.

Leading strand template

2 Helicase unwinds the double helix.

Leading strand

Okazaki fragment

Lagging strand

Lagging strand template

Parent DNA

3 Single-stranded DNA-binding proteins make the templates available to RNA primase and DNA polymerase III.

4 RNA primase makes primer.

5 DNA polymerase III elongates the lagging strand.

11.16 Many Proteins Collaborate at the Replication Fork
In addition to DNA polymerase III, several proteins are involved in DNA replication.

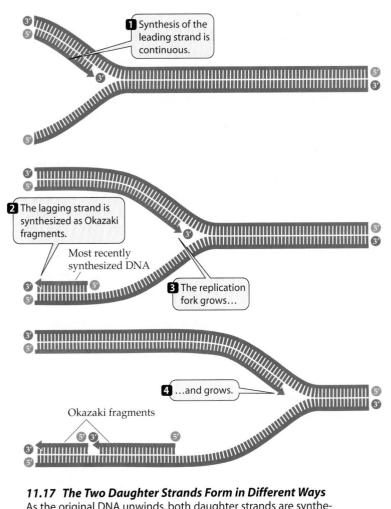

11.17 The Two Daughter Strands Form in Different Ways
As the original DNA unwinds, both daughter strands are synthesized in the 5'-to-3' direction, although their template strands are antiparallel. The leading strand grows continuously forward, but the lagging strand grows in shorter, backward stretches called Okazaki fragments. Eukaryotic Okazaki fragments are hundreds of nucleotides long, with gaps between them.

bond between the adjacent Okazaki fragments is missing (Figure 11.18). Another enzyme, **DNA ligase**, catalyzes the formation of that bond, linking the fragments and making the lagging strand whole.

Working together, DNA helicase, the two DNA polymerases, RNA primase, DNA ligase, and the other proteins do the complex job of DNA synthesis with a speed and accuracy that are almost unimaginable. In *E. coli*, the replication complex makes new DNA at a rate in excess of 1,000 base pairs per second, committing errors in fewer than one base in 10^6; or one in a million.

DNA Proofreading and Repair

DNA must be faithfully replicated and maintained. The price of failure can be great; the transmission of genetic information is at stake, as is the functioning and even the life of a cell or multicellular organism. Yet the replication of DNA is *not* perfectly accurate, and the DNA of nondividing

cells is subject to damage by environmental agents. In the face of these threats, how has life gone on so long?

The preservers of life are DNA repair mechanisms. DNA polymerases initially make a significant number of mistakes in assembling polynucleotide strands. In *E. coli*, for example, the observed error rate every 10^6 bases replicated would result in flaws in approximately one out of every three genes each time the cell divided. In humans, about 1,000 genes in every cell would be affected each time the cell divided.

11.18 The Lagging Strand Story
DNA polymerase I and DNA ligase join DNA polymerase III to complete the complex task of synthesizing the lagging strand.

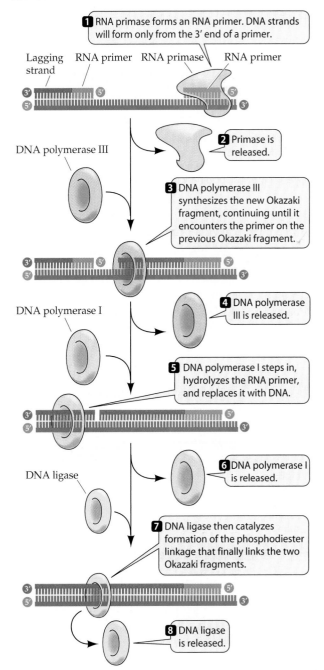

Fortunately, our cells normally have at least three DNA repair mechanisms at their disposal:

▶ A *proofreading mechanism* corrects errors as DNA polymerase makes them.
▶ A *mismatch repair mechanism* scans DNA after it has been made and corrects any base-pairing mismatches.
▶ An *excision repair mechanism* removes abnormal bases that have formed because of chemical damage and replaces them with functional bases.

Proofreading and repair mechanisms ensure that DNA replication is accurate

After introducing a new nucleotide into a growing polynucleotide strand, the DNA polymerases perform a **proofreading** function (Figure 11.19*a*). When a DNA polymerase recognizes a mispairing of bases, it removes the improperly introduced nucleotide and tries again. (DNA helicase, DNA ligase, and other proteins of the replication complex also play roles in this key mechanism.) The polymerase is likely to be successful in inserting the correct monomer the second time, because the error rate for this process is only about 1 in 10,000 base pairs. This proofreading mechanism greatly lowers the overall error rate for replication, to about 1 base in every 10^9 bases replicated.

After DNA has been replicated and during genetic recombination, a second mechanism surveys the newly replicated molecule and looks for mismatched base pairs (Figure 11.19*b*). For example, this **mismatch repair** system might detect an AC base pair instead of an AT pair. Since both AT and GC pairs obey the base-pairing rules, how does the repair mechanism "know" whether the AC pair should be repaired by removing the C and replacing it with T, for instance, or by removing the A and replacing it with G?

The repair mechanism can detect the "wrong" base because a newly synthesized DNA strand is chemically modified some time after replication. In eukaryotes, methyl groups ($-CH_3$) are added to some cytosines to form 5-methylcytosine. In prokaryotes, guanine is methylated. Right after replication, methylation has not yet occurred, so the newly replicated strand is "marked" by being unmethylated, as the one in which errors should be corrected. When mismatch repair fails, DNA sequences are altered. One form of colon cancer arises in part from a failure of mismatch repair.

DNA molecules can also be damaged during the life of a cell (e.g., when it is in G1). Some cells live and play important roles in the organism for many years, even though their DNA is constantly at risk from hazards such as high-energy radiation, chemicals that induce mutations, and random spontaneous chemical reactions. Cells owe their lives

11.19 *DNA Repair Mechanisms*
The proteins of DNA replication also play roles in the life-preserving DNA repair mechanisms, helping to ensure the exact replication of template DNA.

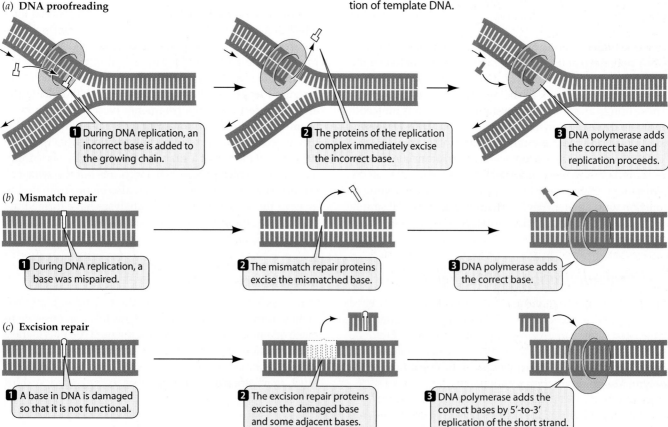

(*a*) **DNA proofreading**

1 During DNA replication, an incorrect base is added to the growing chain.

2 The proteins of the replication complex immediately excise the incorrect base.

3 DNA polymerase adds the correct base and replication proceeds.

(*b*) **Mismatch repair**

1 During DNA replication, a base was mispaired.

2 The mismatch repair proteins excise the mismatched base.

3 DNA polymerase adds the correct base.

(*c*) **Excision repair**

1 A base in DNA is damaged so that it is not functional.

2 The excision repair proteins excise the damaged base and some adjacent bases.

3 DNA polymerase adds the correct bases by 5'-to-3' replication of the short strand.

to DNA repair mechanisms. For example, in **excision repair**, certain enzymes "inspect" the cell's DNA (Figure 11.19c). When they find mispaired bases, chemically modified bases, or points at which one strand has more bases than the other (with the result that one or more bases of one strand form an unpaired loop), these enzymes cut the defective strand. Another enzyme cuts away the bases adjacent to and including the offending base, and DNA polymerase and DNA ligase synthesize and seal up a new (usually correct) piece to replace the excised one.

Our dependence on this repair mechanism is underscored by our susceptibility to various diseases that arise from excision repair defects. One example is the skin disease xeroderma pigmentosum. People with this disease lack a mechanism that normally repairs damage caused by the ultraviolet radiation in sunlight. Without this mechanism, a person exposed to sunlight develops skin cancer.

DNA repair requires energy

What does it cost the cell to keep its DNA accurate and ensure that it replicates properly? At first glance, you might expect DNA polymerization to be fairly "neutral" energetically: Adding a new monomer to the chain requires the formation of a new phosphodiester bond, but is supported by the hydrolysis of one of the high-energy bonds in the deoxyribonucleoside triphosphate (see Figure 11.11). Overall, however, this reaction is slightly endergonic. But help is available in the form of the pyrophosphate ion released in the polymerization reaction. The enzyme pyrophosphatase cleaves the high-energy bond in the pyrophosphate. Coupling this reaction to the polymerization gives it a big boost.

Noncovalent bonds also play a major role in favoring DNA polymerization. Hydrogen bonds form between the complementarily paired bases, and other weak interactions form as the bases stack in the middle of the double helix. These bonds and interactions stabilize the DNA molecule and help drive the polymerization reaction. Thus DNA synthesis itself does not take a tremendous toll in energy.

DNA repair processes, however, are far from cheap energetically. Some are very inefficient. Nonetheless, the cell deploys many DNA repair mechanisms, some overlapping in function with others. Why? Perhaps because the cell simply can't afford to leave its genetic information unprotected, regardless of the cost.

Practical Applications of DNA Replication

The principles that underlie DNA replication in cells have been applied to develop two laboratory techniques that have been vital in analyzing genes and genomes. First, the nucleotide sequence of a DNA molecule can be determined, and second, short DNA sequences can be copied using the polymerase chain reaction technique.

The nucleotide sequence of DNA can be determined

As we saw earlier in this chapter, the deoxyribonucleoside triphosphates (dNTP's) that are the normal substrates for DNA replication contain the sugar 2-deoxyribose. If instead of this, the sugar used is 2,3-dideoxyribose, the resulting nucleoside triphosphates (ddNTP's) will still be picked up by DNA polymerase and added to a growing DNA chain. However, because ddNTP's lack a hydroxyl group at the 3' position, the next nucleotide cannot be added (Figure 11.20a). Thus synthesis stops at the place where ddNTP is incorporated into the growing end of a DNA strand.

In the sequencing technique, a molecule of DNA (usually no more than 500 base pairs long) is denatured. The resulting single-strands of DNA are mixed with:

▸ DNA polymerase to synthesize the complementary strand,
▸ short primers appropriate for that sequence,
▸ the four dNTPs (dATP, dGTP, dCTP, and dTTP), and
▸ small amounts of the four ddNTPs, each with a fluorescent "tag" that emits a different color of light.

The reaction mixture soon contains mostly a DNA mixture of the template DNA strands and shorter, new complementary strands. The latter, each ending with a ddNTP, are of varying lengths. For example, each time a T is reached on the template strand, the growing complementary strand adds either dATP or ddATP. If dATP is added, the strand continues to grow. If ddATP is added, chain growth stops.

After DNA replication has been allowed to proceed for a while in a test tube, the numerous short fragments are denatured from their templates and separated by electrophoresis (see Figure 17.2). This technique measures the length of the DNA fragments, and can detect differences in fragment length as short as one base in 500. During the electrophoresis run, the fragments pass through a laser beam that excites the fluorescent tags. The light emitted is then detected, and the resulting information—that is, which ddNTP is at the end of a strand of which length—is fed into a computer, which processes it and prints out the sequence (Figure 11.20b). The computer can also be programmed to analyze the sequence, and these analyses have formed the basis of the new science of genomics, as we will describe in Chapters 13, 14, and 18.

The polymerase chain reaction makes multiple copies of DNA

Since DNA can be replicated in the test tube, using enzymes from *E. coli* and simple substrates, it is possible to make quantities of a single DNA sequence. The **polymerase chain reaction** (**PCR**) technique is an extension of this early work, which essentially automates the process and makes it much more efficient. PCR is not very complicated: A short

RESEARCH METHOD

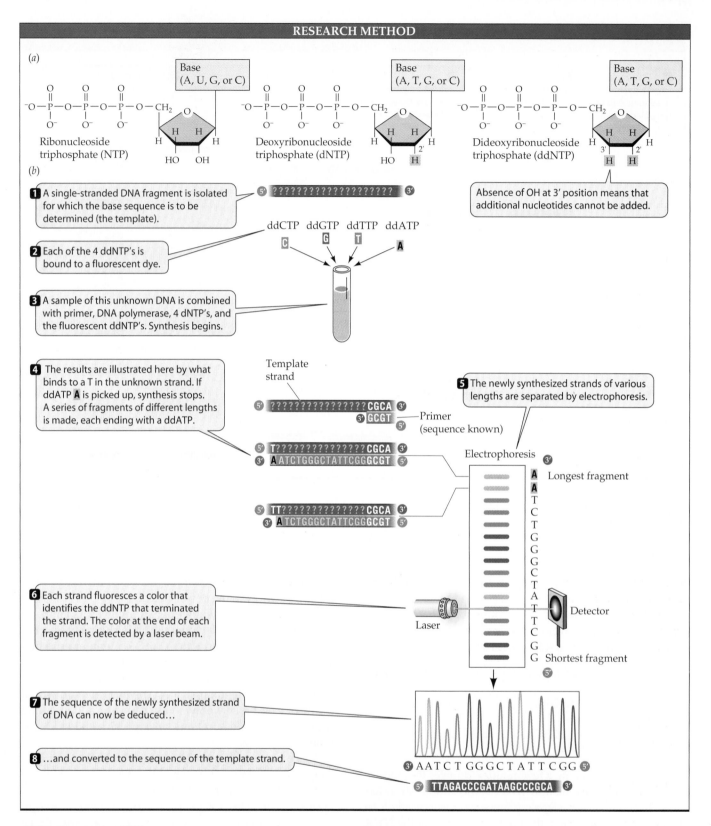

(a)

Ribonucleoside triphosphate (NTP)
Base (A, U, G, or C)

Deoxyribonucleoside triphosphate (dNTP)
Base (A, T, G, or C)

Dideoxyribonucleoside triphosphate (ddNTP)
Base (A, T, G, or C)

Absence of OH at 3′ position means that additional nucleotides cannot be added.

(b)

1 A single-stranded DNA fragment is isolated for which the base sequence is to be determined (the template).

5′ ??????????????????? 3′

2 Each of the 4 ddNTP's is bound to a fluorescent dye.

ddCTP ddGTP ddTTP ddATP
C   G   T   A

3 A sample of this unknown DNA is combined with primer, DNA polymerase, 4 dNTP's, and the fluorescent ddNTP's. Synthesis begins.

4 The results are illustrated here by what binds to a T in the unknown strand. If ddATP **A** is picked up, synthesis stops. A series of fragments of different lengths is made, each ending with a ddATP.

Template strand

5 The newly synthesized strands of various lengths are separated by electrophoresis.

5′ ??????????????CGCA 3′
                3′ GCGT 5′ — Primer (sequence known)

5′ T??????????????CGCA 3′
3′ AATCTGGGCTATTCGGGCGT 5′

5′ TT?????????????CGCA 3′
3′ ATCTGGGCTATTCGGGCGT 5′

Electrophoresis

3′
A Longest fragment
A
T
C
T
G
G
G
C
T
A
T
T
C
G
G Shortest fragment
5′

6 Each strand fluoresces a color that identifies the ddNTP that terminated the strand. The color at the end of each fragment is detected by a laser beam.

Laser

Detector

7 The sequence of the newly synthesized strand of DNA can now be deduced…

8 …and converted to the sequence of the template strand.

3′ A A T C T G G G C T A T T C G G 5′

5′ TTAGACCCGATAAGCCCGCA 3′

11.20 Sequencing DNA
(a) The normal substrates for DNA replication are dNTP's. The slightly different structure of ddNTP's can cause DNA synthesis to stop. (b) When labeled ddNTP's are incorporated into a mixture containing a single-stranded DNA template of unknown sequence, the result is an electrophoresis of fragments of varying lengths.

RESEARCH METHOD

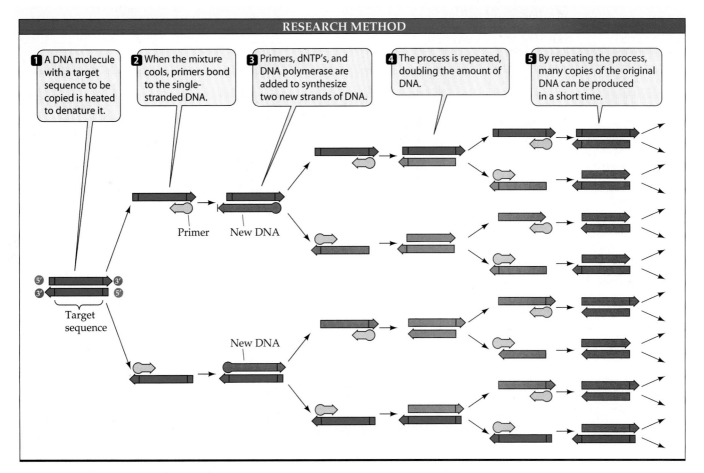

1. A DNA molecule with a target sequence to be copied is heated to denature it.

2. When the mixture cools, primers bond to the single-stranded DNA.

3. Primers, dNTP's, and DNA polymerase are added to synthesize two new strands of DNA.

4. The process is repeated, doubling the amount of DNA.

5. By repeating the process, many copies of the original DNA can be produced in a short time.

Primer   New DNA

5′ 3′
3′ 5′

Target sequence

New DNA

11.21 The Polymerase Chain Reaction
The steps in this cyclic process are repeated many times to produce multiple copies of a DNA sequence.

region of DNA is copied many times in the test tube by DNA polymerase.

PCR is a cyclic process in which the following sequence of steps is repeated over and over again (Figure 11.21). It begins with the same first two steps as DNA sequencing:

▶ Double-stranded DNA is denatured by heat into single strands.

▶ Short primers for DNA replication are added to the mixture.

▶ DNA polymerase catalyzes the production of complementary new strands.

A single cycle, taking a few minutes, doubles the amount of DNA and leaves the new DNA in the double-stranded state. Repeating the cycle many times can theoretically lead to a geometric increase in the number of copies of the DNA sequence.

The PCR technique requires that the base sequences at the 3′ end of each strand of the target DNA be known so that a complementary primer, usually 15 to 20 bases long, can be made in the laboratory. Because of the uniqueness of DNA sequences, usually two primers of this length will bind to only one region of DNA in an organism's genome.

This specificity in the face of the incredible diversity of target DNA is a key to the power of PCR.

One potential problem with PCR involves its temperature requirements. To denature the DNA during each cycle, it must be heated to more than 90°C—a temperature that destroys most DNA polymerases. Then it must be cooled to about 55°C to allow the primer to hydrogen-bond to the single strands of template DNA. The PCR method would not be practical if new polymerase had to be added during each cycle after denaturation—an expensive and laborious proposition.

This problem was solved by nature: In the hot springs at Yellowstone National Park, as well as other locations, there live bacteria called, appropriately, *Thermus aquaticus*. The means by which these organisms survive temperatures up to 95°C was investigated by bacteriologist Thomas Brock and his colleagues. They discovered that *T. aquaticus* has an entire metabolic machinery that is heat-resistant, including DNA polymerase that does not denature at this high temperature.

Scientists pondering the problem of amplifying DNA by PCR read Brock's basic research articles and got a clever idea: Why not use *T. aquaticus* DNA polymerase in the PCR

reaction? It would not be denatured, and thus would not have to be added during each cycle. The idea worked, and earned biochemist Kerry Mullis a Nobel prize. PCR has had an enormous impact on genetic research. Some of its most striking applications will be described in Chapters 13 through 17.

Chapter Summary

DNA: The Genetic Material

▶ In addition to circumstantial evidence (the location and quantity of DNA in the cell), two experiments provided a convincing demonstration that DNA is the genetic material.

▶ In one experiment, DNA from a virulent strain of pneumococcus bacteria genetically transformed nonvirulent bacteria into virulent bacteria. **Review Figure 11.1**

▶ In a second set of experiments, labeled viruses were incubated with host bacteria. Labeled viral DNA entered the host cells, where it produced hundreds of new viruses bearing the label. **Review Figures 11.2, 11.3**

The Structure of DNA

▶ X-ray crystallography showed that the DNA molecule is a helix. **Review Figure 11.4**

▶ DNA is composed of nucleotides, each containing one of four bases—adenine, cytosine, thymine, or guanine. Biochemical analysis revealed that the amount of adenine equals the amount of thymine and the amount of guanine equals the amount of cytosine. **Review Figure 11.5**

▶ Putting the accumulated data together, Watson and Crick built a model of the DNA molecule. They proposed that DNA is a double-stranded helix in which the strands are antiparallel and the bases are held together by hydrogen bonding. This model accounts for the genetic information, mutation, and replication functions of DNA. **Review Figures 11.6, 11.7**

DNA Replication

▶ Three possible models for DNA replication were hypothesized: semiconservative, conservative, and dispersive. **Review Figure 11.8**

▶ An experiment by Meselson and Stahl proved the replication of DNA to be semiconservative. Each parent strand acts as a template for the synthesis of a new strand; thus the two replicated DNA helices contain one parent strand and one newly synthesized strand each. **Review Figures 11.9, 11.10**

The Mechanism of DNA Replication

▶ In DNA replication, the enzyme DNA polymerase catalyzes the addition of nucleotides to the 3′ end of each strand. Nucleotides are added by complementary base pairing with the template strand of DNA. The substrates are deoxyribonucleoside triphosphates, which are hydrolyzed as they are added to the growing chain, releasing energy that fuels the synthesis of DNA. **Review Figure 11.11**

▶ The DNA replication complex is in a fixed location and DNA is threaded through it for replication. **Review Figure 11.12**

▶ Many proteins assist in DNA replication. DNA helicases unwind the double helix, and the template strands are stabilized by other proteins.

▶ Prokaryotes have a single origin of replication; eukaryotes have many. Replication in both cases proceeds in both directions from an origin of replication. **Review Figure 11.13**

▶ An RNA primase catalyzes the synthesis of short RNA primers, to which nucleotides are added as the chain grows. **Review Figure 11.15**

▶ Through the action of DNA polymerase, the leading strand grows continuously in the 5′-to-3′ direction until the replication of that section of DNA has been completed. Then the RNA primer is degraded and DNA is added in its place.

▶ On the lagging strand, which grows in the other direction, DNA is still made in the 5′-to-3′ direction (away from the replication fork). But synthesis of the lagging strand is discontinuous: The DNA is added as short fragments to primers, then the polymerase skips past the 5′ end to make the next fragment. **Review Figures 11.16, 11.17, 11.18**

DNA Proofreading and Repair

▶ The machinery of DNA replication makes about one error in 10^6 nucleotides bases added. These errors are repaired by three different mechanisms: proofreading, mismatch repair, and excision repair. DNA repair mechanisms lower the overall error rate of replication to about one base in 10^9. **Review Figure 11.19**

▶ Although energetically costly and somewhat redundant, DNA repair is crucial to the survival of the cell.

Practical Applications of DNA Replication

▶ The principles of DNA replication can be used to determine the nucleotide sequence of DNA. **Review Figure 11.20**

▶ The polymerase chain reaction technique uses DNA polymerases to repeatedly replicate DNA in the test tube. **Review Figure 11.21**

For Discussion

1. Outline a series of experiments using radioactive isotopes to show that bacterial DNA and not protein enters the host cell and is responsible for bacterial transformation.

2. Suppose that Meselson and Stahl had continued their experiment on DNA replication for another ten bacterial generations. Would there still have been any ^{14}N–^{15}N hybrid DNA present? Would it still have appeared in the centrifuge tube? Explain.

3. If DNA replication were conservative rather than semiconservative, what results would Meselson and Stahl have observed? Diagram the results using the conventions of Figure 11.10.

4. Using the following information, calculate the number of origins of DNA replication on a human chromosome: DNA polymerase adds nucleotides at 3,000 base pairs per minute in one direction; replication is bidirectional; the S phase lasts 300 minutes; there are 120 million base pairs per chromosome. With a typical chromosome 3 cm long, how many origins are there per micrometer?

5. The drug dideoxycytidine (used to treat certain viral infections) is a nucleoside made with 2′-3′-dideoxyribose. This sugar lacks —OH groups at both the 2′ and the 3′ positions. Explain why this drug would stop the growth of a DNA chain if it was added to DNA.

12 *From DNA to Protein: Genotype to Phenotype*

THE OCEAN IS LITERALLY ALIVE WITH LIGHT. About 90 percent of all marine animals living at depths between 200 and 1,000 meters are bioluminescent—they can use chemical reactions to produce light. In some cases, it helps them find a mate or food. In others, it acts like a burglar alarm, startling a predator. The jellyfish in the photograph is in the latter category. Clearly, this is a genetically transmitted characteristic that is important to the animal's survival.

This jellyfish creates light by means of two proteins acting in sequence. The first, aequorin, absorbs light and transmits some of its excitation energy to a second protein, green fluorescent protein (GFP). It is GFP that gives off a green glow. As you will see in Chapter 17, GFP has become an invaluable tool in biological research, as it can be attached to other molecules, allowing them to be observed in real time in living cells and tissues because of the green glow. Here, however, the important conclusion is that bioluminescence—a phenotype—is essentially produced by the actions of two proteins. And these proteins are coded for by DNA sequences.

This chapter deals with the mechanisms by which genes are expressed as proteins. We begin with evidence for the relationship between genes and proteins, and then fill in some of the details of the processes of transcription and translation. Finally, we define mutations and their phenotypes in specific molecular terms.

One Gene, One Polypeptide

There are many steps between genotype and phenotype. Genes cannot, all by themselves, directly produce a phenotypic result, such as a particular eye color, a specific seed shape, or a cleft chin, any more than a compact disk can play a symphony without the help of a CD player.

With the gene defined as a DNA sequence, the first step in relating genes to phenotypes was to define phenotypes in molecular terms. The molecular basis of phenotypes was actually discovered before the discovery of DNA as the genetic material. Using organisms as diverse as humans and bread molds, scientists studied the chemical differences between organisms carrying wild-type and mutant alleles. They found that the major phenotypic differences were in specific proteins.

In the 1940s, a series of experiments by George W. Beadle and Edward L. Tatum at Stanford University showed that when an altered gene resulted in an altered phenotype, the latter showed up as an altered enzyme protein (Figure 12.1). This finding was critically important in defining the phenotype in chemical terms.

Beadle and Tatum experimented with the bread mold *Neurospora crassa*. The nuclei in the body of the mold are haploid (*n*), as are the reproductive spores. (This fact is important because it means that even recessive mutant alleles are easy to detect in experiments.) Beadle and Tatum grew *Neurospora* on a minimal nutritional medium of sucrose, minerals, and a vitamin. Using this medium, the enzymes of wild-type *Neurospora* could catalyze the metabolic reactions needed to make all the chemical constituents of their cells, including proteins. These wild-type strains are called *prototrophs* ("original eaters").

Beadle and Tatum treated wild-type *Neurospora* with X-rays, a mutagen (something known to cause mutations). When they examined the treated molds, they found some

Light in the Depths of the Ocean
The bioluminescent Pacific jellyfish, *Aequorea victoria*, "lights up" when startled by a predator.

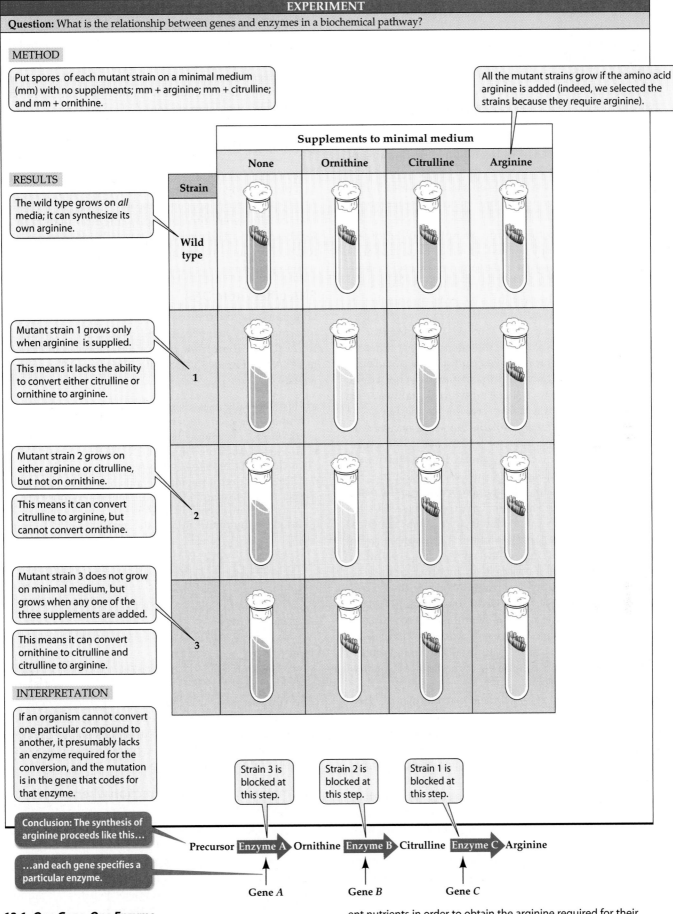

EXPERIMENT

Question: What is the relationship between genes and enzymes in a biochemical pathway?

METHOD

Put spores of each mutant strain on a minimal medium (mm) with no supplements; mm + arginine; mm + citrulline; and mm + ornithine.

All the mutant strains grow if the amino acid arginine is added (indeed, we selected the strains because they require arginine).

Supplements to minimal medium

Strain	None	Ornithine	Citrulline	Arginine

RESULTS

The wild type grows on *all* media; it can synthesize its own arginine.

Wild type

Mutant strain 1 grows only when arginine is supplied.

This means it lacks the ability to convert either citrulline or ornithine to arginine.

1

Mutant strain 2 grows on either arginine or citrulline, but not on ornithine.

This means it can convert citrulline to arginine, but cannot convert ornithine.

2

Mutant strain 3 does not grow on minimal medium, but grows when any one of the three supplements are added.

This means it can convert ornithine to citrulline and citrulline to arginine.

3

INTERPRETATION

If an organism cannot convert one particular compound to another, it presumably lacks an enzyme required for the conversion, and the mutation is in the gene that codes for that enzyme.

Strain 3 is blocked at this step.

Strain 2 is blocked at this step.

Strain 1 is blocked at this step.

Conclusion: The synthesis of arginine proceeds like this…

…and each gene specifies a particular enzyme.

Precursor → Enzyme A → Ornithine → Enzyme B → Citrulline → Enzyme C → Arginine

Gene *A*   Gene *B*   Gene *C*

12.1 One Gene, One Enzyme
Beadle and Tatum studied several auxotrophic mutants of *Neurospora*. Different auxotrophic mutants required the addition of different nutrients in order to obtain the arginine required for their growth. Step through the figure to follow the reasoning that upheld the "one-gene, one-enzyme" hypothesis.

mutant strains could no longer grow on minimal medium, but needed additional nutrients that they could not make on their own. The scientists proposed that these *auxotrophs* ("increased eaters") must have suffered mutations in genes that coded for enzymes necessary for the synthesis of the nutrients they now needed to ingest. For each auxotrophic strain, Beadle and Tatum were able to find a single compound that, when added to the minimal medium, supported the growth of that strain. This result supported the idea that mutations have simple effects, and that each mutation causes a defect in only one enzyme in a metabolic pathway.

One group of auxotrophs, for example, could grow on minimal medium supplemented with the amino acid arginine. (Wild-type *Neurospora* makes arginine by itself.) These mutant strains were classified as *arg* mutants. Were their mutations different alleles of the same gene, or were they in different genes, each coding for an enzyme along the biochemical pathway for arginine synthesis? Mapping studies established that some of the *arg* mutations were at different loci, or on different chromosomes, and so were not alleles. Beadle and Tatum concluded that different genes could participate in governing a single biosynthetic pathway—in this case, the pathway leading to arginine synthesis.

By growing different *arg* mutants in the presence of various compounds suspected to be intermediates in the synthetic metabolic pathway for arginine, Beadle and Tatum were able to classify each mutation as affecting one enzyme or another, and to order the compounds on the pathway (see Figure 12.1). Then they broke open the wild-type and mutant cells and examined them for enzyme activities. The results confirmed their hypothesis: Each mutant strain was indeed missing a single active enzyme in the pathway.

If a genetic mutation results in an abnormal or missing enzyme, then the wild-type gene must code for the normal enzyme. This conclusion led Beadle and Tatum to formulate the one-gene, one-enzyme hypothesis. According to this hypothesis, the *function of a gene is to control the production of a single, specific enzyme.* This proposal strongly influenced the subsequent development of the sciences of genetics and molecular biology.

The English physician Archibald Garrod, who studied the inherited disease alkaptonuria, had made this proposal 40 years before. He linked the biochemical phenotype of the disease to an abnormal gene and a missing enzyme. There are hundreds of examples of such hereditary diseases, and we will return to them in Chapter 18.

The gene–enzyme relationship requires modification when we consider that many enzymes are composed of more than one polypeptide chain, or subunit (that is, they have a quaternary structure). In this case, each polypeptide chain is specified by its own separate gene. Thus, it is more correct to speak of a **one-gene, one-polypeptide hypothesis**: *The function of a gene is to control the production of a single, specific polypeptide.*

Much later, it was discovered that some genes code for forms of RNA that do not become translated into polypeptides, and that still other genes are involved in controlling which other DNA sequences are expressed. While these discoveries have overthrown the idea that all genes code for proteins, they did not invalidate the relationship between genes and polypeptides.

DNA, RNA, and the Flow of Information

Now let us turn our attention to *how* a gene expresses itself as a polypeptide. This expression occurs in two steps. The first step, transcription, copies the information of a DNA sequence (the gene) into corresponding information in an RNA sequence. The second step, translation, converts this RNA information into an appropriate amino acid sequence in a polypeptide.

RNA differs from DNA

To understand the transcription and translation of genetic information, you need to understand the structure of RNA. **RNA (ribonucleic acid)** is a polynucleotide similar to DNA (see Figure 3.17), but it differs from DNA in three ways:

▶ RNA generally consists of only one polynucleotide strand [thus Chargaff's rule, G = C and A = T (see Figure 11.5), is usually true for DNA and not for RNA].

▶ The sugar molecule found in RNA is ribose, rather than the deoxyribose found in DNA.

▶ Although three of the nitrogenous bases (adenine, guanine, and cytosine) in RNA are identical to those in DNA, the fourth base in RNA is uracil (U), which is similar to thymine but lacks the methyl (—CH_3) group.

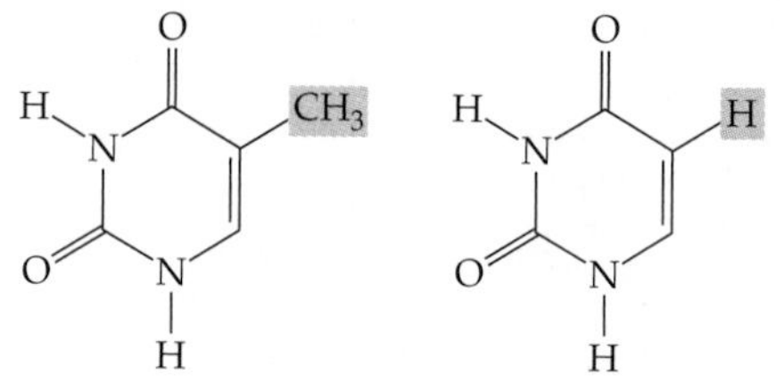

Thymine   Uracil

RNA can base-pair with single-stranded DNA, and this pairing obeys the same hydrogen-bonding rules as in DNA, except that adenine pairs wth uracil instead of thymine. RNA can also fold over and base-pair with itself, as we will see with tRNA later in this chapter.

Information flows in one direction when genes are expressed

Francis Crick (of the Watson–Crick model) proposed what he called the **central dogma** of molecular biology. The central dogma is, simply, that DNA codes for the production of RNA (transcription), RNA codes for the production of protein (translation), and protein does *not* code for the production of protein, RNA, or DNA (Figure 12.2*a*). In Crick's words, "once 'information' has passed into protein *it cannot get out again.*"

(a)

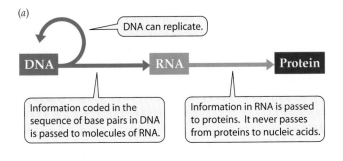

(b)

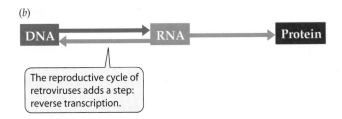

12.2 The Central Dogma
(a) Information flows from DNA to proteins, as indicated by the arrows. In certain viruses, RNA can replicate to RNA. *(b)* The reproductive cycle of retroviruses adds a step, reverse transcription, to the central dogma.

The central dogma posed two questions:

▶ How does genetic information get from the nucleus to the cytoplasm? (As you know, most of the DNA of a eukaryotic cell is confined to the nucleus, but proteins are synthesized in the cytoplasm.)

▶ What is the relationship between a specific nucleotide sequence (in DNA) and a specific amino acid sequence (in a protein)?

To answer the first question, Crick and his colleagues developed the messenger hypothesis, according to which an RNA molecule forms as a complementary copy of one DNA strand of a particular gene. The process by which this RNA forms is called **transcription**. If each such RNA molecule contains the information from a gene, there should be as many different kinds of RNA molecules as there are genes. This **messenger RNA**, or **mRNA**, then travels from the nucleus to the cytoplasm, where it serves as a template for the synthesis of proteins.

To answer the second question, Crick proposed the adapter hypothesis: there must be an adapter molecule that can bind a specific amino acid at one end and recognize a sequence of nucleotides with another region. In due course, these adapters, called **transfer RNA**, or **tRNA**, were identified. Because they recognize the genetic message of mRNA and simultaneously carry specific amino acids, tRNA's can translate the language of DNA into the language of proteins. The tRNA adapters line up on the mRNA so that the amino acids are in the proper sequence for a growing polypeptide chain—a process called **translation** (Figure 12.3).

Summarizing the main features of the central dogma, the messenger hypothesis, and the adapter hypothesis, we may say that *a given gene is transcribed to produce a messenger RNA (mRNA) complementary to one of the DNA strands*, and that *transfer RNA (tRNA) molecules translate the sequence of bases in the mRNA into the appropriate sequence of amino acids*.

RNA viruses modify the central dogma

According to the central dogma, DNA codes for RNA, and RNA codes for protein. All cellular organisms have DNA as their hereditary material. Only among viruses (and certain DNA sequences) are variations on the central dogma found.

Many viruses, such as the tobacco mosaic virus, influenza virus, and poliovirus, have RNA rather than DNA as their genetic material. With its nucleotide sequence, RNA could potentially act as an information carrier and be expressed as proteins. But since RNA is usually single-stranded, its replication is a problem. The viruses generally solve this problem by transcribing from RNA to RNA, making an RNA strand that is complementary to the genome. This "opposite" strand is used to make to make more copies of the genome by transcription.

HIV and certain tumor viruses also have RNA as their genome, but do not replicate it as RNA-to-RNA. Instead, after infecting a host cell, they make a DNA copy of their genome, and use it to make more RNA. This RNA is then used both as genomes for more copies of the virus and as mRNA (see Figure 12.2*b*). Synthesis of DNA from RNA is called reverse transcription. Not surprisingly, such viruses are called *retroviruses*. We will examine tboth types of RNA viruses in detail in the next chapter.

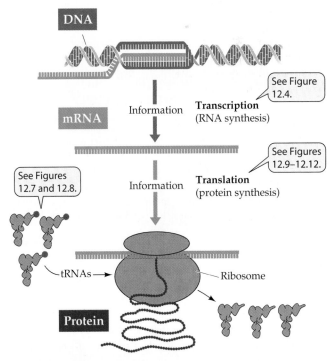

12.3 From Gene to Protein
This figure summarizes the processes of gene expression in prokaryotes. In eukaryotes, the processes are somewhat more complex.

Transcription: DNA-Directed RNA Synthesis

Transcription—the formation of a specific RNA from a specific DNA, requires the enzyme **RNA polymerase**. It also requires the appropriate ribonucleoside triphosphates (ATP, GTP, CTP, and UTP) and a DNA template. Within each gene, only *one* of the strands—the **template strand**—is transcribed. The other, complementary DNA strand remains untranscribed. For different genes in the same DNA molecule, different strands may be transcribed. That is, the strand that is the complementary strand in one gene may be the template strand in another.

Not only mRNA is produced by transcription. The same process is responsible for the synthesis of tRNA and **ribosomal RNA (rRNA)**, which constitutes a major fraction of the ribosomes. Like mRNA, these other forms of RNA are encoded by specific genes.

In DNA replication, as we know, the two strands of the parent molecule unwind, and each strand serves as the template for a new strand. In transcription, DNA partly unwinds so that it can serve as a template for RNA synthesis. As the RNA transcript forms, it peels away, allowing the DNA that has already been transcribed to be rewound into the double helix (Figure 12.4).

Transcription can be divided into three distinct processes: initiation, elongation, and termination. Let's consider each of these in turn.

Initiation of transcription requires a promoter and an RNA polymerase

The transcription of a gene begins at a **promoter**, a special sequence of DNA to which RNA polymerase binds very tightly. There is at least one promoter for each gene (or, in prokaryotes, each set of genes) to be transcribed into

12.4 DNA Is Transcribed into RNA
DNA partially unwinds to serve as a template for RNA synthesis. The RNA transcript forms and then peels away, allowing the DNA that has already been transcribed to rewind into a double helix. Three distinct processes—initiation, elongation, and termination—constitute DNA transcription. RNA polymerase is much larger in reality than indicated here, covering about 50 base pairs.

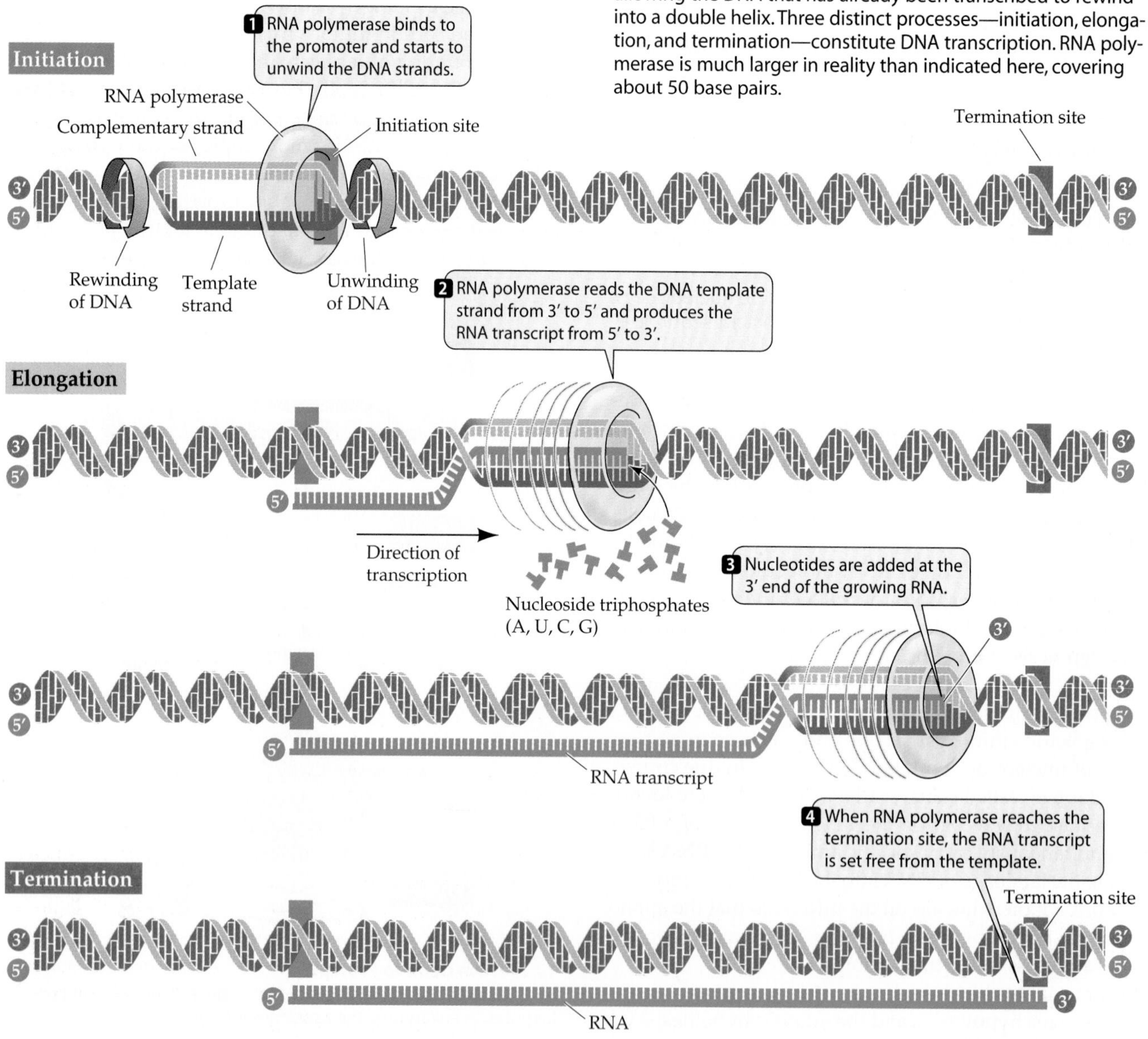

Initiation

RNA polymerase
Complementary strand
Initiation site

1 RNA polymerase binds to the promoter and starts to unwind the DNA strands.

Rewinding of DNA   Template strand   Unwinding of DNA

Termination site

2 RNA polymerase reads the DNA template strand from 3′ to 5′ and produces the RNA transcript from 5′ to 3′.

Elongation

Direction of transcription

Nucleoside triphosphates (A, U, C, G)

3 Nucleotides are added at the 3′ end of the growing RNA.

RNA transcript

4 When RNA polymerase reaches the termination site, the RNA transcript is set free from the template.

Termination site

Termination

RNA

mRNA. Promoters serve as punctuation marks, telling the RNA polymerase where to start, which strand of DNA to read and the direction to take from the start.

A promoter, which is a specific sequence in the DNA that reads in a particular direction, orients the RNA polymerase and thus "aims" it at the correct strand to use as a template. Part of each promoter is the *initiation site*, where transcription begins. Farther toward the 3' end of the promoter lie groups of nucleotides that help the RNA polymerase bind. RNA polymerase moves in a 3'-to-5' direction along the template strand (see Figure 12.4).

Promoter sequences can be identified by scientists in the laboratory in several ways. For example, DNA can be chopped up into short stretches and presented to RNA polymerase in the test tube; sequences to which RNA polymerase binds tightly are promoters. Or, a promoter can be identified by a loss of function due to a change in its DNA sequence; in some cases, a single base-pair change results in a promoter that can no longer bind tightly to the RNA polymerase.

Although every gene has a promoter, not all promoters are identical. One promoter may have a DNA sequence that binds RNA polymerase very effectively and therefore triggers frequent transcription of its gene; in other words, it competes effectively for the available RNA polymerase. Another promoter may bind the polymerase less well, and its genes will rarely be transcribed. We will consider prokaryotic promoters in more detail in Chapter 13 and eukaryotic promoters in Chapter 14.

In addition, not all RNA polymerases are identical. Prokaryotes have a single RNA polymerase that produces mRNA, tRNA, and rRNA. Eukaryotes have three different RNA polymerases with distinct roles. Of these, *RNA polymerase II* is responsible for mRNA production.

Initiation of transcription in eukaryotes differs from that in prokaryotes in another respect. As we will see in more detail in Chapter 14, eukaryotic RNA polymerases cannot bind to a promoter and start the process of transcription until other proteins bind to sites in the promoter and prepare a "docking site" for the RNA polymerase. The requirement for these proteins affords a way to regulate the transcription of particular genes.

RNA polymerase elongates the transcript

Once RNA polymerase has bound to the promoter, it unwinds the DNA about 20 base pairs at a time and reads the template strand in the 3'-to-5' direction (see Figure 12.4). Like DNA polymerase, RNA polymerase adds new nucleotides to the 3' end of the growing strand. That is, the new RNA grows from its own 5' end to its 3' end. The RNA transcript is thus *antiparallel* to the DNA template strand.

We indicated in the previous chapter that DNA replication occurs at fixed locations in the nucleus, with the DNA moving through a replication complex. While there is some evidence that this may be the case in DNA transcription to RNA as well, it is not as convincing as for DNA replication,

so we will present the model of a "moving transcription complex" rather than a moving DNA molecule.

Transcription, like DNA replication, requires a lot of energy. Like replication, transcription draws on energy released by both the removal and the breakdown of the pyrophosphate group from each nucleotide added.

Unlike DNA polymerases, RNA polymerases do not inspect and correct their work. Transcription errors occur at a rate of one mistake for every 10^4 to 10^5 bases. Because many copies of RNA are made, these errors are not as potentially harmful as mutations in DNA.

Transcription terminates at particular base sequences

What tells RNA polymerase to stop adding nucleotides to a growing transcript? Just as initiation sites specify the start of transcription, particular base sequences in the DNA specify its termination. The mechanisms of termination are complex and of more than one kind. For some genes, the newly formed transcript simply falls away from the DNA template and the RNA polymerase. For others, a helper protein pulls the transcript away.

In prokaryotes, the translation of mRNA often begins (at the 5' end of the mRNA) before transcription of the mRNA molecule is complete. In eukaryotes, the situation is more complicated. First, there is a spatial separation of transcription (in the nucleus) and translation (in the cytoplasm). Second, the first product of transcription is a pre-mRNA that is longer than the final mRNA and must undergo considerable processing before it becomes the mRNA that can be translated. The advantages of this processing, and its mechanisms, will be discussed in Chapter 14.

The Genetic Code

The genetic code provides the specificity for protein synthesis. You can think of the genetic information in an mRNA molecule as a series of sequential, non-overlapping three-letter "words." Each sequence of three nucleotides (the three "letters") along the chain specifies a particular amino acid. Each three-letter "word" is called a **codon**. Each codon is complementary to the corresponding triplet in the DNA molecule from which it was transcribed.

The complete genetic code is shown in Figure 12.5. Notice that there are many more codons than there are different amino acids in proteins. Combinations of the four available "letters" (the bases) give 64 (4^3) different three-letter codons, yet these codons determine only 20 amino acids. AUG, which codes for methionine, is also the **start codon**, the initiation signal for translation. Three of the codons (UAA, UAG, UGA) are **stop codons**, or *chain terminators*; when the translation machinery reaches one of these codons, translation stops, and the polypeptide is released from the translation complex.

After describing the properties of the genetic code, we will examine some of the scientific thinking and experimentation that went into deciphering it.

Second letter

		U		C		A		G		
U		UUU UUC	Phenyl-alanine	UCU UCC UCA UCG	Serine	UAU UAC	Tyrosine	UGU UGC	Cysteine	U C
		UUA UUG	Leucine			UAA UAG	Stop codon Stop codon	UGA UGG	Stop codon Tryptophan	A G
C		CUU CUC CUA CUG	Leucine	CCU CCC CCA CCG	Proline	CAU CAC	Histidine	CGU CGC CGA CGG	Arginine	U C A G
						CAA CAG	Glutamine			
A		AUU AUC AUA	Isoleucine	ACU ACC ACA ACG	Threonine	AAU AAC	Asparagine	AGU AGC	Serine	U C
		AUG	Methionine; start codon			AAA AAG	Lysine	AGA AGG	Arginine	A G
G		GUU GUC GUA GUG	Valine	GCU GCC GCA GCG	Alanine	GAU GAC	Aspartic acid	GGU GGC GGA GGG	Glycine	U C A G
						GAA GAG	Glutamic acid			

First letter (left) / *Third letter* (right)

12.5 **The Universal Genetic Code**

12.5 *The Universal Genetic Code*

Genetic information is encoded in mRNA in three-letter units—codons—made up of the bases uracil (U), cytosine (C), adenine (A), and guanine (G). To decode a codon, find its first letter in the left column, then read across the top to its second letter, then read down the right column to its third letter. The amino acid the codon specifies is given in the corresponding row. For example, AUG codes for methionine, and GUA codes for valine.

The genetic code is redundant but not ambiguous

After the start and stop codons, the remaining 60 codons are far more than enough to code for the other 19 amino acids—and indeed there are repeats. Thus we say that the genetic code is **redundant**; that is, an amino acid may be represented by more than one codon. The redundancy is not evenly divided among the amino acids. For example, methionine and tryptophan are represented by only one codon each, whereas leucine is represented by six different codons (see Figure 12.5).

The term *redundancy* should not be confused with *ambiguity*. To say that the code was ambiguous would mean that a single codon could specify either of two (or more) different amino acids; there would then be doubt whether to put in, say, leucine or something else. The genetic code is *not* ambiguous. Redundancy in the code means that there is more than one clear way to say, "Put leucine here." In other words, a given amino acid may be encoded by more than one codon, but a codon can code for only one amino acid. But just as people in different places prefer different ways of saying the same thing—"Good-bye!" "See you!" "Ciao!" and "So long!" have the same meaning—different organisms prefer one or another of the redundant codons.

The genetic code appears to be nearly universal, applying to all the species on our planet. Thus the code must be an ancient one that has been maintained intact throughout the evolution of living things. Exceptions are known: Within mitochondria and chloroplasts, the code differs slightly from that in prokaryotes and in the nuclei of eukaryotic cells; in one group of protists, UAA and UAG code for glutamine rather than functioning as stop codons. The significance of these differences is not yet clear. What is clear is that the exceptions are few and slight.

The common genetic code means that there is also a common language for evolution. As natural selection has resulted in one species replacing another, the raw material of genetic variation has remained the same: DNA sequences with the same "meaning." The common code also has great implications in genetic engineering, as we will see in Chapter 17, since it means that a human gene is in the same language as a bacterial gene. So the protein transcription and translation machinery of a bacterium can utilize genes from a human as well as its own genes.

The codons in Figure 12.5 are *mRNA codons*. The master codons on the DNA strand that was transcribed to produce the mRNA are complementary and antiparallel to these codons. Thus, for example, 3'-AAA-5' in the template DNA strand corresponds to phenylalanine (which is coded for by the mRNA codon 5'-UUU-3'), and 3'-ACC-5' in the template DNA corresponds to tryptophan (which is coded for by the mRNA codon 5'-UGG-3').

Biologists broke the genetic code by using artificial messengers

Molecular biologists broke the genetic code in the early 1960s. The problem seemed difficult: How could more than 20 "code words" be written with an "alphabet" consisting of only four "letters"? How, in other words, could four bases code for 20 or so different amino acids?

That the code was a triplet code, based on three-letter codons, was considered likely. Since there are only four letters (A, G, C, U), a one-letter code clearly could not unambiguously encode 20 amino acids; it could encode only four of them. A two-letter code could contain only $4 \times 4 = 16$ codons—still not enough. But a triplet code could contain up to $4 \times 4 \times 4 = 64$ codons.

Marshall W. Nirenberg and J. H. Matthaei, at the National Institutes of Health, made the first decoding breakthrough in 1961 when they realized that they could use a very simple artificial polynucleotide instead of a complex natural mRNA as a messenger. They could then identify the polypeptide that the artificial messenger encoded.

Nirenberg prepared an artificial mRNA in which all the bases were uracil (poly U). When poly U was added to a test tube containing all the ingredients necessary for protein synthesis (ribosomes, amino acids, activating enzymes,

12.6 Deciphering the Genetic Code
Nirenberg and Matthaei used a test-tube protein synthesis system to determine the amino acids specified by synthetic mRNA's of known codon composition.

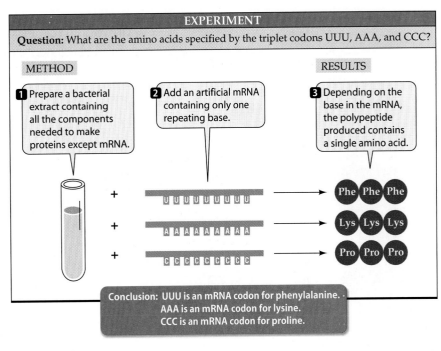

tRNA's, and other factors), a polypeptide formed. This polypeptide contained only one kind of amino acid: phenylalanine (Phe). Poly U coded for poly Phe! Accordingly, UUU appeared to be the mRNA code word—the codon—for phenylalanine. Following up on this success, Nirenberg and Matthaei soon showed that CCC codes for proline and AAA for lysine (Figure 12.6). (Poly G presented some chemical problems and was not tested initially.) UUU, CCC, and AAA were three of the easiest codons; different approaches were required to work out the rest.

Other scientists later found that simple artificial mRNA's only three nucleotides long—each amounting to a codon—could bind to a ribosome, and that the resulting complex could then cause the binding of the corresponding tRNA covalently bound to its specific amino acid. Thus, for example, simple UUU causes the tRNA charged with phenylalanine to bind to the ribosome. After this discovery, complete deciphering of the genetic code was relatively simple. To find the "translation" of a codon, Nirenberg could use a sample of that codon as an artificial mRNA and see which amino acid became bound to it.

Preparation for Translation: Linking RNA's, Amino Acids, and Ribosomes

How is the information contained in mRNA translated into proteins? Recall that Crick proposed that there is an adapter, something that can recognize the information in an mRNA codon and also bind the amino acid specified by the codon. That adapter is tRNA.

Translation occurs at **ribosomes**, which are molecular protein-synthesizing machines that hold the mRNA and tRNA's in place. In prokaryotes, ribosomes bind to mRNA as it is made, coupling the processes of transcription and translation. In eukaryotes, mRNA leaves the nucleus and is bound to ribosomes in the cytoplasm.

Two key events must take place to ensure that the protein made is the one specified by mRNA:

▶ tRNA must read mRNA correctly.
▶ tRNA must carry the amino acid that is correct for its reading of the mRNA.

At first glance, these two events seem similar, but they are quite distinct, as you will see.

Transfer RNA's carry specific amino acids and bind to specific codons

The codon in mRNA and the amino acid in a protein are related by way of an adapter—a specific tRNA. For each of the 20 amino acids, there is at least one specific type of tRNA molecule.

The structure of the tRNA molecule relates clearly to its functions: It carries an amino acid, it associates with mRNA molecules, and it interacts with ribosomes. A tRNA molecule has about 75 to 80 nucleotides. It has a three-dimensional shape that is maintained by complementary base pairing (hydrogen bonding) within its own sequence (Figure 12.7).

At the 3' end of every tRNA molecule is a site to which its specific amino acid binds covalently. At about the midpoint is a group of three bases, called the **anticodon**, that constitutes the point of contact with mRNA. Each tRNA species has a unique anticodon, which is complementary to the mRNA codon for that tRNA's amino acid. The codon and the anticodon unite by complementary base pairing (hydrogen bonding). At contact, they are antiparallel to each other.

As an example of this process, consider the amino acid arginine:

▶ The DNA coding region for arginine is 3'-GCC-5', which is transcribed, by base pairing, to…
▶ the mRNA codon 5'-CGG-3', which binds to…
▶ a tRNA with the anticodon 3'-GCC-5'.

Recall that 61 different codons encode the 20 amino acids in proteins (see Figure 12.5). Does this mean that the cell must produce 61 different tRNA species, each with a different anticodon? No. The cell gets by with about two-thirds that number of tRNA species, because the specificity for the

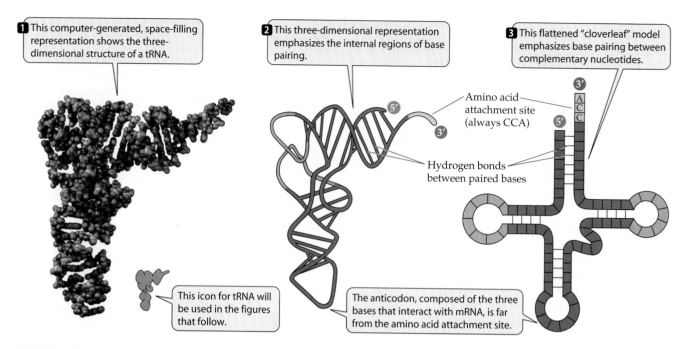

1 This computer-generated, space-filling representation shows the three-dimensional structure of a tRNA.

2 This three-dimensional representation emphasizes the internal regions of base pairing.

3 This flattened "cloverleaf" model emphasizes base pairing between complementary nucleotides.

Amino acid attachment site (always CCA)

Hydrogen bonds between paired bases

This icon for tRNA will be used in the figures that follow.

The anticodon, composed of the three bases that interact with mRNA, is far from the amino acid attachment site.

12.7 Transfer RNA: Crick's Adapter
The tRNA molecule carries amino acids, associates with mRNA molecules, and interacts with ribosomes. There is at least one specific tRNA molecule for each of the amino acids.

base at the 3′ end of the codon (and the 5′ end of the anticodon) is not always strictly observed. This phenomenon, called *wobble*, allows the alanine codons GCA, GCC, and GCU all to be recognized by the same tRNA. Wobble is allowed in some matches but not in others; of most importance, it does not allow the genetic code to be ambiguous!

The three-dimensional shape of tRNAs (see Figure 12.7) allows them to combine specifically with binding sites on ribosomes. The structure of tRNA molecules relates clearly to their functions: They carry amino acids, associate with mRNA molecules, and interact with ribosomes.

Activating enzymes link the right tRNA's and amino acids

How does a tRNA molecule combine with the correct amino acid for the codon it recognizes? A family of **activating enzymes**, known more formally as *aminoacyl-tRNA synthetases*, accomplishes this task (Figure 12.8). Each activating enzyme is specific for one amino acid and for its tRNA. The enzyme has a three-part active site that recognizes three smaller molecules: a specific amino acid, ATP, and a specific tRNA.

The enzyme reacts with tRNA and an amino acid in two steps:

$$\text{enzyme} + \text{ATP} + \text{AA} \rightarrow \text{enzyme—AMP—AA} + \text{PP}_i$$

$$\text{enzyme—AMP—AA} + \text{tRNA} \rightarrow \text{enzyme} + \text{AMP} + \text{tRNA—AA}$$

The amino acid is attached to the 3′ end (a free OH group on the ribose) of tRNA with a high-energy bond, forming

charged tRNA. This bond will provide the energy for the synthesis of the peptide bond that will join adjacent amino acids.

A clever experiment by Seymour Benzer and his colleagues at the California Institute of Technology showed the importance of the specificity of the attachment of tRNA to its amino acid. The amino acid cysteine, already properly attached to its tRNA, was chemically modified to become a different amino acid, alanine. Which component—the amino acid or the tRNA—would be recognized when this hybrid charged tRNA was put into a protein-synthesizing system? The answer was: the tRNA. Everywhere in the synthesized protein where cysteine was supposed to be, alanine appeared instead. The cysteine-specific tRNA delivered its cargo (alanine) to every address where cysteine was called for. This experiment showed that the protein synthesis machinery recognizes the tRNA part of charged tRNA, not the amino acid part.

If activating enzymes in nature did what Benzer did in the laboratory and put the wrong amino acids on tRNA's, those amino acids would be inserted into proteins at inappropriate places, leading to alterations in protein shape and function. The fact that the activating enzymes are highly specific has led to the process of linking tRNA and amino acid being called the "second genetic code."

The ribosome is the staging area for translation

Ribosomes are required for the translation of the genetic information in mRNA into a polypeptide chain. Although ribosomes are the smallest cellular organelles, their mass of several million daltons makes them large in comparison with charged tRNA's.

Each ribosome consists of two subunits, a large one and a small one (Figure 12.9). In eukaryotes, the large subunit consists of three different molecules of rRNA and about 45 different protein molecules, arranged in a precise pattern.

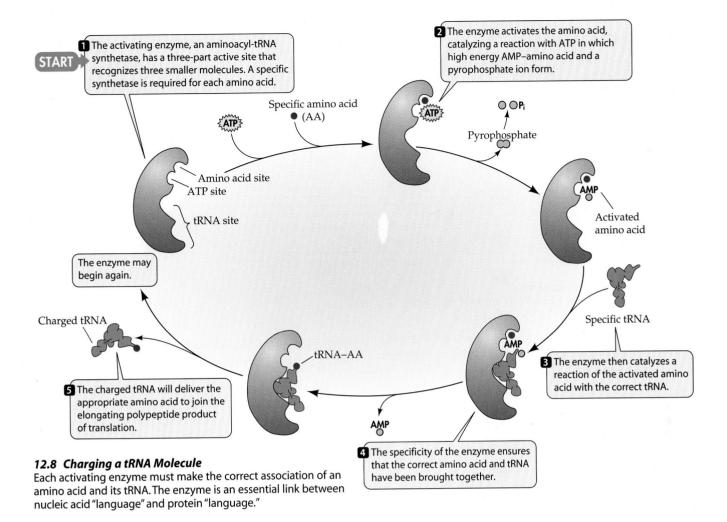

START

1 The activating enzyme, an aminoacyl-tRNA synthetase, has a three-part active site that recognizes three smaller molecules. A specific synthetase is required for each amino acid.

ATP

Specific amino acid (AA)

Amino acid site
ATP site
tRNA site

The enzyme may begin again.

Charged tRNA

5 The charged tRNA will deliver the appropriate amino acid to join the elongating polypeptide product of translation.

tRNA–AA

2 The enzyme activates the amino acid, catalyzing a reaction with ATP in which high energy AMP–amino acid and a pyrophosphate ion form.

ATP

P_i

Pyrophosphate

AMP

Activated amino acid

Specific tRNA

AMP

3 The enzyme then catalyzes a reaction of the activated amino acid with the correct tRNA.

AMP

4 The specificity of the enzyme ensures that the correct amino acid and tRNA have been brought together.

12.8 Charging a tRNA Molecule
Each activating enzyme must make the correct association of an amino acid and its tRNA. The enzyme is an essential link between nucleic acid "language" and protein "language."

The small subunit consists of one rRNA molecule and 33 different protein molecules. When not active in the translation of mRNA, the ribosomes exist as separated subunits.

The ribosomes of prokaryotes are somewhat smaller than those of eukaryotes, and their ribosomal proteins and RNA's are different. Mitochondria and chloroplasts also contain ribosomes, some of which are even smaller than those of prokaryotes.

The different proteins and RNA's in a subunit are held together by ionic and hydrophobic forces, not covalent bonds. If these forces are disrupted by detergents, for example, the proteins and rRNA's separate from each other. When the detergent is removed, the entire complex structure *self-assembles*. This is like separating the pieces of a jigsaw puzzle and having them fit together again without human hands to guide them!

A given ribosome is not specifically adapted to produce just one kind of protein. A ribosome can combine with any mRNA and all tRNA's, and thus can be used to make many different polypeptide products. The mRNA contains the information that specifies the polypeptide sequence; the ribosome is simply the molecular factory where the task is accomplished. Its structure enables it to hold the mRNA and tRNA's in the right positions, thus allowing the growing polypeptide to be assembled efficiently.

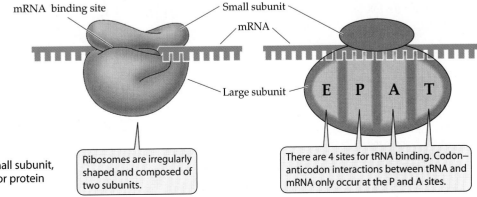

mRNA binding site

Small subunit

mRNA

Large subunit

E P A T

Ribosomes are irregularly shaped and composed of two subunits.

There are 4 sites for tRNA binding. Codon–anticodon interactions between tRNA and mRNA only occur at the P and A sites.

12.9 Ribosome Structure
Each ribosome consists of a large and a small subunit, which separate when they are not in use for protein synthesis.

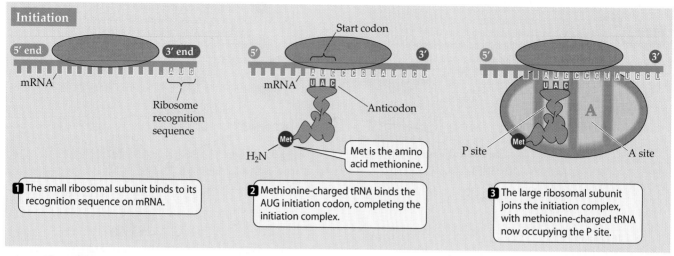

12.10 The Initiation of Translation
Translation begins with the formation of an initiation complex.

On the large subunit of the ribosome there are four sites to which tRNA binds (see Figure 12.9). A tRNA traverses these four sites in order:

▶ The T (transfer) site is where a tRNA carrying an amino acid first lands on the ribosome, accompanied by a special protein "escort" called the *T*, or *transfer, factor*.

▶ The A (amino acid) site is where the tRNA anticodon binds to mRNA codon, thus lining up the correct amino acid to be added to the growing polypeptide chain.

▶ The P (polypeptide) site is where the tRNA adds its amino acid to the growing polypeptide chain.

▶ The E (exit) site is where the tRNA, having given up its amino acid, resides before leaving the ribosome and going back to the cytosol to pick up another amino acid and begin the process again.

Because codon–anticodon interactions and peptide bond formation occur at the A and P sites, we will describe their function in detail in the next section.

Translation: RNA-Directed Polypeptide Synthesis

We have been working our way through the steps by which the sequence of bases in the template strand of a DNA molecule specifies the sequence of amino acids in a protein (see Figure 12.3). We are now at the last step: translation, the RNA-directed assembly of a protein. Like transcription, translation occurs in three steps: initiation, elongation, and termination.

Translation begins with an initiation complex

The translation of mRNA begins with the formation of an **initiation complex**, which consists of a charged tRNA bearing what will be the first amino acid of the polypeptide chain and a small ribosomal subunit, both bound to the mRNA (Figure 12.10). The small ribosomal subunit binds to a sequence that it recognizes on the mRNA. This sequence

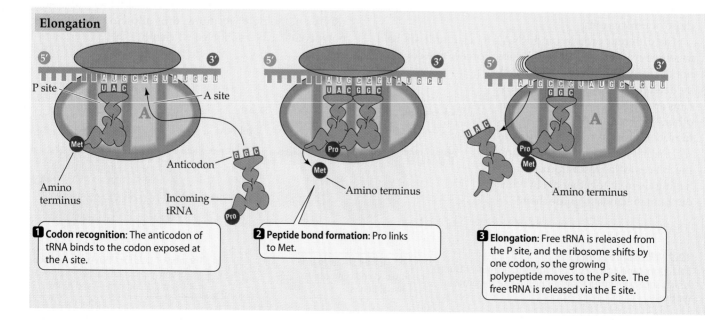

is "upstream" (toward the 5' end) of the actual start codon that begins translation.

Recall that the start codon in the genetic code is AUG (see Figure 12.5). Thus the first amino acid in the chain is always methionine. The anticodon of a methionine-charged tRNA binds to the mRNA start codon by complementary base pairing. (Not all mature proteins have methionine as their N-terminal amino acid, however. In many cases, the initiator methionine is later removed by an enzyme.)

After the methionine-charged tRNA has bound to the mRNA, the large subunit of the ribosome joins the complex. The charged tRNA, bearing methionine, now lies in the P site of the ribosome, and the A site is aligned with the second codon.

How are all these ingredients—mRNA, two ribosomal subunits, and methionine-charged tRNA—put together properly? A group of proteins called *initiation factors* help direct the process, using GTP as an energy supply.

The polypeptide elongates from the N terminus

During translation, the ribosome moves along the mRNA in the 5'-to-3' direction (Figure 12.11). A charged tRNA whose anticodon is complementary to the second codon on the mRNA enters the open A site of the large ribosomal subunit. The large subunit then catalyzes two reactions, collectively called *peptidyl transferase activity*:

▶ breakage of the bond between the tRNA in the P site and its amino acid

▶ peptide bond formation between this amino acid and the one attached to the tRNA in the A site

In this way, methionine (the amino acid in the P site) becomes the N terminus of the new protein. The second amino acid is now bound to methionine, but remains attached to its tRNA by its carboxyl group (—COOH) in the A site.

What catalyzes this binding? In 1992, Harry Noller and his colleagues at the University of California at Santa Cruz found that if they removed almost all the proteins in the large ribosomal subunit, it still catalyzed peptide bond formation. But if the rRNA was destroyed, so was peptidyl transferase activity. Part of the rRNA in the large subunit interacts with the end of the charged tRNA where the amino acid is attached. Thus rRNA appears to be the catalyst.

The idea that RNA—instead of the usual enzymes—can act as a catalyst, or **ribozyme**, was surprising, but is not so far-fetched. Because of its base-pairing ability, RNA can fold into three-dimensional shapes (see Figure 12.7) and bind substrates, just as protein-based enzymes do.

Elongation continues and the polypeptide grows

After the first tRNA releases its methionine, it dissociates from the ribosome, returning to the cytosol to become charged with another methionine. The second tRNA, now bearing a *dipeptide*, shifts to the P site of the ribosome, which moves along the mRNA by another codon. Energy for this movement comes from the hydrolysis of another molecule of GTP.

The elongation process continues, and the polypeptide chain grows, as the steps are repeated:

▶ The next charged tRNA enters the open A site.

▶ Its amino acid forms a peptide bond with the amino acid chain in the P site, so that it picks up the growing polypeptide chain from the tRNA in the P site.

▶ The tRNA in the P site is released. The ribosome shifts one codon, so that the entire tRNA–polypeptide complex, along with its codon, moves to the newly vacated P site.

All these steps are assisted by proteins called *elongation factors*.

A release factor terminates translation

How does the elongation cycle end? When a stop codon—UAA, UAG, or UGA—enters the A site, translation terminates (Figure 12.12). These codons encode no amino acids, nor do they bind any tRNA. Rather, they bind a protein *release factor*, which causes a water molecule instead of an amino acid to attach to the forming protein.

The newly completed protein thereupon separates from the ribosome. Its C terminus is the last amino acid to join the chain. Its N terminus, at least initially, is methionine, as a consequence of the AUG start codon. In its amino acid sequence, it contains information for its three-dimensional shape, as well as its ultimate cellular destination.

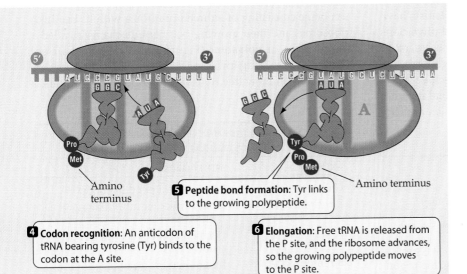

4 **Codon recognition:** An anticodon of tRNA bearing tyrosine (Tyr) binds to the codon at the A site.

5 **Peptide bond formation:** Tyr links to the growing polypeptide.

6 **Elongation:** Free tRNA is released from the P site, and the ribosome advances, so the growing polypeptide moves to the P site.

12.11 Translation: The Elongation Stage The polypeptide chain elongates as the mRNA is translated.

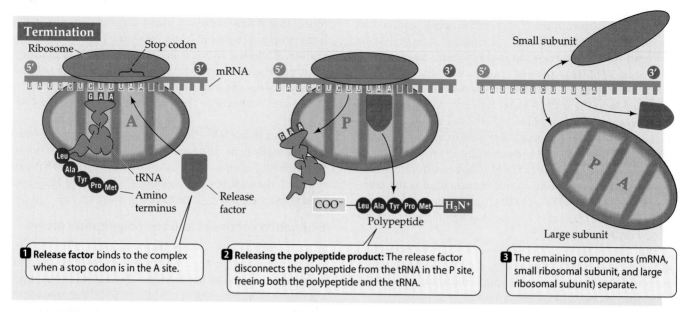

12.12 The Termination of Translation
Translation terminates when the ribosome encounters a stop codon on the mRNA.

Table 12.1 summarizes the initiation and termination of transcription and translation.

Regulation of Translation

Like any factory, the machinery of translation can work at varying rates. For example, externally applied chemicals, such as some antibiotics, can stop translation. Conversely, the presence of more than one ribosome on an mRNA can speed up protein synthesis.

Some antibiotics work by inhibiting translation

Antibiotics are defense molecules produced by microorganisms such as certain bacteria and fungi. These substances often destroy other microbes, which might compete with the defender for nutrients. Since the 1940s, scientists have isolated increasing numbers of antibiotics, and physicians use them to treat a great variety of infectious diseases, ranging from bacterial meningitis to pneumonia to gonorrhea.

The key to the medical use of antibiotics is specificity: An antibiotic must work to destroy the microbial invader, but not harm the human host. One way in which antibacterial antibiotics accomplish this task is to block the synthesis of the bacterial cell wall, something that is essential to the microbe but that is not part of human biochemistry. Penicillin works in this fashion.

Another way in which antibiotics work is to inhibit bacterial protein synthesis. Recall that the bacterial ribosome is smaller, and has a different collection of proteins, than the eukaryotic ribosome. Some antibiotics bind only to bacterial ribosomal proteins that are important in protein synthesis (Table 12.2). Without the ability to make proteins, the bacterial invaders die, and the infection is stemmed.

Polysome formation increases the rate of protein synthesis

Several ribosomes can work simultaneously at translating a single mRNA molecule, producing multiple molecules of the protein at the same time. As soon as the first ribosome has moved far enough from the initiation point, a second initiation complex can form, then a third, and so on. An assemblage consisting of a thread of mRNA with its beadlike ribosomes and their growing polypeptide chains is called a *polyribosome*, or **polysome** (Figure 12.13). Cells that are actively synthesizing proteins contain large numbers of polysomes and few free ribosomes or ribosomal subunits.

12.1	Signals that Start and Stop Transcription and Translation	
	TRANSCRIPTION	TRANSLATION
Initiation	Promoter sequence in DNA	AUG start codon
Termination	Terminator sequence in DNA	UAA, UAG, or UGA stop codon in mRNA

12.2	Antibiotics that Inhibit Bacterial Protein Synthesis
ANTIBIOTIC	STEP INHIBITED
Chloromycetin	Formation of peptide bonds
Erythromycin	Translocation of mRNA along ribosome
Neomycin	Interactions between tRNA and mRNA
Streptomycin	Initiation of translation
Tetracycline	Binding of tRNA to ribosome

(a)

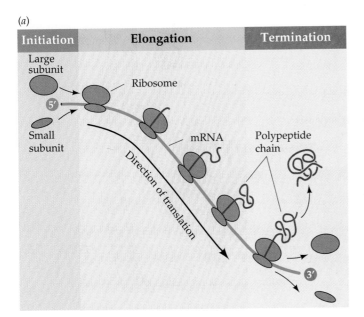

(b)

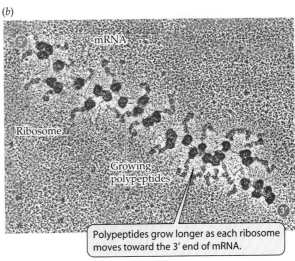

Polypeptides grow longer as each ribosome moves toward the 3' end of mRNA.

12.13 A Polysome
(a) A polysome consists of ribosomes and their growing polypeptide chains moving in single file along an mRNA molecule. (b) An electron microscopic view of a polysome.

A polysome is like a cafeteria line, in which patrons follow each other, adding items to their trays. At any moment, the person at the start has a little food (a newly initiated protein); the person at the end has a complete meal (a completed protein). However, in the polysome cafeteria, everyone gets the same meal: Many copies of the same protein are made from a single mRNA.

Posttranslational Events

The functional protein that results from protein synthesis is not necessarily the same as the polypeptide chain that is released from the ribosome. Especially in eukaryotic cells, the polypeptide may need to be moved far from the site of synthesis in the cytoplasm, to an organelle, or even secreted from the cell. In addition, the polypeptide is often modified by the addition of new chemical groups that have functional significance. In this section, we examine these two posttranslational aspects of protein synthesis.

Chemical signals in proteins direct them to their cellular destinations

As a polypeptide chain forms on the ribosome, it spontaneously folds into its three-dimensional shape. As described in Chapter 3, this shape is determined by the particular sequence of the amino acids that make up the protein, as well as factors such as the polarity and charge of their R groups. Ultimately, this shape allows the polypeptide to interact with other molecules in the cell, such as a substrate or another polypeptide. In addition to this structural information, the amino acid sequence contains an "address label" indicating where in the cell the polypeptide belongs.

All protein synthesis begins on free ribosomes in the cytoplasm. As a polypeptide chain is made, the information contained in its amino acid sequence gives it one of two sets of instructions (Figure 12.14):

▶ *Finish translation and be released to the cytoplasm.* Such proteins are sent to the nucleus, mitochondria, plastids, or peroxisomes, depending on the address in their instructions; or, lacking such specific instructions, they remain in the cytoplasm.

▶ *Stop translation, go to the endoplasmic reticulum (ER), and finish synthesis there.* Such proteins are sent to the lysosomes (via the Golgi apparatus) or the plasma membrane, are instructed to remain in the ER; or, lacking such specific instructions, they are secreted from the cell.

DESTINATION: CYTOPLASM. After translation, some folded polypeptides have a short exposed sequence of amino acids that acts like a postal "zip code," directing them to an organelle. These *signal sequences* are either at the N terminus or in the interior of the amino acid chain. For example, the following sequence is a nuclear localization signal:

—Pro—Pro—Lys—Lys—Lys—Arg—Lys—Val—

This amino acid sequence occurs, for example, in histone proteins, but not in citric acid cycle enzymes, which are addressed to the mitochondria.

The signal sequences have a conformation that allows them to bind to specific receptor proteins, appropriately called *docking proteins,* on the outer membrane of the appropriate organelle. Once the protein has bound to it, the receptor forms a channel in the membrane, allowing the protein to pass through to its organelle destination (it is usually unfolded by a chaperonin so that it can pass through the channel).

DESTINATION: ENDOPLASMIC RETICULUM. If a specific hydrophobic sequence of about 25 amino acids occurs at the beginning of a polypeptide chain, the finished product is

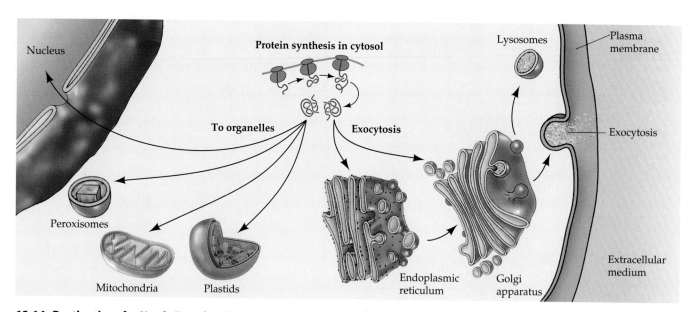

12.14 Destinations for Newly Translated Proteins in a Eukaryotic Cell
Signal sequences on newly synthesized polypeptides bind to specific receptor proteins on the outer membranes of the organelle to which they are "addressed." Once the protein has bound to it, the receptor forms a channel in the membrane and the protein enters the organelle.

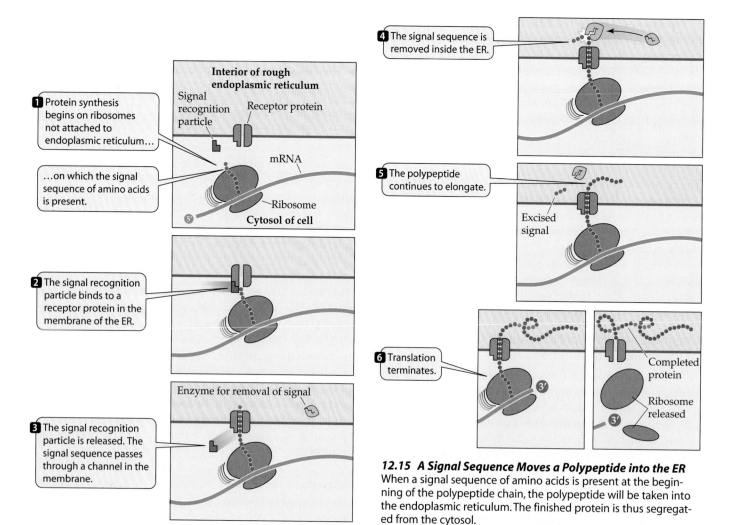

1 Protein synthesis begins on ribosomes not attached to endoplasmic reticulum…

…on which the signal sequence of amino acids is present.

Interior of rough endoplasmic reticulum

Signal recognition particle

Receptor protein

mRNA

Ribosome

Cytosol of cell

2 The signal recognition particle binds to a receptor protein in the membrane of the ER.

3 The signal recognition particle is released. The signal sequence passes through a channel in the membrane.

Enzyme for removal of signal

4 The signal sequence is removed inside the ER.

5 The polypeptide continues to elongate.

Excised signal

6 Translation terminates.

Completed protein

Ribosome released

12.15 A Signal Sequence Moves a Polypeptide into the ER
When a signal sequence of amino acids is present at the beginning of the polypeptide chain, the polypeptide will be taken into the endoplasmic reticulum. The finished protein is thus segregated from the cytosol.

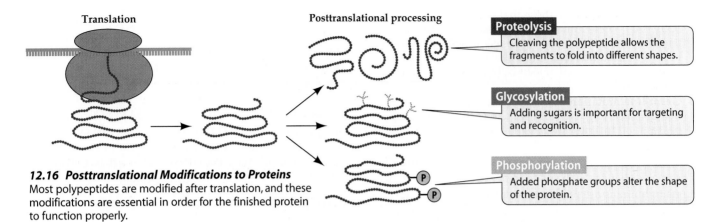

12.16 Posttranslational Modifications to Proteins
Most polypeptides are modified after translation, and these modifications are essential in order for the finished protein to function properly.

Proteolysis
Cleaving the polypeptide allows the fragments to fold into different shapes.

Glycosylation
Adding sugars is important for targeting and recognition.

Phosphorylation
Added phosphate groups alter the shape of the protein.

destined for the ER, lysosomes, and plasma membrane, or for secretion. The signal sequence attaches to a *signal recognition particle* composed of protein and RNA (Figure 12.15). This attachment blocks further protein synthesis until the ribosome can become attached to a specific receptor protein in the membrane of the ER. Once again, the receptor protein becomes a channel through which the growing polypeptide passes, either into the ER membrane itself or into the interior of the ER. An enzyme within the ER interior then removes the signal sequence from the polypeptide chain. At this point, protein synthesis resumes, and the chain grows longer. From the ER, the newly formed protein can be transported to its appropriate location—to other cellular compartments or to the outside of the cell—without mixing with other molecules in the cytoplasm.

Additional signals are needed for sorting the protein further (remember that its ER signal sequence has been removed). These signals are of two kinds. Some are sequences of amino acids that allow the protein's retention within the ER. Others are sugars added in the Golgi apparatus, to which the protein is transferred from the ER; the resulting glycoproteins end up either at the plasma membrane or in a lysosome (or plant vacuole), depending on which sugars are added. Proteins with no signals pass from the ER through the Golgi apparatus and are secreted from the cell.

It is important to emphasize that the addressing of a protein to its destination is a property of its amino acid sequence, and so is genetically determined. An example of what can go wrong if a gene for protein targeting is mutated is *mucoplidosis II*, or *I-cell disease*. People with this disease lack an essential enzyme for the formation of the lysosomal targeting signal. So proteins destined for their lysosomes never get there, and instead either stay in the Golgi (where they form I, or inclusion, bodies) or are secreted from the cell. The inability to perform normal lysosome functions leads to progressive illness and death in childhood.

Many proteins are modified after translation

It is the exception and not the rule that the finished protein product is identical to the polypeptide chain translated from mRNA on the ribosomes. Instead, most polypeptides are modified after translation, both covalently and nonco-

valently. In both cases, the modifications are essential to the final functioning of the protein. Some kinds of covalent changes include the following (Figure 12.16):

▶ **Proteolysis** is the cutting of a polypeptide chain. Cleavage of the signal sequence from the growing polypeptide chain in the ER is an example of proteolysis; the protein might go back out of the ER through the membrane channel if the signal sequence were not cut off. Some proteins are actually made from *polyproteins* (long polypeptides) that are cut into final products by enzymes called proteases. Viruses such as HIV encode such proteases, which are essential because the large viral polyprotein cannot fold properly unless it is cut. Certain drugs used to treat AIDS work by inhibiting HIV's protease.

▶ **Glycosylation** involves the addition of sugars to proteins. In both the ER and the Golgi, resident enzymes catalyze the addition of various sugar residues or short sugar chains to certain amino acid R groups on proteins as they pass through. One type of "sugar coating" is essential for addressing proteins to lysosomes, as we saw above. Other types are important in the three-dimensional structure and recognition of proteins at the cell surface. Still others help in stabilizing proteins stored in storage vacuoles in plant seeds.

▶ **Phosphorylation** is the addition of phosphate groups to proteins and is catalyzed by protein kinases. These charged groups change the three-dimensional structures of targeted proteins, often exposing an active site of an enzyme, or a binding site for another protein—as we will see in Chapter 15.

Mutations: Heritable Changes in Genes

Accurate DNA replication, transcription, and translation all depend on the reliable pairing of complementary bases. Errors occur, though infrequently, in all three processes. Errors happen least often in DNA replication. However, the consequences of those errors can be most severe because only they are heritable.

Mutations are heritable changes in genetic information. In unicellular organisms, any mutations that occur are passed on to the daughter cells when the cell divides. In multicellular organisms, there are two general types of mutations in terms of inheritance:

▶ **Somatic mutations** are those that occur in non-gamete body cells. These mutations are passed on to the daughter cells after mitosis, and to the offspring of these cells in turn. A mutation in a single skin cell, for example, could result in a patch of skin cells, all with the same DNA alteration.

▶ **Germ-line mutations** are those that occur in the cells that give rise to gametes. A gamete with the mutation passes it on to a new organism at fertilization.

Very small changes in the genetic material often lead to easily observable changes in the phenotype. Some effects of mutations in humans are readily detectable—dwarfism, for instance, or the presence of more than five fingers on each hand. A mutant genotype in a microorganism may be obvious if, for example, it results in a change in nutritional requirements, as we described for *Neurospora* earlier (see Figure 12.1).

Other mutations may be unobservable. In humans, for example, a particular mutation drastically lowers the level of an enzyme called glucose 6-phosphate dehydrogenase that is present in many tissues, including red blood cells. The red blood cells of a person carrying the mutant allele are abnormally sensitive to an antimalarial drug called primaquine; when such people are treated with this drug, their red blood cells rupture, causing serious medical problems. People with the normal allele have no such problem. Before the drug came into use, no one was aware that such a mutation existed. Similarly, distinguishing a mutant bacterium from a normal bacterium may require sophisticated chemical methods, not just visual inspection.

Some mutations cause their phenotypes only under certain *restrictive* conditions and are not detectable under other, *permissive* conditions. We call organisms carrying such mutations *conditional mutants*. Many conditional mutants are temperature-sensitive, able to grow normally at a permissive temperature—say, 30°C—but unable to grow at a higher restrictive temperature—say, 37°C. The mutant allele in such an organism may code for an enzyme with an unstable tertiary structure that is altered at the restrictive temperature.

All mutations are alterations in the nucleotide sequence of DNA. We divide mutations into two categories:

▶ **Point mutations** are mutations of single genes: One allele becomes another because of small alterations in the sequence or number of nucleotides—even as small as the substitution of one nucleotide for another.

▶ **Chromosomal mutations** are more extensive alterations. They may change the position or direction of a DNA segment without actually removing any genetic information, or they may cause a segment of DNA to be irretrievably lost.

Point mutations are changes in single bases

Point mutations result from the addition or subtraction of a nucleotide base, or the substitution of one base for another, in the DNA, and hence in the mRNA. Point mutations can be caused by errors in chromosome replication that are not corrected in proofreading, or by environmental mutagens such as chemicals and radiation.

Because of the redundancy of the genetic code, some point mutations result in no change in amino acids when the altered mRNA is translated; for this reason they are called **silent** or **synonymous mutations**. For example, four mRNA codons code for proline: CCA, CCC, CCU, and CCG (see Figure 12.5). If the template strand of DNA has the sequence CGG, it will be transcribed to CCG in mRNA, and proline-charged tRNA will bind at the ribosome. But if there is a mutation such that the codon in the template DNA now reads AGG, the mRNA codon will be CCU—and the tRNA that binds will still carry proline:

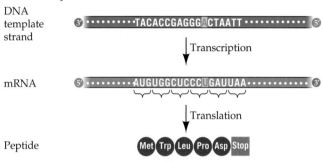

Synonymous mutation

Mutation at position 12 in DNA: C→A

Result: No change in amino acid sequence

Silent mutations are quite common, and account for genetic diversity that is not expressed as phenotypic differences.

In contrast to silent mutations, some base substitution mutations may change the genetic message such that one amino acid substitutes for another in the protein. These are **missense mutations**:

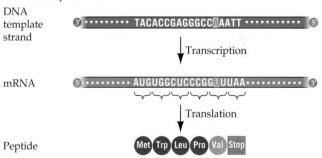

Missense mutation

Mutation at position 14 in DNA: T→A

Result: Amino acid change at position 5: Asp → Val

A specific example of a missense mutation is the sickle-cell allele for human β-globin. *Sickle-cell disease* results from a defect in hemoglobin, a protein that carries oxygen. The gene for β-globin, one of the polypeptides in hemoglobin, differs by one amino acid between the normal and the sickle-cell allele. Persons who are homozygous for this recessive allele have defective red blood cells. Where oxygen is abundant, as in the lungs, the cells are normal in structure and function. But at the low oxygen levels characteristic of working muscles, the red blood cells collapse into the shape of a sickle (Figure 12.17).

A missense mutation may sometimes cause a protein not to function, but often the effect is only to reduce the functional *efficiency* of the protein. Therefore, individuals carrying missense mutations may survive, even though the affected protein is essential to life. Through evolution, some missense mutations even improve functional efficiency.

Nonsense mutations, another type of mutation in which bases are substituted, are more often disruptive than are missense mutations. In a nonsense mutation, the base substitution causes a chain terminator (stop) codon, such as UAG, to form in the mRNA product:

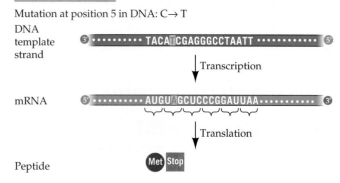

Nonsense mutation

Mutation at position 5 in DNA: C→ T

DNA template strand

Transcription

mRNA

Translation

Peptide

Result: Only one amino acid translated; no protein made

The result is a shortened protein product, since translation does not proceed beyond the point where the mutation occurred.

Not all point mutations are base substitutions. Single base pairs may be inserted into or deleted from DNA. Such mutations are known as **frame-shift mutations** because they interfere with the decoding of the genetic message by throwing it out of register:

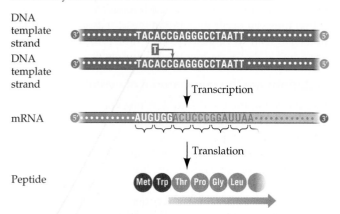

Frame-shift mutation

Mutation by insertion of T between bases 6 and 7 in DNA

DNA template strand

DNA template strand

Transcription

mRNA

Translation

Peptide

Result: All amino acids changed beyond the insertion

Think again of codons as three-letter words, each corresponding to a particular amino acid. Translation proceeds codon by codon; if a base is added to the message or subtracted from it, translation proceeds perfectly until it comes to the one-base insertion or deletion. From that point on, the three-letter words in the message are one letter out of regis-

Sickle-cell phenotype   Normal phenotype

12.17 Sickled and Normal Red Blood Cells
The misshapen red blood cell on the left is caused by a mutation that substitutes an incorrect amino acid in one of the two polypeptides of hemoglobin.

ter. In other words, such mutations shift the "reading frame" of the genetic message. Frame-shift mutations almost always lead to the production of nonfunctional proteins.

Chromosomal mutations are extensive changes in the genetic material

DNA molecules can break and rejoin, grossly disrupting the sequence of genetic information. There are four types of such chromosomal mutations: deletions, duplications, inversions, and translocations (Figure 12.18). These mutations can be caused by severe damage to chromosomes resulting from chemical or radiation exposure or by drastic errors in chromosome replication.

Deletions remove part of the genetic material (Figure 12.18*a*). Like frame-shift point mutations, their consequences can be severe unless they affect unnecessary genes or are masked by the presence, in the same cell, of normal alleles of the deleted genes. It is easy to imagine one mechanism that could produce deletions: A DNA molecule might break at two points, and the two end pieces might rejoin, leaving out the DNA between the breaks.

Another mechanism by which deletion mutations might arise would lead simultaneously to the production of a second kind of chromosomal mutation: a **duplication** (Figure 12.18*b*). Duplication would arise if homologous chromosomes broke at different positions and then reconnected to the wrong partners. One of the two molecules produced by this mechanism would lack a segment of DNA (it would have a deletion), and the other would have two copies (a duplication) of the segment that was deleted from the first.

Breaking and rejoining can also lead to **inversion**—the removal of a segment of DNA and its reinsertion into the same location, but "flipped" end over end so that it runs in the opposite direction (Figure 12.18*c*). If an inversion includes part of a segment of DNA that codes for a protein, the resulting protein will be drastically altered and almost certainly nonfunctional.

12.18 Chromosomal Mutations

Chromosomes may break during replication, and parts of chromosomes may then rejoin incorrectly. Letters on the colored chromosomes distinguish segments and identify consequences of duplications, deletions, inversions, and reciprocal translocations.

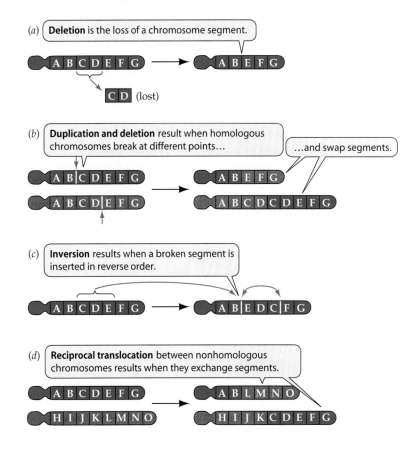

(a) **Deletion** is the loss of a chromosome segment.

A B C D E F G → A B E F G

C D (lost)

(b) **Duplication and deletion** result when homologous chromosomes break at different points… …and swap segments.

A B C D E F G → A B E F G
A B C D E F G → A B C D C D E F G

(c) **Inversion** results when a broken segment is inserted in reverse order.

A B C D E F G → A B E D C F G

(d) **Reciprocal translocation** between nonhomologous chromosomes results when they exchange segments.

A B C D E F G → A B L M N O
H I J K L M N O → H I J K C D E F G

The fourth type of chromosomal mutation, called **translocation**, results when a segment of DNA breaks off, moves from a chromosome, and is inserted into a different chromosome. Translocations may be reciprocal, as in Figure 12.18*d*, or nonreciprocal, as the mutation involving duplication and deletion in Figure 12.18*b* illustrates. Translocations can make synapsis in meiosis difficult and thus sometimes lead to aneuploidy (a lack or excess of chromosomes; see Chapter 9).

Mutations can be spontaneous or induced

Spontaneous mutations are permanent changes in the genome that occur without any outside influence. In other words, they occur simply because the machinery of the cell is imperfect. They may occur by several mechanisms:

▶ *The four nucleotide bases are somewhat unstable* and can exist in two different forms (called *tautomers*), one of which is common and one rare. When a base temporarily forms its unusual tautomer, it can pair with a different base. For example, C normally pairs with G. But if C is in its rare form at the time of DNA replication, it pairs with (and DNA polymerase will insert) A. So there is a mutation of G → A (Figure 12.19*a, c*).

▶ *DNA polymerase makes errors in replication* (see Chapter 11); for example, inserting a T opposite a G. Most of these errors are repaired by the proofreading function of the replication complex, but some errors escape and become permanent.

▶ *Meiosis is not perfect*. Nondisjunction can occur, leading to one too many or one too few chromosomes (aneuploidy; see Figure 9.19). Random chromosome breaks and rejoining among nonhomologous chromatids can occur, leading to translocations.

Induced mutations occur when some outside agent causes a permanent change in DNA:

▶ *Some chemicals can covalently alter the nucleotide bases.* For example, nitrous acid and its relatives can turn cytosine in DNA into uracil by deamination: It converts an amino group on cytosine (—NH$_2$) into a keto group (—C=O). This changes the base pairing properties: C still pairs with G, but when U is present, DNA polymerase inserts an A. (Figure 12.19*b, c*).

▶ *Other chemicals add groups to the bases.* For instance, benzpyrene, a component of cigarette smoke (see page 199), adds a large chemical group to guanine, making it unavailable for base pairing. When DNA polymerase reaches such a modified guanine, it inserts *any* of the four bases; of course, three-fourths of the time the base will not be cytosine, and a mutation results.

▶ *Radiation damages the genetic material in two ways.* Ionizing radiation (X rays) produces highly reactive chemical species called *free radicals*, which can change bases in DNA to unrecognizable (by DNA polymerase) forms or break the sugar–phosphate backbone, causing chromosomal abnormalities. Ultraviolet radiation from the sun (or a tanning lamp) is absorbed by thymine in DNA, causing it to form interbase covalent bonds with adjacent nucleotides. This too creates havoc with DNA replication.

Mutations have both benefits and costs. Mutations provide genetic diversity for evolution to work on, as we will see below. But they can also produce an organism that does poorly in its environment. An additional cost of mutations is that they can occur in non-gametes. Such somatic mutations can lead to cancer. We will return to the effects of germ-line and somatic mutations in humans in Chapter 18.

Mutations are the raw material of evolution

Without mutation, there would be no evolution. As we will see in Part Three of this book, mutation does not drive evolution, but it provides the genetic diversity on which natural selection and other agents of evolution act.

All mutations are rare events, but mutation frequencies vary from organism to organism and from gene to gene within a given organism. The frequency of mutation is usually much lower than one mutation per 10^4 base pairs per DNA duplication, and sometimes as low as one mutation per 10^9 base pairs per duplication. Most mutations are point mutations in which one nucleotide is substituted for another during the synthesis of a new DNA strand.

Mutations can harm the organism that carries them, or be neutral (have no effect on the organism's ability to sur-

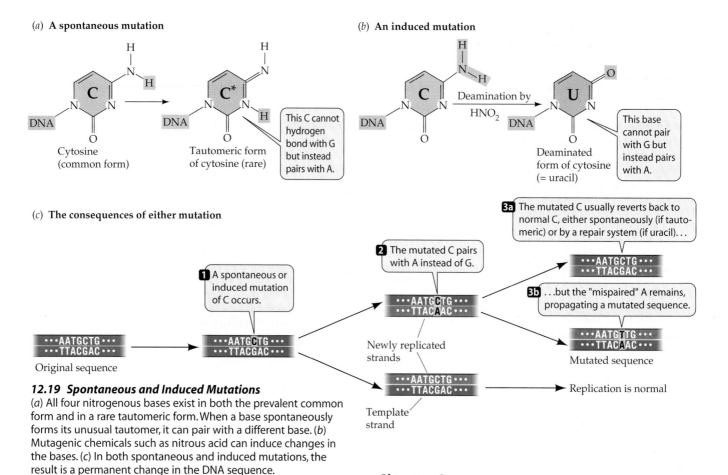

12.19 Spontaneous and Induced Mutations
(*a*) All four nitrogenous bases exist in both the prevalent common form and in a rare tautomeric form. When a base spontaneously forms its unusual tautomer, it can pair with a different base. (*b*) Mutagenic chemicals such as nitrous acid can induce changes in the bases. (*c*) In both spontaneous and induced mutations, the result is a permanent change in the DNA sequence.

vive or produce offspring). Once in a while, however, a mutation improves an organism's adaptation to its environment, or becomes favorable when environmental variables change.

Most of the complex creatures living on Earth have more DNA, and therefore more genes, than the simpler creatures do. Humans, for example, have 1,000 times more genetic material per cell than prokaryotes have. How did these new genes arise? If whole genes were sometimes duplicated by the mechanisms described in the previous section, the bearer of the duplication would have a surplus of genetic information that might be turned to good use. Subsequent mutations in one of the two copies of the gene might not have an adverse effect on survival, because the other copy of the gene would continue to produce functional protein. The extra gene might mutate over and over again without ill effect because its original function would be fulfilled by the original copy.

If the random accumulation of mutations in the extra gene led to the production of a useful protein (for example, an enzyme with an altered specificity for the substrates it binds, allowing it to catalyze different—but related—reactions), natural selection would tend to perpetuate the existence of this new gene. New copies of genes may also arise through the activity of transposable elements, which are discussed in Chapters 13 and 14.

Chapter Summary

One Gene, One Polypeptide

▶ Genes are made up of DNA and are expressed in the phenotype as polypeptides (proteins).

▶ Beadle and Tatum's experiments with the bread mold *Neurospora* resulted in mutant strains lacking a specific enzyme in a biochemical pathway. These results led to the one-gene, one-polypeptide hypothesis. **Review Figure 12.1**

▶ Certain hereditary diseases in humans had been found to be caused by the absence of certain enzymes. These observations supported the one-gene, one-polypeptide hypothesis.

DNA, RNA, and the Flow of Information

▶ RNA differs from DNA in three ways: It is single-stranded, its sugar molecule is ribose rather than deoxyribose, and its fourth base is uracil rather than thymine.

▶ The central dogma of molecular biology is DNA → RNA → protein. **Review Figure 12.2**

▶ A gene is expressed in two steps: First, DNA is transcribed to RNA; then RNA is translated into protein. **Review Figure 12.3**

▶ In retroviruses, the rule for transcription is reversed: RNA → DNA. Other RNA viruses exclude DNA altogether, going directly from RNA to protein. **Review Figure 12.2**

Transcription: DNA-Directed RNA Synthesis

▶ RNA is transcribed from a DNA template after the bases of DNA are exposed by unwinding of the double helix.

▶ In a given region of DNA, only one of the two strands (the template strand) can act as a template for transcription.

▶ RNA polymerase catalyzes transcription from the template strand of DNA.

▶ The initiation of transcription requires that RNA polymerase recognize and bind tightly to a promoter sequence on DNA.

▶ RNA elongates in a 5'-to-3' direction, antiparallel to the template DNA. Special sequences and protein helpers terminate transcription. **Review Figure 12.4**

The Genetic Code

▶ The genetic code consists of triplets of nucleotides (codons). Since there are four bases, there are 64 possible codons.

▶ One mRNA codon indicates the starting point of translation and codes for methionine. Three stop codons indicate the end of translation. The other 60 codons code only for particular amino acids.

▶ Since there are only 20 different amino acids, the genetic code is redundant; that is, there is more than one codon for certain amino acids. But the code is not ambiguous: A single codon does not specify more than one amino acid. **Review Figure 12.5**

Preparation for Translation: Linking RNA's, Amino Acids, and Ribosomes

▶ In prokaryotes, translation begins before the mRNA is completed. In eukaryotes, transcription occurs in the nucleus and translation occurs in the cytoplasm.

▶ Translation requires four components: amino acids, tRNA's, activating enzymes, and ribosomes.

▶ In translation, amino acids are linked in an order specified by the codons in mRNA. This task is achieved by an adapter, transfer RNA (tRNA), which binds the correct amino acid and has an anticodon complementary to the mRNA codon. **Review Figure 12.7**

▶ The aminoacyl-tRNA synthetases, a family of activating enzymes, attach specific amino acids to their appropriate tRNA's, forming charged tRNA's. **Review Figure 12.8**

▶ The mRNA meets the charged tRNA's at a ribosome. **Review Figure 12.9**

Translation: RNA-Directed Polypeptide Synthesis

▶ An initiation complex consisting of an amino acid-charged tRNA and a small ribosomal subunit bound to mRNA triggers the beginning of translation. **Review Figure 12.10**

▶ Polypeptides grow from the N terminus toward the C terminus. The ribosome moves along the mRNA one codon at a time. **Review Figure 12.11**

▶ The presence of a stop codon in the A site of the ribosome causes translation to terminate. **Review Figure 12.12**

Regulation of Translation

▶ Some antibiotics work by blocking events in translation. **Review Table 12.2**

▶ In a polysome, more than one ribosome moves along the mRNA at one time. **Review Figure 12.13**

Posttranslational Events

▶ Signals contained in the amino acid sequences of proteins direct them to their cellular destinations. **Review Figure 12.14**

▶ Protein synthesis begins on free ribosomes in the cytoplasm. Those proteins destined for the nucleus, mitochondria, and plastids are completed there and have signals that allow them to bind to and enter their destined organelles.

▶ Proteins destined for the ER, Golgi apparatus, lysosomes, and outside the cell complete their synthesis on the surface of the ER. They enter the ER by the interaction of a hydrophobic signal sequence with a channel in the membrane. **Review Figure 12.15**

▶ Covalent modifications of proteins after translation include proteolysis, glycosylation, and phosphorylation. **Review Figure 12.16**

Mutations: Heritable Changes in Genes

▶ Mutations in DNA are often expressed as abnormal proteins. However, the result may not be easily observable phenotypic changes. Some mutations appear only under certain conditions, such as exposure to a certain environmental agent (such as a drug) or condition (such as temperature).

▶ Point mutations (silent, missense, nonsense, or frame-shift) result from alterations in single base pairs of DNA. **Review Pages 234–235**

▶ Chromosomal mutations (deletions, duplications, inversions, or translocations) involve large regions of a chromosome. **Review Figure 12.18**

▶ Mutations can be spontaneous or induced. Spontaneous mutations occur because of instabilities in DNA or chromosomes. Induced mutations occur when an outside agent, such as a chemical or radiation, damages DNA. **Review Figure 12.19**

For Discussion

1. The genetic code is described as redundant. What does this mean? How is it possible that a point mutation, consisting of the replacement of a single nitrogenous base in DNA by a different base, might not result in an error in protein production?

2. Har Gobind Khorana, at the University of Wisconsin, synthesized artificial mRNA's such as poly CA (CACACA...) and poly CAA (CAACAACAA...). He found that poly CA codes for a polypeptide consisting of threonine (Thr) and histidine (His) in alternation (His—Thr—His—Thr...). There are two possible codons in poly CA, CAC and ACA. One of these must code for histidine and the other for threonine—but which is which? The answer comes from results with poly CAA, which produces three different polypeptides: poly Thr, poly Gln (glutamine), and poly Asn (asparagine). (An artificial messenger can be read, inefficiently, beginning at any point in the chain; there is no specific start codon.) Thus poly CAA can be read as a polymer of CAA, of ACA, or of AAC. Compare the results of the poly CA and poly CAA experiments, and determine which codon codes for threonine and which for histidine.

3. Look back at Question 2. Using the genetic code (Figure 12.5) as a guide, deduce what results Khorana would have obtained had he used poly UG and poly UGG as artificial messengers. In fact, very few such artificial messengers would have given useful results. For an example of what could happen, consider poly CG and poly CGG. If poly C were the messenger, a mixed polypeptide of arginine and alanine (Arg—Ala—Ala—Arg...) would be obtained; poly CGG would give three polypeptides: poly Arg, poly Ala, and poly Gly (glycine). Can any codons be determined from only these data? Explain.

4. Errors in transcription occur about 100,000 times as often as do errors in DNA replication. Why can this high rate be tolerated in RNA synthesis but not in DNA synthesis?

13 The Genetics of Viruses and Prokaryotes

JANET, A MEMBER OF HER UNIVERSITY'S CROSS-country team, entered the hospital just after final exams for some long-delayed surgery on a tendon in her knee. The tendon repair went well, but she left the hospital with something new: bacteria called *Pseudomonas aeruginosa* had infected the surgical wounds. The antibiotics typically used to kill these bacteria did not work. She ended up back in the hospital two weeks later, where she received intensive antibiotic therapy and ultimately recovered.

Janet developed what is called a *nosocomial infection*—an infection acquired as a result of a hospital stay. Why would a hospital, which we think of as a place to get better, sometimes—in fact, for about 10 percent of all patients—be a place where we get sick? Of course, the stresses of Janet's surgery could have reduced her immunity to the bacteria that are common everywhere in our environment. Increasingly, however, the heavy use of antibiotics in hospitals makes them breeding grounds for bacteria that have genes for resistance to those antibiotics.

How have bacteria acquired antibiotic resistance so rapidly, and how do they pass that acquired resistance along to other bacteria? The answer involves some DNA sequences called R factors. But before we can discuss these remarkable pieces of DNA, we must introduce the genetics of prokaryotes. Prokaryotes usually reproduce asexually by cell division, but can acquire new genes in several ways. These range from simple recombination in a sexual process to using infective viruses as carriers for prokaryotic genes. We also describe how the expression of prokaryotic genes is regulated, and what DNA sequencing has revealed about the prokaryotic genome.

Viruses are not prokaryotes. In fact, they are not even cells, but intracellular parasites that can reproduce only within living cells. We begin this chapter by examining the structures, classification, reproduction, and genetics of viruses.

Using Prokaryotes and Viruses to Probe the Nature of Genes

Prokaryotes such as *Escherichia coli* and the viruses that infect them have been important tools in discovering the structure, function, and transmission of genes, as we saw in Chapter 11. What are the advantages of working with prokaryotes and viruses?

First, it is easier to work with small amounts of DNA than with large amounts. A typical bacterium contains about a thousandth as much DNA as a single human cell, and a typical bacterial virus contains about a hundredth as much DNA as a bacterium. Second, data on large numbers of individuals can be obtained easily from prokaryotes. A single milliliter of medium can contain more than 10^9 *E. coli* cells or 10^{11} bacteriophages. In addition, most prokaryotes grow rapidly. A culture of *E. coli* can be grown under conditions that allow their numbers to double every 20 minutes. By contrast, 10^9 mice would cost more than 10^9 dollars, would require a cage that would cover about 3 square miles, and growing a generation of mice takes about 3 months, not 20 minutes. Third, prokaryotes and viruses are usually haploid, which makes genetic analyses easier.

The ease of growing and handling bacteria and their viruses permitted the explosion of genetics and molecular biology that began shortly after the mid-twentieth century. Their relative biological simplicity contributed immeasurably to discoveries about the genetic material, the replication of DNA, and the mechanisms of gene expression. Later, they were the first subjects of recombinant DNA technology (see Chapter 16).

Questions of interest to all biologists continue to be studied in prokaryotes, and prokaryotes continue to be important tools for biotechnology and for research on eukaryotes. Prokaryotes are important players in the environment, performing much of the cycling of elements in the atmosphere

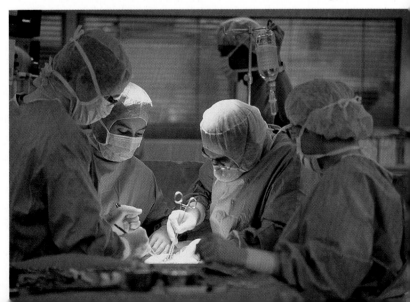

Are There Uninvited Guests Here?
A masked team performs surgery on a patient. But have harmful, drug-resistant bacteria invaded the surgical suite?

and water. And, as we saw at the opening of this chapter, infectious diseases caused by prokaryotes and viruses continue to challenge humankind.

Viruses: Reproduction and Recombination

Although there are many kinds of viruses, most of them are composed of nothing but nucleic acid and a few proteins. Most viruses have relatively simple means of infecting their host cells. Some can infect a cell but postpone reproduction, lying low in the host chromosome until conditions are favorable. The simplest infective agents of all are viroids, which are made up only of genetic material.

Scientists studied viruses before they could see them

Most viruses are much smaller than most bacteria (Table 13.1). Viruses have become well understood only within the last half century, but the first step on this path of discovery was taken by the Russian botanist Dmitri Ivanovsky in 1892. He was trying to find the cause of *tobacco mosaic disease*, which results in the destruction of photosynthetic tissues and can devastate a tobacco crop. Ivanovsky passed an extract of diseased tobacco leaves through a fine porcelain filter, a technique that had been used previously by physicians and veterinarians to isolate disease-causing bacteria.

To Ivanovsky's surprise, the disease agent in this case did not stick to the filter: It passed through, and the liquid filtrate still caused tobacco mosaic disease. But instead of concluding that the agent was smaller than a bacterium, he assumed that his filter was faulty. Pasteur's recent demonstration that bacteria could cause disease was the dominant idea at the time, and Ivanovsky chose not to challenge it. But, as often happens in science, someone soon came along who did. In 1898, Martinus Beijerinck repeated Ivanovsky's experiment, and also showed that the tiny tobacco mosaic agent could diffuse through an agar gel. He called the agent *contagium vivum fluidum*, which later became shortened to **virus**.

Almost 40 years later, the infective agent was crystallized by Wendell Stanley (who won the Nobel Prize for his efforts). The crystalline viral preparation became infectious again when it was dissolved. It was soon shown that crys-

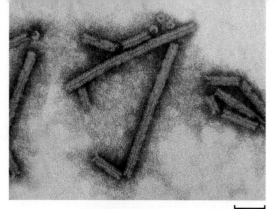

(a)

75 nm

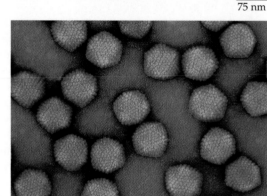

(b)

75 nm

(c)

20 nm

13.1 Virions Come in Various Shapes
(a) The tobacco mosaic virus (a plant virus) consists of an inner helix of RNA covered with a helical array of protein molecules. (b) Many animal viruses, such as this adenovirus, have an icosahedral (20-sided) capsid as an outer shell. Inside the shell is a spherical mass of proteins and DNA. (c) Not all virions are regularly shaped. Wormlike virions of the influenza A virus infect humans, causing chills, fever and sometimes, death.

tallized viral preparations consist of protein and nucleic acid. Finally, direct observation of viruses with electron microscopes in the 1950s showed clearly how much they differ from bacteria and other organisms

Viruses reproduce only with the help of living cells

Unlike the organisms that make up the six taxonomic kingdoms of the living world, viruses are *acellular*; that is, they are not cells and do not consist of cells. Unlike cellular creatures, viruses do not metabolize energy—they neither produce ATP nor conduct fermentation, cellular respiration, or photosynthesis.

Whole viruses never arise directly from preexisting viruses. Viruses are *obligate intracellular parasites*; that is, they

13.1 **Common Sizes of Microorganisms**

MICROORGANISM	TYPE	TYPICAL SIZE RANGE (μm³)
Protists	Eukaryote	5,000–50,000
Photosynthetic bacteria	Prokaryote	5–50
Spirochetes	Prokaryote	0.1–2.0
Mycoplasmas	Prokaryote	0.01–0.1
Poxviruses	Virus	0.01
Influenza virus	Virus	0.0005
Poliovirus	Virus	0.00001

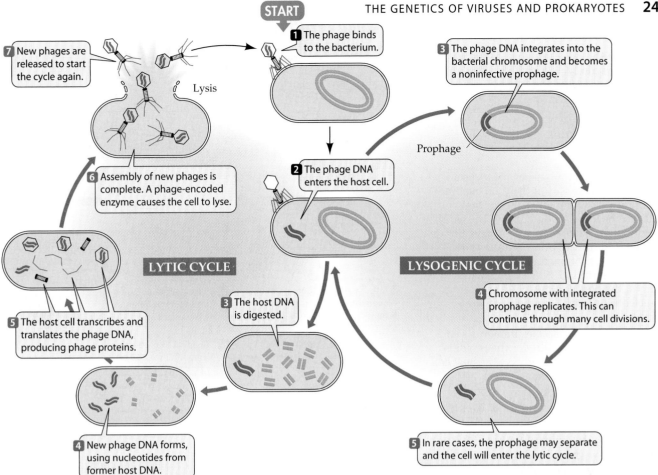

START
1 The phage binds to the bacterium.

7 New phages are released to start the cycle again.

Lysis

6 Assembly of new phages is complete. A phage-encoded enzyme causes the cell to lyse.

2 The phage DNA enters the host cell.

3 The phage DNA integrates into the bacterial chromosome and becomes a noninfective prophage.

Prophage

LYTIC CYCLE

LYSOGENIC CYCLE

4 Chromosome with integrated prophage replicates. This can continue through many cell divisions.

5 The host cell transcribes and translates the phage DNA, producing phage proteins.

3 The host DNA is digested.

4 New phage DNA forms, using nucleotides from former host DNA.

5 In rare cases, the prophage may separate and the cell will enter the lytic cycle.

13.2 The Lytic and Lysogenic Cycles of Bacteriophages
In the lytic cycle, infection by viral DNA leads directly to the multiplication of the virus and lysis of the host bacterial cell. In the lysogenic cycle, a prophage is replicated as part of the host's chromosome.

develop and reproduce only within the cells of specific hosts. The cells of animals, plants, fungi, protists, and prokaryotes (both bacteria and archaea) serve as hosts to viruses. When they reproduce, viruses usually destroy the host cell, releasing progeny viruses that then seek new hosts.

Many diseases of humans, animals, and plants are caused by viruses. Because they lack the distinctive cell wall and ribosomal biochemistry of bacteria, viruses are not affected by antibiotics.

Viruses outside of host cells exist as individual particles called **virions**. The virion, the basic unit of a virus, consists of a central core of either DNA or RNA (but not both) surrounded by a **capsid**, or coat, composed of one or more proteins. The way in which these proteins assemble gives each type of virion a characteristic shape (Figure 13.1). In addition, many animal viruses have a lipid and protein membrane acquired from host cell plasma membranes.

There are many kinds of viruses

A common way to classify viruses separates them by whether their genetic material is DNA or RNA, and then by whether their nucleic acid is single-stranded or double-stranded. Some RNA viruses have more than one molecule of RNA, and the DNA of one virus family is circular. Further levels of classification depend on factors such as the overall shape of the virus and the symmetry of the capsid (see Figure 13.1). Another level of classification is based on the presence or absence of a membranous envelope around the virion; still further subdivision is based on capsid size.

One way to classify viruses is based on the type of host. Let's see how reproductive cycles and other properties vary among the major groups of viruses: those that infect bacteria, animals, and plants.

Bacteriophages reproduce by a lytic cycle or a lysogenic cycle

Viruses that infect bacteria are known as **bacteriophages**. Bacteriophages recognize their hosts by means of specific binding between proteins in the capsid and receptor proteins in the host's cell wall. The virions, which must penetrate the cell wall, are often equipped with tail assemblies that insert their nucleic acid through the cell wall into the host bacterium. After the phage has injected its nucleic acid into the host, one of two things happens, depending on the kind of phage.

We saw one type of viral reproductive cycle when we studied the Hershey–Chase experiment (see Figure 11.3). That was the **lytic cycle**, so named because the infected bacterium *lyses* (bursts), releasing progeny phages. In the **lysogenic cycle**, the infected bacterium does not lyse, but instead harbors the viral nucleic acid for many generations. Some viruses reproduce only by the lytic cycle; others undergo both types of reproductive cycles (Figure 13.2).

THE LYTIC CYCLE. A phage that reproduces only by the lytic cycle is called a **virulent virus**. Once the phage has injected its nucleic acid into the host cell, the phage nucleic acid takes over the host's synthetic machinery. It does so in two stages (Figure 13.3):

▶ The early stage transcribes the virus's *early genes*. This part of the viral genome contains the promoter sequence that attracts host RNA polymerase. The early genes often include proteins that shut down host transcription, stimulate viral genome replication, and stimulate late gene transcription. Nuclease enzymes digest the host chromosome, providing nucleotides for the synthesis of viral genomes.

▶ The late stage transcribes the virus's *late genes*. These genes code for the proteins that package virions and lyse the host cell to release the new virions.

This sequence of transcriptional events is carefully controlled: Premature lysis of the host cell before virus particles are ready for release would stop the infection. The whole process—from binding and infection to lysis of the host—takes about half an hour.

Rarely, two viruses infect a cell at the same time. This is an unusual event, as once an infection cycle is under way, there is usually not enough time for an additional infection. In addition, an early gene product prevents further infections in some cases. The presence of two viral genomes in the same host cell affords the opportunity for genetic recombination by crossing over, as in prophase of meiosis. This enables genetically different viruses of the same kind to swap genes and create new strains.

THE LYSOGENIC CYCLE. Phage infection does not always result in lysis of the host cell. Some phages seem to disappear from a bacterial culture, leaving the bacteria immune to further attack by the same strain of phage. In such cultures, however, a few free phages are always present. Bacteria harboring phages that are not lytic are called *lysogenic*, and the phages are called **temperate viruses**.

Lysogenic bacteria contain a noninfective entity called a **prophage**: a molecule of phage DNA that has been integrated into the bacterial chromosome (see Figure 13.2). As part of the bacterial genome, the prophage can remain quiet within the bacteria through many cell divisions. However, an occasional lysogenic bacterium can be induced to activate its prophage. This results in a lytic cycle, releasing a large number of free phages, which can then infect other uninfected bacteria and renew the reproductive cycle.

This capacity to switch between the lysogenic and the lytic cycle is very useful to the phage, whose purpose is to reproduce as many offspring as possible. When its host cell is growing slowly and is low on energy, the phage becomes

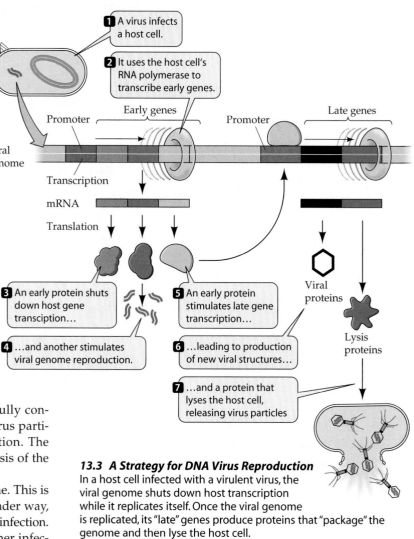

13.3 A Strategy for DNA Virus Reproduction
In a host cell infected with a virulent virus, the viral genome shuts down host transcription while it replicates itself. Once the viral genome is replicated, its "late" genes produce proteins that "package" the genome and then lyse the host cell.

lysogenic. Then, when the host's health is restored to a level that provides maximal resources for phage reproduction, the prophage is released from its dormant state, and the lytic cycle proceeds. We will see how this switch works later in the chapter when we discuss control of gene expression.

Animal viruses have diverse reproductive cycles

Almost all vertebrates are susceptible to viral infections, but among invertebrates, such infections are common only in arthropods (the group that includes insects and crustaceans). One group of viruses, called *arboviruses* (short for "*arthropod-borne viruses*"), is transmitted to a mammalian host through an insect bite. Although carried within the arthropod host's cells, arboviruses apparently do not affect that host severely; they affect only the bitten and infected mammal. The arthropod acts as a **vector**—an intermediate carrier—transmitting the disease organism from one host to another.

Animal viruses are very diverse. Some are just particles of proteins surrounding a nucleic acid core. Others have a membrane derived from the host cell's plasma membrane. Some animal viruses have DNA as their genetic material; others have RNA. In all cases, the small viral genome has limited coding capacity, making only a few proteins.

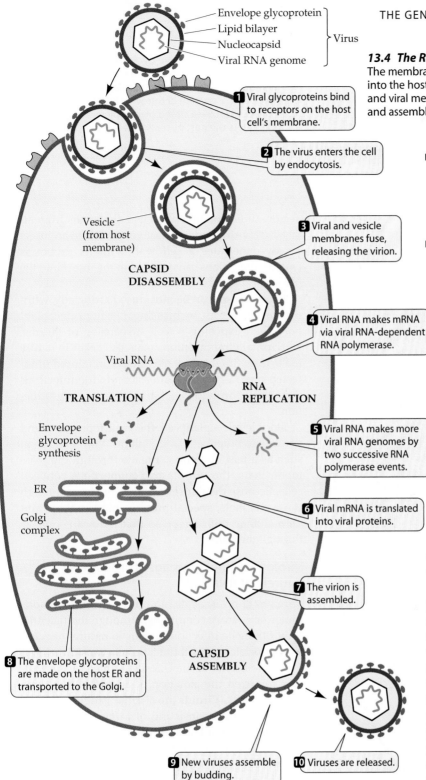

Envelope glycoprotein ⎤
Lipid bilayer ⎥ Virus
Nucleocapsid ⎥
Viral RNA genome ⎦

1 Viral glycoproteins bind to receptors on the host cell's membrane.

2 The virus enters the cell by endocytosis.

Vesicle (from host membrane)

3 Viral and vesicle membranes fuse, releasing the virion.

CAPSID DISASSEMBLY

4 Viral RNA makes mRNA via viral RNA-dependent RNA polymerase.

Viral RNA

TRANSLATION   **RNA REPLICATION**

Envelope glycoprotein synthesis

5 Viral RNA makes more viral RNA genomes by two successive RNA polymerase events.

ER

Golgi complex

6 Viral mRNA is translated into viral proteins.

7 The virion is assembled.

8 The envelope glycoproteins are made on the host ER and transported to the Golgi.

CAPSID ASSEMBLY

9 New viruses assemble by budding.

10 Viruses are released.

13.4 The Reproductive Cycle of the Influenza Virus
The membrane-enclosed, or enveloped, influenza virus is taken into the host cell by endocytosis. Once inside, fusion of the vesicle and viral membranes releases the RNA genome, which replicates and assembles new virions.

▶ Viruses with membranes (called *enveloped viruses*) may also be taken up by endocytosis (see Figure 13.4), and released from a vesicle. In these viruses, the viral membrane is studded with glycoproteins that bind to receptors on the host cell plasma membrane.

▶ More commonly, the host and viral membranes fuse, releasing the rest of the virion into the cell (see Figure 13.5).

Enveloped viruses usually escape from the host cell by budding through virus-modified areas of the host's plasma membrane. During this process, the completed virions acquire a membrane similar to that of the host cell.

The life cycles of influenza virus and HIV, two important RNA viruses, illustrate the two different styles of infection by enveloped viruses. Influenza virus is endocytosed into a membrane vesicle (Figure 13.4). Fusion of the viral and vesicle membranes releases the virion into the cell. The virus carries its own enzyme to replicate its RNA genome into a complementary strand. The latter is then used as mRNA to make, by complementary base pairing, more copies of the viral genome.

Retroviruses such as HIV have a more complex reproductive cycle (Figure 13.5). The virus enters a host cell by direct fusion of viral and cellular membranes. A major feature of the retroviral life cycle is the reverse transcription of retroviral RNA to produce a DNA *provirus* (cDNA), which is the form of the viral genome that gets integrated into the host DNA. The provirus may reside in the host chromosome permanently, occasionally being expressed to produce new virions. Almost every step in this complex cycle can, in principle, be attacked by therapeutic drugs; this fact is used by researchers in their quest to conquer AIDS, the deadly condition caused by HIV infection in humans. This medical battle will be discussed further in Chapter 19.

Many plant viruses spread with the help of vectors

Viral diseases of flowering plants are very common. Plant viruses can be transmitted *horizontally*, from one plant to another, or *vertically*, from parent to offspring. To infect a plant cell, viruses must pass through a cell wall and through the host plasma membrane. Most plant viruses accomplish this penetration through their association with

Like that of bacteriophages, the life cycle of animal viruses can be divided into early and late stages (see Figure 13.3). Animal viruses enter cells in one of three ways:

▶ A naked virion (without a membrane) is taken up by endocytosis, which traps it within a membranous vesicle inside the host cell. The membrane of the vesicle breaks down, releasing the virion into the cytoplasm, and the host cell digests the protein capsid, liberating the viral nucleic acid, which takes charge of the host cell.

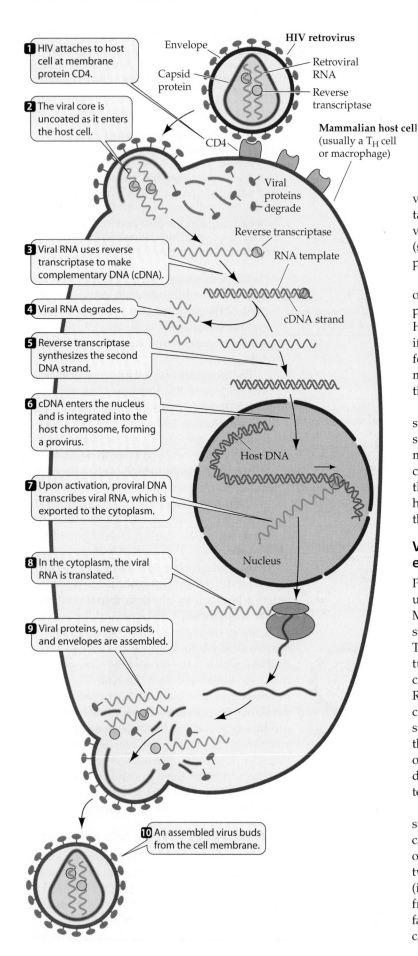

1 HIV attaches to host cell at membrane protein CD4.

2 The viral core is uncoated as it enters the host cell.

3 Viral RNA uses reverse transcriptase to make complementary DNA (cDNA).

4 Viral RNA degrades.

5 Reverse transcriptase synthesizes the second DNA strand.

6 cDNA enters the nucleus and is integrated into the host chromosome, forming a provirus.

7 Upon activation, proviral DNA transcribes viral RNA, which is exported to the cytoplasm.

8 In the cytoplasm, the viral RNA is translated.

9 Viral proteins, new capsids, and envelopes are assembled.

10 An assembled virus buds from the cell membrane.

Envelope
HIV retrovirus
Capsid protein
Retroviral RNA
Reverse transcriptase
CD4
Mammalian host cell (usually a T_H cell or macrophage)
Viral proteins degrade
Reverse transcriptase
RNA template
cDNA strand
Host DNA
Nucleus

13.5 The Reproductive Cycle of HIV
The retrovirus HIV enters a host cell via fusion of its membranes with the host's plasma membrane. Reverse transcription of retroviral RNA then produces a DNA provirus—a strand of complementary DNA that enters the host nucleus, where it transcribes viral RNA.

vectors. Infection of a plant usually results from attack by a virion-laden insect vector. When an insect vector penetrates a cell wall with its proboscis (snout), the virions can move from the insect into the plant.

Plant viruses can be introduced artificially, without insect vectors, by bruising a leaf or other plant part, then exposing it to a suspension of virions. Horizontal viral infections may also occur in nature if a bruised infected plant contacts an injured uninfected one. Vertical transmission of viral infections may occur through vegetative or sexual reproduction.

Once inside a plant cell, the virus reproduces and spreads to other cells in the plant. Within an organ such as a leaf, the virus spreads through the plasmodesmata, the cytoplasmic connections between cells. Because the viruses are too large to go through these channels, special proteins bind to them and help change their shape so that they can squeeze through the pores.

Viroids are infectious agents consisting entirely of RNA

Pure viral nucleic acids can produce viral infections under laboratory conditions, although inefficiently. Might there be infectious agents in nature that consist of nucleic acid without a protein capsid? In 1971, Theodore Diener of the U.S. Department of Agriculture reported the isolation of agents of this type, called viroids. **Viroids** are circular, single-stranded RNA molecules consisting of a few hundred nucleotides. They are one-thousandth the size of the smallest viruses. These RNA's are most abundant in the nuclei of infected cells. Viroids have been found only in plants, in which they produce a variety of diseases. Like plant viruses, viroids can be transmitted horizontally or vertically.

There is no evidence that viroids are translated to synthesize proteins, and it is not known how they cause disease. Viroids are replicated by the enzymes of their plant hosts. Similarities in base sequences between viroids and certain nontranslated sequences (introns) of plant genes suggest that viroids evolved from introns. This conclusion is supported by the fact that viroids, although made of RNA, are catalytically active in the way that some introns are.

Prokaryotes: Reproduction, Mutation, and Recombination

In contrast to viruses, bacteria and archaea are living cells. Prokaryotes carry out all the functions required for their own reproduction. They harvest and use energy, and they produce and use the molecular equipment that synthesizes their components and replicates their genes.

Prokaryotes usually reproduce asexually, but nonetheless have several ways of recombining their genes. Whereas in eukaryotes, genetic recombination occurs between the genomes of two parents, in prokaryotes it results from the interaction of the genome of one cell with a much smaller sample of genes from another cell.

The reproduction of prokaryotes gives rise to clones

Most prokaryotes reproduce by the division of single cells into two identical offspring (see Figure 9.3). In this way, a single cell gives rise to a clone—a population of genetically identical individuals. Prokaryotes reproduce very rapidly. A population of *E. coli*, for example, can double every 20 minutes, as long as conditions remain favorable.

Pure cultures of *E. coli* or other bacteria can be grown on the surface of a solid medium that contain a sugar, minerals, a nitrogen source such as ammonium chloride (NH_4Cl), and a solidifying agent such as agar (Figure 13.6). If the number of cells spread on the medium is small, each cell will give rise to a small, rapidly growing bacterial *colony*. If a large number of cells is poured onto the solid medium, their growth will produce one continuous colony—a bacterial "lawn." Bacteria can also be grown in a liquid nutrient medium. We'll see examples of all these techniques in this chapter.

Some bacteria conjugate, recombining their genes

The existence and heritability of mutations in bacteria attracted the attention of geneticists to these microbes. But if there were no form of exchange of genetic information between individuals, bacteria would not be useful for genetic analysis. Luckily, in 1946, Joshua Lederberg and Edward Tatum demonstrated that such exchanges do occur, although they are rare events.

Lederberg and Tatum grew two nutrient-requiring, or *auxotrophic*, strains of *E. coli*. Like the *Neurospora* in Figure 12.1, these strains will not grow on a minimal medium, but require supplementation with a nutrient that they cannot synthesize for themselves because of an enzyme defect. *E. coli* strain 1 requires the amino acid methionine and the vitamin biotin for growth, and its genotype is symbolized as *met⁻bio⁻*. Strain 2 requires neither of these substances, but cannot grow without the amino acids threonine and leucine. Considering all four factors, we say that strain 1 is *met⁻bio⁻thr⁺leu⁺* and strain 2 is *met⁺bio⁺thr⁻leu⁻*.

Lederberg and Tatum mixed these two mutant strains and cultured them together for several hours on a medium supplemented with methionine, biotin, threonine, and leucine, so that both strains could grow. The bacteria were then removed from the medium by centrifugation, washed, and transferred to minimal medium, which lacked all four supplements. Neither strain could grow on this medium because of their nutritional requirements. However, a few colonies *did* appear on the plates (Figure 13.7). Because they grew in the minimal medium, these colonies must have consisted of bacteria that were *met⁺bio⁺thr⁺leu⁺*; that is, they must have been *prototrophic*. These colonies appeared at a rate of approximately 1 for every 10 million cells originally put on the plates ($1/10^7$).

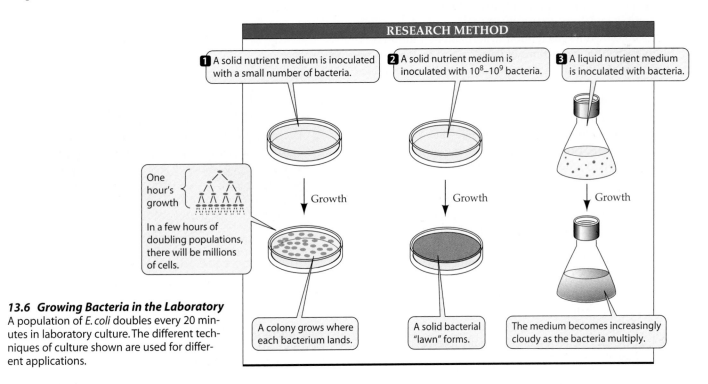

RESEARCH METHOD

1 A solid nutrient medium is inoculated with a small number of bacteria.

2 A solid nutrient medium is inoculated with 10^8–10^9 bacteria.

3 A liquid nutrient medium is inoculated with bacteria.

One hour's growth

In a few hours of doubling populations, there will be millions of cells.

Growth

Growth

Growth

A colony grows where each bacterium lands.

A solid bacterial "lawn" forms.

The medium becomes increasingly cloudy as the bacteria multiply.

13.6 Growing Bacteria in the Laboratory
A population of *E. coli* doubles every 20 minutes in laboratory culture. The different techniques of culture shown are used for different applications.

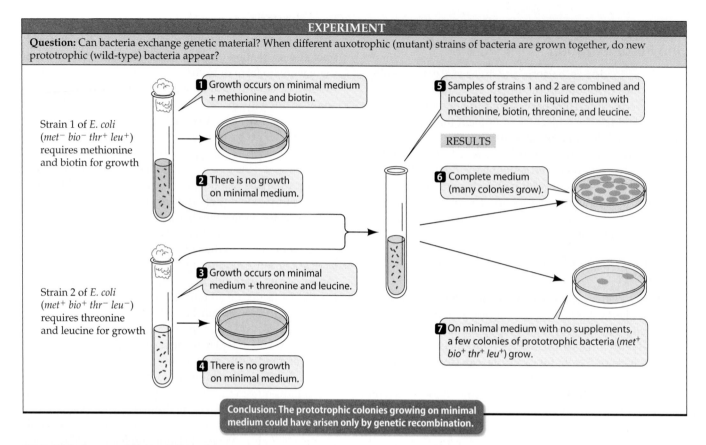

EXPERIMENT

Question: Can bacteria exchange genetic material? When different auxotrophic (mutant) strains of bacteria are grown together, do new prototrophic (wild-type) bacteria appear?

Strain 1 of *E. coli* (met^- bio^- thr^+ leu^+) requires methionine and biotin for growth

1 Growth occurs on minimal medium + methionine and biotin.

2 There is no growth on minimal medium.

Strain 2 of *E. coli* (met^+ bio^+ thr^- leu^-) requires threonine and leucine for growth

3 Growth occurs on minimal medium + threonine and leucine.

4 There is no growth on minimal medium.

5 Samples of strains 1 and 2 are combined and incubated together in liquid medium with methionine, biotin, threonine, and leucine.

RESULTS

6 Complete medium (many colonies grow).

7 On minimal medium with no supplements, a few colonies of prototrophic bacteria (met^+ bio^+ thr^+ leu^+) grow.

Conclusion: The prototrophic colonies growing on minimal medium could have arisen only by genetic recombination.

13.7 Lederberg and Tatum's Experiment
After growing together, a mixture of complementary auxotrophic strains of *E. coli* contained a few cells that gave rise to new prototrophic colonies. This experiment proved that genetic recombination takes place in prokaryotes.

Where did these prototrophic colonies come from? Lederberg and Tatum were able to rule out mutation, and other investigators ruled out transformation. A third possibility is that the two strains of bacteria had exchanged genetic material, allowing it to mix and recombine to produce cells containing met^+ and bio^+ alleles from strain 2 and thr^+ and leu^+ alleles from strain 1 (see Figure 13.7). Later experiments showed that such an exchange, called **conjugation**, had indeed occurred. One bacterial cell—the *recipient*—had received DNA from the other cell—the *donor*—that included the two wild-type alleles that were missing in the recipient. Recombination then created a genotype with four wild-type alleles.

The physical contact required for conjugation can be observed under the electron microscope (Figure 13.8). It is initiated by a thin projection called a *pilus*. Then the actual transfer of DNA from one cell to another occurs by a thin conjugation tube. Since bacterial DNA is circular, it must be made linear (broken) so that it can pass through the tube. Contact between the cells is brief—certainly not long enough for all of the donor genome to enter the recipient cell. Therefore, the recipient cell usually receives only a portion of the donor DNA.

Once the donor fragment is inside the recipient cell, recombination can occur. In much the same way that chromosomes pair up, gene for gene, in prophase of meiosis, the donor DNA can line up beside its homologous gene in the recipient. Enzymes that can cut and rejoin DNA molecules are active in bacteria, and so gene(s) of the donor can end up integrated into the genome of the recipient, thus changing its genetic constitution (Figure 13.9).

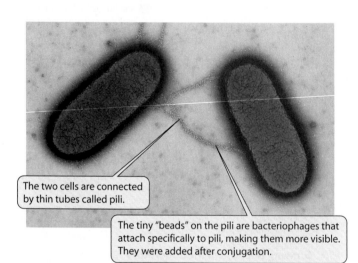

The two cells are connected by thin tubes called pili.

The tiny "beads" on the pili are bacteriophages that attach specifically to pili, making them more visible. They were added after conjugation.

13.8 Bacterial Conjugation
Pili draw two bacteria into close contact, and DNA is transferred from one cell to the other via a conjugation tube.

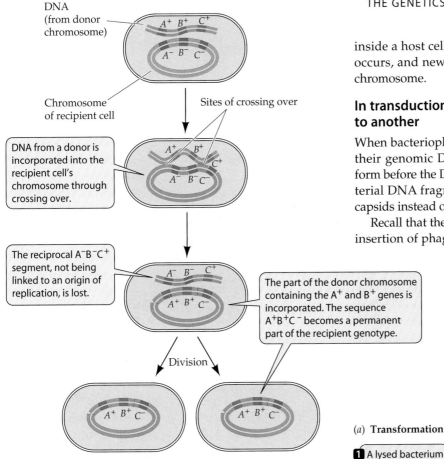

DNA (from donor chromosome)

Chromosome of recipient cell

Sites of crossing over

DNA from a donor is incorporated into the recipient cell's chromosome through crossing over.

The reciprocal $A^-B^-C^+$ segment, not being linked to an origin of replication, is lost.

The part of the donor chromosome containing the A^+ and B^+ genes is incorporated. The sequence $A^+B^+C^-$ becomes a permanent part of the recipient genotype.

Division

13.9 Recombination Following Conjugation
DNA from a donor cell may become incorporated into the recipient cell's chromosome through crossing over. Only about half the transferred genes become integrated in this way.

In transformation, cells pick up genes from their environment

Frederick Griffith obtained the first evidence for the transfer of prokaryotic genes more than 75 years ago when he discovered the transforming principle (see Figure 11.1). We now know the reason for his results: DNA had leaked from dead cells of virulent pneumococci and was taken up as free DNA by living nonvirulent pneumococci, which became virulent as a result. This phenomenon, called **transformation**, occurs in nature in some species of bacteria when cells die and their DNA leaks out (Figure 13.10a). Once transforming DNA is

inside a host cell, an event very similar to recombination occurs, and new genes can be incorporated into the host chromosome.

In transduction, viruses carry genes from one cell to another

When bacteriophages undergo a lytic cycle, they package their genomic DNA in capsids. These capsids generally form before the DNA is inserted into them. Sometimes, bacterial DNA fragments get inserted into the empty phage capsids instead of the phage DNA. (Figure 13.10b)

Recall that the binding of a phage to its host cell and the insertion of phage DNA are carried out by the capsid. So, when a phage capsid carries a piece of *bacterial* DNA, the latter is injected into the "infected" bacterium. This mechanism of DNA transfer is called **transduction**. Needless to say, it does not result in a productive viral infection. Instead, the incoming DNA fragment can recombine with the host chromosome.

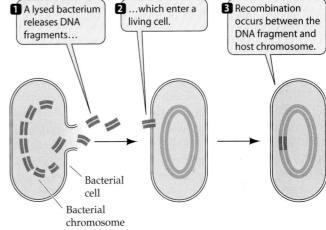

(a) Transformation

1 A lysed bacterium releases DNA fragments…

2 …which enter a living cell.

3 Recombination occurs between the DNA fragment and host chromosome.

Bacterial cell

Bacterial chromosome

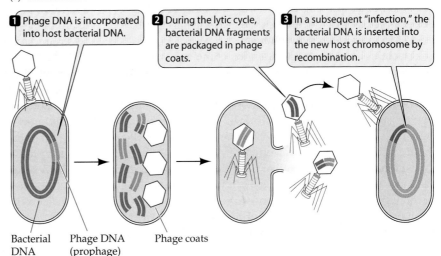

(b) Transduction

1 Phage DNA is incorporated into host bacterial DNA.

2 During the lytic cycle, bacterial DNA fragments are packaged in phage coats.

3 In a subsequent "infection," the bacterial DNA is inserted into the new host chromosome by recombination.

Bacterial DNA

Phage DNA (prophage)

Phage coats

13.10 Transformation and Transduction
After a new DNA fragment enters the host cell, recombination can occur. (a) Transforming DNA can leak from dead bacterial cells and be taken up by a living host bacterium, which may incorporate new genes into its chromosome. (b) In transduction, viruses carry DNA fragments from one cell to another.

Plasmids are extra chromosomes in bacteria

In addition to their main chromosome, many bacteria harbor additional smaller, circular chromosomes. These chromosomes, called **plasmids**, contain at most a few dozen genes, and, importantly, an origin sequence where DNA replication starts, which defines them as chromosomes. Usually, plasmids replicate at the same time as the host chromosome during the bacterial cell cycle, but this is not necessarily the case.

Plasmids are not viruses. They do not take over the cell's molecular machinery or make a protein coat to help them move from cell to cell. Instead, they can move between cells during conjugation, thereby adding some new genes to the recipient bacterium (Figure 13.11). Since plasmids exist independently of the main chromosome (the term *episomes* is sometimes used), they do not need to recombine with the main chromosome to add their genes to the cell's genome.

There are several types of plasmids, classified according to the kinds of genes they carry.

METABOLIC FACTORS CARRY GENES FOR UNUSUAL METABOLIC FUNCTIONS. Some plasmids, called *metabolic factors*, have genes that allow their recipients to carry out unusual metabolic functions. For example, there are many unusual hydrocarbons in oil spills. Some bacteria can actually thrive on these molecules, using them as a carbon source. The genes for the enzymes involved in these degradative pathways are carried on plasmids.

F FACTORS CARRY GENES FOR CONJUGATION. The "F" in F *factors* stands for fertility. Their approximately 25 genes include the ones that make both the pilus for attachment and the conjugation tube for DNA transfer to a recipient bacterium. A cell harboring the F factor is called F$^+$. It can transfer a copy of the F factor to an F$^-$ cell, making the recipient F$^+$. Sometimes the factor integrates into the main chromosome (at which point it is no longer a plasmid), and when it does, it can bring along some bacterial genes when it moves through the conjugation tube from one cell to another.

R FACTORS ARE RESISTANCE FACTORS. *R factors* may carry genes coding for proteins that destroy or modify antibiotics. Other R factors provide resistance to heavy metals that bacteria encounter in their environment.

R factors first came to the attention of biologists in 1957 during an epidemic in Japan, when it was discovered that some strains of the *Shigella* bacterium, which causes dysentery, were resistant to several antibiotics. Researchers found that resistance to the entire spectrum of antibiotics could be transferred by conjugation even when no genes on the main chromosome were transferred.

Eventually it was shown that the genes for antibiotic resistance are carried on plasmids. Each R factor carries one or more genes conferring resistance to particular antibiotics, as well as genes that code for proteins involved in the transfer of DNA to a recipient bacterium. As far as biologists can determine, R factors appeared long before antibi-

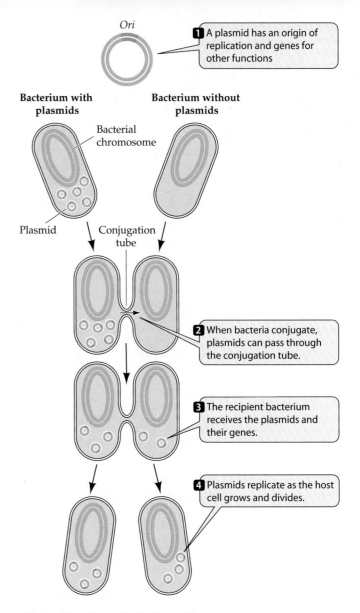

13.11 Gene Transfer by Plasmids
When plasmids enter a cell via conjugation, their genes can be expressed in the new cell.

otics were discovered, but they seem to have become more abundant in modern times, possibly because the heavy use of antibiotics in the hospital environment selects for bacterial strains bearing them.

R factors also pose a threat to people in the general clinical setting if antibiotics are used inappropriately. You probably have gone to see a physician because of a sore throat, which can have either a viral or a bacterial cause. The best way to know is for the doctor to take a small sample from your inflamed throat, culture it, and identify any bacteria that are present. But perhaps you cannot wait another day for the results. Impatient, you ask the doctor to give you something to make you feel better. She prescribes an antibiotic, which you take. The sore throat gradually gets better, and you think that the antibiotic did the job.

But suppose the infection was viral? In that case, the antibiotic did nothing to combat the disease, which just ran its normal course. However, it may have done something harmful: By killing many normal bacteria in your body, it may have exerted selection for bacteria harboring R factors. These bacteria may reproduce in the presence of the antibiotic, and may soon become quite numerous. The next time you got a *bacterial* infection, there would be a ready supply of R factors to be transferred to the invading bacteria, and antibiotics might be ineffective.

Transposable elements move genes among plasmids and chromosomes

As we have seen, plasmids, viruses, and even phage capsids (in the case of transduction) can transport genes from one bacterial cell to another. There is another type of "gene transport" that occurs *within* the individual cell. It relies on segments of chromosomal or plasmid DNA that can be inserted either at new locations on the same chromosome, or into another chromosome. These DNA sequences are called **transposable elements**. Their insertion often produces phenotypic effects by disrupting the genes into which they are inserted (Figure 13.12*a*).

The first transposable elements to be discovered in prokaryotes were large pieces of DNA, typically 1,000 to 2,000 base pairs long, found in many places in the *E. coli*

main chromosome. In one mechanism of transposition, the sequence of a transposable element replicates independently of the rest of the chromosome. The copy then inserts itself at other, seemingly random places in the chromosome. The genes encoding the enzymes necessary for this insertion are found within the transposable element itself. Some other transposable elements are cut from their original sites and inserted elsewhere without replication. Many transposable elements discovered later were longer (about 5,000 base pairs) and carried one or more additional genes with them. These longer elements with additional genes are called **transposons** (Figure 13.12*b*).

What do transposons and other transposable elements have to do with the genetics of prokaryotes—or with hospitals? Transposable elements have contributed to the evolution of plasmids. R plasmids probably originally gained their genes for antibiotic resistance through the activity of transposable elements. One piece of evidence for this conclusion is that each resistance gene in an R plasmid is part of a transposon.

Regulation of Gene Expression in Prokaryotes

Except for mutations, all cells of a bacterial species have the same DNA, and thus the capacity to make the same proteins. Yet the protein content of a bacterium can change rapidly when conditions warrant. For example, there are two ways for a bacterium to get the amino groups that it needs to make amino acids and proteins. One way involves taking N_2 from the air and "fixing" it into ammonia (NH_3), then using the ammonia as a source of amino groups. This reaction requires several enzymes and a lot of energy.

The other way to obtain amino groups is to take them from glutamine (see Table 3.2) and use them directly. This reaction requires only one enzyme and is not as endergonic. If there is a lot of glutamine around, the cell takes the easy way out, using the glutamine rather than the N_2 pathway. In fact, the enzymes that are involved in the N_2 pathway are not even made when glutamine is present.

There are several ways in which a prokaryotic cell could shut off the synthesis or activity of an unneeded protein:

▶ The cell could block the transcription of mRNA for that protein.
▶ The cell could hydrolyze the mRNA after it was made.
▶ The cell could prevent translation of the mRNA at the ribosome.
▶ The cell could hydrolyze the protein after it was made.

These methods would all have to be selective, responding to some biochemical signal. In the case of our two pathways for obtaining amino groups, the signal might be an increased concentration of glutamine.

Clearly, the earlier the cell intervenes in the process, the less energy it has to expend. Selective inhibition of transcription is far more efficient than transcribing the gene, translating the message, and then degrading the protein. While there are examples of all four methods of control of

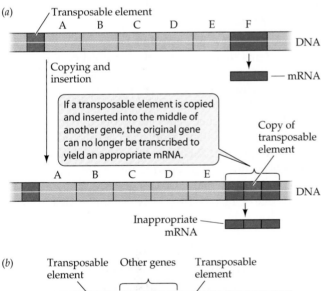

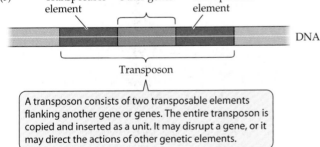

13.12 Transposable Elements and Transposons
(*a*) Transposable elements are segments of DNA that can be inserted at new locations, either on the same chromosome or on a different chromosome. (*b*) Transposons consist of transposable elements combined with other genes.

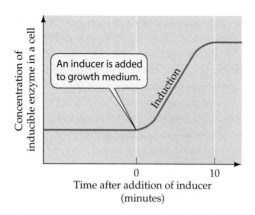

13.13 An Inducer Stimulates the Synthesis of an Enzyme
It is most efficient for a cell to produce an enzyme only when it is needed. Some enzymes are induced by the presence of the substance they act upon (for example, β-galactosidase is induced by the presence of lactose).

protein levels in nature, prokaryotes generally use the most efficient one, transcriptional control.

Regulation of transcription conserves energy

As a normal inhabitant of the human intestine, *E. coli* has to adjust to sudden changes in its chemical environment. Its host may present it with one foodstuff one hour and another the next. This variability presents the bacterium with a metabolic challenge. Glucose is its preferred energy source, and is the easiest sugar to metabolize, but not all of its host's foods contain an abundant supply of glucose. For example, the bacteria may suddenly be deluged with milk, the main carbohydrate of which is the sugar lactose. Lactose is a β-galactoside—a disaccharide containing galactose β linked to glucose (see Chapter 3). Before lactose can be of any use to the bacteria, it must first be taken into the cells by a membrane transport carrier called β-galactoside permease. Then it must be hydrolyzed to glucose and galactose by the enzyme β-galactosidase. A third protein, the enzyme thiogalactoside transacetylase, is also required for lactose metabolism.

When *E. coli* is grown on a medium that does not contain lactose or other β-galactosides, the levels of all three of these enzymes within the bacterial cell are extremely low—the cell does not waste energy and materials making the unneeded proteins. If, however, the environment changes such that lactose is the predominant sugar and very little glucose is

present, the synthesis of all three of these enzymes begins promptly, and they increase rapidly in abundance. For example, there are only two molecules of β-galactosidase present in an *E. coli* cell when glucose is present in the medium. But when it is absent, lactose can induce the synthesis of 3,000 molecules of β-galactosidase per cell!

If lactose is removed from *E. coli*'s environment, synthesis of the three enzymes that process it stops almost immediately. The enzyme molecules that have already formed do not disappear; they are merely diluted during subsequent growth and reproduction until their concentration falls to the original low level within each bacterium.

Compounds that stimulate the synthesis of an enzyme (such as lactose in our example) are called **inducers** (Figure 13.13). The enzymes that are produced are called **inducible enzymes**, whereas enzymes that are made all the time at a constant rate are called **constitutive enzymes**.

We have now seen two basic ways to regulate the rates of metabolic pathways. Chapter 6 described allosteric regulation of enzyme *activity* (the rate of enzyme-catalyzed reactions); this mechanism allows rapid fine-tuning of metabolism. Regulation of protein synthesis—that is, regulation of the *concentration* of enzymes—is slower, but produces a greater savings of energy. Figure 13.14 compares these two modes of regulation.

A single promoter controls the transcription of adjacent genes

The genes that serve as blueprints for the synthesis of the three proteins that process lactose are called **structural genes**, indicating that they specify the primary structure (the amino acid sequence) of a protein molecule. In other words, structural genes are those that can be transcribed into mRNA. Three such genes are involved in the metabolism of lactose, and they lie adjacent to each other on the *E. coli* chromosome. This is no coincidence. Their DNA is transcribed into a single, continuous molecule of mRNA. Because this particular messenger governs the synthesis of all three lactose-metabolizing enzymes, either all or none of

13.14 Two Ways to Regulate a Metabolic Pathway
Feedback from the end product can block enzyme activity, or it can stop the transcription of genes that code for the enzyme.

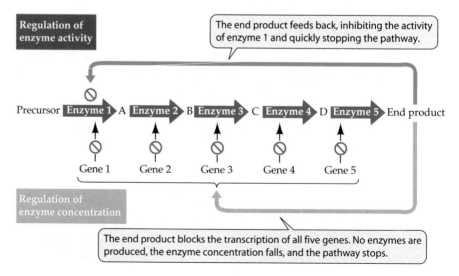

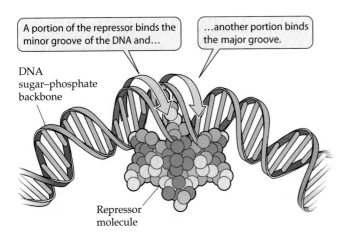

A portion of the repressor binds the minor groove of the DNA and...

...another portion binds the major groove.

DNA sugar–phosphate backbone

Repressor molecule

13.15 Repressor Bound to Operator Blocks Transcription
Portions of the repressor bind to the major and minor grooves in the DNA helix, preventing transcription.

the enzymes are made, depending on whether their common message—their mRNA—is present in the cell.

The three genes share a single promoter. Recall from Chapter 12 that a promoter is a site on DNA where RNA polymerase binds to initiate transcription. The promoter for these three structural genes is very efficient, so that the maximum rate of mRNA synthesis can be high, but there must also be a way to shut down mRNA synthesis when the enzymes are not needed.

Operons are units of transcription in prokaryotes

Prokaryotes shut down transcription by placing an obstacle between the promoter and its structural genes. A short stretch of DNA called the **operator** lies in this position. It can bind very tightly to a special type of protein molecule, called a **repressor**, to create such an obstacle. When the repressor protein is bound to the operator region of DNA, it blocks the transcription of mRNA (Figure 13.15). When the repressor is not attached to the operator, mRNA synthesis proceeds rapidly.

The whole unit, consisting of the closely linked structural genes and the stretches of DNA that control their transcription, is called an **operon** (Figure 13.16). An operon al-

ways consists of a promoter, an operator, and two or more structural genes. The promoter and operator are binding sites on DNA and are not transcribed.

E. coli has numerous ways to control the transcription of operons, and we will focus on three of them. Two ways depend on interactions of the repressor protein with the operator, and the third depends on interactions of other proteins with the promoter. Let's consider each of these three control systems in turn.

Operator–repressor control that induces transcription: The *lac* operon

The operon that controls and contains the structural genes for the three *lac*tose-metabolizing enzymes is called the *lac* operon (see Figure 13.16). As we have just learned, RNA polymerase can bind to the promoter, and a repressor can bind to the operator.

How is the operon controlled? The key lies in the repressor and its binding to the operator. The repressor protein has two binding sites: one for the operator and the other for inducers. The inducers of the *lac* operon, as we know, are molecules of lactose and certain other β-galactosides. Binding of an inducer changes the shape of the repressor (by allosteric modification; see Chapter 6). This change in shape prevents the repressor from binding to the operator (Figure 13.17). As a result, RNA polymerase can bind to the promoter and start transcribing the structural genes of the *lac* operon. The mRNA transcribed from these genes is translated on ribosomes, synthesizing the three proteins required for metabolizing lactose.

What happens if the concentration of lactose drops? As the lactose concentration decreases, the inducer (lactose) molecules separate from the repressor. Free of lactose molecules, the repressor returns to its original shape and binds to the operator, and transcription of the *lac* operon stops. Translation stops soon thereafter, because the mRNA that is already present breaks down quickly. Thus, it is the pres-

13.16 The lac Operon of E. coli and Its Regulator
The *lac* operon of *E. coli* is a segment of DNA that includes a promoter, an operator, and the three structural genes that code for the lactose-metabolizing enzymes.

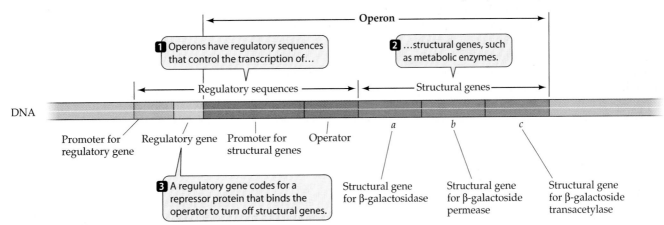

Operon

1 Operons have regulatory sequences that control the transcription of...

2 ...structural genes, such as metabolic enzymes.

Regulatory sequences

Structural genes

DNA

Promoter for regulatory gene

Regulatory gene

Promoter for structural genes

Operator

a *b* *c*

3 A regulatory gene codes for a repressor protein that binds the operator to turn off structural genes.

Structural gene for β-galactosidase

Structural gene for β-galactoside permease

Structural gene for β-galactoside transacetylase

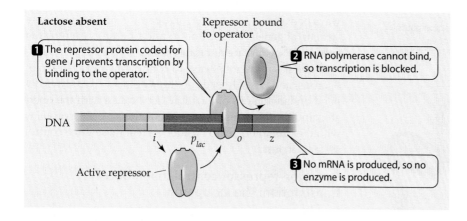

Lactose absent

1 The repressor protein coded for gene *i* prevents transcription by binding to the operator.

Repressor bound to operator

2 RNA polymerase cannot bind, so transcription is blocked.

DNA

i p_{lac} *o*   *z*

Active repressor

3 No mRNA is produced, so no enzyme is produced.

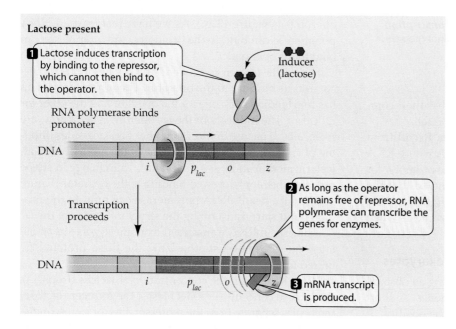

Lactose present

1 Lactose induces transcription by binding to the repressor, which cannot then bind to the operator.

Inducer (lactose)

RNA polymerase binds promoter

DNA

i p_{lac} *o*   *z*

Transcription proceeds

2 As long as the operator remains free of repressor, RNA polymerase can transcribe the genes for enzymes.

DNA

i p_{lac} *o*   *z*

3 mRNA transcript is produced.

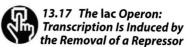

13.17 The lac Operon: Transcription Is Induced by the Removal of a Repressor
Lactose (the inducer) leads to enzyme synthesis by preventing the repressor protein (which would have stopped transcription) from binding to the operator.

► Control is exerted by a regulatory protein—the repressor—that turns the operon *off*.

► Some genes, such as *i*, produce proteins whose sole function is to regulate the expression of other genes.

► Certain other DNA sequences (operators and promoters) do not code for proteins but are binding sites for regulatory or other proteins.

► Adding inducer turns the operon *on*.

Operator–repressor control that represses transcription: The *trp* operon

We have seen that *E. coli* benefits from having an inducible system for lactose metabolism. Only when lactose is present does the system switch on. Equally valuable to a bacterium is the ability to switch *off* the synthesis of certain enzymes in response to the excessive accumulation of their end products. For example, if the amino acid tryptophan, an essential constituent of proteins, is present in ample concentration, it is advantageous to stop making the enzymes for tryptophan synthesis. When the synthesis of an enzyme can be turned off in response to such a biochemical cue, it is said to be **repressible**.

The French biochemist Jacques Monod, who had been part of the team that deciphered the *lac* operon, realized that repressible systems, such as the *trp* operon for *trp*tophan synthesis, could work by mechanisms similar to those of inducible systems. In repressible systems, the repressor protein cannot shut off its operon unless it first binds to a **corepressor**, which may be either the metabolite itself (tryptophan in this case) or an analog of it (Figure 13.18). If the metabolite is absent, the operon is transcribed at a maximum rate. If the metabolite is present, the operon is turned off.

The difference between inducible and repressible systems is small, but significant. In inducible systems, a substance in the environment (the inducer) interacts with the regulatory gene product (the repressor), rendering it *incapable* of binding to the operator and thus incapable of blocking transcription. In repressible systems, a substance in the cell (the corepressor) interacts with the regulatory gene product to make it *capable* of binding to the operator and

ence or absence of lactose—the inducer—that regulates the binding of the repressor to the operator, and therefore the synthesis of the proteins needed to metabolize it.

Repressor proteins are coded by **regulatory genes**. The regulatory gene that codes for the repressor of the *lac* operon is called the *i* (inducibility) gene. The *i* gene happens to lie close to the operon that it controls, but some other regulatory genes are distant from their operons. Like all other genes, the *i* gene itself has a promoter, which can be designated p_i. Because this promoter does not bind RNA polymerase very effectively, only enough mRNA to synthesize about ten molecules of repressor protein per cell per generation is produced. This quantity of the repressor is enough to regulate the operon effectively—to produce more would be a waste of energy. There is no operator between p_i and the *i* gene. Therefore, the repressor of the *lac* operon is constitutive; that is, it is made at a constant rate that is not subject to environmental control.

Let's review the important features of inducible systems such as the *lac* operon:

► In the absence of inducer, the *lac* operon is turned *off*.

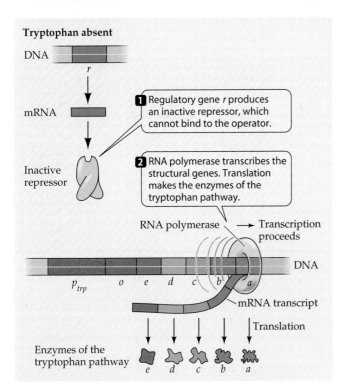

Tryptophan absent

DNA

r

mRNA

Inactive repressor

1 Regulatory gene *r* produces an inactive repressor, which cannot bind to the operator.

2 RNA polymerase transcribes the structural genes. Translation makes the enzymes of the tryptophan pathway.

RNA polymerase → Transcription proceeds

DNA

p_{trp} *o*   *e*   *d*   *c*   *b*   *a*

mRNA transcript

Translation

Enzymes of the tryptophan pathway
e   *d*   *c*   *b*   *a*

13.18 The trp Operon: Transcription Is Repressed by the Binding of a Repressor
Because tryptophan activates an otherwise inactive repressor, it is called a corepressor.

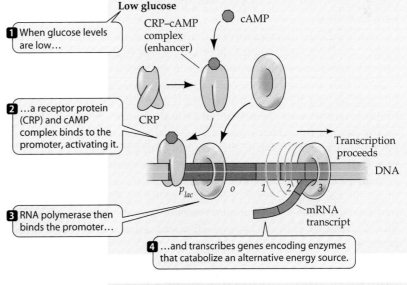

Tryptophan present

DNA

r

mRNA

Inactive repressor → Active repressor

Corepressor (tryptophan)

1 Tryptophan binds the repressor…

2 …which then binds to the operator.

DNA

p_{trp} *o*   *e*   *d*   *c*   *b*   *a*

3 Tryptophan blocks RNA polymerase from binding and transcribing the structural genes, preventing synthesis of tryptophan pathway enzymes.

blocking transcription. Although the effects of the substances are exactly opposite, the systems as a whole are strikingly similar.

In both the inducible lactose system and the repressible tryptophan system, the regulatory molecule functions by binding the operator. Next we'll consider an example of control by binding the *promoter*.

Protein synthesis can be controlled by increasing promoter efficiency

The mechanisms of transcriptional regulation that we have discussed thus far involve repressor–operator interactions that turn the operon on or off. Another way to regulate transcription is to make the promoter work more efficiently.

Suppose that a bacterial cell lacks a supply of glucose, its preferred energy source, but instead has access to another sugar (e.g., lactose or maltose) that can be broken down to enter an energy pathway. In operons such as the *lac* operon that have genes for enzymes that catabolize such alternative energy sources, the promoters bind RNA polymerase in a series of steps (Figure 13.19). First, a protein called **CRP** (short for

13.19 Transcription Is Enhanced by the Binding of the CRP–cAMP Complex to the Promoter
The structural genes of this operon encode enzymes that break down a food source other than glucose.

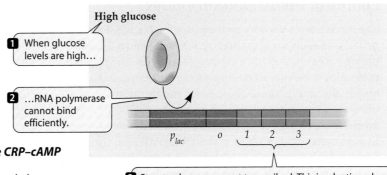

Low glucose

1 When glucose levels are low…

CRP–cAMP complex (enhancer)

cAMP

CRP

2 …a receptor protein (CRP) and cAMP complex binds to the promoter, activating it.

Transcription proceeds

DNA

p_{lac} *o*   *1*   *2*   *3*

mRNA transcript

3 RNA polymerase then binds the promoter…

4 …and transcribes genes encoding enzymes that catabolize an alternative energy source.

High glucose

1 When glucose levels are high…

2 …RNA polymerase cannot bind efficiently.

p_{lac} *o*   *1*   *2*   *3*

3 Structural genes are not transcribed. This is adaptive when the cell does not require an alternative energy source.

13.2 The Relationships Between Positive and Negative Control in the lac *Operon*

GLUCOSE	cAMP LEVELS	RNA POLYMERASE BINDING TO PROMOTER	LACTOSE	*LAC* REPRESSOR	TRANSCRIPTION OF *LAC* GENES?	LACTOSE USED BY CELLS?
Present	Low	Absent	Absent	Active and bound to operator	No	No
Present	Low	Absent	Present	Inactive and not bound to operator	No	No
Absent	High	Present	Present	Inactive and not bound to operator	Yes	Yes
Absent	High	Absent	Absent	Active and bound to operator	No	No

cAMP receptor protein) binds the low-molecular-weight compound adenosine 3′,5′-cyclic monophosphate, better known as **cyclic AMP**, or **cAMP**. Next, the CRP–cAMP complex binds to DNA just upstream of the promoter. This binding results in more efficient binding of RNA polymerase to the promoter, and an elevated level of transcription of the structural genes.

When glucose becomes abundant in the medium, the bacteria do not need to break down alternative food molecules, so the cell diminishes or ceases synthesizing the enzymes that catabolize these alternative sources. Glucose decreases the synthesis of these enzymes—a phenomenon called *catabolite repression*—by lowering the cellular concentration of cAMP.

As you will see in later chapters of this book, cAMP is a widely used signaling molecule in eukaryotes, as well as in prokaryotes. The use of this nucleotide in such widely diverse situations as a bacterium sensing glucose levels and humans sensing hunger demonstrates the prevalence of common themes in biochemistry and natural selection.

The *lac* and *trp* systems—the two operator–repressor systems—are examples of *negative* control of transcription because the regulatory molecule (the repressor) in each case *prevents* transcription. The promoter system is an example of *positive* control of transcription because the regulatory molecule (the CRP–cAMP complex) *enhances* transcription. The relationships between these positive and negative systems are summarized in Table 13.2.

Control of Transcription in Viruses

The mechanisms used by prokaryotes for the regulation of gene expression are also used by viruses. Even a "simple" biological agent such as a virus is faced with complicated molecular decisions when its genome enters a cell. For example, the viral genome must direct the shutdown of host transcription and translation, then redirect the host protein synthesis machinery to virus production. Genes must be activated in the right order; it makes little sense, for example, for the viral genome to transcribe and translate proteins that lyse the host cell membrane before the virus particles are assembled, ready for release. In temperate viruses,

which insert their genome (or a DNA copy) into the host chromosome, another issue arises: When should the provirus leave the host chromosome and undertake a lytic cycle?

Bacteriophage lambda is a temperate phage, which can undergo either a lytic or a lysogenic cycle (see Figure 13.2). When there is a rich medium and its host bacteria are growing, the phage takes advantage of its favorable environment (lots of resources for the phage in the host cell cytoplasm) and undergoes a lytic cycle. When the host bacteria are not as healthy, the phage senses this, and "lays low" as a lysogenic prophage. When things improve, the prophage leaves the host chromosome and becomes lytic.

The phage makes this decision by means of a "genetic switch": *Two* regulatory proteins compete for *two* operator/promoter sites on phage DNA. The two operators control the transcription of genes involved in the lytic and the lysogenic cycles, respectively, and the two regulatory proteins have *opposite* effects on the two operators (Figure 13.20):

PROTEIN	LYTIC OPERATOR/ PROMOTER	LYSOGENIC OPERATOR/PROMOTER
cI	Represses	Activates
Cro	Activates	Represses

So phage infection is a "race" between these two regulatory proteins. If cI "wins," which occurs when Cro synthesis is low in an unhealthy *E. coli* host cell, the phage enters a lysogenic cycle. If the host cell is healthy, a lot of Cro is made, lysogeny is blocked, and lysis ensues. These regulatory proteins, made very early in phage infection, both have binding domains for recognition of specific phage DNA sequences.

The life cycle of phage lambda, which has been greatly simplified here, is a paradigm for viral infections throughout the biological world. The lessons learned from transcriptional controls in this system have been applied again and again to other viruses, including HIV. The control of gene activity in eukaryotic cells is somewhat different, as we will see in the next chapter, but nevertheless usually involves protein–DNA interactions.

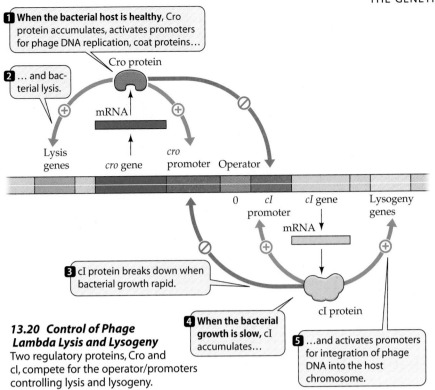

1 When the bacterial host is healthy, Cro protein accumulates, activates promoters for phage DNA replication, coat proteins...

2 ...and bacterial lysis.

Cro protein

mRNA

Lysis genes

cro gene

cro promoter   Operator

0   *cI* promoter   *cI* gene   Lysogeny genes

mRNA

3 cI protein breaks down when bacterial growth rapid.

cI protein

4 When the bacterial growth is slow, cI accumulates...

5 ...and activates promoters for integration of phage DNA into the host chromosome.

13.20 Control of Phage Lambda Lysis and Lysogeny
Two regulatory proteins, Cro and cI, compete for the operator/promoters controlling lysis and lysogeny.

Prokaryotic Genomes

When DNA sequencing first became possible in the late 1970s, the first biological agents to be sequenced were the simplest viruses.

Soon, over 150 viral genomes, including those of important animal and plant pathogens, were sequenced. Information on how they infected and reproduced came quickly as a result.

But manual sequencing was not up to the task of elucidating the genomes of prokaryotes and eukaryotes, the smallest of which are 100 times larger than those of a bacteriophage. In the past 6 years, however, the automation of sequencing has rapidly added many prokaryotic sequences to the biologists' store of knowledge.

In 1995, a team led by Craig Venter and Hamilton Smith determined the first sequence of a free-living organism, the bacterium *Haemophilus influenzae*. Many more sequences have followed, and they have revealed not only how these organisms apportion their genes to perform different cellular roles, but how their specialized functions are carried out. A beginning has even been made on the provocative question of what the minimal requirements for a living cell might be.

Three types of information can be obtained from a genomic sequence:

▶ *Open reading frames*, which are the coding regions of genes. For protein-coding genes, these regions can be recognized by the start and stop codons for translation.

▶ *Amino acid sequences*. For proteins, these can be deduced from the DNA sequence by looking up the genetic code.

▶ *Gene control sequences*, such as promoters and terminators for transcription.

Functional genomics relates gene sequences to functions

The only host for the bacterium *Haemophilus influenzae* is humans. It lives in the upper respiratory tract and can cause ear infections or, more seriously, meningitis in children. Its 1,830,137 base pairs are in a single circular chromosome (Figure 13.21). In addition to its origin of DNA replication and genes coding for rRNA's and tRNA's, this bacterial chromosome has 1,743 regions containing amino acid codons along with the transcriptional (promoter) and translational (start and stop codons) information needed for protein synthesis.

When this sequence was announced, only 1,007 (58%) of its genes had amino acid sequences that corresponded to proteins with known functions—in other words, were genes that researchers, based on their knowledge of the functions of bacteria, expected to find. Roles for most of the unknown proteins have been identified since then, a process known as *annotation*. **Functional genomics**, the assignment of roles to genes and the description of how they work in the organism, has become the major occupation of many biologists.

Of the genes and proteins with known roles, most confirm a century of biochemical description of bacterial enzymatic pathways. For example, there are genes for the entire pathways of glycolysis, fermentation, and electron transport.

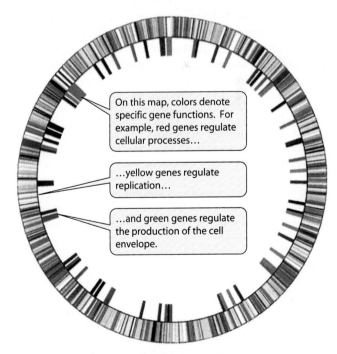

On this map, colors denote specific gene functions. For example, red genes regulate cellular processes...

...yellow genes regulate replication...

...and green genes regulate the production of the cell envelope.

13.21 Functional Organization of the Genome of H. influenzae
The entire DNA sequence has 1,830,137 base pairs.

Some of the gene sequences for unknown proteins may code for membrane proteins, possibly those involved in active transport. Another important finding is that highly infective strains of this bacterium have genes coding for surface proteins that attach them to the human respiratory tract, while noninfective strains lack those genes.

Soon after the sequence of *H. influenzae* was announced, smaller (*Mycoplasma genitalium*, 580,070 base pairs) and larger (*E. coli*, 4,639,221 base pairs) prokaryotic sequences were completed. Thus began a new era in biology, the era of **comparative genomics**, in which genome sequences are compared to see what genes one organism has or is missing, in order to relate the results to physiology.

M. genitalium, for example, lacks the enzymes needed to synthesize amino acids, which the other two organisms possess. This finding reveals that *M. genitalium* is a parasite, which must obtain all its amino acids from its environment, the human urogenital tract. *E. coli* has 55 genes coding for transcriptional activation and 58 repressors; *M. genitalium* has only 3 activators. Comparisons such as these have led to the formulation of specific questions about how an organism lives the way it does.

The sequencing of prokaryotic genomes has medical applications

Sequencing has important ramifications for the study of prokaryotes that cause human diseases. Indeed, most of the early efforts in sequencing have focused on human pathogens.

▶ *Chlamydia trachomatis* causes the most common sexually transmitted disease in the United States. Because it is an intracellular parasite, it has been very hard to study. Among its 900 genes are ones for ATP synthesis—something scientists used to think this bacterium could not do:

▶ *Rickettsai prowazekii* causes typhus; it infects people bitten by lice vectors. Of its 634 genes, 6 code for proteins that are essential for its virulence. These genes are being used to develop vaccines.

▶ *Mycobacterium tuberculosis* causes tuberculosis. It has a large (for a prokaryote) genome, coding for 4,000 proteins. Over 250 of these are used to metabolize lipids, so this may be the main way that the bacterium gets its energy. Some genes coding for previously unidentified cell surface proteins are targets for potential vaccines. The sequence has enough similarities to those of mitochondria to lead to the proposal that this bacterium's ancestor was the one that colonized a cell to ultimately produce that organelle.

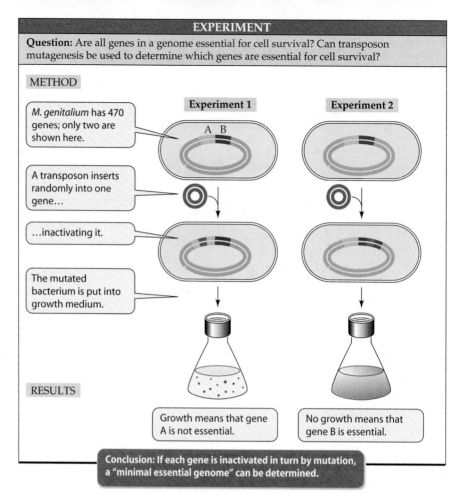

EXPERIMENT

Question: Are all genes in a genome essential for cell survival? Can transposon mutagenesis be used to determine which genes are essential for cell survival?

METHOD

M. genitalium has 470 genes; only two are shown here.

A transposon inserts randomly into one gene...

...inactivating it.

The mutated bacterium is put into growth medium.

RESULTS

Experiment 1

Experiment 2

Growth means that gene A is not essential.

No growth means that gene B is essential.

Conclusion: If each gene is inactivated in turn by mutation, a "minimal essential genome" can be determined.

13.22 Using Transposon Mutagenesis to Determine the Minimal Genome
By inactivating genes one by one, scientists can determine which ones are essential for the cell's survival.

Sequencing has also provided the necessary information for the design of primers and hybridization probes used to detect these and other pathogens.

What genes are required for cellular life?

One striking conclusion arising from comparing the genomes of prokaryotes and eukaryotes is that there are some truly universal genes, present in all organisms. There are also some universal gene segments—coding for an ATP binding site, for example—that are present in many genes in many organisms. These findings suggest that there is some ancient, minimal set of DNA sequences that all cells must have. One way to identify these sequences is to look for them (or, more realistically, to have a computer look for them).

Another way to define a minimal genome is to take the simplest genome and deliberately mutate one gene at a time and see what happens. *M. genitalium*, with only 470 genes, has the smallest known genome of any organism. Some of its genes are dispensable under some circumstances: There are genes for using both glucose and fructose, and in the laboratory, the organism could survive on only one of those sugars, making the genes for the other unnecessary. But what about other genes? Experiments using transposons as mutagens have been performed to address this question. The transposons insert themselves into a gene at random, mutating and inactivating it (Figure 13.22). The resulting mutated cell is sequenced to determine which gene was mutated, and then examined for growth and survival.

The astonishing result of these studies is that *M. genitalium* can survive in the laboratory without the services of 133 of its genes, leaving a minimum of 337 genes! Putting it another way, these 337 genes could theoretically be spliced together (or even synthesized in the laboratory) to make totally artificial life. It is not surprising that the scientists involved in this research have convened a panel of theologians, philosophers and lawyers to advise them.

 Chapter Summary

Using Prokaryotes and Viruses to Probe the Nature of Genes

▶ Prokaryotes and viruses are useful for the study of genetics and molecular biology because they contain much less DNA than eukaryotes, they grow and reproduce rapidly, and they are haploid.

Viruses: Reproduction and Recombination

▶ Viruses were discovered as disease-causing agents small enough to pass through a filter that retains bacteria. They consist of a nucleic acid genome, which codes for a few proteins, and a protein coat. Some viruses also have a lipid membrane derived from host membranes.

▶ Viruses are obligate intracellular parasites, needing the biochemical machinery of living cells to reproduce.

▶ There are many types of viruses, classified by their size and shape, by their genetic material (RNA or DNA), or by their host organism. **Review Figure 13.1**

▶ Bacteriophages are viruses that infect bacteria. In the lytic cycle, the host cell breaks open, releasing many new phage particles. Some phages can also undergo a lysogenic cycle, in which their DNA is inserted into the host chromosome, where it replicates for generations. When conditions are appropriate, the lysogenic DNA exits the host chromosome and enters a lytic cycle. **Review Figure 13.2**

▶ Some viruses have promoters for host RNA polymerase, which they use to transcribe their own genes. **Review Figure 13.3**

▶ Most of the many types of RNA and DNA viruses that infect animals cause diseases. Some animal viruses are surrounded by membranes derived from the host's plasma membrane. Retroviruses, such as HIV, have RNA genomes that they reproduce through a DNA intermediate. Other RNA viruses use their RNA as mRNA to code for enzymes and replicate their genomes without using DNA. **Review Figures 13.4, 13.5, and Table 13.2**

▶ Many plant viruses are spread by other organisms, such as insects.

▶ Viroids are made only of RNA and infect plants, where they are replicated by the plant's enzymes.

Prokaryotes: Reproduction, Mutation, and Recombination

▶ When bacteria divide, they form clones of identical cells that can be observed as colonies when grown on solid media. **Review Figure 13.6**

▶ A bacterium can transfer its genes to another bacterium by conjugation, transformation, or transduction.

▶ In conjugation, a bacterium attaches to another bacterium and passes a partial copy of its DNA to the adjacent cell. **Review Figures 13.7, 13.8, 13.9**

▶ In transformation, genes are transferred between cells when fragments of bacterial DNA are taken up by a cell from the medium. These genetic fragments may recombine with the host chromosome, thereby permanently adding new genes. **Review Figure 13.10**

▶ In transduction, phage capsids carry bacterial DNA from one bacterium to another. **Review Figure 13.10**

▶ Plasmids are small bacterial chromosomes that are independent of the main chromosome. R factors, which are plasmids that carry genes for antibiotic resistance, are a serious public health threat. **Review Figure 13.11**

▶ Transposable elements are movable stretches of DNA that can jump from one place to another on the bacterial chromosome—either by actually moving or by making a new copy, which is inserted at a new location. **Review Figure 13.12**

Regulation of Gene Expression in Prokaryotes

▶ In prokaryotes, the expression of some genes is regulated; their products are made only when they are needed. Other genes, called constitutive genes, whose products are essential to the cell at all times, are constantly expressed. A compound that stimulates the synthesis of an enzyme needed to process it is called an inducer. **Review Figures 13.13, 13.14**

▶ An operon consists of a promoter, an operator, and a number of structural genes. Promoters and operators do not code for proteins, but serve as binding sites for regulatory proteins. When a repressor protein binds to the operator, transcription of the structural genes is inhibited. **Review Figures 13.15, 13.16**

▶ The expression of prokaryotic genes is regulated by three different mechanisms: inducible operator–repressor systems, repressible operator–repressor systems, and systems that increase the efficiency of a promoter.

▶ The *lac* operon is an example of an inducible system whose proteins allow bacteria to metabolize lactose. When lactose is absent, a repressor protein binds tightly to the operator. The repressor prevents RNA polymerase from binding to the promoter, turning transcription off. When glucose is absent and lactose is present, lactose acts as an inducer by binding to the repressor. This changes the repressor's shape so that it no longer recognizes the operator. With the operator unbound, RNA polymerase binds to the promoter, and transcription is turned on.

▶ Repressor proteins are coded by constitutive regulatory genes.

▶ The *trp* operon is a repressible system in which the presence of the end product of a biochemical pathway, tryptophan, represses the synthesis of enzymes involved in its own synthesis. Tryptophan acts as a corepressor by binding to an inactive repressor protein and making it active. When the activated repressor binds to the operator, transcription is turned off. **Review Figure 13.18**

▶ The efficiency of RNA polymerase can be increased by regulation of the level of cyclic AMP, which binds to a protein called CRP. The CRP–cAMP complex then binds to a site near the promoter of a target gene, enhancing the binding of RNA polymerase and hence transcription. **Review Figure 13.19**

Control of Transcription in Viruses

▶ In bacteriophages that can undergo a lytic or a lysogenic cycle, the decision as to which pathway to take is made by operator–regulatory protein interactions. **Review Figure 13.20**

Prokaryotic Genomes

▶ Functional genomics relates gene sequences to functions. **Review Figure 13.21**

▶ By mutating individual genes in a small genome, scientists can determine the minimal genome required for a prokaryote. **Review Figure 13.22**

For Discussion

1. Viruses sometimes carry DNA from one cell to another by transduction. Sometimes a segment of bacterial DNA is incorporated into a phage capsid without any phage DNA. These particles can infect a new host. Would the new host become lysogenic if the phage originally came from a lysogenic host? Why or why not?

2. Compare the life cycles of the viruses that cause influenza (Figure 13.4) and AIDS (Figure 13.5) with respect to how the virus enters the cell; how the virion is released into the cell; how the viral genome is replicated; and how new virus particles are produced.

3. Lederberg, Tatum, and colleagues were able to rule out new mutation and transformation as explanations for the prototrophic colonies that appeared when they mixed cultures of different auxotrophic *E. coli*. Propose experiments to rule out each of these alternatives.

4. Compare promoters adjacent to "early" and "late" genes in the viral life cycle.

5. The repressor protein that turns off the *lac* operon of *E. coli* is encoded by a regulatory gene. The repressor molecules are made in very small quantities and at a constant rate per cell. Would you surmise that the promoter for these repressor molecules is efficient or inefficient? Is synthesis of the repressor constitutive, or is it under environmental control?

6. A key characteristic of a repressible enzyme system is that the repressor molecule must react with a corepressor (typically, the end product of a metabolic pathway) before it can combine with the operator of an operon to shut the operon off. How is this different from an inducible enzyme system?

14 The Eukaryotic Genome and Its Expression

When Tom was diagnosed with leukemia—cancer of the blood cells—his initial treatment included chemotherapy. Combinations of powerful antimitotic drugs were administered to kill the rapidly dividing cancer cells that were spreading throughout his body. But the dosages his physicians prescribed were not up to the task, and the cells continued to spread. Higher dosages of these chemotherapeutic drugs would be lethal; they would kill not only the cancer cells, but the healthy and essential cells in the bone marrow that divide by the hundreds of millions to form blood cells. Without these cells, Tom's bone marrow would no longer produce red blood cells with their vital oxygen-carrying protein hemoglobin, nor would he be able to produce white blood cells, which make the proteins that combat infectious diseases as well as some tumors.

Tom's doctors tried a new approach. They extracted some of his bone marrow and removed the cancer cells from it, then stored the marrow in a refrigerator. Then they gave Tom extremely high doses of the chemotherapeutic drugs, which killed the cancer cells. Finally, the stored bone marrow was replaced in Tom's body. The healthy bone marrow cells began to divide, and after a few weeks they were forming populations of normal red and white blood cells. Tom's leukemia had disappeared.

The success of Tom's bone marrow transplant depended on many things, but the principle behind it is based on the specificity of gene expression during cell differentiation. What are the genetic mechanisms that ensure that healthy red blood cells will contain hemoglobin, and that white blood cells are able to create the vital antibody proteins of the immune system? What features of the DNA sequences of eukaryotic genes determine these mechanisms, and how do they differ from the genes that code for proteins in prokaryotes?

In this chapter, you will see that, although both prokaryotes and eukaryotes use DNA as genetic material, eukaryotic DNA differs from prokaryotic DNA in both its content and its

organization. In addition to the genes for metabolism that prokaryotes have, eukaryotes have genes that mark them as complex cells: genes for addressing, or *targeting*, proteins to organelles, and genes for cell–cell interaction and cell differentiation.

Unlike prokaryotes, eukaryotes have repetitive sequences of DNA, many of which do not code for proteins. In addition, the transcription and later tailoring of mRNA is more complicated in eukaryotes than in prokaryotes. Elegant molecular machinery allows the precise regulation of gene expression needed for all the cells of these complex organisms to develop and function.

The Eukaryotic Genome

As biologists unraveled the intricacies of gene structure and expression in prokaryotes, they tried to generalize their findings by saying, "What's true for *E. coli* is also true for elephants." Although much of prokaryotic biochemistry does apply to eukaryotes, the old saying has its limitations. Table 14.1 lists some of the differences between prokaryotic and eukaryotic genomes.

The eukaryotic genome is larger and more complex than the prokaryotic genome

The fact that the genome of eukaryotes (in terms of haploid DNA content) is larger than that of prokaryotes might be

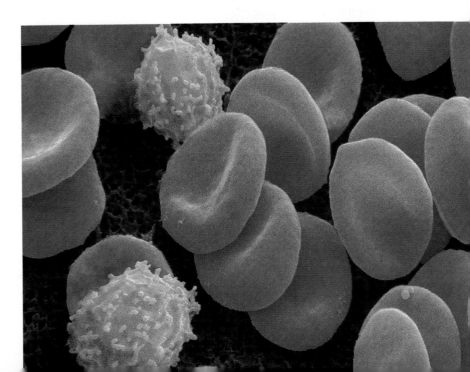

Two Cells, Two Different Protein Products
The red blood cells—erythrocytes—contain abundant hemoglobin, while the white blood cells synthesize proteins of the immune system.

14.1 A Comparison of Prokaryotic and Eukaryotic Genes and Genomes

CHARACTERISTIC	PROKARYOTES	EUKARYOTES
Genome size (base pairs)	10^4–10^7	10^8–10^{11}
Repeated sequences	Few	Many
Noncoding DNA within coding sequences	Rare	Common
Transcription and translation separated in cell	No	Yes
DNA segregated within a nucleus	No	Yes
DNA bound to proteins	Some	Extensive
Promoter	Yes	Yes
Enhancer/silencer	Rare	Common
Capping and tailing of mRNA	No	Yes
RNA splicing required	Rare	Common
Number of chromosomes in genome	One	Many

expected, given that in multicellular organisms there are many cell types, many jobs to do, and many proteins—all coded for by DNA—to do those jobs. A typical virus contains only enough DNA to code for a few proteins—about 10,000 base pairs (bp). The most thoroughly studied prokaryote, *E. coli*, has sufficient DNA (about 4.5 million bp) to make several thousand different proteins and regulate their synthesis. Humans have considerably more genes and regulators: Nearly 6 billion bp (2 meters of DNA) are crammed into each human cell. However, the idea of a more complex organism needing more DNA seems to break down with some plants. For example, the lily (which produces beautiful flowers each spring, but produces fewer proteins than a human does) has 18 times more DNA than humans have (Figure 14.1).

As we will see, the organization of the nuclear eukaryotic genome is fundamentally about regulation. The great complexity of eukaryotes requires a great deal of regulation, and this fact is evident in the many processes and points of control associated with the expression of the eukaryotic genome.

Unlike prokaryotic DNA, most eukaryotic DNA is noncoding. Interspersed throughout the eukaryotic genome are various kinds of repeated DNA sequences that are not transcribed into proteins. Even the coding regions of genes contain sequences that do not end up in mature mRNA.

Some of this noncoding DNA maintains structural integrity at the ends of chromosomes, and some regulates gene expression. But the presence of much of this noncoding DNA remains an enigma.

In contrast to the single main chromosome of most prokaryotes, the eukaryotic genome is partitioned into several separate chromosomes. In humans, each chromosome contains a double helix of DNA with 20 million to 100 million bp. This separation of genomic encyclopedia into multiple volumes requires that each chromosome have, at a minimum, three defining DNA sequences: *Recognition sequences* for the DNA replication machinery, a *centromere region* that holds the replicated sequences together before mitosis, and a *telomeric sequence* at each end of the chromosome. We have described the roles of the first two types of sequences in previous chapters, and will discuss telomeres later in this chapter.

In eukaryotes, the nuclear envelope separates DNA and its transcription (inside the nucleus) from the cytoplasmic sites where mRNA is translated into protein. This separation allows for many points of control in the synthesis, processing, and transport of mRNA to the cytoplasm.

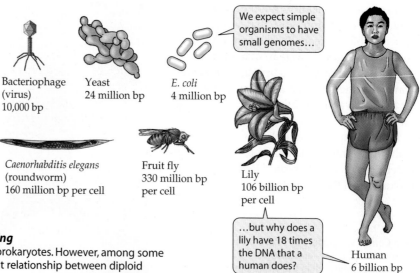

14.1 Amounts of Genomic DNA Can Be Deceiving
Eukaryotes have more DNA in their genomes than prokaryotes. However, among some eukaryotes—especially plants—there is no apparent relationship between diploid genome size and organism complexity.

Bacteriophage (virus) 10,000 bp

Yeast 24 million bp

E. coli 4 million bp

Caenorhabditis elegans (roundworm) 160 million bp per cell

Fruit fly 330 million bp per cell

Lily 106 billion bp per cell

We expect simple organisms to have small genomes...

...but why does a lily have 18 times the DNA that a human does?

Human 6 billion bp per cell

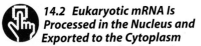

14.2 Eukaryotic mRNA Is Processed in the Nucleus and Exported to the Cytoplasm
Compare this "road map" to the prokaryotic one shown in Figure 12.3.

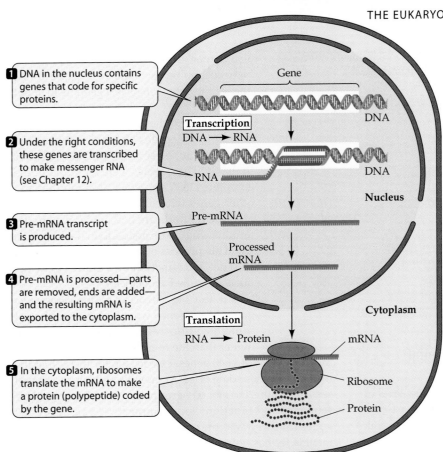

1 DNA in the nucleus contains genes that code for specific proteins.

2 Under the right conditions, these genes are transcribed to make messenger RNA (see Chapter 12).

3 Pre-mRNA transcript is produced.

4 Pre-mRNA is processed—parts are removed, ends are added—and the resulting mRNA is exported to the cytoplasm.

5 In the cytoplasm, ribosomes translate the mRNA to make a protein (polypeptide) coded by the gene.

entire sequences of many prokaryotic genomes. A next step in size and complexity is the sequencing of simple eukaryotes. This has been achieved for a single-celled organism, yeast, as well as for the multicellular roundworm *Caenorhabditis elegans*. Further breakthroughs in the speed and sophistication of the equipment used to sequence DNA accelerated the work on complex eukaryotic genomes, and complete base sequencing of the fruit fly *Drosophila melanogaster* was completed in late 1999. And summer of the year 2000 saw the announcement, attended by frenzied media coverage, of the complete base sequencing of the human genome.

The yeast genome adds some eukaryotic functions onto a prokaryotic model

In comparison with *E. coli*, whose genome has about 4,500,000 bp on a single chromosome, the genome of budding yeast, *Saccharomyces cerevisiae*, has 16 chromosomes and a haploid content of more than 12,068,000 bp. More than 600 scientists around the world collaborated in mapping and sequencing the yeast genome. When they began, they knew of about 1,000 yeast genes coding for RNA or protein. The final sequence revealed 6,200 genes; sequence analyses have assigned probable roles to about 70 percent of them. The functions of the other 30 percent are being investigated by gene inactivation studies similar to those performed on prokaryotes (see Figure 13.22).

It is now possible to estimate what proportions of the yeast genome code for specific metabolic roles. Apparently, 11 percent of yeast proteins are for general metabolism, 3 percent for energy production and storage, 3 percent for DNA replication and repair, 12 percent for protein synthesis, and 6 percent for protein targeting and secretion. Many of the others are involved in cell structure, cell division, and the regulation of gene expression.

The most striking difference between the yeast genome and that of *E. coli* is in the genes for protein targeting (Table 14.2). Both of these single-celled organisms appear to use about the same number of genes to perform the basic functions of cell survival. It is the compartmentalization of the eukaryotic cell into organelles that requires it to have many more genes. This finding is direct, quantitative confirmation of something we have known for a century: The eukaryotic cell is structurally more complex than the prokaryotic cell.

Most eukaryotic DNA is not even fully exposed to the nuclear environment. Instead, it is extensively packaged by proteins into nucleosomes, chromatin fibers, and ultimately chromosomes (see Figure 9.7). This extensive compaction is a means of restricting access of the RNA synthesis machinery to the DNA, as well as a way of segregating replicated DNA's during mitosis and meiosis.

Like the genes of prokaryotes that code for proteins, eukaryotic genes are flanked by noncoding sequences that regulate their transcription. These include the promoter region, where RNA polymerase binds to begin transcription. Of equal importance in eukaryotes, but rare in prokaryotes, is a second set of regulatory DNA sequences, the enhancers and silencers. These sequences are often located quite far from the promoter, and they act by binding proteins that then stimulate or inhibit transcription.

The noncoding DNA sequences found within protein-coding genes present a special problem: How do cells ensure that transcripts of these noncoding regions do not end up in the mRNA that exits the nucleus? The answer lies in an elaborate cutting and splicing mechanism within the nucleus that modifies the initial transcript, called **pre-mRNA**, by cutting out the noncoding regions after transcription (Figure 14.2). Thus, in contrast to the "what is transcribed is what is translated" scheme of most prokaryotic genes, the mature mRNA that is translated at the eukaryotic ribosome is a modified and much smaller molecule than the one initially made in the nucleus.

As we described in Chapter 13, advances in the automation of DNA sequencing have made it possible to obtain the

14.2 *Comparison of the Genomes of E. coli and Yeast*

	E. COLI	YEAST
Genome length (base pairs)	4,640,000	12,068,000
Number of proteins	4,300	6,200
Proteins with roles in:		
Metabolism	650	650
Energy production/storage	240	175
Membrane transporters	280	250
DNA replication/repair/recombination	120	175
Transcription	230	400
Translation	180	350
Protein targeting/secretion	35	430
Cell structure	180	250

The nematode genome adds developmental complexity

The presence of more than a single cell adds a new level of complexity to the genome. *Caenorhabditis elegans* is a 1–mm long nematode (roundworm) that normally lives in the soil. But it also lives in the laboratory, where it is a favorite organism of developmental biologists (see Chapter 16). It has a transparent body, through which scientists can watch over 3 days as its fertilized egg divides and forms an adult worm of 1,000 cells. In spite of its small number of cells, the adult worm has a nervous system, digests food, reproduces sexually, and ages. So it is not surprising that an intense effort was made to sequence the genome of this organism.

Just as with yeast and *E. coli*, the computer-based science of comparative genomics has given us much information on the *C. elegans* genome. It is eight times larger than that of yeast (97 million base pairs) and has four times more protein-coding genes (19,099). Once again, sequencing revealed far more than expected: When the sequencing effort began, researchers estimated that the worm would have about 6,000 proteins.

About 3,000 genes in the worm have direct homologs in yeast; these genes code for basic eukaryotic cell functions. What do the rest of the genes—the bulk of the worm genome—do? In addition to surviving, growing, and dividing, as single-celled organisms do, multicellular organisms must have genes for holding cells together to form tissues, for cell differentiation to divide up tasks in the organism, and for intercellular communication to coordinate its activities (Table 14.3). Many of the genes so far identified in *C. elegans* that are not present in yeast perform these roles, which will be described in the remainder of this chapter and the next one.

The fruit fly genome has surprisingly few genes

The fruit fly *Drosophila melanogaster* is a much larger organism than *C. elegans*, both in size (the fly has 10 times more cells) and complexity. Not surprisingly, the fly genome is also larger, about 180,000,000 base pairs. New technologies made it possible to sequence the entire *Drosophila* genome in about a year.

Even before the complete sequence was announced, decades of genetic studies had already identified some 2,500 different genes in the fly. These genes were all found in the complete DNA sequence, along with many other genes whose functions are as yet unidentified. Efforts are now under way to determine what these genes do in the life of the fly. (This process of discovering the protein product and function of a known gene sequence is called *annotation*.)

But the big surprise of the *Drosophila* genome sequence was the total number of protein-coding regions. Instead of being higher than the roundworm's (18,000 genes), the fly has only 13,600 genes. One reason for this is that the roundworm has some large *gene families*, which, as we will see later in this chapter, are groups of genes related in their sequence and function. For example, *C. elegans* has 1,100 genes involved in either nerve cell signaling or development; a fly has only 160 genes for these two functions. Another major expansion in the worm is in its genes coding for proteins that sense chemicals in its environment.

Many genes that are present in the worm genome have homologs with similar sequences in fly DNA, accounting for a third of the fly genes. And about half of the fly genes have mammalian homologs. An important medical contribution of comparative genomics has resulted from finding genes that are implicated in human diseases in other organisms. Often the roles of such genes can be elucidated in the simpler organism, providing a clue to how the gene might function in human disease. The fly genome contains 177 genes whose sequences are known to be directly involved in human diseases, including cancer and neurological conditions.

14.3 *C. elegans Genes Essential to Multicellularity*

FUNCTION	PROTEIN/DOMAIN	GENES
Transcription control	Zinc finger; homeobox	540
RNA processing	RNA binding domains	100
Nerve impulse transmission	Gated ion channels	80
Tissue formation	Collagens	170
Cell interactions	Extracellular domains; glycotransferases	330
Cell–cell signaling	G protein-linked receptors; protein kinases; protein phosphatases	1,290

Gene sequences for other organisms are rapidly becoming known

A "rough" human genome sequence is already available, with a more detailed one just a few years away. The human genome sequence and its myriad implications will be discussed in Chapter 18. Meanwhile, sequencing is proceeding rapidly for another model organism: the weedy plant *Arabidopsis thaliana* (130 million base pairs). These eukaryotic sequences will pose great challenges and opportunities for biologists in the next decades. "The sequence is not the end of the day," says Sydney Brenner, a leader of this effort. "It's the beginning of the day."

Repetitive Sequences in the Eukaryotic Genome

As we have mentioned, and as you have seen in the genome sequences we have examined, the eukaryotic genome has some base sequences that are repeated many times. Some of these sequences are present in millions of copies in a single genome. In this section, we will examine the organization and possible roles of these repetitive sequences.

Highly repetitive sequences are present in large numbers of copies

Three types of *highly repetitive sequences* are found in eukaryotes:

▶ *Satellites* are 5–50 base pairs long, repeated side by side up to a million times. For example, in guinea pigs, the satellite sequence is CCCTAA.* Satellites are usually present at the centromeres of chromosomes. Their role is not known.

▶ *Minisatellites* are 12–100 base pairs long and are repeated several thousand times. Because DNA polymerase tends to slip and make errors in copying these sequences, they are variable in the numbers of copies. For

*When a DNA sequence such as CCCTAA is written, the complementary bases on the other strand are assumed.

example, one person might have 300, and another, 500. This variation provides a set of molecular genetic markers for identifying an individual.

▶ *Microsatellites* are very short (1–5 base pairs) sequences, present in small clusters of 10–50 copies. They are scattered all over the genome, and have been used in human gene sequencing.

While laboratory scientists have made use of these sequences in genetic studies, their roles in eukaryotes are not clear.

Telomeres are repetitive sequences at the ends of chromosomes

There are several types of *moderately repetitive sequences* in the eukaryotic genome. One type is important in maintaining the ends of chromosomes when DNA is replicated. Recall from Chapter 11 that replication proceeds differently on the two strands of a DNA molecule. Both new strands form in the 5′-to-3′ direction, but one strand (the leading strand) grows continuously from one end to the other, while the other (the lagging strand) grows as a series of short Okazaki fragments (see Figure 11.17).

In a eukaryotic chromosome, replication must begin with an RNA primer at the 5′ end of the forming strand. The leading strand can grow without interruption to the very end, but on the lagging strand there is nothing beyond the primer in the 5′ direction to replace the RNA. So the new chromosome formed after DNA replication lacks a bit of double-stranded DNA at each end. This situation signals DNA repair mechanisms in the cell, and the single-stranded regions, along with some of the intact double-stranded end, is cut off. In this way, the chromosome becomes shorter with each cell division.

In many eukaryotes, there are moderately repetitive sequences at the ends of chromosomes called **telomeres**. In humans, the sequence is TTAGGG, and it is repeated about 2,500 times (Figure 14.3*a*). These repeats bind special proteins that maintain the stability of chromosome ends. Otherwise, the DNA rapidly breaks down.

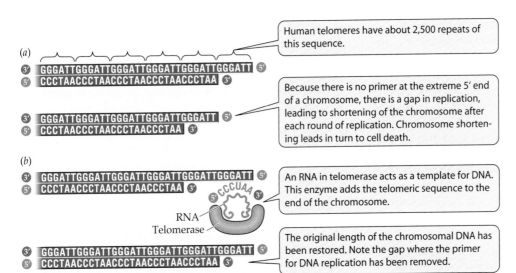

Human telomeres have about 2,500 repeats of this sequence.

(*a*)

Because there is no primer at the extreme 5′ end of a chromosome, there is a gap in replication, leading to shortening of the chromosome after each round of replication. Chromosome shortening leads in turn to cell death.

(*b*)

An RNA in telomerase acts as a template for DNA. This enzyme adds the telomeric sequence to the end of the chromosome.

RNA Telomerase

The original length of the chromosomal DNA has been restored. Note the gap where the primer for DNA replication has been removed.

14.3 Telomeres and Telomerase
(*a*) The loss of moderately repetitive sequences from the telomere leads to cell death. (*b*) In cells that divide continuously (such as germ line cells), the enzyme telomerase prevents the loss of telomeric ends.

When human cells are removed from the body and put in a nutritious medium in the laboratory, they will grow and divide. But each chromosome can lose 50–200 bp of telomeric DNA after each round of DNA replication and cell division. This shortening compromises the stability of the chromosomes. After 20–30 divisions, chromosomes are unable to take part in cell division, and the cell dies. The same thing happens in the body, and explains in part why cells do not last the entire lifetime of the organism: Their telomeres shorten.

Yet constantly dividing cells, such as bone marrow cells and germ line cells, manage to maintain their moderately repetitive telomeric DNA. An enzyme, appropriately called *telomerase*, prevents the loss of this DNA by catalyzing the addition of any lost telomeric sequences (Figure 14.3*b*). Telomerase is made up not only of proteins, but also of an RNA sequence that acts as a template for the telomeric sequence addition.

Considerable interest has been generated by the finding that telomerase is expressed in more than 90 percent of human cancers. Telomerase may be an important factor in the ability of cancer cells to divide continuously. Since most normal cells do not have this ability, telomerase is an attractive target for drugs designed to attack tumors specifically.

There is also interest in telomerase and aging. When a gene expressing high levels of telomerase is added to human cells in culture, their telomeres do not shorten, and instead of dying after 20–30 cell generations, the cells become immortal. It remains to be seen how this finding relates to the aging of a large organism.

Some moderately repetitive sequences are transcribed

Some moderately repetitive DNA sequences code for tRNA's and rRNA's, which are used in protein synthesis (see Chapter 12). These RNA's are constantly being made, but even at the maximum rate of transcription, single copies of these sequences would be inadequate to supply the large amounts of these molecules needed by most cells; hence there are multiple copies of the DNA sequences coding for them. Since these moderately repetitive sequences are transcribed into RNA, they are properly termed "genes," and we can speak of rRNA genes and tRNA genes.

In mammals, there are four different rRNA molecules that make up the ribosome—the 18S, 5.8S, 28S, and 5S rRNA's.* The 18S, 5.8S, and 28S rRNA's are transcribed from a repeated sequence of DNA as a single precursor, which is twice the size of the three ultimate products (Figure 14.4). Several posttranscriptional steps cut this precursor into its final three rRNA's and discard the nonuseful, or "spacer," RNA. The DNA coding for these RNA's is moder-

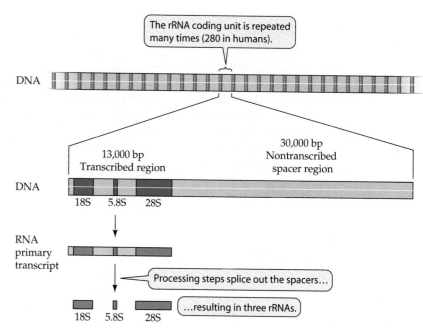

14.4 A Moderately Repetitive Sequence Codes for rRNA
This rRNA gene, along with its nontranscribed spacer, is repeated 280 times in the human genome.

ately repetitive in humans: A total of 280 copies of the sequence are located in clusters on five different chromosomes.

Other moderately repetitive sequences in mammals are not clustered, but instead are scattered throughout the genome. These DNA's usually are not transcribed and usually are short, about 300 bp long. In humans, half of these DNA's are of a single type, called the *Alu* family (because they contain a sequence that is recognized by a nuclease enzyme, Alu I). There are 300,000 copies of the *Alu* family in the genome, and they may act as multiple origins for DNA replication.

Transposable elements move about the genome

Most of the remaining scattered moderately repetitive DNA is not stably integrated into the genome. Instead, these DNA sequences can move from place to place in the genome. Such sequences are called **transposable elements**, or **transposons**.

There are four main types of transposable elements in eukaryotes:

▶ *SINEs* (short *in*terspersed *e*lements) are up to 500 bp long and are transcribed, but not translated.

▶ *LINEs* (long *in*terspersed *e*lements) are up to 7,000 bp long, and some are transcribed and translated into proteins. They constitute about 15 percent of the human genome.

Both of these elements are present in more than 100,000 copies. They move about the genome in a distinctive way: They make an RNA copy, which acts as a template for the

---

*The measure "S" refers to the movement of a molecule in a centrifuge: In general, larger molecules have a higher S value.

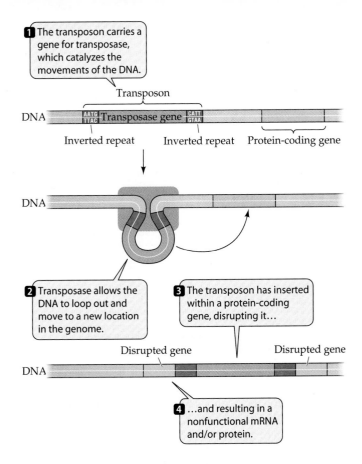

1 The transposon carries a gene for transposase, which catalyzes the movements of the DNA.

Transposon

DNA — Transposase gene — Inverted repeat — Inverted repeat — Protein-coding gene

DNA

2 Transposase allows the DNA to loop out and move to a new location in the genome.

3 The transposon has inserted within a protein-coding gene, disrupting it…

Disrupted gene   Disrupted gene

DNA

4 …and resulting in a nonfunctional mRNA and/or protein.

14.5 Transposons and Transposition
At the end of each transposable element is an inverted repeat sequence that helps in the transposition process.

new DNA, which then inserts itself at a new location in the genome.

▶ *Retrotransposons* also make an RNA copy when they move. They are rare in mammals, but are more common in other animals and yeasts. The genetic organization of *viral retrotransposons* resembles that of retroviruses such as HIV, but these segments lack the genes for protein coats and thus cannot produce viruses.

▶ *DNA transposons* are similar to their prokaryotic counterparts. They do not use an RNA intermediate, but actually move to a new spot in the genome without replicating (Figure 14.5).

What role do these moving sequences play in the cell? There are few answers to this question. The best answer so far seems to be that transposons are cellular parasites that simply replicate themselves. But these replications can lead to the insertion of a transposon at a new location, and this event has important consequences. For example, insertion of a transposon into the coding region of a gene causes a mutation because of the addition of new base pairs. This has been found in rare forms of several human genetic diseases, including hemophilia and muscular dystrophy.

If the insertion of a transposon takes place in the germ line, a gamete with a new mutation results. If the insertion takes place in a somatic cell, cancer may result. If a transposon replicates not just itself but also an adjacent gene, the result may be a gene duplication. A transposon can carry a

gene, or a part of it, to another location on a chromosome, shuffling genetic material and creating new genes. Clearly, transposition stirs the genetic pot in the eukaryotic genome and thus contributes to genetic variability.

In Chapter 4, we described the endosymbiosis theory of the origin of chloroplasts and mitochondria, which proposes that these organelles are the descendants of once free-living prokaryotes. Transposable elements may have played a role in this process. In living eukaryotes, although the organelles have some DNA, the nucleus contains most of the genes that encode the organelle proteins. If the organelles were once independent, they must originally have contained all of these genes. How did the genes move to the nucleus? The answer may lie in DNA transposition. Genes once in the organelle may have moved to the nucleus by well-known molecular events that still occur today. The DNA that remains in the organelles may be the remnants of more complete prokaryotic genomes.

The Structures of Protein-Coding Genes

Like their prokaryotic counterparts, many protein-coding genes in eukaryotes are single-copy DNA sequences. But eukaryotic genes have two distinctive characteristics that are uncommon among prokaryotes. First, they contain noncoding internal sequences, and second, they form gene families with structurally and functionally related cousins in the genome.

Protein-coding genes contain noncoding internal and flanking sequences

Preceding the coding region of a eukaryotic gene is a *promoter*, where RNA polymerase begins the transcription process. Unlike the prokaryotic enzyme, eukaryotic RNA polymerase does not recognize the promoter sequence by itself, but requires help from other molecules, as we'll see later. At the other end of the gene, after the coding region, is a DNA sequence appropriately called the *terminator*, which RNA polymerase recognizes as the end point for transcription (Figure 14.6). Neither the promoter nor the terminator sequence is transcribed into RNA.

Eukaryotic protein-coding genes also contain noncoding base sequences, called **introns**. One or more introns are interspersed with the coding regions—called **exons**—in most eukaryotic genes. Transcripts of the introns appear in the primary transcript of RNA—the pre-mRNA—within the nucleus, but by the time the mature mRNA exits the organelle, they have been removed. The transcripts of the introns are cut out of the pre-mRNA, and the transcripts of the exons are spliced together.

The locations of the introns can be determined by comparing the base sequences of a gene (DNA) with those of its final mRNA. Although direct sequencing of the DNA that codes for an mRNA is the easiest way to map the locations of introns within a gene, **nucleic acid hybridization** is the method that originally revealed the existence of introns in protein-coding genes. This method, outlined in

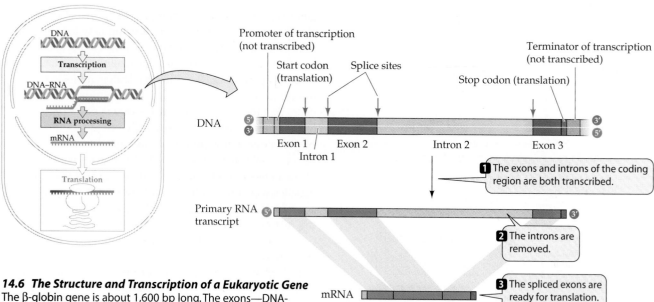

Promoter of transcription (not transcribed)

Start codon (translation)

Splice sites

Terminator of transcription (not transcribed)

Stop codon (translation)

DNA

Exon 1   Exon 2   Intron 2   Exon 3

Intron 1

1 The exons and introns of the coding region are both transcribed.

Primary RNA transcript

2 The introns are removed.

mRNA

3 The spliced exons are ready for translation.

14.6 The Structure and Transcription of a Eukaryotic Gene
The β-globin gene is about 1,600 bp long. The exons—DNA-coding sequences— contain 441 base pairs (triplet codons for 146 amino acids plus a triplet stop codon). Noncoding sequences of DNA—introns—are initially transcribed between codons 30 and 31 (130 bp long) and 104 and 105 (850 bp long), but are spliced out of the final transcript.

Figure 14.7, has been crucial to genetic research; in later chapters we will see its use in localizing genes, testing for alleles, localizing mRNA's during development, and many other applications.

To examine the relationship between a gene and its transcript, biologists used nucleic hybridization to examine the gene for one of the globin proteins that make up hemoglobin (Figure 14.8). They first denatured the globin DNA by heating it, then added mature globin mRNA. As expected, the mRNA bound to the DNA by complementary base pairing. The researchers expected to obtain a linear matchup of the mRNA to the globin-coding DNA. They got their wish, in part: There were indeed stretches of RNA–DNA hybridization. But some looped structures were also visible. These loops were the introns, stretches of DNA that did not have complementary bases on the mRNA. Later studies showed that hybridization to the gene using pre-mRNA was complete, and that the introns were indeed transcribed. Somewhere on the path from transcript to mature mRNA, the introns had been removed, and the exons had been spliced together. We will examine this splicing process later in the chapter.

Most (but not all) vertebrate genes contain introns, as do many other eukaryotic genes (and even a few prokaryotic ones). Introns interrupt, but do not scramble, the DNA sequence that codes for a polypeptide chain. The base sequence of the exons, taken in order, is exactly complementary to that of the mature mRNA product. The introns, therefore, separate a gene's protein-coding region into distinct parts—the exons. In some cases, the separated exons code for different functional regions, or *domains*, of the protein. For example, the globin proteins that make up hemoglobin have two domains: one for binding to heme, and another for binding to the other globin chains. These two domains are coded for by different exons in the globin gene.

RESEARCH METHOD

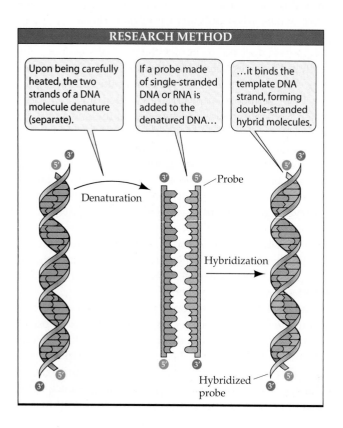

Upon being carefully heated, the two strands of a DNA molecule denature (separate).

If a probe made of single-stranded DNA or RNA is added to the denatured DNA…

…it binds the template DNA strand, forming double-stranded hybrid molecules.

Denaturation

Probe

Hybridization

Hybridized probe

14.7 Nucleic Acid Hybridization
Base pairing permits the detection of a sequence complementary to the probe.

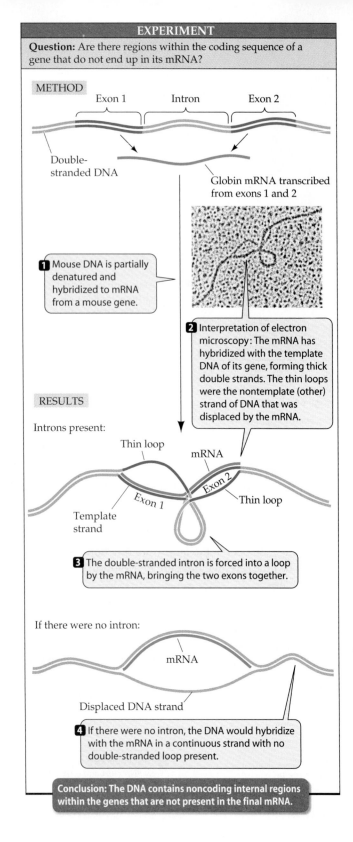

EXPERIMENT

Question: Are there regions within the coding sequence of a gene that do not end up in its mRNA?

METHOD

Exon 1 — Intron — Exon 2

Double-stranded DNA

Globin mRNA transcribed from exons 1 and 2

1 Mouse DNA is partially denatured and hybridized to mRNA from a mouse gene.

2 Interpretation of electron microscopy: The mRNA has hybridized with the template DNA of its gene, forming thick double strands. The thin loops were the nontemplate (other) strand of DNA that was displaced by the mRNA.

RESULTS

Introns present:

Thin loop

mRNA

Exon 1   Exon 2

Thin loop

Template strand

3 The double-stranded intron is forced into a loop by the mRNA, bringing the two exons together.

If there were no intron:

mRNA

Displaced DNA strand

4 If there were no intron, the DNA would hybridize with the mRNA in a continuous strand with no double-stranded loop present.

Conclusion: The DNA contains noncoding internal regions within the genes that are not present in the final mRNA.

Many eukaryotic genes are members of gene families

About half of all eukaryotic protein-coding genes are present in only one copy in the haploid genome. The rest have multiple copies. Often, inexact, nonfunctional copies of a particular gene, called **pseudogenes**, are located near it on a chromosome. These duplicates may have arisen by an ab-

14.8 Hybridization Revealed Noncoding DNA
When an mRNA transcript was experimentally hybridized to the double-stranded DNA of a gene, the introns from the DNA "looped out," demonstrating that the coding region of a eukaryotic gene can contain noncoding DNA that is not present in the mRNA transcript.

normal event in chromosomal crossing over during meiosis or by the action of retrotransposons.

In other cases, however, the genome contains slightly altered copies of a gene that are functional. A set of duplicated or related genes is called a **gene family**. Some families, such as the β-globins that are part of hemoglobin, contain only a few members; other families, such as the immunoglobulins that make up antibodies, have hundreds of members.

Like the members of any family, the DNA sequences in a gene family are usually different from one another to a certain extent. As long as one member retains the original DNA sequence and thus codes for the proper protein, the other members can mutate slightly, extensively, or not at all. The availability of such extra genes is important for "experiments" in evolution: If the mutated gene is useful, it may be selected for in succeeding generations. If the gene is a total loss (a pseudogene), the functional copy is still there to save the day.

The gene family for the *globins* is a good example of the gene families found in vertebrates. These proteins are found in hemoglobin, as well as in myoglobin (an oxygen-binding protein present in muscle). The globin genes probably all arose from a single common ancestor gene long ago. In humans, there are three functional members of the alpha-globin (α-globin) cluster and five in the beta-globin (β-globin) cluster (Figure 14.9). In a human adult, each hemoglobin molecule is a tetramer containing the heme pigments (each held inside a globin polypeptide), two identical α-globins, and two identical β-globins.

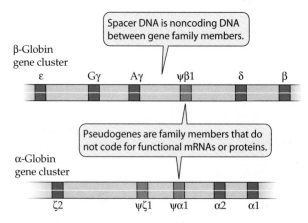

β-Globin gene cluster

Spacer DNA is noncoding DNA between gene family members.

ε   Gγ   Aγ   ψβ1   δ   β

Pseudogenes are family members that do not code for functional mRNAs or proteins.

α-Globin gene cluster

ζ2   ψζ1   ψα1   α2   α1

14.9 Gene Families
The human α-globin and β-globin gene clusters are located on different chromosomes. Each family is organized into a cluster of genes separated by noncoding "spacer" DNA. The nonfunctional pseudogenes are indicated by the Greek letter psi (ψ).

14.10 Differential Expression in the β-Globin Gene Family

During human development, different members of the β-globin gene family are expressed at different times and in different tissues.

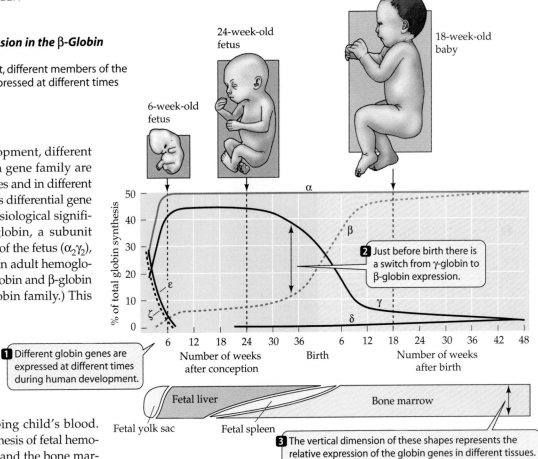

1 Different globin genes are expressed at different times during human development.

2 Just before birth there is a switch from γ-globin to β-globin expression.

3 The vertical dimension of these shapes represents the relative expression of the globin genes in different tissues.

During human development, different members of the β-globin gene family are expressed at different times and in different tissues (Figure 14.10). This differential gene expression has great physiological significance. For example, γ-globin, a subunit found in the hemoglobin of the fetus ($\alpha_2\gamma_2$), binds O_2 more tightly than adult hemoglobin ($\alpha_2\beta_2$) does. (Both γ-globin and β-globin are members of the β-globin family.) This specialized form of hemoglobin ensures that in the placenta, where the maternal and fetal circulation come near each other, O_2 will be transferred from the mother's to the developing child's blood. Just before birth, the synthesis of fetal hemoglobin in the liver stops, and the bone marrow cells take over, making the adult form.

In addition to genes that encode proteins, the globin family includes nonfunctional pseudogenes, designated with the Greek letter psi (φ). These pseudogenes are the "black sheep" of any gene family: They result from mutations that cause a *loss* of function rather than an enhanced or new function.

The DNA sequence of a pseudogene may not differ vastly from that of other family members. It may just lack a promoter, for example, and thus cannot be transcribed. Or it may lack the recognition sites for the removal of introns, and thus will be transcribed into pre-mRNA, but not correctly processed into a useful mRNA. In some gene families, pseudogenes outnumber functional genes. However,

since some members of the family are functional, there appears to be little selective pressure in evolution to eliminate pseudogenes.

RNA Processing

As we have seen, the primary RNA transcript (pre-mRNA) of a eukaryotic gene is not the same as the mature mRNA. To produce the mRNA, the primary transcript is processed by the addition of bases at both ends, and by the removal of introns and the joining of exons.

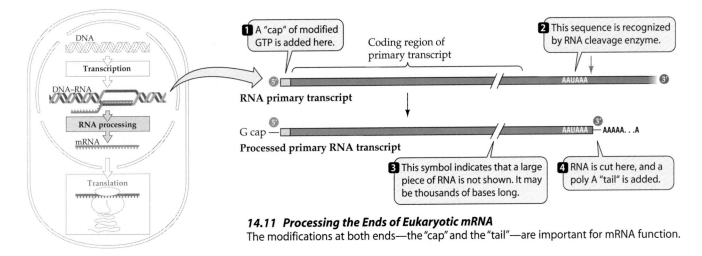

1 A "cap" of modified GTP is added here.

2 This sequence is recognized by RNA cleavage enzyme.

3 This symbol indicates that a large piece of RNA is not shown. It may be thousands of bases long.

4 RNA is cut here, and a poly A "tail" is added.

14.11 Processing the Ends of Eukaryotic mRNA

The modifications at both ends—the "cap" and the "tail"—are important for mRNA function.

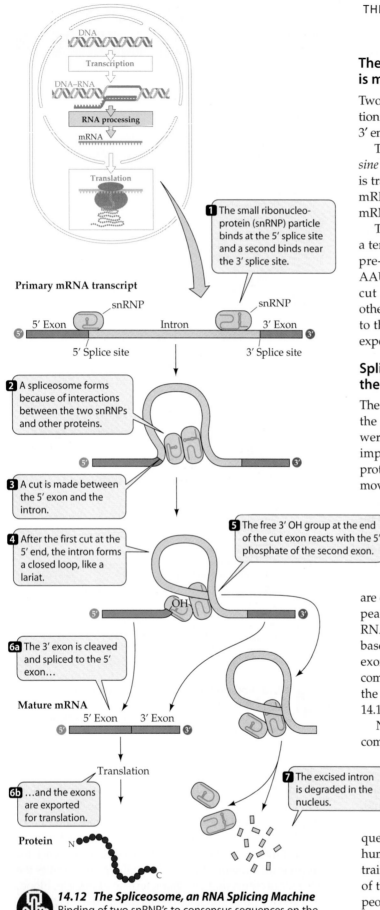

1 The small ribonucleo-protein (snRNP) particle binds at the 5' splice site and a second binds near the 3' splice site.

Primary mRNA transcript

snRNP   snRNP

5' Exon   Intron   3' Exon

5' Splice site   3' Splice site

2 A spliceosome forms because of interactions between the two snRNPs and other proteins.

3 A cut is made between the 5' exon and the intron.

5 The free 3' OH group at the end of the cut exon reacts with the 5' phosphate of the second exon.

4 After the first cut at the 5' end, the intron forms a closed loop, like a lariat.

OH

6a The 3' exon is cleaved and spliced to the 5' exon…

Mature mRNA

5' Exon   3' Exon

Translation

6b …and the exons are exported for translation.

7 The excised intron is degraded in the nucleus.

Protein N   C

14.12 The Spliceosome, an RNA Splicing Machine
Binding of two snRNP's to consensus sequences on the pre-mRNA lines up the splicing machinery. After the snRNP's bind to pre-mRNA, other proteins join the complex to form a spliceosome.

The primary transcript of a protein-coding gene is modified at both ends

Two early steps in the processing of pre-mRNA are the addition of a "cap" at the 5' end and the addition of a "tail" at the 3' end (Figure 14.11).

The **G cap** is a chemically modified molecule of *guanosine triphosphate (GTP)*. It is added to the 5' end as the RNA is transcribed. The cap apparently facilitates the binding of mRNA to the ribosome for translation and protects the mRNA from breaking down.

The **poly A tail** is added to the 3' end of pre-mRNA after a terminal sequence has been removed. Near the 3' end of pre-mRNA, and after the last codon, is the sequence AAUAAA. This sequence acts as a signal for an enzyme to cut the pre-mRNA. Immediately after this cleavage, another enzyme adds 100 to 300 residues of adenine (poly A) to the 3' end of the pre-mRNA. This tail may assist in the export of the mRNA from the nucleus.

Splicing removes introns from the primary transcript

The next step in processing of eukaryotic pre-mRNA within the nucleus is deleting the introns. If these RNA regions were not removed, a nonfunctional mRNA, producing an improper amino acid sequence and thus a nonfunctional protein, would result. The process called RNA **splicing** removes the introns and splices the exons together.

As soon as the pre-mRNA is transcribed, it is quickly bound to several **small nuclear ribonucleo-protein particles** (snRNP's, commonly pronounced "snurps"). There are several types of these RNA–protein particles in the nucleus.

At the boundaries between introns and exons are **consensus sequences**—short stretches of DNA that appear, with little variation, in many different genes. The RNA in one of the snRNP's (called U1) has a stretch of bases complementary to the consensus sequence at the 5' exon–intron boundary, and binds to the pre-mRNA by complementary base pairing. Another snRNP (U2) binds to the pre-mRNA near the 3' intron–exon boundary (Figure 14.12).

Next, other proteins bind, forming a large RNA–protein complex called a **spliceosome**. The spliceosome uses energy from ATP for its assembly. It cuts the RNA, releases the introns, and joins the ends of the exons together to produce mature mRNA.

Molecular studies of human diseases have been valuable tools in the investigation of consensus sequences and splicing machinery. Beta thalassemia is a human genetic disease inherited as an autosomal recessive trait. People with this disease make an inadequate amount of the β-globin subunit that is part of hemoglobin. These people suffer from severe anemia because they have an inadequate supply of red blood cells. In some cases, the genetic mutation that causes the disease occurs at a consensus sequence in the β-globin gene. Consequently, the pre-

mRNA cannot be spliced correctly, and nonfunctional β-globin mRNA is made.

This finding is an excellent example of the use of mutations in determining a cause-and-effect relationship in biology. In the logic of science, merely linking two phenomena (for example, consensus sequences and splicing) does not prove that one is necessary for the other. In an experiment, the scientist alters one phenomenon (for example, the base sequence at the consensus region) to see whether the other event (for example, splicing) occurs. In beta thalassemia, nature has done the experiment for us.

After the processing events are completed in the nucleus, the mRNA exits the organelle, apparently through the nuclear pores (see Figure 4.9). A receptor at the nuclear pore recognizes the processed mRNA (or a protein bound to it). Unprocessed or incompletely processed pre-mRNA's remain in the nucleus.

Transcriptional Control

In a multicellular organism with specialized cells and tissues, every cell contains every gene in the organism's genome. For development to proceed normally, and for each cell to acquire and maintain its proper function, certain proteins must be synthesized at just the right times and in just the right cells. Thus, the expression of eukaryotic genes must precisely regulated.

Regulation of gene expression can occur at many points (Figure 14.13). This section describes the mechanisms that control the transcription of specific genes. These often involve nuclear proteins that alter chromosome function or structure. In some cases, the regulation of transcription involves changes in the DNA itself: Genes are selectively replicated to give more templates to transcribe, or even rearranged on the chromosome.

Posttranscriptional events can also regulate gene expression. As we have seen, the processing of pre-mRNA can be controlled after transcription. The transport of the mRNA into the cytoplasm, and how long it remains there, can also be controlled. The translation of mRNA into protein can also be regulated. Finally, once the protein itself is made, its structure can be modified, or it can be broken down and destroyed.

Specific genes can be selectively transcribed

The brain cells and the liver cells of a mouse have some proteins in common and some that are distinctive for each cell type. Yet both cells have the same DNA sequences and, therefore, the same genes. Are the differences in protein content due to *differential transcription* of genes? Or is it that all the genes are transcribed in both cell types, and a *posttranscriptional mechanism* is responsible for the differences in proteins?

These two alternatives—transcriptional or posttranscriptional control—can be distinguished by examination of the actual RNA sequences made within the nucleus of each cell type. Such analyses indicate that for some proteins, the

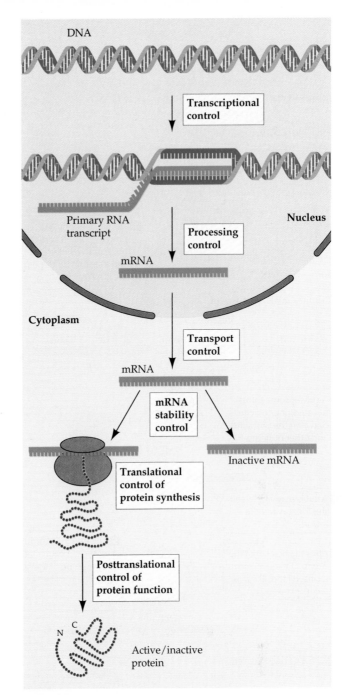

14.13 Potential Points for the Regulation of Gene Expression
Gene expression can be regulated at three levels: at transcription, at translation, or after translation.

mechanism of control is differential gene transcription. Both brain and liver cells, for example, transcribe "housekeeping" genes, such as those for glycolysis enzymes and ribosomal RNA's. But liver cells transcribe some genes for liver-specific proteins, and brain cells transcribe some genes for brain-specific proteins. And neither cell type transcribes the genes for proteins that are characteristic of muscle, blood, bone, and the other specialized cell types in the body.

CONTRASTING EUKARYOTES AND PROKARYOTES. Unlike prokaryotes, in which related genes are transcribed as a unit in

operons, eukaryotes tend to have solitary genes. Thus, regulating several genes at once requires common control elements in each gene, which allow all of the genes to respond to the same signal.

In contrast to the single RNA polymerase in bacteria, eukaryotes have three different RNA polymerases. Each eukaryotic polymerase catalyzes the transcription of a specific type of gene. Only one (RNA polymerase II) transcribes protein-coding genes to mRNA. The other two transcribe the DNA that codes for rRNA (polymerase I) and for tRNA and small nuclear RNA's (polymerase III). The diversity of eukaryotic polymerases is reflected in the diversity of eukaryotic promoters, which tend to be much more variable than prokaryotic promoters. In addition, most eukaryotic genes have regulator, enhancer, and silencer elements (which we will discuss shortly) that can control the rate of transcription. Whether a eukaryotic gene is transcribed depends on the sum total of the effects of all of these DNA and protein elements; thus there are many points of possible control.

Finally, the transcription complex in eukaryotes is very different from that of prokaryotes, in which a single peptide subunit can cause RNA polymerase to recognize the promoter. In eukaryotes, many proteins are involved in initiating transcription. We will confine the following discussion to RNA polymerase II, which catalyzes the transcription of most protein-coding genes, but the mechanisms for the other two polymerases are similar.

TRANSCRIPTION FACTORS. In prokaryotes, the promoter is a sequence of DNA near the 5′ end of the coding region of a gene where RNA polymerase begins transcription. A prokaryotic promoter has two essential regions. One is the *recognition sequence*—the sequence recognized by RNA polymerase. The second, closer to the initiation point, is the *TATA box* (so called because it is rich in AT base pairs), where DNA begins to denature so that its templates can be exposed. In eukaryotes, there is a TATA box about 25 bp away from the initiation site for transcription, and one or two recognition sequences of about 50 to 70 bp 5′ from the TATA box.

Eukaryotic RNA polymerase II cannot simply bind tightly to the promoter and initiate transcription. Rather, it binds and acts only after various regulatory proteins, called **transcription factors**, have assembled on the chromosome (Figure 14.14). First, the protein *TFIID* ("TF" stands for *transcription factor*) binds to the TATA box. Its binding changes both its own shape and that of the DNA, presenting a new surface that attracts the binding of other transcription factors. RNA polymerase II does not bind until several other proteins have already bound to this complex.

Some DNA sequences, such as the TATA box, are common to the promoters of many genes and are recognized by transcription factors that are found in all the cells of an organism. Other sequences in promoters are specific to only a few genes and are recognized by transcription factors found only in certain tissues. These specific transcription

factors play an important role in *differentiation*, the specialization of cells during development.

REGULATORS, ENHANCERS, AND SILENCERS IN DNA. In addition to the promoter, two other regions of DNA bind proteins that activate RNA polymerase. The recently discovered **regulator** regions are clustered just upstream of the promoter. Various regulator proteins (seven in the β-globin

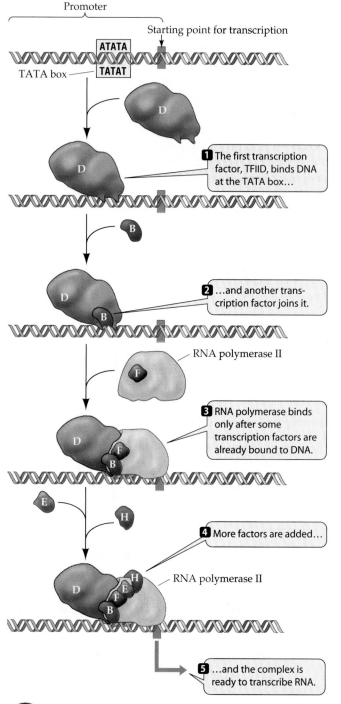

14.14 The Initiation of Transcription in Eukaryotes
Except for TFIID, which also binds to the TATA box, each transcription factor has binding sites only for the other proteins and does not bind directly to DNA.

14.15 The Roles of Transcription Factors, Regulators, and Activators

The actions of many proteins determine whether and where RNA polymerase II will transcribe DNA.

gene) may bind to these regions (Figure 14.15). Their net effect is to bind to the adjacent transcription complex and activate it.

Much farther away—up to 20,000 bp away—are the **enhancer** regions. Enhancer regions bind *activator* proteins, and this binding strongly stimulates the transcription complex. How enhancers can exert this influence is not clear. In one proposed model, the DNA bends—it is known to do so—so that the activator is in contact with the transcription complex (see Figure 14.15).

Finally, there are negative regulatory regions on DNA called **silencers**, which have the reverse effect of enhancers. Silencers turn off transcription by binding proteins appropriately called *repressors*.

DNA bending can bring an activator protein, bound to an enhancer element far from the promoter in linear DNA, to interact with the transcription–initiation complex.

A long stretch of DNA lies between the activator binding site and the transcription complex.

How do these proteins and DNA sequences—transcription factors, activators, repressors, regulators, enhancers, and silencers—regulate transcription? Apparently, all genes in most tissues can transcribe a small amount of RNA. But the right *combination* of the factors is what determines the maximum rate of transcription. In the immature red blood cells of bone marrow, for example, which make a large amount of β-globin, the transcription of globin genes is stimulated by the binding of 7 regulators and 6 activators. But in white blood cells in the same bone marrow, these 13 proteins are not made, and they do not bind to their sites adjacent to the β-globin genes; consequently, these genes are hardly transcribed at all.

COORDINATING THE EXPRESSION OF GENES. How do eukaryotic cells coordinate the regulation of several genes whose transcription must be turned on at the same time? In prokaryotes, in which related genes are linked together in an operon, a single regulatory system can regulate several adjacent genes. But in eukaryotes, the several genes whose regulation requires coordination may be on different chromosomes.

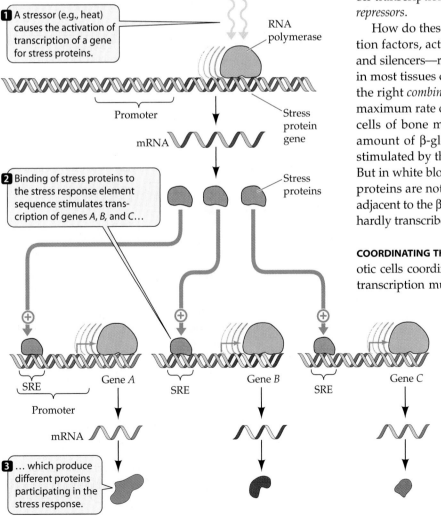

1 A stressor (e.g., heat) causes the activation of transcription of a gene for stress proteins.

2 Binding of stress proteins to the stress response element sequence stimulates transcription of genes *A*, *B*, and *C*...

3 ... which produce different proteins participating in the stress response.

14.16 Coordinating Gene Expression

A single signal, for example heat stress, causes the synthesis of a transcriptional regulator for many genes.

In such a case, regulation can be achieved if the various genes all have the same regulatory sequences near them, which bind to the same activators and regulators. One of the many examples of this phenomenon is provided by the response of organisms to a stressor—for example, drought in plants. Under conditions of drought stress, a plant must synthesize various proteins, but the genes for these proteins are scattered throughout the genome. However, each of these genes has a specific regulatory sequence near its promoter called the *stress response element* (SRE). The binding of a regulator protein to this element stimulates RNA synthesis (Figure 14.16). In the drought example, the proteins made are involved not only in water conservation, but also in protecting the plant against excess salt in the soil and against freezing. This finding has considerable importance for agriculture, in which crops are often grown under less than optimal conditions.

THE BINDING OF PROTEINS TO DNA. A key to transcriptional control in eukaryotes is that transcription factors, regulators, activators, and repressors all bind to specific DNA sequences. In these proteins, there are four common structural themes in the domains that bind to DNA. These themes are called *motifs* and consist of combinations of structures and special components.

The *helix-turn-helix* motif involves several α-helices, one of which makes contact with DNA; the others stabilize the structure. This motif appears in the proteins that activate genes involved in embryonic development (homeobox proteins; see Chapter 16) and in the proteins that regulate the development of the immune and central nervous systems.

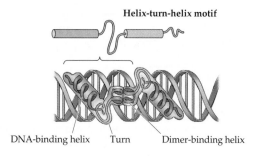

Helix-turn-helix motif

DNA-binding helix   Turn   Dimer-binding helix

The *zinc finger* motif has loops that form when a zinc ion is held by the amino acids cysteine and histidine. It occurs most notably in the receptors for steroid hormones (see Figure 15.9).

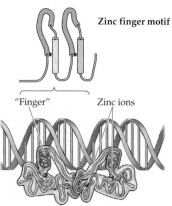

Zinc finger motif

"Finger"   Zinc ions

The *leucine zipper* motif places hydrophobic leucine residues on one side of a polypeptide. Their presence allows two polypeptide chains to interact (zipper) hydrophobically, setting up the positively charged residues just past the zipper to bind to DNA. This motif occurs in many DNA-binding proteins—for example, the transcription factor AP-1, which binds near promoters of genes involved in mammalian cell growth and division. Overactivity of AP-1 has been linked to several types of cancer.

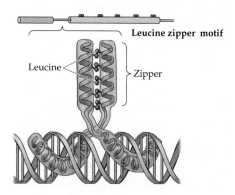

Leucine zipper motif

Leucine   Zipper

The *helix-loop-helix* motif is two helices separated by a loop. This region is adjacent to a stretch of amino acids that interact with DNA. This motif occurs in the activator proteins that bind to enhancers for the immunoglobulin genes that synthesize antibodies, as well as in the transcription factors involved in muscle protein synthesis.

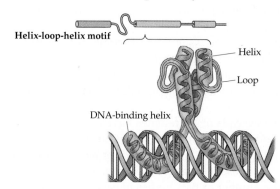

Helix-loop-helix motif

Helix

Loop

DNA-binding helix

Genes can be inactivated by chromatin structure

Chromatin contains nucleosomes and many other chromosomal proteins (see Chapter 9). The packaging of DNA by these nuclear proteins can make DNA physically inaccessible to RNA polymerase and the rest of the transcription apparatus, much as the binding of a repressor to the operator in the prokaryotic *lac* operon prevents transcription. Both local and global chromatin structure affect transcription.

LOCAL EFFECTS. Nucleosomes inhibit both the initiation and elongation of transcription. To alleviate these blocks, cells recruit two protein complexes. One binds upstream of the initiation site, disaggregating the nucleosomes so that the large initiation complex can bind and begin transcription. The other binds once transcription is under way,

allowing the transcription complex to move through these nucleosomes. These processes are called *chromatin remodeling* (Figure 14.17).

GLOBAL EFFECTS. Two kinds of chromatin can be distinguished by staining of the interphase nucleus: euchromatin and heterochromatin. *Euchromatin* is diffuse and stains lightly; it contains the DNA that is transcribed into mRNA. *Heterochromatin* stains densely and is generally not transcribed; any genes that it contains are thus inactivated. Perhaps the most dramatic example of heterochromatin is the inactive X chromosome of mammals.

A normal female mammal has two X chromosomes; a normal male, an X and a Y. The Y chromosome has only a few genes that are also present on the X, and is largely transcriptionally inactive in most cells. So there is a great difference between females and males in the "dosage" of X chromosome genes. In other words, each female cell has two copies of the genes on the X chromosome, and therefore has the potential to produce twice as much protein product of these genes as a male has. Yet X-linked gene expression is generally the same in males and females. How can this happen?

The answer was found in 1961 independently by Mary Lyon, Liane Russell, and Ernest Beutler. They suggested that one of the X chromosomes in each cell of an XX female is transcriptionally inactivated early in embryonic development. That copy of the X remains inactive in that cell, and in all the cells arising from it. In a given cell, the "choice" of which X in the pair of Xs to inactivate is usually random. Recall that one of the Xs in a female comes from her father and one from her mother. Thus, in one embryonic cell, the paternal X might be the one remaining active in mRNA synthesis, but in a neighboring cell, the maternal X might be active.

Interphase cells of XX females have a single, stainable nuclear body called a **Barr body**, after its discoverer, Murray Barr (Figure 14.18). This clump of heterochromatin, which is not present in males, is the inactivated X chromosome. The number of Barr bodies in each nucleus is equal to the number of X chromosomes minus one (the one represents the X chromosome that remains transcriptionally active). So a female with the normal two X chromosomes will have one Barr body, one with three X's will have two, an XXXX female will have three, and an XXY male will have one. We may infer that the interphase cells of each person, male or female, have a single *active* X chromosome, making the dosage of the expressed X chromosome genes constant across both sexes.

The mechanism of X inactivation involves chromosome condensation that makes the DNA sequences physically unavailable to the transcription machinery. One method may be the addition of a methyl group (—CH$_3$) to the 5' position of cytosine on DNA. Such *methylation* seems to be most prevalent in transcriptionally inactive genes. For example, most of the DNA of the inactive X chromosome has many cytosines methylated, while few of them on the ac-

INITIATION OF TRANSCRIPTION

1 DNA wraps around histone proteins, forming a nucleosome.

DNA

2 Nucleosomes block transcription.

3 Remodeling proteins bind, disaggregating the nucleosome.

Remodeling protein

4 Now the initiation complex can bind to begin transcription.

Histone protein

Initiation complex

mRNA

ELONGATION

5 A second remodeling complex can bind to the DNA–histone complex…

Remodeling protein

mRNA

6 …allowing transcription without disaggregation.

14.17 Local Remodeling of Chromatin for Transcription
Initiation of transcription requires that nucleosomes disaggregate. During elongation, however, they remain intact.

tive X are methylated. Methylated DNA appears to bind certain chromosomal proteins, which may be responsible for heterochromatin formation. But this seems to occur after the actual inactivation event, making methylation a way to keep genes turned off.

The otherwise inactive X chromosome has one gene that is only lightly methylated and *is* transcriptionally active. This gene is called *XIST* (for *X* inactivation specific transcript), and it is heavily methylated on, and *not* transcribed from, the other, "active" X chromosome. The RNA transcribed from *XIST* does not leave the nucleus and is not an mRNA. Instead, it appears to bind to the X chromosome that transcribes it, and this binding somehow leads to a spreading of inactivation along the chromosome.

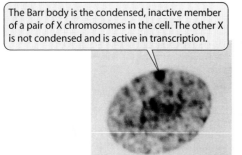

The Barr body is the condensed, inactive member of a pair of X chromosomes in the cell. The other X is not condensed and is active in transcription.

14.18 Barr Bodies in the Nuclei of Female Cells
The number of Barr bodies per nucleus is equal to the number of X chromosomes minus one. Thus males (XY) have no Barr body, whereas females (XX) have one.

A DNA sequence can move to a new location to activate transcription

In some instances, gene expression is regulated by the movement of a gene to a new location on the chromosome. An example of this mechanism is found in the yeast *Saccharomyces cerevisiae*. The haploid single cells of this fungus exist in two *mating types*, *a* and α, which fuse to form a diploid zygote. Although all yeast cells have an allele for each of these types, the allele that is expressed determines the mating type of the cell. In some yeasts, the mating type changes with almost every cell division cycle. How does it change so rapidly?

The yeast cell keeps the two different alleles (coding for type α and type *a*) at separate locations on its chromosome, away from a third site, the *MAT locus*. The two mating type alleles are usually transcriptionally silent because a repressor protein binds to them. However, when a copy of the α or *a* allele is inserted at the MAT region, the gene for proteins of the appropriate mating type is transcribed.

A change in mating type requires three steps:

▶ First, a new DNA copy of the nonexpressed allele is made (if the cell is now α, the new copy will be the *a* allele).

▶ Second, the current occupant of the MAT region (in this case, the α DNA) is removed by an enzyme.

▶ Third, the new allele (*a*) is inserted at the MAT region and transcribed. The *a* proteins are now made, and the mating type is changed.

DNA rearrangement is also important in producing the highly variable proteins that make up the human repertoire of antibodies, and in cancer, when inactive genes move to be adjacent to active promoters.

Selective gene amplification results in more templates for transcription

Another way for one cell to make more of a certain gene product than another cell does is to have more copies of the appropriate gene and to transcribe them all. The process of creating more copies of a specific gene in order to increase transcription is called **gene amplification**.

As described earlier, the genes that code for three of the four human ribosomal RNA's are linked together in a unit, and this unit is repeated several hundred times in the genome to provide multiple templates for rRNA synthesis (rRNA is the most abundant kind of RNA in the cell). In some circumstances, however, even this moderate repetition is not enough to satisfy the demands of the cell. For example, the mature eggs of frogs and fishes have up to a trillion ribosomes. These ribosomes are used for the massive protein synthesis that follows fertilization. The cell that differentiates into the egg contains fewer than 1,000 copies of the rRNA gene cluster, and would take 50 years to make a trillion ribosomes if it transcribed those rRNA genes at peak efficiency. How does the egg end up with so many ribosomes (and so much rRNA)?

The egg cell solves this problem by selectively amplifying its rRNA gene clusters until there are more than a million copies. In fact, this gene complex goes from being 0.2 percent of the total genome DNA to being 68 percent. These million copies transcribed at maximum rate (Figure 14.19) are just enough to make the necessary trillion ribosomes in a few days.

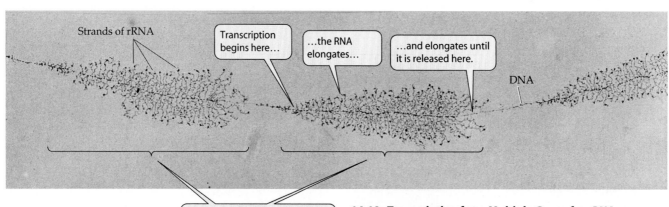

Strands of rRNA

Transcription begins here…

…the RNA elongates…

…and elongates until it is released here.

DNA

Multiple rRNA genes are actively transcribing rRNA precursors.

14.19 Transcription from Multiple Genes for rRNA
Elongating strands of rRNA transcripts form arrowhead-shaped regions, each centered on a strand of DNA that codes for rRNA.

The mechanism for selective overreplication of a single gene is not clearly understood, but it has important medical implications. As Chapter 18 will show, in some cancers, a cancer-causing gene called an oncogene becomes amplified. Also, in some tumors treated with a drug that targets a single protein, amplification of the gene for the target protein leads to an excess of that protein, and the cell becomes resistant to the prescribed dose of the drug.

Posttranscriptional Control

There are many ways to regulate the presence of mature mRNA in a cell even after a precursor has been transcribed. As we saw earlier, pre-mRNA can be processed by cutting out the introns and splicing the exons together. If the exons of the pre-mRNA are recombined in different ways by alternate splicing, different proteins can be synthesized. The longevity of mRNA in the cytoplasm can also be regulated. The longer an mRNA exists in the cytoplasm, the more of its coded protein can be made.

Different mRNA's can be made from the same gene by alternate splicing

Most primary transcripts contain several introns (see Figure 14.6). We have seen how the splicing mechanism recognizes the boundaries between exons and introns. What would happen if the β-globin pre-mRNA, which has two introns, was spliced from the start of the first intron to the end of the second? Not only the two introns but also the middle exon would be spliced out. An entirely new protein (certainly not a β-globin) would be made, and the functions of normal β-globin would be lost.

Alternate splicing can be a deliberate mechanism for generating a family of different proteins from a single gene. For example, a single pre-mRNA for the structural protein tropomyosin is alternatively spliced to give five different mRNA's and five different forms of tropomyosin found in five different tissues: skeletal muscle, smooth muscle, fibroblast, liver, and brain (Figure 14.20).

The stability of mRNA can be regulated

DNA, as the genetic material, must remain stable, and there are elaborate mechanisms for repairing it if it becomes damaged. RNA has no such repair system. After it arrives in the cytoplasm, mRNA is subject to breakdown catalyzed by ribonucleases, which exist both in the cytoplasm and in lysosomes. But not all eukaryotic mRNA's have the same life span. Differences in the stabilities of mRNA's provide another mechanism for posttranscriptional control of protein synthesis. The less time an mRNA spends in the cytoplasm, the less of its protein can be translated.

Tubulin is a protein that polymerizes to form microtubules, a component of the cytoskeleton (see Chapter 4). When a large pool of free tubulin is available in the cytoplasm, there is no particular need for the cell to make more of it. Under these conditions, some tubulin molecules bind to tubulin mRNA. This binding makes the mRNA especially susceptible to breakdown, and less tubulin is made.

Translational and Posttranslational Control

Is the amount of a protein in a cell determined by the amount of its mRNA? Recently, a survey was made of the relationships between mRNA's and proteins in yeast cells. Dozens of genes were surveyed. For about a third of them, the relationship between mRNA and protein held: More of one led to more of the other. But for two-thirds of the proteins, there was no apparent relationship. Their concentration in the cell must be determined by factors acting after the mRNA is made.

Just as proteins can control the synthesis of mRNA by binding to DNA, they can also control the translation of mRNA by binding to mRNA in the cytoplasm. This mode of control is especially important in long-lived mRNA's. A cell must not continue to make proteins that it does not

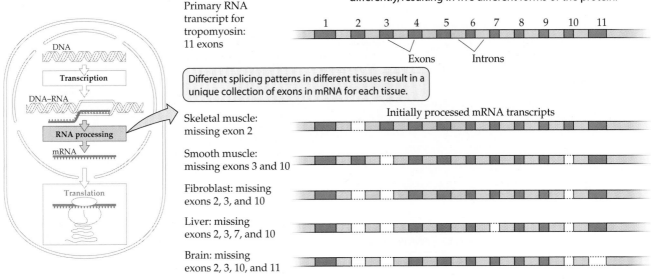

14.20 Alternate Splicing Results in Different mRNA's and Proteins
In mammals, the protein tropomyosin is coded for by a gene that has 11 exons. Different tissues splice tropomyosin pre-mRNA differently, resulting in five different forms of the protein.

Primary RNA transcript for tropomyosin: 11 exons

Exons   Introns

Different splicing patterns in different tissues result in a unique collection of exons in mRNA for each tissue.

Initially processed mRNA transcripts

Skeletal muscle: missing exon 2

Smooth muscle: missing exons 3 and 10

Fibroblast: missing exons 2, 3, and 10

Liver: missing exons 2, 3, 7, and 10

Brain: missing exons 2, 3, 10, and 11

DNA

Transcription

DNA–RNA

RNA processing

mRNA

Translation

need. For example, mammalian cells respond to certain stimuli by making cyclins, which stimulate the events of the cell cycle. If the mRNA for a cyclin is still in the cytoplasm and available for translation long after the cyclin is needed, cyclin will be made and released inappropriately. Its presence might cause a target cell population to divide inappropriately, forming a tumor.

The translation of mRNA can be controlled

One way to control translation is by the capping mechanism on mRNA. As already noted, mRNA is capped at its 5′ end by a modified guanosine molecule (see Figure 14.11). Messenger RNA's that have unmodified caps are not translated. For example, stored mRNA in the oocyte of the tobacco hornworm moth has the guanosine added to its 5′ end, but the G is not modified. Hence, this stored mRNA is not translated. However, after fertilization, the cap is modified, allowing the mRNA to be translated to produce proteins needed for early embryogenesis.

Free iron ions (Fe^{2+}) within a mammalian cell are bound by a storage protein, *ferritin*. When iron is in excess, ferritin synthesis rises dramatically. Yet the amount of ferritin mRNA remains constant. The increase in ferritin synthesis is due to an increased rate of mRNA translation. When the iron level in the cell is low, a *translational repressor* protein binds to ferritin mRNA and prevents its translation by blocking its attachment to a ribosome. When iron levels rise, the excess iron binds to the repressor and alters its three-dimensional structure, causing it to detach from the mRNA, and translation of ferritin proceeds.

Translational control also acts in the synthesis of hemoglobin. As we described earlier, hemoglobin consists of four polypeptide chains and a nonprotein pigment, heme. If heme synthesis does not equal globin synthesis, some polypeptide chains stay free in the cell, waiting for a heme partner. Excess heme in the cell increases the rate of translation of globin mRNA by removing a block to the initiation of translation at the ribosome, helping to maintain the balance.

The proteasome controls the longevity of proteins after translation

We have considered how gene expression may be regulated by the control of transcription, RNA processing, and translation. However, the story does not end here, because most gene products—proteins—are modified after translation. Some of these changes are permanent, such as the addition of sugars (glycosylation), the addition of phosphate groups, or the removal of a signal sequence after a protein has crossed a membrane (see Figure 12.14).

An important way to regulate the action of a protein in a cell is to regulate its *lifetime* in the cell. Proteins involved in cell division (e.g., the cyclins) are hydrolyzed at just the right moment to time the sequence of events. Proteins identified for breakdown are often covalently linked to a 76-amino acid protein called *ubiquitin* (so called because it is ubiquitous, or widespread). The protein–ubiquitin complex then binds to a huge complex of several dozen polypeptide chains called a **proteasome** (Figure 14.21). The entryway to this "molecular chamber of doom" is a hollow cylinder, with ATPase activity, that cuts off the ubiquitin for recycling and unfolds its targeted protein victim. The protein then passes by three different proteases (thus the name of the complex) that digest it into small peptides and amino acids.

The cellular concentrations of many proteins are determined not by differential transcription of their genes, but by their degradation in proteasomes. For example, cyclins are degraded at just the right time during the cell cycle (see Figure 9.5). Transcription factors are broken down after they are used, lest the affected genes be always "on." Abnormal proteins are often targeted for destruction by a quality control mechanism. Human papillomavirus, which causes cervical cancer, targets the cell division inhibitory protein p53 for proteasomal degradation, so that unregulated cell division—and cancer—results.

14.21 The Proteasome Breaks Down Proteins
Proteins targeted for breakdown are bound to ubiquitin, which "leads" them to the proteasome, a complex composed of many polypeptides.

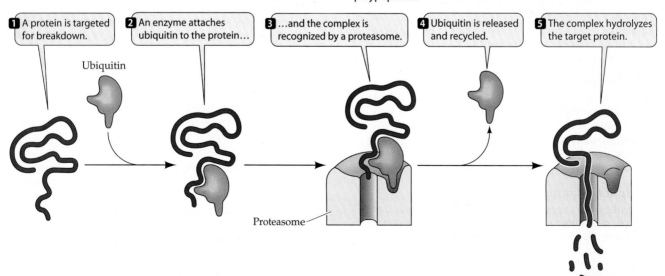

1 A protein is targeted for breakdown.

2 An enzyme attaches ubiquitin to the protein…

3 …and the complex is recognized by a proteasome.

4 Ubiquitin is released and recycled.

5 The complex hydrolyzes the target protein.

Ubiquitin

Proteasome

Chapter Summary

The Eukaryotic Genome

▶ Although eukaryotes have more DNA in their genomes than prokaryotes, in some cases there is no apparent relationship between genome size and organism complexity. **Review Figure 14.1**

▶ Unlike prokaryotic DNA, eukaryotic DNA is separated from the cytoplasm by being contained within a nucleus. The initial mRNA transcript of the DNA may be modified before it is exported from the cytoplasm. **Review Figure 14.2**

▶ The genome of the single-celled budding yeast contains genes for the same metabolic machinery as bacteria, with the addition of genes for protein targeting in the cell. **Review Table 14.2**

▶ The genome of the multicellular roundworm *Caenorhabditis elegans* contains genes required for intercellular interactions. **Review Table 14.3**

▶ The genome of the fruit fly has fewer genes than that of the roundworm. Many of its sequences are homologs of sequences on roundworm and mammalian genes.

Repetitive Sequences in the Eukaryotic Genome

▶ Highly repetitive DNA is present in up to millions of copies of short sequences. It is not transcribed. Its role is unknown.

▶ Telomeric DNA is found at the ends of chromosomes. Some telomeric DNA may be lost during each DNA replication, eventually leading to chromosome instability and cell death. The enzyme telomerase catalyzes the restoration of the lost telomeric DNA. Most somatic cells lack telomerase and thus have limited life spans. **Review Figure 14.3**

▶ Some moderately repetitive DNA sequences, such as those coding for rRNA's, are transcribed. **Review Figure 14.4**

▶ Some moderately repetitive DNA sequences are transposable, or able to move about the genome. **Review Figure 14.5**

The Structures of Protein-Coding Genes

▶ A typical protein-coding gene has noncoding internal sequences (introns) as well as flanking sequences that are involved in the machinery of transcription and translation. **Review Figures 14.6, 14.8**

▶ Nucleic acid hybridization is an important technique for analyzing eukaryotic genes. **Review Figure 14.7**

▶ Some eukaryotic genes form families of related genes that have similar sequences and code for similar proteins. These related proteins may be made at different times and in different tissues. Some sequences in gene families are pseudogenes, which code for nonfunctional mRNA's or proteins. **Review Figure 14.9**

▶ Differential expression of different genes in the β-globin family ensures important physiological changes during human development. **Review Figure 14.10**

RNA Processing

▶ After transcription, the pre-mRNA is altered by the addition of a G cap at the 5′ end and a poly A tail at the 3′ end. **Review Figure 14.11**

▶ The introns are removed from the mRNA precursor by the spliceosome, a complex of RNA's and proteins. **Review Figure 14.12**

Transcriptional Control

▶ Eukaryotic gene expression can be controlled at the transcriptional, posttranscriptional, translational, and posttranslational levels. **Review Figure 14.13**

▶ The major method of control of eukaryotic gene expression is selective transcription, which results from specific proteins binding to regulatory regions on DNA.

▶ A series of transcription factors must bind to the promoter before RNA polymerase can bind. Whether RNA polymerase will initiate transcription also depends on the binding of regulatory proteins, activator proteins (which are bound by enhancers and stimulate transcription), and repressor proteins (which are bound by silencers and inhibit transcription). **Review Figures 14.14, 14.15**

▶ The simultaneous control of widely separated genes is possible through proteins that bind to common sequences in their promoters. **Review Figure 14.16**

▶ The DNA-binding domains of most DNA-binding proteins have one of four structural motifs: helix-turn-helix, zinc finger, leucine zipper, or helix-loop-helix.

▶ Remodeling of chromatin occurs during transcription. **Review Figure 14.17**

▶ Heterochromatin is a condensed form of DNA that cannot be transcribed. It is found in the inactive X chromosome of female mammals. **Review Figure 14.18**

▶ The movement of a gene to a new location on a chromosome may alter its ability to be transcribed, as in the change from one mating type to another in yeast.

▶ Some genes become selectively amplified, and the extra copies result in increased transcription of their protein product. **Review Figure 14.19**

Posttranscriptional Control

▶ Because eukaryotic genes have several exons, alternate splicing can be used to produce different proteins. **Review Figure 14.20**

▶ The stability of mRNA in the cytoplasm can be regulated by the binding of proteins.

Translational and Posttranslational Control

▶ Translational repressors can inhibit the translation of mRNA.

▶ Proteasomes degrade proteins targeted for breakdown. **Review Figure 14.21**

For Discussion

1. In rats, a gene 1,440 bp long codes for an enzyme made up of 192 amino acid units. Discuss this apparent discrepancy. How long would the initial and final mRNA transcripts be?

2. The activity of the enzyme dihydrofolate reductase (DHFR) is high in some tumor cells. This activity makes the cells resistant to the anticancer drug methotrexate, which targets DHFR. Assuming that you had the complementary DNA for the gene that encodes DHFR, how would you show whether this increased activity was due to increased transcription of the single-copy DHFR gene or to amplification of the gene?

3. Describe the steps in the production of a mature, translatable mRNA from a eukaryotic gene that contains introns. Compare this to the situation in prokaryotes (see Chapter 13).

4. A certain protein-coding gene has three introns. How many different proteins can be made from alternate splicing of the pre-mRNA transcribed from this gene?

5. Most somatic cells in mammals do not express telomerase. Yet the germ line cells that produce gametes by meiosis do express this enzyme. Explain.

15

Cell Signaling and Communication

Carol had intended to complete the report summarizing her semester's biology lab project long before it was due; it would, after all, count for significant points toward her grade. But between her other courses and a few interesting distractions, she kept putting off the report until "next week." Finally, at 9:30 the night before it was due, she sat down to create the 20-page report. By 11:00 she realized she still had hours of work to do. She made a frantic call to her lab partner, asking him to provide her with some data from one of their experiments. And Carol filled the coffee-brewing machine in her dorm room, knowing she would need a "caffeine jolt" to keep her awake so she could finish the job.

To understand how a "caffeine jolt" works, we must understand the pathways by which the body's cells respond to certain signals in their environment. The signals might be chemicals traveling between brain cells, or hormones produced in response to an outside event. There are three sequential processes involved in the cell's response to any signal. First, the signal binds to a receptor protein. Second, the binding of the signal causes a message to be conveyed to the cell's cytoplasm and amplified. Third, the cell changes its activity in response to the signal.

Caffeine acts in different ways in different tissues. First, a tired person's brain produces adenosine molecules that bind to specific receptor proteins, resulting in decreased brain activity and increased drowsiness. Caffeine's molecular structure is similar to that of adenosine, so it occupies the adenosine receptors without inhibiting brain cell function, and alertness is restored. Then, in the heart and liver, caffeine stimulates a pathway inside cells so that they do not need hormonal stimulation. In the heart, the result is an increased rate of beating; the liver is stimulated to release glucose into the bloodstream.

We begin this chapter with a discussion of signals that affect cells. As you will see, these range from chemicals such as hormones to physical entities such as light. Whatever the signal, it will affect a cell only if that cell has a receptor protein that binds to the signal. In addition to binding the signal, the receptor must somehow communicate to the rest of the cell that binding has occurred. This process of signal transduction often involves special small molecules called second messengers, which initiate a series of events that amplify the signal. As a result, the third phase—alteration of cell function—may involve many instances of the same event, such as the opening of an ion channel in the plasma membrane or increased transcription of a number of genes.

We close with a description of how cells communicate with one another directly through specialized channels in their adjacent plasma membranes.

Signals

Both prokaryotic and eukaryotic cells process information from their environment. This information can be in the form of a physical stimulus, such as the light reaching your eye as you read this book, or chemicals that bathe a cell, such as lactose in the medium surrounding *E. coli*. It may come from outside the organism, such as the ions dissolved in the water that bathes plant roots, or from a neighboring cell within the organism, as occurs in the heart, where thousands of cells contract in unison by transmitting signals to one another.

Multiple Signals at Work
The computer, the telephone, and the coffee all send signals to the brain. All may be necessary for the successful completion of a research project.

Of course, the mere presence of a signal does not mean that a cell will respond, just as you do not pay close attention to every sound in your environment as you study. To respond, the cell must have a specific **receptor protein** that can bind to the signal. In this section, we describe some of the signals different cells respond to and look at one model system of signal transduction.

Cells receive signals from the physical environment and from other cells

The physical environment is full of signals. Our sense organs allow us to respond to light, odors, touch, and sound. Bacteria and protists respond to even minute chemical changes in their environment. Plants respond to light as a signal. For example, at sunset, at night, or in the shade, not only the amount of sunlight but also the spectrum of light reaching the surface of Earth differs from that of full sunlight in the daytime. These variations are signals that affect plant growth and reproduction. Even magnetism can be a signal: Some bacteria and birds orient themselves to the Earth's magnetic poles, like a needle on a compass.

But a cell inside a large organism is far from the exterior environment. Instead, its environment is other cells and extracellular fluids. Cells receive their nutrients from, and

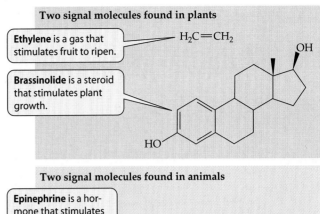

Two signal molecules found in plants

Ethylene is a gas that stimulates fruit to ripen.

Brassinolide is a steroid that stimulates plant growth.

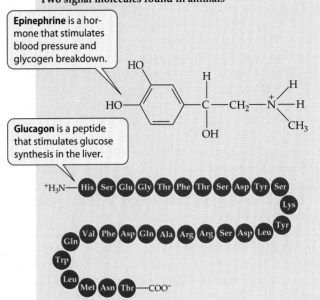

Two signal molecules found in animals

Epinephrine is a hormone that stimulates blood pressure and glycogen breakdown.

Glucagon is a peptide that stimulates glucose synthesis in the liver.

15.2 A Variety of Biological Signals
Many different kinds of molecules can serve as biological signals. The structure of glucagon is simplified so that a "bead" represents an amino acid, each about the size of the epinephrine molecule whose chemical formula is shown.

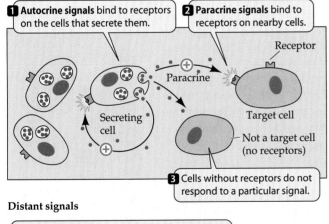

Local signals

1 Autocrine signals bind to receptors on the cells that secrete them.

2 Paracrine signals bind to receptors on nearby cells.

Receptor

Paracrine

Secreting cell

Target cell

Not a target cell (no receptors)

3 Cells without receptors do not respond to a particular signal.

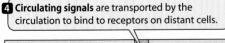

Distant signals

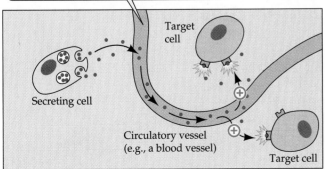

4 Circulating signals are transported by the circulation to bind to receptors on distant cells.

Target cell

Secreting cell

Circulatory vessel (e.g., a blood vessel)

Target cell

15.1 Chemical Signaling Systems
A signal molecule can act on the same cell that produces it, or on a nearby cell. Most signals act on distant cells, to which they are transported by the organism's circulatory system.

pass their wastes into, extracellular fluids. Cells also receive signals—mostly chemical signals—from their extracellular fluid environment. Most of these chemical signals come from other cells. In animal cells, they include hormones (Chapter 41), neurotransmitters (Chapter 44), and chemical messages from the immune system (Chapter 19). Cells also respond to chemical signals coming from the environment via the digestive and respiratory systems. And cells can respond to chemicals, such as CO_2 and H^+, whose presence in the extracellular fluids results from the metabolic activities of other cells.

Inside a large organism, chemical signals reach a target cell by local diffusion or by circulation within the blood. *Autocrine* signals affect the cells that make them. *Paracrine* signals diffuse to nearby cells. Signals to distant cells usually travel through the circulatory system (Figure 15.1).

The biological signals cells receive are diverse (Figure 15.2). In each case, the cell must be able to receive or sense the signal and respond to it. Depending on the cell and the signal, the responses range from entering the cell division

cycle to heal a wound, to moving to a new location in the embryo to form a tissue, to releasing enzymes that digest food, to sending messages to the brain about the book you are reading. Clearly, signaling underlies a lot of biology. The whole process, from signal detection to final response, is called a **signal transduction pathway**.

Signaling involves a receptor, transduction, and effects

In Chapter 13, we saw that bacteria respond to changes in nutrients in their environment by altering their transcription of genes, as in the *lac* operon. In addition to responding to such changes, these same bacteria must be able to sense and respond to non-nutritive changes in their environment, such as changes in osmotic concentration. For example, if the solute concentration around *E. coli* rises far above that inside the cell, the law of diffusion tells us that water will diffuse out of the cell and solutes into the cell. Since the cell must maintain homeostasis in its cytoplasm, it must perceive and respond to this environmental change. The way in which this one-celled organism responds to such signals has much in common with signaling in more complex animals and plants (Figure 15.3).

The receptor protein in *E. coli* for osmotic changes is called EnvZ. It is a transmembrane protein that extends through the bacterium's plasma membrane into the space between the plasma membrane and a highly porous outer membrane that forms a complex with the cell wall. When the solute concentration of the extracellular environment rises, so does the concentration in the environment between the two membranes. This change in its aqueous medium causes the part of the receptor protein sticking into the intermembrane space to undergo a conformational change.

We saw in Chapter 6 that changing the tertiary structure of one part of a protein often leads to changes in distant parts of the protein. In the case of the bacterial EnvZ receptor, the conformational change in the intermembrane region of the protein is transmitted to the region that lies in the bacterium's cytoplasm. This change initiates the events of signal transduction. EnvZ becomes an active *protein kinase*, which catalyzes the addition of a phosphate group from ATP to one of EnvZ's own histidine residues. In other words, EnvZ phosphorylates itself.

The charged phosphate group added to the histidine causes the cytoplasmic tail of the EnvZ protein to change its shape again. It now binds to a second protein, OmpR,

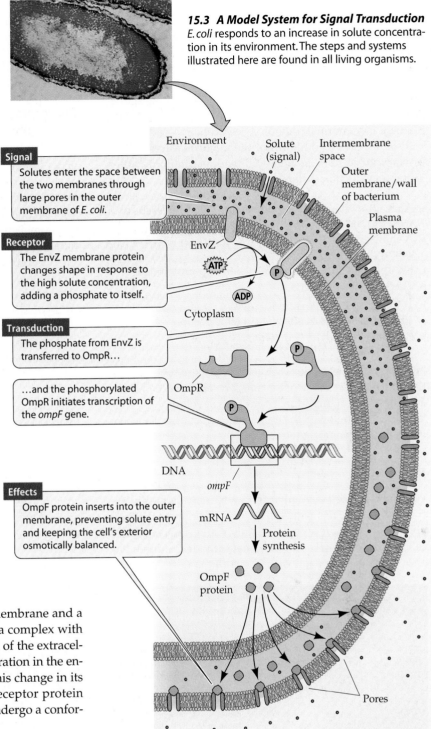

15.3 A Model System for Signal Transduction
E. coli responds to an increase in solute concentration in its environment. The steps and systems illustrated here are found in all living organisms.

Signal
Solutes enter the space between the two membranes through large pores in the outer membrane of *E. coli*.

Receptor
The EnvZ membrane protein changes shape in response to the high solute concentration, adding a phosphate to itself.

Transduction
The phosphate from EnvZ is transferred to OmpR...

...and the phosphorylated OmpR initiates transcription of the *ompF* gene.

Effects
OmpF protein inserts into the outer membrane, preventing solute entry and keeping the cell's exterior osmotically balanced.

Environment · Solute (signal) · Intermembrane space · Outer membrane/wall of bacterium · Plasma membrane · EnvZ · ATP · ADP · Cytoplasm · OmpR · DNA · *ompF* · mRNA · Protein synthesis · OmpF protein · Pores

which takes the phosphate group from EnvZ. OmpR also changes its structure due to this phosphorylation. This change is a key event in signaling for three reasons. First, the signal on the outside of the cell has now been *transduced* to a protein totally within the cell's cytoplasm. Second, OmpR can *do* something, and that is to bind to a promoter on *E. coli* DNA adjacent to the DNA coding for the protein OmpF. This binding begins the final phase of this signaling pathway: the *effect* of the signal, which is an alteration in cell function. Third, the signal has been *amplified*. EnvZ can alter the structure of many OmpR molecules.

Phosphorylated OmpR has the correct three-dimensional structure to bind to DNA at the *ompF* promoter, resulting in an increase in transcription of that gene. Translation of *ompF* mRNA results in the production of OmpF protein, which is inserted into the outer membrane and prevents solutes from entering the intermembrane space. Thus the *E. coli* cell can go on behaving just as if the environment has a normal osmotic concentration.

It is important to highlight the major features of this prokaryotic system, as they will reappear in many other signal transduction systems in animals and plants:

▶ A receptor changes its conformation upon interacting with the signal.
▶ A conformational change exposes a protein kinase activity.
▶ Phosphorylation alters the function of a protein.
▶ The signal is amplified.
▶ Transcription factors are activated.
▶ Altered synthesis of specific proteins occurs.
▶ Protein action alters cell activity.

Signal transduction pathways featuring these seven activities occur in all types of organisms. The emergence of these activities was an important event in the evolution of cellular life, as they allowed the organism to react to and survive in a rapidly changing environment.

Receptors

While a given cell is bombarded with many signals, it responds to only a few of them. The reason for this is that any given cell makes receptors for only some signals. Which cells make which receptors is genetically determined: If a cell transcribes the gene encoding a particular receptor and the resulting mRNA is translated, the cell will have that receptor. A receptor protein binds to a signal in much the same way as an enzyme binds to a substrate or a membrane transport protein binds to the molecule it is transporting.

Receptors have specific binding sites for their signals

A signaling molecule, usually called a **ligand**, fits into a site on its receptor much like a substrate fits into the active site of an enzyme (Figure 15.4). Whether the receptor protrudes from the plasma membrane surface or is located in the cytoplasm, the result of ligand binding is the same: The receptor protein changes its three-dimensional structure and initiates a cellular response. The ligand does not contribute further to this response. In fact, the ligand usually is not metabolized into useful products. Its role is purely to "knock on the door." This is in sharp contrast to enzyme–substrate interactions, in which the whole purpose is to change the substrate into a useful product.

Receptors bind to their ligands according to chemistry's law of mass action:

$$R + L \rightleftharpoons RL$$

This means that the binding is reversible, although for most ligand/receptor complexes, the equilibrium point is far to the right—that is, favoring binding. Release of the ligand is important because if it does not happen, the receptor will be continuously stimulated.

Just as with enzymes, *inhibitors* can bind to the ligand site on a receptor protein. Both natural and artificial inhibitors of receptor binding are important in medicine.

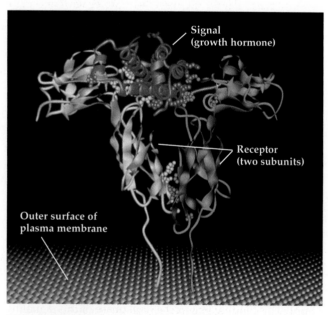

15.4 A Signal Bound to Its Receptor
Only the extracellular regions of the human growth hormone receptor are shown.

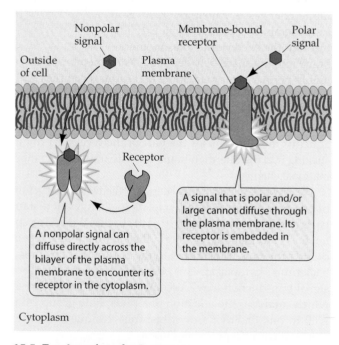

15.5 Two Locations for Receptors
Receptors can be located on the plasma membrane or in the cytoplasm of the cell.

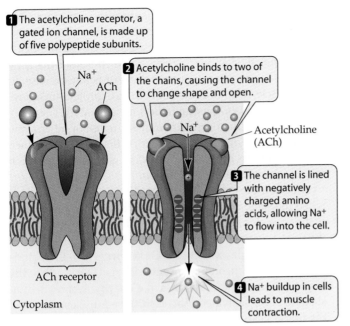

1 The acetylcholine receptor, a gated ion channel, is made up of five polypeptide subunits.

2 Acetylcholine binds to two of the chains, causing the channel to change shape and open.

Acetylcholine (ACh)

3 The channel is lined with negatively charged amino acids, allowing Na$^+$ to flow into the cell.

4 Na$^+$ buildup in cells leads to muscle contraction.

ACh receptor

Cytoplasm

15.6 An Ion Channel Receptor
The acetylcholine receptor is a channel for sodium ions that resembles a gate. The gate opens when its ligand, acetylcholine, binds to it, allowing Na$^+$ to flow into the cell.

There are several types of receptors

A major division among receptors is in their cellular location, which largely depends on the nature of their ligands. The chemistry of signals is quite variable, but they can be divided into two classes: those that are nonpolar, and can cross the plasma membrane and enter the cell, and those that are large or polar, and cannot cross the membrane (Figure 15.5). Estrogen, for example, is a steroid and can easily diffuse across the plasma membrane and enter the cell; it binds to a receptor inside the cytoplasm. Insulin, on the other hand, is

1 The α subunit binds insulin (the signal).

2 The β subunit transmits a signal from bound insulin to the cytoplasm.

Insulin

Outside of cell

Membrane

Phosphate groups

3 The insulin signal activates the receptor's protein kinase domain in the cytoplasm.

Insulin receptor

Cytoplasm

Insulin response substrate (IRS)

4 Protein kinases from the receptor phosphorylate insulin-response substrates, triggering other chemical responses inside the cell.

Cellular responses

a protein hormone that cannot diffuse through the plasma membrane; instead, it binds to a receptor that is a transmembrane protein with an extracellular binding region.

In more complex eukaryotes, there are three well-studied types of receptors on plasma membranes: ion channels, protein kinases, and G protein-linked receptors.

ION CHANNELS. In the plasma membranes of many types of cells, there are channel proteins that can be open or closed. These **ion channels** act as "gates," allowing ions such as Na$^+$, K$^+$, Ca^{2+}, or Cl$^-$ to enter or leave the cell. The gate-opening mechanism is an alteration in the three-dimensional structure of the channel protein upon ligand binding. Each ion channel has its own signal. These signals include sensory stimuli, such as light and sound, voltage differences across the plasma membrane, and chemical ligands such as small molecules and hormones.

An example of a gated ion channel is the acetylcholine receptor (Figure 15.6). This receptor is located at the plasma membranes of vertebrate skeletal muscle cells and binds the ligand acetylcholine, which is released from nerve cells. When two molecules of acetylcholine bind to the receptor, it opens for about a thousandth of a second. This is enough time for Na$^+$, which is more concentrated outside the cell than inside, to rush into the cell. The change in Na$^+$ concentration in the cell ultimately results in muscle contraction. Right after the channel opens, the ligand is released from the receptor and then degraded. This makes the receptor (and the cell) responsive to the next signal, so that the muscle can contract again.

PROTEIN KINASES. Like the activated EnvZ protein of *E. coli*, some eukaryotic receptor proteins become kinases when they are activated: That is, they catalyze the transfer of a phosphate group from ATP to a protein. The targets for the protein kinase activity are both the receptor itself and cytoplasmic molecules, which alter their shape and then act to change the cell's activities. While histidine is phosphorylated by EnvZ in *E. coli*, the amino acid targeted by the protein kinase receptors of animal cells is usually tyrosine. In plants, either serine or threonine is phosphorylated.

Insulin is a protein hormone made by the mammalian pancreas. The receptor for insulin is a protein consisting of two copies each of two different polypeptide subunits (Figure 15.7). As with acetylcholine, two molecules of insulin must bind to the receptor. After binding insulin on its extracellular surface, the receptor changes

15.7 A Protein Kinase Receptor
The mammalian hormone insulin does not enter the cell, but is bound by a membrane receptor protein with four subunits (two α and two β). The β subunits transmit a signal that changes the cytoplasmic end of the receptor protein, activating a protein kinase domain and triggering further responses by the cell, eventually resulting in the transport of glucose across the membrane.

its shape to expose a cytoplasmic protein kinase active site. Like the EnvZ receptor described above, the insulin receptor self-phosphorylates. Then, as a protein kinase signal, it targets certain cytoplasmic proteins, appropriately called insulin response substrates. These proteins then initiate many cell responses, including the insertion of glucose transporters into the plasma membrane.

G PROTEIN-LINKED RECEPTORS. A third category of eukaryotic plasma membrane receptor is the seven-spanning G protein-linked receptors. This long name identifies a fascinating group of receptors, all of which are composed of a single protein with seven regions that pass through the lipid bilayer, separated by short loops that extend either outside or inside the cell. Ligand binding on the extracellular side changes the shape of the receptor's cytoplasmic region, opening up a binding site for a mobile membrane protein.

This membrane protein, known as a **G protein**, has two important binding sites: one for the G protein-linked receptor, and the other for the nucleotide GDP/GTP (Figure 15.8). G proteins have several polypeptide subunits. When the G protein binds to the activated receptor, it also binds GTP to one of its subunits. At the same time, the ligand is released from the extracellular side of the receptor. The GTP-bound subunit of the G protein now separates from the parent G protein, diffusing in the plane of the lipid bilayer until it encounters an *effector protein* to which it can bind. Effector proteins are what their name implies: They cause an effect. The binding of the GTP-bearing G protein subunit activates the effector—which may be an ion channel or an enzyme—thereby causing changes in cell function.

After binding to the effector protein, the GTP on the G protein is hydrolyzed to GDP. The now inactive G protein subunit separates from the effector protein. The G protein subunit must form a complex with another subunit before binding to yet another activated receptor. When this activated receptor is bound, the G protein exchanges its GDP for GTP, and the cycle begins again.

By means of their diffusing subunits, G proteins can either activate or inhibit an effector. An example of an *activating* response involves the receptor for epinephrine (adrenaline), the famous "fight-or-flight" hormone made by the adrenal gland in response to stress or heavy exercise. In heart muscle, this hormone binds to its G protein-linked receptor, causing a G protein to become activated. The GTP-bound subunit then activates a membrane-bound enzyme to produce a small molecule, cyclic AMP (see below), which has many effects on the cell, including glucose mobilization for energy and muscle contraction.

An example of G protein-mediated *inhibition* occurs when the same hormone, epinephrine, binds to its receptor in the smooth muscle cells surrounding blood vessels lining the digestive tract. Again, the epinephrine-bound receptor changes its shape and binds a G protein, which then binds GTP, and the G protein subunit with its GTP binds to the target enzyme. But in this case, the enzyme is inhibited instead of being activated. As a result, the muscles relax, and the blood vessel diameter increases, allowing more nutrients to be carried away from the digestive system to the rest of the body. Thus the same signal and initial signaling mechanism can have different consequences in different cells, depending on the nature of the responding cell.

CYTOPLASMIC RECEPTORS. Not all signals act at the plasma membrane. Some signals diffuse across the lipid bilayer of the plasma membrane and enter the cytoplasm. In these cases, the receptor protein lies inside the cytoplasm. Steroid hormones in animals, for example, enter the cytoplasm and bind to steroid hormone receptors. Binding to the ligand

15.8 A G Protein-Linked Receptor

Binding of an extracellular signal—in this case, a hormone—causes the activation of a seven-segmented G protein. The G protein then activates an enzyme that catalyzes a reaction in the cytoplasm, amplifying production of the product. The figure is a generalized diagram that could apply to any of the large family of G proteins and the signals they react with.

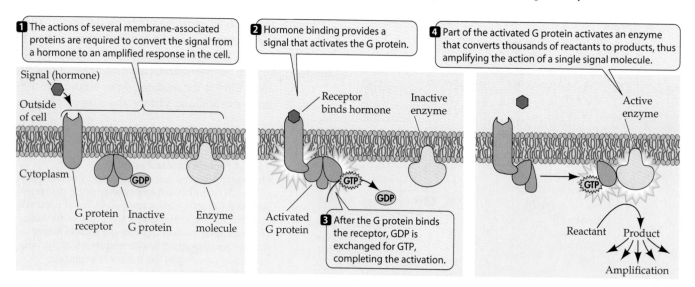

1 The actions of several membrane-associated proteins are required to convert the signal from a hormone to an amplified response in the cell.

2 Hormone binding provides a signal that activates the G protein.

4 Part of the activated G protein activates an enzyme that converts thousands of reactants to products, thus amplifying the action of a single signal molecule.

Signal (hormone)

Outside of cell

Cytoplasm

G protein receptor Inactive G protein Enzyme molecule

Receptor binds hormone Inactive enzyme

Activated G protein

3 After the G protein binds the receptor, GDP is exchanged for GTP, completing the activation.

Active enzyme

Reactant Product

Amplification

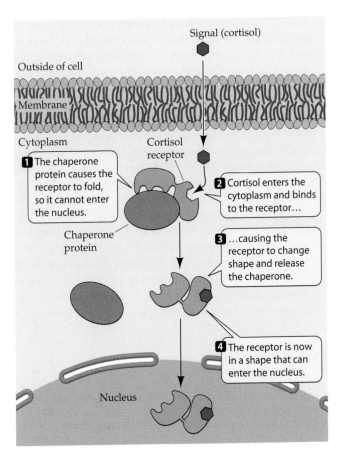

15.9 A Cytoplasmic Receptor
The receptor for cortisol is bound to a chaperone protein. Binding of the signal (which diffuses directly through the membrane) releases the chaperone and allows the receptor protein to enter the cell's nucleus, where it functions as a transcription factor.

causes the receptor to change its shape so that it can enter the cell nucleus, where it acts as a transcription factor (Figure 15.9). But this general view is somewhat simplified. The receptor for the hormone cortisol, for example, is bound to a chaperone protein, which blocks it from entering the nucleus. Binding of the hormone causes the receptor to change its shape so that the chaperone is released. This allows the receptor, which is a transcription factor, to fold into an appropriate configuration for entering the nucleus and initiating transcription.

Transducers

As previously mentioned, the same signal may produce different responses in different tissues. Acetylcholine, for example, can bind to receptors on skeletal muscle cells, where it stimulates muscle contraction, but on heart muscle cells, it slows contraction. These different responses to the same ligand/receptor complex are mediated by the events of signal transduction. These events, which are critical to the cell's response, may be either direct or indirect.

Direct transduction is a property of the receptor itself and occurs at the plasma membrane. In *indirect transduction*, which is more common, another molecule, termed a second messenger, mediates the interaction between receptor binding and cellular reaction. In neither case is transduction a single event. Rather, the signal initiates a cascade of events, in which proteins interact with other proteins until the final responses are achieved (Figure 15.10).

15.10 Direct and Indirect Signal Transduction
(a) All the events of direct transduction occur on the plasma membrane, at or near the receptor protein. (b) In indirect transduction, the binding of the signal to the receptor triggers formation of a "second messenger" molecule that works in the cytoplasm. It is the second messenger that sets off the necessary biochemical reactions.

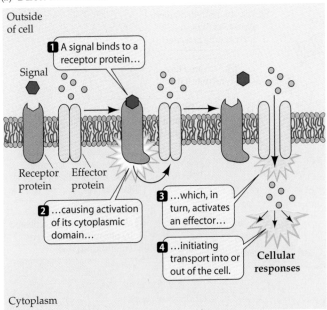

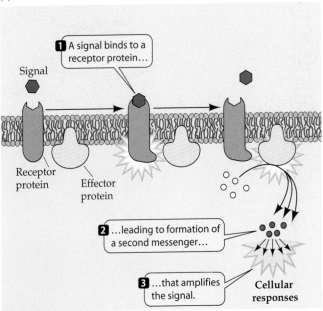

Protein kinase cascades amplify a response to receptor binding

We have seen that when a signal binds to a protein kinase receptor, the receptor changes its structure to expose a protein kinase active site, which catalyzes the phosphorylation of target proteins. This process is an example of direct signal transduction. Protein kinase receptors are important in binding ligands that stimulate cell division in both plants and animals. In Chapter 9, we described growth factors that were external inducers of the cell cycle. These factors stimulate cell division by binding to protein kinase receptors.

The complete signal transduction pathway that occurs after a protein kinase receptor is activated was worked out from studies on a cell that went wrong. Many human bladder cancers contain an abnormal protein called Ras (so named because it was first isolated from a *ra*t sarcoma tumor). Investigations of these bladder cancers showed that the ras protein was a G protein, but was always active because it was permanently bound to GTP. So the abnormal ras protein caused continuous tumor cell division. If the cancer cells' Ras protein was inhibited, the tumor cells stopped dividing. This discovery has led to a major effort to develop specific Ras inhibitors for cancer treatment.

What does Ras do in normal, noncancerous cells? When scientists treated cells in a laboratory dish with both a Ras inhibitor and a growth factor, the expected cell division did

not occur. Since growth factor binding is the first event in stimulating cell division, these results meant that Ras, like other G proteins, must be an intermediary between signal (growth factor) and response (cell division). After this discovery, the challenge was to work out what the activated growth factor receptor did to Ras, and what Ras did to stimulate further events in signal transduction.

This signaling pathway has been worked out, and it is an excellent example of a more general phenomenon, a **protein kinase cascade** (Figure 15.11). Such cascades are key to the external regulation of many cellular activities. Indeed, as we saw in Chapter 14, the eukaryotic genome codes for hundreds, even thousands, of such kinases. The unbound receptors for growth factors exist in the plasma membrane as separate polypeptide chains (subunits). When the growth factor signal binds to a subunit, it associates with another subunit to form a dimer, which changes shape to expose a protein kinase active site. The kinase activity sets off a series of events, activating several other protein ki-

15.11 A Protein Kinase Cascade
In a protein kinase cascade, a series of proteins becomes sequentially activated. In this example, the growth factor receptor protein stimulates the G protein Ras, which mediates a cascading series of reactions. The final product of the cascade, MAP kinase (MAPk), enters the nucleus and causes changes in transcription. Inactive forms of the proteins are on the left, activated forms are on the right.

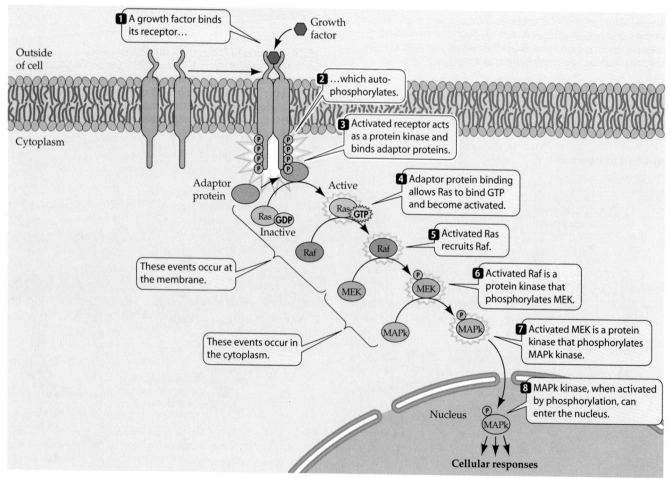

1. A growth factor binds its receptor…
2. …which auto-phosphorylates.
3. Activated receptor acts as a protein kinase and binds adaptor proteins.
4. Adaptor protein binding allows Ras to bind GTP and become activated.
5. Activated Ras recruits Raf.
6. Activated Raf is a protein kinase that phosphorylates MEK.
7. Activated MEK is a protein kinase that phosphorylates MAPk kinase.
8. MAPk kinase, when activated by phosphorylation, can enter the nucleus.

Outside of cell

Growth factor

Cytoplasm

Adaptor protein

Ras GDP Inactive

Active

Ras GTP

Raf

Raf

These events occur at the membrane.

MEK

MEK

MAPk

MAPk

These events occur in the cytoplasm.

Nucleus

MAPk

Cellular responses

nases in turn. The final phosphorylated, activated protein—MAP kinase—moves into the nucleus and phosphorylates target proteins necessary for cell division.

Is the protein kinase cascade pathway universal in eukaryotes? Genome sequencing of the plant *Arabidopsis* has revealed proteins with strong homologies to many of the proteins in the mammalian pathway. A number of proteins that resemble tyrosine kinase receptors are also present. There is even a Ras-like protein. The functions of this pathway in *Arabidopsis* are under investigation.

Protein kinase cascades are very useful for signal transduction for three reasons:

▶ At each step in the cascade of events, the signal is amplified. Because each newly activated protein kinase is an enzyme, each can catalyze the phosphorylation of many target proteins.

▶ The information from the signal that was originally at the plasma membrane is communicated to the nucleus.

▶ The multitude of steps provides some specificity to the process. As we have seen with epinephrine, signal binding and receptor activation do not result in the same response in all cells. Different target proteins at every step in the cascade can provide variability of response.

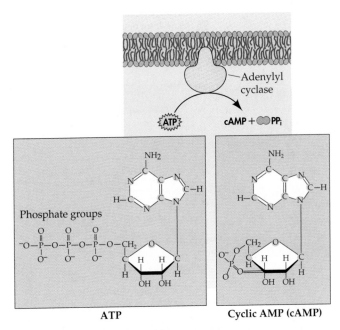

15.12 The Formation of Cyclic AMP
The formation of cAMP from ATP is catalyzed by adenylyl cyclase, an enzyme that is activated by G proteins.

Cyclic AMP is a common second messenger

As we have seen, protein kinase receptors stimulate the protein kinase cascade right at the plasma membrane. However, the stimulation of events in the cell is more often indirect. In a series of clever experiments, Earl Sutherland, Edwin Krebs, and Edmond Fischer showed that in many cases, there is a small, water-soluble chemical messenger between the membrane receptor and cytoplasmic events. These researchers were investigating the activation of the liver enzyme phosphorylase by the hormone epinephrine. Phosphorylase catalyzes the breakdown of glycogen stored in the liver so that carbohydrate can be released to the blood to fuel the fight-or-flight response.

The researchers found that phosphorylase could be activated in liver cells that had been broken open, but only if the entire cell contents, including the plasma membrane, were present. Epinephrine had bound to the plasma membrane, and phosphorylase was present in the cytoplasm. So they tried the steps of this experiment in sequence. First, they incubated membranes of broken liver cells with epinephrine. Then they removed the membranes, but kept the solution in which the membranes had been bathed. When they added this solution to the cytoplasm, the phosphorylase enzyme present became activated! Hormone binding to the membrane receptor had caused the production of a small, water-soluble molecule that then diffused to the cytoplasm, where it activated the enzyme.

This small molecule was identified as **cyclic AMP (cAMP)**, which we also encountered in the *lac* operon regulatory system in *E. coli*, where cAMP was working as a second messenger. **Second messengers** are substances re-

leased into the cytoplasm after the first messenger—the signal—binds its receptor.

In contrast to the uniqueness of receptor binding, *second messengers affect many processes in the cell, and allow a cell to respond to a single event at the plasma membrane with many events inside the cell.* Like the kinase cascade, second messengers amplify the signal—a single epinephrine molecule leads to the production of several dozen molecules of cAMP, which then activate many enzyme targets.

Adenylyl cyclase, the enzyme that catalyzes the formation of cAMP from ATP, is located on the cytoplasmic surface of the plasma membrane of target cells (Figure 15.12). Usually, it is activated by the binding of G proteins, themselves activated by receptors. Second messengers do not have enzymatic activity; rather, they act as cofactors or allosteric regulators of target proteins. In the case of cAMP, there are two major target types. In many kinds of sensory cells, cAMP binds to ion channels to cause them to open. A second major target type is cytoplasmic. Cyclic AMP binds to an enzyme such as a protein kinase, whose active site gets exposed as a result. The sequential activation of yet another kinase ensues, leading to the final effects in the cell.

Two second messengers are derived from lipids

Membrane phospholipids are involved in signal transduction in addition to their roles as structural components of the plasma membrane. When certain phospholipids are hydrolyzed into their component parts (see Figure 3.21) by enzymes called phospholipases, second messengers are formed. The best-studied of these come from hydrolysis of the lipid phosphatidyl inositol-bisphosphate (PTI), which

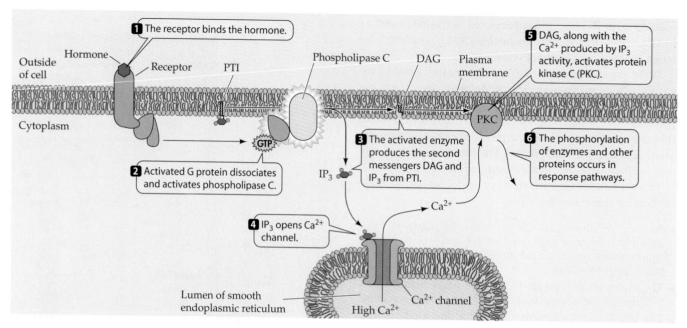

15.13 The IP₃ and DAG Second Messenger System
Phospholipase C hydrolyzes the lipid phosphatidyl inositol-bis-phosphate (PTI) into its components IP₃ and DAG, both of which are second messengers. IP₃ and DAG act separately but in concert, ultimately producing a wide range of responses in the cell.

has two fatty acid chains (diacylglycerol, or DAG) embedded in the plasma membrane, and a hydrophilic inositol group (inositol triphosphate, or IP_3) projecting into the cytoplasm. There are over two dozen signals whose actions are mediated by the products of PTI hydrolysis. Once again, the receptors involved are often linked to G proteins. The activated G proteins diffuse through the plasma membrane and activate an enzyme, phospholipase C. This enzyme cleaves off the IP_3 from PTI, leaving the glycerol and the two attached fatty acids (DAG) embedded in the lipid bilayer:

$$PTI \xrightarrow{\text{Phospholipase C}} IP_3 + DAG$$

IP_3 and DAG are both second messengers and have different modes of action that build on each other (Figure 15.13). DAG activates a membrane-bound enzyme, protein kinase C (PKC), in much the same way that cAMP activates protein kinase A. PKC is dependent on Ca^{2+} (hence the "C"), and this is where IP_3 comes in. IP_3 is charged and diffuses through the cytoplasm to the endoplasmic reticulum, where it causes the release of Ca^{2+} into the cytoplasm. There, in combination with DAG, the Ca^{2+} causes PKC to become active. PKC then phosphorylates a wide variety of proteins, leading to the ultimate response of the cell (Figure 15.13).

Calcium ions are involved in many transduction pathways

Calcium ions can also act as a second messenger. They are scarce in most cells, with a cytoplasmic concentration of only about 0.1 μM, while the concentrations of Ca^{2+} outside

the cell and within the ER are usually much higher. This difference is maintained by active transport proteins at the plasma and ER membranes that pump the ion out of the cytoplasm. In contrast to cAMP and the lipid second messengers, the level of intracellular Ca^{2+} cannot be increased by making more of it. Instead, the opening and closing of channels and the action of membrane pumps regulate levels of the ion in a cellular compartment.

There are many signals that can cause Ca^{2+} channels to open, including IP_3 (see the previous section) and the entry of a sperm into an egg cell (Figure 15.14). Whatever the signal, the open channels result in a dramatic increase in cyto-

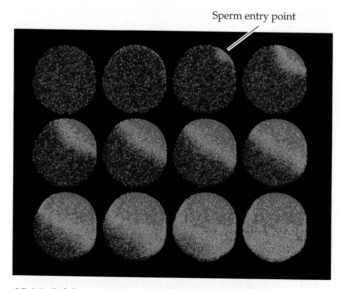

15.14 Calcium Ions as an Intracellular Messenger
The concentration of Ca^{2+} can be measured by a dye that fluoresces and turns red when it binds the ion. Here, photographed at 5-second intervals, fertilization causes a wave of Ca^{2+} to pass through the egg of a starfish. The message that fertilization is complete and development can begin is thus delivered.

plasmic Ca^{2+} concentration, up to a hundredfold within a fraction of a second. As we saw earlier, this ion activates protein kinase C. In addition, Ca^{2+} controls other ion channels and stimulates secretion by exocytosis.

A distinctive aspect of Ca^{2+} signaling is that the ion can stimulate its own release from intracellular stores. For example, in some plant leaf cells, the hormone abscisic acid binds to gated Ca^{2+} channels and opens them, causing the ion to rush into the cells. This influx is not enough to trigger the cell's response, however. The ion binds to Ca^{2+} channels in the endoplasmic reticulum and vacuolar membranes, causing those organelles to release their Ca^{2+} stores as well.

In some cases, Ca^{2+} ions act via a calcium-binding protein called *calmodulin*, and it is the Ca^{2+}–calmodulin complex that performs cellular functions by binding to target proteins. Calmodulin, which is present in many cells, has four binding sites for Ca^{2+}. When the cytoplasmic Ca^{2+} concentration is low, calmodulin does not bind enough of it to become activated. But when the cell is stimulated by a signal and the Ca^{2+} level rises, all four binding sites are filled. Then the calmodulin changes shape and binds to a number of cellular targets, activating them in turn. One such target is a protein kinase in smooth muscle cells that phosphorylates the muscle protein myosin, initiating muscle contraction.

Nitric oxide is a gas that can act as a second messenger

Pharmacologist Robert Furchgott, at the State University of New York in Brooklyn, was investigating how acetylcholine causes smooth muscles lining blood vessels to relax, thus allowing more blood to flow to certain organs. Acetylcholine appeared to stimulate the IP_3 signal transduction system to produce an influx of Ca^{2+}, which led to an increase in the level of an unusual second messenger, cyclic GMP (cGMP). This nucleotide bound to a protein kinase, which then stimulated a kinase cascade leading to muscle relaxation. So far, the pathway seemed straightforward.

But while this pathway seemed to work in intact animals, it did not work on isolated strips of artery tissue. When Furchgott switched to tubular sections of artery, however, signal transduction did occur. There turned out to be a crucial difference between these two tissue preparations: In the strips, the delicate inner layer of cells that lines the blood vessel had been lost. Furchgott hypothesized that this layer, the endothelium, was making something that diffused into the muscle cells and was needed for their response to acetylcholine. The substance was not easy to isolate. It seemed to break down quickly, with a half-life (the time in which half of it disappeared) of 5 seconds in living tissues. It turned out to be a gas, nitric oxide (NO), that had always been thought of as a toxic air pollutant!

In the body, NO is made via an enzyme, NO synthase. This enzyme is activated by Ca^{2+}, which enters the endothelial cell through a channel opened by IP_3, released after acetylcholine binds to its receptor. The NO formed is chem-

ically very unstable and although it diffuses readily, it does not get too far. Conveniently, the endothelial cells are close to the smooth muscle cells, where NO acts as a second messenger, stimulating the formation of cGMP (Figure 15.15).

The spectacular discovery of NO as a second messenger explained the action of nitroglycerin, a drug that has been used for over a century to treat angina, the chest pain caused by insufficient blood flow to the heart. Nitroglycerin releases NO, which results in relaxation of the blood vessels and increased blood flow.

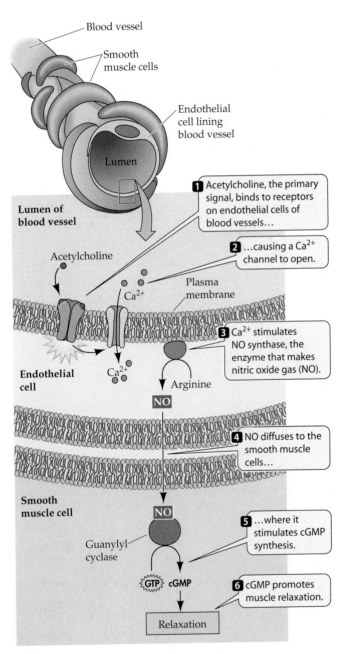

15.15 Nitric Oxide as a Second Messenger
Nitric oxide (NO) is an unstable gas, which nevertheless serves as a second messenger between a signal, acetylcholine, and its effect, the relaxation of smooth muscles. The endothelial tissue of blood vessels is a crucial intermediary in this communication between three types of tissue.

Signal transduction is highly regulated

There are several ways in which cells can regulate the activity of a signal transduction mechanism. The concentration of NO, which breaks down quickly, can be regulated only by how much of it is made. The level of Ca^{2+}, on the other hand, is determined by both membrane pumps and ion channels. For protein kinase cascades, G proteins, and cAMP, there are enzymes that convert the activated form back to its inactivated precursor:

▸ Protein phosphatases remove the phosphate groups from phosphorylated proteins.
▸ GTPases convert the GTP on an active G protein back to GDP, inactivating the protein.
▸ cAMP phosphodiesterase converts cAMP into its precursor, AMP, which has no second messenger activity.

These three inactivation systems are themselves under controls. For example, the major protein phosphatase can be inhibited by a protein whose activity is determined by phosphorylation by protein kinase A, which itself is under control of cAMP. So the cAMP pathway can intersect with the protein kinase cascade. On the other hand, some cAMP phosphodiesterases are stimulated by Ca^{2+}, thus showing an interaction between these two signaling pathways. The caffeine in coffee acts as a stimulant in part because it inhibits cAMP phosphodiesterase.

Effects

We have seen that the binding of a signal to its receptor initiates the response of a cell to an environmental signal, and how the direct or indirect transduction of this signal to the cytoplasm of the cell amplifies the stimulus. In this section, we consider the third and final step in the process, the actual effects of the signal on cell function. These effects are primarily the opening of membrane channels, changes in the activities of enzymes, and differential gene transcription.

Membrane channels are opened

The opening of ion channels is of great importance when the nervous system responds to a signal. Sensory nerve cells of the sense organs, for example, become stimulated through the opening of ion channels. We will focus here on one such signal transduction pathway, that for the sense of smell (Figure 15.16).

The sense of smell is well developed in mammals, some of which have an amazing *1,000 genes* for odorant receptors, the largest gene family known. Each of the thousands of nerve cells in the nose expresses one of these receptors. The identification of which chemical signal, or odorant, activates which receptor is just getting under way.

15.16 A Signal Transduction Pathway Leads to the Opening of Membrane Channels
In the signal transduction pathway for the sense of smell, the final effect is the opening of Na^+ channels. The resulting influx of Na^+ stimulates the transmission of a scent message to a specific region of the brain.

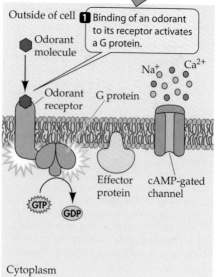
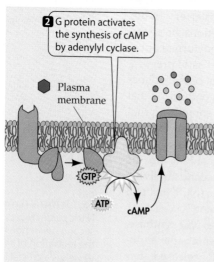
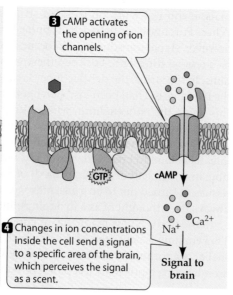

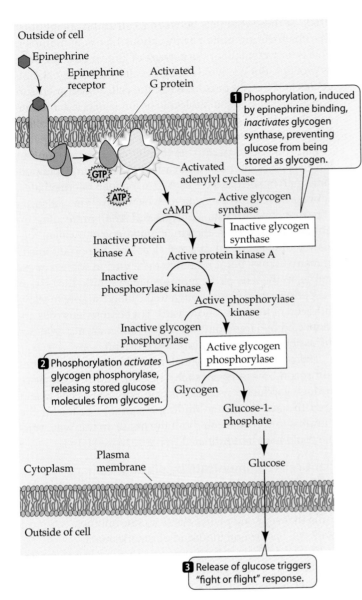

1 Phosphorylation, induced by epinephrine binding, *inactivates* glycogen synthase, preventing glucose from being stored as glycogen.

2 Phosphorylation *activates* glycogen phosphorylase, releasing stored glucose molecules from glycogen.

3 Release of glucose triggers "fight or flight" response.

15.17 A Cascade of Reactions Leads to Altered Enzyme Activity
Liver cells respond to epinephrine by activating G proteins, which in turn activate cAMP synthesis. The second messenger initiates a series of kinase reactions. The cascade both inhibits the continued storage of glucose molecules and stimulates the release of previously stored glucose.

The G protein-mediated protein kinase cascade stimulated by epinephrine in liver cells results in the phosphorylation of two key enzymes in glycogen metabolism (Figure 15.17). One of them, glycogen synthase, catalyzes the joining of glucose molecules to synthesize the energy-storing molecule glycogen, but it is inactivated by phosphorylation. Thus the epinephrine signal *prevents* glucose from being stored in glycogen. On the other hand, phosphorylase kinase becomes activated when a phosphate group is added to it, and goes on to stimulate a protein kinase cascade that ultimately leads to the activation by phosphorylation of glycogen phosphorylase, the other key enzyme in glucose metabolism. This enzyme *liberates* glucose molecules from glycogen. Thus the same signaling pathway inhibits the storage of glucose as glycogen (by inhibiting glycogen synthase) and promotes the release of glucose through glycogen breakdown (by activating glycogen phosphorylase). As we mentioned earlier, the released glucose fuels the ATP-requiring fight-or-flight response to epinephrine.

Phosphorylation by activated protein kinase A alters the activities of many other proteins, including enzymes involved in glycolysis, a ribosomal protein, and a receptor for a neurotransmitter. Likewise, Ca^{2+} binds to many proteins in the cell, changing their activities. In addition to protein kinase C, Ca^{2+}-activated targets include proteins that bind to and organize actin microfilaments and microtubules, as well as troponin, a modulator of muscle contraction.

Different genes are transcribed

Cell surface receptors are involved in activating a broad range of gene expression responses. For example, the Ras signaling pathway ends in the nucleus (see Figure 15.11). The final protein kinase enters the nucleus and phosphorylates a leucine zipper protein called AP-1. This activated protein stimulates the transcription of a number of genes involved in cell proliferation.

As we described earlier in this chapter, lipid-soluble hormones can diffuse directly through the plasma membrane and meet their receptors in the cytoplasm. Binding of the ligand allows the ligand/receptor complex to enter the nucleus, where it binds to hormone-responsive elements at the promoters of a number of genes. In some cases, transcription is stimulated, and in others it is inhibited.

In plants, light acts as a signal to initiate the formation of chloroplasts. Between this signal and response is a transcription-mediated signaling pathway. In bright sunlight, red wavelengths are absorbed by a receptor protein called phytochrome. We will say more about this important recep-

When an odorant molecule binds to its receptor, a G protein becomes activated, which in turns activates adenylyl cyclase to form the second messenger cAMP. This molecule then binds to an ion channel, causing it to open. The resulting influx of Na^+ causes the nerve cell to become stimulated so it sends a signal back to the brain that a particular odor is present.

Enzyme activities are changed

Proteins will change their shape, and their functioning, if they are modified either covalently or noncovalently. We have seen examples of both types of modificaton in signal transduction. Protein kinases add phosphate groups to a target protein, and the covalent change alters the protein's shape. Cyclic AMP binds to target proteins allosterically, and this noncovalent interaction changes the protein's shape. In both cases, previously inaccessible active sites are exposed, and the target protein goes on to perform a cellular role.

tor later in the book, but for now it is important to note that it is activated by red light. The activated phytochrome binds to cytoplasmic regulatory proteins, which enter the nucleus and bind to promoters for genes involved in the synthesis of important chloroplast proteins. Synthesis of these proteins is the key to the plant "greening".

Direct Intercellular Communication

Up to now, we have described how signals from the environment can influence a cell. But the environment of a cell in a multicellular organism is more than the extracellular medium. Most cells are in contact with their neighbors. In Chapter 5, we described how cells adhere to one another by the noncovalent interactions of recognition proteins protruding from the cell surface. There are also specialized cell junctions, such as tight junctions and desmosomes, that help "cement" the cells together (see Figure 5.7).

However, as we know from our own neighbors (and roommates), just being in proximity does not necessarily mean that there is functional communication. In this section, we look at the specialized junctions between cells that allow them to signal one another. In animals, these are gap junctions; in plants, they are plasmodesmata.

Animal cells communicate by gap junctions

Gap junctions are channels between adjacent cells that occur in many multicellular animals, occupying up to 25 percent of the area of the plasma membrane (Figure 15.18). Gap junctions traverse the 2-nm space between the plasma membranes of two cells (the "gap") by means of thin molecular channels called connexons. The walls of these tubes are composed of six subunits of an integral membrane protein appropriately named *connexin*. In two cells close to each another, two connexons come together, forming a channel that links the two cytoplasms. There may be hun-

dreds of these channels between a cell and its neighbors. The channels are quite narrow, about 1.5 nm in diameter. This is far too small for the passage of large molecules, such as proteins. But it is wide enough to allow small signal molecules and ions to pass between the cells. Experiments in which a labeled signal molecule or ion is injected into one cell show that it can readily pass into the adjacent cells if they are connected by gap junctions.

Gap junctions permit metabolic cooperation among linked cells. Such cooperation assures the sharing of important small molecules such as ATP, metabolic intermediates, amino acids, and coenzymes between cells. It may also assure that concentrations of ions and small molecules are similar in linked cells, thereby maintaining equivalent regulation of metabolism. It is not clear how important this is in many tissues, but it is known to be vital in some. For example, in the lens of the mammalian eye, only the cells at the periphery are close enough to the blood supply to allow diffusion of nutrients and wastes. But because lens cells are connected by large numbers of gap junctions, material can diffuse between them rapidly and efficiently.

There is evidence that signal molecules such as hormones and second messengers such as cAMP and IP_3 can move through gap junctions. If this is true, only a few cells would need to have receptors binding a signal in order for the stimulus to spread throughout the tissue. In this way, a tissue could have a coordinated response to the signal.

Plant cells communicate by plasmodesmata

Instead of gap junctions, plants have **plasmodesmata**, which are membrane-lined bridges spanning the thick cell walls that separate plant cells from one another. A typical plant cell has several thousand plasmodesmata.

Plasmodesmata differ from gap junctions in one fundamental way: Unlike gap junctions, in which the wall of the channel is made of integral proteins from the adjacent plasma membranes, plasmodesmata are lined by the fused plasma membranes themselves. Plant biologists are so familiar with the notion of a tissue as cells interconnected in this way that they refer to these continuous cytoplasms as a *symplast* (see Chapter 35).

The diameter of a plasmodesma is about 6 nm, far larger than the gap junction channel. But the actual space available for diffusion is about the same—1.5 nm—as with gap junctions. A look at the interior of the plasmodesma gives the reason for this reduction in pore size: A tubule called the *desmotubule*, apparently derived from the endoplasmic reticulum, fills up most of the opening of the plasmodesma (Figure 15.19). So, typically, only small metabolites and ions move between plant cells. This fact is important physiologically to plants, which lack the tiny circulatory vessels (capillaries) many animals use to bring gases and nutrients near enough to every cell.

Diffusion from cell to cell through plasma membranes is probably inadequate for hormonal responses in plants. Instead, they rely on more rapid diffusion through plasmo-

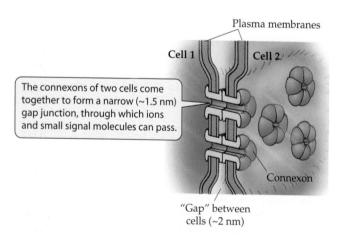

Plasma membranes

Cell 1   Cell 2

The connexons of two cells come together to form a narrow (~1.5 nm) gap junction, through which ions and small signal molecules can pass.

Connexon

"Gap" between cells (~2 nm)

15.18 Gap Junctions Connect Animal Cells
An animal cell may contain hundreds of gap junctions connecting it to neighboring cells. Gap junctions are too small for proteins to pass through, but small molecules such as ATP, metabolic intermediates, amino acids, and coenzymes can be shared in this way.

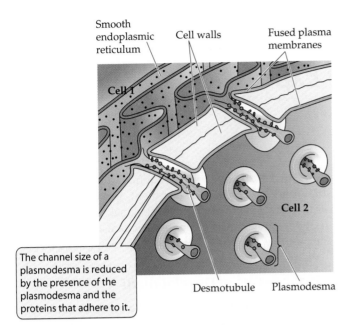

Smooth
endoplasmic
reticulum   Cell walls   Fused plasma
                               membranes

Cell 1

Cell 2

The channel size of a
plasmodesma is reduced
by the presence of the
plasmodesma and the
proteins that adhere to it.

Desmotubule   Plasmodesma

15.19 Plasmodesmata Connect Plant Cells
The desmotubule, derived from the endoplasmic reticulum, fills up most of the space inside plasmodesmata, leaving a tiny gap through which small metabolites and ions can pass.

desmata to ensure that all cells of a tissue respond to a signal at the same time. In C_4 plants (see Chapter 8), there are abundant plasmodesmata between the mesophyll and bundle sheath cells, helping to rapidly move the fixed carbon in the former cell type to the latter. A similar transport system, found at the junctions of nonvascular tissues and phloem, conducts organic solutes throughout the plant.

Plasmodesmata are not merely passive channels, but can be regulated. Plant viruses may infect cells at one location, then spread rapidly through a plant organ by plasmodesmata until they reach the plant's vascular tissue (circulatory system). These viruses, and even their RNA, would appear to be many times too large to pass through the plasmodesmal channel. But they get through, apparently by making "movement proteins" that increase the pore size temporarily while attached to the viral genome. Similar movement proteins are involved in transporting mRNAs between plant cells. This finding opens up the possibility of long-distance regulation of translation.

 ## Chapter Summary

Signals
▶ Cells receive many signals from both the physical environment and other cells. **Review Figures 15.1, 15.2**
▶ Signaling involves three steps: the binding of a signal by a receptor, the transduction of the signal within the cell, and the ultimate cellular response. **Review Figure 15.3**

Receptors
▶ Cells respond to signals only if they have specific receptor proteins that can bind to those signals. **Review Figure 15.4**
▶ Depending on the nature of the signal, the receptor may be at the plasma membrane or in the cytoplasm of the target cell. **Review Figure 15.5**
▶ Membrane receptors include ion channels, protein kinases. and G protein-linked receptors. **Review Figures 15.6, 15.7, 15.8**

Transducers
▶ The events of signal transduction may be direct, occurring at the plasma membrane, or indirect, involving the formation of a second messenger. **Review Figure 15.10**
▶ Protein kinase cascades amplify a response to receptor binding. **Review Figure 15.11**
▶ Second messengers include cyclic AMP, the lipid-derived substances phosphatidylinositol and diacylglycerol, calcium ions, and the gas nitric oxide. **Review Figures 15.12, 15.13, 15.14, 15.15**

Effects
▶ The ultimate cell response to a signal may be the opening of membrane channels, the alteration of enzyme activities, or changes in gene transcription. **Review Figures 15.16, 15.17**

Direct Intercellular Communication
▶ Animal cells can communicate directly, through small pores in their plasma membranes called gap junctions. Small molecules and ions can pass through these channels. **Review Figure 15.18**
▶ Plant cells are connected by somewhat larger pores called plasmodesmata, which traverse both membranes and cell walls. **Review Figure 15.19**

For Discussion

1. Like Ras itself, the various components of the Ras signaling pathway (see Figure 15.11) were discovered when tumors showed mutations in one or another of the components. What might be the biochemical consequences of mutations in the genes for (*a*) Raf and (*b*) MAP kinase that result in rapid cell division?

2. Cyclic AMP is a second messenger in many different responses, ranging from the sense of smell to the breakdown of glycogen. How can one messenger act in different ways in different cells?

3. Compare direct communication via plasmodesmata or gap junctions with ligand/receptor-mediated communication between cells. What are the advantages of one method over the other?

4. The tiny invertebrate *Hydra* has an apical region, which has tentacles, and a long, slender body. *Hydra* can reproduce asexually when cells on the body wall differentiate and form a bud, which breaks off as a new organism. Buds form only at a certain distance from the apex, leading to the idea that the apex releases a molecule that diffuses down the body and, at high concentrations (near the apex), inhibits bud formation. *Hydra* lacks a circulatory system, so the inhibitor must diffuse from cell to cell. If you had an antibody that binds to connexin and plugs up gap junctions, how would you show that the inhibitory factor passes through them?

16 Development: Differential Gene Expression

IT IS A DAY IN THE NOT-TOO-DISTANT FUTURE. Decades of eating fatty foods, combined with a genetically based tendency to deposit cholesterol in his arteries, finally catch up with 60-year-old Don: A blood clot closes off the blood flow to part of his heart, leading to a heart attack and irreversible tissue damage. Today, Don would be faced with a long period of rehabilitation, taking medications to manage his weakened heart. Instead, his physicians inject undifferentiated embryonic stem cells directly into his heart. The cells differentiate into cardiac muscle cells, replacing the ones that were lost to oxygen starvation, and full heart function is restored.

Embryonic stem cells, which are cells from a very young mammalian embryo, are able to form an entire organism if separated from one another. These cells can be removed from an embryo and maintained indefinitely in the laboratory. If they could be genetically altered to make them acceptable for transplants, these cells could be a source of tissue replacement not only for damaged hearts, but for the pancreas in people with diabetes and the brain in people with Alzheimer's disease.

While the application of stem cells to medicine is not yet possible, considerable new knowledge about the molecular biology of development has emerged. Much of this knowledge has come from studies on organisms such as the fruit fly *Drosophila*, the roundworm *Caenorhabditis elegans*, frogs, sea urchins, and a flowering plant, *Arabidopsis thaliana*. As we saw in Chapter 14, the genomes of eukaryotes are surprisingly similar, and the cellular and molecular principles underlying their development also turn out to be similar. Thus discoveries from one organism aid us in understanding other organisms, including ourselves.

Two major conclusions have emerged from studies of development. The first is that all types of somatic cells—all cells except gametes—in an organism retain all of the genes that were present in the fertilized egg. In other words, cell differentiation does not result from a loss of DNA. The second is that cellular changes during development and cell differentiation result from differential expression of genes. During development, the various mechanisms of transcriptional and translational control described in Chapter 14 and the signaling mechanisms described in Chapter 15 work together to produce a complex organism.

The Processes of Development

Development is a process in which an organism undergoes a series of progressive changes, taking on the successive forms that characterize its life cycle (Figure 16.1). In its earliest stages of development, a plant or animal is called an **embryo**. Sometimes the embryo is contained within a protective structure such as a seed coat, an eggshell, or a uterus. An embryo does not photosynthesize or feed actively; instead, it obtains its food from its mother directly or indirectly (by way of nutrients stored in the egg, for example). A series of embryonic stages may precede the birth of the new, independent organism. Most individual organisms continue to develop throughout their life cycle; development ceases only with death.

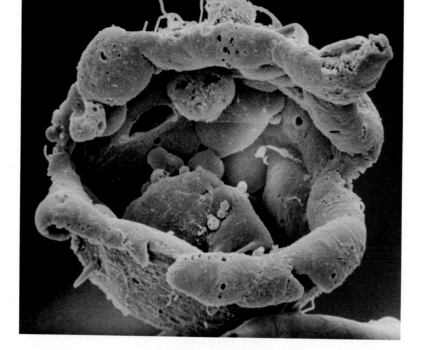

An Early Embryo
This mammalian embryo has been opened up to show the undifferentiated stem cells.

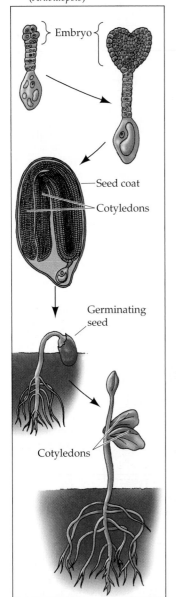

(a) **Flowering plant** (*Arabidopsis*)

Embryo

Seed coat

Cotyledons

Germinating seed

Cotyledons

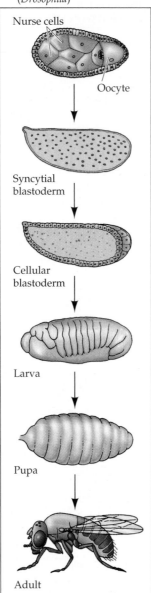

(b) **Insect** (*Drosophila*)

Nurse cells

Oocyte

Syncytial blastoderm

Cellular blastoderm

Larva

Pupa

Adult

16.1 Stages of Development
Stages from embryo to adult are shown for a plant and an animal. Cell division and expansion, growth, cell differentiation, and the creation of the organs and tissues of the adult body are all part of the complex process of development.

Development consists of growth, differentiation, and morphogenesis

Growth (increase in size) occurs through cell division and cell expansion. It continues throughout the individual's life in some species, but reaches a more or less stable end point in others. Repeated mitotic divisions generate the multicellular body. In plants, unless the daughter cells become longer (expand) after they form, the embryo does not grow very much; thus in plant development, cell expansion begins shortly after the first divisions of the fertilized egg. In animal development, on the other hand, cell expansion is often slow to begin: The animal embryo may consist of

thousands of cells before it becomes larger than the fertilized egg.

Differentiation is the generation of cellular specializations; that is, *differentiation defines the specific structure and function of a cell*. Mitosis, as we have seen, produces daughter nuclei that are chromosomally and genetically identical to the nucleus that divides to produce them. However, the cells of an animal or plant body are obviously not all identical in structure or function. The human body, with its approximately 100 trillion (10^{14}) cells, consists of about 200 functionally distinct cell types—for example, muscle cells, blood cells, and nerve cells. This apparent contradiction results from regulation of the expression of various parts of the genome. When the embryo consists of only a few cells, each cell has the potential to develop in many different ways. As development proceeds, however, the possibilities available to individual cells gradually narrow, until each cell's *fate* is fully determined and the cell has differentiated.

Whereas differentiation gives rise to cells of different kinds, **morphogenesis** (literally, "creation of form") gives rise to the shape of the multicellular body and its organs. Morphogenesis results from *pattern formation*, the organization of differentiated tissues into specific structures. In plant development, cells are constrained by cell walls and do not move around, so organized division and expansion of cells are the major processes that build the plant body. In animals, cell movements are very important in morphogenesis. And in both plants and animals, programmed cell death is essential to orderly development.

In plants and animals alike, differentiation and morphogenesis result ultimately from the regulated activities of genes and their products, as well as the interplay of extracellular signals and their transduction in target cells.

As development proceeds, cells become more and more specialized

Marking specific cells of an early embryo with stains reveals which adult structures are derived from which part of the embryo. For instance, the shaded area of the frog embryo shown in Figure 16.2 normally becomes part of the skin of the tadpole larva. However, if we cut out a piece from this region and transplant it to another location on an early embryo, the type of tissue it becomes is determined by its new environment. The *developmental potential* of such cells—that is, their range of possible development—is thus greater than their actual *fate*, which is limited to what normally develops.

Does embryonic tissue retain its broad developmental potential? Generally speaking, the answer is no. The developmental potential of cells becomes restricted fairly early in normal development. Tissue from a later-stage frog embryo, for example, if taken from a region fated to develop into brain, becomes brain tissue even if transplanted to parts of an early-stage embryo destined to become other structures. The tissue of the later-stage embryo is thus said to be *determined*: Its fate has been sealed, regardless of its surroundings. By contrast, the younger transplant tissue in Figure 16.2 has not yet become determined.

Determination precedes differentiation

Determination, the commitment of a cell to a particular fate, is a process influenced by the extracellular environment and the contents of the cell acting on the cell's genome. *Determination* is not something that is visible under the microscope—cells do not change appearance when they become determined. Determination is followed by differentiation, the actual changes in biochemistry, structure, and function that result in cells of different types. *Differentiation* often involves a change in appearance as well as function. Determination precedes differentiation. Determination is a commitment; the final realization of this commitment is differentiation.

The Role of Differential Gene Expression in Cell Differentiation

Differentiated cells are recognizably different from one another, sometimes visually as well as in their protein products. Certain cells in our hair follicles continuously produce keratin, the protein that makes up hair, nails, feathers, and porcupine quills. Other cell types in the body do not produce keratin. In the hair follicle cells, the gene that encodes keratin is transcribed; in most other cells in our body, that gene is not transcribed. Activation of the keratin gene is a key step in the differentiation of the hair follicle cells.

Generalizing from examples like this one, we may say that *differentiation results from differential gene expression*—that is, from the differential regulation of transcription, posttranscriptional events such as mRNA splicing, and translation (see Chapter 14).

Because the fertilized egg, or zygote, has the ability to give rise to every type of cell in the adult body, we say it is **totipotent**. Its genome contains instructions for all of the structures and functions that will arise throughout the life cycle. Later in the development of animals (and probably to a lesser extent in plants), the cellular descendants of the zygote lose their totipotency and become determined—that is, committed to form only certain parts of the embryo. Determined cells differentiate into specific types of specialized cells, such as nerve cells or muscle cells. When a cell becomes specialized, it is said to have differentiated.

Differentiation usually does not include an irreversible change in the genome

Differentiation is irreversible in certain types of cells. Examples include the mammalian red blood cell, which loses its nucleus during development, and the tracheid, a water-conducting cell in vascular plants. Tracheid development culminates in the death of the cell, leaving only the pitted cell walls that formed while the cell was alive (see Chapter 34). In both of these extreme cases, the irreversibility of differentiation can be explained by the absence of a nucleus.

Generalizing about mature cells that retain functional nuclei is more difficult. We tend to think of plant differentiation as reversible and of animal differentiation as irreversible, but this is not a hard-and-fast rule. Why is differentiation apparently reversible in some cells but not in

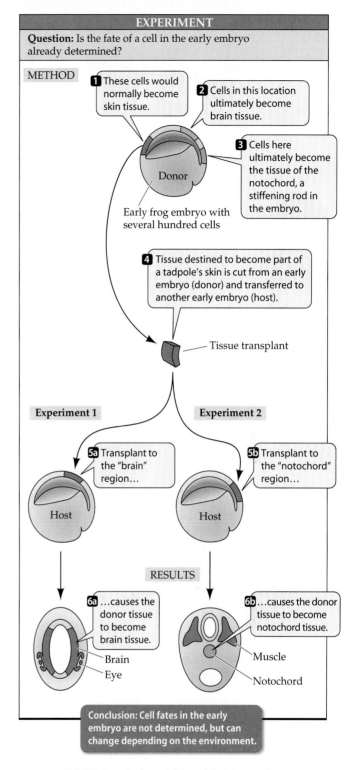

EXPERIMENT

Question: Is the fate of a cell in the early embryo already determined?

METHOD

1 These cells would normally become skin tissue.

2 Cells in this location ultimately become brain tissue.

3 Cells here ultimately become the tissue of the notochord, a stiffening rod in the embryo.

Donor

Early frog embryo with several hundred cells

4 Tissue destined to become part of a tadpole's skin is cut from an early embryo (donor) and transferred to another early embryo (host).

Tissue transplant

Experiment 1

Experiment 2

5a Transplant to the "brain" region…

Host

5b Transplant to the "notochord" region…

Host

RESULTS

6a …causes the donor tissue to become brain tissue.

Brain
Eye

6b …causes the donor tissue to become notochord tissue.

Muscle
Notochord

Conclusion: Cell fates in the early embryo are not determined, but can change depending on the environment.

16.2 Developmental Potential in Early Frog Embryos
Cells that would be expected to form one kind of tissue can form completely different tissues when they are experimentally moved within the early embryo.

others? At some stage of development, do changes within the nucleus permanently commit a cell to specialization? For both higher plants and animals, the answer appears to be no. Under the right environmental circumstances, differentiation is reversible.

A food storage cell in a carrot root faces a dark future. It cannot photosynthesize or give rise to new carrot plants.

However, if we isolate that cell from the root, maintain it in a suitable nutrient medium, and provide it with appropriate chemical cues, we can "fool" the cell into acting as if it were a fertilized egg. It can divide and give rise to a normal carrot embryo and, eventually, a complete plant (Figure 16.3). Since the new plant is genetically identical to the somatic cell from which it came, we call the plant a *clone*.

The ability to clone an entire carrot plant from a differentiated root cell indicates that the cell contains the entire carrot genome and that it can express the appropriate genes in the right sequence. Many cells from other plant species show similar behavior in the laboratory, and this ability to generate a whole plant from a single cell has been invaluable in agricultural biotechnology (see Chapter 17).

These experiments with plants establish that a somatic *cell* is totipotent. A more direct demonstration that all the genetic material is present in differentiated cells has come from *nuclear* transplant experiments. Such experiments were first done on frogs by Robert Briggs and Thomas King, who asked whether the nuclei of early frog embryos had lost the ability to do what the totipotent zygote nucleus could do. They first removed the nucleus from an unfertilized egg, thus forming an *enucleated* egg. Then, with a very fine glass tube, they punctured a cell from an early embryo and drew up part of its contents, including the nucleus, which they injected into the enucleated egg.

More than 80 percent of these nuclear transplant operations resulted in the formation, from the egg and its new nucleus, of a normal early embryo. Of these embryos, more than half developed into normal tadpoles and, ultimately, normal adult frogs. These experiments showed that no information is lost from the nuclei of cells as they pass through the early stages of embryonic development, and that the cytoplasmic environment around a nucleus can modify its fate.

Similar experiments have been performed on rhesus monkeys, in which a single cell can be removed from an 8-celled embryo and fused with an enucleated egg, allowing the nucleus of the embryonic cell to enter the egg cytoplasm. The resulting cell acts like a zygote, forming an embryo, which can be implanted into a foster mother, who ultimately gives birth to a normal monkey. Each of the remaining 7 cells from the original embryo can similarly give rise to offspring by the same cell fusion technique.

In humans, the totipotency of early embryonic cells permits both genetic screening and in vitro fertilization. An 8-cell human embryo can be isolated in the laboratory and a single cell removed to determine whether a harmful genetic condition is present (Chapter 18). Each remaining cell, being totipotent, can be stimulated to divide and form an embryo, which can be implanted into the mother, where it develops into an infant (Chapter 42).

Later frog nuclear transplant experiments by John Gurdon showed that a donor nucleus obtained from later stages in the frog's life could occasionally give rise to a normal tadpole, again showing totipotency. But successful cloning of animals was very difficult until the late 1990s,

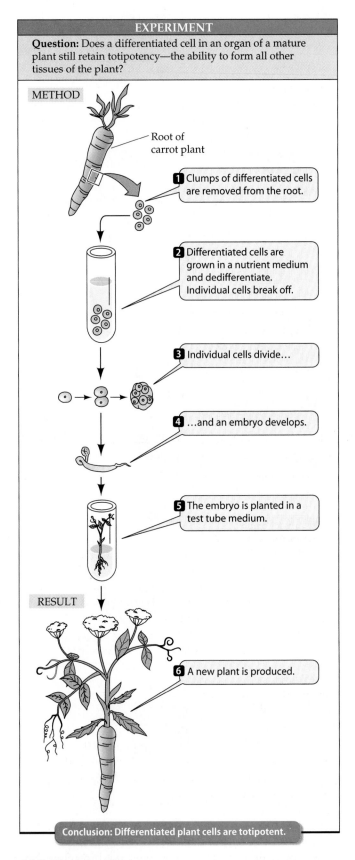

EXPERIMENT

Question: Does a differentiated cell in an organ of a mature plant still retain totipotency—the ability to form all other tissues of the plant?

METHOD

Root of carrot plant

1 Clumps of differentiated cells are removed from the root.

2 Differentiated cells are grown in a nutrient medium and dedifferentiate. Individual cells break off.

3 Individual cells divide...

4 ...and an embryo develops.

5 The embryo is planted in a test tube medium.

RESULT

6 A new plant is produced.

Conclusion: Differentiated plant cells are totipotent.

16.3 Cloning a Plant
Differentiated, specialized food storage cells from the root of a carrot can be induced to dedifferentiate, act like embryonic cells, and form a new plant.

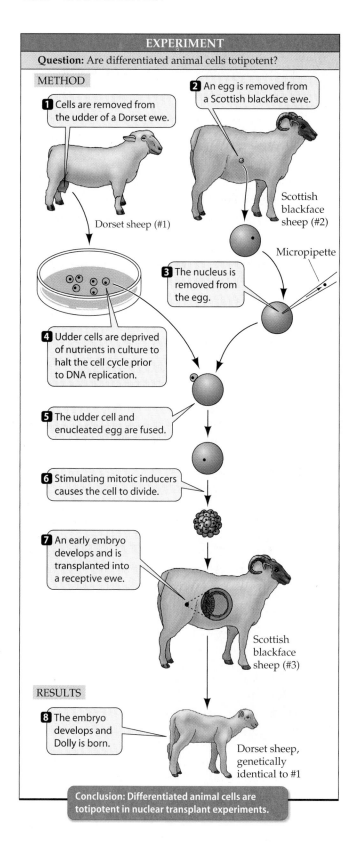

EXPERIMENT

Question: Are differentiated animal cells totipotent?

METHOD

1 Cells are removed from the udder of a Dorset ewe.

2 An egg is removed from a Scottish blackface ewe.

Dorset sheep (#1)

Scottish blackface sheep (#2)

Micropipette

3 The nucleus is removed from the egg.

4 Udder cells are deprived of nutrients in culture to halt the cell cycle prior to DNA replication.

5 The udder cell and enucleated egg are fused.

6 Stimulating mitotic inducers causes the cell to divide.

7 An early embryo develops and is transplanted into a receptive ewe.

Scottish blackface sheep (#3)

RESULTS

8 The embryo develops and Dolly is born.

Dorset sheep, genetically identical to #1

Conclusion: Differentiated animal cells are totipotent in nuclear transplant experiments.

16.4 A Clone and Her Offspring
Although Dolly herself (right) is a clone with only one parent, she has mated and given birth to "normal" offspring (the lamb on the left), proving the genetic viability of cloned mammals.

embryo. Apparently, when mammalian donor cells were in the G2 phase of the cell cycle and were fused with egg cytoplasm also in G2, some extra DNA replication took place that created havoc with the cell cycle in the egg when it attempted to divide.

Wilmut took differentiated cells from a ewe's udder and starved them of nutrients for a week, thus halting the cells in G1. After one of these cells was fused with an enucleated egg from a different breed of ewe, mitotic inducers in the egg cytoplasm (see Chapter 9) were stimulated and the donor nucleus entered S phase; the rest of the cell cycle proceeded normally. After several cell divisions, the resulting early embryo was transplanted into the womb of a surrogate mother. Out of 277 successful attempts to fuse adult cells with enucleated eggs, one lamb, named Dolly, survived to be born. DNA analyses confirmed that Dolly was genetically identical to the ewe from whose udder the donor nucleus had been obtained.

The purpose of Wilmut's experiment was to clone sheep that have been genetically programmed to produce products such as pharmaceuticals in their milk. The cloning procedure could make multiple, identical copies of sheep that are reliable producers of a drug such as α-1-antitrypsin, which is used to treat people with cystic fibrosis (see Figure 17.15).

The trick of starving donor cells for cloning has been applied to other mammals. Mice have been cloned using the cells lining the egg as a source of donor nuclei (Figure 16.5). Cows have been cloned to preserve a rare breed in New Zealand. Genetically engineered goats have been cloned to produce several useful proteins in their milk. This flurry of cloning has touched off a flurry of controversy, but cloning

when Ian Wilmut and his colleagues at a biotechnology company in Scotland used the cell fusion procedure to clone sheep (Figure 16.4). Previous attempts to produce mammals by this method had worked, as in the rhesus monkey case, only if the donor nucleus was from an early

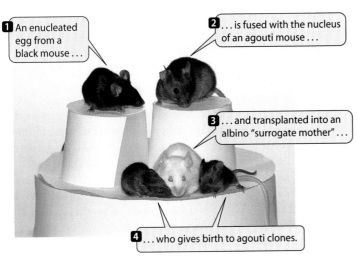

1 An enucleated egg from a black mouse . . .

2 . . . is fused with the nucleus of an agouti mouse . . .

3 . . . and transplanted into an albino "surrogate mother" . . .

4 . . . who gives birth to agouti clones.

16.5 Cloned Mice
Because so much is known about mouse genetics and molecular biology, cloned mice may be useful in studies of basic biology.

is not a new scientific concept. The idea of totipotency was accepted long before Dolly was born, but achieving it is an impressive technical achievement.

An example of nuclear totipotency gone awry occurs in a human tumor called a *teratocarcinoma*. Here, a differentiated cell *dedifferentiates* to form an unspecialized cell. Then it divides, forming a tumor, as occurs in most cancers. But some cells in the tumor redifferentiate to form specialized tissue arrangements. So the tumor can be a single mass of cells inside the abdomen, with some of the cells forming kidney tubes, others hair, and still others teeth! How this occurs is not clear.

Stem cells can be induced to differentiate by environmental signals

Totipotency implies that a differentiated cell stays that way because of its environment, and that appropriate environmental changes could result in a new pattern of differentiation. In normal development, a complex series of signals and their transduction results in the patterns of differentiation we see in a newborn organism. If these signals could be described in enough detail, we should be able to understand how any cell type becomes any other.

Stem cells are undifferentiated, dividing cells that are found in adult animal tissues that need frequent cell replacement, such as skin, the inner lining of the intestine, and the blood system. As they divide, stem cells produce cells that differentiate to replace dead cells and maintain tissue homeostasis. In the body, stem cells have limited abilities to differentiate. The stem cells in bone marrow, for example, produce the various types of red and white blood cells, while stem cells in the nervous system produce the various differentiated nerve cells.

Can one kind of stem cell be manipulated by its environment to produce cells that differentiate into cells of another

tissue? The answer appears to be yes. For example, when stem cells of the brain were transplanted into the bone marrow of mice whose bone marrow stem cells had been depleted, they proceeded to act like bone marrow stem cells, producing blood cells. In the reverse experiment, bone marrow stem cells were implanted into the brains of mice, where they formed brain cells. These experiments indicate that the environment—presumably intercellular signals—determines what a stem cell will do.

The stem cell populations closest to totipotency are not the ones in adults, but those of the early embryo. In mice, these embryonic stem cells can be removed from an early embryo (called a *blastocyst*) and then induced to differentiate in some particular way. Normally, these cells are formed a few days after fertilization, and soon become determined as to what their fate will be in the developing embryo. Before then, however, they are virtually totipotent. Such cells can be grown indefinitely in the laboratory and, when injected back into a mouse blastocyst, will mix with the resident cells and differentiate to form all the cells of the mouse. This kind of experiment shows that they have not lost any of their developmental potential while growing in the laboratory.

Embryonic stem cells growing in the laboratory can be induced to differentiate if the right signal is provided (Figure 16.6). For example, treatment of mouse embryonic stem cells with a derivative of vitamin A causes them to form nerve cells, while other growth factors induce them to form blood cells, again demonstrating their developmental potential and the roles of environmental signals. This finding raises the possibility of using stem cell cultures as sources of differentiated cells for clinical medicine. A key advance toward this use has been the ability to grow human embryonic stem cells in the laboratory. The age of custom-made cells to replace ones lost to disease or injury is rapidly approaching.

Genes are differentially expressed in cell differentiation

Experiments such as nuclear transplants in frogs and sheep—as well as plant cell cloning—point to the conclusion of genome constancy or equivalence in all somatic cells of an organism. Molecular experiments have provided even more convincing evidence. For example, the gene for β-globin, one of the protein components of hemoglobin, is present and expressed in red blood cells as they form in the bone marrow of mammals. Is the same gene also present—but unexpressed—in nerve cells in the brain, which do not make hemoglobin?

Nucleic acid hybridization (see Figure 14.7) can provide an answer. A probe for the β-globin gene can be applied to DNA from both brain cells and immature red blood cells (recall that mature red blood cells lose their nuclei and DNA). In both cases, the probe finds its complement, showing that the β-globin gene is present in both types of cells.

On the other hand, if the probe is applied to cellular mRNA rather than cellular DNA, it finds β-globin mRNA

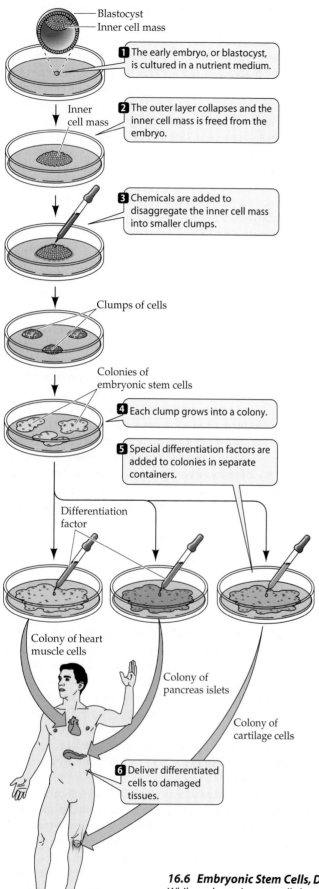

Blastocyst
Inner cell mass

1 The early embryo, or blastocyst, is cultured in a nutrient medium.

Inner cell mass

2 The outer layer collapses and the inner cell mass is freed from the embryo.

3 Chemicals are added to disaggregate the inner cell mass into smaller clumps.

Clumps of cells

Colonies of embryonic stem cells

4 Each clump grows into a colony.

5 Special differentiation factors are added to colonies in separate containers.

Differentiation factor

Colony of heart muscle cells

Colony of pancreas islets

Colony of cartilage cells

6 Deliver differentiated cells to damaged tissues.

16.6 Embryonic Stem Cells, Differentiation, and Medicine
While embryonic stem cells have been cultured from humans, their potential use in medicine was suggested by experiments in mice. This technique is under intensive investigation.

only in the red blood cells, and not in the brain cells. This result shows that the gene is expressed in only one of the two tissues. Many similar experiments have shown convincingly that differentiated cells lose none of the genes that were present in the fertilized egg.

What leads to this differential gene expression? One well-studied example is the conversion of undifferentiated muscle precursor cells, called *myoblasts*, into the large, multinucleated cells that make up mammalian skeletal muscle cells (called *muscle fibers*). The key event that starts this conversion is the expression of *MyoD1* (*My*oblast *D*etermining Gene *1*). The protein product of this gene is a transcription factor (MyoD1) with a helix-loop-helix domain (see Chapter 14) that not only binds to promoters of the muscle-determining genes to stimulate their transcription, but also acts on its own promoter to keep its levels high in the myoblasts and in their descendants.

Strong evidence for the controlling role of MyoD1 comes from experiments in which a gene containing an active promoter adjacent to *MyoD1* DNA is injected into the precursors of other cell types. For example, if *MyoD1* DNA is put into fat cell precursors, they are reprogrammed to become muscle cells. Genes such as *MyoD1*, which code for proteins that direct fundamental decisions in development, often by regulating genes on other chromosomes, usually code for transcription factors.

The Role of Polarity in Cell Determination

What initially stimulates the *MyoD1* promoter to begin transcription is not clear, but a chemical signal clearly is involved. In general, two overall mechanisms for producing such signals have been found. In **cytoplasmic segregation**, a factor within eggs, zygotes, or precursor cells is unequally distributed in the cytoplasm. After cell division, the factor ends up in some cells or regions of cells and not others. In **induction**, a factor is actively produced and secreted by certain cells to induce other cells to differentiate.

Polarity—the difference between one end of an embryo and the other—is obvious in development. Our heads are distinct from our feet, and the distal ends of our arms (wrists and fingers) differ from the proximal ends (shoulders). An animal's polarity develops early, even in the egg itself. Yolk may be distributed asymmetrically in the egg and embryo. In addition, other chemical substances may be confined to specific parts of the egg, or may be more concentrated at one pole than at the other.

In some animals, the original polar distribution of materials in the egg's cytoplasm changes as a result of fertilization. As cell division proceeds, the resulting cells contain unequal amounts of the materials that were not distributed uniformly in the zygote. As we learned from the work on

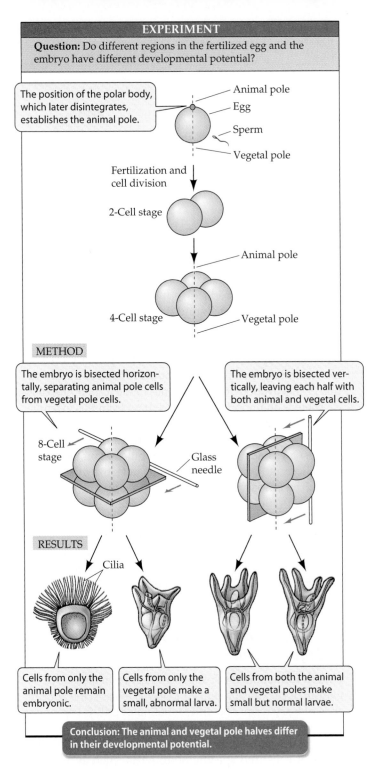

EXPERIMENT

Question: Do different regions in the fertilized egg and the embryo have different developmental potential?

The position of the polar body, which later disintegrates, establishes the animal pole.

Animal pole
Egg
Sperm
Vegetal pole

Fertilization and cell division

2-Cell stage

Animal pole

4-Cell stage — Vegetal pole

METHOD

The embryo is bisected horizontally, separating animal pole cells from vegetal pole cells.

The embryo is bisected vertically, leaving each half with both animal and vegetal cells.

8-Cell stage

Glass needle

RESULTS

Cilia

Cells from only the animal pole remain embryonic.

Cells from only the vegetal pole make a small, abnormal larva.

Cells from both the animal and vegetal poles make small but normal larvae.

Conclusion: The animal and vegetal pole halves differ in their developmental potential.

16.7 Early Asymmetry in the Embryo
The upper (animal pole) and lower (vegetal pole) halves of very early sea urchin embryos differ in their developmental potential. Cells from both halves are necessary to produce a normal larva.

cloning, cell nuclei do not always undergo irreversible changes during early development; thus we can explain some embryological events on the basis of the *cytoplasmic* differences in cells.

Even a structure as apparently uniform as a sea urchin egg has polarity. A striking difference between cells can be demonstrated very early in embryonic sea urchin development. The development of embryos that have been divided in half at the 8-cell stage depends on how they are separated. If the embryo is split into "left" and "right" halves, with each half containing cells from both the upper and the lower halves, normal-shaped but dwarfed larvae develop. If, however, the cut separates the upper four cells from the lower four, the result is different. The upper four cells develop into an abnormal early embryo with large cilia at one end that cannot form a larva. The lower four cells develop into small, somewhat misshapen larvae with an oversized gut (Figure 16.7). These results shows that for fully normal development, factors from both the upper and lower parts of the embryo are necessary.

This and many other experiments established that certain materials, called **cytoplasmic determinants**, are distributed unequally in the egg cytoplasm, and that these materials play a role in directing the embryonic development of many organisms.

The Role of Embryonic Induction in Cell Determination

Experimental work on developing embryos has clearly established that in many cases, the fates of particular tissues are determined by interactions with other specific tissues in the embryo. In developing animal embryos there are many such instances of induction, in which one tissue causes an adjacent tissue to develop in a particular manner. These effects are mediated by intercellular biochemical communication—that is, signal transduction mechanisms. We will describe two examples of such induction: one in the developing vertebrate eye, and the other in a developing reproductive structure in the nematode *C. elegans*.

Tissues direct the development of their neighbors by secreting inducers

The development of the lens in the vertebrate eye is a classic example of induction. In a frog embryo, the developing forebrain bulges out at both sides to form the *optic vesicles*, which expand until they come into contact with the cells at the surface of the head (Figure 16.8). The surface tissue in the region of contact with the optic vesicles thickens, forming a *lens placode*. The lens placode bends inward, folds over on itself, and ultimately detaches from the surface to produce a structure that will develop into the lens.

If the growing optic vesicle is cut away before it contacts the surface cells, no lens forms. Placing an impermeable barrier between the optic vesicle and the surface cells also prevents the lens from forming. These observations suggest that the surface tissue begins to develop into a lens when it receives a signal—an embryonic **inducer**—from the optic vesicle.

The interaction of tissues in eye development is a two-way street: There is a "dialogue" between the developing optic vesicle and the surface tissue. The optic vesicle induces lens development, and the developing lens determines the

16.8 Inducers in the Vertebrate Eye
The eye of a frog develops as inducers take their turns.

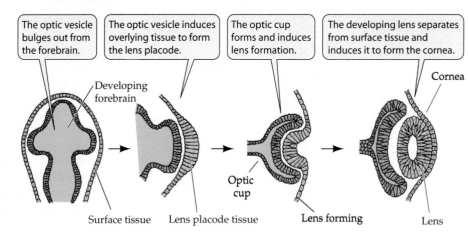

The optic vesicle bulges out from the forebrain.

The optic vesicle induces overlying tissue to form the lens placode.

The optic cup forms and induces lens formation.

The developing lens separates from surface tissue and induces it to form the cornea.

Developing forebrain

Cornea

Surface tissue   Lens placode tissue   Optic cup   Lens forming   Lens

size of the optic cup that forms from the optic vesicle. If head surface tissue from a frog species with small eyes is grafted over the optic vesicle of one with large eyes, both lens and optic cup are of intermediate size.

The developing lens also induces the surface tissue over it to develop into a *cornea*, a specialized layer that allows light to pass through and enter the eye. Thus a chain of inductive interactions participates in the development of the parts required to make an eye. Induction triggers a sequence of gene expression in the responding cells. Tissues do not induce themselves; rather, different tissues interact and molecularly induce each other. We will return to embryonic induction in Chapter 43.

Single cells can induce changes in their neighbors

The tiny nematode roundworm *Caenorhabditis elegans* is used as a model organism in many biological studies, but it is especially useful for studying development. It normally lives in the soil, where it feeds on bacteria, but can also grow in the laboratory if supplied with its food source. The entire process of development from egg to larva takes about 8 hours. It is easily observed using a low-magnification dissecting microscope because the body covering is transparent (Figure 16.9a). Because *C. elegans* is easy to culture, develops rapidly, and is easily observed, it is a favorite experimental organism. The development of *C. elegans* does not vary, so it has been possible to identify the source of each of the 959 somatic cells of the adult form.

The adult nematode is hermaphroditic, containing both male and female reproductive organs. It lays eggs through a pore called the *vulva* on the ventral surface. During development, a single cell, called the *anchor cell*, induces the vulva to form. If the anchor cell is destroyed by laser surgery, no vulva forms. The eggs develop inside the parent, and a "bag of worms" that eventually consume the parent results.

The anchor cell controls the fates of six cells on the animal's ventral surface through two molecular switches. Each of these cells has three possible fates: It may become a primary vulval precursor, a secondary vulval precursor, or simply part of the worm's surface—an epidermal cell (Figure 16.9b).

The anchor cell produces an inducer that diffuses out of the cell and interacts with adjacent cells. Cells that receive enough of the inducer become vulval precursors; cells slightly farther from the anchor cell become epidermis. The first molecular switch, controlled by the inducer from the anchor cell, determines whether a cell takes the "track" toward becoming part of the vulva or the track toward becoming epidermis.

The cell closest to the anchor cell, having received the most inducer, differentiates into the primary vulval precursor and apparently produces its own inducer, which acts on the two neighboring cells and directs them to become secondary vulval precursors. Thus the primary vulval precursor cell controls a second molecular switch, determining whether a vulval precursor will take the primary track or the secondary track. The two inducers control the activation or inactivation of specific genes in the responding cells.

There is an important lesson to draw from this example: Much of development is controlled by molecular switches that allow a cell to proceed down one of two alternative tracks. One challenge for the developmental biologist is to find these molecular switches and determine how they work. The primary inducer for the *C. elegans* vulva appears to be a growth factor homologous to the mammalian epidermal growth factor (EGF). The nematode growth factor, called LIN-3, binds to a receptor on the surface of a vulval precursor cell. This binding sets in motion a signal transduction cascade involving the ras protein and MAP kinases (see Figure 15.11). The end result is increased transcription of genes involved in the differentiation of vulval cells.

The Role of Pattern Formation in Organ Development

Pattern formation, the spatial organization of a tissue or organism, is inextricably linked to *morphogenesis*, the appearance of body form. The differentiation of cells is beginning to be understood in terms of molecular events, but how do molecular events contribute to the organization of multitudes of cells into specific body parts, such as a leaf, a flower, a shoulder blade, or a tear duct?

Some cells are programmed to die

Apoptosis is programmed cell death, a series of events caused by the expression of certain genes (see Figure 9.20). Some of these "death genes" have been pinpointed, and related ones have been found in organisms as diverse as nematodes and humans.

Apoptosis is vital to the normal development of all animals. For example, the nematode *C. elegans* produces precisely 1,090 somatic cells as it develops from a fertilized egg to an adult (see Figure 16.9). But 131 of these cells die. The sequential expression of two genes called *ced-4* and *ced-3*

(a) The nematode *C. elegans*

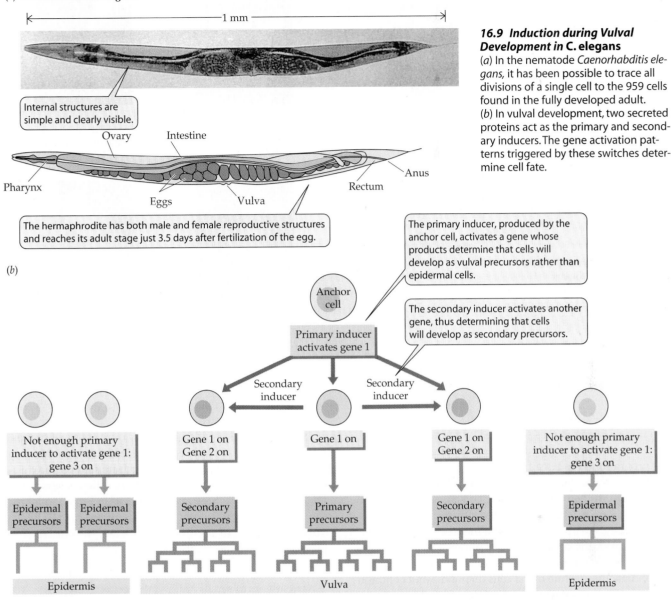

Internal structures are simple and clearly visible.

Ovary Intestine

Pharynx

Anus

Eggs Vulva Rectum

The hermaphrodite has both male and female reproductive structures and reaches its adult stage just 3.5 days after fertilization of the egg.

16.9 Induction during Vulval Development in C. elegans
(a) In the nematode *Caenorhabditis elegans*, it has been possible to trace all divisions of a single cell to the 959 cells found in the fully developed adult.
(b) In vulval development, two secreted proteins act as the primary and secondary inducers. The gene activation patterns triggered by these switches determine cell fate.

(b)

The primary inducer, produced by the anchor cell, activates a gene whose products determine that cells will develop as vulval precursors rather than epidermal cells.

The secondary inducer activates another gene, thus determining that cells will develop as secondary precursors.

Anchor cell

Primary inducer activates gene 1

Secondary inducer

Secondary inducer

Not enough primary inducer to activate gene 1: gene 3 on

Gene 1 on
Gene 2 on

Gene 1 on

Gene 1 on
Gene 2 on

Not enough primary inducer to activate gene 1: gene 3 on

Epidermal precursors | Epidermal precursors

Secondary precursors

Primary precursors

Secondary precursors

Epidermal precursors

Epidermis

Vulva

Epidermis

(for *c*ell *d*eath) appears to control this process. In the nervous system, for example, there are 302 nerve cells that come from 405 precursors; thus 103 cells undergo apoptosis. If the protein coded for by either *ced-3* or *ced-4* is nonfunctional, all 405 cells form neurons, and disorganization results. A third gene, *ced-9*, codes for an inhibitor of apoptosis: that is, its protein blocks the function of the *ced-4* gene. So, where cell death is required, *ced-3* and *ced-4* are active and *ced-9* is inactive; where cell death does not occur, the reverse is true.

Remarkably, a similar system of cell death genes acts in humans. During early development, human hands and feet look like tiny paddles—the fingers and toes are linked by webbing. Between days 41 and 56 of development, the cells in the webbing die, freeing the individual fingers and toes (Figure 16.10). The protein—an enzyme called *caspase*—that stimulates this apoptosis is similar in amino acid sequence to the protein encoded by *ced-3*, and a human protein (*bcl-2*) that inhibits apoptosis is similar to *ced-9*. So humans and

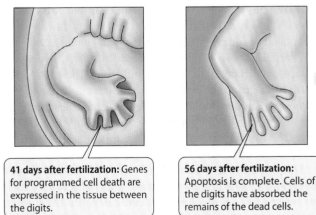

41 days after fertilization: Genes for programmed cell death are expressed in the tissue between the digits.

56 days after fertilization: Apoptosis is complete. Cells of the digits have absorbed the remains of the dead cells.

16.10 Apoptosis Removes the Webbing between Fingers
Early in the second month of human development, the webbing connecting the fingers is removed by apoptosis, freeing the individual fingers.

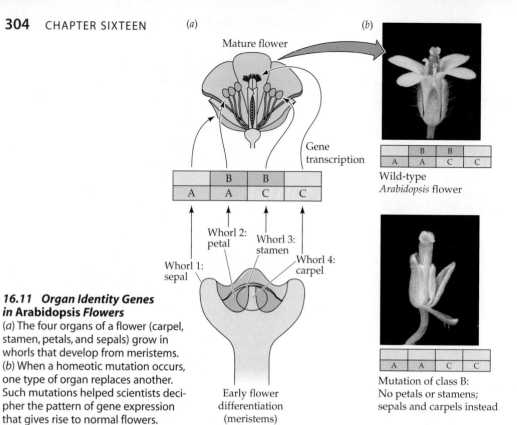

16.11 Organ Identity Genes in Arabidopsis Flowers
(a) The four organs of a flower (carpel, stamen, petals, and sepals) grow in whorls that develop from meristems. (b) When a homeotic mutation occurs, one type of organ replaces another. Such mutations helped scientists decipher the pattern of gene expression that gives rise to normal flowers.

nematodes, two creatures separated by more than 600 million years of evolutionary time, have similar genes controlling programmed cell death.

Apoptosis plays many other roles in your life. The dead cells that form the outermost layer of your skin and those from the uterine wall that are lost during menstruation have undergone apoptosis. White blood cells live only a few months in the circulation, then undergo apoptosis (see Figure 9.20). In a form of cancer called follicular large-cell lymphoma, these white blood cells do not die, but continue to divide. The reason is a mutation that causes the overexpression of *bcl-2*, the gene that inhibits cell death.

Plants have organ identity genes

Like animals, plants have organs—for example, leaves and roots. Many plants form flowers, and many flowers are composed of four types of organs: sepals, petals, stamens, and carpels. These floral organs occur in *whorls*, which are groups of each organ around a central axis. The whorls develop from groups of cells called *meristems* in the shape of domes, which develop at growing points on the stem (Figure 16.11a). How is the identity of a particular whorl determined? The answer appears to lie in the activities of a group of genes.

These genes have been best described in *Arabidopsis thaliana*, also called mouse-ear cress. This plant is very useful for studies of development because of its small size, abundant seed production (over 1,000 per plant), rapid development (from seed to plant to seed in 6 weeks), and small genome (10 chromosomes with 80 million base pairs of DNA). Finally, it is easy to produce mutations in this plant by treating the seeds with mutagens.

Normal *Arabidopsis* flowers have four whorls of organs, but there are mutant strains that have the wrong organs in particular whorls (Figure 16.11b). Such mutations, in which one organ is replaced with another, are called *homeotic mutations*. Studies on three mutant strains led to a model for the determination of floral organ identity. This model involves three organ identity genes, each of which is expressed in *two* of the whorls:

▶ A class A gene is expressed in whorls 1 and 2 (which normally form sepals and petals, respectively).
▶ A class B gene is expressed in whorls 2 and 3 (which normally form petals and stamens, respectively).
▶ A class C gene is expressed in whorls 3 and 4 (which normally form stamens and carpels, respectively).

Nucleic acid hybridization confirmed these locations for the mRNA transcripts of the three genes.

Not surprisingly, these three genes code for transcription factors, which are active as dimers. Possibly, gene regulation in these cases is *combinatorial*—that is, the composition of the dimer determines which other genes will be activated by the transcription factor. For example, a dimer made up only of transcription factor A would activate transcription of the genes that make sepals; a dimer made up of A and B would result in petals, and so forth.

A gene called *leafy* appears to control the transcription of the ABC genes. Plants with the *leafy* mutation are just that—they don't make flowers. The protein product of this gene acts as a transcription factor stimulating genes A, B, and C so that they produce the flower (Figure 16.12).

In addition to being fascinating to biologists, organ identity genes have caught the eye of horticultural and agricultural scientists. Flowers filled with petals instead of stamens and carpels often have mutations of the C genes. Many of the foods that make up the human diet come from fruits and seeds, such as the grains of wheat, rice, and corn. These fruits and seeds form from the carpels (the female reproductive organs) of the flower. Genetically modifying the number of these organs on a particular plant could increase the amount of grain a crop could produce.

Plants and animals use positional information

Certain cells in both plants and animals appear to "know" where they are with respect to the body as a whole. This spatial "sense" is called **positional information**. In plants, the pattern of development of two major types of conducting tissue suggested to scientists long ago that distance from the body surface may play a role in their formation. Cells destined to become water conductors are farther from the body surface than are those destined to become conductors of sucrose, a product of photosynthesis. Thus those cells destined to become water conductors are exposed to lower concentrations of O_2 and higher concentrations of CO_2. These differences may help determine which genes are expressed in which parts of the stem and root.

Recently it has been suggested that cells on the surface of the stem secrete a protein or other signal that is more concentrated close to the surface than deeper in the stem. Other signals may diffuse from the stem tip and root tip, establishing positional information along the plant's axis. These signals are called *morphogens*.

The wing of a chick develops from a round bud. The cells that become the bones and muscles of the wing must receive positional information. If they do not, the limbs will be totally disorganized (imagine fingers growing out of your shoulders). Three groups of cells—one at the junction of the bud and the body, a second at the tip of the bud, and a third on the surface of the bud—produce different morphogens that diffuse through the bud. Each cell in the bud receives unique concentrations of each morphogen. The first morphogen determines the *proximal–distal* ("shoulder to fingertip") axis of the wing, the second determines the *anterior–posterior* ("thumb to little finger") axis, and the third determines the *dorsal–ventral* ("palm to knuckles") axis.

The signaling pathways involved in limb development have been conserved through animal evolution. Comparative genomic studies have revealed developmental signaling pathways using homologous morphogen proteins in organisms ranging from nematodes to humans.

The Role of Differential Gene Expression in Establishing Body Segmentation

Another experimental subject that developmental biologists have used to study pattern formation is the fruit fly, *Drosophila*. Insects (and many other animals) develop a highly modular body composed of different types of *segments*. Complex interactions of different sets of genes underlie the pattern formation of segmented bodies.

Unlike the body segments of segmented worms such as earthworms, the segments of the *Drosophila* body are clearly different from one another. The adult fly has a head (composed of several fused segments), 3 different thoracic segments, 8 abdominal segments, and a terminal segment at the posterior end. The 13 seemingly identical segments of the *Drosophila* larva correspond to these specialized adult segments. Several types of genes are expressed sequentially in the embryo to define these segments. The first step in this process is to establish the polarity of the embryo.

Maternal effect genes determine polarity

In *Drosophila* eggs and larvae, polarity is based on the distribution of morphogens, of which some are mRNA's and some are proteins. These molecules are products of specific **maternal effect genes** in the mother and are distributed to the eggs, often in a nonuniform manner. The maternal effect genes are transcribed in the *nurse cells*, which surround and nurture the developing egg and are localized at certain specific regions of the egg as it forms. Maternal effect genes produce their effects on the embryo regardless of the genotype of the father. Their products determine the dorsal–ventral (back–belly) and anterior–posterior (head–tail) axes of the embryo.

The fact that these morphogens specify these axes has been established by the results of experiments in which cytoplasm was transferred from one egg to another. Females homozygous for a particular mutation of the maternal effect gene *bicoid* produce larvae with no head and or no thorax. However, if eggs of homozygous *bicoid* mutant females are inoculated at the anterior end with cytoplasm from the anterior region of a wild-type egg, the treated eggs develop into normal larvae, with heads developing from the part of the egg that receives the wild-type cytoplasm. Conversely, removal of 5 percent or more of the cytoplasm from the anterior of a wild-type egg results in an abnormal larva that looks like a *bicoid* mutant larva.

Wild-type   *Leafy* mutant

16.12 A Nonflowering Mutant
Mutations in the *leafy* gene of *Arabidopsis* prevent the transcription of the ABC genes, and the resulting plant does not produce any flower.

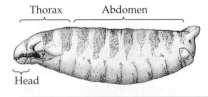

Thorax　Abdomen

Head

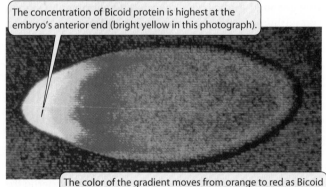

The concentration of Bicoid protein is highest at the embryo's anterior end (bright yellow in this photograph).

The color of the gradient moves from orange to red as Bicoid concentration decreases into the dark blue posterior end.

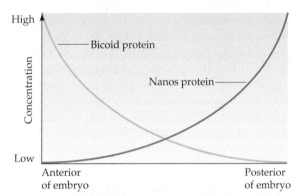

16.13 Bicoid and Nanos Protein Gradients Provide Positional Information
The anterior–posterior axis of *Drosophila* arises from morphogens produced by the maternal effect genes *bicoid* and *nanos*. The mRNA's are translated at the ends of the larva, and the resulting gradient controls the developing body's polarity.

Another maternal effect gene, *nanos*, plays a comparable role in the development of the posterior end of the larva. Eggs from homozygous *nanos* mutant females develop into larvae with missing abdominal segments. Injecting a *nanos* egg with cytoplasm from the posterior region of a wild-type egg allows normal development. These findings show that, in wild-type larvae, the overall framework of the anterior–posterior axis is laid down by the activity of these two maternal effect genes (Figure 16.13).

After the axes of the embryo are determined, the next step in pattern formation is to determine the larval segments.

Segmentation and homeotic genes act after the maternal effect genes

The number, boundaries, and polarity of the larval segments are determined by proteins encoded by the **segmentation genes**. These genes are expressed when there are

about 6,000 nuclei in the embryo. The nuclei all look the same, but in terms of gene expression, they are not.

The maternal effect genes set the segmentation genes in motion. Three classes of segmentation genes act, one after the other, to regulate finer and finer details of the segmentation pattern (Figure 16.14):

▶ First, **gap genes** organize large areas along the anterior–posterior axis. Mutations in gap genes result in gaps in the body plan—the omission of several larval segments.

▶ Second, **pair rule genes** divide the embryo into units of two segments each. Mutations in pair rule genes result in embryos missing every other segment.

▶ Third, **segment polarity genes** determine the boundaries and anterior–posterior organization of the segments. Mutations in segment polarity genes result in segments in which posterior structures are replaced by reversed (mirror-image) anterior structures.

▶ Finally, after the basic pattern of segmentation has been established by the segmentation genes, differences between the segments are mediated by the activities of **homeotic genes**. These genes are expressed in different combinations along the length of the body and tell each segment what to become. Homeotic genes are analogous to the organ identity genes of plants.

The maternal effect, segmentation, and homeotic genes interact to "build" a *Drosophila* larva step by step, beginning with the unfertilized egg.

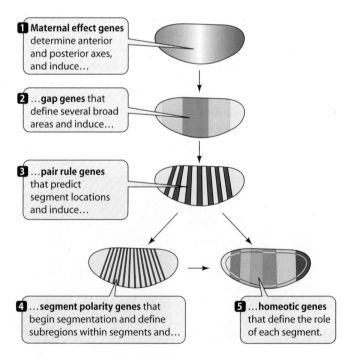

1 Maternal effect genes determine anterior and posterior axes, and induce…

2 …gap genes that define several broad areas and induce…

3 …pair rule genes that predict segment locations and induce…

4 …segment polarity genes that begin segmentation and define subregions within segments and…

5 …homeotic genes that define the role of each segment.

16.14 A Gene Cascade Controls Pattern Formation in the Drosophila Embryo
Gap, pair rule, and segment polarity genes are collectively referred to as the segmentation genes. The shading shows the locations of their gene products in the embryo.

Drosophila development results from a transcriptionally controlled cascade

One of the most striking and important observations about development in *Drosophila*—and in other animals—is that it results from a sequence of changes, with each change triggering the next. The sequence, or *cascade*, is largely controlled at the levels of transcription and translation.

In general, unfertilized eggs are storehouses of mRNA's, which are made prior to fertilization to support protein synthesis during the early stages of embryo development. Indeed, early embryos do not carry out transcription. After several cell divisions, mRNA production begins, forming the mRNA's needed for later development.

Some of the prefabricated mRNA's in the egg provide positional information. Before the egg is fertilized, mRNA for the Bicoid protein is localized at the end destined to become the anterior end of the fly. After the egg is fertilized and laid, nuclear divisions begin. (In *Drosophila*, cell divisions do not begin right away; until the thirteenth cell division, the embryo is a single, multinucleated cell called a *syncytium*.) At this early point, *bicoid* mRNA is translated, forming Bicoid protein, which diffuses away from the anterior end, establishing a gradient. At the posterior end, Nanos forms a gradient in the other direction. Thus each nucleus in the developing embryo is exposed to a different concentration ratio of Bicoid and Nanos proteins.

The two morphogens regulate the expression of the gap genes, although in different ways. Bicoid protein affects transcription, while Nanos affects translation. The high concentrations of Bicoid protein in the anterior portion of the egg turn on a gap gene called *hunchback*, while simultaneously turning off another gap gene, *Krüppel*. Nanos at the posterior end reduces the translation of *hunchback*, so a difference in concentration of these two gap gene products at the two ends is established.

The proteins encoded by the gap genes control the expression of the pair rule genes. Many pair rule genes in turn encode transcription factors that control the expression of the segment polarity genes, giving rise to a complex, striped pattern (see Figure 16.14) that foreshadows the segmented body plan of *Drosophila*.

By this point, each nucleus of the embryo is exposed to a distinct set of transcription factors. The segmented body pattern of the larva is established even before any sign of segmentation is visible. When the segments do appear, they are not all identical, because the homeotic genes specify the different structural and functional properties of each segment. Each homeotic gene is expressed over a characteristic portion of the embryo.

Let's turn now to the homeotic genes and see how their mutation can alter the course of development.

Homeotic mutations produce large-scale effects

Two bizarre homeotic mutations in *Drosophila* are the *Antennapedia* mutation, in which legs grow in place of antennae (Figure 16.15), and the *bithorax* mutation, in which an extra pair of wings grows in a thoracic segment (see Figure 24.9).

Edward Lewis at the California Institute of Technology found that *bithorax* was not a single gene, but a cluster of genes, each one determining the functional identity of a segment. Moreover, the genes were lined up along the chromosome in the order of the segments they determined: Genes at the beginning of the cluster determined thoracic segments, then the next genes determined the upper abdomen, and so on. A similar cluster of genes—the ones that are mutated in the *Antennapedia* flies—was found to determine the identities of the segments at the front of the fly. Lewis predicted that all of these genes might have come from the duplication of a single gene in an ancestral, unsegmented organism.

Molecular biologists confirmed Lewis's prediction. Using a nucleic acid hybridization probe for part of one of the genes of the cluster, several scientists found the probe binding not only to its own gene, but also to adjacent genes in its cluster and to genes in the other homeotic cluster. In

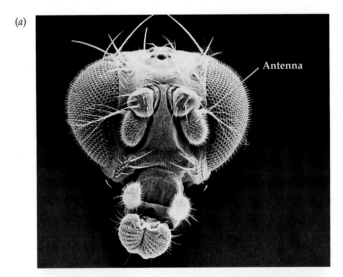

(a)

Antenna

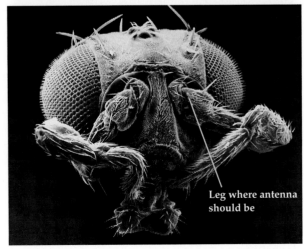

(b)

Leg where antenna should be

16.15 *A Homeotic Mutation in* Drosophila
Mutations of the homeotic genes cause body parts to form on inappropriate segments. (a) A wild-type fruit fly. (b) An *Antennapedia* mutant fruit fly.

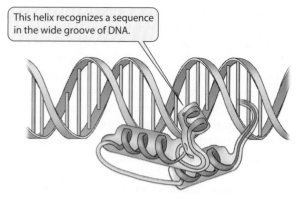

This helix recognizes a sequence in the wide groove of DNA.

16.16 A Homeodomain Binds to DNA
The homeodomain region has a helix-turn-helix motif. There are three α-helices, one of which is involved in recognition of a DNA sequence.

other words, there are DNA sequences that are common to all the homeotic genes in both clusters.

Homeobox-containing genes encode transcription factors

The 180-base-pair DNA sequence that is common to the *bithorax* and *Antennapedia* gene clusters is called the **homeobox**. It encodes a 60-amino-acid sequence, called the **homeodomain**, which binds to DNA (Figure 16.16). This sequence turns out to be present in other proteins involved in *Drosophila* pattern formation, such as *bicoid*. In all cases, the DNA-binding region of the protein has a helix-turn-helix motif (see Chapter 14). Each type of homeodomain recognizes a specific DNA sequence in target genes. The *bicoid* homeodomain, for example, recognizes TCCTAATCCC.

What do these proteins do when they recognize their target sequence in DNA? Not surprisingly, they are transcription factors, affecting genes involved in development. For example, Bicoid protein binds to promoters for the gap gene *hunchback*, activating its transcription. Hunchback protein is also a transcription factor, binding to enhancers and activating genes involved in head and thorax formation. In this way, the homeodomain proteins produce the cascade of events that controls *Drosophila* development.

Evolution and Development

Once DNA probes had identified the homeobox sequences involved in fruit

fly development, those probes were applied to other organisms. Soon, homeoboxes were found in over a hundred proteins from organisms as diverse as nematodes, frogs, mice, chickens, plants, and humans. As developmental biologists have described more and more intricate developmental systems at the molecular level, they have found that these systems are present in other organisms as well. These systems are an example of gene conservation. Comparative genomic studies have shown that, like the genes involved in biochemical pathways, the genes controlling development have much in common among different organisms. There may be only small differences that have turned one species into another, or a fin into a limb. This comparative approach has spawned a new discipline called *evolutionary developmental biology*, or "evo-devo."

HOMEOTIC GENE CLUSTERS. As Lewis predicted, homeobox genes are involved in many developmental pathways. In the mouse, for example, there are 38 such genes in four clusters, each located on a different chromosome. As do the homeobox genes in *Drosophila*, these **Hox genes** control the development of specific regions of the mouse embryo, and are arranged in the same order on the chromosome as they are expressed, from anterior to posterior, in the developing animal (Figure 16.17). These four clusters are present in all vertebrate animals, and apparently arose from repeated duplications, followed by small mutations, of a single ancestral gene. This gene was recently found in the lancelet, a tiny marine organism thought to be the simplest living member of the chordates, a lineage that includes the vertebrates. As we will see in Chapter 24, gene duplication is a common mechanism of evolutionary change. Here, it allowed the formation of gene clusters that could determine increasingly complex organisms.

16.17 Homeobox-Containing Genes Are Common to Vastly Different Organisms
In both the fruit fly and the mouse—two very dissimilar animals—the homeobox-containing genes are lined up on the chromosomes in the same order, and produce their effects in the same order as the genes along the anterior–posterior axis.

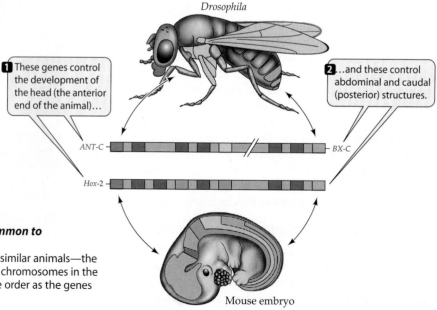

Drosophila

1 These genes control the development of the head (the anterior end of the animal)…

2 …and these control abdominal and caudal (posterior) structures.

ANT-C   BX-C

Hox-2

Mouse embryo

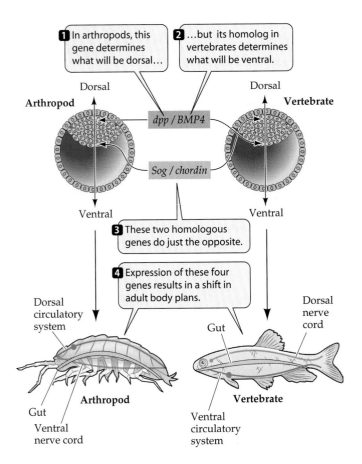

1 In arthropods, this gene determines what will be dorsal...

2 ...but its homolog in vertebrates determines what will be ventral.

3 These two homologous genes do just the opposite.

4 Expression of these four genes results in a shift in adult body plans.

16.18 Shifts of Homologous Gene Expression and Shift in Body Plans of Arthropods and Vertebrates
Although arthropods and vertebrates have similar and analogous genes governing development, these genes determine opposite locations for the nervous and circulatory systems.

ings! So the protein chordin determines dorsal identity in vertebrates, but its homolog, Sog, specifies the ventral region of arthropods. Likewise, BMP4 is a ventral determiner in vertebrates, but its relative, Dpp, is a dorsal determiner in arthropods. These findings suggest that the difference in body plan between arthropods and vertebrates involved an inversion of the axis of expression of these genes during evolution.

Chapter Summary

The Processes of Development

▶ A multicellular organism develops through a series of embryonic stages and eventually into an adult. Development continues until death. **Review Figure 16.1**

▶ Growth results from a combination of cell division and cell expansion.

▶ Differentiation produces specialized cell types.

▶ Morphogenesis—the creation of the overall form of the multicellular organism—is the result of pattern formation.

▶ In many organisms, the fates of the earliest embryonic cells have not yet been decided. These early embryonic cells may develop into different tissues if transplanted to other parts of an embryo. **Review Figure 16.2**

▶ As the embryo develops, its cells gradually become determined—committed to developing into particular parts of the embryo and particular adult structures. Following determination, cells eventually differentiate into their final, often specialized, forms.

The Role of Differential Gene Expression in Cell Differentiation

▶ The zygote is totipotent; it contains the entire genetic constitution of the organism and is capable of forming all adult tissues.

▶ Two lines of evidence show that differentiation does not involve permanent changes in the genome. First, nuclear transplant and cloning experiments show that the nucleus of a differentiated cell retains the ability to act like a zygote nucleus and stimulate the production of an entire organism. Second, molecular investigations have shown directly that all cells contain all genes for the organism, but that only certain genes are expressed in a given tissue. **Review Figures 16.3, 16.4, 16.5**

▶ Embryonic stem cells are totipotent, and can be cultured in the laboratory. With suitable environmental stimulation, these cells can be induced to form cells that differentiate. **Review Figure 16.6**

The Role of Polarity in Cell Determination

▶ Unequal distribution of cytoplasmic determinants in the egg, zygote, or embryo leads to cell determination in normal development. Experimentally altering this distribution can alter gene expression and produce abnormal or nonfunctional organisms. **Review Figure 16.7**

CONSERVATION OF A GENE FOR EYE FORMATION. The eye of an insect is very different in form and function from the eye of a mammal. Yet similar genes are involved in their formation. Mutations of a *Drosophila* gene called *eyeless* result in the reduction or absence of eyes in the adult fly. The protein encoded by the wild-type gene is a transcription factor that binds to promoters in over 1,500 different genes involved in eye formation. The homolog of *eyeless* in vertebrates, *Pax6*, is also involved in eye formation. In fact, mutations in the human version of *Pax6* result in a disease called aniridia—the partial or total absence of the iris of the eye. Remarkably, the *Pax6* and *eyeless* genes can be swapped for one another and yet remain functional in the other organism's development. This conservation of a global role for a gene for eye formation suggests that these "general instructions" evolved early and have been used repeatedly as different types of eyes have evolved.

DORSAL AND VENTRAL. In 1822, Geoffrey Saint-Hillaire noticed that in arthropods, such as the lobster, the nerve cord is on the same side as the mouth (ventral), with the circulatory system on the opposite side (dorsal) and the digestive system between the two. This pattern is the inverse of the one in vertebrates, in which the circulatory system is ventral and the nerve cord is dorsal (Figure 16.18). Molecular biologists have now described this "flip" in terms of genes that act during development. The signals for dorsal–ventral determination in arthropods and in vertebrates are the same; it's just that they have opposite mean-

The Role of Embryonic Induction in Cell Determination

▶ Some embryonic animal tissues direct the development of their neighbors by secreting inducers.

▶ Induction is often reciprocal: One tissue induces a neighbor to change, and the neighbor, in turn, induces the first tissue to change, as in eye formation in vertebrate embryos. **Review Figure 16.8**

▶ The nematode *Caenorhabditis elegans* provides a striking example of induction. The adult consists of 959 cells that develop from the fertilized egg by a precise pattern of cell divisions and other events. **Review Figure 16.9**

▶ Induction in *C. elegans* can be very precise, with individual cells producing specific effects in just two or three neighboring cells. **Review Figure 16.9**

The Role of Pattern Formation in Organ Development

▶ In plants and animals, programmed cell death (apoptosis) is important in pattern formation. Some genes whose protein products regulate apoptosis have been identified. **Review Figure 16.10**

▶ Plants have organ identity genes that interact to cause the formation of sepals, petals, stamens, and carpels. Mutations of these genes may cause undifferentiated cells to form a different organ. **Review Figures 16.11, 16.12**

▶ Both plants and animals use positional information as a basis for pattern formation. Gradients of morphogens provide this information.

The Role of Differential Gene Expression in Establishing Body Segmentation

▶ The fruit fly *Drosophila melanogaster* has provided much information about the development of body segmentation. Some of this information applies to other animals.

▶ The first genes to act in determining *Drosophila* segmentation are maternal effect genes, such as *bicoid* and *nanos*, which encode morphogens that form gradients in the egg. These morphogens act on segmentation genes to define the anterior–posterior organization of the embryo. **Review Figures 16.13, 16.14, 16.15**

▶ Segmentation develops as the result of a transcriptionally controlled cascade, with the product of one gene promoting or repressing the expression of another gene. There are three kinds of segmentation genes, each responsible for a different step in segmentation. Gap genes organize large areas along the anterior–posterior axis, pair rule genes divide the axis into pairs of segments, and segment polarity genes see to it that each segment has an appropriate anterior–posterior axis. **Review Figure 16.14**

▶ The Bicoid and Nanos proteins act as a transcription factor and translation regulator, respectively, to control the level of expression of gap genes. Gap genes encode transcription factors that regulate the expression of pair rule genes. The products of the pair rule genes are transcription factors that regulate the segment polarity genes.

▶ Activation of the segmentation genes leads to the activation of the appropriate homeotic genes in different segments. The homeotic genes define the functional characteristics of the segments.

▶ Mutations in homeotic genes often have bizarre effects, causing structures to form in inappropriate parts of the body. **Review Figure 16.16**

▶ Homeotic genes contain the homeobox, which encodes an amino acid sequence that is part of many transcription factors. **Review Figure 16.17**

Evolution and Development

▶ The homeobox is found in key genes of distantly related species; thus numerous regulatory mechanisms may trace back to a single evolutionary precursor. **Review Figure 16.18**

▶ The activities of similar genes result in similar programs of development in different organisms. These genes include homeotic genes, genes involved in eye formation, and genes that determine dorsal and ventral surfaces. **Review Figure 16.19**

For Discussion

1. Molecular biologists can insert genes attached to high-level promoters into cells (see Chapter 16). What would happen if the following were inserted and overexpressed? Explain your answers.
   a. *ced-9* in embryonic nerve cell precursors in *C. elegans*
   b. *MyoD1* in undifferentiated myoblasts
   c. *nanos* at the anterior end of the *Drosophila* embryo

2. A powerful method to test for the function of a gene in development is to generate a "knockout" organism, in which the gene in question is inactivated. What do you think would happen in each of the following?
   a. *C. elegans* with *ced-9* knocked out
   b. *Drosophila* with *nanos* knocked out
   c. *Drosophila* with *bithorax* knocked out

3. During development, the potential of a tissue becomes ever more limited until, in the normal course of events, the potential is the same as the original fate. On the basis of what you have learned here and in Chapter 14, discuss possible mechanisms for the progressive limitation of developmental potential.

4. How were biologists able to obtain such a complete accounting of all the cells in *C. elegans*? Why can't we reason directly from studies of *C. elegans* to comparable problems in our own species?

17 Recombinant DNA and Biotechnology

At the beginning of Chapter 3, we introduced you to spider silk, a family of proteins that show a fascinating combination of strength and elasticity. There is great interest in studying this biomaterial, not only to find out how it meets the structural challenges of the spider's web, but also because it has potential uses to humans, ranging from replacing expensive Kevlar in bulletproof vests to being used for surgical sutures and even to snag airplanes landing on aircraft carriers. The problem with spider silk is one of supply. "Milking" spiders, as has been done for centuries to get silk for fabrics from moth larvae, does not work on a commercial scale.

Because silk is made of a protein, one approach to the mass production of silk has been to use applied biology—biotechnology—to produce silk proteins in quantity and then develop ways to spin them into fibers, just as the spider does in its abdomen. This process has two parts. First, the gene for the silk protein is isolated from the spider genome. Then, the gene is put into a system that can express it in quantity. Silk genes turn out to be similar to the proteins they code for in that they are composed of repetitive domains. Through cutting and splicing of DNA, silk genes have been inserted into bacteria and into yeast, a relatively simple eukaryote. In both cases, an active promoter region was fused to the silk gene so that it would be expressed. Unfortunately, both of these types of host cells for this new DNA—called recombinant DNA—make insoluble silk that remains inside the cells.

Spider silk glands are similar in structure and function to animal mammary glands, in that both of them have epithelial cells that manufacture and secrete water-soluble, complex proteins in large amounts. Exploiting these similarities, scientists at a biotechnology company inserted the spider silk gene, with an accompanying mammary gland promoter for tissue-specific expression, into a goat, which produced abundant (10 g/L), soluble, easily purified silk in its milk. Creating such a transgenic goat is a tricky procedure, so a reliable silk producer goat has now been cloned, using the method described in the previous chapter. The next step, currently under way, is to develop a way to spin this silk protein into fibers.

This story—from problem to solution, from protein to expressed gene—has been repeated many times in the past two decades. The products of biotechnology range from life-saving drugs that there is no other way to make in adequate amounts to crop plants with improved agricultural characteristics. Although the basic techniques of DNA manipulation have been called revolutionary, most of them come from the knowledge of DNA transcription and translation that we described in earlier chapters.

We begin this chapter with a description of how DNA molecules can be cut into smaller fragments, and the fragments from different sources covalently linked to create recombinant DNA in the test tube. Recombinant (or any other) DNA can be introduced into a suitable prokaryotic or eukaryotic host cell. Sometimes, the purpose of adding new gene(s) to a host cell or organism is to ask an experimental question about the role of that gene, which can be answered by placing it in a new environment. In other instances, the purpose is to coax the host cell to make a new gene product.

Cleaving and Rejoining DNA

Scientists have long realized that the chemical reactions used in living cells for one purpose may be applied in the laboratory for other, novel purposes. Recombinant DNA technology—the manipulation and combination of DNA molecules from different sources—is based on this realization, and on an understanding of the properties of certain enzymes and of DNA itself.

A Factory for Spider Silk
These goats belong to a strain that is being genetically engineered to make spider silk in their milk.

As we saw in previous chapters, the nucleic acid base-pairing rules underlie many fundamental processes of molecular biology. The mechanisms of DNA replication, transcription, and translation rely on complementary base pairing. Similarly, all the key techniques in recombinant DNA technology—sequencing, rejoining, amplifying, and locating DNA fragments—make use of the complementary base pairing of A with T (or U) and of G with C.

In this section we will identify some of the numerous naturally occurring enzymes that cleave DNA, help it replicate, and repair it. Many of these enzymes have been isolated and purified, and are now used in the laboratory to manipulate and combine DNA. Then we will see how fragments of DNA can be separated and covalently linked to other fragments.

Restriction endonucleases cleave DNA at specific sequences

All organisms must have mechanisms to deal with their enemies. As we saw in Chapter 13, bacteria are attacked by viruses called bacteriophages that inject their genetic material into the host cell. Some bacteria defend themselves against such invasions by first altering their own DNA and then producing enzymes called **restriction endonucleases**, which catalyze the cleavage of double-stranded DNA molecules—such as those injected by phages—into smaller, noninfectious fragments (Figure 17.1). The bonds cut are between the 3′ hydroxyl of one nucleotide and the 5′ phosphate of the next one.

There are many such restriction enzymes, each of which cleaves DNA at a specific site defined by a sequence of bases called a **recognition site** or **restriction site**. The DNA of the host cell is not cleaved by its own restriction enzymes, because specific modifying enzymes called methylases add methyl ($-CH_3$) groups to certain bases at the restriction sites of the host's DNA when it is being replicated. The methylation of the host's bases makes the recognition sequence unrecognizable to the restriction endonuclease. But the unmethylated phage DNA is efficiently recognized and cleaved.

A specific sequence of bases defines each recognition site. For example, the enzyme *Eco*RI (named after its source, a strain of the bacterium *E. coli*) cuts DNA only where it encounters the following paired sequence in the DNA double helix:

5′ ... GAATTC ... 3′
3′ ... CTTAAG ... 5′

Notice that this sequence reads the same in the 5′-to-3′ direction on both strands. It is palindromic, like the word "mom," in the sense that it is the same in both directions from the 5′ end. *Eco*RI has two identical subunits that cleave the two strands between the G and the A.

This recognition sequence occurs on average about once in 4,000 base pairs in a typical prokaryotic genome—or about once per four prokaryotic genes. So *Eco*RI can chop a large piece of DNA into smaller pieces containing, on average, just a few genes. Using *Eco*RI in the laboratory to cut

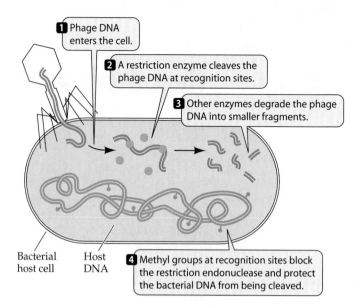

17.1 Bacteria Fight Invading Viruses with Restriction Enzymes
Bacteria produce restriction enzymes that cleave and degrade phage DNA. Other enzymes protect the bacteria's own DNA from being cleaved.

small genomes, such as those of viruses that have only a few thousand base pairs, may result in a few fragments. For a huge eukaryotic chromosome with tens of millions of base pairs, the number of fragments will be very large.

Of course, "on average" does not mean that the enzyme cuts all stretches of DNA at regular intervals. The *Eco*RI recognition sequence does not occur even once in the 40,000 base pairs of the genome of a phage called T7—a fact that is crucial to the survival of this virus, since its host is *E. coli*. Fortunately for the *E. coli* that make *Eco*RI, the DNA of other phages does contain the recognition sequence.

Hundreds of restriction enzymes have been purified from various microorganisms. In the test tube, different restriction enzymes that recognize different recognition sequences can be used to cut the same sample of DNA. Thus cutting a sample of DNA in many different, specific places is an easy task, and restriction enzymes can be used as "knives" for genetic "surgery."

Gel electrophoresis identifies the sizes of DNA fragments

After a laboratory sample of DNA has been cut with a restriction enzyme, the DNA is in fragments, each of which is bounded at its ends by the recognition sequence. As we noted, these fragments are not all the same size, and this property provides a way to separate them from one another. Separating the fragments is necessary to determine the number and sizes (in base pairs) of fragments produced, or to identify and purify an individual fragment of particular interest.

The best way to separate DNA fragments is by **gel electrophoresis** (Figure 17.2). Because of its phosphate groups,

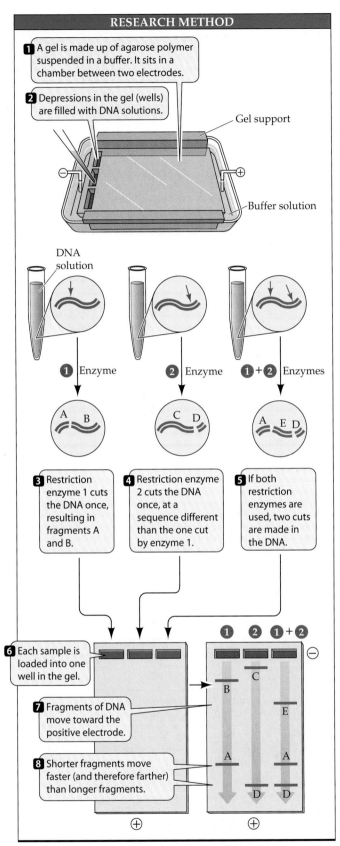

RESEARCH METHOD

1 A gel is made up of agarose polymer suspended in a buffer. It sits in a chamber between two electrodes.

2 Depressions in the gel (wells) are filled with DNA solutions.

Gel support

Buffer solution

DNA solution

1 Enzyme   **2** Enzyme   **1** + **2** Enzymes

3 Restriction enzyme 1 cuts the DNA once, resulting in fragments A and B.

4 Restriction enzyme 2 cuts the DNA once, at a sequence different than the one cut by enzyme 1.

5 If both restriction enzymes are used, two cuts are made in the DNA.

1   **2**   **1** + **2**

6 Each sample is loaded into one well in the gel.

7 Fragments of DNA move toward the positive electrode.

8 Shorter fragments move faster (and therefore farther) than longer fragments.

17.2 Separating Fragments of DNA by Gel Electrophoresis
A mixture of DNA fragments is placed in a gel and an electric field is applied across the gel. The negatively charged DNA moves toward the positive end of the field, with smaller molecules moving faster than larger ones. When the electric power is shut off, the separate fragments can be analyzed.

DNA is negatively charged at neutral pH. A mixture of DNA fragments is placed in a porous gel, and an electric field (with positive and negative ends) is applied across the gel. Because opposite charges attract, the DNA moves toward the positive end of the field. Since the porous gel acts as a sieve, the smaller molecules move faster than the larger ones. After a fixed time, and while all fragments are still on the gel, the electric power is shut off. The separated fragments can then be examined or removed individually.

Different samples of fragmented DNA can be analyzed side by side on a gel. DNA fragments of known molecular size are often run in a lane on the gel next to the sample to provide a size reference. The separated DNA fragments can be visualized by staining them with a dye that fluoresces under ultraviolet light. Or a specific DNA sequence can be located by denaturing the DNA in the gel, affixing the denatured DNA to a nylon membrane to make a "blot" of the gel, and exposing the fragments to a single-stranded DNA *probe* with a sequence complementary to the one that is being sought (Figure 17.3). The probe can be labeled in some way—for example, with radioactive phosphorus (P^{32}). Therefore, after hybridization, the presence of radioactivity on the membrane indicates that the probe has hybridized to its target at that location. The gel region containing a desired fragment can be removed when the gel is sliced, and then the pure DNA fragment can be removed from the gel.

Recombinant DNA can be made in a test tube

Some restriction enzymes cut the DNA backbone cleanly, cutting both strands exactly opposite one another. Others make two staggered cuts, cutting one strand of the double helix several bases away from where they cut the other. Fragments cut in this manner are particularly useful in biotechnology.

*Eco*RI, for example, cuts DNA within its recognition sequence in a staggered manner, as shown at the top of Figure 17.4. After the two cuts in the opposing strands are made, the strands are held together only by the hydrogen bonding between four base pairs. The hydrogen bonds of these few base pairs are too weak to persist at warm temperatures (above room temperature), so the two strands of DNA separate, or denature. As a result, there are single-stranded

17.3 Analyzing DNA Fragments

A hybridization probe can be used to locate a specific DNA fragment on an electrophoresis gel.

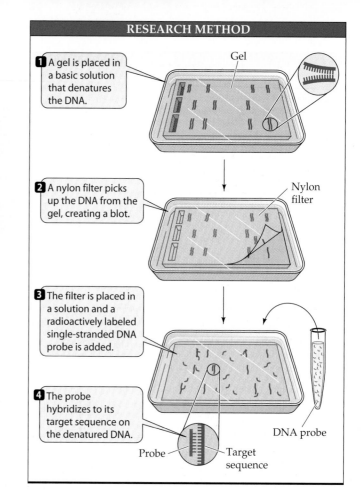

RESEARCH METHOD

1 A gel is placed in a basic solution that denatures the DNA.

Gel

2 A nylon filter picks up the DNA from the gel, creating a blot.

Nylon filter

3 The filter is placed in a solution and a radioactively labeled single-stranded DNA probe is added.

4 The probe hybridizes to its target sequence on the denatured DNA.

DNA probe

Probe   Target sequence

"tails" at the location of each cut. These tails are called *sticky ends* because they have a specific base sequence that can bind by base pairing with complementary sticky ends. If more than one recognition site for a given restriction enzyme is present in a DNA sample, numerous fragments can be made, all with the same sequence at their sticky ends.

After a DNA molecule has been cut with a restriction enzyme, the complementary sticky ends can form hydrogen bonds with one another. The original ends may rejoin, or an end may pair with a complementary end from another fragment. Furthermore, because all *Eco*RI ends are the same, fragments from one source, such as a human, can be joined to fragments from another, such as a bacterium.

When the temperature is lowered, the fragments anneal (come together by hydrogen bonding) at random, but these associations are unstable because they are held together by only a few pairs of hydrogen bonds. The associated sticky ends can be permanently united by a second enzyme, **DNA ligase**, which forms a covalent bond to "seal" each DNA strand. In the cell, this enzyme unites the Okazaki fragments and mends breaks in DNA (see Chapter 11).

Many restriction enzymes do not produce sticky ends. Instead, they cut both DNA strands at the same base pair within the recognition sequence, making "blunt" ends. DNA ligase can also connect blunt-ended fragments, but it does so with reduced efficiency.

With these two enzyme tools—restriction endonucleases and DNA ligases—scientists can cut and rejoin different DNA molecules to form **recombinant DNA** (see Figure 17.4). These simple techniques have revolutionized biological science in the past 25 years.

Cloning Genes

The goal of recombinant DNA work is to produce many copies (clones) of a particular gene, either for purposes of analysis or to produce its protein product in quantity. If the DNA is to make its protein, it must be inserted, or *transfected*, into a host cell. The choice of a host cell—prokaryotic or eukaryotic—is important. Once the host species is selected, the recombinant DNA is brought together with the host cells and, under specific conditions, can enter some of them.

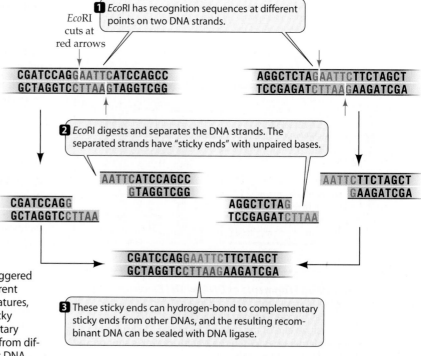

1 *Eco*RI has recognition sequences at different points on two DNA strands.

*Eco*RI cuts at red arrows

```
CGATCCAGGAATTCATCCAGCC
GCTAGGTCCTTAAGTAGGTCGG
```

```
AGGCTCTAGAATTCTTCTAGCT
TCCGAGATCTTAAGAAGATCGA
```

2 *Eco*RI digests and separates the DNA strands. The separated strands have "sticky ends" with unpaired bases.

```
AATTCATCCAGCC
    GTAGGTCGG
```

```
AATTCTTCTAGCT
    GAAGATCGA
```

```
CGATCCAGG
GCTAGGTCCTTAA
```

```
AGGCTCTAG
TCCGAGATCTTAA
```

```
CGATCCAGGAATTCTTCTAGCT
GCTAGGTCCTTAAGAAGATCGA
```

3 These sticky ends can hydrogen-bond to complementary sticky ends from other DNAs, and the resulting recombinant DNA can be sealed with DNA ligase.

17.4 Cutting and Splicing DNA

Some restriction enzymes (*Eco*RI is shown here) make staggered cuts in DNA. *Eco*RI can be used to cut DNA from two different sources (shown here in blue and green). At warm temperatures, the two DNA strands will separate (denature), leaving "sticky ends," exposed bases that can hybridize with complementary fragments. When the temperature is lowered, sticky ends from different DNAs can bind to each other, forming recombinant DNA.

Because all the host cells proliferate—not just the few that receive the recombinant DNA—the scientist must be able to determine which cells actually contain the sequence of interest. One common method of identifying cells with recombinant DNA is to tag the inserted sequence with genetic markers, called reporter genes, whose phenotypes are easily observed.

Genes can be inserted into prokaryotic or eukaryotic cells

The initial successes of recombinant DNA technology were achieved using bacteria as hosts. As noted in preceding chapters, bacterial cells are easily grown and manipulated in the laboratory. Much of their molecular biology is known, especially for certain bacteria, such as *E. coli*, and numerous genetic markers can be used to select for cells harboring the recombinant DNA. Bacteria also contain small circular chromosomes called plasmids, which, as we will see, can carry recombinant DNA into the cell.

In some important ways, however, bacteria are not ideal organisms for studying and expressing eukaryotic genes. Bacteria lack the splicing machinery to excise introns from the initial RNA transcript of eukaryotic genes. In addition, many eukaryotic proteins are extensively modified after translation by reactions such as glycosylation and phosphorylation. Often these modifications are essential for the protein's activity. Finally, in some instances, the expression of the new gene in a eukaryote is the point of the experiment. That is, the aim is to produce a **transgenic** organism, defined as an organism to which a new gene has been added. In these cases, the host for the new DNA may be a mouse, a wheat plant, a yeast, or a human, to name a few examples.

Yeasts, such as *Saccharomyces*, the baker's and brewer's yeasts, are common eukaryotic hosts for recombinant DNA studies. Advantages of using yeasts include rapid cell division (a life cycle completed in 2 to 8 hours), ease of growth in the laboratory, and a relatively small genome size (about 20 million base pairs). The yeast genome is several times larger than that of *E. coli*, yet only 1/150 the size of the mammalian one. Nevertheless, yeast has most of the characteristics of a eukaryote, except for those involved with multicellularity.

Plant cells can also be used as hosts, especially if the desired result is a transgenic plant. The property that makes plant cells good hosts is their *totipotency*—that is, the ability of a differentiated cell to act like a fertilized egg and produce an entire new organism. Isolated plant cells grown in culture can take up recombinant DNA, and by manipulation of the growth medium, these transgenic cells can be induced to form an entire new plant. The transgenic plant can then be reproduced naturally in the field and will carry and express the gene carried on the recombinant DNA.

Vectors can carry new DNA into host cells

In natural environments, DNA released from one bacterium can sometimes be taken up by another bacterium and genetically transform that bacterium (see Chapter 11),

but this is not common. The challenge of inserting new DNA into a cell is not just getting it into the host cell, but getting it to replicate in the host cell as it divides. As you know, DNA polymerase, the enzyme that catalyzes replication, does not bind to just any sequence of DNA to begin the replication. Rather, like any DNA-binding protein, it recognizes a specific sequence, the *origin of replication* (see Chapter 11).

There are two general ways in which the newly introduced DNA can become part of a *replicon*, or replication unit. First, it can insert into the host chromosome after entering the cell. Although this insertion is often a random event, it is nevertheless a common method of integrating a new gene into the host cell. Alternatively, the new DNA can enter the host cell as part of a carrier DNA sequence that already has the appropriate origin of replication. This carrier DNA, targeted at the host cell, is called a **vector**.

In addition to its ability to replicate independently in the host cell, a vector must have two other properties:

▶ A vector must have a recognition sequence for a restriction enzyme, permitting it to form recombinant DNA.
▶ A vector must have a **genetic marker** that will announce its presence in the host cell.

For ease of isolation and manipulation, a vector should also be small in comparison to host chromosomes.

PLASMIDS AS VECTORS. The properties of plasmids make them ideal vectors for the introduction of recombinant DNA into bacteria. Each plasmid is a naturally occurring bacterial chromosome, and has an origin of replication. An *E. coli* plasmid is small, usually 2,000 to 6,000 base pairs, as compared to the main *E. coli* chromosome, which has more than 4 million base pairs. Because it is so small, a plasmid often has only a single site for a given restriction enzyme (Figure 17.5a). This fact is essential because it allows for insertion of new DNA at only that location (see Figure 17.4). When the plasmid is cut with a restriction enzyme, it is transformed into a linear molecule with sticky ends. The sticky ends of another DNA fragment cut with the same restriction enzyme can pair with the sticky ends of the plasmid, resulting in a circular plasmid containing the new DNA.

Two other characteristics make plasmids good vectors. First, many plasmids contain genes for enzymes that confer resistance to antibiotics. This characteristic provides a genetic marker for host cells carrying the recombinant plasmid. The second useful property of plasmids is their capacity to replicate independently of the host chromosome. It is not uncommon for a bacterial cell with a single main chromosome to have hundreds of copies of a recombinant plasmid.

The plasmids commonly used as vectors in the laboratory have been extensively altered, and most are combinations of genes and other sequences from several sources. Many of these plasmids have a single marker for antibiotic resistance.

(*a*) Plasmid pBR322
Host: *E. coli*

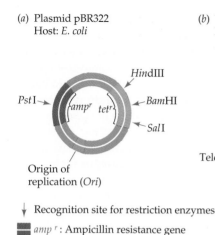

Pst I

*amp*ʳ *tet*ʳ

Hind III

Bam HI

Sal I

Origin of
replication (*Ori*)

↓ Recognition site for restriction enzymes

■ *amp*ʳ: Ampicillin resistance gene

▬ *tet*ʳ: Tetracycline resistance gene

(*b*) Yeast artificial chromosome
Host: yeast

Centromere

Eco RI

Ori

Selectable
marker

Telomere

Telomere

Bam HI

(*c*) Ti plasmid
Hosts: *Agrobacterium tumefaciens* (plasmid)
and infected plants (Ti DNA)

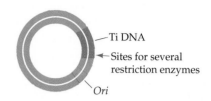

Ti DNA

Sites for several
restriction enzymes

Ori

17.5 Vectors for Carrying DNA into Cells
(*a*) A plasmid with genes for antibiotic resistance can be incorporated into an *E. coli* cell. (*b*) A DNA molecule synthesized in the laboratory becomes a chromosome that can carry its inserted DNA into yeasts. (*c*) The Ti plasmid, isolated from the bacterium *Agrobacterium tumefaciens*, is an important vector for inserting DNA into many types of plants.

VIRUSES AS VECTORS. Constraints on plasmid replication limit the size of the new DNA that can be spliced into a plasmid to about 5,000 base pairs. Although a prokaryotic gene may be this small, 5,000 base pairs is much smaller than most eukaryotic genes, with their introns and extensive flanking sequences. So, a vector that accommodates larger DNA inserts is needed.

Both prokaryotic and eukaryotic viruses are often used as vectors for eukaryotic DNA. Bacteriophage lambda, which infects *E. coli*, has a DNA genome of more than 45,000 base pairs. If the genes that cause the host cell to die and lyse—about 20,000 base pairs—are eliminated, the virus can still infect a host cell and inject its DNA. The deleted 20,000 base pairs can be replaced with DNA from another organism, thereby creating recombinant viral DNA.

Because viruses infect cells naturally, they offer a great advantage as vectors over plasmids, which often require artificial means to coax them to enter cells. As we will see in Chapter 18, viruses are important vectors for delivering new genes to people in gene therapy.

ARTIFICIAL CHROMOSOMES AS VECTORS. Bacterial plasmids are not good vectors for yeast hosts, because prokaryotic and eukaryotic DNA sequences use different origins of replication. Thus a recombinant bacterial plasmid will not replicate in yeast. To remedy this problem, scientists have created in the laboratory a "minimalist chromosome" called the *yeast artificial chromosome*, or *YAC* (Figure 17.5*b*). This DNA molecule contains not only the yeast origin of replication, but sequences for the yeast centromere and telomeres as well, making it a true eukaryotic chromosome. YACs also contain artificially synthesized single restriction sites and useful marker genes (for yeast nutritional requirements). YACs are only about 10,000 base pairs in size, but can accommodate 50,000 to 1.5 million base pairs of inserted DNA.

There has been considerable progress in creating a *human* artificial chromosome (HAC), which could someday be used as a gene therapy vector. Instead of yeast cen-

tromere and telomere sequences, their human counterparts have been used. The vector acts as a separate minichromosome in human cells, and can be maintained there for months.

PLASMID VECTORS FOR PLANTS. An important vector for carrying new DNA into many types of plants is a plasmid that is found in *Agrobacterium tumefaciens*. This bacterium lives in the soil and causes a plant disease called crown gall, which is characterized by the presence of growths, or tumors, in the plant. *A. tumefaciens* contains a plasmid called Ti (for *tumor-inducing*) (Figure 17.5*c*).

Part of the Ti plasmid is T DNA, a transposon that produces copies of itself in the chromosomes of infected plant cells. The T DNA has recognition sites for restriction enzymes, and new DNA can be spliced into the T DNA region of the plasmid. When the T DNA is thus replaced, the plasmid no longer produces tumors, but the transposon, with the new DNA, is inserted into the host cell's chromosomes. The plant cell containing this DNA can then be grown in culture or induced to form a new, transgenic, plant.

There are many ways to insert recombinant DNA into host cells

Although some vectors, such as viruses, can enter host cells on their own, most vectors require help to do so. A major barrier to DNA entry is that the exterior surface of the plasma membrane, with its phospholipid heads, is negatively charged, as is DNA. The resulting charge repulsion can be alleviated if the exterior of the cells and the DNA are both neutralized with Ca^{2+} (calcium) salts. The salts reduce the charge effect, and the plasma membrane becomes permeable to DNA. In this way, almost any cell, prokaryotic or eukaryotic, can take up a DNA molecule from its environment. In plants and fungi, the cell wall must first be removed by hydrolysis with fungal enzymes; the resulting wall-less plant cells are called *protoplasts*.

In addition to this "naked" DNA approach, DNA can be introduced into host cells by a variety of mechanical methods:

▶ In *electroporation*, host cells are exposed to rapid pulses of high-voltage current. This treatment temporarily renders the plasma membrane permeable to DNA in the surrounding medium.

▶ In *injection*, a very fine pipette is used to insert DNA into cells. This method is especially useful on large cells such as eggs.

▶ In *lipofection*, DNA is coated with lipid, which allows it to pass through the plasma membrane. For example, DNA can be encased in liposomes, bubbles of lipid that fuse with the membranes of the host cell.

▶ In *particle bombardment*, tiny high-velocity particles of tungsten or gold are coated with DNA and then shot into host cells. This "gene gun" approach must be undertaken with great care to prevent the cell contents from being damaged.

Genetic markers identify host cells that contain recombinant DNA

Even when a population of host cells is allowed to interact with an appropriate vector, only a small percentage of the cells actually take up the vector. Also, since the process of cutting the vector and inserting the new DNA to make recombinant DNA is far from perfect, only a few of the vectors that have moved into the host cells will actually contain the new DNA sequence. How can we select only the host cells that contain the recombinant DNA?

The experiment we are about to describe illustrates an elegant, commonly used approach to this problem. In this example, we use *E. coli* bacteria as hosts and a plasmid vector (see Figure 17.5*a*) that carries the genes for resistance to the antibiotics ampicillin and tetracycline.

When the plasmid is incubated with the restriction enzyme *Bam*HI, the enzyme encounters its recognition sequence, GGATCC, only once, at a site within the gene for tetracycline resistance. If foreign DNA is inserted into this restriction site, the presence of these "extra" base pairs within the tetracycline resistance gene inactivates it. So plasmids containing the inserted DNA will carry an intact gene for ampicillin resistance, but *not* an intact gene for tetracycline resistance. This is the key to the selection of host bacteria that contain the recombinant plasmid (Figure 17.6).

The cutting and splicing process results in three types of DNA, all of which can be taken up by the host bacteria:

▶ The recombinant plasmid—the one we want—turns out to be the rarest type of DNA. Its uptake confers on host *E. coli* resistance only to ampicillin.

▶ More common are bacteria that take up plasmids that have sealed their own ends back together. These plasmids retain intact genes for resistance to both ampicillin and tetracycline.

▶ Even more common are bacteria that take up the foreign DNA sequence alone, without the plasmid; since it is not part of a replicon, it does not survive as the bacteria divide. These host cells will remain susceptible to both antibiotics, as will the vast majority (more than 99.9 percent) of cells that take up no DNA at all.

So the unique drug resistance phenotype of the cells with recombinant DNA (tetracycline-sensitive and ampicillin-resistant) marks them in a way that can be detected by simply adding ampicillin and/or tetracycline to the medium surrounding the cells.

In addition to genes for antibiotic resistance, several other marker genes are used to detect recombinant DNA in host cells. Scientists have created several artificial vectors in the laboratory that include sites for restriction enzymes within the *lac* operon (see Chapter 13). When this gene is inactivated by the insertion of foreign DNA, the vector no longer carries this operon's function into the host cell. Other *reporter genes* that have been used in vectors include the gene for luciferase, the enzyme that causes fireflies to glow in the dark; this enzyme causes host cells to glow when supplied with its substrate. Green fluorescent protein, which normally occurs in the jellyfish *Aequopora victo-*

17.6 Marking Recombinant DNA by Inactivating a Gene
Scientists manipulate marker genes within plasmids so they will know which host cells have incorporated the recombinant genes. The host bacteria in this experiment could display any of the phenotypes indicated in the table. Assuming we wish to select only those that have taken up the recombinant plasmid, we can do so by adding antibiotics to the medium surrounding the cells.

1 A plasmid has genes for resistance to both ampicillin (*amp^r*) and tetracycline (*tet^r*).

2 Foreign DNA is inserted at the *Bam*HI recognition site, which is within the *tet^r* gene.

3 The resulting recombinant DNA has an intact functional gene for ampicillin resistance but not tetracycline resistance.

Detection of Recombinant DNA in E. coli		
DNA TAKEN UP BY AMP^S AND TET^S *E. COLI*	PHENOTYPE FOR AMPICILLIN	PHENOTYPE FOR TETRACYCLINE
None	Sensitive	Sensitive
Foreign DNA only	Sensitive	Sensitive
pBR322 plasmid	Resistant	Resistant
pBR322 recombinant plasmid	Resistant	Sensitive

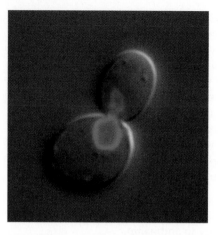

17.7 A Reporter Gene Announces the Presence of a Vector in Eukaryotic Cells
These cells have taken up a vector that expresses a gene producing green fluorescent protein.

riana, does not require a substrate to glow, and is now widely used (Figure 17.7).

Many vectors in common use contain only a *single* marker for antibiotic resistance, outside of the sites for foreign DNA insertion. In this case, the recombinant DNA will have the same antibiotic resistance gene that the non-recombinant plasmid does. The formation of recombinant DNA in the ligase reaction is favored if there is a high concentration of foreign DNA fragments compared to the cut plasmid. So there will be a preponderance of colonies containing recombinant DNA among those that grow in the presence of the antibiotic.

After DNA uptake (or not), host cells are usually first grown on a solid medium. If the concentration of cells dispersed on the solid medium is low, each cell will divide and grow into a distinct colony (see Chapter 13). The colonies that contain recombinant DNA can be identified and removed from the medium, and then grown in large amounts in liquid culture. A quick examination of a plasmid can confirm whether the plasmids in the cells of the colony actually have the recombinant DNA. The power of bacterial transfection to amplify a gene is indicated by the fact that a 1-liter culture of bacteria harboring the human β-globin gene in the pBR322 plasmid has as many copies of the gene as the sum total of all the cells in a typical human being.

Sources of Genes for Cloning

The genes or DNA fragments used in recombinant DNA work are obtained from three principal sources. One source is random pieces of chromosomes maintained as gene libraries. The second source is complementary DNA, obtained by reverse transcription from specific mRNA's. The third source is DNA synthesized by organic chemists in the laboratory. Specific fragments can be deliberately modified to create mutations or to change a mutant sequence back to the wild type.

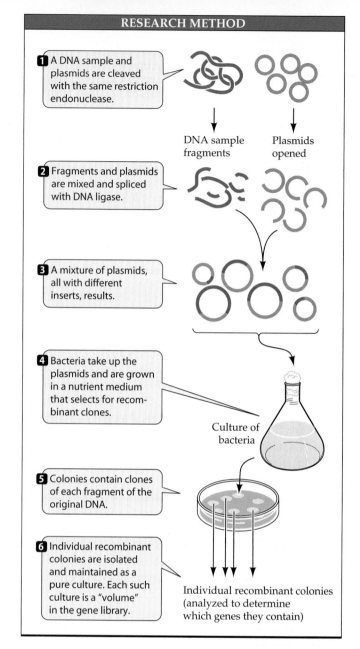

RESEARCH METHOD

1 A DNA sample and plasmids are cleaved with the same restriction endonuclease.

DNA sample fragments   Plasmids opened

2 Fragments and plasmids are mixed and spliced with DNA ligase.

3 A mixture of plasmids, all with different inserts, results.

4 Bacteria take up the plasmids and are grown in a nutrient medium that selects for recombinant clones.

Culture of bacteria

5 Colonies contain clones of each fragment of the original DNA.

6 Individual recombinant colonies are isolated and maintained as a pure culture. Each such culture is a "volume" in the gene library.

Individual recombinant colonies (analyzed to determine which genes they contain)

17.8 Constructing a Gene Library
Human chromosomes are broken up into fragments of DNA that are inserted into vectors (plasmids are shown here) and taken up by host bacterial cells, each of which harbors a single fragment of the human DNA. The information in these bacterial colonies constitutes a gene library.

Gene libraries contain pieces of a genome

The 23 pairs of human chromosomes can be thought of as a library that contains the entire genome of our species. Each chromosome, or "volume" in the library, contains, on average, 80 million base pairs of DNA, encoding several thousand genes. Such a huge molecule is not very useful for studying genome organization or for isolating a specific gene. To address this problem, researchers can break each chromosome into smaller pieces using restriction enzymes, and then analyze each piece. These smaller fragments still represent a **gene library** (Figure 17.8); however, the information is now in many more volumes than 23. Each of the fragments can be inserted into a vector, which can then be taken up by a host bacterial cell. Each host cell colony, then, harbors a single fragment of human DNA.

Using plasmids, which are able to insert only a few thousand base pairs of foreign DNA into a bacterium, about a million separate fragments are required to make a library of the human genome. By using phage lambda, which can carry four times as much DNA as a plasmid, the number of volumes is reduced to about 250,000. Although this seems like a large number, a single growth plate can hold up to 80,000 phage colonies, or *plaques*, and is easily screened for the presence of a particular DNA sequence by denaturing the phage DNA and applying a particular probe for hybridization.

A DNA copy of mRNA can be made

A much smaller DNA library—one that includes only genes transcribed in a particular tissue—can be made from **complementary DNA**, or **cDNA** (Figure 17.9). Recall that most eukaryotic mRNA's have a poly A tail—a string of adenine residues at their 3' end (see Figure 14.11). The first step in cDNA production is to extract mRNA from a tissue and allow it to hybridize with a molecule called oligo dT (the "d" indicates *deoxyribose*), which consists of a string of thymine residues. After the oligo dT hybridizes with the poly A tail of the mRNA, it serves as a primer, and the mRNA as a template, for the enzyme reverse transcriptase, which synthesizes DNA from RNA. In this way, a cDNA strand complementary to the mRNA is formed.

A collection of cDNA's from a particular tissue at a particular time in the life cycle of an organism is called a **cDNA library**. mRNA's do not last long in the cytoplasm and are often present in small amounts, so a cDNA library is a "snapshot" that preserves the transcription pattern of the cell. cDNA libraries have been invaluable in comparisons of gene expression in different tissues at different stages of development. For example, their use has shown that up to one-third of all the genes of an animal are ex-

pressed only during prenatal development. Complementary DNA is also a good starting point for the cloning of eukaryotic genes. It is especially useful for genes expressed at low levels in only a few cell types.

DNA can be synthesized chemically in the laboratory

When we know the amino acid sequence of a protein, we can obtain the DNA that codes for it by simply making it in the laboratory, using organic chemistry techniques. DNA synthesis has even been automated, and at many institutions, a special service laboratory can make short-to-medium-length sequences overnight for any number of investigators.

How do we design a synthetic gene? Using the genetic code and the known amino acid sequence, we can figure out the most likely base sequence for the gene. With this sequence as a starting point, we can add other sequences, such as codons for translation initiation and termination and flanking sequences for transcription initiation, termination, and regulation. Of course, these noncoding DNA sequences must be the ones actually recognized by the host cell if the synthetic gene is to be transcribed. It does no good to have a prokaryotic promoter sequence near a gene if that gene is to be inserted into a yeast cell for expression. Codon usage is also important: Many amino acids are encoded by more than one codon, and different organisms stress the use of different synonymous codons.

DNA can be mutated in the laboratory

Mutations that occur in nature have been important in proving cause-and-effect relationships in biology. For ex-

17.9 Synthesizing Complementary DNA
Gene libraries that include only genes transcribed in a particular tissue at a particular time can be made from complementary DNA. cDNA synthesis is especially useful for identifying genes that are present only in a few copies, and is often a starting point for gene cloning.

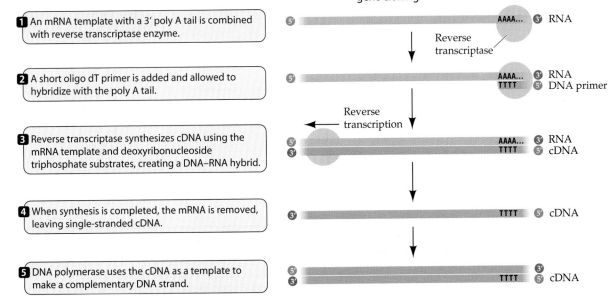

1 An mRNA template with a 3' poly A tail is combined with reverse transcriptase enzyme.

2 A short oligo dT primer is added and allowed to hybridize with the poly A tail.

3 Reverse transcriptase synthesizes cDNA using the mRNA template and deoxyribonucleoside triphosphate substrates, creating a DNA–RNA hybrid.

4 When synthesis is completed, the mRNA is removed, leaving single-stranded cDNA.

5 DNA polymerase uses the cDNA as a template to make a complementary DNA strand.

ample, in Chapter 14, we learned that some people with the disease beta thalassemia have a mutation at the consensus sequence for intron removal and so cannot make proper β-globin mRNA. This example shows the importance of the consensus sequence.

Recombinant DNA technology has allowed us to ask "What if?" questions without having to look for mutations in nature. Because synthetic DNA can be made in any sequence desired, we can manipulate DNA to create specific mutations and then see what happens when the mutant DNA expresses itself in a host cell. Additions, deletions, and base-pair substitutions are all possible with isolated or synthetic DNA.

These mutagenesis techniques have allowed scientists to bypass the search for naturally occurring mutant strains, leading to many cause-and-effect proofs. For example, it was proposed that the signal sequence at the beginning of a secreted protein is essential to its passage through the endoplasmic reticulum membrane. So, a gene coding for such a protein, but with the codons for the signal sequence deleted, was made. Sure enough, when this gene was expressed in yeast cells, the protein did not cross the ER membrane. When the signal sequence codons were added to an unrelated gene encoding a soluble cytoplasmic protein, that protein crossed the ER membrane.

Mutagenesis has also begun to be useful in the design of specific drugs. The advent of a new branch of biology called *computational biology* has led to sophisticated studies of the three-dimensional shapes and chemical properties of enzymes, substrates, and their possible regulators. Attempts are being made to devise rules to predict the tertiary structure of a protein from its primary structure. For example, if we know the structure of an enzyme, the three-dimensional design of a polypeptide regulating that enzyme might be proposed. Mutant bacterial strains with genes coding for variants of this polypeptide could be made. Then, the variant polypeptides could be isolated and used to test the relationship between structure and activity.

Some Additional Tools for DNA Manipulation

Biological methods are not the only ways of manipulating DNA managed in the laboratory. In Chapter 11, we described DNA sequencing and the polymerase chain reaction, two applications of DNA replication techniques. Here, we examine three additional techniques. One is the use of genetic recombination to create an inactive, or "knocked-out," gene. The second is the use of "DNA chips" to detect the presence of many different sequences simultaneously. The third is the use of antisense RNA to block the translation of specific mRNA's.

Genes can be inactivated by homologous recombination

As we have seen, laboratory-created mutations are an excellent way to ask the "what if" questions about the role of a gene in cell function. *Homologous recombination* is used to ask these questions at the organism level (Figure 17.10). The

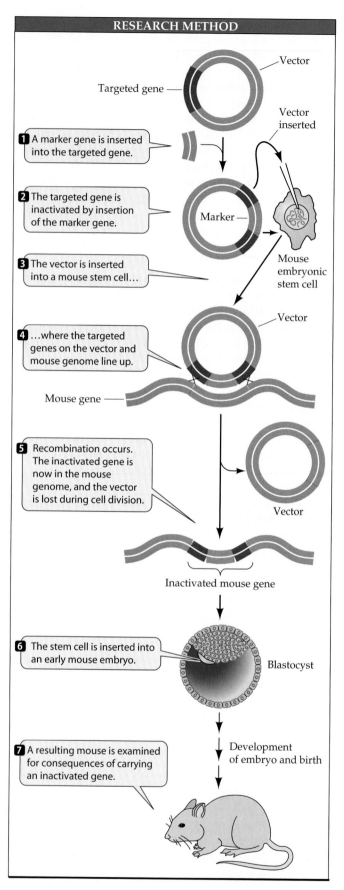

RESEARCH METHOD

1 A marker gene is inserted into the targeted gene.

2 The targeted gene is inactivated by insertion of the marker gene.

3 The vector is inserted into a mouse stem cell...

4 ...where the targeted genes on the vector and mouse genome line up.

5 Recombination occurs. The inactivated gene is now in the mouse genome, and the vector is lost during cell division.

6 The stem cell is inserted into an early mouse embryo.

7 A resulting mouse is examined for consequences of carrying an inactivated gene.

Vector

Targeted gene

Vector inserted

Marker

Mouse embryonic stem cell

Vector

Mouse gene

Vector

Inactivated mouse gene

Blastocyst

Development of embryo and birth

17.10 Making a Knockout Mouse
Homologous recombination is used to replace a normal mouse gene with an inactivated copy, thus "knocking out" the active gene. Discovering what happens to a mouse with an inactive gene tells us much about the role of that gene.

aim of this technique is to replace a gene inside a cell with an inactivated form of that gene, and then see what happens when the inactive gene is part of an organism. Such a manipulation is called a *knockout* experiment.

Mice are frequently used in knockout experiments. The mouse gene to be tested is inserted into a plasmid. Restriction enzymes are then used to insert another fragment, containing a genetic marker, in the middle of this gene. Addition of extra DNA to the gene creates havoc with its transcription and translation; a functional mRNA is seldom made from such an interrupted gene. Next, the plasmid is transfected into a mouse embryonic stem cell (see Chapter 16). Because much of the targeted gene is still present in the plasmid (although in two separated regions), there is DNA sequence recognition between the gene on the plasmid and the homologous gene in the mouse genome. As in prophase I of meiosis, the plasmid and the mouse chromosomes line up, and, sometimes, a genetic exchange occurs in which the plasmid's inactive gene is swapped for the functional gene in the host cell. The genetic marker in the insert is used to identify those stem cells carrying the inactivated gene. The transfected stem cell is now inserted into an early mouse embryo, and through some clever tricks, a knockout mouse carrying the inactivated gene in homozygous form can be produced. The phenotype of the mutant mouse is an indication of the role of the gene in the normal, wild-type animal. The knockout technique has been very important in assessing the roles of genes during development.

DNA chips can reveal DNA mutations and RNA expression

The emerging science of genomics deals with two major quantitative circumstances: First, there are a large number of genes in eukaryotic genomes. Second, the pattern of gene expression in different tissues at different times is quite distinctive. For example, a skin cancer cell at its early stage may have a unique mRNA "fingerprint" that differs from that of both normal skin cells and more advanced skin cancer cells.

To find these patterns, scientists could isolate total cell mRNA and test it by hybridization with each gene in the genome, one gene at a time. But it would be far better to do these hybridizations all in one step. For this, one needs some way to arrange all the DNA sequences in a genome in an array on some solid support.

DNA chip technology has been developed to provide these large arrays of sequences for hybridization. The chips were developed by modifications of methods that have been used for several decades in the semiconductor industry. You may be familiar with the silicon microchip, in which an array of microscopic electric circuits is etched onto a tiny chip. In the same way, DNA chips are glass slides onto which are attached, in precise order, pre-established sequences of DNA (Figure 17.11). Typically, the slide is divided into 24×24 μM squares, each of which contains about 10 million copies of a particular sequence, up to 20 nucleotides long. A computer controls the addition of nucleotides in a predetermined pattern. Up to 60,000 different sequences can be put on a single chip.

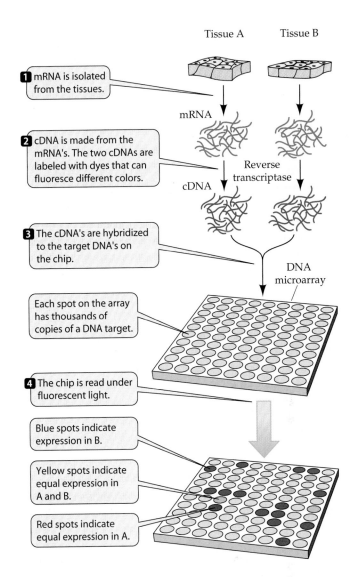

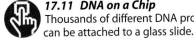

17.11 DNA on a Chip
Thousands of different DNA probes of known sequence can be attached to a glass slide.

If cellular mRNA is to be analyzed, it is usually incubated with reverse transcriptase (RT) to make cDNA (see Figure 17.9), and the cDNA is amplified by PCR prior to hybridization. This technique is called "RT-PCR," and it ensures that mRNA sequences present in only a few copies (or in a small sample such as a cancer biopsy) will be numerous enough to form a signal when hybridized. The amplified cDNA's are coupled to a fluorescent dye and then allowed to hybridize with DNA on a chip. Those DNA sequences that form a hybrid can be located by a sensitive scanner. With the number of genes on a chip approaching that of the largest genomes, these chips will result in an information explosion on mRNA transcription patterns in cells in different physiological states.

Another use for DNA chips is in detecting genetic variants. Suppose one wants to find out if a particular gene, which is 5,500 base pairs long, has any mutations in a particular individual. One way would be to sequence the entire gene, but this would be difficult to do, and would require a large tissue sample. On the other hand, chip technology can be used to make 20-nucleotide fragments along the gene in every possible mutant sequence. Then, probing with the individual's DNA might reveal a particular mutation if it hybridized to a mutant sequence on the chip. This method may provide a rapid way to detect mutations in people.

Antisense RNA and ribozymes can prevent the expression of specific genes

The base-pairing rules not only can be used to make genes; they can also be employed to stop the translation of mRNA. As is often the case, this technique is an example of scientists imitating nature. In normal cells, a rare method of controlling gene expression is the production of an RNA molecule that is complementary to mRNA. This complementary molecule is called **antisense RNA** because it binds by base pairing to the "sense" bases on the mRNA that code for a protein. The formation of a double-stranded RNA hybrid prevents tRNA from binding to the mRNA, and the hybrid tends to be broken down rapidly in the cytoplasm. So, although the gene continues to be transcribed, translation does not take place.

In the laboratory, after determining the sequence of a gene and its mRNA, scientists can add antisense RNA to a cell to prevent translation of the mRNA (Figure 17.12). The antisense RNA can be added as itself—RNA can be inserted into cells in the same way that DNA is—or it can be made in the cell by transcription from a DNA molecule introduced as a part of a vector.

Without this technique, repressing the synthesis of a specific protein would be very difficult. It is especially useful if a tissue-specific promoter is used to prime transcription of

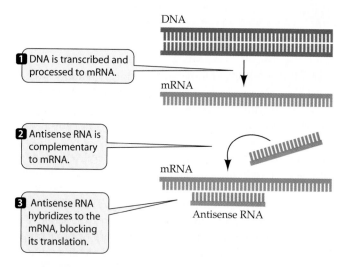

1 DNA is transcribed and processed to mRNA.

2 Antisense RNA is complementary to mRNA.

3 Antisense RNA hybridizes to the mRNA, blocking its translation.

DNA

mRNA

mRNA

Antisense RNA

17.12 Using Antisense RNA to Block Translation of an mRNA
Once a gene's sequence is determined in the laboratory, the synthesis of its protein can be prevented using antisense RNA that is complementary to its mRNA.

the antisense RNA, so that its expression occurs only in a targeted tissue. An even more effective way to ensure that antisense RNA works is to couple the antisense sequence to a special RNA sequence—a *ribozyme*—that catalyzes the cleavage of its target RNA.

Antisense RNA (with or without a ribozyme) has been widely used to test cause-and-effect relationships. For example, when antisense RNA was used to block the synthesis of a protein essential for the growth of cancer cells, the cells reverted to a normal state.

Biotechnology: Applications of DNA Manipulation

Biotechnology is the use of microbial, plant, and animal cells to produce materials useful to people. These products include foods, medicines, and chemicals. We have been making some of them for a long time. For example, the use of yeasts to brew beer and wine dates back at least 8,000 years in human history, and the use of bacterial cultures to make cheese and yogurt is a technique many centuries old.

For a long time, people were not aware of the cellular bases of these biochemical transformations. About 100 years ago, thanks largely to Pasteur's work, it became clear that specific bacteria, yeasts, and other microbes could be used as biological converters to make certain products. Alexander Fleming's discovery that the mold *Penicillium* makes the antibiotic penicillin led to the large-scale commercial culture of microbes to produce antibiotics as well as other useful chemicals. Today, microbes are grown in vast quantities to make much of the industrial-grade alcohol, glycerol, butyric acid, and citric acid that are used by themselves or as starting materials in the manufacture of other products.

In the past, the list of such products was limited to those that were naturally made by microbes. The many products that eukaryotes make, such as hormones and certain enzymes, had to be extracted from those complex organisms. Yields were low, and purification was difficult and costly. All this has changed with the advent of gene cloning. The ability to insert almost any gene into bacteria or yeast, along with methods to induce the gene to make its product, has turned these microbes into versatile factories for important products.

Expression vectors can turn cells into protein factories

If a eukaryotic gene is inserted into a typical plasmid (see Figure 17.5a) and cloned into *E. coli*, little, if any, of the product of the gene will be made by the host cell. The reason is that the eukaryotic gene lacks the bacterial promoter for RNA polymerase binding, the terminator for transcription, and a special sequence on mRNA for ribosome binding. All of these are necessary for the gene to be expressed and its products synthesized in the bacterial cell.

Expression vectors can be made that have all the characteristics of typical vectors, as well as the extra sequences

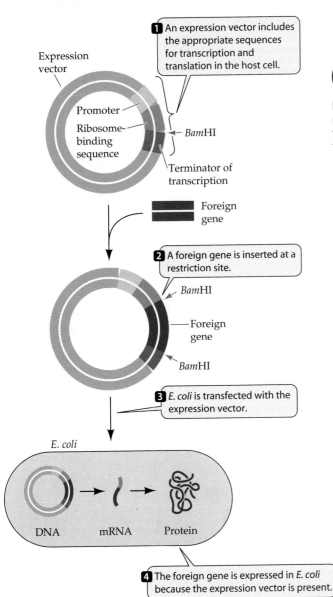

1 An expression vector includes the appropriate sequences for transcription and translation in the host cell.

Expression vector

Promoter

Ribosome-binding sequence

*Bam*HI

Terminator of transcription

Foreign gene

2 A foreign gene is inserted at a restriction site.

*Bam*HI

Foreign gene

*Bam*HI

3 *E. coli* is transfected with the expression vector.

E. coli

DNA → mRNA → Protein

4 The foreign gene is expressed in *E. coli* because the expression vector is present.

17.13 An Expression Vector Allows a Foreign Gene to Be Expressed in a Host Cell
An inserted eukaryotic gene may not be expressed in *E. coli* because it lacks the necessary bacterial sequences for promotion, termination, and ribosome binding. Expression vectors contain these additional sequences, enabling the eukaryotic protein to be synthesized in the prokaryotic cell.

hormone is added. An enhancer that responds to hormonal stimulation might also be added so that transcription and protein production will occur at very high rates—a goal of obvious importance in the manufacture of an industrial product.

A *tissue-specific promoter*, which is expressed only in a certain tissue at a certain time, can be used if localized expression in an organism is desired. For example, many seed proteins are expressed only in the plant embryo. So coupling a gene to a seed-specific promoter will allow the gene to be expressed only as a seed protein.

Targeting sequences can be added to the gene in the expression vector so that the protein product is directed to an appropriate destination. For example, in a large vessel containing yeast cells making a protein, it might be useful for the protein to be secreted into the extracellular medium for easier recovery.

Medically useful proteins can be made by DNA technology

Many medically useful products have been made by recombinant DNA technology (Table 17.1), and hundreds more are in various stages of development. We will describe three such products to illustrate the techniques that have been used in their development.

needed for the foreign gene to be expressed in the host cell. For bacteria, these additional sequences include the bacterial promoter, the transcription terminator, and the sequence for ribosome binding (Figure 17.13). For eukaryotes, expression vectors would include the poly A addition site, transcription factor binding sites, and enhancers. Once these sequences are placed at the appropriate location on the vector, the gene will be expressed in the host cell.

An expression vector can be refined in various ways. An *inducible promoter*, which responds to a specific signal, can be made part of an expression vector. For example, a specific promoter can be used that responds to hormonal stimulation so that the foreign gene can be induced to transcribe its mRNA when the

17.1 Some Medically Useful Products of Biotechnology

PRODUCT	USE
Brain-derived neurotropic factor	Stimulates regrowth of brain tissue in patients with Lou Gehrig's disease
Colony-stimulating factor	Stimulates production of white blood cells in patients with cancer and AIDS
Erythropoietin	Prevents anemia in patients undergoing kidney dialysis
Factor VIII	Replaces clotting factor missing in patients with hemophilia A
Growth hormone	Replaces missing hormone in people of short stature
Insulin	Stimulates glucose uptake from blood in some people with diabetes
Platelet-derived growth factor	Stimulates wound healing
Tissue plasminogen activator	Dissolves blood clots after heart attacks and strokes
Vaccine proteins: Hepatitis B, herpes, influenza, Lyme disease, meningitis, pertussis, etc.	Prevent and treat infectious diseases

TISSUE PLASMINOGEN ACTIVATOR. In most people, when a wound begins bleeding, a blood clot soon forms to stop the flow. Later, as the wound heals, the clot dissolves. How does the blood perform these conflicting functions at the right times? Mammalian blood contains an enzyme called *plasmin* that catalyzes the dissolution of the clotting proteins. But plasmin is not always active; if it were, a blood clot would dissolve as soon as it formed! Instead, plasmin is "stored" in the blood in an inactive form called *plasminogen*. The conversion of plasminogen to plasmin is activated by an enzyme appropriately called *tissue plasminogen activator* (TPA), which is produced by cells lining the blood vessels. Thus, the reaction is

$$\text{plasminogen} \xrightarrow{\text{TPA}} \text{plasmin}$$
$$\text{(inactive)} \qquad\qquad \text{(active)}$$

Heart attacks and many strokes are caused by blood clots that form in important blood vessels leading to the heart or the brain, respectively. During the 1970s, a bacterial enzyme, streptokinase, was found to stimulate the quick dissolution of clots in some patients with these afflictions. Treating people with this enzyme saved lives, but there were side effects. The drug was a protein foreign to the body, so patients' immune systems reacted against it. More important, the drug sometimes prevented clotting throughout the circulatory system, leading to an almost hemophilia-like condition in some patients.

The discovery of TPA and its isolation from human tissues led to the hope that this enzyme could be used to bind specifically to clots, and that it would not provoke an immune reaction. But the amounts of TPA available from human tissues were tiny, certainly not enough to inject at the site of a clot in a patient in the emergency room.

Recombinant DNA technology solved the problem. TPA mRNA was isolated and used to make a cDNA copy, which was then inserted into an expression vector and introduced into *E. coli* (Figure 17.14). The transfected bacteria made the protein in quantity, and it soon became available commercially. This protein has had considerable success in dissolving blood clots in people undergoing heart attacks and, especially, strokes.

ERYTHROPOIETIN. Another protein made through recombinant DNA methods and widely used in medicine is *erythropoietin* (EPO). The kidneys produce this hormone, which travels through the blood to the bone marrow, where it stimulates the division of stem cells to produce red blood cells. People who have suffered kidney failure often require a procedure called kidney dialysis to remove toxins from the blood. However, because dialysis also removes EPO, these patients can become severely anemic (depleted of red blood cells).

As with TPA, the amounts of EPO that can be obtained from healthy people to give to people undergoing dialysis are extremely small, but once again, biotechnology has come to the rescue. The gene for EPO was isolated, inserted in an expression vector, and introduced into bacteria. Large

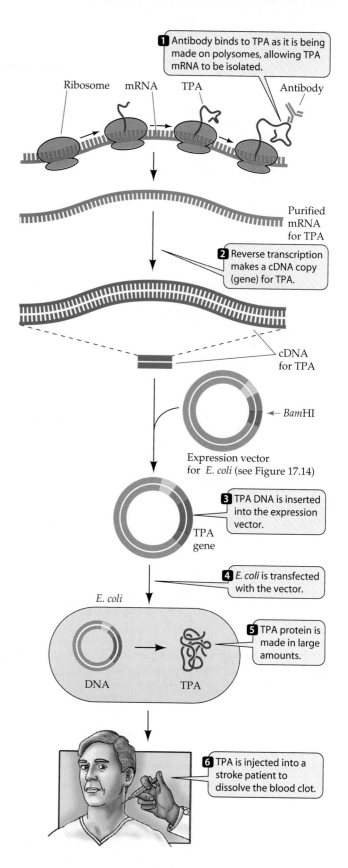

17.14 Tissue Plasminogen Activator: From Protein to Gene to Pharmaceutical
TPA is a naturally occurring human protein that prevents blood from clotting. Its isolation and use as a pharmaceutical agent for treating patients suffering from blood clotting in the brain or heart—in other words, strokes and heart attacks—was made possible by recombinant DNA technology.

amounts of the protein are now produced by bacteria and given to tens of thousands of people undergoing dialysis, with great success at reducing anemia.

HUMAN INSULIN.   One of the first important medications made by recombinant DNA methods was human insulin. This hormone, normally made by the pancreas, stimulates cells to take up glucose from the blood. People who have certain forms of diabetes mellitus have a deficiency of pancreatic insulin. Injections of the hormone can compensate for this deficiency.

In the past, the injected insulin was obtained from the pancreases of cows and pigs, which caused two problems. First, animal insulin is laborious to purify; second, it is slightly different in its amino acid sequence from human insulin. Some diabetics' immune systems detect these differences and react against the foreign protein.

The ideal solution is to use human insulin, but until the advent of recombinant DNA technology, it was available only in minuscule amounts. Since insulin is made up of only 51 amino acids, scientists were able to synthesize a gene for this protein in the laboratory. This gene (there were actually two of them, one for each polypeptide chain of the protein) was inserted into *E. coli* via an expression vector. Production of human insulin by the bacteria made widespread use of the human hormone by diabetics feasible.

DNA manipulation is changing agriculture

The cultivation of plants and husbanding of animals that constitute agriculture give us the world's oldest examples of biotechnology, dating back more than 8,000 years in human history. Over the centuries, people have adapted crops and farm animals to their needs. Through cultivation and breeding (artificial selection), desirable characteristics, such as ease of cooking the seeds or quality of the meat, have been imparted and improved. In addition, people have developed crops with desirable growth characteristics, such as a reliable ripening season and resistance to diseases.

Until recently, the most common way to improve crop plants and farm animals was to select and breed varieties with the desired phenotypes that existed in nature through mutational variation. The advent of genetics in the past century was followed by its application to plant and animal breeding. A crop plant or animal with a desirable gene could be identified, and through deliberate crosses, a single gene or, more usually, many genes could be introduced into a widely used variety of that crop.

Despite spectacular successes, such as the breeding of "supercrops" of wheat, rice, and corn, such deliberate crossing remains a hit-or-miss affair. Many desirable characteristics are complex in their genetics, and it is hard to predict accurately the results of a cross. Moreover, traditional crop plant breeding takes a long time: Many plants can reproduce only once or twice a year—a far cry from the rapid reproduction of bacteria or fruit flies.

Modern recombinant DNA technology has two advantages over traditional methods of breeding. First, the molecular approach allows a breeder to choose specific genes, making the process more precise and less likely to fail as a result of the incorporation of unforeseen genes. The ability to work with cells in the laboratory and then regenerate a whole plant by cloning makes the process much faster than the years needed for traditional breeding. The second advantage—and it is truly an amazing one—is that these molecular methods allow breeders to introduce any gene from any organism into a plant or animal species. This ability, combined with mutagenesis techniques, expands the range of possible new characteristics to an almost limitless horizon.

Biotechnology has found many applications in agriculture (Table 17.2), ranging from improving the nutritional properties of crops, to using animals as gene product factories, to using edible crops to make oral vaccines. We will describe a few examples here to demonstrate the approaches that have been used.

PLANTS THAT MAKE THEIR OWN INSECTICIDES.   Humans are not the only species that consumes crop plants. Plants are subject to infections by viruses, bacteria, and fungi, but probably the most important crop pests are herbivorous insects. From the locusts of biblical (and modern) times to the cotton boll weevil, insects have continually eaten the crops people grow.

The development of insecticides has improved the situation somewhat, but insecticides have their problems. Most, such as the organophosphates, are relatively nonspecific, killing not only the pests in the field but beneficial insects in the ecosystem as well. Some even have toxic effects on other organisms, including people. What's more, insecticides are applied to the surface of crop plants and tend to be blown away to adjacent areas, where they may have unforeseen effects.

17.2 *Agricultural Applications of Biotechnology under Development*	
PROBLEM	**TECHNOLOGY/GENES**
Improving the environmental adaptations of plants	Genes for drought tolerance, salt tolerance
Improving breeding	Male sterility for hybrid seeds
Improving nutritional traits	High-lysine seeds
Improving crops after harvest	Delay of fruit ripening; high-solids tomatoes; sweeter vegetables
Using plants as bioreactors	Plastics, oils, and drugs produced in plants
Controlling crop pests	Herbicide tolerance; resistance to viruses, bacteria, fungi, insects

Some bacteria have solved their own pest problem by producing proteins that kill insect larvae that eat them. For example, there are dozens of strains of *Bacillus thuringiensis*, each of which produces a protein toxic to the insect larvae that prey on it. The toxicity of this protein is 80,000 times that of the usual commercial insecticides. When a hapless larva eats the bacteria, the toxin becomes activated, binding specifically to the insect's gut to produce holes. The insect starves to death.

Dried preparations of *B. thuringiensis* have been sold for decades as a safe, biodegradable insecticide. But biodegradation is their limitation, because it means that the dried bacteria must be applied repeatedly during the growing season. A more permanent approach would be to have the crop plants make the toxin themselves.

The toxin genes from different strains of *B. thuringiensis* have been isolated and cloned. They have been extensively modified by the addition of plant promoters and terminators, plant poly A signals, plant codon usage, and plant regulatory elements on DNA. These modified genes have been introduced into plant cells in the laboratory using the Ti plasmid vector (see Figure 17.5*c*), and transgenic plants have been grown and tested for insect resistance in the field. So far, transgenic tomato, corn, potato, and cotton crops have been successfully shown to have considerable resistance to their insect predators.

CLONED ANIMALS THAT EXPRESS USEFUL GENES. As we described in Chapter 16, the cloning of Dolly the sheep was not done only out of scientific curiosity. One of the main objectives of the biotechnology company associated with this experiment is to make useful products in the milk of transgenic dairy animals.

This transgenic strategy is the one that was described in the opening of this chapter for making spider silk. It can also be used to make pharmaceutical products, such as human α-1-antitrypsin (α-1AT). This protein inhibits elastase, an enzyme that breaks down connective tissue. Elastase is found in excess on the surfaces of the lungs of people with cystic fibrosis, and is partly responsible for their severe breathing problems. Thus, using an inhibitor of elastase could alleviate these symptoms in these patients.

The problem is that it has been hard to get enough α-1AT from human serum. To overcome this problem, the gene for human α-1AT was introduced into the fertilized eggs of sheep, next to the promoter for lactoglobulin, a protein made in large amounts in milk (Figure 17.15). The resulting transgenic sheep made large amount of α-1AT in its milk. Since milk is produced in large amounts all year, this natural "bioreactor" produced a large supply of α-1AT, which was easily purified from the other components of the milk.

The production of animals with reliably integrated transgenes is difficult, however, so another approach is to make transgenic clones. In this case, the human gene (with its promoter) is inserted into sheep somatic cells. Those sheep cells that incorporate the transgene can then be used

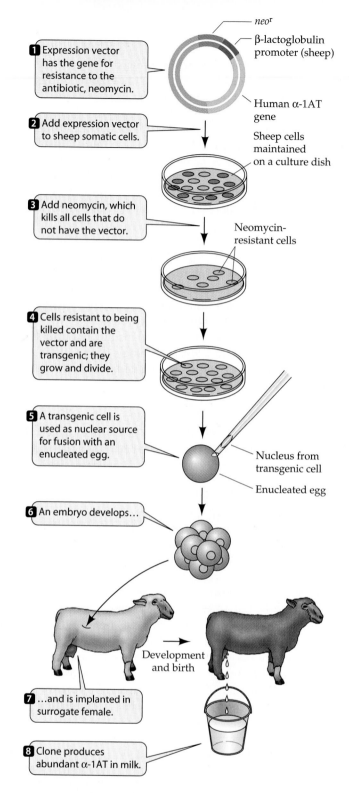

1 Expression vector has the gene for resistance to the antibiotic, neomycin.

*neo*ʳ

β-lactoglobulin promoter (sheep)

Human α-1AT gene

2 Add expression vector to sheep somatic cells.

Sheep cells maintained on a culture dish

3 Add neomycin, which kills all cells that do not have the vector.

Neomycin-resistant cells

4 Cells resistant to being killed contain the vector and are transgenic; they grow and divide.

5 A transgenic cell is used as nuclear source for fusion with an enucleated egg.

Nucleus from transgenic cell

Enucleated egg

6 An embryo develops…

Development and birth

7 …and is implanted in surrogate female.

8 Clone produces abundant α-1AT in milk.

17.15 Production of Transgenic Clones for "Pharming"
The production of transgenic animals involves a combination of DNA technology and reproductive technology.

as the donor nucleus source for cloning (see Figure 16.4). This was the motivation behind the creation of Dolly.

Goats, sheep, and cows are all being used for what has come to be called "pharming," the production of medically useful products in milk. These products include blood clotting factors for treating hemophilia and antibodies for treating colon cancer.

CROPS THAT ARE RESISTANT TO HERBICIDES. Herbivorous insects are not the only threat to agriculture. Weeds may grow in fields and compete with crop plants for water and soil nutrients. Glyphosate (which is known by the trade name Roundup) is a widely used and effective weed killer, or herbicide. It works only on plants, by inhibiting an enzyme system in the chloroplast that is involved in the synthesis of amino acids. Glyphosate is truly a "miracle herbicide," killing 76 of the world's 78 most prevalent weeds. Unfortunately, it also kills crop plants, so great care must be taken with its use. In fact, it is best used to rid a field of weeds before the crop plant starts to grow. But as any gardener knows, when the crop begins to grow, the weeds reappear. So it would be advantageous if the crop were not affected by the herbicide. Then, the herbicide could be applied to the field at any time, and would kill only the weeds.

Fortunately, some soil bacteria have mutated to develop an enzyme that breaks down glyphosate. Scientists have isolated the gene for this enzyme, cloned it, and added plant sequences for transcription, translation, and targeting to the chloroplast. The gene has been inserted into corn, cotton, and soybean plants, making them resistant to glyphosate. In the late 1990s, this technology expanded so rapidly that half of the U.S. crops of these three plants are now transgenic in this way.

GRAINS WITH IMPROVED NUTRITIONAL CHARACTERISTICS. Humans must eat foods (or supplements) containing an adequate amount of β-carotene, which the body converts into vitamin A. About 400 million people worldwide suffer from vitamin A deficiency, which makes them susceptible to infections and blindness. One reason is that they eat rice grains, which do not contain β-carotene, but have only a precursor molecule for it. However, other organisms, such as the bacterium *Erwinia* and daffodil plants, have enzymes that can convert the precursor into β-carotene. The genes for this biochemical pathway are present in the bacterial and daffodil genomes, but not in the rice genome.

Scientists isolated two of the genes for the β-carotene pathway from the bacterium and the other two from daffodil plants. They added promoter signals for expression in the developing rice grain, and then added each gene to rice plants by using the vector *Agrobacterium tumefaciens* (see Figure 17.5c). The resulting rice plants produce grains that look yellow because of their high content of β-carotene (Figure 17.16). About 300 grams of this cooked rice a day can supply all the β-carotene a person needs. This new transgenic strain is now being crossed with more locally adapted strains, and it is hoped that the diets of millions of people will be improved soon.

There is public concern about biotechnology

When the initial experiments creating recombinant DNA in the laboratory were done in the 1970s, there was considerable concern, especially by the scientists involved, over the

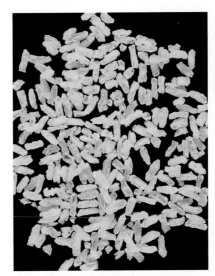

17.16 Grains From Transgenic Rice Rich in β-Carotene
The grains from this transgenic strain (right) are yellow because they make the pigment β-carotene, which is converted by humans into vitamin A. Normal rice (left) does not contain β-carotene.

safety of recombinant DNA. After all, the bacterium they used, *E. coli*, normally lives in the human intestine. What would happen if the laboratory strain shared its new genes with the bacteria living in humans? In response to this concern, the scientists involved initially stopped their research, took stock of the implications of what they were doing, and then took elaborate safety precautions to prevent accidental release of the recombinant organisms and their genes. For example, the strains of *E. coli* used in the lab have a number of mutations that make their survival in the human intestine impossible.

As biotechnology developed, it became apparent that these fears for safety were exaggerated. Accidental release of organisms and transfer of genes has not been a problem. Medical products made by DNA technology are widely used and accepted.

However, with the rapid expansion of genetically modified crops, new concerns have been raised. The issue now is a different one in that genetically modified organisms are being designed to be introduced into the natural environment. Indeed, some countries have banned foods that come from genetically modified crops. These concerns are centered on three claims:

▶ Genetic manipulation is an unnatural interference with nature.

▶ Genetically altered foods are unsafe to eat.

▶ Genetically altered plants are dangerous to the environment.

Advocates of biotechnology tend to agree with the first claim. However, they point out that all major crops are unnatural in the sense that they come from highly bred plants growing in a manipulated environment (a farmer's field). The new technology just adds another level of sophistication.

The concern about safety for humans is countered by the facts that only single genes are added, and that these genes are specific for plant function. For example, the *B. thuringiensis* toxin produced by transgenic plants does not have any effects on people. However, as plant biotechnology moves from adding genes to improve plant growth to adding genes that affect human nutrition, such concerns will become more pressing.

The third concern, about environmental effects, involves the possible "escape" of transgenes from crops to other species. If the gene for herbicide resistance, for example, was inadvertently transferred from a crop to a nearby weed, the latter could thrive in herbicide-treated areas. Or beneficial insects could eat plant materials containing *B. thuringiensis* toxin and die. Transgenic plants undergo extensive field testing before they are approved for use, but the complexity of the biological world makes it impossible to predict all potential environmental effects of transgenic organisms. Because of the potential benefits of agricultural biotechnology (see Table 17.1), scientists believe that it is wise to "proceed with caution."

DNA fingerprinting uses the polymerase chain reaction

"Everyone is unique." This old saying applies not only to human behavior, but also to the human genome. Mutations and recombination through sexual reproduction ensure that each member of a species (except identical twins) has a unique DNA sequence. An individual can be definitively characterized ("fingerprinted") by his or her DNA base sequence.

The ideal way to distinguish an individual from all the other people on Earth would be to describe his or her entire genomic DNA sequence. But since the human genome contains more than 3 billion nucleotides, this idea is clearly not practical. Instead, scientists have looked for genes that are highly polymorphic—that is, genes that have multiple alleles in the human population and are therefore different in different individuals.

One easily analyzed genetic system consists of short moderately repetitive DNA sequences that occur side by side in the chromosomes. These repeat patterns are inherited. For example, an individual might inherit a chromosome 15 with a short sequence repeated six times from her mother, and the same sequence repeated two times from her father. These repeats, called VNTRs (*variable number of tandem repeats*), are easily detectable if they lie between two recognition sites for a restriction enzyme. If the DNA from this individual is cut with the restriction enzyme, it will form two different-sized fragments: one larger (the one from the mother) and the other smaller (the one from the father). These patterns are easily seen by use of gel electrophoresis (Figure 17.17). With several different repeated sequences (as many as eight are used, each with numerous alleles), an individual's unique pattern becomes apparent.

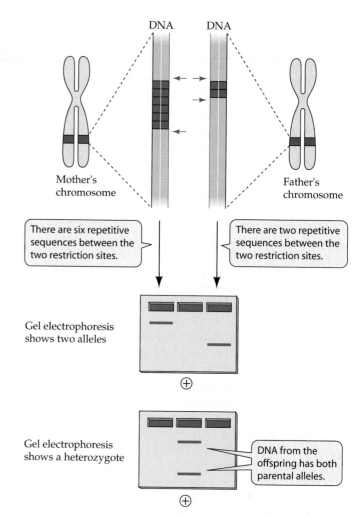

17.17 DNA Fingerprinting
The number of VNTRs inherited by an individual can be used to make a DNA fingerprint.

Typically, these methods require 1 μg of DNA, or the DNA content of about 100,000 human cells, but this amount is not always available. The power of the polymerase chain reaction (see Figure 11.21) permits the DNA from a single cell to be amplified, producing in a few hours the necessary 1 μg for restriction and gel analysis.

DNA fingerprints are used in forensics (crime investigation) to help prove the innocence or guilt of a suspect. For example, in a rape case, DNA can be extracted from dried semen or hair left by the attacker and compared with DNA from a suspect. So far, this method has been used to prove innocence (the DNA patterns are different) more often than guilt (the DNA patterns are the same). It is easy to exclude someone on the basis of these tests, but two people could theoretically have the same patterns, since what is being tested is just a small sample of the genome. Therefore, proving that a suspect is guilty cannot rest on DNA fingerprinting alone.

Two fascinating examples demonstrate the use of DNA fingerprinting in the analysis of historical events. Three hundred years of rule by the Romanov dynasty in Russia

ended on July 16, 1918, when Tsar Nicholas II, his wife, and their five children were executed by a firing squad during the Communist revolution. A report that the bodies had been burned to ashes was never questioned until 1989, when a shallow grave with several skeletons was discovered several miles from the presumed execution site. Recent DNA fingerprinting of bone fragments found in this grave indicated that they came from an older man and woman and three female children, who were clearly related to each other (Figure 17.18) and were also related to several living descendants of the Tsar.

The other example involves Thomas Jefferson, the third president of the United States. In 1802, Jefferson was alleged to have fathered a son by his female slave, Sally Hemmings. Jefferson denied this, and his denial was accepted by many historians because of his vocal opposition to mixed-race relationships. But descendants of Hemmings' two oldest sons (the second was named Eston Jefferson) pressed their case. DNA fingerprinting was done using Y chromosome markers from descendants of these two sons as well as the president's paternal uncle (the president had no acknowledged sons). The results show that Thomas Jefferson may have been the father of the second son, but was not the father of the first son.

In addition to such highly publicized cases, there are many other applications of PCR-based DNA fingerprinting. In 1992, the California condor was extinct in the wild. There were only 52 California condors on Earth, all cared for by the San Diego and Los Angeles zoos. Scientists made DNA fingerprints of all these birds so that the geneticists at the zoos could select unrelated individuals for mating in order to increase genetic variation and increase the viability of the offspring. A number of these young birds have now been returned to the wild. A similar program is under way for the threatened Galápagos tortoises (see Chapter 58).

Plant scientists have found in nature or produced by artificial selection thousands of varieties of crops such as rice, wheat, corn, and grapes. The seeds of many of these varieties are kept in cold storage in "seed banks." Samples of these plants are being DNA-fingerprinted to determine which varieties are genetically the same and which are the most diverse, as a guide to future breeding programs.

A related use of PCR is in the diagnosis of infections. In this case, the test shows whether the DNA of an infectious agent is present in a blood or tissue sample. A primer strand matching the pathogen's DNA is added to the sample. If the pathogen is present, its DNA will serve as a template for the primer, and will be amplified. Because so little of the target sequence is needed, and because primers can be made to bind only to a specific viral or bacterial genome, the PCR-based test is extremely sensitive. If an organism is present in small amounts, PCR testing will detect it.

Finally, the isolation and characterization of genes for various human diseases, such as sickle-cell anemia and cystic fibrosis, has made PCR-based genetic testing a reality. We will discuss this subject in depth in the next chapter.

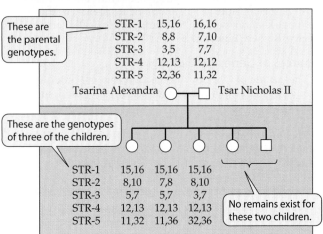

These are the parental genotypes.

STR-1	15,16	16,16
STR-2	8,8	7,10
STR-3	3,5	7,7
STR-4	12,13	12,12
STR-5	32,36	11,32

Tsarina Alexandra ◯━☐ Tsar Nicholas II

These are the genotypes of three of the children.

STR-1	15,16	15,16	15,16
STR-2	8,10	7,8	8,10
STR-3	5,7	5,7	3,7
STR-4	12,13	12,13	12,13
STR-5	11,32	11,36	32,36

No remains exist for these two children.

17.18 DNA Fingerprinting the Russian Royal Family
The skeletal remains of Tsar Nicholas II, his wife Alexandra, and three of their children were found in 1989 and subjected to DNA fingerprinting. Five VNTRs were tested. The results can be interpreted as follows. Using the VNTR STR-2 as an example, the parents had genotypes 8,8 (homozygous) and 7,10 (heterozygous). The three children all inherited type 8 from the Tsarina and either type 7 or type 10 from the Tsar.

Chapter Summary

Cleaving and Rejoining DNA

▶ Knowledge of DNA transcription, translation, and replication has been used to create recombinant DNA molecules, made up of sequences from different organisms.

▶ Restriction enzymes, which are made by microbes as a defense mechanism against viruses, bind to DNA at specific sequences and cut it. **Review Figure 17.1**

▶ DNA fragments generated from cleavage by restriction enzymes can be separated by size using gel electrophoresis. The sequences of these fragments can be further identified by hybridization with a probe. **Review Figures 17.2, 17.3**

▶ Many restriction enzymes make staggered cuts in the two strands of DNA, creating "sticky ends" with unpaired bases. The sticky ends can be used to create recombinant DNA if

DNA molecules from different species are cut with the same restriction enzyme. **Review Figure 17.4**

Cloning Genes

▶ Bacteria, yeasts, and cultured plant cells are commonly used as hosts for recombinant DNA experiments.

▶ Newly introduced DNA must be part of a replicon if it is to be propagated in host cells. One way to make sure that the introduced DNA is part of a replicon is to introduce it as part of a carrier DNA, or vector, that has a replicon.

▶ There are specialized vectors to transfect bacteria, yeasts, and plant cells. These vectors must contain a replicon, recognition sequences for restriction enzymes, and genetic markers to identify their presence in the host cells. **Review Figure 17.5**

▶ Naked DNA may be introduced into a host cell by chemical or mechanical means. In this case, the DNA must integrate into the host DNA by itself.

▶ When vectors carrying recombinant DNA are incubated with host cells, nutritional, antibiotic resistance, or fluorescent markers can be used to identify which cells contain the vector. **Review Figure 17.6**

Sources of Genes for Cloning

▶ The cutting of DNA by a restriction enzyme produces many fragments that can be individually and randomly combined with a vector and inserted into a host to create a gene library. **Review Figure 17.8**

▶ The mRNA's produced in a certain tissue at a certain time can be extracted and used to create complementary DNA (cDNA) by reverse transcription. This cDNA is then used to make a library. **Review Figure 17.9**

▶ A third source of DNA is synthetic DNA made by chemists in the laboratory. The methods of organic chemistry can be used to create specific, mutated DNA sequences.

Some Additional Tools for DNA Manipulation

▶ Homologous recombination can be used to "knock out" a gene in an organism. **Review Figure 17.10**

▶ DNA chip technology permits the screening of thousands of sequences at the same time. **Review Figure 17.11**

▶ An antisense RNA complementary to a specific mRNA can prevent its translation by hybridizing to the mRNA. **Review Figure 17.12**

Biotechnology: Applications of DNA Manipulation

▶ The ability to clone genes has made possible many new applications of biotechnology, such as the large-scale production of eukaryotic gene products.

▶ For a vector carrying a gene of interest to be expressed in a host cell, the gene must be adjacent to appropriate sequences for its transcription and translation in the host cell. **Review Figure 17.13**

▶ Recombinant DNA and expression vectors have been used to make medically useful proteins that would otherwise have been difficult to obtain in necessary quantities. **Review Figure 17.14, Table 17.1**

▶ Because plant cells can be cloned to produce adult plants, the introduction of new genes into plants via vectors has been advancing rapidly. The result is crop plants that carry new, useful genes. **Review Table 17.2**

▶ "Pharming" uses trangenic dairy animals that produce useful products in their milk. **Review Figure 17.15**

▶ There is public concern about the applications of biotechnology to food production.

▶ Because the DNA of an individual is unique, the polymerase chain reaction can be used to identify an organism from a small sample of its cells—that is, to create a DNA fingerprint. **Review Figures 17.17, 17.18**

For Discussion

1. Compare PCR and cloning as methods to amplify a gene. What are the requirements, benefits, and drawbacks of each method?

2. As specifically as you can, outline the steps you would take to (a) insert and express the gene for a new, nutritious seed protein in wheat; (b) insert and express a gene for a human enzyme in sheep's milk.

3. The *E. coli* plasmid pSCI carries genes for resistance to the antibiotics tetracycline and kanamycin. The *tet^r* gene has a single restriction site for the enzyme *Hin*dIII. Suppose that both the plasmid and the gene for corn glutein protein are cleaved with *Hin*dIII and incubated to create recombinant DNA. The reaction mixture is then incubated with *E. coli* that are sensitive to both antibiotics. What would be the characteristics, with respect to antibiotic sensitivity or resistance, of colonies of *E. coli* containing, in addition to its own genome: (a) no new DNA; (b) native pSCI DNA; (c) recombinant pSCI DNA; (d) corn DNA only? How would you detect these colonies?

18 *Molecular Biology and Medicine*

THE MOTHER BROUGHT HER TWO CHILDREN to Dr. Asbjørn Følling in 1934 as a last resort. Since their births, she had watched the conditions of her 6-year-old daughter and 4-year-old son deteriorate. Now both were severely mentally retarded. So far, all of the doctors who had examined the children had expressed sympathy but could do nothing. The mother had noticed a peculiar smell clinging to her children, and she had heard that Dr. Følling was trained as both a chemist and a physician. Could he help? It turned out he couldn't, because their retardation was irreversible. But while examining these children, Dr. Følling made a major discovery.

As part of his examination, Dr. Følling tested the children's urine by adding a brown solution of ferric chloride to look for ketones, which are often excreted by diabetics. This solution normally stays brown, but in diabetics it turns purple. To his surprise, the urine of these children turned the solution dark green. He had never seen this color before, and it was not described in any of his reference books. At first, he suspected that the children were taking a medication that ended up in the urine and reacted with ferric chloride. So he asked the mother to refrain from giving her children any medications for a week and then to bring him two new urine samples. Once again, the samples turned green. Clearly, a substance unique to the bodies of these two children was responsible for the strange color.

Følling's chemistry training served him well. Using analytic chemistry, he purified the substance from the children's urine and identified it as phenylpyruvic acid. Because of the similarity between this substance and the amino acid phenylalanine, Følling hypothesized that the children were unable to metabolize phenylalanine, and that the excess was being converted to phenylpyruvic acid.

Følling soon found other mentally retarded people who excreted this substance, and among the first ten were three pairs of siblings. The parents of these children were mentally normal and did not excrete phenylpyruvic acid. All of these observations fit the idea of an autosomal recessive inherited condition.

Dr. Følling had discovered the genetic disease *phenylketonuria*. But it was not the first such disease to be described in biochemical terms. In 1909, Dr. Archibald Garrod had found the cause of *alkaptonuria*—an inherited disorder in which the patient's urine turns black. Garrod coined the term "inborn errors of metabolism" as a general description of diseases in which genetics and biochemistry are clearly linked. Later, the phenotypes of both phenylketonuria and alkaptonuria were identified as abnormalities of specific enzymes in the same biochemical pathway.

Today the causes of hundreds of such single-gene, single-enzyme diseases are known. In some cases, these discoveries have led to the design of specific therapies and ways to screen for the abnormal proteins in people who do not overtly show the disease. As we will see in this chapter, more precision in describing these abnormalities at the DNA level has come from molecular biology. Even cancer, it turns out, is caused in most cases by abnormalities in genes. The rise of "molecular medicine" is most dramatically shown by undertakings such as gene therapy and the Human Genome Project, which we will discuss at the end of this chapter.

Treatment for Phenylketouria
These siblings both suffer from the genetic condition phenylketonuria. The 11-year-old boy was not diagnosed or treated and is severely affected; his younger sister was treated from early infancy with a low-phenylalanine diet; her development and intelligence are normal. Today there is hope of correcting the defect in the gene itself.

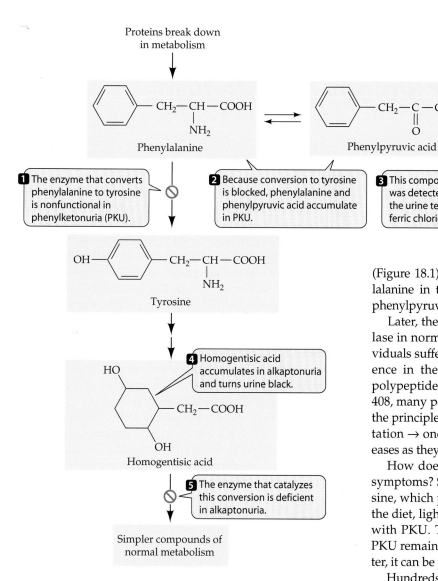

Proteins break down in metabolism

Phenylalanine

Phenylpyruvic acid

1 The enzyme that converts phenylalanine to tyrosine is nonfunctional in phenylketonuria (PKU).

2 Because conversion to tyrosine is blocked, phenylalanine and phenylpyruvic acid accumulate in PKU.

3 This compound was detected in the urine test with ferric chloride.

Tyrosine

4 Homogentisic acid accumulates in alkaptonuria and turns urine black.

Homogentisic acid

5 The enzyme that catalyzes this conversion is deficient in alkaptonuria.

Simpler compounds of normal metabolism

18.1 One Gene, One Enzyme in Humans
Phenylketonuria and alkaptonuria are both caused by abnormalities in a specific enzyme. Knowing the causes of such single-gene, single-enzyme metabolic diseases can aid in the development of screening tests and treatments.

Protein as Phenotype

As we saw in Chapters 11 and 12, genetic mutations are often expressed phenotypically as proteins that differ from the normal wild type. In the first section of this chapter, we identify and discuss the kinds of abnormal proteins that can result from inheritance of an abnormal allele or its origin by mutation. Then we will consider the role of the environment and of patterns of inheritance resulting from autosomal recessives, autosomal dominants, X linkages, and chromosomal abnormalities.

Many genetic diseases result from abnormal or missing proteins

Proteins have many roles in eukaryotic cells, and the genes that code for them can be mutated to cause genetic diseases. Enzymes, receptors, transport proteins, structural pro-

teins, and carriers such as hemoglobin have all been implicated in genetic diseases.

ENZYMES. Although Dr. Følling made his discovery in 1934, it was not until 1957 that the complex clinical phenotype of phenylketonuria (PKU) were traced back to its *primary phenotype*: a single abnormal protein. As Følling had predicted, phenylalanine hydroxylase, the enzyme that catalyzes the conversion of dietary phenylalanine to tyrosine, was not active in patients' livers (Figure 18.1). Lack of this conversion led to excess phenylalanine in the blood and explained the accumulation of phenylpyruvic acid.

Later, the protein sequences of phenylalanine hydroxylase in normal people were compared with those in individuals suffering from PKU. In many cases, the only difference in the 451 amino acids that constitute this long polypeptide chain was that instead of arginine at position 408, many people with PKU have tryptophan. Once again, the principles of one gene → one polypeptide and one mutation → one amino acid change hold true in human diseases as they do in studies of so many other organisms.

How does the abnormality in PKU lead to its clinical symptoms? Since the pigment melanin is made from tyrosine, which patients cannot synthesize but must obtain in the diet, lighter skin and hair color are observed in people with PKU. The exact cause of the mental retardation in PKU remains elusive, but as we will see later in this chapter, it can be prevented.

Hundreds of human genetic diseases that result from enzyme abnormalities have been discovered, many of which lead to mental retardation and premature death. Most of these diseases are rare; PKU, for example, shows up in one newborn infant out of every 12,000. But this is just the tip of the mutational iceberg. Undoubtedly, some mutations result in altered proteins that have no obvious clinical effects. For example, there could be many amino acid changes in phenylalanine hydroxylase that do not affect its catalytic activity.

Analysis of the same protein in different people often shows variations that have no functional significance. In fact, at least 30 percent of all proteins whose sequences are known show detectable amino acid differences among individuals. If one protein variant exists in less than 99 percent of a population (that is, if the protein has another variant at least 1 percent of the time), the protein is said to be *polymorphic*. The key point is that polymorphism does not necessarily mean disease.

HEMOGLOBIN. The first human genetic disease for which an amino acid abnormality was tracked down as the cause was not PKU. It was the blood disease *sickle-cell anemia*, which most often afflicts people whose ancestors came from

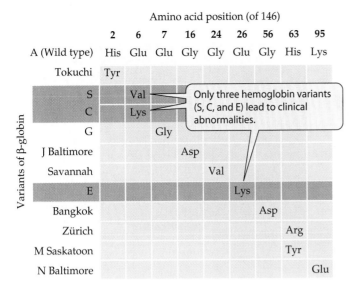

				Amino acid position (of 146)					
	2	6	7	16	24	26	56	63	95
A (Wild type)	His	Glu	Glu	Gly	Gly	Glu	Gly	His	Lys
Tokuchi	Tyr								
S		Val							
C		Lys							
G			Gly						
J Baltimore				Asp					
Savannah					Val				
E						Lys			
Bangkok							Asp		
Zürich								Arg	
M Saskatoon								Tyr	
N Baltimore									Glu

Variants of β-globin

Only three hemoglobin variants (S, C, and E) lead to clinical abnormalities.

18.2 Hemoglobin Polymorphism
Only three of the many variants of hemoglobin are known to lead to clinical abnormalities.

the Tropics or from the Mediterranean. Among African-Americans, about 1 in 655 are homozygous for the sickle allele and have the disease. The abnormal allele produces an abnormal protein that leads to sickled red blood cells (see Figure 12.17). These cells tend to block narrow blood capillaries, especially when the oxygen concentration of the blood is low, and the result is tissue damage.

Human hemoglobin is a protein with quaternary structure, containing four globin chains—two α chains and two β chains—as well as the pigment heme (see Figure 3.7). In sickle-cell anemia, one of the 146 amino acids in the β–globin chain is abnormal: At position 6, the normal glutamic acid has been replaced by valine. This replacement changes the charge of the protein (glutamic acid is negatively charged and valine is neutral; see Table 3.2), causing the protein to form long aggregates in the red blood cells. The result is anemia, a deficiency of normal red blood cells.

Because hemoglobin is easy to isolate and study, its variations in the human population have been extensively documented (Figure 18.2). Hundreds of single amino acid alterations in β-globin have been reported. Some of these polymorphisms are even at the same amino acid position. For example, at the same position that is mutated in sickle-cell anemia, the normal glutamic acid may be replaced by lysine, causing *hemoglobin C disease*. In this case, the anemia is usually not severe. Many alterations of hemoglobin have no effect on the protein's function, and thus no clinical phenotype. This is fortunate, because about 5 percent of all humans are carriers for one of these variants.

RECEPTORS AND TRANSPORT PROTEINS. Some of the most common human genetic diseases show their primary phenotype as altered membrane proteins. About one person in 500 is born with *familial hypercholesterolemia* (FH), in which levels of cholesterol in the blood are several times higher than normal. The excess cholesterol can accumulate on the

inner walls of blood vessels, leading to complete blockage if a blood clot forms. If a blood clot forms in a major vessel serving the heart, the heart becomes starved of oxygen, and a heart attack results. If a blood clot forms in the brain, the result is a stroke. People with FH often die of heart attacks before the age of 45, and in severe cases, before they are 20 years old.

Unlike PKU, which is characterized by the inability to convert phenylalanine to tyrosine, the problem in FH is not an inability to convert cholesterol to other products. People with FH have all the machinery needed to metabolize cholesterol. The problem is that they are unable to transport the cholesterol into the liver cells that use it.

Cholesterol travels in the bloodstream in protein-containing particles called lipoproteins. One type of lipoprotein, *low-density lipoprotein*, carries cholesterol to the liver cells (Figure 18.3a). After binding to a specific receptor on the membrane of a liver cell, the lipoprotein is taken up by endocytosis and delivers its cholesterol to the interior of the cell. People with FH lack a functional version of

(a) **Hypercholesterolemia**

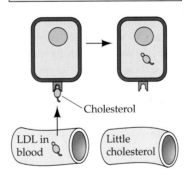

Normal liver cell: Cholesterol, as part of low-density lipoprotein (LDL), enters the cell after LDL binds to a receptor.

Familial hypercholesterolemia: Absence of an LDL receptor prevents cholesterol from entering the cells, and it accumulates in the blood.

(b) **Cystic fibrosis**

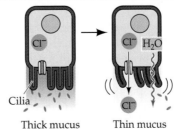

Normal cell lining the airway: Cl⁻ leaves the cell through a channel. Water follows by osmosis, and moist thin mucus allows cilia to beat and sweep away foreign particles, including bacteria.

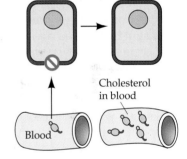

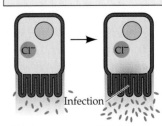

Cystic fibrosis: Lack of a Cl⁻ channel causes a thick viscous mucus to form. Protective cilia cannot beat properly and remove bacteria; infections can easily take hold.

18.3 Genetic Diseases of Membrane Proteins
The left two panels illustrate normal cell function, while the two right panels show the abnormalities caused by (*a*) hypercholesterolemia and (*b*) cystic fibrosis.

the receptor protein. Of the 840 amino acids that make up the receptor, often only one is abnormal in FH, but this is enough to change its structure so that it cannot bind to the lipoprotein.

Among Caucasians, about one baby in 2,500 is born with *cystic fibrosis*. The clinical phenotype of this genetic disease is an unusually thick and dry mucus that lines organs such as the tubes that serve the respiratory system. Its dryness prevents cilia on the surfaces of the epithelial cells from working efficiently to clear out the bacteria and fungal spores that we take in with every breath. The results are recurrent and serious infections, as well as liver, pancreatic, and digestive failures. Patients often die in their twenties or thirties.

The reason for the thick mucus in patients with cystic fibrosis is a defective version of a membrane protein, the *chloride transporter* (Figure 18.3b). In normal cells, this membrane channel opens to release Cl⁻ to the outside of an epithelial cell. The imbalance of Cl⁻ ions (more are now on the outside of the cell than on the inside) causes water to leave the cell by osmosis, resulting in a moist mucus outside the cell. A single amino acid change in the transporter renders it nonfunctional, which leads to dry mucus and the consequent clinical problems.

STRUCTURAL PROTEINS. About one boy in 3,000 is born with *Duchenne's muscular dystrophy*. In this genetic disease, the problem is not an enzyme or receptor, but a protein involved in biological structure. People with this disease show progressively weaker muscles and are wheelchair-bound by their teenage years. Patients usually die in their twenties, when the muscles that serve their respiratory system fail. Most people have a protein in their skeletal muscles called dystrophin, which may bind the major muscle protein actin to the plasma membrane of the muscle cells. Patients with Duchenne's muscular dystrophy do not have a working copy of dystrophin, so their muscles do not work.

Coagulation proteins are involved in the clotting of blood at a wound. As we saw in Chapter 17, inactive clotting proteins are always present in the blood, and become active only at a wound. People with the genetic disease *hemophilia* lack one of the coagulation proteins. Some people with this disease risk death from even minor cuts, since they cannot stop bleeding.

Prion diseases are disorders of protein conformation

Transmissible spongiform encephalopathies (TSE's) are degenerative brain diseases that occur in many mammals, including humans. In these diseases, the brain gradually develops holes, leaving it looking like a sponge. Scrapie, a TSE that causes affected sheep and goats to rub the wool off their bodies, has been known for 250 years. In the 1980s, a TSE appeared in many cows in Britain and was traced to the cows eating products from sheep that had scrapie. Then, in the 1990s, some people who had eaten beef from cows with

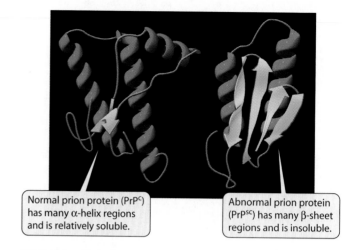

Normal prion protein (PrPᶜ) has many α-helix regions and is relatively soluble.

Abnormal prion protein (PrPˢᶜ) has many β-sheet regions and is insoluble.

18.4 Prion Proteins
Normal prion proteins (PrPᶜ, left) can be converted to the disease-causing form (PrPˢᶜ, right), which has a different three-dimensional structure.

TSE got a human version of the disease (dubbed "mad cow disease" by the media), again suggesting that the causative agent could cross species lines.

Another instance of humans consuming an infective agent and getting TSE was kuru, a disease resulting in dementia that occurred among the Fore tribe of New Guinea. In the 1950s, it was discovered that people with kuru had consumed the brains of people who had died of it. When this ritual cannibalism stopped, so did the epidemic of kuru.

Researchers found that TSE's could be transmitted from one animal to another via brain extracts from a diseased animal. But when Tikva Alper treated these extracts with high doses of ultraviolet light to inactivate nucleic acids, they still caused TSE's. She proposed that the causative agent for TSE's was a protein and not a virus, as had been suspected. Later, Stanley Prusiner purified the protein responsible and showed it to be free of DNA or RNA. He called it a **proteinaceous infective particle (prion)**.

Normal brain cells have a membrane protein called PrPᶜ. A protein with the same amino acid sequence is present in TSE-affected brain tissues, but this protein has an altered shape and is called PrPˢᶜ (Figure 18.4). So TSE is not caused by a mutated gene (the primary structures of the two proteins are the same), but is somehow caused by altered protein conformation. The altered three-dimensional structure has profound effects on the protein's function in the cell. Insoluble PrPˢᶜ piles up as fibers in brain tissue, causing cell death.

How can the exposure of a normal cell to material containing PrPˢᶜ result in a TSE? The abnormal PrPˢᶜ protein seems to induce a conformational change in the normal PrPᶜ protein so that it too becomes abnormal, just as one rotten apple results in a whole barrel full of rotten apples. Just how the conversion occurs and how it causes TSE are unclear.

Most diseases are caused by both heredity and environment

The human diseases for which clinical phenotypes can be traced to a single altered protein may number in the thousands, and in most cases they are dramatic evidence of the one-gene, one-polypeptide principle. Taken together, these diseases have a frequency of about 1 percent in the total population.

Far more common, however, are diseases that are *multifactorial*; that is, they are caused by many genes and proteins interacting with the environment. Although we tend to call individuals either normal (wild-type) or abnormal (mutant), the sum total of our genes is what determines which of us who eat a high-fat diet will die of a heart attack, or which of us exposed to infectious bacteria will come down with a disease. Estimates suggest that up to 60 percent of all people have diseases that are genetically influenced.

Human genetic diseases have several patterns of inheritance

As in any human genetic system, the alleles that cause diseases are inherited as dominants or recessives, and are carried on autosomes or sex chromosomes (see Chapter 10). In addition, some human diseases are caused by more extensive chromosomal abnormalities.

AUTOSOMAL RECESSIVES.   PKU, sickle-cell anemia, and cystic fibrosis are all caused by autosomal recessive mutant alleles. Typically, both parents of an affected person are normal, heterozygous carriers of the abnormal allele. The parents have a 25 percent (one in four) chance of having an affected (homozygous) son or daughter. Because of this low probability and the fact that many families in Western societies now have fewer than four children, it is unusual for more than one child in a family to have an autosomal recessive disease.

In the cells of a person who is homozygous for an autosomal recessive mutant allele, only the nonfunctional, mutant version of the protein it encodes is made. Thus a biochemical pathway or important cell function is disrupted, and disease results. Not unexpectedly, heterozygotes, with one normal and one mutant allele, often have 50 percent of the normal level of functional protein. For example, people who are heterozygous for the allele for PKU have half the number of active molecules of phenylalanine hydroxylase in their liver cells as individuals who carry two normal alleles for this enzyme. But by one mechanism or another, this 50 percent suffices for relatively normal cellular function.

AUTOSOMAL DOMINANTS.   An example of a disease caused by abnormal autosomal dominant alleles is familial hypercholesterolemia. In autosomal dominance, the presence of only one mutant allele is enough to produce the clinical phenotype. In people who are heterozygous for familial hypercholesterolemia having half the normal number of functional receptors for low-density lipoprotein on the sur-

face of liver cells is simply not enough to clear cholesterol from the blood. In autosomal dominance, direct transmission from parent to offspring is the rule.

X-LINKED INHERITANCE.   Both hemophilia and Duchenne's muscular dystrophy are inherited as X-linked recessives; that is, the mutant alleles responsible are located on the X chromosome. Thus a son who inherits a mutant allele on the X chromosome from his mother will have the disease, because the Y chromosome does not contain a normal allele. However, a daughter who inherits one mutant allele will be a heterozygous carrier, since she has two X chromosomes, and hence two alleles. Because, until recently, few males with these diseases lived to reproduce, the most common pattern of inheritance has been from carrier mother to son, and these diseases are much more common in males than in females.

CHROMOSOMAL ABNORMALITIES.   Chromosomal abnormalities also cause human diseases. Such abnormalities include an excess or loss of one or two chromosomes (*aneuploidy*), loss of a piece of a chromosome (*deletion*), and transfer of a piece of one chromosome to another chromosome (*translocation*). About one newborn in 200 is born with a chromosomal abnormality. While some of them are inherited, many are the result of meiotic problems such as nondisjunction (see Chapter 9).

Many zygotes that have abnormal chromosomes do not survive development and are spontaneously aborted. Of the 20 percent of pregnancies that are spontaneously aborted during the first 3 months of human development, an estimated half of them have chromosomal aberrations. For example, more than 90 percent of human zygotes that have only one X chromosome and no Y (Turner syndrome) do not live beyond the fourth month of pregnancy.

A common cause of mental retardation is *fragile-X syndrome* (Figure 18.5). About one male in 1,500 and one female in 2,000 are affected. Near the tip of the abnormal X chromosome is a constriction that tends to break during preparation for microscopy, giving the name for this syndrome. Although the basic pattern of inheritance is that of an X-linked recessive trait, there are departures from this pat-

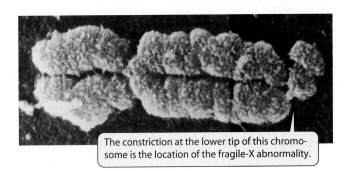

The constriction at the lower tip of this chromosome is the location of the fragile-X abnormality.

18.5 A Fragile-X Chromosome at Metaphase
The chromosomal abnormality that causes the mental retardation symptomatic of fragile-X syndrome shows up physically as a constriction.

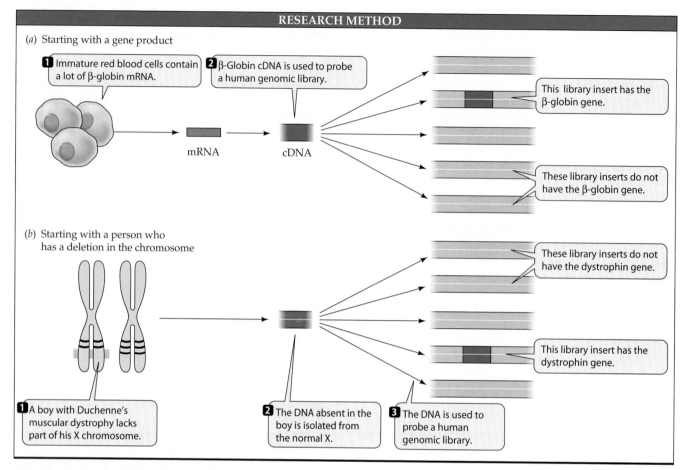

RESEARCH METHOD

(a) Starting with a gene product

1 Immature red blood cells contain a lot of β-globin mRNA.

2 β-Globin cDNA is used to probe a human genomic library.

mRNA

cDNA

This library insert has the β-globin gene.

These library inserts do not have the β-globin gene.

(b) Starting with a person who has a deletion in the chromosome

These library inserts do not have the dystrophin gene.

This library insert has the dystrophin gene.

1 A boy with Duchenne's muscular dystrophy lacks part of his X chromosome.

2 The DNA absent in the boy is isolated from the normal X.

3 The DNA is used to probe a human genomic library.

18.6 Strategies for Isolating Human Genes
(a) Once the sequence for the normal β-globin gene was established by cloning from from the isolated protein, it could be compared to the gene sequence of patients with sickle-cell anemia. *(b)* When an abnormality is caused by a missing gene, as in Duchenne's muscular dystrophy, researchers can compare the affected chromosome with a normal chromosome and isolate the DNA that is missing, then determine the protein for which this DNA codes.

tern. Not all people with the fragile-X abnormality, however, are mentally retarded; we will describe the reason for this variation later in the chapter.

Mutations and Human Diseases

The isolation and description of human mutations has proceeded rapidly since the development of molecular biology techniques (see Chapter 17). When the primary phenotype was known, as in the case of abnormal hemoglobins, cloning the gene responsible was straightforward, although time-consuming. In other cases, such as Duchenne's muscular dystrophy, a chromosome deletion associated with the disease pointed the way to the missing gene. In still other cases, such as cystic fibrosis, only a subtle molecular marker was available to lead investigators to the gene. In both of the latter examples, the primary phenotype—the defective protein—was unknown; only when the gene was isolated was the protein found.

In the discussions that follow, we will examine how mRNA, chromosome deletions, and DNA markers are used to identify both mutant genes and abnormal proteins involved in genetic diseases. We close this discussion by considering the role of triplet repeats in several genetic diseases.

The logical way to identify a gene is to start with its protein

The primary phenotype for sickle-cell anemia was described in the 1950s as a single amino acid change in β-globin. On the basis of the clinical picture of sickled red blood cells, β-globin was certainly the right protein to examine. By the 1970s, researchers were able to isolate β-globin mRNA from immature red blood cells, which transcribe the globins as their major gene product. A cDNA copy of this mRNA was made and used to probe a human DNA library to find the β-globin gene (Figure 18.6a). DNA sequencing was then used to compare the normal gene with the gene from patients with sickle-cell anemia, and as previously described, a single-gene mutation was found.

Chromosome deletions can lead to gene and then protein isolation

The inheritance pattern of Duchenne's muscular dystrophy is consistent with an X-linked recessive trait. But until the late 1980s, neither the abnormal protein involved nor its gene had been described. This failure was not from lack of effort: Almost every muscle protein that could be isolated had been compared for affected people and those without the disease, and had shown no differences.

Then several boys with the disease were found to have a small deletion in their X chromosome. Comparison of the affected chromosomes with normal X chromosomes made possible the isolation of the gene that was missing in the boys (Figure 18.6b).

DNA markers can point the way to important genes

In cases in which no candidate protein or visible chromosome deletion is available to help scientists in isolating a gene responsible for a disease, a technique called **positional cloning** has been invaluable. To understand this method, imagine an astronaut looking down from space, trying to find her son on a park bench on Chicago's North Shore. Unable to spot the boy with her naked eye, the astronaut picks out landmarks that will lead her to the park. She recognizes the shape of North America, then moves to Lake Michigan, the Sears Tower, and so on. Once she has zeroed in on the North Shore park, she can use advanced optical instruments to find her son.

The reference points for positional cloning are genetic markers on the DNA. These markers can be located within protein-coding regions, within introns, or in spacer DNA between genes. The only requirement is that they be polymorphic, with more than one allele.

As we described in Chapter 17, restriction enzymes cut DNA molecules at specific recognition sequences. On a particular human chromosome, a given restriction enzyme may make hundreds of cuts, producing many DNA fragments that can be probed using gel electrophoresis.

The enzyme *Eco*RI, for example, cuts DNA at 5′… GAATTC … 3′. Suppose this recognition site exists in a stretch of human chromosome 7. The enzyme will cut this stretch once and make two fragments of DNA. Now, suppose in some people this sequence is mutated as follows: 5′… GAGTTC … 3′. This sequence will not be recognized by the restriction enzyme; thus it will remain intact and yield one larger fragment of DNA.

Such DNA differences are called **restriction fragment length polymorphisms**, or **RFLP's** (Figure 18.7). A RFLP band pattern is inherited in a Mendelian fashion and can be followed through a pedigree. More than 1,000 such markers have been described for the human genome.

Genetic markers such as RFLP's can be used as landmarks to find genes of interest if they, too, are polymorphic. The key to this method is the well-known observation that if two genes are located near each other on the same chromosome, they are usually passed on together from parent to offspring. The same holds true for any pair of genetic markers.

So, in order to narrow down the location of a gene, a scientist must find a marker and gene that are always inherited together. To do this, family medical histories are taken

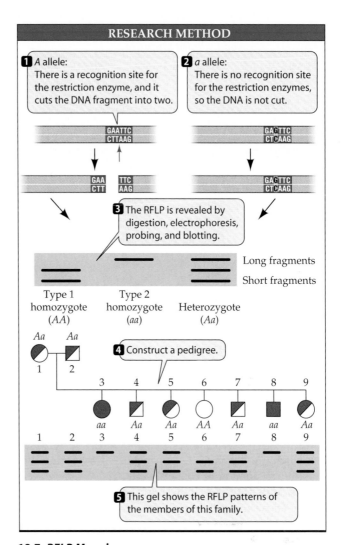

18.7 RFLP Mapping
Restriction fragment length polymorphisms are differences in DNA sequences that serve as genetic markers. More than 1,000 such markers have been described for the human genome.

and pedigrees are constructed. If the DNA marker and genetic disease are inherited together, then they must be near each other on the chromosome. Unfortunately, "near each other" might be as much as a million base pairs away. The process of locating the gene is thus similar to the astronaut focusing on Chicago: The first landmarks lead to only an approximate location.

How can the gene be isolated? Many sophisticated methods are available for narrowing the search. For example, the neighborhood around the RFLP can be screened for further RFLP's involving other restriction enzymes. With luck, one of them might be linked to the disease-causing gene. Then, DNA fragments from this region can be used to probe for sequences that are expressed and therefore encode proteins. Finally, the candidate gene is sequenced from normal people and from people who have the disease in question. If appropriate mutations are found, the gene of interest has been isolated.

18.1 Comparison of Two Genetic Diseases Caused by Point Mutations		
VARIABLE	SICKLE-CELL ANEMIA	PHENYLKETONURIA
Protein in phenotype	β-globin	Phenylalanine hydroxylase
Length of chain	146 amino acids	451 amino acids
Normal protein	Glutamic acid at position 6	Arginine at position 408
Disease protein	Valine at position 6	Tryptophan at position 408
Length of gene	1,512 base pairs	2,448 base pairs
Normal allele	CGG at codon 6	GAA at codon 408
Disease allele	TGG at codon 6	GTA at codon 408

The isolation of genes responsible for hereditary diseases has led to spectacular advances in the understanding of human biology. Before the genes, and then the proteins, for Duchenne's muscular dystrophy and for cystic fibrosis were isolated, dystrophin and the chloride transporter had never been described. This identification of mutant genes has opened up new vistas in our understanding of how the human body works.

Human gene mutations come in many sizes

Phenylketonuria and sickle-cell anemia are caused by single base-pair point mutations (Table 18.1). As we have seen, some variants of the β-globin gene cause disease, but others do not. Those single base-pair mutations that alter a protein's function usually affect its three-dimensional structure; for example, such a mutation may alter the shape at the active site of an enzyme.

Some mutations lead to not much of a protein at all. For example, some people with cystic fibrosis have a nonsense mutation such that a codon for an amino acid near the beginning of the long chloride transporter protein chain has been changed to a stop codon. Protein translation stops at that point, and a very short peptide is made. As we noted in Chapter 12, other point mutations affect RNA processing, leading to nonfunctional mRNA and no protein synthesis.

DNA sequencing has revealed that mutations occur most often at certain base pairs. These "hot spots" are often located where cytosine residues have been methylated to 5-methylcytosine (see Chapter 14). This phenomenon is a result of the natural instability of the bases in DNA. Either spontaneously, or with chemical prodding, unmethylated cytosine residues can lose their amino group and form uracil (Figure 18.8a). But the cell nucleus has a repair system that recognizes this uracil as being inappropriate for DNA: After all, uracil occurs only in RNA! So, the uracil is removed and replaced with cytosine.

The fate of 5-methylcytosine that loses its amino group is rather different, since the result of that loss is thymine, a natural base for DNA. The uracil repair system ignores this

thymine (Figure 18.8b). However, since the GC pair is now a mismatched GT pair, a different type of repair system comes in and tries to fix the mismatch. Half the time, the mismatch repair system matches a new C to the G, but the other half of the time, it matches a new A to the T, resulting in a mutation.

Larger mutations may involve many base pairs of DNA. For example, some of the deletions in the X chromosome that result in Duchenne's muscular dystrophy cover only part of the dystrophin gene, leading to an incomplete protein and a mild form of the muscle disease. Others cover all of the gene, and thus the protein is missing entirely from muscle, resulting in the severe form of the disease. Still other deletions involve millions of base pairs, and cover not only the dystrophin gene but adjacent genes as well; the result may be several diseases simultaneously.

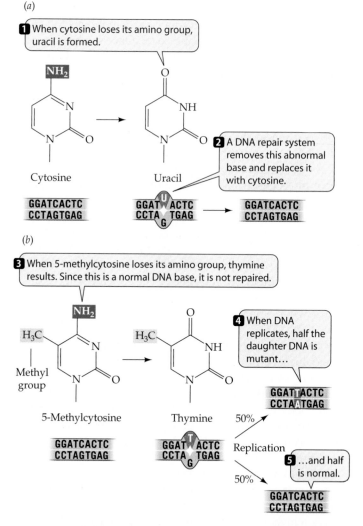

18.8 5-Methylcytosine in DNA Is a Hot Spot for Mutagenesis Cytosine can lose an amino group either spontaneously or because of exposure to certain chemical mutagens. The abnormality is usually repaired, unless the cytosine residue has been methylated to 5-methylcytosine, in which case a mutation is likely to occur.

Expanding triplet repeats demonstrate the fragility of some human genes

About one-fifth of all males that have a fragile-X chromosome are phenotypically normal, as are most of their daughters. But many of those daughters' sons are mentally retarded. In a family in which the fragile-X syndrome appears, later generations tend to show earlier onset and more severe symptoms of the disease. It is almost as if the abnormal allele itself is changing—and getting worse. And that's exactly what is happening.

The gene responsible for fragile-X syndrome (*FMR1*) contains a repeated triplet, CGG, at a certain point in the 5′ untranslated region. In normal people, this triplet is repeated 6 to 54 times (average: 29). In the alleles of mentally retarded people with fragile-X syndrome, the CGG sequence is repeated 200 to 1,300 times.

In "premutated" males—those who show no symptoms but have affected offspring—the repeats are fewer—52 to 200 times. These repeats become more numerous as the daughters of these males pass the chromosome on to their children (Figure 18.9). Expansion to more than 200 repeats leads to the increased methylation of cytosines in the CGG units, accompanied by transcriptional inactivation of the *FMR1* gene, which somehow causes mental retardation.

Expanding triplet repeats have been found in over a dozen other diseases, such as myotonic dystrophy (involving repeated CTG triplets) and Huntington's disease (in which CAG is repeated). Many non-disease-causing genes also appear to have these repeats, which can be found within a protein-coding region or outside of it. Because these repeats are so common, they are assumed to play some important role in the genome. How they expand is not known, but may involve DNA polymerase slipping after copying a repeat and then falling back to copy it again.

Genomic imprinting shows that mammals need both a mother and father

Just after fertilization in a mammalian egg, there are two haploid *pronuclei*—one from the sperm and the other from the egg—in the zygote. They can be distinguished from each other, and they can be removed with a pipette and placed in other eggs. So it is possible to make mouse zygotes in the laboratory with two male or two female pronuclei. These diploid cells should go on to develop into mice, but they don't. Invariably, if the two sets of chromosomes come from only one sex, development begins but is quickly aborted. The same happens in those rare instances when this occurs in humans—for instance, if two sperm enter an empty egg. Again, a fetus never develops.

In addition to showing the obvious need for two sexes, these observations raise the possibility that the male and female genomes are not functionally equivalent. In fact, there are groups of genes that differ in their phenotypic effects depending on which parent they came from. This phenomenon is called **genomic imprinting**.

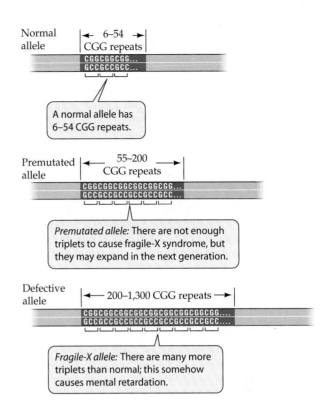

18.9 The CGG Repeat in the Fragile-X Gene Expands with Each Generation
The genetic defect in fragile-X syndrome is caused by excessive repetitions of the CGG triplet.

A dramatic example of genomic imprinting is the inheritance and phenotypic pattern of a certain small deletion on human chromosome 15. If the deletion is on the mother's chromosome 15, the result is a thin child with a wide mouth and prominent jaw (Angelman syndrome). If the deletion is on the father's chromosome 15, the child is short and obese, with small hands and feet (Prader-Willi syndrome). The remaining functional alleles in this region of chromosome 15 must be imprinted in very different ways in the two sexes to result in such different phenotypes. How this happens is not clear.

Detecting Human Genetic Variations

The determination of the precise molecular phenotypes and genotypes for human genetic diseases has had three consequences:

▶ Normal cell physiology has been illuminated by studying mutations.

▶ Specific biochemical treatments and, potentially, cures have been suggested.

▶ Diagnoses may be possible before symptoms first appear, thus making medical intervention possible.

Here we consider the third consequence: the ability to screen for genetic diseases. We will return to the potential for treatments and cures later in the chapter.

Genetic screening is the application of a test to identify people who have, are predisposed to, or are carriers of a certain disease. It can be applied at many times of life and used for many purposes. Prenatal testing can identify an embryo or fetus with a disease so that medical intervention can be applied or decisions about continuing the pregnancy can be made. Newborn babies can be tested so that proper medical intervention can be initiated. Asymptomatic people who have a relative with a genetic disease can be tested to determine whether they are carriers. The goal of any screening is not just to provide information; it is to provide information that can be used to reduce an individual's burden resulting from the disease. However, the existence of genetic screening techniques poses ethical questions concerning the uses of this information, as we will see later in the chapter.

Screening for abnormal phenotypes can make use of protein expression

Screening for phenylketonuria in newborns is legally mandatory in many countries, including all of the United States and Canada. Babies who are homozygous for this genetic disease are born with a normal phenotype, because excess phenylalanine in their blood before birth diffuses across the placenta to the mother's circulation. Since the mother is almost always heterozygous, and therefore has adequate phenylalanine hydroxylase activity, her body metabolizes the excess phenylalanine from the fetus. Thus at birth the baby has not yet accumulated abnormal levels of phenylalanine.

After birth, however, the situation changes. The baby begins to consume protein-rich food (milk) and to break down some of its own proteins. Phenylalanine enters the baby's blood, and without the mother's phenylalanine hydroxylase to help, accumulates there. After a few days, the phenylalanine level in the baby's blood may be ten times higher than normal. Within days, the developing brain is damaged, and as Dr. Følling saw, untreated children with PKU become profoundly mentally retarded.

If PKU is detected early, it can be treated with a special diet low in phenylalanine, and the brain damage avoided. Thus, early detection is imperative. At first, physicians used Følling's ferric chloride test for phenylpyruvic acid in the urine. Unfortunately, babies with PKU do not start excreting large quantities of this substance until they are 4 to 8 weeks old, which can be too late to prevent brain damage. In 1963, Robert Guthrie described a simple screening test for PKU in newborns that today is used almost universally (Figure 18.10). This elegant application of auxotrophic bacteria can be automated so that a screening laboratory can process many samples in a day.*

If an infant tests positive for PKU in this screening, he or she must be re-tested using a more accurate chemical

---

*Guthrie refused to patent the screening test he developed, or to accept any royalties or payment for it. Its immediate and widespread use was at least in part a result of his generosity in allowing the test to be available to all hospitals at low cost.

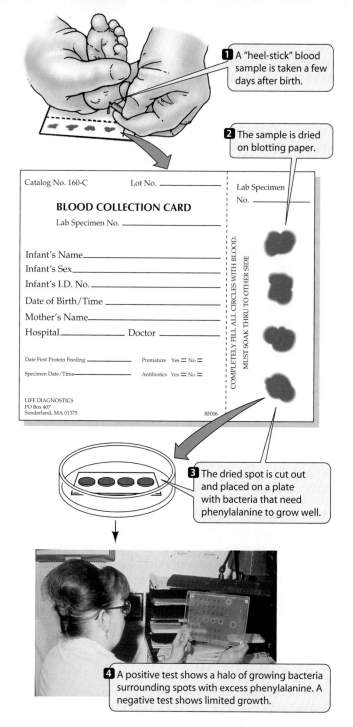

1 A "heel-stick" blood sample is taken a few days after birth.

2 The sample is dried on blotting paper.

3 The dried spot is cut out and placed on a plate with bacteria that need phenylalanine to grow well.

4 A positive test shows a halo of growing bacteria surrounding spots with excess phenylalanine. A negative test shows limited growth.

18.10 Genetic Screening of Newborns for Phenylketonuria
A simple test devised by Robert Guthrie in 1963 is used today to screen newborns for phenylketonuria. Early detection means that the symptoms of the condition can be prevented by following a therapeutic diet.

assay for phenylalanine. If this test also shows a very high level in the blood, dietary intervention is begun. The whole process—the newborn genetic screen, confirmatory test, diagnosis, and initiation of treatment—must be completed by the end of the second week of life. Since the screening test is inexpensive (about a dollar per test), and since babies with PKU who receive early medical intervention develop practically normally, the benefit of screening is significant.

Various other conditions are screened for in newborns. The effects of screening on a target population are seen in the example of Tay-Sachs disease, an autosomal recessive disorder that is lethal in childhood. The missing enzyme can be detected with a simple blood test.

Although they do not exhibit any symptoms of the condition, heterozygous carriers of the Tay-Sachs allele are identifiable by their intermediate level of enzyme activity, when compared with that of homozygous wild-type and mutant (affected) persons. If both potential parents know that they carry the mutated allele, the couple can make informed (if difficult) choices about whether to have a child who is likely to be born with a lethal disease.

In the United States, Jewish people of eastern European origin (Ashkenazic Jews) have a Tay-Sachs carrier frequency of 1 in 27, much higher than the overall frequency of 1 in 200 in the general population. Since the 1970s, massive publicity campaigns in the Jewish community have led to widespread screening. As a result, new diagnoses of this disease among Jews have fallen from 65 a year to fewer than 5. In the non-Jewish population, the number of newborns with this disease has remained constant at 15 per year.

There are several ways to screen for abnormal genes

Screening tests based on enzyme activity or protein structure, such as those for PKU and sickle-cell anemia, must be performed on the tissues in which the relevant gene is expressed. For example, the blood level of phenylalanine is an indirect measure of phenylalanine hydroxylase activity in the liver, and hemoglobin electrophoresis shows the presence of sickle β-globin. But what if blood is difficult to test, as it is in a fetus? What about diseases that are expressed only in the liver or brain and are not reflected in blood? What about proteins whose expression is under cellular control, such that low levels might be the result of a simple dietary factor? Finally, since tissues in heterozygotes often compensate for having just one functional gene by raising the activity of the remaining proteins to near normal levels, how can heterozygotes be identified?

These problems are overcome by DNA testing, which is the most accurate way to test for an abnormal gene. With the description of the genetic mutations responsible for human diseases, it has become possible to directly examine any cell in the body at any time during the life span for mutations. However, these methods work best for diseases caused by only one or a few different mutations. If there are dozens of possible mutations of the gene in the population, simple tests such as the ones we will describe are much less informative.

The polymerase chain reaction (PCR) allows testing of the DNA from even a single cell. Consider, for example, two parents that are both heterozygous for the cystic fibrosis allele, have had a child with the disease, and want a normal child. If the woman is treated with the appropriate hormones, she can be induced to "superovulate," releasing several eggs. One of them can be injected with a single sperm from her husband and the resulting zygote allowed to divide to the 8-cell stage. If one of these cells is removed, it can be tested by PCR for the presence of the cystic fibrosis allele(s). The remaining 7-cell embryo can be implanted in the mother's womb and go on to develop normally.

Such pre-implantation screening is unusual. More typical are analyses of fetal cells after implantation in the womb. Fetal cells can be analyzed at about the tenth week of pregnancy (by chorionic villus sampling) or during the thirteenth to seventeenth weeks (by amniocentesis). These two sampling methods are described in Chapter 43. In either case, only a few fetal cells are required.

Newborns can also be screened for genetic mutations. The blood samples used for screening for PKU and other disorders contain enough of the baby's blood cells to permit extraction of the DNA, its amplification by PCR, and testing. Pilot studies are under way for testing for sickle-cell disease and cystic fibrosis, and other genes will surely follow.

DNA testing is also now widely used to test adults for heterozygosity. For example, a sister or female cousin of a boy with Duchenne's muscular dystrophy may want to know if she is a carrier of the X chromosome that contains the dystrophin gene mutation. Similarly, the relatives of children with cystic fibrosis can determine their carrier status via DNA testing.

Of the numerous methods of genetic analysis, two are the most widespread. We will describe their use to detect the mutation in the β-globin gene that results in sickle-cell anemia.

ALLELE-SPECIFIC CLEAVAGE DIFFERENCES. There is a difference between the normal and sickle alleles of the β-globin gene with respect to a restriction enzyme recognition site. Around position 6 in the normal gene is the sequence 5′... CCTGAGGAG... 3′. This sequence is recognized by the restriction enzyme *Mst*II, which will cleave DNA at 5′... CCTNAGG... 3′, where "N" is any base. In the sickle mutation (see Table 18.1), the DNA sequence is changed to 5′... CCTGTGGAG... 3′. The single base-pair alteration makes this sequence unrecognizable by *Mst*II. So, when *Mst*II fails to make the cut in the mutant gene, gel electrophoresis detects a larger DNA fragment (Figure 18.11). This analysis is similar to the use of RFLP's (see Figure 18.7). Thus, the method works only if a restriction enzyme exists that can recognize either the sequence at the mutation or the original sequence that is altered by that mutation.

ALLELE-SPECIFIC OLIGONUCLEOTIDE HYBRIDIZATION. The allele-specific oligonucleotide hybridization method uses oligonucleotides made in the laboratory that will hybridize either to the denatured normal β-globin DNA sequence around position 6 or to the sickle mutant sequence. Usually, a probe of at least a dozen bases is needed to form a stable double helix with the target DNA. If the probe is labeled with radioactivity or with a colored or fluorescent substrate, hybridization is readily detected (Figure 18.12). This method is easier and faster than allele-specific cleavage, and will work no matter what the sequence of the normal or mutant allele.

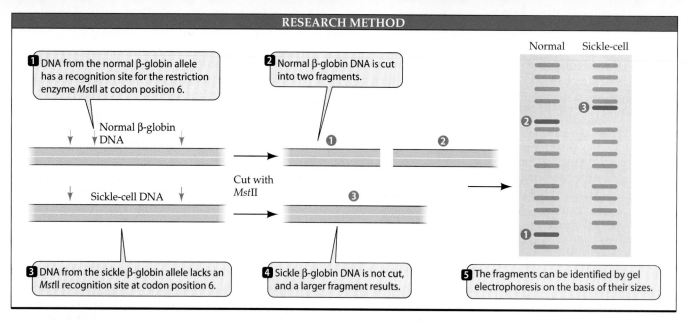

18.11 DNA Testing by Allele-Specific Cleavage
Allele-specific cleavage, a technique similar to RFLP analysis, can be used to detect mutations such as the one that causes sickle-cell anemia.

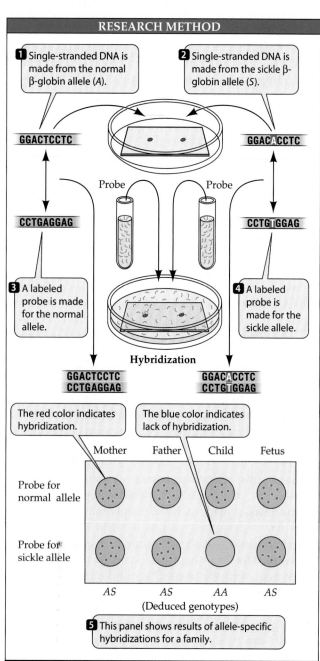

18.12 DNA Testing by Allele-Specific Oligonucleotide Hybridization
Testing of this family reveals that three of them are heterozygous carriers of the sickle-cell allele. The first child, however, has inherited two normal alleles and is neither affected by the disease nor a carrier.

Cancer: A Disease of Genetic Changes

Perhaps no malady affecting people in the industrialized world instills more fear than cancer. One in three Americans will have some form of cancer in their lifetime, and at present, one in four will die of it. With a million new cases and half a million deaths in the United States annually, cancer ranks second only to heart disease as a killer. Cancer was less common a century ago; then, as now in many poor regions of the world, people died of infectious diseases and did not live long enough to get cancer. Cancer tends to be a disease of the later years of life; children are less frequently afflicted.

Since the U.S. government declared a "war on cancer" in 1970, a tremendous amount of information on cancer cells—their growth and spread and their molecular changes—has been obtained. Perhaps the most remarkable discovery is that cancer is a disease caused primarily by genetic changes. These changes are mostly alterations in the DNA of somatic cells that are propagated by mitosis.

Cancer cells differ from their normal counterparts

Cancer cells differ from the normal cells from which they originate in two major ways. First, a cancer cell loses control over cell division. Most cells in the body divide only if they are exposed to outside influences, such as growth factors and hormones. Cancer cells do not respond to these controls, and instead divide more or less continuously, ultimately forming *tumors* (large masses of cells). By the time a physician can feel a tumor or see one on an X-ray or CAT scan, it already contains millions of cells.

Benign tumors resemble the tissue they came from, grow slowly, and remain localized where they develop. A lipoma, for example, is a benign tumor of fat cells that may arise in the armpit and remain there. Benign tumors are usually not a problem, but they must be removed if they impinge on an important organ, such as the brain.

Malignant tumors, on the other hand, do not look like their parent tissue at all. A flat, specialized lung epithelial cell in the lung wall may turn into a relatively featureless, round, malignant lung cancer cell. Malignant cells often have irregular structures, such as variable sizes and shapes of nuclei. Many malignant cells express the gene for telomerase and thus do not shorten the ends of their chromosomes after each DNA replication.

The second, and most fearsome, characteristic of cancer cells is their ability to invade surrounding tissues and spread to other parts of the body. This spreading of cancer, called **metastasis**, occurs in several stages. First, the malignant tumor secretes chemical signals that cause blood vessels to grow to the tumor and supply it with oxygen and nutrients. This process is called *angiogenesis*. Then, the cancer cells extend into the tissue that surrounds them by actively secreting digestive enzymes to disintegrate the surrounding cells and extracellular materials, working their way toward a blood vessel. Finally, some of the cancer cells enter either the bloodstream or the lymphatic system (Figure 18.13).

The journey of malignant cells through these vessels is perilous, and few of them survive—perhaps one in 10,000. When by chance a cancer cell arrives at an organ suitable for its further growth, it expresses cell surface proteins that allow it to bind to and invade the new host tissue.

Different forms of cancer affect different parts of the body. About 85 percent of all human tumors are *carcinomas*—cancers that arise in surface tissues such as the skin

18.2	**Human Cancers Known to Be Caused by Viruses**
CANCER	**ASSOCIATED VIRUS**
Liver cancer	Hepatitis B virus
Lymphoma, nasopharyngeal cancer	Epstein–Barr virus
T cell leukemia	Human T cell leukemia virus (HTLV-I)
Anogenital cancers	Papillomavirus
Kaposi's sarcoma	Kaposi's sarcoma herpesvirus

and the epithelial cells that line the organs. Lung cancer, breast cancer, colon cancer, and liver cancer are all carcinomas. *Sarcomas* are cancers of tissues such as bone, blood vessels, and muscle. *Leukemias* and *lymphomas* affect the cells that give rise to blood cells.

Some cancers are caused by viruses

Peyton Rous's discovery in 1910 that a sarcoma in chickens is caused by a virus that is transmitted from one bird to another spawned an intensive search for cancer-causing viruses in humans. At least five types of human cancer are probably caused by viruses (Table 18.2).

Hepatitis B, a liver disease that affects people all over the world, is caused by the hepatitis B virus, which contaminates blood or is carried from mother to child during birth. The viral infection can be long-lasting and may flare up numerous times. This virus is associated with liver cancer, especially in Asia and Africa, where millions of people are infected. But it does not act to cause cancer by itself. Some gene mutations that are necessary for tumor formation occur in the infected cells of Asians and Africans, although apparently not in Europeans and North Americans.

An important group of virus-caused cancers in North Americans and Europeans are the various anogenital cancers caused by papillomaviruses. The genital and anal warts that these viruses cause often develop into tumors. These viruses seem to be able to act on their own, not needing mutations in the host tissue for tumors to arise. Sexual transmission of these papillomaviruses is unfortunately widespread.

Most cancers are caused by genetic mutations

Worldwide, no more than 15 percent of all cancers may be caused by viruses. What causes the other 85 percent? Because most cancers develop in older people, it is reasonable to assume that one must live long enough for a series of events to occur. This assumption turns out to be correct, and the events are genetic mutations. But these are usually not the germ-line mutations found in genetic diseases. Instead, the mutations in cancer cells are usually *somatic* mutations—alterations in the genes of non-gamete-producing cells.

DNA can become damaged in many ways. As we have seen, spontaneous mutations arise because of chemical

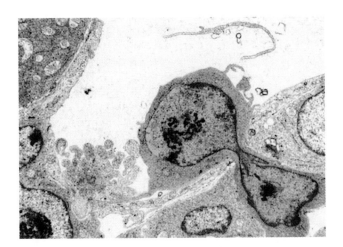

18.13 The Spread of Cancer
A cancer cell squeezes into a small blood vessel through the vessel's wall. The cancer cell is then carried through the blood and, if it survives the journey, may spread into other tissue.

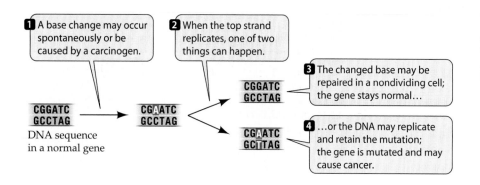

1 A base change may occur spontaneously or be caused by a carcinogen.

2 When the top strand replicates, one of two things can happen.

3 The changed base may be repaired in a nondividing cell; the gene stays normal…

4 …or the DNA may replicate and retain the mutation; the gene is mutated and may cause cancer.

CGGATC
GCCTAG
DNA sequence
in a normal gene

CG**A**ATC
GCCTAG

CGGATC
GCCTAG

CG**A**ATC
GC**T**TAG

18.14 Dividing Cells Are Especially Susceptible to Genetic Damage and Cancer
A base change is more likely to be repaired in a nondividing cell.

changes in the nucleotides. In addition, certain substances called *carcinogens* cause mutations that lead to cancer. Familiar carcinogens include the chemicals that are present in tobacco smoke and meat preservatives, ultraviolet light from the sun, and ionizing radiation from sources of radioactivity. Less familiar, but just as harmful, are thousands of chemicals present naturally in the foods people eat. According to one estimate, these "natural" carcinogens account for well over 80 percent of the human exposure to agents that cause cancer.

The common theme in natural and human-made carcinogens is that almost all of them damage DNA, usually by causing changes from one base to another (Figure 18.14). In somatic cells that divide often, such as epithelial and bone marrow cells, there is less time for DNA repair mechanisms to work before replication occurs again. Therefore, such cells are especially susceptible to genetic damage.

Two kinds of genes are changed in many cancers

The changes in the control of cell division that lie at the heart of cancer can be likened to the controls of an automobile. To make a car move, two things must happen: The gas pedal must be pressed, and the brake must be released. In the human genome, some genes act as **oncogenes**, which act as the "gas pedal" to stimulate cell division, and some as **tumor suppressor genes**, which "put the brake on" to inhibit it.

ONCOGENES. The first hint that oncogenes (from the Greek *onco-*, "mass") were necessary for cells to become cancerous came with the identification of virally induced cancers in animals. In many cases, these viruses bring a new gene into their host cells, one that acts to stimulate cell division when it is expressed in the viral genome. It soon became apparent that these viral oncogenes had counterparts in the genomes of the host cells, called **proto-oncogenes**, that were not actively transcribed. So the search for genes that are damaged by carcinogens quickly zeroed in on the proto-oncogenes.

Proto-oncogenes are genes that have the capacity to stimulate cell division, but are normally turned "off" in differentiated, nondividing cells. Many are involved in growth factor stimulation (Figure 18.15). Some remarkable proto-oncogenes control apoptosis (programmed cell death). Acti-

vation of these genes by mutation causes them to prevent apoptosis, allowing cells that normally die to continue dividing.

Some proto-oncogenes can be activated by point mutations, others by chromosome changes such as translocations, and still others by gene amplification. Whatever the mechanism, the result is the same: The proto-oncogene becomes activated, and the "gas pedal" for cell division is pressed.

TUMOR SUPPRESSOR GENES. About 10 percent of all cancer is clearly inherited. Often the inherited form of the cancer is clinically similar to the noninherited form that occurs later in life, called the *sporadic form*. The major differences are that the inherited form strikes a person much earlier in life and usually shows up as multiple tumors.

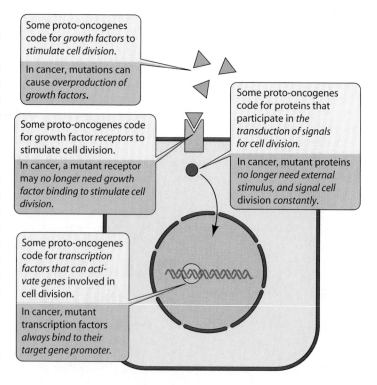

Some proto-oncogenes code for *growth factors* to *stimulate cell division*.

In cancer, mutations can cause *overproduction of growth factors*.

Some proto-oncogenes code for growth factor *receptors* to stimulate cell division.

In cancer, a mutant receptor may *no longer need growth factor binding to stimulate cell division*.

Some proto-oncogenes code for *transcription factors that can activate genes* involved in cell division.

In cancer, mutant transcription factors *always bind to their target gene promoter*.

Some proto-oncogenes code for proteins that participate in *the transduction of signals for cell division*.

In cancer, mutant proteins *no longer need external stimulus, and signal cell division constantly*.

18.15 Proto-Oncogene Products Stimulate Cell Division
Mutations can affect any of the several ways proto-oncogenes normally stimulate cell division, thus causing cancer.

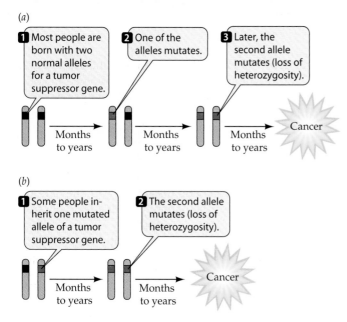

(a)

1 Most people are born with two normal alleles for a tumor suppressor gene.

2 One of the alleles mutates.

3 Later, the second allele mutates (loss of heterozygosity).

Months to years → Months to years → Months to years → Cancer

(b)

1 Some people inherit one mutated allele of a tumor suppressor gene.

2 The second allele mutates (loss of heterozygosity).

Months to years → Months to years → Cancer

18.16 The "Two-Hit" Hypothesis for Cancer
(a) Although a single mutation can activate a proto-oncogene, two mutations are needed to inactivate a tumor suppressor gene. (b) An inherited predisposition to cancer occurs in people born with one allele already mutated.

In 1971, Alfred Knudson used these observations to predict that for a tumor to occur, a tumor suppressor gene must be inactivated. But unlike oncogenes, in which one mutated allele is all that is needed for activation, the full inactivation of a tumor suppressor gene requires that both alleles be turned off, which requires two mutational events. It takes a long time for both alleles in a single cell to mutate and cause sporadic cancer. But in inherited cancer, people are born with one mutant allele for the tumor suppressor, and need just one more mutational event for full inactivation of the "brakes" (Figure 18.16).

Various tumor suppressor genes have been isolated and confirm Knudson's "two-hit" hypothesis. Some of these genes are involved in inherited forms of rare childhood cancers such as retinoblastoma (a tumor of the eye) and Wilms' tumor of the kidney, as well as in inherited breast and prostate cancers.

An inherited form of breast cancer demonstrates the effect of tumor suppressor genes. The 9 percent of women who inherit one mutated copy of the gene *BRCA1* have a 60 percent chance of having breast cancer by age 50 and an 82 percent chance of developing it by age 70. The comparable figures for women who inherit two normal copies of the gene are 2 percent and 7 percent, respectively.

How do tumor suppressor genes act in the cell? Like the proto-oncogenes, they are normally involved in vital cell functions (Figure 18.17). Some control progress through the cell cycle. The protein encoded by *Rb*, a gene that was first described for its contribution to retinoblastoma, is active during G1. In the active form, it encodes a protein that binds to and inactivates transcription factors that are necessary for progress to the S phase and the rest of the cell cycle. In nondividing cells, *Rb* remains active, preventing

cell division until the proper growth factor signals are present. When the Rb protein is inactivated by mutation, the cell cycle moves forward independent of growth factors.

The protein product of another widespread tumor suppressor gene, *p53*, also stops cells during G1. It does this by acting as a transcription factor, stimulating the production of (among other things) a protein that blocks the interaction of a cyclin and protein kinase needed for moving the cell cycle beyond G1. This gene is mutated in many types of cancers, including lung cancer and colon cancer.

The pathway from normal cell to cancerous cell is complex

The "gas pedal" and "brake" analogies for oncogenes and tumor suppressor genes, respectively, are elegant but simplified. There are many oncogenes and tumor suppressor genes, some of which act only in certain cells at certain times. Therefore, a sequence of events must occur before a normal cell becomes malignant.

Because colon cancers progress to full malignancy slowly, it is possible to describe the oncogenes and tumor suppressor genes at each stage in great molecular detail. Figure 18.18 outlines the progress of this form of cancer. At least three tumor suppressor genes and one oncogene must be mutated in sequence for an epithelial cell in the colon to become metastatic. Although the occurrence of all these events in a single cell might appear unlikely, remember that the colon has millions of cells, that the cells giving rise to epithelial cells are constantly dividing, and that these changes take

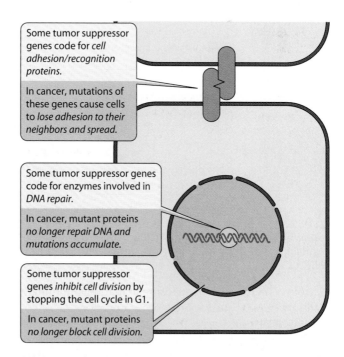

Some tumor suppressor genes code for *cell adhesion/recognition proteins.*

In cancer, mutations of these genes cause cells to *lose adhesion to their neighbors and spread.*

Some tumor suppressor genes code for enzymes involved in *DNA repair.*

In cancer, mutant proteins *no longer repair DNA and mutations accumulate.*

Some tumor suppressor genes *inhibit cell division* by stopping the cell cycle in G1.

In cancer, mutant proteins *no longer block cell division.*

18.17 Tumor Suppressor Gene Products Inhibit Cell Division and Cancer
Mutations can affect any of the several ways that tumor suppressor genes inhibit cell division, causing cells to divide and form a tumor.

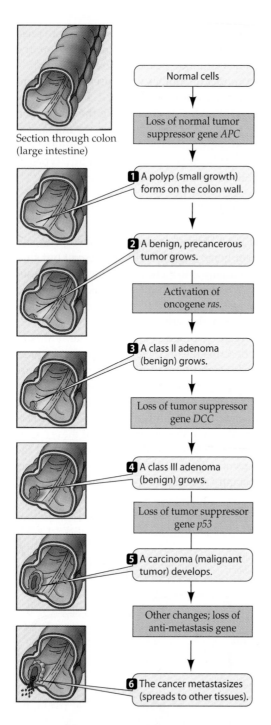

Section through colon (large intestine)

Normal cells

↓

Loss of normal tumor suppressor gene *APC*

↓

1 A polyp (small growth) forms on the colon wall.

↓

2 A benign, precancerous tumor grows.

↓

Activation of oncogene *ras*.

↓

3 A class II adenoma (benign) grows.

↓

Loss of tumor suppressor gene *DCC*

↓

4 A class III adenoma (benign) grows.

↓

Loss of tumor suppressor gene *p53*

↓

5 A carcinoma (malignant tumor) develops.

↓

Other changes; loss of anti-metastasis gene

↓

6 The cancer metastasizes (spreads to other tissues).

18.18 Multiple Mutations Transform a Normal Colon Epithelial Cell into a Cancerous Cell
In this form of cancer, at least five genes are mutated in a single cell.

place over many years of exposure to natural and human-made carcinogens and to spontaneous mutations.

The characterization of the molecular changes in tumor cells has opened up the possibility of genetic diagnosis and screening for cancer, as is done for genetic diseases. Many tumors are now commonly diagnosed in part by specific oligonucleotide probes for oncogene and/or tumor suppressor gene alterations. It is also possible to detect early in life whether an individual has inherited a mutated tumor sup-

pressor gene. For example, a person who inherits mutated copies of the tumor suppressor genes involved in colon cancer normally would have a high probability of developing this cancer by age 40. Surgical removal of the colon would prevent this metastatic tumor from arising.

Treating Genetic Diseases

Most treatments of genetic diseases try to alleviate the symptoms that affect the patient. But to effectively treat diseases caused by genes—whether they affect all cells, as in inherited disorders such as PKU, or only somatic cells, as in cancer—physicians must be able to diagnose the disease accurately, must know how the disease works at the biochemical level, and must be able to intervene early, before the disease ravages or kills the individual.

Basic research has provided the knowledge needed for accurate diagnostic tests, as well as a beginning at understanding *pathogenesis* (the cause of diseases) at the molecular level. Physicians are now applying this knowledge to treat genetic diseases. In this section, we will see that these treatments range from specifically modifying the mutant phenotype to supplying the normal version of a mutant gene.

One approach to treatment is to modify the phenotype

There are three ways to alter the phenotype of a genetic disease so that it no longer harms an individual.

RESTRICTING THE SUBSTRATE. Restricting the substrate of a deficient enzyme is the approach taken when a newborn is diagnosed with PKU. In this case, the deficient enzyme is phenylalanine hydroxylase, and the substrate is phenylalanine. The infant's inability to break down the phenylalanine in food leads to a buildup of the substrate, which causes the clinical symptoms. So the infant is immediately put on a special diet that contains only enough phenylalanine for immediate use.

Lofenelac, a milk-based product that is low in phenylalanine, is fed to these infants just like formula. Later, certain fruits, vegetables, cereals, and noodles low in phenylalanine can be added to the diet. Meat, fish, eggs, dairy products, and bread, which contain high amounts of phenylalanine, must be avoided, especially during childhood, when brain development is most rapid. The artificial sweetener aspartame must also be avoided, because it is made of two amino acids, one of which is phenylalanine.

People with PKU are generally advised to stay on a low-phenylalanine diet for life. Although maintaining these dietary restrictions may be difficult, it is effective. Numerous follow-up studies since newborn screening was initiated have shown that people with PKU who stay on the diet are no different from the rest of the population in terms of mental ability. This is an impressive achievement in public health, given the extent of mental retardation in untreated patients.

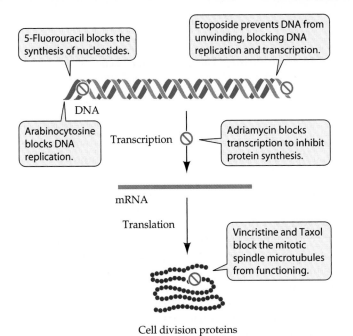

5-Fluorouracil blocks the synthesis of nucleotides.

Etoposide prevents DNA from unwinding, blocking DNA replication and transcription.

DNA

Arabinocytosine blocks DNA replication.

Transcription

Adriamycin blocks transcription to inhibit protein synthesis.

mRNA

Translation

Vincristine and Taxol block the mitotic spindle microtubules from functioning.

Cell division proteins

18.19 Drug Strategies for Killing Cancer Cells
The medications used against cancer attack rapidly dividing cancer cells in several ways. Unfortunately, most of them also affect noncancerous dividing cells.

METABOLIC INHIBITORS. As we described earlier, people with familial hypercholesterolemia accumulate dangerous levels of cholesterol in their blood. Not only are these people unable to metabolize dietary cholesterol, they also synthesize a lot of it. One effective treatment for people with this disease is the drug mevinolin, which blocks the patients' own cholesterol synthesis. Patients who receive this drug need only worry about cholesterol in their diet, and not about what their cells are making.

Metabolic inhibitors also form the basis of cancer therapy with drugs. The strategy is to kill rapidly dividing cells, since rapid cell division is the hallmark of malignancy. Many drugs kill dividing cells (Figure 18.19), but most of these drugs are given in the blood and thus also damage other, noncancerous, dividing cells in the body. Therefore, it is not surprising that people undergoing chemotherapy suffer side effects such as loss of hair (due to damage to the skin epithelium), digestive upsets (gut epithelial cells), and anemia (bone marrow stem cells). The effective dose of these highly toxic drugs for treating the cancer is often just below the dose that would kill the patient, so they must be used with utmost care. Often they can control the spread of cancer, but not cure it.

SUPPLYING THE MISSING PROTEIN. An obvious way to treat a disease phenotype in which a functional protein is missing is to supply the missing protein. This approach is the basis of treatment of hemophilia A, in which the missing blood clotting factor is supplied in pure form. The production of human clotting protein by biotechnology has made it possible for a pure protein to be given instead of crude blood products, which could be contaminated with the AIDS virus or other pathogens.

Unfortunately, the phenotypes of many diseases caused by genetic mutations are very complex. Simple interventions, such as some of those we have described, do not work for most such diseases. Indeed, a recent survey showed that current therapies for 351 diseases caused by single-gene mutations improved the patients' life span by only 15 percent.

Gene therapy offers the hope of specific treatments

Perhaps the most obvious thing to do when a cell lacks a functional allele is to provide one. Such **gene therapy** approaches are under intensive investigation for diseases ranging from the rare inherited disorders caused by single mutations, to cancer, AIDS, and atherosclerosis.

Gene therapy in humans seeks to insert a new gene that will be expressed in the host. Thus, the new DNA is often attached to a promoter that will be active in human cells. Presenting the DNA for cellular uptake follows the methods used in biotechnology: Calcium salts, liposomes, and viral vectors are used to get the "good gene" into the human cells. The physicians who are developing this "molecular medicine" are confronted by all the challenges of genetic engineering: effective vectors, efficient uptake, appropriate expression and processing of mRNA and protein, and selection within the body for the cells that contain the recombinant DNA.

Which human cells should be the targets of gene therapy? The best approach would be to replace the nonfunctional allele with a functional one in every cell of the body. But vectors to do this are simply not available, and delivery to every cell poses a formidable challenge. Until recently, the major attempts at gene therapy have been *ex vivo*. That is, physicians have taken cells from the patient's body, added the new gene to the cells in the laboratory, and then returned the cells to the patient in the hope that the correct gene product would be made (Figure 18.20).

A widely publicized example of this approach was the introduction of a functional gene for the enzyme adenosine deaminase into the white blood cells of a girl with a genetic deficiency of this enzyme. Unfortunately, these were mature white blood cells, and although they survived for a time in the girl, and provided some therapeutic benefit, they eventually died, as is the normal fate of such cells. It would be more effective to insert the functional gene into *stem cells*—the bone marrow cells that constantly divide to produce white blood cells. The use of stem cells is a major thrust of many current clinical experiments on gene therapy.

The other approach to gene therapy is to insert the gene directly into cells in the body of the patient. This *in vivo* approach is being attempted for various types of cancer. For example, lung cancer cells are accessible if the DNA or vector is given as an aerosol through the respiratory system. Vectors carrying functional alleles of the tumor suppressor genes that are mutated in the tumors, as well as vectors expressing antisense RNA's against oncogene mRNA's, have been successfully introduced in this way to patients with lung cancer, with some clinical improvement.

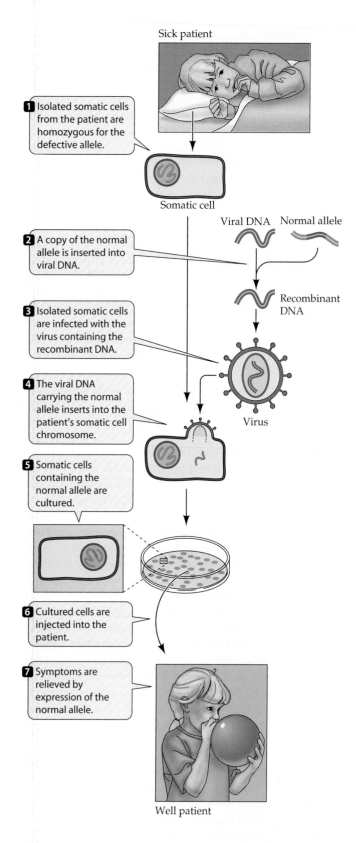

Sick patient

1 Isolated somatic cells from the patient are homozygous for the defective allele.

Somatic cell

Viral DNA Normal allele

2 A copy of the normal allele is inserted into viral DNA.

Recombinant DNA

3 Isolated somatic cells are infected with the virus containing the recombinant DNA.

Virus

4 The viral DNA carrying the normal allele inserts into the patient's somatic cell chromosome.

5 Somatic cells containing the normal allele are cultured.

6 Cultured cells are injected into the patient.

7 Symptoms are relieved by expression of the normal allele.

Well patient

18.20 Gene Therapy: The Ex Vivo Approach
New genes are added to somatic cells taken from a patient's body, then returned to the body to make the missing gene product.

larger number of patients receive the therapy with the hope that their disease will disappear, or at least improve.

Sequencing the Human Genome

In 1984, the United States government sponsored a conference to examine the problem of detecting DNA damage in people exposed to low levels of radiation, such as those who had survived the atomic bomb in Japan 39 years earlier. Scientists attending this conference quickly realized that the ability to detect such damage would also be useful in evaluating environmental mutagens. But in order to detect *changes* in the human genome, scientists first needed to know its *normal* sequence.

In 1986, Renato Dulbecco, who won the Nobel prize for his pioneering work on cancer viruses, suggested that determining the sequence of human DNA could also be a boon to cancer research. He proposed that the scientific community be mobilized for the task. The result was the publicly funded Human Genome Project, an international effort. In the 1990s, private industry launched their own sequencing effort. By the summer of 2000, "draft" sequences of the human genome were ready. The final sequence is expected to be completed by 2003.

There are two approaches to genome sequencing

Each human chromosome consists of one double-stranded molecule of DNA. Because of their differing sizes, the 46 human chromosomes can be separated from each other and identified (see Figure 9.14). So it is possible to carefully isolate the DNA of each chromosome for sequencing. The straightforward approach would be to start at one end and simply sequence the entire 50 million base pairs of the chromosome. Unfortunately, this approach does not work.

The DNA of a molecule that is 50 million base pairs long cannot be sequenced all at once; only about 700 base pairs at a time can be sequenced. (See Figure 11.21 to review the DNA sequencing technique.) To sequence the entire genome, chromosomal DNA is first cut into fragments about 500 base pairs long, then each fragment is sequenced. For the human genome, which has about 3 billion base pairs, there are more than 6 million such fragments.

The problem then becomes putting these millions of pieces of the jigsaw puzzle back together. This problem can be overcome by breaking up the DNA in several ways into "sub-jigsaws" that overlap, and aligning the overlapping fragments. There are two ways to do this.

HIERARCHICAL SEQUENCING. The publicly funded effort first systematically identified short marker sequences along the chromosomes, so that every small sequenced section of DNA would contain a marker. Then sequences with the common

Several thousand patients, over half of them with cancer, have undergone gene therapy. Most of these clinical trials have been at a preliminary level, in which people are given the therapy to see whether it has any toxicity, and whether the new gene is actually incorporated into the genome of the patient. More ambitious trials are under way, in which a

RESEARCH METHOD

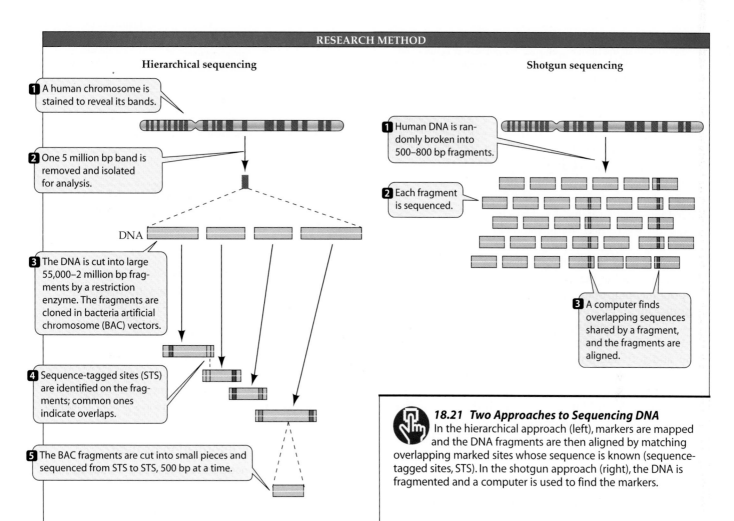

Hierarchical sequencing

1 A human chromosome is stained to reveal its bands.

2 One 5 million bp band is removed and isolated for analysis.

DNA

3 The DNA is cut into large 55,000–2 million bp fragments by a restriction enzyme. The fragments are cloned in bacteria artificial chromosome (BAC) vectors.

4 Sequence-tagged sites (STS) are identified on the fragments; common ones indicate overlaps.

5 The BAC fragments are cut into small pieces and sequenced from STS to STS, 500 bp at a time.

Shotgun sequencing

1 Human DNA is randomly broken into 500–800 bp fragments.

2 Each fragment is sequenced.

3 A computer finds overlapping sequences shared by a fragment, and the fragments are aligned.

18.21 Two Approaches to Sequencing DNA
In the hierarchical approach (left), markers are mapped and the DNA fragments are then aligned by matching overlapping marked sites whose sequence is known (sequence-tagged sites, STS). In the shotgun approach (right), the DNA is fragmented and a computer is used to find the markers.

markers could then be aligned (Figure 18.21, left panel). There are two types of mapped markers along the chromosome:

▶ *Physical map markers* are chromosomal landmarks that can be ordered and the distances between them determined. The result can be compared to a road map showing towns with the mileage separating them. The "towns" are DNA markers, and the "mileage" is in base pairs.

▶ *Genetic map markers* are specific DNA sequences, such as RFLPs or short simple sequence repeats, whose locations are determined genetically.

The simplest of the physical map markers are the recognition sites for restriction enzymes. Since there are hundreds of these recognition site–restriction enzyme pairs, there are hundreds of ways to cut the DNA and then generate a restriction map. Restriction mapping has been useful in generating maps for relatively small chromosome regions of thousands of base pairs.

Some restriction enzymes recognize 8–12 base pairs in DNA, not just the usual 4–6 base pairs. A DNA molecule with several million base pairs will have relatively few of these larger sites, and thus the enzyme will generate a small number of relatively large fragments. These large

fragments can be put into a vector called a *bacterial artificial chromosome* (BAC), which can carry about 250,000 base pairs of inserted DNA in a single copy and cloned in bacteria to create a DNA library.

The books (fragments) in this library can be arranged in the proper order by using *sequence-tagged sites* (STS's). These are unique stretches of DNA, 60 to 100 base pairs long, whose sequences are known. About 41,000 STS's have been precisely mapped on human chromosomes, meaning that each is less than 100,000 base pairs (or just a few genes) away from the next STS.

To arrange DNA fragments on a map, libraries made from different restriction enzyme cuts are compared. If two large fragments of DNA cut with different enzymes have the same STS, they must overlap.

Genetic map markers are also useful tools in analyzing the genome. As we described earlier, linkage studies with markers have been very important in tracking down disease-causing genes by positional cloning. The genetic and physical map markers can complement each other— one providing new markers for the other.

SHOTGUN SEQUENCING. Instead of the "top-down" approach—getting map makers, then fragmenting the DNA and sequencing it—the "shotgun" approach cuts the DNA at random into small, sequence-ready fragments and lets

18.22 Is This the Future of Medicine?
The elucidation of the human genome sequence may result in an approach to medicine that is oriented to the genetic and functional individuality of each patient.

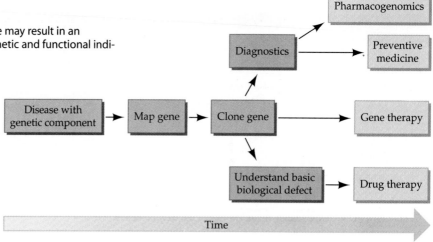

powerful computers determine markers that overlap (Figure 18.21, right panel). The fragments can then be aligned.

The shotgun method, which has been used by private industry, is much faster than the hierarchical approach because there is no need for physical or genetic maps. At first there was considerable skepticism about the method. There were concerns that without rigorous prior mapping of marker sites on the chromosomes, the computer might pick out repetitive sequences common to many DNA fragments and line the fragments up incorrectly. But the rapid rate of development of sophisticated computer and software power has allowed the technique to quickly be refined to a point where inaccurate alignment is not a major problem.

The entire 180 million-base-pair fruit fly genome (see Chapter 14) was sequenced by the shotgun method in little over a year. This proved that the rapid method might work for the much larger human genome, and in fact it did.

The human genome is more than just a sequence

Reading the human "book of life" is an achievement that ranks with other recent great events in exploration, such as landing on the moon. But gene sequencing and the tools developed to carry it out, are changing biology in many other ways as well:

▶ The sequences of other organisms have provided insights and practical information on both prokaryotic (Chapter 13) and eukaryotic (Chapter 14) genomes. Many genes sequenced and identified in "simpler" organisms have counterparts in humans, so these findings have facilitated the identification of human genes.

▶ Mapping technology has made the isolation of human genes by positional cloning much easier, because of the many chromosome markers available. Over 100 disease-related genes have been identified in this way.

▶ Genetic variability in drug metabolism has been a medical problem for a long time. The identification of the genes responsible is leading to tests that predict who will react best to which medications. This is the emerging science of *pharmacogenomics.*

▶ DNA chips (see Figure 17.12) are being used to analyze the specific expression of thousands of genes in cells in different biochemical states. For example, a Cancer Genome Anatomy Project seeks to make an mRNA "fingerprint" of a tumor at each stage of its development. Finding out which sequences are expressed at what stage will be important not only in diagnosis, but also to identify targets for gene therapy.

▶ The Human Genome Diversity Project is looking for important polymorphisms in specific human popula-

tions. For example, the Pima Indians in Arizona have a high frequency of extreme obesity and diabetes. A search of their genomes might reveal genes predisposing them to these diseases.

The end result of all of this knowledge of the human genome may lead to a new approach to medical care, in which each person's genes will be used to prescribe lifestyles and treatments that can maximize that person's genetic potential (Figure 18.22).

How should genetic information be used?

After the primary genetic defect that causes cystic fibrosis was discovered, many people predicted a "tidal wave" of genetic testing for heterozygous carriers. Everyone would want the test, it was thought—especially the relatives of patients with the genetic disease. But the tidal wave has not developed. To find out why, a team of psychologists, ethicists, and geneticists interviewed 20,000 people in the United States. What the researchers found surprised them: Most people are simply not very interested in their genetic makeup, unless they have a close relative with a genetic disease and are involved in a decision about pregnancy.

There are other people, however, who might be very interested in the results of genetic testing. People who test positive for genetic abnormalities, from hypercholesterolemia to cancer, might be denied employment or health insurance. The linking of genetic abnormalities to behavioral characteristics, such as manic depression and schizophrenia, has led to the potential for screening and then social manipulations of those at risk. Consequently, many legislative bodies are considering and passing bills that prohibit genetic discrimination.

The Human Genome Diversity Project has raised many concerns about exploitation and commercialization of people's DNA sequences. Is a gene that confers resistance to cancer, for example, the property of an individual, an ethnic group in which it may be frequent, a pharmaceutical company that finds it, or humanity at large? This issue of ownership is being tested worldwide, perhaps no more acutely than in Iceland, most of whose 270,000 people trace their ancestry back to the first settlers that arrived on the island

1,000 years ago. Tissues from the entire population have been sampled and stored for several generations. This tissue bank is a potential gold mine for genetic prospectors, and a single company has been set up, with government support, to sell the knowledge that comes out of analyzing the DNA's of Iceland's people.

Chapter Summary

Protein as Phenotype

▶ In many human genetic diseases, a single protein is missing or nonfunctional. Therefore, the one-gene, one-polypeptide relationship applies to human genetic diseases. **Review Figure 18.1**

▶ A mutation in a single gene causes alterations in its protein product that may lead to clinical abnormalities or have no effect. **Review Figure 18.2**

▶ Some diseases are caused by mutations that affect structural proteins; examples include Duchenne's muscular dystrophy and hemophilia.

▶ The genes that code for receptors and membrane transport proteins can also be mutated and cause diseases such as familial hypercholesterolemia and cystic fibrosis. **Review Figure 18.3**

▶ Prion diseases are caused by a protein with an altered shape that is transmitted from one person to another and alters the same protein in the second person. **Review Figure 18.4**

▶ Few human diseases are caused by a single-gene mutation. Most are caused by the interactions of many genes and proteins with the environment.

▶ Human genetic diseases show different patterns of inheritance. Mutant alleles may be inherited as autosomal recessives, autosomal dominants, X-linked conditions, or chromosomal abnormalities.

Mutations and Human Diseases

▶ Molecular biology techniques have made possible the isolation of many genes responsible for human diseases.

▶ One method of identifying the gene responsible for a disease is to isolate the mRNA for the protein in question and then use the mRNA to isolate the gene from a gene library. DNA from a patient that lacks a piece of a chromosome can be compared with DNA from a person who does not show this deletion to isolate a missing gene. **Review Figure 18.6**

▶ In positional cloning, DNA markers are used as guides to point the way to a gene. These markers may be restriction fragment length polymorphisms that are linked to a mutant gene. **Review Figure 18.7**

▶ Human mutations range from single point mutations to large deletions. Some of the most common mutations occur where the modified base 5-methylcytosine is converted to thymine. **Review Figure 18.8**

▶ The effects of the fragile-X chromosome worsen with each generation. This pattern is caused by a triplet repeat that tends to expand with each new generation. **Review Figure 18.9**

▶ Genomic imprinting results in a gene being differentially expressed depending on which parent it comes from.

Detecting Human Genetic Variations

▶ Genetic screening detects human gene mutations. Some protein abnormalities can be detected by simple tests, such as tests for the presence of excess substrate or lack of product. **Review Figure 18.10**

▶ The advantage of testing DNA for mutations directly is that any cell can be tested at any time in the life cycle.

▶ There are two methods of DNA testing: allele-specific cleavage and allele-specific oligonucleotide hybridization. **Review Figures 18. 11, 18.12**

Cancer: A Disease of Genetic Changes

▶ Tumors may be benign, growing only to a certain extent and then stopping, or malignant, spreading through organs and to other parts of the body.

▶ At least five types of human cancers are caused by viruses, which account for about 15 percent of all cancers. **Review Table 18.2**

▶ Eighty-five percent of human cancers are caused by genetic mutations of somatic cells. These mutations occur most commonly in dividing cells. **Review Figure 18.14**

▶ Normal cells contain proto-oncogenes, which, when mutated, can become activated and cause cancer by stimulating cell division or preventing cell death. **Review Figure 18.15**

▶ About 10 percent of all cancer is inherited as a result of the mutation of tumor suppressor genes, which normally act to slow down the cell cycle. For cancer to develop, both alleles of a tumor suppressor gene must be mutated.

▶ In inherited cancer, an individual inherits one mutant allele and then somatic mutation occurs in the second one. In sporadic cancer, two normal alleles are inherited, so two mutational events must occur in the same somatic cell. **Review Figures 18.16, 18.17**

▶ Mutations must activate several oncogenes and inactivate several tumor suppressor genes for a cell to produce a malignant tumor. **Review Figure 18.18**

Treating Genetic Diseases

▶ Most genetic diseases are treated symptomatically. However, as more knowledge is accumulated, specific treatments are being devised.

▶ One treatment approach is to modify the phenotype—for example, by manipulating the diet, providing specific metabolic inhibitors to prevent the accumulation of a harmful substrate, or supplying a missing protein. **Review Figure 18.19**

▶ In gene therapy, a mutant gene is replaced with a normal gene. Either the affected cells can be removed, the new gene added, and the cells returned to the body, or the new gene can be inserted via a vector directly into the patient. **Review Figure 18.20**

Sequencing the Human Genome

▶ Human genome sequencing is determining the entire human DNA sequence, which requires sequencing many 500-base-pair fragments and then fitting the sequences back together.

▶ In hierarchical gene sequencing, marker sequences are identified and mapped. The markers are then sought in sequenced fragments and are used to align the fragments. In the shotgun approach, the fragments are sequenced and then common markers are identified by computer. **Review Figure 18.21**

▶ The identification of more than 30,000 human genes may lead to a new molecular medicine. **Review Figure 18.22**

▶ As more genes relevant to human health are described, concerns about how such information is used are growing.

For Discussion

1. Compare the roles of proto-oncogenes and tumor sup-pressor genes in normal cells. How do these genes and their functions change in tumor cells? Propose targets for cancer therapy involving these gene products.

2. In the past, it was common for people with phenylke-tonuria (PKU) who were placed on a low-phenylalanine diet after birth to be allowed to return to a normal diet during their teenage years. Although the levels of pheny-lalanine in their blood were high, their brains were thought to be beyond the stage of being harmed. If a woman with PKU becomes pregnant, however, a problem arises. Typically, the fetus is heterozygous, but is unable at early stages of development to metabolize the high levels of phenylalanine that arrive from the mother's blood. Why is the fetus heterozygous? What do you think would happen to the fetus during this "maternal PKU" situa-tion? What would be your advice to a woman with PKU who wants to have a child?

3. A "knockout" mouse has been made that has deletions in both copies of the gene coding for Pr^c. Would you expect this mouse to develop a prion disease if infected with Pr^{sc}? Explain your answer.

4. Cystic fibrosis is an autosomal recessive disease in which thick mucus is produced in the lungs and airway. The gene responsible for this disease codes for a protein com-posed of 1,480 amino acids. In most patients with cystic fibrosis, the protein has 1,479 amino acids: A phenylala-nine is missing at position 508. How would you test the older brother of a baby with cystic fibrosis to determine whether he is a carrier for the disease? How would you design a gene therapy protocol to "cure" the cells in the lung and airway?

19 *Natural Defenses against Disease*

ON JANUARY 6, 1777, GEORGE WASHINGTON, commanding the Revolutionary army of the fledgling United States, made a fateful medical/military decision. As he wrote to his chief physician, "Finding smallpox to be spreading much, and fearing that no precaution can prevent it from running through the whole of our army, I have determined that the troops shall be inoculated. Should the disease rage with its usual virulence, we should have more to dread from it than the sword of the enemy."

Washington was speaking from experience. He himself had survived the disease in 1751. During 1776 his army lost 1,000 men in battle, but 10,000 to smallpox. This highly virulent disease, which killed up to 1 in 4 people who were exposed to it, had already figured prominently in American history. A century before, it had decimated the Native Americans, making colonization by Europeans easier. Two years previously at Quebec, it had laid waste to an American army that was trying to annex Canada by force.

The death rate due to smallpox in the Revolutionary army plummeted after Washington's order was carried out. How did inoculation, a practice that was learned from the people in the Near East and from African slaves, save the soldiers? And why was Washington himself immune to the disease as it ravaged his army?

The answers to these questions lie in the cells and molecules of the immune system. When Washington caught smallpox as a teenager, cells called macrophages engulfed some of the smallpox viruses by phagocytosis and partly digested them. The macrophages displayed fragments of the viruses on their cell surfaces. Specialized white blood cells called T cells recognized those fragments and were activated to divide and differentiate further. The descendants of these activated T cells then attacked Washington's virus-infected cells, preventing the lethal spread of the disease. Other descendants of the T cells persisted in his body as "memory cells" and rapidly defended him when he was exposed to the disease as an adult. Inoculation of his soldiers with powdered scabs containing dead viruses from smallpox patients stimulated the formation of these memory cells in their bodies, once again preventing the virus from spreading following infection. This practice had been used for centuries, and was finally put on a more scientific basis by Edward Jenner two decades after Washington's army was inoculated.

These defensive events in the bodies of Washington and his soldiers required the participation of many kinds of proteins. Some cellular proteins functioned as specific receptors, such as the markers identifying Washington's cells, some as signals triggering events in the macrophages and T cells, and others as weapons leading to the breakdown of infected cells.

Animal defense systems are based on the distinction between *self* —the animal's own molecules—and *nonself*, or foreign, molecules. In this chapter we consider the mechanisms by which animals recognize nonself molecules and combat infection and disease. Many of these mechanisms are based on the principles of genetics and molecular biology that have been discussed in earlier chapters.

In general, there are two types of defenses. **Innate**, or nonspecific, defenses are general mechanisms that protect the body from many pathogens. An example is the skin, which acts as a barrier to stop potentially invading viruses

George Washington
Washington's decision to immunize his army against smallpox saved many lives and probably helped him win the Revolutionary War.

from entering the body. Most animals and plants have innate defenses. **Specific** defenses are aimed at a single target. For example, an antibody protein can be made that binds to a virus if that virus ever enters the bloodstream, and this binding results in the virus being destroyed. Specific defense mechanisms are present in vertebrate animals. In animals that have both, *innate and specific defenses operate together and offer a coordinated defense.*

Defensive Cells and Proteins

The components of the defense system are dispersed throughout the body and interact with almost all of its other tissues and organs. The *lymphoid tissues*, which include the thymus, bone marrow, spleen, and lymph nodes, are essential parts of our defense system (Figure 19.1), but central to their functioning are the blood and lymph.

Blood and lymph are fluid tissues that consist of water, dissolved solutes, and cells. *Blood plasma* is a yellowish solution containing ions, small molecular solutes, and soluble proteins. Suspended in the plasma are red blood cells, white blood cells, and platelets (cell fragments essential to

clotting) (Figure 19.2). While red blood cells are normally confined to the *closed circulatory system* of (the heart, arteries, capillaries, and veins), white blood cells and platelets are also found in the lymph.

Lymph is a fluid derived from blood and other tissues. It accumulates in spaces outside the blood vessels and contains many of the components of blood, but not red blood cells. From the spaces around body cells, the lymph moves slowly into the lymphatic system. Tiny lymph capillaries conduct fluid to larger vessels that join together, forming one large lymph duct that joins a major vein (the left subclavian vein) near the heart. By this system of vessels, the lymphatic fluid is eventually returned to the blood and the circulatory system.

At many sites along the lymph vessels are small, roundish structures called **lymph nodes**, which contain a variety of white blood cells. As fluid passes through the node, it is filtered and "inspected" for foreign materials by these defensive cells.

White blood cells play many defensive roles

One milliliter of blood typically contains about 5 billion red blood cells and 7 million of the larger white blood cells. White blood cells have nuclei and are colorless, whereas mammalian red blood cells lose their nuclei during development. White blood cells can leave the circulatory system and enter extracellular spaces where foreign cells or substances are present. In response to invading pathogens, the number of white blood cells in the blood and lymph may rise sharply, providing medical professionals with a useful clue for detecting an infection.

Several types of white blood cells are important in the body's defenses. But they are all members of two broad groups, phagocytes and lymphocytes.

Phagocytes engulf and digest nonself materials. The most important phagocytes are the neutrophils and the macrophages. In addition to phagocytosis of nonself materials, **macrophages** have the important additional function of presenting partly digested nonself materials to the T cells.

Lymphocytes are the most abundant white cells. A healthy person has about a trillion lymphocytes, making them as numerous as brain cells. There are two types of lymphocytes, **B cells** and **T cells**. Both originate from cells in the bone marrow. Immature T cells migrate via the blood to the thymus, where they mature. They participate in specific defenses against foreign or altered cells, such as virus-infected cells and tumor cells. The B cells leave the bone marrow and circulate through the blood and lymph vessels. B cells make specialized proteins called **antibodies** that enter the blood and bind to nonself substances.

Immune system proteins bind pathogens or signal other cells

The cells that defend our bodies work together like cast members in a drama, interacting with one another and with the cells of invading pathogens. These cell–cell interactions are accomplished by a variety of key proteins, including re-

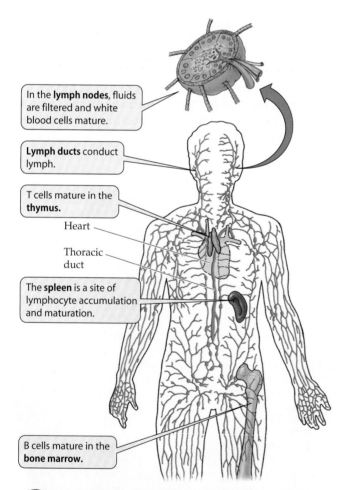

In the **lymph nodes**, fluids are filtered and white blood cells mature.

Lymph ducts conduct lymph.

T cells mature in the **thymus.**

Heart

Thoracic duct

The **spleen** is a site of lymphocyte accumulation and maturation.

B cells mature in the **bone marrow.**

19.1 The Human Defense System
A network of ducts collects lymph from the body's tissues and carries it toward the heart, where it mixes with blood to be pumped back to the tissues. The thymus, spleen, and bone marrow are essential to the body's defensive network.

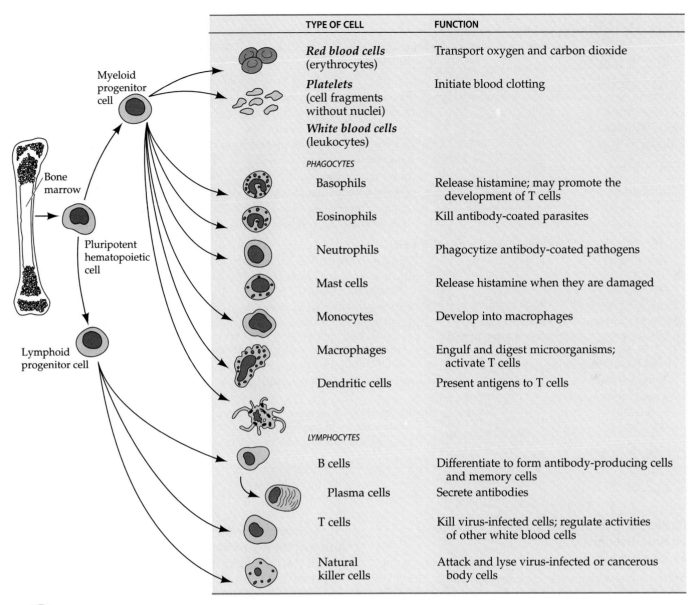

	TYPE OF CELL	FUNCTION
	Red blood cells (erythrocytes)	Transport oxygen and carbon dioxide
	Platelets (cell fragments without nuclei)	Initiate blood clotting
	White blood cells (leukocytes)	
	PHAGOCYTES	
	Basophils	Release histamine; may promote the development of T cells
	Eosinophils	Kill antibody-coated parasites
	Neutrophils	Phagocytize antibody-coated pathogens
	Mast cells	Release histamine when they are damaged
	Monocytes	Develop into macrophages
	Macrophages	Engulf and digest microorganisms; activate T cells
	Dendritic cells	Present antigens to T cells
	LYMPHOCYTES	
	B cells	Differentiate to form antibody-producing cells and memory cells
	Plasma cells	Secrete antibodies
	T cells	Kill virus-infected cells; regulate activities of other white blood cells
	Natural killer cells	Attack and lyse virus-infected or cancerous body cells

19.2 Blood Cells
Pluripotent stem cells in the bone marrow can differentiate into red blood cells, platelets, and the various types of white blood cells.

ceptors, surface markers, signaling molecules, and toxins. They will be discussed later in the chapter, as they appear in the context of our story. However, let's take a brief look at four of the major players here.

▶ **Antibodies** are proteins that bind specifically to certain substances identified by the immune system as nonself or altered self. B cells secrete antibodies as defensive weapons.

▶ **T cell receptors** are proteins on the surfaces of T cells. They recognize and bind to nonself substances on the surfaces of other cells.

▶ **Major histocompatibility complex (MHC) proteins** protrude from the surfaces of most cells in the mammalian body. They are important "self"-identifying

labels and play major parts in coordinating interactions among lymphocytes and macrophages.

▶ **Cytokines** are soluble signal proteins released by T cells, macrophages, and other cells. They bind to and alter the behavior of their target cells. Different cytokines activate or inactivate B cells, macrophages, and T cells. Some cytokines limit tumor growth by killing tumor cells.

Innate Defenses

Innate defenses are general protection mechanisms to stop *pathogens*—harmful organisms that can cause disease—from invading the body. In humans, these defenses include physical barriers as well as cellular and chemical defenses. Table 19.1 provides a summary of the innate defense mechanisms.

19.1 Human Innate Defenses

DEFENSIVE AGENT	FUNCTION
Surface barriers	
Skin	Prevents entry of pathogens and foreign substances
Acid secretions	Inhibit bacterial growth on skin
Mucus membranes	Prevent entry of pathogens
Mucus secretions	Trap bacteria and other pathogens in digestive and respiratory tracts
Nasal hairs	Filter bacteria in nasal passages
Cilia	Move mucus and trapped materials away from respiratory passages
Gastric juice	Concentrated HCl and proteases destroy pathogens in stomach
Acid in vagina	Limits growth of fungi and bacteria in female reproductive tract
Tears, saliva	Lubricate and cleanse; contain lysozyme, which destroys bacteria
Nonspecific cellular, chemical, and coordinated defenses	
Normal flora	Compete with pathogens; may produce substances toxic to pathogens
Fever	Body-wide response inhibits microbial multiplication and speeds body repair processes
Coughing, sneezing	Expels pathogens from upper respiratory passages
Inflammatory response (involves leakage of blood plasma and phagocytes from capillaries)	Limits spread of pathogens to neighboring tissues; concentrates defenses; digests pathogens and dead tissue cells; released chemical mediators attract phagocytes and specific defense lymphocytes to site
Phagocytes (macrophages and neutrophils)	Engulf and destroy pathogens that enter body
Natural killer cells	Attack and lyse virus-infected or cancerous body cells
Antimicrobial proteins	
Interferon	Released by virus-infected cells to protect healthy tissue from viral infection; mobilizes specific defenses
Complement proteins	Lyse microorganisms, enhance phagocytosis, and assist in inflammatory response

Barriers and local agents defend the body against invaders

Skin is a primary innate defense against invasion. Fungi, bacteria, and viruses rarely penetrate healthy, unbroken skin. But damaged skin or internal surface tissue greatly increase the risk of infection by pathogenic agents.

The bacteria and fungi that normally live and reproduce in great numbers on our body surfaces without causing disease are referred to as *normal flora*. These natural occupants of our bodies compete with pathogens for space and nutrients, so normal flora are a form of innate defense.

The mucus-secreting tissues found in parts of the visual, respiratory, digestive, excretory, and reproductive systems have other defenses against pathogens. Tears, nasal mucus, and saliva possess an enzyme called *lysozyme* that attacks the cell walls of many bacteria. Mucus in the nose traps airborne microorganisms, and most of those that get past this filter end up trapped in mucus deeper in the respiratory tract. Mucus and trapped pathogens are removed by the beating of cilia in the respiratory passageway, which moves a sheet of mucus and the debris it contains up toward the nose and mouth. Sneezing is another way to remove microorganisms from the respiratory tract.

Pathogens that reach the digestive tract (stomach, small intestine, and large intestine) are met by other defenses.

The stomach is a deadly environment for many bacteria because of the hydrochloric acid and protein-digesting enzymes that are secreted into it. The intact lining of the small intestine is not normally penetrated by bacteria, and some pathogens are killed by bile salts secreted into this part of the tract. The large intestine harbors many bacteria, which multiply freely; however, these are usually removed quickly with the feces. Most of the bacteria in the large intestine are normal flora that provide benefits to their host.

All of these barriers and secretions are *nonspecific* defenses because they act on all invading pathogens in the same way. But there are more complicated nonspecific cellular chemical defenses.

Innate defenses include chemical and cellular processes

Pathogens that manage to penetrate the body's outer and inner surfaces encounter additional nonspecific defenses. These defenses include the secretion of various defensive proteins as well as cellular defenses involving phagocytosis.

COMPLEMENT PROTEINS. Vertebrate blood contains about 20 different antimicrobial proteins that make up the **complement system**. These proteins, in different combinations, provide three types of defenses. In each type, the comple-

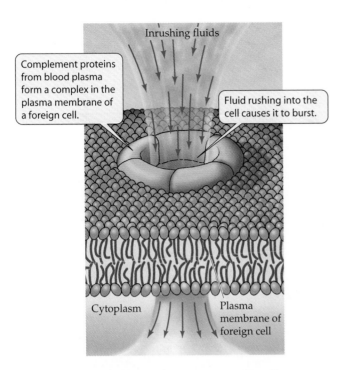

Complement proteins from blood plasma form a complex in the plasma membrane of a foreign cell.

Inrushing fluids

Fluid rushing into the cell causes it to burst.

Cytoplasm

Plasma membrane of foreign cell

19.3 Complement Proteins Destroy a Foreign Cell
Some complement proteins attach to foreign cells such as bacteria after antibodies have bound to them. The porelike structure of the protein allows fluids to pour into the foreign cell until it bursts.

ment proteins act in a characteristic sequence, or cascade, with each protein activating the next.

▶ Complement proteins attach to microbes, which helps phagocytes destroy them.
▶ Complement proteins activate the inflammatory response and attract phagocytes to sites of infection.
▶ Complement proteins, acting with antibodies, lyse (burst) invading cells such as bacteria (Figure 19.3).

INTERFERONS.   When cells are infected by a virus, they produce small amounts of antimicrobial proteins called **interferons** that increase the resistance of neighboring cells to infection by the same *or other* viruses. Interferons have been found in many vertebrates and are one of the body's lines of nonspecific defense against viral infection.

Interferons differ from species to species, and each vertebrate species produces at least three different interferons. All interferons are glycoproteins (proteins with a carbohydrate component) consisting of about 160 amino acids. By binding to receptors in the plasma membranes of their target cells, interferons inhibit viral replication.

PHAGOCYTOSIS AND OTHER CELLULAR ASSAULTS.   Phagocytes provide an important nonspecific defense against pathogens that penetrate the surface of the host. Some phagocytes travel freely in the circulatory system; others can move out of blood vessels and adhere to certain tissues. Pathogens such as large molecules, cells, and viruses become attached to the membrane of a phagocyte (Figure 19.4), which ingests them by endocytosis. After lysosomes fuse with the phagosome,

the pathogens are degraded by lysosomal enzymes (see Figure 4.15).

There are three types of phagocytes:

▶ **Neutrophils** are the most abundant phagocytes, but they are relatively short-lived. They recognize and attack pathogens in infected tissue.
▶ **Monocytes** mature into **macrophages**, which live longer than neutrophils and can consume large numbers of pathogens. Some macrophages roam through the body; others reside permanently in lymph nodes, the spleen, and certain other lymphoid tissues, "inspecting" the lymph for pathogens.
▶ **Eosinophils** are weakly phagocytic. Their primary function is to kill parasites, such as worms, that have been coated by antibodies.

A class of nonphagocytic white blood cells, known as **natural killer cells**, can distinguish virus-infected cells and some tumor cells from their normal counterparts and initiate the lysis of these target cells. In addition to this nonspecific action, natural killer cells are part of the specific defenses, as we will describe later in this chapter.

INFLAMMATION.   The body employs the inflammation response in dealing with infection or with any other process that causes tissue injury either on the surface of the body or internally. You have experienced the symptoms of inflammation: redness and swelling, accompanied by heat and pain. The damaged body cells cause the inflammation by releasing various substances, such as the chemical signal **histamine**. Cells adhering to the skin and linings of organs, called **mast cells**, release histamine when they are damaged, as do white blood cells called **basophils**.

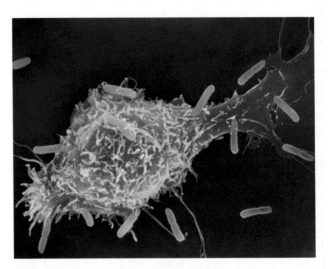

19.4 A Phagocyte and Its Bacterial Prey
Some bacteria (appearing yellow in this artificially colored scanning electron micrograph) have become attached to the surface of a phagocyte in the human bloodstream. Many of these bacteria will be engulfed by the phagocyte and destroyed before they can multiply and damage the human host. A single phagocyte can digest about 20 bacteria.

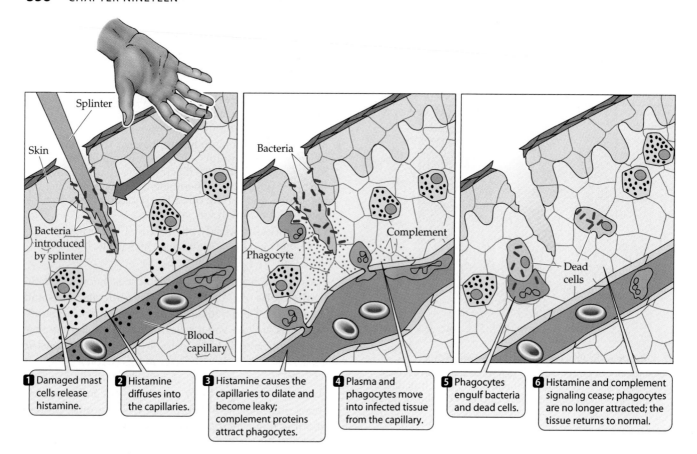

1 Damaged mast cells release histamine.

2 Histamine diffuses into the capillaries.

3 Histamine causes the capillaries to dilate and become leaky; complement proteins attract phagocytes.

4 Plasma and phagocytes move into infected tissue from the capillary.

5 Phagocytes engulf bacteria and dead cells.

6 Histamine and complement signaling cease; phagocytes are no longer attracted; the tissue returns to normal.

19.5 Interactions of Cells and Chemical Signals in Inflammation

The histamine-induced swelling of the inflammation reaction is accompanied by redness, heat, and pain. The chemical signals associated with the reaction attract the phagocytes that are largely responsible for healing the wound.

The redness and heat of inflammation result from histamine-induced dilation of blood vessels in the infected or injured area (Figure 19.5). Histamine also causes the capillaries (the smallest blood vessels) to become leaky, allowing blood plasma and phagocytes to escape into the tissue, causing the characteristic swelling. The pain of inflammation results from increased pressure (from the swelling) and from the action of leaked enzymes.

In damaged or infected tissue, complement proteins and other chemical signals attract phagocytes—neutrophils first, and then monocytes, which become macrophages. The macrophages engulf the invaders and debris and are responsible for most of the healing associated with inflammation. They produce several cytokines, which, among other functions, signal the brain to produce a fever, which inhibits the growth of the invading pathogen. Cytokines may also attract phagocytic cells to the site of injury and initiate a specific immune response to the pathogen.

Following inflammation, pus may accumulate. It is composed of dead cells (neutrophils and the damaged body cells) and leaked fluid. A normal result of inflammation, pus is gradually consumed and digested by macrophages.

Specific Defenses: The Immune Response

Nonspecific defenses are numerous and effective, but some invaders elude them and must be dealt with by defenses targeted against specific threats. The recognition and destruction of specific nonself substances is an important function of an animal's immune system.

Four features characterize the immune response

The characteristic features of the immune system are specificity, the ability to respond to an enormous diversity of foreign molecules and organisms, the ability to distinguish self from nonself, and immunological memory.

SPECIFICITY. **Antigens** are organisms or molecules that are recognized by and/or interact with the immune system to initiate an immune response. The specific sites on antigens that the immune system recognizes are called **antigenic determinants** (Figure 19.6). Chemically, an antigenic determinant is a specific portion of a large molecule, such as a sequence of amino acids that may be present in several proteins. A large antigen, such as a whole cell, may have many different antigenic determinants on its surface, each capable of being bound by a specific antibody or T cell. Even a single protein has multiple, different antigenic determinants. The host responds to the presence of an antigen by producing highly specific defenses—T cells or antibodies that correspond to the antigenic determinants of the antigen. Each T cell and each antibody is specific for a single antigenic determinant.

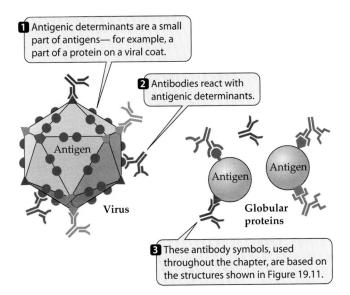

1 Antigenic determinants are a small part of antigens— for example, a part of a protein on a viral coat.

2 Antibodies react with antigenic determinants.

3 These antibody symbols, used throughout the chapter, are based on the structures shown in Figure 19.11.

19.6 Each Antibody Matches an Antigenic Determinant
Each antigen has many different antigenic determinants that are recognized by specific antibodies. An antibody recognizes and binds to its antigenic determinant to initiate defensive measures against the antigen.

DIVERSITY. Challenges to the immune system are legion: individual molecules, viruses, bacteria, protists, and multicellular parasites. Each of these types of potential pathogens includes many species; each species includes many subtly differing genetic strains; each strain possesses multiple surface features, each of which is presented to the immune system. Estimates vary, but a reasonable guess is that humans can respond *specifically* to 10 million different antigenic determinants. Upon recognition of an antigenic determinant, the immune system responds by activating lymphocytes of the appropriate specificity.

DISTINGUISHING SELF FROM NONSELF. The human body contains tens of thousands of different proteins, each with a specific three-dimensional structure capable of generating an immune response. Every cell in the body bears a tremendous number of antigenic determinants. A crucial attribute of an individual's immune system is that it recognizes the body's own antigenic determinants and does not attack them. Failure to make this distinction may lead to an *autoimmune disease*—an attack on one's own body. Such diseases include rheumatoid arthritis and lupus.

IMMUNOLOGICAL MEMORY. After responding to a particular type of pathogen once, the immune system "remembers" that pathogen and can usually respond more rapidly and powerfully to the same threat in the future. This immunological memory usually saves us from repeats of childhood diseases such as chicken pox. Vaccination and inoculation against disease work because the immune system "remembers" the antigenic determinants that are inoculated into the body.

There are two interactive immune responses

The immune system has two responses against invaders: the humoral immune response and the cellular immune response. The two responses operate in concert—simultaneously and cooperatively, sharing mechanisms.

In the **humoral immune response** (from the Latin *humor*, "fluid"), antibodies react with antigenic determinants on foreign invaders in blood, lymph, and tissue fluids. An animal produces a vast diversity of antibodies that, among them, can react with almost any conceivable antigen encountered. Some antibodies are soluble and travel free in the blood and lymph; others exist as integral membrane proteins on specialized lymphocytes called B cells.

The first time a specific antigen invades the body, it may be detected by and bind to a B cell whose membrane antibody recognizes one of its antegenic determinants. This activated B cell forms a **plasma cell** that makes multiple soluble copies of an antibody with the same specificity as the membrane antibody.

The **cellular immune response** is directed against an antigen that has become established within a cell of the host animal. It detects and destroys virus-infected or mutated cells.

Unlike the humoral response, the cellular immune response does not use antibodies. Instead, it is carried out by T cells within the lymph nodes, the bloodstream, and the intercellular spaces. The T cells have integral membrane proteins called T cell receptors—surface glycoproteins that recognize and bind to antigenic determinants while remaining part of the cell's plasma membrane. Like antibodies, T cell receptors have specific molecular configurations that bind to specific antigenic determinants. Once a T cell is bound to a determinant, it initiates an immune response.

Clonal selection accounts for the characteristic features of the immune response

Each person possesses an enormous number of different B cells and T cells, apparently capable of dealing with almost any antigen ever likely to be encountered. How does this diversity arise? How do B and T cells specific for certain antigens proliferate? And why don't our antibodies and T cells attack and destroy our own bodies? The versatility of immune responses, the proliferation of specific cells, the ability to distinguish between self and nonself, and immunological memory can all be explained by the theory of **clonal selection**.

According to clonal selection theory, each individual human contains an enormous variety of different B cells, and each type of B cell is able to produce *only one kind of antibody*. Thus there are millions of different B cells, each one producing a particular antibody and displaying it on the cell surface. When an antigen that fits this surface antibody binds to it, the B cell is activated. It divides to form a clone of cells, all of them producing that particular antibody. Thus the antigen "selects" a particular antibody-producing B cell for proliferation (Figure 19.7). In the same way, an antigenic cell "selects" a T cell expressing a particular T cell receptor on its surface for proliferation.

The clonal selection theory accounts nicely for the body's ability to respond rapidly to any of a vast number of different antigens. In the extreme case, even a single B cell might be sufficient for an immunological response by the body, provided that it encounters its antigen and then proliferates into a large clone rapidly enough to combat the invasion.

Immunological memory and immunity result from clonal selection

According to clonal selection theory, an activated lymphocyte (B cell or T cell) produces two types of daughter cells, effector cells and memory cells.

▶ **Effector cells** carry out the attack on the antigen. They are either plasma cells that produce antibodies, or T cells that, upon binding an antigenic determinant, release messenger molecules called *cytokines*. Effector cells live only a few days.

▶ **Memory cells** are long-lived cells that retain the ability to start dividing on short notice to produce more effector and more memory cells. Memory B and possibly T cells may survive for decades.

When the body first encounters a particular antigen, a *primary immune response* is activated, and lymphocytes produce clones of effector and memory cells. The effector cells destroy the invaders at hand and then die, but one or more clones of memory cells have now been added to the immune system and provide **immunological memory**.

After the body's first immune response to a particular antigen, subsequent encounters with the same antigen will result in a greater and more rapid production of antigen-specific antibody or T cells. This response is called the *secondary immune response*. The first time a vertebrate animal is exposed to a particular antigen, there is a time lag (usually several days) before the number of antibody molecules and T cells slowly increases (Figure 19.8). But for years afterward—sometimes for life—the immune system "remembers" that particular antigen. The secondary immune response has a shorter lag time, a greater rate of antibody production, and a larger production of total antibody or T cells than the primary response.

Thanks to immunological memory, recovery from many diseases, such as chicken pox, provides a *natural immunity* to those diseases. However, it is possible to protect against many life-threatening diseases, such as typhoid or tetanus, by *artificial immunity*. Artificial immunity can be acquired by the introduction of antigenic proteins or other molecular antigens into the body in a process called **immunization**, or by the introduction of whole pathogens, live or rendered harmless, which is called **vaccination**.

Immunization or vaccination initiates a primary immune response, generating memory cells without making the person ill. Later, if the same or very similar disease organisms attack, B memory cells already exist. They recognize the antigen and quickly overwhelm the invaders with a massive production of lymphocytes and antibodies.

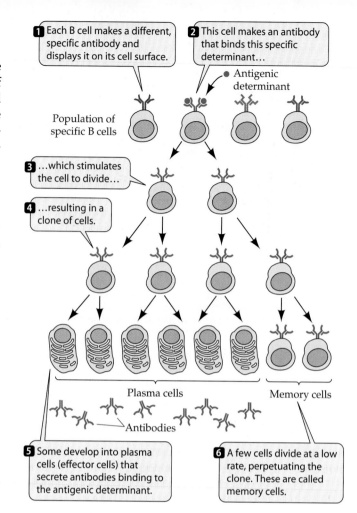

1 Each B cell makes a different, specific antibody and displays it on its cell surface.

2 This cell makes an antibody that binds this specific determinant...

Antigenic determinant

Population of specific B cells

3 ...which stimulates the cell to divide...

4 ...resulting in a clone of cells.

Plasma cells

Memory cells

Antibodies

5 Some develop into plasma cells (effector cells) that secrete antibodies binding to the antigenic determinant.

6 A few cells divide at a low rate, perpetuating the clone. These are called memory cells.

19.7 Clonal Selection in B Cells
The binding of an antigenic determinant to a specific antibody on the surface of a B cell stimulates the cell to divide, rapidly producing a clone of cells to fight the invader.

The ability of the human body to remember a specific antigen explains why immunization has almost completely wiped out deadly diseases such as diphtheria and polio in industrialized countries. In fact, smallpox has been eliminated worldwide, thanks to an international effort by the World Health Organization. As far as we know, the only remaining smallpox viruses on Earth are those kept in some laboratories.

Animals distinguish self from nonself and tolerate their own antigens

Given the presence of lymphocytes directed against so many antigens, why doesn't a healthy human produce self-destructive immune responses? The body is tolerant of its own molecules—the same molecules that would generate an immune response in another individual. Self-tolerance seems to be based on two mechanisms: clonal deletion and clonal anergy.

Clonal deletion physically removes B or T cells from the immune system at some point during their differentiation. For example, immature B cells in bone marrow may encounter self antigens. Any of these cells that shows the potential to mount an immune response against self antigens

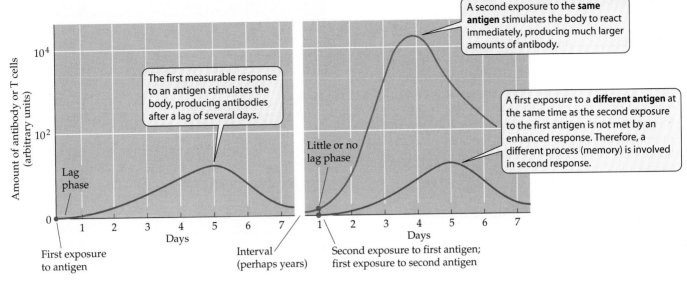

19.8 Immunological Memory

The ability of the body to remember an antigen to which it has been exposed is the basis for natural and artificial immunity against a disease.

undergoes programmed cell death (apoptosis) within a short time, and never differentiates enough to make antibodies. Thus, no clones of antiself lymphocytes normally appear in the bloodstream. Clonal deletion eliminates about 90 percent of all the B cells made in the bone marrow. A similar process occurs with T cells in the thymus.

Clonal anergy is the suppression of the immune response. For example, a mature T cell may encounter and recognize a self antigen on the surface of a body cell. But it does not send out the cytokines that signal the initiation of an immune response. Before it does so, the T cell must encounter not only an antigen, but also a second molecule called CD28 on the cell surface. This *co-stimulatory signal* is expressed only on certain cells called antigen-presenting

cells. So most body cells, lacking CD28, will not be attacked by the cellular immune system.

The phenomenon of **immunological tolerance** (Figure 19.9) was discovered through the observation that some *nonidentical* twin cattle with different blood types contained some of each other's red blood cells. Why did these "foreign" blood cells not cause immune responses resulting in their elimination? The hypothesis suggested was that the blood cells had passed between the fetal animals in the womb before the lymphocytes had matured. Thus each calf regarded the other's red blood cells as self. This hypothesis was confirmed when it was shown that injecting foreign

19.9 Making Nonself Seem Like Self

The ability of adult mice to recognize and reject grafts of foreign skin can be overcome by earlier exposure to nonself antigens. Both of the mouse strains used in this experiment are highly inbred, and so each member of a strain is genetically identical to the others in that strain. Strain B mice tolerate grafts from other strain B mice.

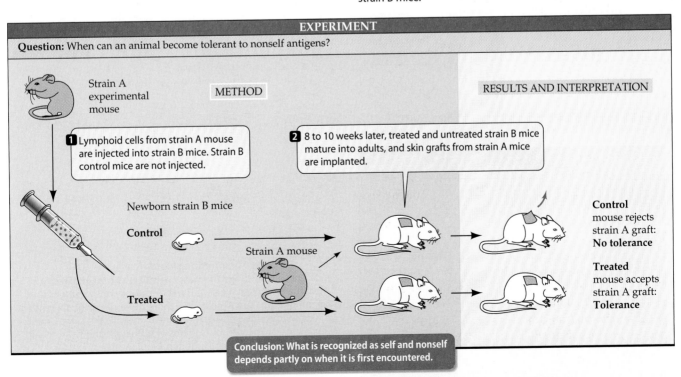

antigen into an animal early in fetal development caused that animal henceforth to recognize that antigen as self.

Tolerance must be established repeatedly throughout the life of the animal because lymphocytes are produced constantly. Continued exposure to self antigen helps maintain tolerance. For unknown reasons, tolerance to self antigens may be lost. When this happens, the body produces antibodies or T cells against its own proteins, resulting in an autoimmune disease.

B Cells: The Humoral Immune Response

Every day, billions of B cells survive the test of clonal deletion and are released from the bone marrow to enter the circulation. The B cells are the basis for the humoral immune response. Since each B cell expresses on its surface an antibody that is specific for a particular antigen, that antigen can bind to and activate the B cell, causing it to form a clone.

Some B cells develop into plasma cells

As described above, the activation of a B cell involves the binding of a particular antigenic determinant to the antibody protein on the surface of the B cell. Normally, for such

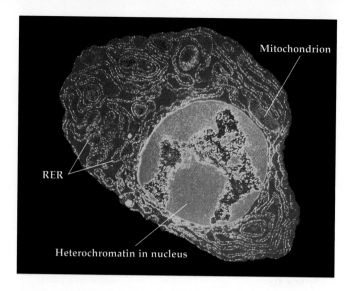

19.10 A Plasma Cell
The prominent nucleus with large amounts of heterochomatin (orange) and the cytoplasm (bright blue) crowded with rough endoplasmic reticulum are features of a cell that is actively synthesizing and exporting proteins—in this case, antibodies.

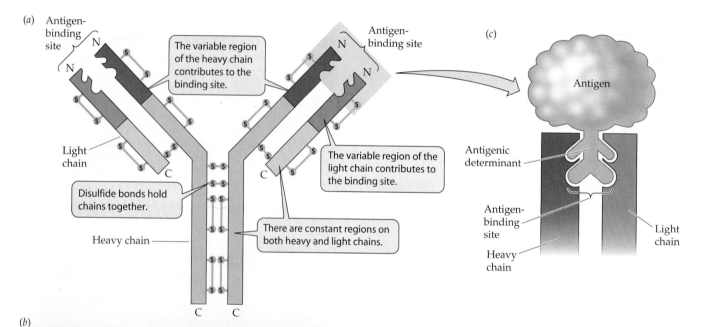

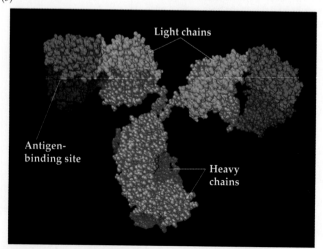

19.11 Structure of Immunoglobulins
In all these renderings, the light chains are shown in green and the heavy chains in blue. (*a*) The four polypeptide chains of an immunoglobulin molecule. (*b*) A more three-dimensional rendering of an immunoglobulin molecule in roughly the same orientation as (*a*). (*c*) The variable regions of both light and heavy chains participate in defining the antigen-binding site.

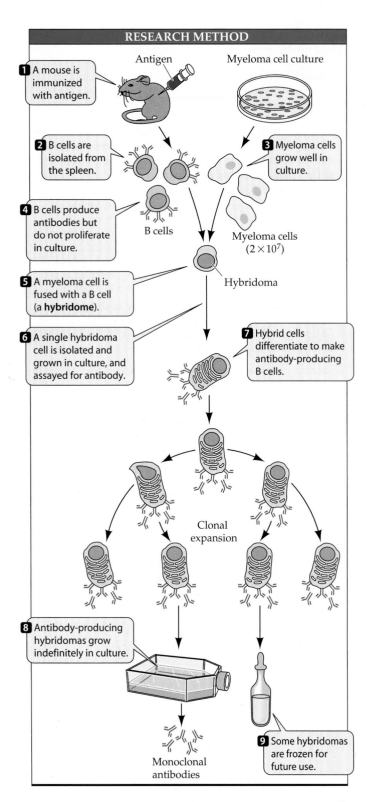

RESEARCH METHOD

1 A mouse is immunized with antigen.

Antigen   Myeloma cell culture

2 B cells are isolated from the spleen.

3 Myeloma cells grow well in culture.

4 B cells produce antibodies but do not proliferate in culture.

B cells

Myeloma cells (2×10^7)

5 A myeloma cell is fused with a B cell (a **hybridome**).

Hybridoma

6 A single hybridoma cell is isolated and grown in culture, and assayed for antibody.

7 Hybrid cells differentiate to make antibody-producing B cells.

Clonal expansion

8 Antibody-producing hybridomas grow indefinitely in culture.

Monoclonal antibodies

9 Some hybridomas are frozen for future use.

19.13 Creating Hybridomas for the Production of Monoclonal Antibodies
Cancerous myeloma cells and normal lymphocytes can be hybridized so that the proliferative properties of the myeloma cells are merged with the properties of of the antibody-producing lymphocytes.

quires special proteins encoded by the MHC (major histocompatibility complex) genes. These proteins underlie the immune system's tolerance for the cells of its own body and are responsible for the rejection of foreign tissues by the body.

T cell receptors are found on two types of T cells

Like B cells, T cells possess specific surface receptors. T cell receptors are not immunoglobulins, but glycoproteins with molecular weights about half that of an IgG. They are made up of two polypeptide chains, each encoded by a separate gene (Figure 19.14).

The genes that code for T cell receptors are similar to those for immunoglobulins, suggesting that both are derived from a single, evolutionarily more ancient group of genes. Like the immunoglobulins, T cell receptors include both variable and constant regions. The variable regions provide the specificity for reaction with a single antigenic determinant.

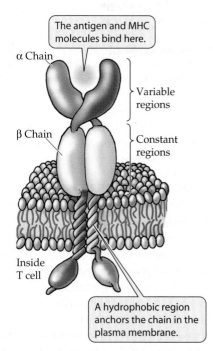

The antigen and MHC molecules bind here.

α Chain

β Chain

Variable regions

Constant regions

Inside T cell

A hydrophobic region anchors the chain in the plasma membrane.

19.14 A T Cell Surface Receptor
T cell receptors are glycoproteins, not immunoglobulins, although the structures of the two molecules are similar. The receptors are bound to the plasma membrane of the T cell that produces them.

vated B cells. T cells are the effectors of the cellular immune response, which is directed against any factor, such as a virus or mutation, that changes a normal cell into an abnormal cell.

In this section, we will describe two types of T cells (helper T cells and cytotoxic T cells). We will discover that the binding of a T cell receptor to an antigenic determinant re-

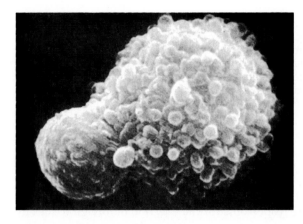

19.15 A Cytotoxic T Cell in Action
A cytotoxic T cell (the smaller sphere) has come into contact with a virus-infected cell, causing the infected cell to lyse. The blisters on the infected cell's surface indicate that it is beginning to break up.

There is a major difference between antibodies and T cell receptors: *While antibodies bind to an intact antigen, T cell receptors bind to a piece of the antigen displayed on the surface of an antigen-presenting cell.*

When T cells are activated by contact with a specific antigenic determinant, they proliferate and give rise to two types of effector cells:

▶ **Cytotoxic T cells**, or T_C **cells**, recognize virus-infected cells and kill them by causing them to lyse (Figure 19.15).

▶ **Helper T cells**, or T_H **cells**, assist both the cellular and humoral immune systems.

As mentioned already, a specific T_H cell must bind an antigen presented on a B cell before that B cell can become activated. The helper cell becomes the "conductor" of the "immunological orchestra," as it sends out chemical signals that not only result in its own clonal expansion, but also set in motion the actions of cytotoxic T cells as well as B cells.

Now that we are familiar with the major types of T cells, we can address the question of how T cells meet the antigenic determinants.

The major histocompatibility complex encodes proteins that present antigens to the immune system

We have seen that a body's defenses recognize its own cells as self—that proteins on our own cell surfaces are tolerated by our immune systems. There are several types of mammalian cell surface proteins, but we will focus on one very important group, the products of a cluster of genes called the major histocompatibility complex, or MHC.

The MHC gene products are plasma membrane glycoproteins—proteins with attached carbohydrate groups. In humans, the MHC molecules are called *human leukocyte antigens* (*HLA*), while in mice they are called *H-2 proteins*. Their major role is to "present" antigens on the cell surface to a T cell receptor.

There are three classes of MHC proteins:

▶ **Class I MHC proteins** are present on the surface of every nucleated cell in the animal. When cellular proteins are degraded to small peptide fragments in the proteasome (see Chapter 14), an MHC I protein may bind to a fragment and travel to the plasma membrane. There, the MHC I protein "presents" the cellular peptide to T_C cells. The T_C cells have a surface protein called CD8 that recognizes MHC I.

▶ **Class II MHC proteins** are found mostly on the surfaces of B cells, macrophages, and other antigen-presenting cells. When an antigen-presenting cell ingests an antigen, such as a virus, it is broken down in an endosome. An MHC II molecule may bind to one of the fragments and carry it to the cell surface, where it is presented to a T_H cell (Figure 19.16). T_H cells have a surface protein called CD4 that recognizes MHC II.

▶ **Class III MHC proteins** include some of the proteins of the complement system that interact with antigen–antibody complexes and result in the lysis of foreign cells (see Figure 19.4).

To accomplish their roles in antigen binding and presentation, both MHC I and MHC II have an *antigen-binding groove*, which can hold a peptide of about 10–20 amino acids (Figure 19.17). The T cell receptor recognizes not just the antigenic fragment, but the fragment *bound to an MHC I or MHC II molecule*. The table in Figure 19.17 summarizes the relationships of T cells and antigen-presenting cells.

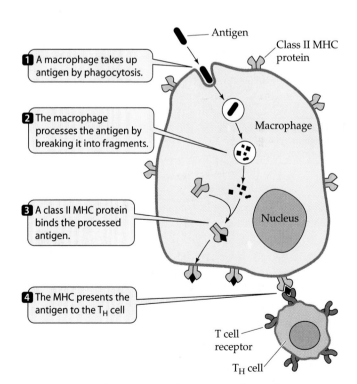

1 A macrophage takes up antigen by phagocytosis.

2 The macrophage processes the antigen by breaking it into fragments.

3 A class II MHC protein binds the processed antigen.

4 The MHC presents the antigen to the T_H cell

Antigen

Class II MHC protein

Macrophage

Nucleus

T cell receptor

T_H cell

19.16 Macrophages Are Antigen-Presenting Cells
Processed antigen is displayed by MHC II protein on the surface of a macrophage. Receptors on the helper T cell can then interact with the processed antigen/MHC II protein complex.

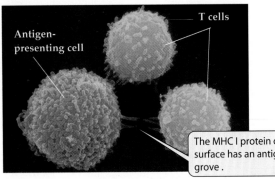

19.17 The Interaction between T Cells and Antigen-Presenting Cells
A groove in the MHC 1 protein holds an antigen, which it "presents" to a cytotoxic T cell. CD8 surface proteins on the T cells insure the specificity of interaction.

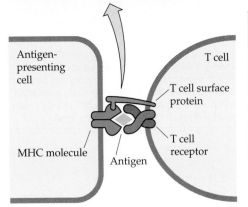

The MHC I protein on the cell's surface has an antigen-binding grove .

Antigen-Presenting and T Cell Types

PRESENTING CELL TYPE	ANTIGEN PRESENTED	MHC CLASS	T CELL TYPE	T CELL SURFACE PROTEIN
Any cell	Intracellular protein fragment	Class I	Cytotoxic T cell (T_C)	CD8
Macrophages and B cells	Fragments from extracellular proteins	Class II	T helper cell (T_H)	CD4

There are three genetic loci each for MHC I and MHC II, and all six loci have as many as 100 different alleles. With so many possible allelic combinations, it is not surprising that different individuals are very likely to have different MHC genotypes. Similarities in base sequences between MHC genes and the genes coding for antibodies and T cell receptors suggest that all three may have descended from the same ancestral genes and are part of a "superfamily." Major aspects of the immune systems seem to be woven together by a common evolutionary thread.

Helper T cells and MHC II proteins contribute to the humoral immune response

When a T_H cell binds to an antigen-presenting macrophage, the T_H cell releases cytokines, which activate it to produce a clone of differentiated cells capable of interacting with B cells. The steps to this point constitute the **activation phase** of the response, and they occur in lymphatic tissue. Next comes the **effector phase**, in which B cells are activated to produce antibodies (Figure 19.18a).

B cells are also antigen-presenting cells. B cells take up by endocytosis antigen bound to their immunoglobulin receptors, process it, and display it on class II MHC proteins. When a T_H cell binds to the displayed antigen–MHC II complex, it releases cytokines that cause the B cell to produce a clone of plasma cells. Finally, the plasma cells secrete antibody, completing the effector phase of the humoral immune response.

Cytotoxic T cells and MHC I proteins contribute to the cellular immune response

Class I MHC molecules play a role in the cellular immune response that is similar to the role played by class II MHC molecules in the humoral immune response. In a virus-infected or mutated cell, "foreign" proteins or peptide frag-

ments combine with MHC class I molecules. The complex is displayed on the cell surface and presented to T_C cells. When a T_C cell binds to this complex, it is activated to proliferate and differentiate (Figure 19.18b).

In the effector phase of the cellular immune response, T_C cells produce molecules that lyse the target cell. In addition, the T_C cell can bind to a specific receptor (called Fas) on the target cell, that initiates apoptosis in the target cell. These two mechanisms, cell lysis and programmed cell death, work in concert to eliminate the altered host cell.

Because T cell receptors recognize self MHC molecules complexed with *nonself* antigens, they help rid the body of its own virus-infected cells. Because they also recognize MHC molecules complexed with *altered self* antigens (as a result of mutations), they help eliminate tumor cells, since most tumor cells have mutations.

In addition to the binding of an antigen–MHC complex to a cell surface receptor, T cells must receive a second signal for activation. This "co-stimulatory" signal occurs after the initial specific binding, and involves the interaction of additional proteins on the T cell and antigen-presenting cell. This second binding event leads to T cell activation, including cytokine production and cell division. It also sets in motion the production of an *inhibitor* of these events, so that the response is appropriately terminated. This inhibitor, a cell surface protein called CTLA4, is also important for the acquisition of tolerance, the capacity to avoid attacking one's own antigenic determinants.

MHC molecules underlie the tolerance of self

MHC molecules play a key role in establishing tolerance to self, without which an animal would be destroyed by its own immune system. Throughout the animal's life, developing T cells are tested in the thymus. One test question is, Can this cell recognize the body's MHC proteins? A T cell

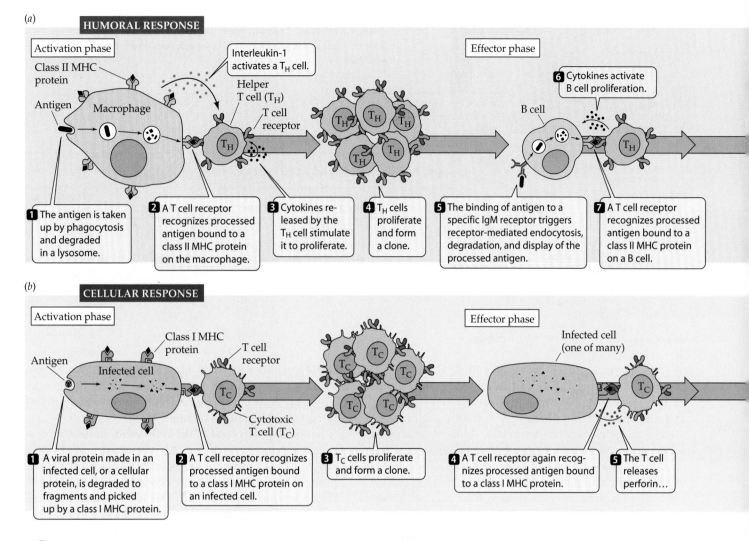

(a)

HUMORAL RESPONSE

Activation phase

Class II MHC protein

Antigen

Macrophage

Interleukin-1 activates a T$_H$ cell.

Helper T cell (T$_H$)

T cell receptor

Effector phase

6 Cytokines activate B cell proliferation.

B cell

T$_H$

1 The antigen is taken up by phagocytosis and degraded in a lysosome.

2 A T cell receptor recognizes processed antigen bound to a class II MHC protein on the macrophage.

3 Cytokines released by the T$_H$ cell stimulate it to proliferate.

4 T$_H$ cells proliferate and form a clone.

5 The binding of antigen to a specific IgM receptor triggers receptor-mediated endocytosis, degradation, and display of the processed antigen.

7 A T cell receptor recognizes processed antigen bound to a class II MHC protein on a B cell.

(b)

CELLULAR RESPONSE

Activation phase

Class I MHC protein

T cell receptor

Antigen

Infected cell

Cytotoxic T cell (T$_C$)

Effector phase

Infected cell (one of many)

T$_C$

1 A viral protein made in an infected cell, or a cellular protein, is degraded to fragments and picked up by a class I MHC protein.

2 A T cell receptor recognizes processed antigen bound to a class I MHC protein on an infected cell.

3 T$_C$ cells proliferate and form a clone.

4 A T cell receptor again recognizes processed antigen bound to a class I MHC protein.

5 The T cell releases perforin…

19.18 Phases of the Humoral and Cellular Immune Responses
Both immune responses have an activation phase and an effector phase.

unable to recognize self MHC proteins would be useless to the animal because it could not participate in any immune reactions. Such a T cell fails the test and dies within about 3 days.

The second question is, Does this cell bind to self MHC protein *and* to one of the body's own antigens? A T cell that satisfied both of these criteria would be harmful or lethal to the animal; it also fails the test and undergoes apoptosis. T cells that survive these tests mature into either T$_C$ cells or T$_H$ cells.

MHC molecules are responsible for transplant rejection

In humans, a major consequence of the MHC molecules became important with the development of organ transplant surgery. Because the proteins produced by the MHC are specific to each individual, they act as antigens if trans-

planted into another individual. An organ or a piece of skin transplanted from one person to another is recognized as nonself and soon provokes an immune response; the tissue then is killed, or "rejected," by the cellular immune system. But if the transplant is performed immediately after birth, or if it comes from a genetically identical person (an identical twin), the material is recognized as self and is not rejected.

The rejection problem can be overcome by treating a patient with drugs, such as cyclosporin, that suppress the immune system. However, this approach compromises the ability of patients to defend themselves against bacteria and viruses. Cyclosporin and some other immunosuppressants interfere with communication between cells of the immune system. Specifically, they inhibit the production of cytokines.

The Genetic Basis of Antibody Diversity

A newborn mammal possesses a full set of genetic information for immunoglobulin synthesis. At each of the loci coding for the heavy and light chains, it has an allele from the mother and one from the father. Throughout the animal's

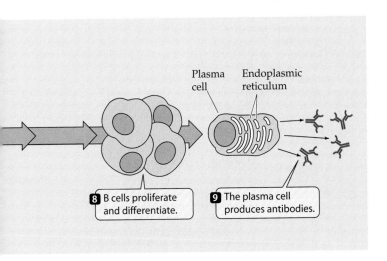

8 B cells proliferate and differentiate.

9 The plasma cell produces antibodies.

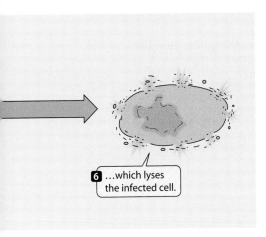

6 ...which lyses the infected cell.

life, each of its cells begins with the same full set of genes. However, as B cells develop, their genomes become modified in such a way that each cell eventually can produce one—and only one—specific type of antibody. In other words, different B cells develop slightly different genomes encoding different antibody specificities. How can a single organism produce millions of different immunoglobulins?

One hypothesis was that we simply have millions of antibody genes. However, a simple calculation (the number of base pairs needed per antibody gene multiplied by millions) shows that if this were true, our entire genome would be taken up by antibody genes. More than 30 years ago, an alternative hypothesis was proposed: A relatively small number of genes recombine at the DNA level to produce many unique combinations, and it is this shuffling of the genetic deck that produces antibody diversity. This is now the accepted theory.

In this section, we will describe the unusual events that generate the enormous antibody diversity normally characterizing each individual mammal. Then we will see how similar events produce the five classes of antibodies by producing slightly different "constant regions" that have special properties.

Antibody diversity results from DNA rearrangement and other mutations

In an unusual genetic process, functional immunoglobulin genes are assembled from DNA segments that initially are spatially separate. Every cell in the body has hundreds of DNA segments potentially capable of participating in the synthesis of the variable regions of the Ig molecule. In most body cells, these DNA sequences remain intact and separate from one another. During B cell development, however, these DNA segments are *rearranged and joined*. Pieces of the DNA are deleted, and DNA segments formerly distant from one another are joined together. In this fashion, an immunoglobulin gene is assembled from randomly selected pieces of DNA. Each B cell precursor in the animal assembles its own unique set of immunoglobulin genes. This remarkable process generates many diverse antibodies from the same starting genome. The same type of process also accounts for the diversity of T cell receptors.

In both humans and mice, the DNA segments coding for immunoglobulin heavy chains are on one chromosome and those for light chains are on others. The variable region of the light chain is encoded by two families of DNA segments, and the variable region of the heavy chain is encoded by three families.

Look at Figure 19.19 for an example of the gene families coding for the constant and variable regions of the heavy chain in mice. There are multiple genes coding for each of four kinds of segments in the polypeptide chain: 100 *V*, 30 *D*, 6 *J*, and 8 *C*. Each B cell that becomes committed to making an antibody randomly selects *one* gene for each of these segments to make the final heavy-chain coding sequence, *VDJC*. So the number of *different* heavy chains that can be made through this random recombination process is quite large.

Now consider that the light chains are similarly constructed, with a similar amount of diversity made possible by random recombination. If we assume that light-chain diversity is the same as heavy-chain diversity (144,000 possibilities), the number of possible combinations of light *and* heavy chains is 144,000 different light chains × 144,000 different heavy chains = 21 *billion* possibilities!

Even if this number is an overestimate by severalfold (and it is), the number of different immunoglobulin molecules that a B cell can make is huge. But there are still other mechanisms that generate even more diversity:

▶ When the DNA sequences for the *V*, *J*, and *C* regions are rearranged so that they are next to one another, the recombination event is not precise, and errors occur at the junctions. This *imprecise recombination* can create new codons at the junctions, with resulting amino acid changes.

▶ After the DNA fragments are cut out and before they are joined, an enzyme, *terminal transferase*, often adds some nucleotides to the free ends of the DNA's. These additional bases create insertion mutations.

▶ Finally, there is a relatively *high mutation rate* in immunoglobulin genes. Once again, this process creates many new alleles and antibody diversity.

19.19 Heavy-Chain Genes

Mouse immunoglobulin heavy chains have four segments, each of which is coded for by one of multiple genes.

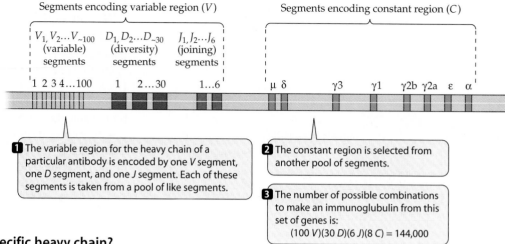

Segments encoding variable region (*V*)   Segments encoding constant region (*C*)

$V_1, V_2 \ldots V_{\sim 100}$ (variable) segments $D_1, D_2 \ldots D_{\sim 30}$ (diversity) segments $J_1, J_2 \ldots J_6$ (joining) segments

1 2 3 4 … 100   1   2 … 30   1 … 6   μ δ   γ3   γ1   γ2b γ2a ε   α

❶ The variable region for the heavy chain of a particular antibody is encoded by one *V* segment, one *D* segment, and one *J* segment. Each of these segments is taken from a pool of like segments.

❷ The constant region is selected from another pool of segments.

❸ The number of possible combinations to make an immunoglubulin from this set of genes is:

(100 *V*)(30 *D*)(6 *J*)(8 *C*) = 144,000

Adding these possibilities to the billions of combinations that can be made by random DNA rearrangements makes it not surprising that the immune system can mount a response to almost any natural or human-made substance.

How does a B cell produce a specific heavy chain?

As an example of how DNA rearrangement generates antibody diversity, let's consider how the heavy chain of IgM is produced in the mouse, a favorite subject for immunology studies.

The gene families governing all heavy-chain synthesis are on mouse chromosome 12, with the members arranged as shown in Figure 19.19. Light chains are produced from similar families, but they lack *D* segments.

How does order emerge from this seeming chaos of DNA segments? Two distinct types of nucleic acid rearrangements contribute to the formation of an antibody:

▸ DNA rearrangements, before transcription, join the *V*, *D*, and *J* segments.

▸ RNA splicing, after transcription, joins the variable region (*VDJ*) to the constant region.

First, substantial chunks of DNA are deleted from the chromosome during rearrangement of the segments. As a result of these deletions, a particular *D* segment ends up directly beside a particular *J* segment, and then the *DJ* segment ends up adjacent to one of the *V* segments. Thus, a single

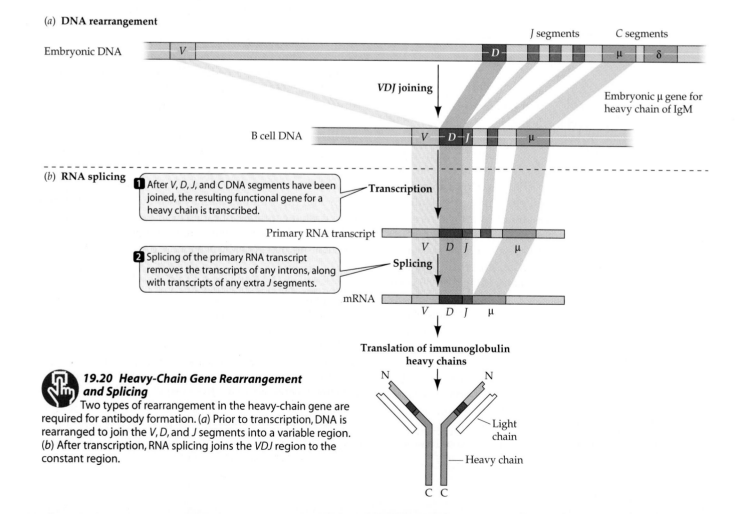

(*a*) **DNA rearrangement**

J segments   *C* segments

Embryonic DNA

VDJ joining

Embryonic μ gene for heavy chain of IgM

B cell DNA

(*b*) **RNA splicing**

❶ After *V*, *D*, *J*, and *C* DNA segments have been joined, the resulting functional gene for a heavy chain is transcribed.

Transcription

Primary RNA transcript

❷ Splicing of the primary RNA transcript removes the transcripts of any introns, along with transcripts of any extra *J* segments.

Splicing

mRNA

Translation of immunoglobulin heavy chains

19.20 Heavy-Chain Gene Rearrangement and Splicing

Two types of rearrangement in the heavy-chain gene are required for antibody formation. (*a*) Prior to transcription, DNA is rearranged to join the *V*, *D*, and *J* segments into a variable region. (*b*) After transcription, RNA splicing joins the *VDJ* region to the constant region.

Light chain

Heavy chain

"new" sequence, consisting of one *V*, one *D*, and one *J* segment, can now code for the variable region of the heavy chain. All the progeny of this cell constitute a clone having the same sequence for the variable region (Figure 19.20*a*).

The second step follows transcription. Splicing of the RNA transcript removes introns and any *J* segments lying between the selected *J* segment and the first constant region segment (Figure 19.20*b*). The result is an mRNA that can be translated, directly yielding the heavy chain of the cell's specific antibody.

The constant region is involved in class switching

In Table 19.2, we described the different classes of antibodies and their functions. Generally, a B cell makes only one class at a time. But **class switching** can occur, in which a B cell changes which antibody class it synthesizes.

Early in its life, a B cell produces IgM molecules, which are responsible for the specific recognition of a particular antigenic determinant. At this time, the constant region of the antibody's heavy chain is encoded by the first constant region segment, the μ segment (see Figure 19.19). If the B cell later becomes a plasma cell during an immunological response, another deletion commonly occurs in the cell's DNA, positioning the heavy-chain variable region gene (consisting of the same *V*, *D*, and *J* segments) next to a constant region segment farther down the original DNA, such as the γ, ε, or α segments (Figure 19.21). Such a DNA deletion results in the production of an antibody with a differ-

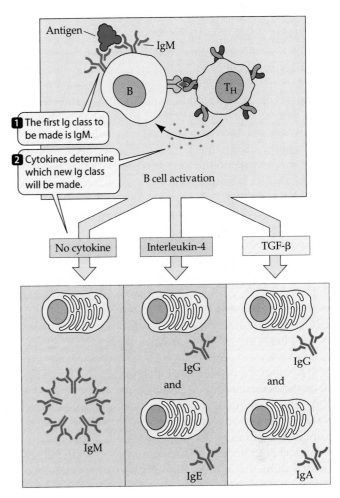

19.22 Cytokines Determine How the Antibody Switches Class
A T_H cell initiates class switching in a B cell by secreting a cytokine. Different cytokines produce different switches.

ent constant region and *function*, but the same *antigen specificity*. The new antibody has *the same variable regions of the light and heavy chains, but a different constant region of the heavy chain*. This new antibody falls into one of the four other immunoglobulin classes (IgA, IgD, IgE, or IgG), depending on which of the constant region segments is placed adjacent to the variable region gene.

After switching classes, the plasma cell cannot go back to making the previous immunoglobulin class, because that part of the DNA has been lost. On the other hand, if additional constant region segments are still present, the cell may switch classes again.

What triggers class switching, and what determines the class to which a given B cell will switch? T_H cells direct the course of an antibody response and determine the nature of the attack on the antigen. These T cells induce class switching by sending cytokine signals (Figure 19.22). These signals bind to receptors on the target B cells, generating a signal transduction cascade and resulting in altered transcription of immunoglobulin genes.

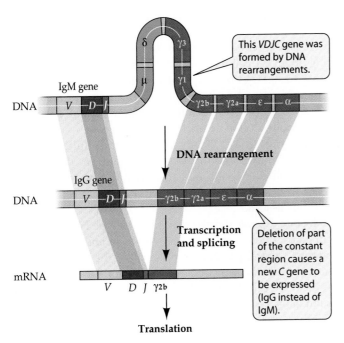

19.21 Class Switching
The gene produced by joining *V*, *D*, *J*, and *C* segments (see Figure19.20) may later be modified, causing a different *C* segment to be transcribed. This modification, known as class switching, is accomplished by deletion of part of the constant region. Shown here is class switching from an IgM gene to an IgG gene.

The Evolution of Animal Defense Systems

The strategy of rearranging DNA to make variable antibodies is used by all vertebrates with jaws, but nowhere else in the animal world. It is an "anticipatory" strategy, since the organism makes not only the defensive proteins that it needs, but also the antibodies and T cell receptors that it *might need*. Therefore, it must have provided an evolutionary advantage to the organisms that first had it.

This anticipatory strategy may first have arisen in an ancient creature resembling today's sharks, when a transposon inserted itself into a gene used for a defensive protein. Over time, this inserted element developed the ability to cut out adjacent DNA sequences and move them elsewhere in the genome.

The invertebrates, in all their diversity, have sturdy innate defense systems, and certain defense system elements are found even in unicellular protists. Many protists carry on phagocytosis, as do our own macrophages, and some protists use phagocytosis as a defense mechanism. Multicellular animals, both invertebrate and vertebrate, employ mobile phagocytic cells to patrol their bodies.

Like vertebrates, invertebrates (and probably some protists) distinguish between self and nonself. Making such distinctions enables invertebrates to reject tissue grafted from other individuals of the same species. Unlike vertebrates, however, invertebrates reject a second graft no more rapidly than a first graft—indicating that invertebrates lack immunological memory. This and other observations show that although immunological functions of invertebrates and vertebrates may be similar, their mechanisms often differ.

Invertebrates do not produce immunoglobulins, lymphocytes, or the complement system. However, they achieve similar defensive goals by analogous methods, and the analogs are probably evolutionary precursors of the systems found in vertebrates. Many invertebrates make proteins very similar to vertebrate cytokines, and those proteins play regulatory roles similar to those in humans.

Disorders of the Immune System

Immune deficiency diseases such as AIDS show us how much we depend on our immune system to protect us from pathogens. However, sometimes the immune system fails us in one way or another. It may overreact, as in an allergy; it may attack self antigens, as in an autoimmune disease; or it may function weakly or not at all, as in an immune deficiency disease. After a look at allergies and autoimmune conditions, we will examine the acquired immune deficiency that characterizes AIDS.

An inappropriately active immune system can cause problems

HYPERSENSITIVITY. A common type of condition can arise when the human immune system overreacts to a dose of antigen (**hypersensitivity**). Although the antigen itself may present no danger to the host, the inappropriate immune response may produce inflammation and other symptoms that can cause serious illness or death. **Allergies** are the most familiar examples of such a problem.

There are two types of allergic reactions. *Immediate hypersensitivity* occurs when an individual makes large amounts of IgE that bind to a molecule or structure in a food, pollen, or the venom of an insect. Mast cells in tissues and basophils in blood bind the IgE, which causes them to release amines such as histamine. The result is symptoms such as dilation of blood vessels, inflammation, and difficulty breathing. If not treated with antihistamines, a severe allergic reaction can lead to death.

Delayed hypersensitivity does not begin until hours after exposure to an antigen. In this case, the antigen is processed by antigen-presenting cells and a T cell response is initiated. This response can be so massive that the cytokines released cause macrophages to become activated and damage tissues. This is what happens when the bacteria that cause tuberculosis colonize the lung.

AUTOIMMUNITY. Sometimes clonal deletion fails, resulting in the appearance of one or more "forbidden clones" of B and T cells directed against self antigens (**autoimmunity**). This failure does not always result in disease, but in some instances it can.

People with *systemic lupus erythematosis* (*SLE*) have antibodies to many cellular components, including DNA and nuclear proteins. These antinuclear antibodies can cause serious damage when they link up with normal tissue antigens to form circulating immune complexes, which become stuck in tissues and provoke inflammation. A person with SLE has hyperactive B cells (thus the excess antibodies).

A person with *rheumatoid arthritis* has difficulty in shutting down a T cell response. We mentioned earlier that the inhibitor CTLA4 blocks T cells from reacting to self antigens. People with rheumatoid arthritis may have low CTLA4 activity, which results in inflammation of joints due to the infiltration of excess white blood cells.

Multiple sclerosis usually affects young adults, causing progressive damage to the nervous system. It involves both T cell- and B cell-mediated attack on two major proteins in myelin, the special membrane that coats some nervous tissues.

Insulin-dependent diabetes mellitus, or type I diabetes, occurs most often in children. It involves an immune reaction against several proteins in the cells of the pancreas that manufacture the protein hormone insulin. This reaction results in the cells being killed. These patients must take insulin daily in order to survive.

The causes of these autoimmune diseases are not known. They tend to "run in families," indicating a genetic component. Some alleles of MHC II are strongly linked to certain of these diseases. In some cases, the underlying reason may be molecular mimicry, in which T cells that recognize a nonself antigen also recognize something on the self that has a similar structure.

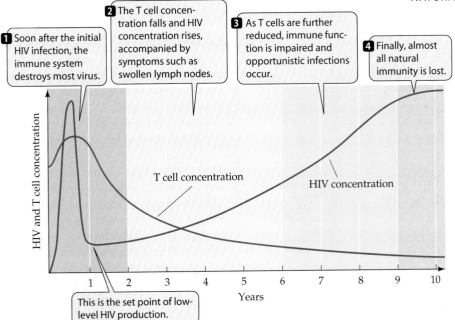

1 Soon after the initial HIV infection, the immune system destroys most virus.

2 The T cell concentration falls and HIV concentration rises, accompanied by symptoms such as swollen lymph nodes.

3 As T cells are further reduced, immune function is impaired and opportunistic infections occur.

4 Finally, almost all natural immunity is lost.

T cell concentration

HIV concentration

This is the set point of low-level HIV production.

Years

19.23 The Course of an HIV Infection
HIV infection may be carried, unsuspected, for many years before the onset of symptoms. This long "dormant" period means the infection is often spread by people who are unaware that they are carrying the virus.

to the nodes and having cells in the nodes already receptive to virus infection—combine to ensure that HIV reproduces vigorously. Up to 10 billion viruses are made every day during this phase. The numbers of T_H cells quickly drop, and people show symptoms similar to mononucleosis, such as enlarged lymph nodes and fever.

These symptoms abate within 3 weeks, however, when T cells recognize infected lymphocytes, an immune response is mounted, and antibodies appear in the blood (Figure 19.23). By this time (several months after initial infection) the patient has a lot of circulating HIV complexed with antibodies that is gradually removed by the action of dendritic cells. But before they are filtered out, these antibody-complexed viruses can still infect T_H cells that come in contact with them. This secondary infection process reaches a low, steady-state level called the "set point." This point varies between people, and is a strong predictor of the rate of progression of the disease. For most people, it takes 8–10 years, even if untreated, for the more severe manifestations of AIDS to develop. In some, it can take as little as a year; in others, 20 years. During this "incubation period," infected people generally feel fine, and their T_H cell levels are adequate for them to mount immune responses.

However, in time, the virus destroys the T_H cells, and their numbers fall to dangerous levels. At this point, the infected patient is considered to have full-blown AIDS, and is susceptible to infections that the T_H cells would normally have been involved in eliminating (Figure 19.24). Most notable are the otherwise rare skin tumor called Kaposi's sarcoma caused by a herpesvirus, pneumonia caused by the fungus *Pneumocystis carinii*, and lymphoma tumors caused by Epstein-Barr virus. These are called **opportunistic infections**, because they take advantage of the crippled immune system of the host. They lead to death within a year or two.

HIV infection and replication occur in T_H cells

As a retrovirus, HIV uses RNA as its genetic material. A central core, with a protein coat, contains two identical molecules of RNA as well as the enzymes reverse transcriptase, integrase, and a protease. An envelope, derived from the plasma membrane of the cell in which the virus was formed, surrounds the core. The envelope is studded with envelope proteins (gp120 and gp41, where "gp" stands for *glycoprotein*) that enable the virus to infect its target cells. The HIV replication cycle has several stages.

AIDS is an immune deficiency disorder

There are various immune deficiency disorders, such as those in which T or B cells never form, and others in which B cells lose the ability to give rise to plasma cells. In either case, the affected individual is unable to mount an immune response and thus lacks a major line of defense against microbial pathogens.

Because of its essential roles in both the humoral and cellular immune responses, the T_H cell is perhaps the most central of all the components of the immune system—a significant cell to lose to an immune deficiency disorder. This cell is the target of HIV (*human immunodeficiency virus*), the virus that eventually results in AIDS (*acquired immune deficiency syndrome*).

HIV is transmitted from person to person several ways:

▶ Through blood, such as by an injection needle contaminated with the virus after being used by an infected individual

▶ Through exposure of broken skin, an open wound, or mucous membranes to body fluids, such as semen, containing HIV

▶ Through the blood of an infected mother to her baby during birth

HIV initially infects macrophages, T_H cells, and dendritic cells in blood and tissues. Dendritic cells are antigen-presenting cells with highly folded plasma membranes that can capture antigens (see Figure 19.2). These infected cells carry the virus to the lymphoid tissues (lymph nodes and spleen) where B cells mature and T cells reside.

Normally in the lymph node, the dendritic cells present their captured antigen to T_H cells, and this causes the T_H cells to divide and form a clone (see Figure 19.18). But HIV preferentially infects activated, and not resting, T_H cells. So the HIV arriving in the lymph nodes proceeds to infect the many activated T_H cells that are already responding to other antigens. These two processes—having cells take the virus

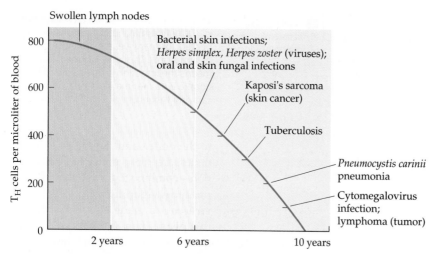

19.24 Relationship Between T_H Cell Count and Opportunistic Infections
As HIV kills more and more T_H cells, the immune system is less and less able to defend the body against various diseases.

VIRUS ENTRY INTO CELLS. HIV attaches to T_H cells and macrophages via CD4, which acts as a receptor for the viral envelope protein gp120. Following binding, the membrane that surrounds the HIV core particle fuses with the host cell plasma membrane, resulting in the entry of the core into the cytoplasm (Figure 19.25; see also Figure 13.7). These events require the participation of at least two other membrane proteins, one from the virus (gp41) and another from the host (appropriately called fusin in T_H cells).

REVERSE TRANSCRIPTION. HIV can insert a cDNA copy of its genome into the host cell's DNA. The process of making a cDNA copy from viral RNA occurs in the viral core particle in an infected cell. It requires the participation of three separate enzymes:

▶ Reverse transcriptase, to make single-stranded DNA from an RNA template
▶ RNAse H, to degrade the viral RNA
▶ DNA polymerase, to make the second strand of cDNA

HIV reverse transcriptase does not have the proofreading activity of many DNA polymerases, so the errors that inevitably creep into the process are not corrected. Up to 10 incorrect bases out of about 8,000 may end up in each cDNA produced. This is a great advantage to the virus, as genomic mutations allow its proteins to escape the host's immune response; however, the mutations present a challenge to scientists trying to design drugs and vaccines to bind to the constantly changing viral proteins.

INTEGRATION OF VIRAL cDNA INTO THE HOST GENOME. The viral core proteins have an amino acid sequence that is recognized by a receptor on the surface of the host cell's nucleus; thus the particle rapidly enters the nucleus. At this point, a viral enzyme called integrase catalyzes the breakage of host DNA and insertion of viral cDNA. This is similar to the way in which bacteriophage DNA becomes incorporated into a bacterial chromosome as a prophage (see Chapter 13). The cDNA thus becomes a permanent part of the T_H cell's DNA, replicating with it at each cell division, and may remain in the T_H cell genome for a decade or more.

VIRUS PRODUCTION. This latent period ends if the HIV-infected T_H cell becomes activated as it responds naturally to an antigen. The expression of viral genes requires the collaboration of host transcription factors that are made in activated T_H cells and a virally encoded protein called Tat. When the T_H cell is activated, the entire viral genome is transcribed into RNA, which can either remain as it is or be spliced. Unspliced RNA's become the genomes of new HIV particles; spliced RNA's make the viral structural proteins. An important activator of splicing is the viral protein called Rev.

The protease encoded by HIV is needed to complete the formation of individual viral proteins from larger products of translation. Packaging domains on viral proteins cause the RNA genomes to fold into them and form core particles. In the meantime, the viral membrane proteins are made on the endoplasmic reticulum of the host cell and transported to the plasma membrane via the Golgi complex. The cytoplasmic tails of the gp120 membrane proteins bind to the core particles, and the viruses bud from the infected cell, surrounding themselves with modified plasma membrane from the host.

Treatments for HIV infection rely on knowledge of its molecular biology

As the AIDS epidemic has grown, so has knowledge of HIV molecular biology. The general therapeutic strategy is to try to block a stage in the viral life cycle and hold HIV infection in check. Potential therapeutic agents that interfere with the major steps of the life cycle are being tested (see Figure 19.25). Of course, it is crucial to block only steps that are unique to the virus, so that drug therapies do not harm the patient by blocking a step in the patient's own metabolism.

*H*ighly *a*ctive *a*ntiretroviral *t*herapy (HAART) was developed in the late 1990s and has had considerable success in delaying the onset of AIDS symptoms in people infected with HIV by 3 years or more, and in prolonging the lives of people with AIDS. The logic of HAART comes from cancer treatment: Employ a combination of drugs acting at different parts of the virus life cycle. Generally, the HAART regimen uses:

▶ A protease inhibitor. These drugs obstruct the active site of the HIV protease.
▶ Two reverse transcriptase inhibitors. These molecules are incorporated into the growing cDNA chain, but no nucleotides can be added to them, so reverse transcription stops.

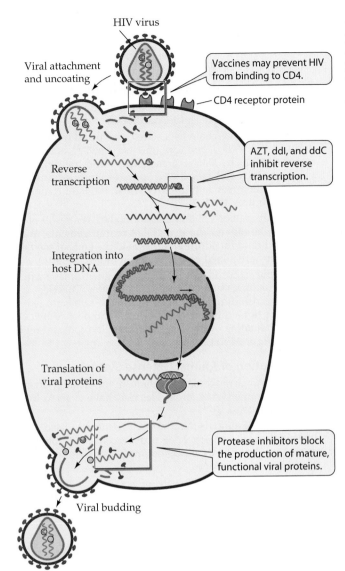

 19.25 Strategies to Combat HIV Reproduction
Several widely used drugs block specific steps in the HIV life cycle.

These drugs have had such dramatic effects on patients that they may eliminate HIV entirely in some people, especially in those treated within the first few days after infection, before the virus has arrived in the lymph nodes. Most patients, however, face a lifetime of anti-HIV therapy.

The many treatments under development include:

▶ Vaccines to inhibit virus entry into cells and to form immune complexes with circulating viruses

▶ Integrase inhibitors, to block cDNA incorporation into the host genome

▶ Tat and Rev inhibitors, to block HIV transcription and splicing

▶ Antisense RNA and ribozymes directed against HIV RNA

What can be done until biomedical science provides the tools to bring the worldwide AIDS epidemic to an end? Above all, people must recognize that they are in danger whenever they have sex with a partner whose total sexual history is not known. The danger rises as the number of sex partners rises, and the danger is much greater if partners participating in sexual intercourse are not protected by a latex condom. The danger that heterosexual intercourse will transmit HIV rises tenfold to a hundredfold if either partner has another sexually transmitted disease.

Chapter Summary

▶ Animals defend themselves against pathogens by both innate (nonspecific) and specific means.

Defensive Cells and Proteins

▶ Many of our defenses are implemented by cells and proteins carried in the bloodstream and in the lymphatic system. **Review Figure 19.1**

▶ White blood cells, including lymphocytes (B and T cells) and phagocytes (such as neutrophils and macrophages), play many defensive roles. **Review Figure 19.2**

▶ Immune system cells produce several kinds of proteins. Antibodies and T cell receptors bind foreign substances, MHC proteins help coordinate the recognition of foreign substances and the activation of defensive cells, and cytokines alter the behavior of other cells.

Innate Defenses

▶ An animal's innate defenses include physical barriers, competing resident microorganisms, and local agents, such as secretions that contain an antibacterial enzyme. **Review Table 19.1**

▶ The complement system, composed of about 20 proteins, assembles itself in a cascade of reactions to cooperate with phagocytes or antibodies to lyse foreign cells. **Review Figure 19.3**

▶ Interferons produced by virus-infected cells inhibit the ability of viruses to replicate in neighboring cells.

▶ Macrophages and neutrophils engulf invading bacteria. Natural killer cells attack tumor cells and virus-infected body cells.

▶ Macrophages play an important role in the inflammation response. Activated mast cells release histamine, which causes blood capillaries to leak. Complement proteins attract macrophages to the site, where they engulf bacteria and dead cells. **Review Figure 19.5**

Specific Defenses: The Immune Response

▶ Four features characterize the immune response: specificity, the ability to respond to an enormous diversity of antigens, the ability to distinguish self from nonself, and memory.

▶ The immune response is directed against antigens that evade the nonspecific defenses. Each antibody or T cell is directed against a particular antigenic determinant. **Review Figure 19.6**

▶ The immune response is highly diverse and can respond specifically to millions of different antigenic determinants.

▶ The immune response distinguishes its own cells from foreign cells, attacking only cells recognized as nonself.

▶ The immune system remembers; it can respond rapidly and effectively to a second exposure to an antigen.

▶ There are two interactive immune responses: the humoral immune response and the cellular immune response. The humoral immune response employs antibodies secreted by plasma cells to target antigens in body fluids. The cellular immune response employs T cells to attack body cells that have been altered by viral infection or mutation or to target antigens that have invaded the body's cells.

▶ Clonal selection theory accounts for the rapidity, specificity, and diversity of the immune response. It also accounts for immunological memory, which is based on the production of both effector and memory cells as T cell and B cell clones expand. **Review Figure 19.7**

▶ Immunological memory plays roles in both natural immunity and artificial immunity based on vaccination. **Review Figure 19.8**

▶ Clonal selection theory also accounts for the immune system's recognition of self. Tolerance of self results from clonal deletion of antiself lymphocytes. **Review Figure 19.9**

B Cells: The Humoral Immune Response

▶ Activated B cells form plasma cells, which synthesize and secrete specific antibodies.

▶ The basic unit of an antibody, or immunoglobulin, is a tetramer of four polypeptides: two identical light chains and two identical heavy chains, each consisting of a constant and a variable region. **Review Figure 19.11**

▶ The variable regions of the light and heavy chains collaborate to form the antigen-binding sites of an antibody. The variable regions determine an antibody's specificity; the constant region determines the destination and function of the antibody.

▶ There are five immunoglobulin classes. IgM, formed first, is a membrane receptor on B cells, as is IgD. IgG is the most abundant antibody class and performs several defensive functions. IgE takes part in inflammation and allergic reactions. IgA is present in various body secretions. **Review Table 19.2**

▶ Monoclonal antibodies consist of identical immunoglobulin molecules directed against a single antigenic determinant. Hybridomas are produced by fusing B cells with myeloma cells (from cancerous tumors of plasma cells). **Review Figure 19.13**

T Cells: The Cellular Immune Response

▶ The cellular immune response is directed against altered or antigen-infected cells of the body. T_C cells attack virus-infected or tumor cells, causing them to lyse. T_H cells activate B cells and influence the development of other T cells and macrophages. **Review Figure 19.14**

▶ T cell receptors in the cellular immune response are analogous to immunoglobulins in the humoral immune response.

▶ The major histocompatibility complex (MHC) encodes many membrane proteins. MHC molecules in macrophages, B cells, or body cells bind processed antigen and present it to T cells. **Review Figure 19.16, 19.17**

▶ The activation of the humoral immune response requires the collaboration of class II MHC molecules, the T cell surface protein CD4, and cytokines. The effector phase of the humoral immune response involves T cells, class II MHC molecules, B cells, and cytokines, and results in the formation of active plasma cells. **Review Figure 19.18**

▶ In the cellular immune response, class I MHC molecules, T_C cells, CD8, and cytokines collaborate to activate T_C cells with the appropriate specificity. **Review Figure 19.18**

▶ Developing T cells undergo two tests: They must be able to recognize self MHC molecules, and they must *not* bind to both self MHC and any of the body's own antigens. T cells that fail either of these tests die.

▶ The rejection of organ transplants results from the genetic diversity of MHC molecules.

The Genetic Basis of Antibody Diversity

▶ Several gene families underlie the incredible diversity of antibody and T cell receptor specificities.

▶ Antibody heavy-chain genes are constructed from one each of numerous *V*, *D*, *J*, and *C* segments. The *V*, *D*, and *J* segments combine by DNA rearrangement, and transcription yields an RNA molecule that is spliced to form a translatable mRNA. Other gene families give rise to the light chains. **Review Figures 19.19, 19.20**

▶ There are millions of possible antibodies as a result of these DNA combinations. Imprecise DNA rearrangements, mutations, and random addition of bases to the ends of the DNA's before they are joined contribute even more diversity.

▶ A plasma cell produces IgM first, but later it may switch to the production of other classes of antibodies. This class switching, resulting in antibodies with the same antigen specificity but a different function, is accomplished by cutting and rejoining of the genes encoding the constant region. **Review Figure 19.21, 19.22**

The Evolution of Animal Defense Systems

▶ The DNA rearrangement mechanism for creating diversity among immune system molecules may have evolved from a DNA transposon.

▶ Invertebrate animals reject nonself tissues but lack immunological memory. They possess cells and molecules analogous, but not identical, to lymphocytes, immunoglobulins, and cytokines.

▶ Even the most evolutionarily ancient groups among today's vertebrates have immune systems more similar to those of humans than to those of invertebrates.

Disorders of the Immune System

▶ Allergies result from an overreaction of the immune system to an antigen.

▶ Autoimmune diseases result from a failure in the immune recognition of self, with the appearance of antiself B and T cells that attack the body's own cells.

▶ Immune deficiency disorders result from failures of one or another part of the immune system. AIDS is an immune deficiency disorder arising from depletion of the body's T_H cells as a result of infection with HIV. Depletion of the T_H cells weakens and eventually destroys the immune system, leaving the host defenseless against "opportunistic" infections. **Review Figures 19.23, 19.24**

▶ HIV inserts a copy of its genome into a chromosome of a macrophage or T_H cell, where it may lie dormant for years. When the viral genome is transcribed and translated, new viruses form. **Review Figure 19.25**

▶ All steps in the reproductive cycle of HIV are under investigation as possible targets for drugs. Currently the most effective drugs are those directed against reverse transcriptase and protease. **Review Figure 19.25**

▶ Some treatments may provide a dramatic reduction in HIV levels, but there is as yet no indication that we can prevent infection with HIV, as by vaccination. The only strategy currently available is for people to avoid behaviors that place them at risk.

For Discussion

1. Describe the part of an antibody molecule that interacts with an antigenic determinant. How is it similar to the active site of an enzyme? How does it differ from the active site of an enzyme?

2. Contrast immunoglobulins and T cell receptors with respect to structure and function.

3. Discuss the diversity of antibody specificities in an individual in relation to the diversity of enzymes. Does every cell in an animal contain genetic information for all the organism's enzymes? Does every cell contain genetic information for all the organism's immunoglobulins?

4. The gene family determining MHC on the cell surface in humans is on a single chromosome. A father's MHC type is A1, A3, B5, B7, D9, D11. His wife's phenotype is A2, A4, B6, B7, D11, D12. They have a child who is A1, A4, B6, B7, D11, D12. What are the parents' haplotypes—that is, which alleles are linked on the diploid chromosomes in each parent? Assuming that there is no recombination among the genes determining the MHC type, can these parents have a child who is A1, A2, B7, B8, D10, D11?

5. Is it true that any child can accept an organ transplant from either parent, but parents cannot accept a graft from a child? Explain your answer.

Appendix:
Some Measurements Used in Biology

QUANTITY	NAME OF UNIT	SYMBOL	DEFINITION
Length	meter (*also* metre)	m	A base unit. 1 m = 100 cm = 39.37 inches
	kilometer	km	1 km = 1000 m = 10^3 m
	centimeter	cm	1 cm = $\frac{1}{100}$ m = 10^{-2} m
	millimeter	mm	1 mm = $\frac{1}{1000}$ m = 10^{-3} m
	micrometer	μm	1 μm = $\frac{1}{1000}$ mm = 10^{-6} m
	nanometer	nm	1 nm = $\frac{1}{1000}$ μm = 10^{-9} m
Area	square meter	m^2	Area encompassed by a square, each side of which is 1 m in length
	hectare	ha	1 ha = 10,000 m^2 = 10^4 m^2 (2.47 acres)
	square centimeter	cm^2	1 cm^2 = $\frac{1}{10,000}$ m^2 = 10^{-4} m^2
Volume	liter (*also* litre)	l	1 l = $\frac{1}{1000}$ m^3 = 10^{-3} m^3 (1.057 qts)
	milliliter	ml	1 ml = $\frac{1}{1000}$ l = 10^{-3} l = 1 cm^3 = 1 cc
	microliter	μl	1 μl = $\frac{1}{1000}$ ml = 10^{-3} ml = 10^{-6} l
Mass	kilogram	kg	A basic unit. 1 kg = 1000 g = 2.20 lbs
	gram	g	1 g = $\frac{1}{1000}$ kg = 10^{-3} kg
	milligram	mg	1 mg = $\frac{1}{1000}$ g = 10^{-3} g = 10^{-6} kg
Time	second	s	A basic unit. 1 s = $\frac{1}{60}$ min
	minute	min	1 min = 60 s
	hour	h	1 h = 60 min = 3,600 s
	day	d	1 d = 24 h = 86,400 s
Temperature	kelvin	K	A basic unit. 0 K = −273.15°C = absolute zero
	degree Celsius	°C	0°C = 273.15 K = melting point of ice
Heat, work	calorie	cal	1 cal = heat necessary to raise 1 gram of pure water from 14.5°C to 15.5°C = 4.184 J
	kilocalorie	kcal	1 kcal = 1000 cal = 10^3 cal = (in nutrition) 1 Calorie
	joule	J	1 J = 0.2389 cal (The joule is now the accepted unit of heat in most sciences.)
Electric potential	volt	V	A unit of potential difference or electromotive force
	millivolt	mV	1 mV = $\frac{1}{1000}$ V = 10^{-3} V

Glossary

Abdomen (ab' duh mun) [L.: belly] • In arthropods, the posterior portion of the body; in mammals, the part of the body containing the intestines and most other internal organs, posterior to the thorax.

Abscisic acid (ab sighs' ik) [L. *abscissio*: breaking off] • A plant growth substance having growth-inhibiting action. Causes stomata to close.

Abscission (ab sizh' un) [L. *abscissio*: breaking off] • The process by which leaves, petals, and fruits separate from a plant.

Absolute temperature scale • Also known as the Kelvin scale. A temperature scale in which zero is the state of no molecular motion. This "absolute zero" is –273° on the Celsius scale.

Absorption • (1) Of light: complete retention, without reflection or transmission. (2) Of liquids: soaking up (taking in through pores or cracks).

Absorption spectrum • A graph of light absorption versus wavelength of light; shows how much light is absorbed at each wavelength.

Abyssal zone (uh biss' ul) [Gr. *abyssos*: bottomless] • That portion of the deep ocean floor where no light penetrates.

Accessory pigments • Pigments that absorb light and transfer energy to chlorophylls for photosynthesis.

Acetylcholine • A neurotransmitter substance that carries information across vertebrate neuromuscular junctions and some other synapses. **Acetylcholinesterase** is an enzyme that breaks down acetylcholine.

Acetyl CoA (acetyl coenzyme A) • Compound that reacts with oxaloacetate to produce citrate at the beginning of the citric acid cycle; a key metabolic intermediate in the formation of many compounds.

Acid [L. *acidus*: sharp, sour] • A substance that can release a proton in solution. (Contrast with base.)

Acid precipitation • Precipitation that has a lower pH than normal as a result of acid-forming precursors introduced into the atmosphere by human activities.

Acidic • Having a pH of less than 7.0 (a hydrogen ion concentration greater than 10^{-7} molar).

Acoelomate • Lacking a coelom.

Acquired Immune Deficiency Syndrome • See AIDS.

Acrosome (a' krow soam) [Gr. *akros*: highest or outermost + *soma*: body] • The structure at the forward tip of an animal sperm which is the first to fuse with the egg membrane and enter the egg cell.

ACTH (adrenocorticotropin) • A pituitary hormone that stimulates the adrenal cortex.

Actin [Gr. *aktis*: a ray] • One of the two major proteins of muscle; it makes up the thin filaments. Forms the microfilaments found in most eukaryotic cells.

Action potential • An impulse in a neuron taking the form of a wave of depolarization or hyperpolarization imposed on a polarized cell surface.

Activating enzymes (also called aminoacyl-tRNA synthetases) • These enzymes catalyze the addition of amino acids to their appropriate tRNAs.

Activation energy (E_a) • The energy barrier that blocks the tendency for a set of chemical substances to react.

Active site • The region on the surface of an enzyme where the substrate binds, and where catalysis occurs.

Active transport • The transport of a substance across a biological membrane against a concentration gradient—that is, from a region of low concentration (of that substance) to a region of high concentration. Active transport requires the expenditure of energy and is a saturable process. (Contrast with facilitated diffusion, free diffusion; see primary active transport, secondary active transport.)

Adaptation (a dap tay' shun) • In evolutionary biology, a particular structure, physiological process, or behavior that makes an organism better able to survive and reproduce. Also, the evolutionary process that leads to the development or persistence of such a trait.

Adenine (a' den een) • A nitrogen-containing base found in nucleic acids, ATP, NAD, etc.

Adenosine triphosphate • See ATP.

Adenylate cyclase • Enzyme catalyzing the formation of cyclic AMP from ATP.

Adrenal (a dree' nal) [L. *ad-*: toward + *renes*: kidneys] • An endocrine gland located near the kidneys of vertebrates, consisting of two glandular parts, the cortex and medulla.

Adrenaline • See epinephrine.

Adrenocorticotropin • See ACTH.

Adsorption • Binding of a gas or a solute to the surface of a solid.

Aerobic (air oh' bic) [Gr. *aer*: air + *bios*: life] • In the presence of oxygen, or requiring oxygen.

Afferent (af' ur unt) [L. *ad*: to + *ferre*: to bear] • To or toward, as in a neuron that carries impulses to the central nervous system, or a blood vessel that carries blood to a structure. (Contrast with efferents.)

AIDS (acquired immune deficiency syndrome) • Condition caused by a virus (HIV) in which the body's helper T lymphocytes are reduced, leaving the victim subject to opportunistic diseases.

Aldehyde (al' duh hide) • A compound with a –CHO functional group. Many sugars are aldehydes. (Contrast with ketone.)

Aldosterone (al dahs' ter own) • A steroid hormone produced in the adrenal cortex of mammals. Promotes secretion of potassium and reabsorption of sodium in the kidney.

Alga (al' gah) (plural: algae) [L.: seaweed] • Any one of a wide diversity of protists belonging to the phyla Pyrrophyta, Chrysophyta, Phaeophyta, Rhodophyta, and Chlorophyta.

Allele (a leel') [Gr. *allos*: other] • The alternate forms of a genetic character found at a given locus on a chromosome.

Allele frequency • The relative proportion of a particular allele in a specific population.

Allergy [Ger. *allergie*: altered reaction] • An overreaction to an antigen in amounts that do not affect most people; often involves IgE antibodies.

Allometric growth • A pattern of growth in which some parts of the body of an organism grow faster than others, resulting in a change in body proportions as the organism grows.

Allopatric speciation (al' lo pat' rick) [Gr. *allos*: other + *patria*: fatherland] • Also called geographical speciation, this is the formation of two species from one when reproductive isolation occurs because of the the interposition of (or crossing of) a physical geographic barrier such as a river. (Contrast with parapatric speciation, sympatric speciation.)

Allopolyploid • A polyploid in which the chromosome sets are derived from more than one species.

Allostery (al' lo steer' y) [Gr. *allos*: other + *stereos*: structure] • Regulation of the activity of a protein by the binding of an effector molecule at a site other than the active site.

Alpha helix • Type of protein secondary structure; a right-handed spiral.

Alternation of generations • The succession of haploid and diploid phases in some sexually reproducing organisms, notably plants.

Altruistism • A behavior whose performance harms the actor but benefits other individuals.

Alveolus (al ve' o lus) (plural: alveoli) [L. *alveus*: cavity] • A small, baglike cavity, especially the blind sacs of the lung.

Amensalism (a men' sul ism) • Interaction in which one animal is harmed and the other is unaffected. (Contrast with commensalism, mutualism.)

Amine • An organic compound with an amino group (see Amino acid).

Amino acid • An organic compound of the general formula $H_2N–CHR–COOH$, where R can be one of 20 or more different side groups. An amino acid is so named because it has both a basic amine group, $–NH_2$, and an acidic carboxyl group, $–COOH$. Proteins are polymers of amino acids.

Ammonotelic (am moan' o teel' ic) [Gr. *telos*: end] • Describes an organism in which the final product of breakdown of nitrogen-containing compounds (primarily proteins) is ammonia. (Contrast with ureotelic, uricotelic.)

Amniocentesis • A medical procedure in which cells from the fetus are obtained from the amniotic fluid. The genetic material of the cells is then examined. (Contrast with chorionic villus sampling.)

Amniote • An organism that lays eggs that can be incubated in air (externally) because the embryo is enclosed by a fluid-filled sac. Birds and reptiles are amniotes.

Amphipathic (am' fi path' ic) [Gr. *amphi*: both + *pathos*: emotion] • Of a molecule, having both hydrophilic and hydrophobic regions.

Amylase (am' ill ase) • Any of a group of enzymes that digest starch.

Anabolism (an ab' uh liz' em) [Gr. *ana*: up, throughout + *ballein*: to throw] • Synthetic reactions of metabolism, in which complex molecules are formed from simpler ones. (Contrast with catabolism.)

Anaerobic (an ur row' bic) [Gr. *an*: not + *aer*: air + *bios*: life] • Occurring without the use of molecular oxygen, O_2.

Anagenesis • Evolutionary change in a single lineage over time.

Analogy (a nal' o jee) [Gr. *analogia*: resembling] • A resemblance in function, and often appearance as well, between two structures which is due to convergence in evolution rather than to common ancestry. (Contrast with homology.)

Anaphase (an' a phase) [Gr. *ana*: indicating upward progress] • The stage in nuclear division at which the first separation of sister chromatids (or, in the first meiotic division, of paired homologues) occurs. Anaphase lasts from the moment of first separation to the time at which the moving chromosomes converge at the poles of the spindle.

Anaphylactic shock • A precipitous drop in blood pressure caused by loss of fluid from capillaries because of an increase in their permeability stimulated by an allergic reaction.

Ancestral trait • Trait shared by a group of organisms as a result of descent from a common ancestor.

Androgens (an' dro jens) • The male sex steroids.

Aneuploidy (an' you ploy dee) • A condition in which one or more chromosomes or pieces of chromosomes are either lacking or present in excess.

Angiosperm (an' jee oh spurm) [Gr. *angion*: vessel + *sperma*: seed] • One of the flowering plants; literally, one whose seed is carried in a "vessel," which is the fruit. (See fruit.)

Angiotensin (an' jee oh ten' sin) • A peptide hormone that raises blood pressure by causing peripheral vessels to constrict; maintains glomerular filtration by constricting efferent glomerular vessels; stimulates thirst; and stimulates the release of aldosterone.

Animal [L. *animus*: breath, soul] • A member of the kingdom Animalia. In general, a multicellular eukaryote that obtains its food by ingestion.

Animal hemisphere • The metabolically active upper portion of some animal eggs, zygotes, and embryos, which does *not* contain the dense nutrient yolk. The **animal pole** refers to the very top of the egg or embryro. (Contrast with vegetal hemisphere.)

Anion (an' eye one) • An ion with one or more negative charges. (Contrast with cation.)

Anisogamy (an' eye sog' a mee) [Gr. *aniso*: unequal + *gamos*: marriage] • The existence of two dissimilar gametes (egg and sperm).

Annual • Referring to a plant whose life cycle is completed in one growing season. (Contrast with biennial, perennial.)

Anterior pituitary • The portion of the vertebrate pituitary gland that derives from gut epithelium and produces tropic hormones.

Anther (an' thur) [Gr. *anthos*: flower] • A pollen-bearing portion of the stamen of a flower.

Antheridium (an' thur id' ee um) (plural: antheridia) [Gr. *antheros*: blooming] • The multicellular structure that produces the sperm in bryophytes and ferns.

Antibody • One of millions of proteins, produced by the immune system, that specifically recognizes a foreign substance and initiates its removal from the body.

Anticodon • A "triplet" of three nucleotides in transfer RNA that is able to pair with a complementary triplet (a codon) in messenger RNA, thus aligning the transfer RNA on the proper place on the messenger. The codon (and, reciprocally, the anticodon) codes for a specific amino acid.

Antidiuretic hormone • A hormone that controls water reabsorption in the mammalian kidney. Also called vasopressin.

Antigen (an' ti jun) • Any substance that stimulates the production of an antibody or antibodies in the body of a vertebrate.

Antigen processing • The breakdown of antigenic proteins into smaller fragments, which are then presented on the cell surface, along with MHC proteins, to T cells.

Antigenic determinant • A specific region of an antigen, which is recognized by and binds to a specific antibody.

Antiport • A membrane transport process that carries one substance in one direction and another in the opposite direction. (Contrast with symport.)

Antisense nucleic acid • A single-stranded RNA or DNA complementary to and thus targeted against the mRNA transcribed from a harmful gene such as an oncogene.

Anus (a' nus) • Opening through which digestive wastes are expelled, located at the posterior end of the gut.

Aorta (a or' tuh) [Gr. *aorte*: aorta] • The main trunk of the arteries leading to the systemic (as opposed to the pulmonary) circulation.

Apex (a' pecks) • The tip or highest point of a structure, as the apex of a growing stem or root.

Apical (a' pi kul) • Pertaining to the apex, or tip, usually in reference to plants.

Apical dominance • Inhibition by the apical bud of the growth of axillary buds.

Apical meristem • The meristem at the tip of a shoot or root; responsible for the plant's primary growth.

Apomixis (ap oh mix' is) [Gr. *apo*: away from + *mixis*: sexual intercourse] • The asexual production of seeds.

Apoplast (ap' oh plast) • in plants, the continuous meshwork of cell walls and extracellular spaces through which material can pass without crossing a plasma membrane. (Contrast with symplast.)

Apoptosis (ay' pu toh sis) • A series of genetically programmed events leading to cell death.

Aquaporin • A transport protein in plant and animals cells through which water passes in osmosis.

Archegonium (ar' ke go' nee um) [Gr. *archegonos*: first of a kind] • The multicellular structure that produces eggs in bryophytes, ferns, and gymnosperms.

Archenteron (ark en' ter on) [Gr. *archos*: beginning + *enteron*: bowel] • The earliest primordial animal digestive tract.

Arteriosclerosis • See atherosclerosis.

Artery • A muscular blood vessel carrying oxygenated blood away from the heart to other parts of the body. (Contrast with vein.)

Ascus (ass' cuss) [Gr. *askos*: bladder] • In fungi belonging to the phylum Ascomycota (the sac fungi), the club-shaped sporangium within which spores (ascospores) are produced by meiosis.

Asexual • Without sex.

Assortative mating • A breeding system in which mates are selected on the basis of a particular trait or group of traits.

Atherosclerosis (ath' er oh sklair oh' sis) • A disease of the lining of the arteries characterized by fatty, cholesterol-rich deposits in the walls of the arteries. When fibroblasts infiltrate these deposits and calcium precipitates in them, the disease become arteriosclerosis, or "hardening of the arteries."

Atmosphere • The gaseous mass surrounding our planet. Also: a unit of pressure, equal to the normal pressure of air at sea level.

Atom [Gr. *atomos*: indivisible] • The smallest unit of a chemical element. Consists of a nucleus and one or more electrons.

Atomic mass (also called atomic weight) • The average mass of an atom of an element on the amu scale. (The average depends upon the relative amounts of different isotopes of an element on Earth.)

Atomic number • The number of protons in the nucleus of an atom, also equal to the number of electrons around the neutral atom. Determines the chemical properties of the atom.

ATP (adenosine triphosphate) • A compound containing adenine, ribose, and three phosphate groups. When it is formed, useful energy is stored; when it is broken down (to ADP or AMP), energy is released to drive endergonic reactions. ATP is an energy storage compound.

ATP synthase • An integral membrane protein that couples the transport of proteins with the formation of ATP.

Atrium (a' tree um) • A body cavity, as in the hearts of vertebrates. The thin-walled chamber(s) entered by blood on its way to the ventricle(s). Also, the outer ear.

Autoimmune disease • A disorder in which the immune system attacks the animal's own antigens.

Autonomic nervous system • The system (which in vertebrates comprises sympathetic and parasympathetic subsystems) that controls such involuntary functions as those of guts and glands.

Autosome • Any chromosome (in a eukaryote) other than a sex chromosome.

Autotroph (au' tow trow' fik) [Gr. *autos*: self + *trophe*: food] • An organism that is capable of living exclusively on inorganic materials, water, and some energy source such as sunlight or chemically reduced matter. (Contrast with heterotroph.)

Auxin (awk' sin) [Gr. *auxein*: increase] • In plants, a substance (indoleacetic acid) that regulates growth and various aspects of development.

Auxotroph (awks' o trofe) [Gr. *auxanein*: to grow + *trophe*: food] • A mutant form of an organism that requires a nutrient or nutrients not required by the wild type, or reference, form of the organism. (Contrast with prototroph.)

Axon [Gr.: axle] • Fiber of a neuron which can carry action potentials. Carries impulses away from the cell body of the neuron; releases a neurotransmitter substance.

Axon hillock • The junction between an axon and its cell body; where action potentials are generated.

Axon terminals • The endings of an axon; they form synapses and release neurotransmitter.

Axoneme (ax' oh neem) • The complex of microtubules and their crossbridges that forms the motile apparatus of a cilium.

Bacillus (buh sil' us) [L.: little rod] • Any of various rod-shaped bacteria.

Bacteriophage (bak teer' ee o fayj) [Gr. *bakterion*: little rod + *phagein*: to eat] • One of a group of viruses that infect bacteria and ultimately cause their disintegration.

Bacteria (bak teer' ee ah) (singular: bacterium) [Gr. *bakterion*: little rod] • Prokaryote in the Domain Bacteria. The chromosomes of bacteria are not contained in nuclear envelopes.

Balanced polymorphism [Gr. *polymorphos*: having many forms] • The maintenance of more than one form, or the maintenance at a given locus of more than one allele, at frequencies of greater than one percent in a population. Often results when heterozygotes are superior to both homozygotes.

Bark • All tissues outside the vascular cambium of a plant.

Baroreceptor [Gr. *baros*: weight] • A pressure-sensing cell or organ.

Barr body • In mammals, an inactivated X chromosome.

Basal body • Centriole found at the base of a eukaryotic flagellum or cilium.

Basal metabolic rate • The minimum rate of energy turnover in an awake (but resting) bird or mammal that is not expending energy for thermoregulation.

Base • (1) A substance which can accept a proton (hydrogen ion; H$^+$) in solution. (Contrast with acid.) (2) In nucleic acids, a nitrogen-containing molecule that is attached to each sugar in the backbone. (See purine; pyrimidine.)

Base pairing • See complementary base pairing.

Basic • having a pH greater than 7.0 (having a hydrogen ion concentration lower than 10^{-7} molar).

Basidium (bass id' ee yum) • In fungi of the class Basidiomycetes, the characteristic sporangium in which four spores are formed by meiosis and then borne externally before being shed.

Batesian mimicry • Mimicry by a relatively harmless kind of organism of a more dangerous one, by which the mimic enjoys protection from predators that mistake it for the dangerous model. (Contrast with Müllerian mimicry.)

B cell • A type of lymphocyte involved in the humoral immune response of vertebrates. Upon recognizing an antigenic determinant, a B cell develops into a plasma cell, which secretes an antibody. (Contrast with a T cell.)

Benefit • An improvement in survival and reproductive success resulting from a behavior. (Contrast with cost.)

Benign (be nine') • A tumor that grows to a certain size and then stops, uaually with a fibrous capsule surrounding the mass of cells. Benign tumors do not spread (metastasize) to other organs.

Benthic zone [Gr. *benthos*: bottom of the sea] • The bottom of the ocean. (Contrast with pelagic zone.)

Beta-pleated sheet • Type of protein secondary structure; results from hydrogen bonding between polypeptide regions running antiparallel to each other.

Biennial • Referring to a plant whose life cycle includes vegetative growth in the first year and flowering and senescence in the second year. (Contrast with annual, perennial.)

Bilateral symmetry • The condition in which only the right and left sides of an organism, divided exactly down the back, are mirror images of each other. (Contrast with biradial symmetry.)

Bile • A secretion of the liver delivered to the small intestine via the common bile duct. In the intestine, bile emulsifies fats.

Binocular cells • Neurons in the visual cortex that respond to input from both retinas; involved in depth perception.

Binomial (bye nome' ee al) • Consisting of two names; for example, the binomial nomenclature of biology which gives the name of the genus followed by the name of the species.

Biodiversity crisis • The current high rate of loss of species, caused primarily by human activities.

Biogeochemical cycles • Movement of elements through living organisms and the physical environment.

Biogeography • The scientific study of the geographic distribution of organisms.

Biogeographic region • A continental-scale part of Earth that has a biota distinct from that of other such regions.

Biological species concept • The view that a species is most usefully defined as a population or series of populations within which there is a significant amount of gene flow under natural conditions, but which is genetically isolated from other populations.

Bioluminescence • The production of light by biochemical processes in an organism.

Biomass • The total weight of all the living organisms, or some designated group of living organisms, in a given area.

Biome (bye' ome) • A major division of the ecological communities of Earth; characterized by distinctive vegetation.

Biota (bye oh' tah) • All of the organisms, including animals, plants, fungi, and microorganisms, found in a given area.

Biotechnology • The use of cells to make medicines, foods and other products useful to humans.

Biradial symmetry • Radial symmetry modified so that only two planes can divide the animal into similar halves.

Blastocoel (blass' toe seal) [Br. *blastos*: sprout + *koilos*: hollow] • The central, hollow cavity of a blastula.

Blastodisc (blass' toe disk) • A disk of cells forming on the surface of a large yolk mass, comparable to a blastula, but occurring in animals such as birds and reptiles, in which the massive yolk restricts cleavage to one side of the egg only.

Blastomere • A cell produced by the division of a fertilized egg.

Blastopore • The opening from the archenteron to the exterior of a gastrula.

Blastula (blass' chu luh) [Gr. *blastos*: sprout] • An early stage in animal embryology; in many species, a hollow sphere of cells surrounding a central cavity, the blastocoel. (Contrast with blastodisc.)

Blood–brain barrier • A property of the blood vessels of the brain that prevents most chemicals from diffusing from the blood into the brain.

Body plan • A basic structural design that includes an entire animal, its organ systems, and the integrated functioning of its parts. Phylogenetic groups of organisms are classified in part on the basis of a shared body plan.

Bowman's capsule • An elaboration of kidney tubule cells that surrounds a knot of capillaries (the glomerulus). Blood is filtered across the walls of these capillaries and the filtrate is collected into Bowman's capsule.

Brain stem • The portion of the vertebrate brain between the spinal cord and the forebrain.

Brassinosteroids • Plant steroid hormones that promote the elongation of stems and pollen tubes.

Bronchus (plural: bronchi) • The major airway(s) branching off the trachea into the vertebrate lung.

Brown fat • Fat tissue in mammals that is specialized to produce heat. It has many mitochondria and capillaries, and a protein that uncouples oxidative phosphorylation.

Browser • An animal that feeds on the tissues of woody plants.

Bryophyte (bri' uh fite') [Gr. *bruon*: moss + *phyton*: plant] • A moss. Formerly was often used to refer to all the nontracheophyte plants.

Budding • Asexual reproduction in which a more or less complete new organism simply grows from the body of the parent organism and eventually detaches itself.

Buffering • A process by which a system resists change—particularly in pH, in which case added acid or base is partially converted to another form.

C_3 **photosynthesis** • The form of photosynthesis in which 3-phosphoglycerate is the first stable product, and ribulose bisphosphate is the CO_2 receptor.

C_4 **photosynthesis** • The form of photosynthesis in which oxaloacetate is the first stable product, and phosphoenolpyruvate is the CO_2 acceptor. C_4 plants also perform the reactions of C_3 photosynthesis.

Calcitonin • A hormone produced by the thyroid gland; it lowers blood calcium and promotes bone formation. (Contrast with parathormone.)

Calmodulin (cal mod' joo lin) • A calcium-binding protein found in all animal and plant cells; mediates many calcium-regulated processes.

calorie [L. *calor*: heat] • The amount of heat required to raise the temperature of one gram of water by one degree Celsius (1°C) from 14.5°C to 15.5°C. In nutrition studies, "Calorie" (spelled with a capital C) refers to the kilocalorie (1 kcal = 1,000 cal).

Calvin–Benson cycle • The stage of photosynthesis in which CO_2 reacts with RuBP to form 3PG, 3PG is reduced to a sugar, and RuBP is regenerated, while other products are released to the rest of the plant.

Calyx (kay' licks) [Gr. *kalyx*: cup] • All of the sepals of a flower, collectively.

CAM • See crassulacean acid metabolism.

Cambium (kam' bee um) [L. *cambiare*: to exchange] • A meristem that gives rise to radial rows of cells in stem and root, increasing them in girth; commonly applied to the vascular cambium which produces wood and phloem, and the cork cambium, which produces bark.

cAMP (cyclic AMP) • A compound, formed from ATP, that mediates the effects of numerous animal hormones. Also needed for the transcription of catabolite-repressible operons in bacteria. Used for communication by cellular slime molds.

Canopy • The leaf-bearing part of a tree. Collectively the aggregate of the leaves and branches of the larger woody plants of an ecological community.

Capillaries [L. *capillaris*: hair] • Very small tubes, especially the smallest blood-carrying vessels of animals between the termination of the arteries and the beginnings of the veins.

Capsid • The protein coat of a virus.

Carbohydrates • Organic compounds with the general formula $C_nH_{2m}O_m$. Common examples are sugars, starch, and cellulose.

Carboxylic acid (kar box sill' ik) • An organic acid containing the carboxyl group, –COOH, which dissociates to the carboxylate ion, –COO⁻.

Carcinogen (car sin' oh jen) • A substance that causes cancer.

Cardiac (kar' dee ak) [Gr. *kardia*: heart] • Pertaining to the heart and its functions.

Carnivore [L. *carn*: flesh + *vovare*: to devour] • An organism that feeds on animal tissue. (Contrast with detritivore, herbivore, omnivore.)

Carotenoid (ka rah' tuh noid) [L. *carota*: carrot] • A yellow, orange, or red lipid pigment commonly found as an accessory pigment in photosynthesis; also found in fungi.

Carpel (kar' pel) [Gr. *karpos*: fruit] • The organ of the flower that contains one or more ovules.

Carrier • (1) In facilitated diffusion, a membrane protein that binds a specific molecule and transports it through the membrane. (2) In respiratory and photosynthetic electron transport, a participating substance such as NAD that exists in both oxidized and reduced forms. (3) In genetics, a person heterozygous for a recessive trait.

Carrying capacity • In ecology, the largest number of organisms of a particular species that can be maintained indefinitely in a given part of the environment.

Cartilage • In vertebrates, a tough connective tissue found in joints, the outer ear, and elsewhere. Forms the entire skeleton in some animal groups.

Casparian strip • A band of cell wall containing suberin and lignin, found in the endodermis. Restricts the movement of water across the endodermis.

Catabolism [Ge. *kata*: down + *ballein*: to throw] • Degradational reactions of metabolism, in which complex molecules are broken down. (Contrast with anabolism.)

Catalyst (cat' a list) [Gr. *kata-*, implying the breaking down of a compound] • A chemical substance that accelerates a reaction without itself being consumed in the overall course of the reaction. Catalysts lower the activation energy of a reaction. Enzymes are biological catalysts.

Cation (cat' eye on) • An ion with one or more positive charges. (Contrast with anion.)

Caudal [L. *cauda*: tail] • Pertaining to the tail, or to the posterior part of the body.

cDNA • See complementary DNA.

Cecum (see' cum) [L. *caecus*: blind] • A blind branch off the large intestine. In many nonruminant mammals, the cecum contains a colony of microorganisms that contribute to the digestion of food.

Cell adhesion molecules • Molecules on animal cell surfaces that affect the selective association of cells during development of the embryo.

Cell cycle • The stages through which a cell passes between one division and the next. Includes all stages of interphase and mitosis.

Cell division • The reproduction of a cell to produce two new cells. In eukaryotes, this process involves nuclear division (mitosis) and cytoplasmic division (cytokinesis).

Cell theory • The theory, well established, that organisms consist of cells, and that all cells come from preexisting cells.

Cell wall • A relatively rigid structure that encloses cells of plants, fungi, many protists, and most bacteria. The cell wall gives these cells their shape and limits their expansion in hypotonic media.

Cellular immune system • That part of the immune system that is based on the activities of T cells. Directed against parasites, fungi, intracellular viruses, and foreign tissues (grafts). (Contrast with humoral immune system.)

Cellular respiration • See respiration.

Cellulose (sell' you lowss) • A straight-chain polymer of glucose molecules, used by plants as a structural supporting material.

Central dogma • The statement that information flows from DNA to RNA to polypeptide (in retroviruses, there is also information flow from RNA to cDNA).

Central nervous system • That part of the nervous system which is condensed and centrally located, e.g., the brain and spinal cord of vertebrates; the chain of cerebral, thoracic and abdominal ganglia of arthropods.

Centrifuge [L. *fugere*: to flee] • A device in which a sample can be spun around a central axis at high speed, creating a centrifugal force that mimics a very strong gravitational force. Used to separate mixtures of suspended materials.

Centriole (sen' tree ole) • A paired organelle that helps organize the microtubules in animal and protist cells during nuclear division.

Centromere (sen' tro meer) [Gr. *centron*: center + *meros*: part] • The region where sister chromatids join.

Centrosome (sen' tro soam) • The major microtubule organizing center of an animal cell.

Cephalization (sef' uh luh zay' shun) [Gr. *kephale*: head] • The evolutionary trend toward increasing concentration of brain and sensory organs at the anterior end of the animal.

Cerebellum (sair' uh bell' um) [L.: diminutive of *cerebrum*: brain] • The brain region that controls muscular coordination; located at the anterior end of the hindbrain.

Cerebral cortex • The thin layer of gray matter (neuronal cell bodies) that overlays the cerebrum.

Cerebrum (su ree' brum) [L.: brain] • The dorsal anterior portion of the forebrain, making up the largest part of the brain of mammals. In mammals, the chief coordination center of the nervous system; consists of two **cerebral hemispheres**.

Cervix (sir' vix) [L.: neck] • The opening of the uterus into the vagina.

cGMP (cyclic guanosine monophosphate) • An intracellular messenger that is part of signal transmission pathways involving G proteins. (See G protein.)

Channel • A membrane protein that forms an aqueous passageway though which specific solutes may pass by simple diffusion; some channels are gated: they open and close in response to binding of specific molecules.

Chaperone protein • A protein that assists a newly forming protein in adopting its appropriate tertiary structure.

Chemical bond • An attractive force stably linking two atoms.

Chemiosmotic mechanism • The formation of ATP in mitochondria and chloroplasts, resulting from a pumping of protons across a membrane (against a gradient of electrical charge and of pH), followed by the return of the protons through a protein channel with ATPase activity.

Chemoautotroph • An organism that uses carbon dioxide as a carbon source and obtains energy by oxidizing inorganic substances from its environment. (Contrast with chemoheterotroph, photoautotroph, photoheterotroph.)

Chemoheterotroph • An organism that must obtain both carbon and energy from organic substances. (Contrast with chemoautotroph, photoautotroph, photoheterotroph.)

Chemoreceptor • A cell or tissue that senses specific substances in its environment.

Chemosynthesis • Synthesis of food substances, using the oxidation of reduced materials from the environment as a source of energy.

Chiasma (kie az' muh) (plural: chiasmata) [Gr.: cross] • An X-shaped connection between paired homologous chromosomes in prophase I of meiosis. A chiasma is the visible manifestation of crossing over between homologous chromosomes.

Chitin (kye' tin) [Gr. *kiton*: tunic] • The characteristic tough but flexible organic component of the exoskeleton of arthropods, consisting of a complex, nitrogen-containing polysaccharide. Also found in cell walls of fungi.

Chlorophyll (klor' o fill) [Gr. *kloros*: green + *phyllon*: leaf] • Any of a few green pigments associated with chloroplasts or with certain bacterial membranes; responsible for trapping light energy for photosynthesis.

Chloroplast [Gr. *kloros*: green + *plast*: a particle] • An organelle bounded by a double membrane containing the enzymes and pigments that perform photosynthesis. Chloroplasts occur only in eukaryotes.

Choanocyte (cho' an oh cite) • The collared, flagellated feeding cells of sponges.

Cholecystokinin (ko' lee sis to kai nin) • A hormone produced and released by the lining of the duodenum when it is stimulated by undigested fats and proteins. It stimulates the gallbladder to release bile and slows stomach activity.

Chorion (kor' ee on) [Gr. *khorion*: afterbirth] • The outermost of the membranes protecting mammal, bird, and reptile embryos; in mammals it forms part of the placenta.

Chorionic villus sampling • A medical procedure that extracts a portion of the chorion from a pregnant woman to enable genetic and biochemical analysis of the embryo. (Contrast with amniocentesis.)

Chromatid (kro' ma tid) • Each of a pair of new sister chromosomes from the time at which the molecular duplication occurs until the time at which the centromeres separate at the anaphase of nuclear division.

Chromatin • The nucleic acid–protein complex found in eukaryotic chromosomes.

Chromatophore (krow mat' o for) [Gr. *kroma*: color + *phoreus*: carrier] • A pigment-bearing cell that expands or contracts to change the color of the organism.

Chromosome (krome' o sowm) [Gr. *kroma*: color + *soma*: body] • In bacteria and viruses, the DNA molecule that contains most or all of the genetic information of the cell or virus. In eukaryotes, a structure composed of DNA and proteins that bears part of the genetic information of the cell.

Chylomicron (ky low my' cron) • Particles of lipid coated with protein, produced in the gut from dietary fats and secreted into the extracellular fluids.

Chyme (kime) [Gr. *kymus*, juice] • Created in the stomach; a mixture of ingested food with the digestive juices secreted by the salivary glands and the stomach lining.

Cilium (sil' ee um) (plural: cilia) [L. *cilium*: eyelash] • Hairlike organelle used for locomotion by many unicellular organisms and for moving water and mucus by many multicellular organisms. Generally shorter than a flagellum.

Circadian rhythm (sir kade' ee an) [L. *circa*: approximately + *dies*: day] • A rhythm in behavior, growth, or some other activity that recurs about every 24 hours under constant conditions.

Circannual rhythm (sir can' you al) [L. *circa*: approximately + *annus*: year] • A rhythm of behavior, growth, or some other activity that recurs on a yearly basis.

Citric acid cycle • A set of chemical reactions in cellular respiration, in which acetyl CoA reacts with oxaloacetate to form citric acid, and oxaloacetate is regenerated. Acetyl CoA is oxidized to carbon dioxide, and hydrogen atoms are stored as NADH and $FADH_2$. Also called the Krebs cycle.

Class • In taxonomy, the category below the phylum and above the order; a group of related, similar orders.

Class I MHC molecules • These cell surface proteins participate in the cellular immune response directed against virus-infected cells.

Class II MHC molecules • These cell surface proteins participate in the cell-cell interactions (of helper T cells, macrophages, and B cells) of the humoral immune response.

Class switching • The process whereby a plasma cell changes the class of immunoglobulin that it synthesizes. This results from the deletion of part of the constant region of DNA, bringing in a new C segment. The variable region is the same as before, so that the new immunoglbulin has the same antigenic specificity.

Clathrin • A fibrous protein on the inner surfaces of animal cell membranes that strengthens coated vesicles and thus participates in receptor-mediated endocytosis.

Clay • A soil constituent comprising particles smaller than 2 micrometers in diameter.

Cleavages • First divisions of the fertilized egg of an animal.

Cline • A gradual change in the traits of a species over a geographical gradient.

Cloaca (klo ay' kuh) [L. *cloaca*: sewer] • In some invertebrates, the posterior part of the gut; in many vertebrates, a cavity receiving material from the digestive, reproductive, and excretory systems.

Clonal anergy • When a naive T cell encounters a self-antigen, the T cell may bind to the antigen but does not receive signals from an antigen-presenting cell. Instead of being activated, the T cell dies (becomes anergic). In this way, we avoid reacting to our own tissue-specific antigens.

Clonal deletion • In immunology, the inactivation or destruction of lymphocyte clones that would produce immune reactions against the animal's own body.

Clonal selection • The mechanism by which exposure to antigen results in the activation of selected T- or B-cell clones, resulting in an immune response.

Clone [Gr. *klon*: twig, shoot] • Genetically identical cells or organisms produced from a common ancestor by asexual means.

Cnidocytes • The feeding cells of cnidarians, within which nematocysts are housed.

Coacervate (ko as' er vate) [L. *coacervare*: to heap up] • An aggregate of colloidal particles in suspension.

Coacervate drop • Drops formed when a mixture of large proteins and polysaccharides is shaken in water. The interiors of these drops, which are often very stable, contain most of the proteins and polysaccharides.

Coated vesicle • Vesicle, sometimes formed from a coated pit, with characteristic "bristly" surface; its membrane contains distinctive proteins, including clathrin.

Coccus (kock' us) [Gr. *kokkos*: berry, pit] • Any of various spherical or spheroidal bacteria.

Cochlea (kock' lee uh) [Gr. *kokhlos*: a land snail] • A spiral tube in the inner ear of vertebrates; it contains the sensory cells involved in hearing.

Codominance • A condition in which two alleles at a locus produce different phenotypic effects and both effects appear in heterozygotes.

Codon • A "triplet" of three nucleotides in messenger RNA that directs the placement of a particular amino acid into a polypeptide chain. (Contrast with anticodon.)

Coefficient of relatedness • The probability that an allele in one individual is an identical copy, by descent, of an allele in another individual.

Coelom (see' lum) [Gr. *koiloma*: cavity] • The body cavity of certain animals, which is lined with cells of mesodermal origin.

Coelomate • Having a coelom.

Coenocyte (seen' a sight) [Gr.: common cell] • A "cell" bounded by a single plasma membrane, but containing many nuclei.

Coenzyme • A nonprotein molecule that plays a role in catalysis by an enzyme. The coenzyme may be part of the enzyme molecule or free in solution. Some coenzymes are oxidizing or reducing agents.

Coevolution • Concurrent evolution of two or more species that are mutually affecting each other's evolution.

Cohort (co' hort) [L. *cohors*: company of soldiers] • A group of similar-age organisms, considered as it passes through time.

Collagen [Gr. *kolla*: glue] • A fibrous protein found extensively in bone and connective tissue.

Collecting duct • In vertebrates, a tubule that receives urine produced in the nephrons of the kidney and delivers that fluid to the ureter for excretion.

Collenchyma (cull eng' kyma) [Gr. *kolla*: glue + *enchyma*: infusion] • A type of plant cell, living at functional maturity, which lends flexible support by virtue of primary cell walls thickened at the corners. (Contrast with parenchyma, sclerenchyma.)

Colon [Gr. *kolon*: large intestine] • The large intestine.

Commensalism • The form of symbiosis in which one species benefits from the association, while the other is neither harmed nor benefited.

Common bile duct • A single duct that delivers bile from the gallbladder and secretions from the pancreas into the small intestine.

Communication • A signal from one organism (or cell) that alters the pattern of behavior in another organism (or cell) in an adaptive fashion.

Community • Any ecologically integrated group of species of microorganisms, plants, and animals inhabiting a given area.

Companion cell • Specialized cell found adjacent to a sieve tube member in flowering plants.

Comparative analysis • An approach to studying evolution in which hypotheses are tested by measuring the distribution of states among a large number of species.

Comparative genomics • Computer-aided comparison of DNA sequences between different organisms to reveal genes with related functions.

Compensation point • The light intensity at which the rates of photosynthesis and of cellular respiration are equal.

Competitive inhibitor • A substance, similar in structure to an enzyme's substrate, that binds the active site and thus inhibits a reaction.

Competition • In ecology, use of the same resource by two or more species, when the resource is present in insufficient supply for the combined needs of the species.

Competitive exclusion • A result of competition between species for a limiting resource in which one species completely eliminates the other.

Competitive inhibitor • A substance, similar in structure to an enzyme's substrate, that binds the active site and inhibits a reaction.

Complement system • A group of eleven proteins that play a role in some reactions of the immune system. The complement proteins are not immunoglobulins.

Complementary base pairing • The A–T (or A–U), T–A (or U–A), C–G and G–C pairing of bases in double-stranded DNA, in transcription, and between tRNA and mRNA.

Complementary DNA (cDNA) • DNA formed by reverse transcriptase acting with an RNA template; essential intermediate in the reproduction of retroviruses; used as a tool in recombinant DNA technology; lacks introns.

Complete metamorphosis • A change of state during the life cycle of an organism in which the body is almost completely rebuilt to produce an individual with a very different body form. Characteristic of insects such as butterflies, moths, beetles, ants, wasps, and flies.

Compound • (1) A substance made up of atoms of more than one element. (2) Made up of many units, as the compound eyes of arthropods (as opposed to the simple eyes of the same group of organisms).

Condensation reaction • A reaction in which two molecules become connected by a covalent bond and a molecule of water is released. ($AH + BOH \rightarrow AB + H_2O$.)

Cones • (1) In the vertebrate retina: photoreceptors responsible for color vision. (2) In gymnosperms: reproductive structures consisting of many sporophylls packed relatively tightly.

Conidium (ko nid' ee um) [Gr. *konis*: dust] • An asexual fungus spore borne singly or in chains either apically or laterally on a hypha.

Conifer (kahn' e fer) [Gr. *konos*: cone + *phero*: carry] • One of the cone-bearing gymnosperms, mostly trees, such as pines and firs.

Conjugation (kahn' jew gay' shun) [L. *conjugare*: yoke together] • The close approximation of two cells during which they exchange genetic material, as in *Paramecium* and other ciliates, or during which DNA passes from one to the other through a tube, as in bacteria.

Connective tissue • An animal tissue that connects or surrounds other tissues; its cells are embedded in a collagen-containing matrix.

Connexon • In a gap junction, a protein channel linking adjacent animal cells.

Consensus sequences • Short stretches of DNA that appear, with little variation, in many different genes.

Constant region • The constant region in an immunoglobulin is encoded by a single exon and determines the function, but not the specificity, of the molecule. The constant region of the T cell receptor anchors the protein to the plasma membrane.

Constitutive enzyme • An enzyme that is present in approximately constant amounts in a system, whether its substrates are present or absent. (Contrast with inducible enzyme.)

Consumer • An organism that eats the tissues of some other organism.

Continental drift • The gradual drifting apart of the world's continents that has occurred over a period of billions of years.

Convergent evolution • The evolution of similar features independently in unrelated taxa from different ancestral structures.

Cooperative act • Behavior in which two or more individuals interact to their mutual benefit. No conscious awareness by the actors of the effects of their behavior is implied.

Cooption • The act of capturing something for a particular use. In ecology refers to the diversion of ecological production for human use. Such production is said to be coopted.

Copulation • Reproductive behavior that results in a male depositing sperm in the reproductive tract of a female.

Corepressor • A low molecular weight compound that unites with a protein (the repressor) to prevent transcription in a repressible operon.

Cork • A waterproofing tissue in plants, with suberin-containing cell walls. Produced by a cork cambium.

Corolla (ko role' lah) [L.: diminutive of *corona*: wreath, crown] • All of the petals of a flower, collectively.

Coronary (kor' oh nair ee) • Referring to the blood vessels of the heart.

Corpus luteum (kor' pus loo' tee um) [L. *corpus*: body + *luteum*: yellow] A structure formed from a follicle after ovulation; it produces hormones important to the maintenance of pregnancy.

Cortex [L.: bark or rind] • (1) In plants, the tissue between the epidermis and the vascular tissue of a stem or root. (2) In animals, the outer tissue of certain organs, such as the adrenal cortex and cerebral cortex.

Corticosteroids • Steroid hormones produced and released by the cortex of the adrenal gland.

Cost • See energetic cost, opportunity cost, risk cost.

Cotyledon (kot' ul lee' dun) [Gr. *kotyledon*: a hollow space] • A "seed leaf." An embryonic organ which stores and digests reserve materials; may expand when seed germinates.

Countercurrent exchange • An adaptation that promotes maximum exchange of heat or any diffusible substance between two fluids by the fluids flow in opposite directions through parallel tubes in close approximation to each other. An example is countercurrent heat exchange between arterioles and venules in the extremities of some animals.

Covalent bond • A chemical bond that arises from the sharing of electrons between two atoms. Usually a strong bond.

Crassulacean acid metabolism (CAM) • A metabolic pathway enabling the plants that possess it to store carbon dioxide at night and then perform photosynthesis during the day with stomata closed.

Crista (plural: cristae) • A small, shelflike projection of the inner membrane of a mitochondrion; the site of oxidative phosphorylation.

Critical night length • In the photoperiodic flowering response of short-day plants, the length of night above which flowering occurs and below which the plant remains vegetative. (The reverse applies in the case of long-day plants.)

Critical period • The age during which some particular type of learning must take place or during which it occurs much more easily than at other times. Typical of song learning among birds.

Cross section (also called a transverse section) • A section taken perpendicular to the longest axis of a structure.

Crossing over • The mechanism by which linked markers undergo recombination. In general, the term refers to the reciprocal exchange of corresponding segments between two homologous chromatids.

CRP • The cAMP receptor protein that interacts with the promoter to enhance transcription; a lowered cAMP concentration results in catabolite repression.

Crustacean (crus tay' see an) • A member of the phylum Crustacea, such as a crab, shrimp, or sowbug.

Cryptic appearance [Gr. *kryptos*: hidden] • The resemblance of an animal to some part of its environment, which helps it to escape detection by predators.

Cryptochromes [Gr. *kryptos*: hidden + *kroma*: color] • Photoreceptors mediating some blue-light effects in plants and animals.

Culture • (1) A laboratory association of organisms under controlled conditions. (2) The collection of knowledge, tools, values, and rules that characterize a human society.

Cuticle • A waxy layer on the outer surface of a plant or an insect, tending to retard water loss.

Cyanobacteria (sigh an' o bacteria) [Gr. *kuanos*: the color blue] • A division of photosynthetic bacteria, formerly referred to as blue-green algae; they lack sexual reproduction, and they use chlorophyll *a* in their photosynthesis.

Cyclic AMP • See cAMP.

Cyclins • Proteins that activate cyclin-dependent kinases, bringing about transitions in the cell cycle.

Cyclin-dependent kinase (cdk) • A kinase is an enzyme that catalzyes the addition of phosphate groups from ATP to target molecules. Cdk's target proteins involved in transitions in the cell cycle and are active only when complexed to additional protein subunits, cyclins.

Cyst (sist) [Gr. *kystis*: pouch] • (1) A resistant, thick-walled cell formed by some protists and other organisms. (2) An abnormal sac, containing a liquid or semisolid substance, produced in response to injury or illness.

Cytochromes (sy' toe chromes) [Gr. *kytos*: container + *chroma*: color] • Iron-containing red proteins, components of the electron-transfer chains in photophosphorylation and respiration.

Cytokinesis (sy' toe kine ee' sis) [Gr. *kytos*: container + *kinein*: to move] • The division of the cytoplasm of a dividing cell. (Contrast with mitosis.)

Cytokinin (sy' toe kine' in) [Gr. *kytos*: container + *kinein*: to move] • A member of a class of plant growth substances playing roles in senescence, cell division, and other phenomena.

Cytoplasm • The contents of the cell, excluding the nucleus.

Cytoplasmic determinants • In animal development, gene products whose spatial distribution may determine such things as embryonic axes.

Cytosine (site' oh seen) • A nitrogen-containing base found in DNA and RNA.

Cytoskeleton • The network of microtubules and microfilaments that gives a eukaryotic cell its shape and its capacity to arrange its organelles and to move.

Cytosol • The fluid portion of the cytoplasm, excluding organelles and other solids.

Cytotoxic T cells • Cells of the cellular immune system that recognize and directly eliminate virus-infected cells. (Contrast with helper T cells, suppressor T cells.)

Decomposer • See detritivore.

Degeneracy • The situation in which a single amino acid may be represented by any of two or more different codons in messenger RNA. Most of the amino acids can be represented by more than one codon.

Degradative succession • Ecological succession occuring on the dead remains of the bodies of plants and animals, as when leaves or animal bodies rot.

Deletion (genetic) • A mutation resulting from the loss of a continuous segment of a gene or chromosome. Such mutations never revert to wild type. (Contrast with duplication, point mutation.)

Deme (deem) [Gr. *demos*: common people] • Any local population of individuals belonging to the same species that interbreed with one another.

Demographic processes • The events—such as births, deaths, immigration, and emigration—that determine the number of individuals in a population.

Demographic stochasticity • Random variations in the factors influencing the size, density, and distribution of a population.

Demography • The study of dynamical changes in the sizes, densities, and distributions of populations.

Denaturation • Loss of activity of an enzyme or nucleic acid molecule as a result of structural changes induced by heat or other means.

Dendrite [Gr. *dendron*: a tree] • A fiber of a neuron which often cannot carry action potentials. Usually much branched and relatively short compared with the axon, and commonly carries information to the cell body of the neuron.

Denitrification • Metabolic activity by which inorganic nitrogen-containing ions are reduced to form nitrogen gas and other products; carried on by certain soil bacteria.

Density dependence • Change in the severity of action of agents affecting birth and death rates within populations that are directly or inversely related to population density.

Density independence • The state where the severity of action of agents affecting birth and death rates within a population does not change with the density of the population.

Deoxyribonucleic acid • See DNA.

Depolarization • A change in the electric potential across a membrane from a condition in which the inside of the cell is more negative than the outside to a condition in which the inside is less negative, or even positive, with reference to the outside of the cell. (Contrast with hyperpolarization.)

Derived trait • A trait found among members of a lineage that was not present in the ancestors of that lineage.

Dermal tissue system • The outer covering of a plant, consisting of epidermis in the young plant and periderm in a plant with extensive secondary growth. (Contrast with ground tissue system and vascular tissue system.)

Desmosome (dez' mo sowm) [Gr. *desmos*: bond + *soma*: body] • An adhering junction between animal cells.

Determination • Process whereby an embryonic cell or group of cells becomes fixed into a predictable developmental pathway.

Detritivore (di try' ti vore) [L. *detritus*: worn away + *vorare*: to devour] • An organism that obtains its energy from the dead bodies and/or waste products of other organisms.

Deuterostome • A major evolutionary lineage in animals, characterized by radial cleavage, enterocoelous development, and other traits. (Compare with protostome.)

Development • Progressive change, as in structure or metabolism; in most kinds of organisms, development continues throughout the life of the organism.

Diaphragm (dye' uh fram) [Gr. *diaphrassein*, to barricade] • (1) A sheet of muscle that separates the thoracic and abdominal cavities in mammals; responsible for the action of breathing. (2) A method of birth control in which a sheet of rubber is fitted over the woman's cervix, blocking the entry of sperm.

Diastole (dye ahs' toll ee) [Gr.: dilation] • The portion of the cardiac cycle when the heart muscle relaxes. (Contrast with systole.)

Dicot (short for dicotyledon) [Gr. *di*: two + *kotyledon*: a hollow space] • This term, not used in this book, formerly referred to all angiosperms other than the monocots. (See eudicot, monocot.)

Differentiation • Process whereby originally similar cells follow different developmental pathways. The actual expression of determination.

Diffusion • Random movement of molecules or other particles, resulting in even distribution of the particles when no barriers are present.

Digestibility-reducing chemicals • Defensive chemicals produced by plants that make the plant's tissued difficult to digest.

Digestion • Enzyme-catalyzed process by which large, usually insoluble, molecules (foods) are hydrolyzed to form smaller molecules of soluble substances.

Dihybrid cross • A mating in which the parents differ with respect to the alleles of two loci of interest.

Dikaryon (di care' ee ahn) [Gr. *dis*: two + *karyon*: kernel] • A cell or organism carrying two genetically distinguishable nuclei. Common in fungi.

Dioecious (die eesh' us) [Gr.: two houses] • Organisms in which the two sexes are "housed" in two different individuals, so that eggs and sperm are not produced in the same individuals. Examples: humans, fruit flies, oak trees, date palms. (Contrast with monoecious.)

Diploblastic • Having two cell layers. (Contrast with triploblastic.)

Diploid (dip' loid) [Gr. *diploos*: double] • Having a chromosome complement consisting of two copies (homologues) of each chromosome. A diploid individual (or cell) usually arises as a result of the fusion of two gametes, each with just one copy of each chromosome. Thus, the two homologues in each chromosome pair in a diploid cell are of separate origin, one derived from the female parent and one from the male parent.

Directional selection • Selection in which phenotypes at one extreme of the population distribution are favored. (Contrast with disruptive selection; stabilizing selection.)

Disaccharide • A carbohydrate made up of two monosaccharides (simple sugars).

Dispersal stage • Stage in its life history at which an organism moves from its birthplace to where it will live as an adult.

Displacement activity • Apparently irrelevant behavior performed by an animal under conflict situations, especially when tendencies to attack and escape are closely balanced.

Display • A behavior that has evolved to influence the actions of other individuals.

Disruptive selection • Selection in which phenotypes at both extremes of the population distribution are favored. (Contrast with directional selection; stabilizing selection.)

Distal • Away from the point of attachment or other reference point. (Contrast with proximal.)

Disturbance • A short-term event that disrupts populations, communities, or ecosystems by changing the environment.

Diverticulum (di ver tic' u lum) [L. *divertere*: turn away] • A small cavity or tube that connects to a major cavity or tube.

Division • A term used by some microbiologists and formerly by botanists, corresponding to the term phylum.

DNA (deoxyribonucleic acid) • The fundamental hereditary material of all living organisms. In eukaryotes, stored primarily in the cell nucleus. A nucleic acid using deoxyribose rather than ribose.

DNA chip • A small glass or plastic square onto which thousands of single-stranded DNA sequences are fixed. Hybridization of cell-derived RNA or DNA to the target sequences can be performed. (See DNA hybridization.)

DNA hybridization • A process by which DNAs from two species are mixed and heated so that interspecific double helixes are formed.

DNA ligase • Enzyme that unites Okazaki fragments of the lagging strand during DNA replication; also mends breaks in DNA strands. It connects pieces of a DNA strand and is used in recombinant DNA technology.

DNA methylation • Addition of methyl groups to DNA; plays role in regulation of gene expression; protects a bacterium's DNA against its restriction endonucleases.

DNA polymerase • Any of a group of enzymes that catalyze the formation of DNA strands from a DNA template.

Domain • The largest unit in the current taxonomic nomenclature. Members of the three domains (Bacteria, Archaea, and Eukarya) are believed to have been evolving independently of each other for at least a billion years.

Dominance • In genetic terminology, the ability of one allelic form of a gene to determine the phenotype of a heterozygous individual, in which the homologous chromosome carries both it and a different allele. For example, if *A* and *a* are two allelic forms of a gene, *A* is said to be dominant to *a* if *AA* diploids and *Aa* diploids are phenotypically identical and are distinguishable from *aa* diploids. The *a* allele is said to be **recessive**.

Dominance hierarchy • In animal behavior, the set of relationships within a group of animals, usually established and maintained by aggression, in which one individual has precedence over all others in eating, mating, and other activities.

Dormancy • A condition in which normal activity is suspended, as in some seeds and buds.

Dorsal [L. *dorsum*: back] • Pertaining to the back or upper surface. (Contrast with ventral.)

Double fertilization • Process virtually unique to angiosperms in which one sperm nucleus combines with the egg to produce a zygote, and the other sperm nucleus combines with the two polar nuclei to produce the first cell of the triploid endosperm.

Double helix • Of DNA: molecular structure in which two complementary polynucleotide strands, antiparallel to each other, form a right-handed spiral.

Duodenum (doo' uh dee' num) • The beginning portion of the vertebrate small intestine. (Contrast with ileum, jejunum.)

Duplication (genetic) • A mutation resulting from the introduction into the genome of an extra copy of a segment of a gene or chromosome. (Contrast with deletion, point mutation.)

Dynein [Gr. *dunamis*: power] • A protein that undergoes conformational changes and thus plays a part in the movement of eukaryotic flagella and cilia.

Ecdysone (eck die' sone) [Gr. *ek*: out of + *dyo*: to clothe] • In insects, a hormone that induces molting.

Ecological biogeography • The study of the distributions of organisms from an ecological perspective, usually concentrating on migration, dispersal, and species interactions.

Ecological community • The species living together at a particular site.

Ecological niche (nitch) [L. *nidus*: nest] • The functioning of a species in relation to other species and its physical environment.

Ecological succession • The sequential replacement of one population assemblage by another in a habitat following some disturbance. Succession sometimes ends in a relatively stable ecosystem.

Ecology [Gr. *oikos*: house + *logos*: discourse, study] • The scientific study of the interaction of organisms with their environment, including both the physical environment and the other organisms that live in it.

Ecoregion • A large geographic unit characterized by a typical climate and a widespread assemblage of similar species.

Ecosystem (eek' oh sis tum) • The organisms of a particular habitat, such as a pond or forest, together with the physical environment in which they live.

Ecto- (eck' toh) [Gr.: outer, outside] • A prefix used to designate a structure on the outer surface of the body. For example, ectoderm. (Contrast with endo- and meso-.)

Ectoderm [Gr. *ektos*: outside + *derma*: skin] • The outermost of the three embryonic tissue layers first delineated during gastrulation. Gives rise to the skin, sense organs, nervous system, etc.

Ectotherm [Gr. *ektos*: outside + *thermos*: heat] • An animal unable to control its body temperature. (Contrast with endotherm.)

Edema (i dee' mah) [Gr. *oidema*: swelling] • Tissue swelling caused by the accumulation of fluid.

Edge effect • The changes in ecological processes in a community caused by physical and biological factors originating in an adjacent community.

Effector • Any organ, cell, or organelle that moves the organism through the environment or else alters the environment to the organism's advantage. Examples include muscle, bone, and a wide variety of exocrine glands.

Effector cell • A lymphocyte that performs a role in the immune system without further differentiation.

Effector phase • In this phase of the immune response, effector T cells called cytotoxic T cells attack virus-infected cells, and effector helper T cells assist B cells to differentiate into plasma cells, which release antibodies.

Efferent [L. *ex*: out + *ferre*: to bear] • Away from, as in neurons that conduct action potentials out from the central nervous system, or arterioles that conduct blood away from a structure. (Contrast with afferent.)

Egg • In all sexually reproducing organisms, the female gamete; in birds, reptiles, and some other vertebrates, a structure witin which early embryonic development occurs.

Elasticity • The property of returning quickly to a former state after a disturbance.

Electrocardiogram (EKG) • A graphic recording of electrical potentials from the heart.

Electroencephalogram (EEG) • A graphic recording of electrical potentials from the brain.

Electromyogram (EMG) • A graphic recording of electrical potentials from muscle.

Electron (e lek' tron) [L. *electrum*: amber (associated with static electricity), from Gr. *slektor*: bright sun (color of amber)] • One of the three most important fundamental particles of matter, with mass approximately 0.00055 amu and charge –1.

Electronegativity • The tendency of an atom to attract electrons when it occurs as part of a compound.

Electrophoresis (e lek' tro fo ree' sis) [L. *electrum*: amber + Gr. *phorein*: to bear] • A separation technique in which substances are separated from one another on the basis of their electric charges and molecular weights.

Electrotonic potential • In neurons, a hyperpolarization or small depolarization of the membrane potential induced by the application of a small electric current. (Contrast with action potential, resting potential.)

Elemental substance • A substance composed of only one type of atom.

Embolus (em' buh lus) [Gr. *embolos*: inserted object; stopper] • A circulating blood clot. Blockage of a blood vessel by an embolus or by a bubble of gas is referred to as an **embolism**. (Contrast with thrombus.)

Embryo [Gr. *en-*: in + *bryein*: to grow] • A young animal, or young plant sporophyte, while it is still contained within a protective structure such as a seed, egg, or uterus.

Embryo sac • In angiosperms, the female gametophyte. Found within the ovule, it consists of eight or fewer cells, membrane bounded, but without cellulose walls between them.

Emergent property • A property of a complex system that is not exhibited by its individual component parts.

Emigration • The deliberate and usually oriented departure of an organism from the habitat in which it has been living.

3′ End (3-prime) • The end of a DNA or RNA strand that has a free hydroxyl group at the 3′-carbon of the sugar (deoxyribose or ribose).

5′ End (5-prime) • The end of a DNA or RNA strand that has a free phosphate group at the 5′-carbon of the sugar (deoxyribose or ribose).

Endemic (en dem' ik) [Gr. *endemos*: dwelling in a place] • Confined to a particular region, thus often having a comparatively restricted distribution.

Endergonic reaction • One for which energy must be supplied. (Contrast with exergonic reaction.)

Endo- [Gr.: within, inside] • A prefix used to designate an innermost structure. For example, endoderm, endocrine. (Contrast with ecto-, meso-.)

Endocrine gland (en' doh krin) [Gr. *endon*: inside + *krinein*: to separate] • Any gland, such as the adrenal or pituitary gland of vertebrates, that secretes certain substances, especially hormones, into the body through the blood.

Endocrinology • The study of hormones and their actions.

Endocytosis • A process by which liquids or solid particles are taken up by a cell through invagination of the plasma membrane. (Contrast with exocytosis.)

Endoderm [Gr. *endon*: within + *derma*: skin] • The innermost of the three embryonic tissue layers first delineated during gastrulation. Gives rise to the digestive and respiratory tracts and structures associated with them.

Endodermis [Gr. *endon*: within + *derma*: skin] • In plants, a specialized cell layer marking the inside of the cortex in roots and some stems. Frequently a barrier to free diffusion of solutes.

Endomembrane system • Endoplasmic reticulum plus Golgi apparatus plus, when present, lysosomes; thus, a system of membranes that exchange material with one another.

Endoplasmic reticulum [Gr. *endon*: within + L. *plasma*: form; L. *reticulum*: little net] • A system of membrane-bounded tubes and flattened sacs found in the cytoplasm of eukaryotes. Exists as rough ER, studded with ribosomes; and smooth ER, lacking ribosomes.

Endorphins • Naturally occurring, opiate-like substances in the mammalian brain.

Endoskeleton [Gr. *endon*: within + *skleros*: hard] • A skeleton covered by other, soft body tissues. (Contrast with exoskeleton.)

Endosperm [Gr. *endon*: within + *sperma*: seed] • A specialized triploid seed tissue found only in angiosperms; contains stored food for the developing embryo.

Endosymbiosis [Gr. *endon*: within + *syn*: together + *bios*: life] • The living together of two species, with one living inside the body (or even the cells) of the other.

Endosymbiotic theory • Theory that the eukaryotic cell evolved from a prokaryote that contained other, endosymbiotic prokaryotes.

Endotherm [Gr. *endon*: within + *thermos*: hot] • An animal that can control its body temperature by the expenditure of its own metabolic energy. (Contrast with ectotherm.)

Endotoxins [Gr. *endon*: within + L. *toxicum*: poison] • Lipopolysaccharides released by the lysis of some Gram-negative bacteria that cause fever and vomiting in a host organism.

Energetic cost • The difference between the energy an animal would have expended had it rested, and that expended in performing a behavior.

Energy • The capacity to do work.

Enhancer • In eukaryotes, a DNA sequence, lying on either side of the gene it regulates, that stimulates a specific promoter.

Enterocoelous development • A pattern of development in which the coelum is formed by an outpocketing of the embryonic gut (enteron).

Enterokinase (ent uh row kine' ase) • An enzyme secreted by the mucosa of the duodenum. It activates the zymogen trypsinogen to create the active digestive enzyme trypsin.

Entrainment • With respect to circadian rhythms, the process whereby the period is adjusted to match the 24-hour environmental cycle.

Entropy (en' tro pee) [Gr. *en*: in + *tropein*: to change] • A measure of the degree of disorder in any system. A perfectly ordered system has zero entropy; increasing disorder is measured by positive entropy. Spontaneous reactions in a closed system are always accompanied by an increase in disorder and entropy.

Environment • An organism's surroundings, both living and nonliving; includes temperature, light intensity, and all other species that influence the focal organism.

Environmental toxicology • The study of the distribution and effects of toxic compounds in the environment.

Enzyme (en' zime) [Gr. *en*: in + *zyme*: yeast] • A protein, on the surface of which are chemical groups so arranged as to make the enzyme a catalyst for a chemical reaction.

Epi- [Gr.: upon, over] • A prefix used to designate a structure located on top of another; for example: epidermis, epiphyte.

Epicotyl (epp' i kot' il) [Gr. *epi*: upon + *kotyle*: something hollow] • That part of a plant embryo or seedling that is above the cotyledons.

Epidermis [Gr. *epi*: upon + *derma*: skin] • In plants and animals, the outermost cell layers. (Only one cell layer thick in plants.)

Epididymis (epuh did' uh mus) [Gr. *epi*: upon + *didymos*: testicle] • Coiled tubules in the testes that store sperm and conduct sperm from the seiminiferous tubules to the vas deferens.

Epinephrine (ep i nef' rin) [Gr. *epi*: upon + *nephros*: a kidney] • The "fight or flight" hormone. Produced by the medulla of the adrenal gland, it also functions as a neurotransmitter. Also known as adrenaline.

Epiphyte (ep' e fyte) [Gr. *epi*: upon + *phyton*: plant] • A specialized plant that grows on

the surface of other plants but does not parasitize them.

Episome • A plasmid that may exist either free or integrated into a chromosome. (See plasmid.)

Epistasis • An interaction between genes, in which the presence of a particular allele of one gene determines whether another gene will be expressed.

Epithelium • In animals, a layer of cells covering or lining an external surface or a cavity.

Equilibrium • (1) In biochemistry, a state in which forward and reverse reactions are proceeding at counterbalancing rates, so there is no observable change in the concentrations of reactants and products. (2) In evolutionary genetics, a condition in which allele and genotype frequencies in a population are constant from generation to generation.

Erythrocyte (ur rith' row sight) [Gr. *erythros*: red + *kytos*: hollow vessel] • A red blood cell.

Esophagus (i soff' i gus) [Gr. *oisophagos*: gullet] • That part of the gut between the pharynx and the stomach.

Ester linkage • A condensation (water-releasing) reaction in which the carboxyl group of a fatty acid reacts with the hydroxyl group of an alcohol. Lipids are formed in this way.

Estivation (ess tuh vay' shun) [L. *aestivalis*: summer] • A state of dormancy and hypometabolism that occurs during the summer; usually a means of surviving drought and/or intense heat. Contrast with hibernation.

Estrogen • Any of several steroid sex hormones, produced chiefly by the ovaries in mammals.

Estrus (es' truss) [L. *oestrus*: frenzy] • The period of heat, or maximum sexual receptivity, in some female mammals. Ordinarily, the estrus is also the time of release of eggs in the female.

Ethylene • One of the plant hormones, the gas $H_2C;h2CH_2$.

Euchromatin • Chromatin that is diffuse and non-staining during interphase; may be transcribed. (Contrast with heterochromatin.)

Eudicots (yew di' kots) [Gr. *eu*: true + *di*: two + *kotyledon*: a cup-shaped hollow] • Members of the angiosperm class Eudicotyledones, flowering plants in which the embryo produces two cotyledons prior to germination. Leaves of most eudicots have major veins arranged in a branched or reticulate pattern.

Eukaryotes (yew car' ry otes) [Gr. *eu*: true + *karyon*: kernel or nucleus] • Organisms whose cells contain their genetic material inside a nucleus. Includes all life other than the viruses, Archaebacteria, and Eubacteria.

Eusocial • Term applied to insects, such as termites, ants, and many bees and wasps, in which individuals cooperate in the care of offspring, there are sterile castes, and generations overlap.

Eutrophication (yoo trofe' ik ay' shun) [Gr. *eu-*: well + *trephein*: to flourish] • The addition of nutrient materials to a body of water, resulting in changes to species composition therein.

Evolution • Any gradual change. Organic evolution, often referred to as evolution, is any genetic and resulting phenotypic change in organisms from generation to generation.

Evolutionary agent • Any factor that influences the direction and rate of evolutionary changes.

Evolutionarily conservative • Traits of organisms that evolve very slowly.

Evolutionary innovations • Major changes in body plans of organisms; these have been very rare during evolutionary history.

Evolutionary radiation • The proliferation of species within a single evolutionary lineage.

Evolutionary reversal • The reappearance of the ancestral state of a trait in a lineage in which that trait had acquired a derived state.

Excision repair • The removal and damaged DNA and its replacement by the appropriate nucleotides.

Excitatory postsynaptic potential (EPSP) • A change in the resting potential of a postsynaptic membrane in a positive (depolarizing) direction. (Contrast with inhibitory postsynaptic potential.)

Excretion • Release of metabolic wastes by an organism.

Exergonic reaction • A reaction in which free energy is released. (Contrast with endergonic reaction.)

Exo- (eks' oh) • Same as ecto-.

Exocrine gland (eks' oh krin) [Gr. *exo*: outside + *krinein*: to separate] • Any gland, such as a salivary gland, that secretes to the outside of the body or into the gut.

Exocytosis • A process by which a vesicle within a cell fuses with the plasma membrane and releases its contents to the outside. (Contrast with endocytosis.)

Exon • A portion of a DNA molecule, in eukaryotes, that codes for part of a polypeptide. (Contrast with intron.)

Exoskeleton (eks' oh skel' e ton) [Gr. *exos*: outside + *skleros*: hard] • A hard covering on the outside of the body to which muscles are attached. (Contrast with endoskeleton.)

Exotoxins • Highly toxic proteins released by living, multiplying bacteria.

Experiment • A scientific method in which particular factors are manipulated while other factors are held constant so that the potential influences of the manipulated factors can be determined.

Exponential growth • Growth, especially in the number of organisms in a population, which is a simple function of the size of the growing entity: the larger the entity, the faster it grows. (Contrast with logistic growth.)

Expression vector • A DNA vector, such as a plasmid, that carries a DNA sequence that includes the adjacent sequences for its expression into mRNA and protein in a host cell.

Expressivity • The degree to which a genotype is expressed in the phenotype— may be affected by the environment.

Extensor • A muscle the extends an appendage.

Extinction • The termination of a lineage of organisms.

Extrinsic protein • A membrane protein found only on the surface of the membrane. (Contrast with intrinsic protein.)

F_1 generation • The immediate progeny of a parental (P) mating; the first filial generation.

F_2 generation • The immediate progeny of a mating between members of the F_1 generation.

Facilitated diffusion • Passive movement through a membrane involving a specific carrier protein; does not proceed against a concentration gradient. (Contrast with active transport, free diffusion.)

Family • In taxonomy, the category below the order and above the genus; a group of related, similar genera.

Fat • A triglyceride that is solid at room temperature. (Contrast with oil.)

Fatty acid • A molecule with a long hydrocarbon tail and a carboxyl group at the other end. Found in many lipids.

Fauna (faw' nah) • All of the animals found in a given area. (Contrast with flora.)

Feces [L. *faeces*: dregs] • Waste excreted from the digestive system.

Feedback control • Control of a particular step of a multistep process, induced by the presence or absence of a product of one of the later steps. A thermostat regulating the flow of heating oil to a furnace in a home is a negative feedback control device.

Fermentation (fur men tay' shun) [L. *fermentum*: yeast] • The degradation of a substance such as glucose to smaller molecules with the extraction of energy, without the use of oxygen (i.e., anaerobically). Involves the glycolytic pathway.

Fertilization • Union of gametes. Also known as syngamy.

Fertilization membrane • A membrane surrounding an animal egg which becomes rapidly raised above the egg surface within seconds after fertilization, serving to prevent entry of a second sperm.

Fetus • The latter stages of an embryo that is still contained in an egg or uterus; in humans, the unborn young from the eighth week of pregnancy to the moment of birth.

Fiber • An elongated and tapering cell of flowering plants, usually with a thick cell wall. Serves a support function.

Fibrin • A protein that polymerizes to form long threads that provide structure to a blood clot.

Filter feeder • An organism that feeds upon much smaller organisms, that are suspended in water or air, by means of a straining device.

Filtration • In the excretory physiology of some animals, the process by which the initial urine is formed; water and most solutes are transferred into the excretory tract, while proteins are retained in the blood or hemolymph.

First law of thermodynamics • Energy can be neither created nor destroyed.

Fission • Reproduction of a prokaryote by division of a cell into two comparable progeny cells.

Fitness • The contribution of a genotype or phenotype to the composition of subsequent generations, relative to the contribution of other genotypes or phenotypes. (See inclusive fitness.)

Fixed action pattern • A behavior that is genetically programmed.

Flagellum (fla jell' um) (plural: flagella) [L. *flagellum*: whip] • Long, whiplike appendage that propels cells. Prokaryotic flagella differ sharply from those found in eukaryotes.

Flexor • A muscle that flexes an appendage.

Flora (flore' ah) • All of the plants found in a given area. (Contrast with fauna.)

Florigen • A plant hormone (not yet isolated) involved in the conversion of a vegetative shoot apex to a flower.

Flower • The total reproductive structure of an angiosperm; its basic parts include the calyx, corolla, stamens, and carpels.

Fluorescence • The emission of a photon of visible light by an excited atom or molecule.

Follicle [L. *folliculus*: little bag] • In female mammals, an immature egg surrounded by nutritive cells.

Follicle-stimulating hormone • A gonadotropic hormone produced by the anterior pituitary.

Food chain • A portion of a food web, most commonly a simple sequence of prey species and the predators that consume them.

Food web • The complete set of food links between species in a community; a diagram indicating which ones are the eaters and which are consumed.

Forb • Any broad-leaved (dicotyledonous), herbaceous plant. Especially applied to such plants growing in grasslands.

Fossil • Any recognizable structure originating from an organism, or any impression from such a structure, that has been preserved over geological time.

Fossil fuel • A fuel (particularly petroleum products) composed of the remains of organisms that lived in the remote past.

Founder effect • Random changes in allele frequencies resulting from establishment of a population by a very small number of individuals.

Fovea [L. *fovea*; a small pit] • The area, in the vertebrate retina, of most distinct vision.

Frame-shift mutation • A mutation resulting from the addition or deletion of a single base pair in the DNA sequence of a gene. As a result of this, mRNA transcribed from such a gene is translated normally until the ribosome reaches the point at which the mutation has occurred. From that point on, codons are read out of proper register and the amino acid sequence bears no resemblance to the normal sequence. (Contrast with missense mutation, nonsense mutation, synonymous mutation.)

Free energy • That energy which is available for doing useful work, after allowance has been made for the increase or decrease of disorder. Designated by the symbol G (for Gibbs free energy), and defined by: $G = H - TS$, where H = heat, S = entropy, and T = absolute (Kelvin) temperature.

Frequency-dependent selection • Selection that changes in intensity with the proportion of individuals having the trait.

Fruit • In angiosperms, a ripened and mature ovary (or group of ovaries) containing the seeds. Sometimes applied to reproductive structures of other groups of plants, and includes any adjacent parts which may be fused with the reproductive structures.

Fruiting body • A structure that bears spores.

Fundamental niche • The range of condition under which an organism could survive if it were the only one in the environment. (Contrast with realized niche.)

Fungus (fung' gus) • A member of the kingdom Fungi, a (usually) multicellular eukaryote with absorptive nutrition.

G_1 phase • In the cell cycle, the gap between the end of mitosis and the onset of the S phase.

G_2 phase • In the cell cycle, the gap between the S (synthesis) phase and the onset of mitosis.

G protein • A membrane protein involved in signal transduction; characterized by binding guanyl nucleotides. The activation of certain receptors activates the G protein, which in turn activates adenylate cyclase. G protein activation involves binding a GTP molecule in place of a GDP molecule.

Gametangium (gam i tan' gee um) [Gr. *gamos*: marriage + *angeion*: vessel or reservoir] • Any plant or fungal structure within which a gamete is formed.

Gamete (gam' eet) [Gr. *gamete*: wife, *gametes*: husband] • The mature sexual reproductive cell: the egg or the sperm.

Gametocyte (ga meet' oh site) [Gr. *gamete*: wife, *gametes*: husband + *kytos*: cell] • The cell that gives rise to sex cells, either the eggs or the sperm. (See oocyte and spermatocyte.)

Gametogenesis (ga meet' oh jen' e sis) [Gr. *gamete*: wife, *gametes*: husband + *genesis*: source] • The specialized series of cellular divisions that leads to the production of sex cells (gametes). (Contrast with oogenesis and spermatogenesis.)

Gametophyte (ga meet' oh fyte) • In plants and photosynthetic protists with alternation of generations, the haploid phase that produces the gametes. (Contrast with sporophyte.)

Ganglion (gang' glee un) [Gr.: tumor] • A group or concentration of neuron cell bodies.

Gap junction • A 2.7-nanometer gap between plasma membranes of two animal cells, spanned by protein channels. Gap junctions allow chemical substances or electrical signals to pass from cell to cell.

Gas exchange • In animals, the process of taking up oxygen from the environment and releasing carbon dioxide to the environment.

Gastrovascular cavity • Serving for both digestion (gastro) and circulation (vascular); in particular, the central cavity of the body of jellyfish and other cnidarians.

Gastrula (gas' true luh) [Gr. *gaster*: stomach] • An embryo forming the characteristic three cell layers (ectoderm, endoderm, and mesoderm) which will give rise to all of the major tissue systems of the adult animal.

Gastrulation • Development of a blastula into a gastrula.

Gated channel • A channel (membrane protein) that opens and closes in response to binding of specific molecules or to changes in membrane potential.

Gel electrophoresis (jel ul lec tro for' eesis) • A semisolid matrix suspended in a salty buffer in which molecules can be separated on the basis of their size and change when current is passed through the gel.

Gene [Gr. *gen*: to produce] • A unit of heredity. Used here as the unit of genetic function which carries the information for a single polypeptide.

Gene amplification • Creation of multiple copies of a particular gene, allowing the production of large amounts of the RNA transcript (as in rRNA synthesis in oocytes).

Gene cloning • Formation of a clone of bacteria or yeast cells containing a particular foreign gene.

Gene family • A set of identical, or once-identical, genes, derived from a single parent gene; need not be on the same chromosomes; classic example is the globin family in vertebrates.

Gene flow • The exchange of genes between different species (an extreme case referred to as hybridization) or between different populations of the same species caused by migration following breeding.

Gene pool • All of the genes in a population.

Gene therapy • Treatment of a genetic disease by providing patients with cells containing wild type alleles for the genes that are nonfunctional in their bodies.

Generative nucleus • In a pollen tube, a haploid nucleus that undergoes mitosis to produce the two sperm nuclei that participate in double fertilization. (Contrast with tube nucleus.)

Genet • The genetic individual of a plant that is composed of a number of nearly identical but repeated units.

Genetic drift • Changes in gene frequencies from generation to generation in a small population as a result of random processes.

Genetic stochasticity • Variation in the frequencies of alleles and genotypes in a population over time.

Genetics • The study of heredity.

Genetic structure • The frequencies of alleles and genotypes in a population.

Genome (jee' nome) • The genes in a complete haploid set of chromosomes.

Genotype (jean' oh type) [Gr. *gen*: to produce + *typos*: impression] • An exact description of the genetic constitution of an individual, either with respect to a single trait or with respect to a larger set of traits. (Contrast with phenotype.)

Genus (jean' us) (plural: genera) [Gr. *genos*: stock, kind] • A group of related, similar species.

Geotropism • See gravitropism.

Germ cell • A reproductive cell or gamete of a multicellular organism.

Germination • The sprouting of a seed or spore.

Gestation (jes tay' shun) [L. *gestare*: to bear] • The period during which the embryo of a mammal develops within the uterus. Also known as **pregnancy**.

Gibberellin (jib er el' lin) [L. *gibberella*: hunchback (refers to shape of a reproductive structure of a fungus that produces gibberellins)] • One of a class of plant growth substances playing roles in stem elongation, seed germination, flowering of certain plants, etc. Named for the fungus *Gibberella*.

Gill • An organ for gas exchange in aquatic organisms.

Gill arch • A skeletal structure that supports gill filaments and the blood vessels that supply them.

Gizzard (giz' erd) [L. *gigeria*: cooked chicken parts] • A very muscular port of the stomach of birds that grinds up food, sometimes with the aid of fragments of stone.

Gland • An organ or group of cells that produces and secretes one or more substances.

Glans penis • Sexually sensitive tissue at the tip of the penis.

Glia (glee' uh) [Gr.: glue] • Cells, found only in the nervous system, which do not conduct action potentials.

Glomerulus (glo mare' yew lus) [L. *glomus*: ball] • Sites in the kidney where blood filtration takes place. Each glomerulus consists of a knot of capillaries served by afferent and efferent arterioles.

Glucocorticoids • Steroid hormones produced by the adrenal cortex. Secreted in response to ACTH, they inhibit glucose uptake by many tissues in addition to mediating other stress responses.

Glucagon • A hormone produced and released by cells in the islets of Langerhans of the pancreas. It stimulates the breakdown of glycogen in liver cells.

Gluconeogenesis • The biochemical synthesis of glucose from other substances, such as amino acids, lactate, and glycerol.

Glucose (glue' kose) [Gr. *gleukos*: sweet wine mash for fermentation] • The most common sugar, one of several monosaccharides with the formula $C_6H_{12}O_6$.

Glycerol (gliss' er ole) • A three-carbon alcohol with three hydroxyl groups, the linking component of phospholipids and triglycerides.

Glycogen (gly' ko jen) • A branched-chain polymer of glucose, similar to starch (which is less branched and may be of lower molecular weight). Exists mostly in liver and muscle; the principal storage carbohydrate of most animals and fungi.

Glycolysis (gly kol' li sis) [from glucose + Gr. *lysis*: loosening] • The enzymatic breakdown of glucose to pyruvic acid. One of the oldest energy-yielding machanisms in living organisms.

Glycosidic linkage • The connection in an oligosaccharide or polysaccharide chain, formed by removal of water during the linking of monosaccharides.by root pressure.

Glyoxysome (gly ox' ee soam) • An organelle found in plants, in which stored lipids are converted to carbohydrates.

Golgi apparatus (goal' jee) • A system of concentrically folded membranes found in the cytoplasm of eukaryotic cells. Plays a role in the production and release of secretory materials such as the digestive enzymes manufactured in the pancreas. First described by Camillo Golgi (1844–1926).

Gonad (go' nad) [Gr. *gone*: seed, that which produces seed] • An organ that produces sex cells in animals: either an ovary (female gonad) or testis (male gonad).

Gonadotropin • A hormone that stimulates the gonads.

Gondwana • The large southern land mass that existed from the Cambrian (540 mya) to the Jurassic (138 mya). Present-day South America, Africa, India, Australia, and Antarctica.

Gram stain • A differential stain useful in characterizing bacteria.

Granum • Within a chloroplast, a stack of thylakoids.

Gravitropism • A directed plant growth response to gravity.

Grazer • An animal that eats the vegetative tissues of herbaceous plants.

Green gland • An excretory organ of crustaceans.

Greenhouse effect • The heating of Earth's atmosphere by gases that are transparent to sunlight but opaque to radiated heat.

Gross primary production • The total energy captured by plants growing in a particular area.

Ground meristem • That part of an apical meristem that gives rise to the ground tissue system of the primary plant body.

Ground tissue system • Those parts of the plant body not included in the dermal or vascular tissue systems. Ground tissues function in storage, photosynthesis, and support.

Group transfer • The exchange of atoms between molecules.

Growth • Irreversible increase in volume (probably the most accurate definition, but at best a dangerous oversimplification).

Growth factors • A group of proteins that circulate in the blood and trigger the normal growth of cells. Each growth factor acts only on certain target cells.

Guanine (gwan'een) • A nitrogen-containing base found in DNA, RNA and GTP.

Guard cells • In plants, paired epidermal cells which surround and control the opening of a stoma (pore).

Gut • An animal's digestive tract.

Guttation • The extrusion of liquid water through openings in leaves, caused by root pressure.

Gymnosperm (jim' no sperm) [Gr. *gymnos*: naked + *sperma*: seed] • A plant, such as a pine or other conifer, whose seeds do not develop within an ovary (hence, the seeds are "naked").

Gyrus (plural: gyri) • The raised or ridged portion of the convoluted surface of the brain. (Contrast to sulcus.)

Habit • The form or pattern of growth characteristic of an organism.

Habitat • The environment in which an organism lives.

Habituation (ha bich' oo ay shun) • The simplest form of learning, in which an animal presented with a stimulus without reward or punishment eventually ceases to respond.

Hair cell • A type of mechanoreceptor in animals.

Half-life • The time required for half of a sample of a radioactive isotope to decay to its stable, nonradioactive form.

Halophyte (hal' oh fyte) [Gr. *halos*: salt + *phyton*: plant] • A plant that grows in a saline (salty) environment.

Haploid (hap' loid) [Gr. *haploeides*: single] • Having a chromosome complement consisting of just one copy of each chromosome. This is the normal "ploidy" of gametes or of asexual spores produced by meiosis or of organisms (such as the gametophyte generation of plants) that grow from such spores without fertilization.

Hardy–Weinberg equilibrium • The percentages of diploid combinations expected from a knowledge of the proportions of alleles in the population if no agents of evolution are acting on the population.

Haustorium (haw stor' ee um) [L. *haustus*: draw up] • A specialized hypha or other structure by which fungi and some parasitic plants draw food from a host plant.

Haversian systems • Units of organization in compact bone that reflect the action of intercommunicating osteoblasts.

Heat-shock proteins • Chaperone proteins expressed in cells exposed to high temperatures or other forms of environmental stress.

Helper T cells • T cells that participate in the activation of B cells and of other T cells; targets of the HIV-I virus, the agent of AIDS. (Contrast with cytotoxic T cells, suppressor T cells.)

Hematocrit (heme at o krit) [Gr. *haima*: blood + *krites*: judge] • The proportion of 100 cc of blood that consists of red blood cells.

Hemizygous (hem' ee zie' gus) [Gr. *hemi*: half + *zygotos*: joined] • In a diploid organism, having only one allele for a given trait, typically the case for X-linked genes in male mammals and Z-linked genes in female birds. (Contrast with homozygous, heterozygous.)

Hemoglobin (hee' mo glow' bin) [Gr. *haima*: blood + L. *globus*: globe] • The colored protein of vertebrate blood (and blood of some invertebrates) which transports oxygen.

Hepatic (heh pat' ik) [Gr. *hepar*: liver] • Pertaining to the liver.

Hepatic duct • The duct that conveys bile from the liver to the gallbladder.

Herbicide (ur' bis ide) • A chemical substance that kills plants.

Herbivore [L. *herba*: plant + *vorare*: to devour] • An animal which eats the tissues of plants. (Contrast with carnivore, detritivore, omnivore.)

Heritable • Able to be inherited; in biology usually refers to genetically determined traits.

Hermaphroditism (her maf' row dite' ism) [Gr. *hermaphroditos*: a person with both male and female traits] • The coexistence of both female and male sex organs in the same organism.

Hertz (abbreviated as Hz) • Cycles per second.

Hetero- [Gr.: other, different] • A prefix used in biology to mean that two or more different conditions are involved; for example, heterotroph, heterozygous.

Heterochromatin • Chromatin that retains its coiling during interphase; generally not transcribed. (Contrast with euchromatin.)

Heterocyst • A large, thick-walled cell in the filaments of certain cyanobacteria; performs nitrogen fixation.

Heterogeneous nuclear RNA (hnRNA) • The product of transcription of a eukaryotic gene, including transcripts of introns.

Heteromorphic (het' er oh more' fik) [Gr. *heteros*: different + *morphe*: form] • having a different form or appearance, as two heteromorphic life stages of a plant. (Contrast with isomorphic.)

Heterosporous (het' er os' por us) • Producing two types of spores, one of which gives rise to a female megaspore and the other to a male microspore. Heterosporous plants produce distinct female and male gametophytes. (Contrast with homosporous.)

Heterotherm • An animal that regulates its body temperature at a constant level at some times but not others, such as a hibernator.

Heterotroph (het' er oh trof) [Gr. *heteros*: different + *trophe*: food] • An organism that requires preformed organic molecules as food. (Contrast with autotroph.)

Heterozygous (het' er oh zie' gus) [Gr. *heteros*: different + *zygotos*: joined] • Of a diploid organism having different alleles of a given gene on the pair of homologues carrying that gene. (Contrast with homozygous.)

Hibernation [L. *hibernus*: winter] • The state of inactivity of some animals during winter; marked by a drop in body temperature and metabolic rate.

Highly repetitive DNA • Short DNA sequences present in millions of copies in the genome, next to each other (in tandem). In a In a reassociation experiment, denatured highly repetitive DNA reanneals very quickly.

Hippocampus • A part of the forebrain that takes part in long-term memory formation.

Histamine (hiss; tah meen) • A substance released within a damaged tissue by a type of white blood cell. Histamines are responsible for aspects of allergice reactions, including the increased vascular permeability that leads to edema (swelling).

Histology • The study of tissues.

Histone • Any one of a group of basic proteins forming the core of a nucleosome, the structural unit of a eukaryotic chromosome. (See nucleosome.)

hnRNA • See heterogeneous nuclear RNA.

Homeobox • A 180-base-pair segment of DNA found in a few genes (called **Hox genes**), perhaps regulating the expression of other genes and thus controlling large-scale developmental processes.

Homeostasis (home' ee o sta' sis) [Gr. *homos*: same + *stasis*: position] • The maintenance of a steady state, such as a constant temperature or a stable social structure, by means of physiological or behavioral feedback responses.

Homeotherm (home' ee o therm) [Gr. *homos*: same + *therme*: heat] • An animal which maintains a constant body temperature by virtue of its own heating and cooling mechanisms. (Contrast with heterotherm, poikilotherm.)

Homeotic genes (home' ee ott' ic) • Genes that determine what entire segments of an animal become. Drastic mutations in these genes cause the transformation of body segments in *Drosophila*. Homeotic genes studied in the plant *Arabidopsis* are called organ identity genes.

Homolog (home' o log') [Gr. *homos*: same + *logos*: word] • One of a pair, or larger set, of chromosomes having the same overall genetic composition and sequence. In diploid organisms, each chromosome inherited from one parent is matched by an identical (except for mutational changes) chromosome—its homolog—from the other parent.

Homology (ho mol' o jee) [Gr. *homologi(a)*: agreement] • A similarity between two structures that is due to inheritance from a common ancestor. The structures are said to be homologous. (Contrast with analogy.)

Homoplasy (home' uh play zee) [Gr. *homos*: same + *plastikos*: to mold] • The presence in several species of a trait not present in their most common ancestor. Can result from convergent evolution, reverse evolution, or parallel evolution.

Homosporous • Producing a single type of spore that gives rise to a single type of gametophyte, bearing both female and male reproductive organs. (Contrast with heterosporous.)

Homozygous (home' o zie' gus) [Gr. *homos*: same + *zygotos*: joined] • Of a diploid organism having identical alleles of a given gene on both homologous chromosomes. An organism may be a "homozygote" with respect to one gene and, at the same time, a "heterozygote" with respect to another. (Contrast with heterozygous.)

Hormone (hore' mone) [Gr. *hormon*: excite, stimulate] • A substance produced in one part of a multicellular organism and transported to another part where it exerts its specific effect on the physiology or biochemistry of the target cells.

Host • An organism that harbors a parasite and provides it with nourishment.

Host–parasite interaction • The dynamic interaction between populations of a host and the parasites that attack it.

Hox genes • See homeobox.

Humoral immune system • The part of the immune system mediated by B cells; it is mediated by circulating antibodies and is active against extracellular bacterial and viral infections.

Humus (hew' muss) • The partly decomposed remains of plants and animals on the surface of a soil. Its characteristics depend primarily upon climate and the species of plants growing on the site.

Hyaluronidase (hill yew ron' uh dase) • An enzyme that digests proteoglycans. Found in sperm cells, it helps digest the coatings surrounding an egg so the sperm can penetrate the egg cell membrane.

Hybrid (high' brid) [L. *hybrida*: mongrel] • The offspring of genetically dissimilar parents. In molecular biology, a double helix formed of nucleic acids from different sources.

Hybridoma • A cell produced by the fusion of an antibody-producing cell with a myeloma cell; it produces monoclonal antibodies.

Hybrid zone • A narrow zone where two populations interbreed, producing hybrid individuals.

Hydrocarbon • A compound containing only carbon and hydrogen atoms.

Hydrogen bond • A chemical bond which arises from the attraction between the slight positive charge on a hydrogen atom and a slight negative charge on a nearby fluorine, oxygen, or nitrogen atom. Weak bonds, but found in great quantities in proteins, nucleic acids, and other biological macromolecules.

Hydrological cycle • The sum total of movement of water from the oceans to the atmosphere, to the soil, and back to the oceans. Some water is cycled many times within compartments of the system before completing one full circuit.

Hydrolyze (hi' dro lize) [Gr. *hydro*: water + *lysis*: cleavage] • To break a chemical bond, as in a peptide linkage, with the insertion of the components of water, –H and –OH, at the cleaved ends of a chain. The digestion of proteins is a hydrolysis.

Hydrophilic [Gr. *hydro*: water + *philia*: love] • Having an affinity for water. (Contrast with hydrophobic.)

Hydrophobic [Gr. *hydro*: water + *phobia*: fear] • Molecules and amino acid side chains, which are mainly hydrocarbons (compounds of C and H with no charged groups or polar groups), have a lower energy when they are clustered together than when they are distributed through an aqueous solution. Because of their attraction for one another and their reluctance to mix with water they are called "hydrophobic." Oil is a hydrophobic substance; phenylalanine is a hydrophobic animo acid in a protein. (Contrast with hydrophilic.)

Hydrostatic skeleton • The incompressible internal liquids of some animals that transfer forces from one part of the body to another when acted upon by the surrounding muscles.

Hydroxyl group • The —OH group, characteristic of alcohols.

Hyperpolarization • A change in the resting potential of a membrane so the inside of a cell becomes more electronegative. (Contrast with depolarization.)

Hypersensitive response • A defensive response of plants to microbial infection; it results in a "dead spot."

Hypertension • High blood pressure.

Hypertonic [Gk. *hyper*: above, over] • Having a greater solute concentration. Said of one solution in comparing it to another. (Contrast with hypotonic, isotonic.)

Hypha (high' fuh) (plural: hyphae) [Gr. *hyphe*: web] • In the fungi, any single filament. May be multinucleate (zygomycetes, ascomycetes) or multicellular (basidiomycetes).

Hypocotyl [Gk. *hypo*: beneath, under + *kotyledon*: hollow space] • That part of the embryonic or seedling plant shoot that is below the cotyledons.

Hypothalamus • The part of the brain lying below the thalamus; it coordinates water balance, reproduction, temperature regulation, and metabolism.

Hypothesis • A tentative answer to a question, from which testable predictions can be generated. (Contrast with theory.)

Hypothetico-deductive method • A method of science in which hypotheses are erected, predictions are made from them, and experiments and observations are performed to test the predictions.

Hypotonic [Gk. *hypo*: beneath, under] • Having a greater solute concentration. Said of one solution in comparing it to another. (Contrast with hypotonic, isotonic.)

Imaginal disc • In insect larvae, groups of cells that develop into specific adult organs.

Immune system [L. *immunis*: exempt] • A system in mammals that recognizes and eliminates or neutralizes either foreign substances or self substances that have been altered to appear foreign.

Immunization • The deliberate introduction of antigen to bring about an immune response.

Immunoglobulins • A class of proteins, with a characteristic structure, active as receptors and effectors in the immune system.

Immunological memory • Certain clones of immune system cells made to respond to an antigen persist. This leads to a more rapid and massive response of the immune system to any subsequenct exposure to that antigen.

Immunological tolerance • A mechanism by which an animal does not mount an immune response to the antigenic determinants of its own macromolecules.

Imprinting • (1) In genetics, the differential modification of a gene depending on whether it is present in a male or a female. (2) In animal behavior, a rapid form of learning in which an animal comes to make a particular response, which is maintained for life, to some object or other organism.

Inclusive fitness • The sum of an individual's own fitness (the effect of producing its own offspring: the individual selection component) plus its influence on fitness in relatives other than direct descendants (the kin selection component).

Incomplete dominance • Condition in which the heterozygous phenotype is intermediate between the two homozygous phenotypes.

Incomplete metamorphosis • Insect development in which changes between instars are gradual.

Incus (in' kus) [L. *incus*: anvil] • The middle of the three bones that conduct movements of the eardrum to the oval window of the inner ear. (See malleus, stapes.)

Independent assortment • The random separation during meiosis of nonhomologous chromosomes and of genes carried on nonhomologous chromosomes.

Individual fitness • That component of inclusive fitness that results from an organism producing its own offspring. (Contrast with kin selection component.)

Indoleacetic acid • See auxin.

Inducer • (1) In enzyme systems, a small molecule which, when added to a growth medium, causes a large increase in the level of some enzyme. (2) In embryology, a substance that causes a group of target cells to differentiate in a particular way.

Inducible enzyme • An enzyme that is present in much larger amounts when a particular compound (the inducer) has been added to the system. (Contrast with constitutive enzyme.)

Inflammation • A nonspecific defense against pathogens; characterized by redness, swelling, pain, and increased temperature.

Inflorescence • A structure composed of several flowers.

Inhibitor • A substance which binds to the surface of an enzyme and interferes with its action on its substrates.

Inhibitory postsynaptic potential • A change in the resting potential of a postsynaptic membrane in the hyperpolarizing (negative) direction.

Initiation complex • Combination of a ribosomal light subunit, an mRNA molecule, and the tRNA charged with the first amino acid coded for by the mRNA; formed at the onset of translation.

Initiation factors • Proteins that assist in forming the translation initiation complex at the ribosome.

Inositol triphosphate (IP3) • An intracellular second messenger derived from membrane phospholipids.

Instar (in' star) [L.: image, form] • An immature stage of an insect between molts.

Insulin (in' su lin) [L. *insula*: island] • A hormone, synthesized in islet cells of the pancreas, that promotes the conversion of glucose to the storage material, glycogen.

Integrase • An enzyme that integrates retroviral cDNA into the genome of the host cell.

Integrated pest management • A method of control of pests in which natural predators and parasites are used in conjunction with sparing use of chemical methods to achieve control of a pest without causing serious adverse environmental side effects.

Integument [L. *integumentum*: covering] • A protective surface structure. In gymnosperms and angiosperms, a layer of tissue around the ovule which will become the seed coat. Gymnosperm ovules have one integument, angiosperm ovules two.

Intercalary meristem • A meristematic region in plants which occurs not apically, but between two regions of mature tissue. Intercalary meristems occur in the nodes of grass stems, for example.

Intercostal muscles • Muscles between the ribs that can augment breathing movements by elevating and suppressing the rib cage.

Interferon • A glycoprotein produced by virus-infected animal cells; increases the resistance of neighboring cells to the virus.

Interkinesis • The phase between the first and second meiotic divisions.

Interleukins • Regulatory proteins, produced by macrophages and lymphocytes, that act upon other lymphocytes and direct their development.

Intermediate filaments • Fibrous proteins that stabilize cell structure and resist tension.

Internode • Section between two nodes of a plant stem.

Interphase • The period between successive nuclear divisions during which the chromosomes are diffuse and the nuclear envelope is intact. It is during this period that the cell is most active in transcribing and translating genetic information.

Interspecific competition • Competition between members of two or more species.

Intertropical convergence zone • The tropical region where the air rises most strongly; moves north and south with the passage of the sun overhead.

Intraspecific competition • Competition among members of a single species.

Intrinsic protein • A membrane protein that is embedded in the phospholipid bilayer of the membrane. (Contrast with extrinsic protein.)

Intrinsic rate of increase • The rate at which a population can grow when its density is low and environmental conditions are highly favorable.

Intron • A portion of a DNA molecule that, because of RNA splicing, is not involved in coding for part of a polypeptide molecule. (Contrast with exon.)

Invagination • An infolding.

Inversion (genetic) • A rare mutational event that leads to the reversal of the order of genes within a segment of a chromosome, as if that segment had been removed from the chromosome, turned 180°, and then reattached.

Invertebrate • Any animal that is not a vertebrate, that is, whose nerve cord is not enclosed in a backbone of bony segments.

In vitro [L.: in glass] • In a test tube, rather than in a living organism. (Contrast with in vivo.)

In vivo [L.: in the living state] • In a living organism. Many processes that occur in vivo can be reproduced in vitro with the right selection of cellular components. (Contrast with in vitro.)

Ion (eye' on) [Gr.: wanderer] • An atom or group of atoms with electrons added or removed, giving it a negative or positive electrical charge.

Ion channel • A membrane protein that can let ions pass across the membrane. The channel can be ion-selective, and it can be voltage-gated or ligand-gated.

Ionic bond • A chemical bond which arises from the electrostatic attraction between positively and negatively charged ions. Usually a strong bond.

Iris (eye' ris) [Gr. *iris*: rainbow] • The round, pigmented membrane that surrounds the pupil of the eye and adjusts its aperture to regulate the amount of light entering the eye.

Irruption • A rapid increase in the density of a population. Often followed by massive emigration.

Islets of Langerhans • Clusters of hormone-producing cells in the pancreas.

Iso- [Gr.: equal] • Prefix used to denote two separate but similar or identical states of a characteristic. (See isomers, isomorphic, isotope.)

Isolating mechanism • Geographical, physiological, ecological, or behavioral mechanisms that lead to a reduction in the frequency of hybrid matings.

Isomers • Molecules consisting of the same numbers and kinds of atoms, but differing in the way in which the atoms are combined.

Isomorphic (eye' so more' fik) [Gr. *isos*: equal + *morphe*: form] • having the same form or appearance, as two isomorphic life stages. (Contrast with heteromorphic.)

Isotonic • Having the same solute concentration; said of two solutions. (Contrast with hypertonic, hypotonic.)

Isotope (eye' so tope) [Gr. *isos*: equal + *topos*: place] • Two isotopes of the same chemical element have the same number of protons in their nuclei, but differ in the number of neutrons.

Jasmonates • Plant hormones that trigger defenses against pathogens and herbivores.

Jejunum (jih jew' num) • The middle division of the small intestine, where most absorption of nutrients occurs. (See duodenum, ileum.)

Joule (jool, or jowl) • A unit of energy, equal to 0.24 calories.

Juvenile hormone • In insects, a hormone maintaining larval growth and preventing maturation or pupation.

Karyotype • The number, forms, and types of chromosomes in a cell.

Kelvin temperature scale • See absolute temperature scale.

Keratin (ker' a tin) [Gr. *keras*: horn] • A protein which contains sulfur and is part of such hard tissues as horn, nail, and the outermost cells of the skin.

Ketone (key' tone) • A compound with a C==O group attached to two other groups, neither of which is an H atom. Many sugars are ketones. (Contrast with aldehyde.)

Keystone species • A species that exerts a major influence on the composition and dynamics of the community in which it lives.

Kidneys • A pair of excretory organs in vertebrates.

Kin selection • The component of inclusive fitness resulting from helping the survival of relatives containing the same alleles by descent from a common ancestor.

Kinase (kye' nase) • An enzyme that transfers a phosphate group from ATP to another molecule. Protein kinases transfer phosphate from ATP to specific proteins, playing important roles in cell regulation.

Kinesis (ki nee' sis) [Gr.: movement] • Orientation behavior in which the organism does not move in a particular direction with reference to a stimulus but instead simply moves at an increasing or decreasing rate until it ends up farther from the object or closer to it. (Contrast with taxis.)

Kinetochore (kin net' oh core) [Gr. *kinetos*: moving + *khorein*: to move] • Specialized structure on a centromere to which microtubules attach.

Koch's posulates • Four rules for establishing that a particular microorganism causes a particular disease.

Krebs cycle • See citric acid cycle.

Lactic acid • The end product of fermentation in vertebrate muscle and some microorganisms.

Lagging strand • In DNA replication, the daughter strand that is synthesized discontinuously.

Lamella • Layer.

Larynx (lar' inks) • A structure between the pharynx and the trachea that includes the vocal cords.

Larva (plural: larvae) [L.: ghost, early stage] • An immature stage of any invertebrate animal that differs dramatically in appearance from the adult.

Lateral • Pertaining to the side.

Lateral gene transfer • The movement of genes from one prokaryotic species to another.

Lateral meristems • The vascular cambium and cork cambium, which give rise to secondary tissue in plants.

Laterization (lat' ur iz ay shun) • The formation of a nutrient-poor soil that is rich in insoluble iron and aluminum compounds.

Law of independent assortment • The random separation during meiosis of nonhomologous chromosomes and of genes carried on nonhomologous chromosomes. Mendel's second law.

Law of segregation • Alleles segregate from one another during gamete formation, Mendel's first law.

Leader sequence • A sequence of amino acids at the N-terminal end of a newly synthesized protein, determining where the protein will be placed in the cell.

Leading strand • In DNA replication, the daughter strand that is synthesized continuously.

Lenticel • Spongy region in a plant's periderm, allowing gas exchange.

Leukocyte (loo' ko sight) [Gr. *leukos*: clear + *kutos*: hollow vessel] • A white blood cell.

Lichen (lie' kun) [Gr. *leikhen*: licker] • An organism resulting from the symbiotic association of a true fungus and either a cyanobacterium or a unicellular alga.

Life cycle • The entire span of the life of an organism from the moment of fertilization (or asexual generation) to the time it reproduces in turn.

Life history • The stages an individual goes through during its life.

Life table • A table showing, for a group of equal-aged individuals, the proportion still alive at different times in the future and the number of offspring they produce during each time interval.

Ligament • A band of connective tissue linking two bones in a joint.

Ligand (lig' and) • A molecule that binds to a receptor site of another molecule.

Lignin • The principal noncarbohydrate component of wood, a polymer that binds together cellulose fibrils in some plant cell walls.

Limbic system • A group of primitive vertebrate forebrain nuclei that form a network and are involved in emotions, drives, instinctive behaviors, learning, and memory.

Limiting resource • The required resource whose supply most strongly influences the size of a population.

Linkage • Association between genetic markers on the same chromosome such that they do not show random assortment and seldom recombine; the closer the markers, the lower the frequency of recombination.

Lipase (lip' ase; lye' pase) • An enzyme that digests fats.

Lipids (lip' ids) [Gr. *lipos*: fat] • Substances in a cell which are easily extracted by organic solvents; fats, oils, waxes, steroids, and other large organic molecules, including those which, with proteins, make up the cell membranes. (See phospholipids.)

Litter • The partly decomposed remains of plants on the surface and in the upper layers of the soil.

Littoral zone • The coastal zone from the upper limits of tidal action down to the depths where the water is thoroughly stirred by wave action.

Liver • A large digestive gland. In vertebrates, it secretes bile and is involved in the formation of blood.

Lobes • Regions of the human cerebral hemispheres; includes the temporal, frontal, parietal, and occipital lobes.

Locus • In genetics, a specific location on a chromosome. May be considered to be synonymous with "gene."

Logistic growth • Growth, especially in the size of an organism or in the number of organisms that constitute a population, which slows steadily as the entity approaches its maximum size. (Contrast with exponential growth.)

Loop of Henle (hen' lee) • Long, hairpin loop of the mammalian renal tubule that runs from the cortex down into the medulla, and back to the cortex. Creates a concentration gradient in the interstitial fluids in the medulla.

Lophophore • A U-shaped fold of the body wall with hollow, ciliated tentacles that encircles the mouth of animals in several different phyla. Used for filtering prey from the surrounding water.

Lordosis (lor doe' sis) [Gk. *lordosis*: curving forward] • A posture assumed by females of some mammalian species (especially rodents) to signal sexual receptivity.

Lumen (loo' men) [L.: light] • The cavity inside any tubular part of an organ, such as a piece of gut or a kidney tubule.

Lungs • A pair of saclike chambers within the bodies of some animals, functioning in gas exchange.

Luteinizing hormone • A gonadotropin produced by the anterior pituitary. It stimulates the gonads to produce sex hormones.

Lymph [L. *lympha*: water] • A clear, watery fluid that is formed as a filtrate of blood; it contains white blood cells; it collects in a series of special vessels and is returned to the bloodstream.

Lymph nodes • Specialized tissue regions that act as filters for cells, bacteria and foreign matter.

Lymphocyte • A major class of white blood cells. Includes T cells, B cells, and other cell types important in the immune response.

Lysis (lie' sis) [Gr.: a loosening] • Bursting of a cell.

Lysogenic • The condition of a bacterium that carries the genome of a virus in a relatively stable form. (Contrast with lytic.)

Lysosome (lie' so soam) [Gr. *lysis*: a loosening + *soma*: body] • A membrane-bounded inclusion found in eukaryotic cells (other than plants). Lysosomes contain a mixture of enzymes that can digest most of the macromolecules found in the rest of the cell.

Lysozyme (lie' so zyme) • An enzyme in saliva, tears, and nasal secretions that attacks bacterial cell walls, as one of the body's nonspecific defense mechanisms.

Lytic • Condition in which a bacterium lyses shortly after infection by a virus; the viral genome does not become stabilized within the bacterial cell. (Contrast with lysogenic.)

Macro- (mack' roh) [Gr. *makros*: large, long] • A prefix commonly used to denote something large. (Contrast with micro-.)

Macroevolution • Evolutionary changes occurring over long time spans and usually involving changes in many traits. (Contrast with microevolution.)

Macromolecule • A giant polymeric molecule. The macromolecules are proteins, polysaccharides, and nucleic acids.

Macronutrient • A mineral element required by plant tissues in concentrations of at least 1 milligram per gram of their dry matter.

Macrophage (mac' roh faj) • A type of white blood cell that endocytoses bacteria and other cells.

Major histocompatibility complex (MHC) • A complex of linked genes, with multiple alleles, that control a number of immunological phenomena; it is important in graft rejection.

Malignant tumor • A tumor whose cells can invade surrounding tissues and spread to other organs.

Malleus (mal' ee us) [L. *malleus*: hammer] • The first of the three bones that conduct movements of the eardrum to the oval window of the inner ear. (See incus, stapes.)

Malpighian tubule (mal pee' gy un) • A type of protonephridium found in insects.

Mammal [L. *mamma*: breast, teat] • Any animal of the class Mammalia, characterized by the production of milk by the female mammary glands and the possession of hair for body covering.

Mantle • A sheet of specialized tissues that covers most of the viscera of mollusks; provides protection to internal organs and secretes the shell.

Map unit • In eukaryotic genetics, one map unit corresponds to a recombinant frequency of 0.01.

Mapping • In genetics, determining the order of genes on a chromosome and the distances between them.

Marine [L. *mare*: sea, ocean] • Pertaining to or living in the ocean. (Contrast with aquatic, terrestrial.)

Marsupial (mar soo' pee al) • A mammal belonging to the subclass Metatheria, such as opossums and kangaroos. Most have a pouch (marsupium) that contains the milk glands and serves as a receptacle for the young.

Mass extinctions • Geological periods during which rates of extinction were much higher than during intervening times.

Mass number • The sum of the number of protons and neutrons in an atom's nucleus.

Mast cells • Typically found in connective tissue, mast cells can be provoked by antigens or inflammation to release histamine.

Maternal effect genes • These genes code for morphogens that determine the polarity of the egg and larva in the fruit fly, *Drosophila melanogaster*.

Maternal inheritance (cytoplasmic inheritance) • Inheritance in which the phenotype of the offspring depends on factors, such as mitochondria or chloroplasts, that are inherited from the female parent through the cytoplasm of the female gamete.

Maturation • The automatic development of a pattern of behavior, which becomes increasingly complex or precise as the animal matures. Unlike learning, the development does not require experience to occur.

Mechanoreceptor • A cell that is sensitive to physical movement and generates action potentials in response.

Medulla (meh dull' luh) [L.: narrow] • (1) The inner, core region of an organ, as in the adrenal medulla (adrenal gland) or the renal medulla (kidneys). (2) The portion of the brain stem that connects to the spinal cord.

Mega- [Gr. *megas*: large, great] • A prefix often used to denote something large. (Contrast with micro-.)

Megaspore [Gr. *megas*: large + *spora*:seed] • In plants, a haploid spore that produces a female gametophyte.

Meiosis (my oh' sis) [Gr.: diminution] • Division of a diploid nucleus to produce four haploid daughter cells. The process consists of two successive nuclear divisions with only one cycle of chromosome replication.

Membrane potential • The difference in electrical charge between the inside and the outside of a cell, caused by a difference in the distribution of ions.

Memory cells • Long-lived lymphocytes produced by exposure to antigen. They persist in the body and are able to mount a rapid response to subsequent exposures to the antigen.

Mendelian population • A local population of individuals belonging to the same species and exchanging genes with one another.

Menopause • The time in a human female's life when the ovarian and menstrual cycles cease.

Menstrual cycle • The monthly sloughing off of the uterine lining if fertilization does not occur in the female. Occurs between puberty and menopause.

Meristem [Gr. *meristos*: divided] • Plant tissue made up of actively dividing cells.

Mesenchyme (mez' en kyme) [Gr. *mesos*: middle + *enchyma*: infusion] • Embryonic or unspecialized cells derived from the mesoderm.

Meso- (mez' oh) [Gr.: middle] • A prefix often used to designate a structure located in the middle, or a stage that appears at some intermediate time. For example, mesoderm, Mesozoic.

Mesoderm [Gr. *mesos*: middle + *derma*: skin] • The middle of the three embryonic tissue layers first delineated during gastrulation. Gives rise to skeleton, circulatory system, muscles, excretory system, and most of the reproductive system.

Mesophyll (mez' a fill) [Gr. *mesos*: middle + *phyllon*: leaf] • Chloroplast-containing, photosynthetic cells in the interior of leaves.

Mesosome (mez' o soam') [Gr. *mesos*: middle + *soma*: body] • A localized infolding of the plasma membrane of a bacterium.

Messenger RNA (mRNA) • A transcript of one of the strands of DNA, it carries information (as a sequence of codons) for the synthesis of one or more proteins.

Meta- [Gr.: between, along with, beyond] • A prefix used in biology to denote a change or a shift to a new form or level; for example, as used in metamorphosis.

Metabolic compensation • Changes in biochemical properties of an organism that render it less sensitive to temperature changes.

Metabolic pathway • A series of enzyme-catalyzed reactions so arranged that the product of one reaction is the substrate of the next.

Metabolism (meh tab' a lizm) [Gr. *metabole*: to change] • The sum total of the chemical reactions that occur in an organism, or some subset of that total (as in "respiratory metabolism").

Metamorphosis (met' a mor' fo sis) [Gr. *meta*: between + *morphe*: form, shape] • A radical change occurring between one developmental stage and another, as for example from a tadpole to a frog or an insect larva to the adult.

Metaphase (met' a phase) [Gr. *meta*: between] • The stage in nuclear division at which the centromeres of the highly supercoiled chromosomes are all lying on a plane (the metaphase plane or plate) perpendicular to a line connecting the division poles.

Metapopulation • A population divided into subpopulations, among which there are occasional exchanges of individuals.

Metastasis (meh tass' tuh sis) • The spread of cancer cells from their original site to other parts of the body.

Methanogen • Any member of a group of Archaebacteria that release methane as a metabolic product. This group is considered to be an extremely ancient one.

MHC • See major histocompatibility complex.

Micro- (mike' roh) [Gr. *mikros*: small] • A prefix often used to denote something small. (Contrast with macro-, mega-.)

Microbiology [Gr. *mikros*: small + *bios*: life + *logos*: discourse] • The scientific study of microscopic organisms, particularly bacteria, unicellular algae, protists, and viruses.

Microevolution • The small evolutionary changes typically occurring over short time spans; generally involving a small number of traits and minor genetic changes. (Contrast with macroevolution.)

Microfilament • Minute fibrous structure generally composed of actin found in the cytoplasm of eukaryotic cells. They play a role in the motion of cells.

Micronutrient • A mineral element required by plant tissues in concentrations of less than 100 micrograms per gram of their dry matter.

Micropyle (mike' roh pile) [Gr. *mikros*: small + *pyle*: gate] • Opening in the integument(s) of a seed plant ovule through which pollen grows to reach the female gametophyte within.

Microspores [Gr. *mikros*: small + *spora*: seed] • In plants, a haploid spore that produces a male gametophyte.

Microtubules • Minute tubular structures found in centrioles, spindle apparatus, cilia, flagella, and other places in the cytoplasm of eukaryotic cells. These tubules play roles in the motion and maintenance of shape of eukaryotic cells.

Microvilli (singular: microvillus) • The projections of epithelial cells, such as the cells lining the small intestine, that increase their surface area.

Middle lamella • A layer of derivative polysaccharides that separates plant cells; a common middle lamella lies outside the primary walls of the two cells.

Migration • The regular, seasonal movements of animals between breeding and nonbreeding ranges.

Mimicry (mim' ik ree) • The resemblance of one kind of organism to another, or to some inanimate object; serves the function of making the organism difficult to find, of discouraging potential enemies or of attracting potential prey. (See Batesian mimicry and Müllerian mimicry.)

Mineral • An inorganic substance other than water.

Mineralocorticoid • A hormone produced by the adrenal cortex that influences mineral ion balance; aldosterone.

Mismatch repair • When a single base in DNA is changed into a different base, or the wrong base inserted during DNA replication, there is a mismatch in base pairing with the base on the opposite strand. A repair system removes the incorrect base and inserts the proper one for pairing with the opposite strand.

Missense mutation • A nonsynonymous mutation, or one that changes a codon for one amino acid to a codon for a different amino acid. (Contrast with frame-shift mutation, nonsense mutation, synonymous mutation.)

Mitochondrial matrix • The fluid interior of the mitochondrion, enclosed by the inner mitochondrial membrane.

Mitochondrion (my' toe kon' dree un) (plural: mitochondria) [Gr. mitos: thread + chondros: cartilage, or grain] • An organelle that occurs in eukaryotic cells and contains the enzymes of the ctric acid cycle, the respiratory chain, and oxidative phosphorylation. A mitochondrion is bounded by a double membrane.

Mitosis (my toe' sis) [Gr. mitos: thread] • Nuclear division in eukaryotes leading to the formation of two daughter nuclei each with a chromosome complement identical to that of the original nucleus.

Mitotic center • Cellular region that organizes the microtubules for mitosis. In animals a centrosome serves as the mitotic center.

Moderately repetitive DNA • DNA sequences that appear hundreds to thousands of times in the genome. They include the DNA sequences coding for rRNAs and tRNAs, as well as the DNA at telomeres.

Modular organism • An organism which grows by producing additional units of body construction (modules) that are very similar to the units of which it is already composed.

Mole • A quantity of a compound whose weight in grams is numerically equal to its molecular weight expressed in atomic mass units. Avogadro's number of molecules: 6.023×10^{23} molecules.

Molecular clock • The theory that macromolecules diverge from one another over evolutionary time at a constant rate, and that discovering this rate gives insight into the phylogenetic relationships of organisms.

Molecular weight • The sum of the atomic weights of the atoms in a molecule.

Molecule • A particle made up of two or more atoms joined by covalent bonds or ionic attractions.

Molting • The process of shedding part or all of an outer covering, as the shedding of feathers by birds or of the entire exoskeleton by arthropods.

Mono- [Gr. monos: one] • Prefix denoting a single entity. (Contrast with poly.)

Monoclonal antibody • Antibody produced in the laboratory from a clone of hybridoma cells, each of which produces the same specific antibody.

Monocot (short for monocotyledon) [Gr. monos: one + kotyledon: a cup-shaped hollow] • Any member of the angiosperm class Monocotyledones, plants in which the embryo produces but a single cotyledon (seed leaf). Leaves of most monocots have their major veins arranged parallel to each other.

Monocytes • White blood cells that produce macrophages.

Monoecious (mo nee' shus) [Gr.: one house] • Organisms in which both sexes are "housed" in a single individual, which produces both eggs and sperm. (In some plants, these are found in different flowers within the same plant.) Examples: corn, peas, earthworms, hydras. (Contrast with dioecious, perfect flower.)

Monohybrid cross • A mating in which the parents differ with respect to the alleles of only one locus of interest.

Monomer [Gr.: one unit] • A small molecule, two or more of which can be combined to form oligomers (consisting of a few monomers) or polymers (consisting of many monomers).

Monophyletic (mon' oh fih leht' ik) [Gk. monos: single + phylon: tribe] • Being descended from a single ancestral stock.

Monosaccharide • A simple sugar. Oligosaccharides and polysaccharides are made up of monosaccharides.

Monosynaptic reflex • A neural reflex that begins in a sensory neuron and makes a single synapse before activating a motor neuron.

Morphogens • Diffusible substances whose concentration gradients determine patterns of development in animals and plants.

Morphogenesis (more' fo jen' e sis) [Gr. morphe: form + genesis: origin] • The development of form. Morphogenesis is the overall consequence of determination, differentiation, and growth.

Morphology (more fol' o jee) [Gr. morphe: form + logos: discourse] • The scientific study of organic form, including both its development and function.

Mosaic development • Pattern of animal embryonic development in which each blastomere contributes a specific part of the adult body. (Contrast with regulative development.)

Motor end plate • The modified area on a muscle cell membrane where a synapse is formed with a motor neuron.

Motor neuron • A neuron carrying information from the central nervous system to an effector such as a muscle fiber.

Motor unit • A motor neuron and the set of muscle fibers it controls.

mRNA • (See messenger RNA.)

Mucosa (mew koh' sah) • An epithelial membrane containing cells that secrete mucus. The inner cell layers of the digestive and respiratory tracts.

Müllerian mimicry • The resemblance of two or more unpleasant or dangerous kinds of organisms to each other.

Multicellular [L. multus: much + cella: chamber] • Consisting of more than one cell, as for example a multicellular organism. (Contrast with unicellular.)

Muscle • Contractile tissue containing actin and myosin organized into polymeric chains called microfilaments. In vertebrates, the tissues are either cardiac muscle, smooth muscle, or striated (skeletal) muscle.

Muscle fiber • A single muscle cell. In the case of striated muscle, a syncitial, multinucleate cell.

Muscle spindle • Modified muscle fibers encased in a connective sheat and functioning as stretch receptors.

Mutagen (mute' ah jen) [L. mutare: change + Gr. genesis: source] • Any agent (e.g., chemicals, radiation) that increases the mutation rate.

Mutation • An inherited change along a very narrow portion of the nucleic acid sequence.

Mutation pressure • Evolution (change in gene proportions) by different mutation rates alone.

Mutualism • The type of symbiosis, such as that exhibited by fungi and algae or cyanobacteria in forming lichens, in which both species profit from the association.

Mycelium (my seel' ee yum) [Gr. mykes: fungus] • In the fungi, a mass of hyphae.

Mycorrhiza (my' ka rye' za) [Gr. mykes: fungus + rhiza: root] • An association of the root of a plant with the mycelium of a fungus.

Myelin (my' a lin) • A material forming a sheath around some axons. It is formed by Schwann cells that wrap themselves about the axon. It serves to insulate the axon electrically and to increase the rate of transmission of a nervous impulse.

Myofibril (my' oh fy' bril) [Gr. mys: muscle + L. fibrilla: small fiber] • A polymeric unit of actin or myosin in a muscle.

Myogenic (my oh jen' ik) [Gr. mys: muscle + genesis: source] • Originating in muscle.

Myoglobin (my' oh globe' in) [Gr. mys: muscle + L. globus: sphere] • An oxygen-binding molecule found in muscle. Consists of a heme unit and a single globiin chain, and carrys less oxygen than hemoglobin.

Myosin [Gr. mys: muscle] • One of the two major proteins of muscle, it makes up the thick filaments. (See actin.)

NAD (nicotinamide adenine dinucleotide) • A compound found in all living cells, existing in two interconvertible forms: the oxidizing agent NAD^+ and the reducing agent NADH.

NADP (nicotinamide adenine dinucleotide phosphate) • Like NAD, but possessing another phosphate group; plays similar roles but is used by different enzymes.

Natural selection • The differential contribution of offspring to the next generation by various genetic types belonging to the same population. The mechanism of evolution proposed by Charles Darwin.

Necrosis (nec roh' sis) • Tissue damage resulting from cell death.

Negative control • The situation in which a regulatory macromolecule (generally a repressor) functions to turn off transcription. In the absence of a regulatory macromolecule, the structural genes are turned on.

Nekton [Gr. *nekhein*: to swim] • Animals, such as fish, that can swim against currents of water. (Contrast with plankton.)

Nematocyst (ne mat' o sist) [Gr. *nema*: thread + *kystis*: cell] • An elaborate, thread-like structure produced by cells of jellyfish and other cnidarians, used chiefly to paralyze and capture prey.

Nephridium (nef rid' ee um) [Gr. *nephros*: kidney] • An organ which is involved in excretion, and often in water balance, involving a tube that opens to the exterior at one end.

Nephron (nef' ron) [Gr. *nephros*: kidney] • The basic component of the kidney, which is made up of numerous nephrons. Its form varies in detail, but it always has at one end a device for receiving a filtrate of blood, and then a tubule that absorbs selected parts of the filtrate back into the bloodstream.

Nephrostome (nef' ro stome) [Gr. *nephros*: kidney + *stoma*: opening] An opening in a nephridium through which body fluids can enter.

Nerve • A structure consisting of many neuronal axons and connective tissue.

Net primary production • Total photosynthesis minus respiration by plants.

Neural plate • A thickened strip of ectoderm along the dorsal side of the early vertebrate embryo; gives rise to the central nervous system.

Neural tube • An early stage in the development of the vertebrate nervous system consisting of a hollow tube created by two opposing folds of the dorsal ectoderm along the anterior–posterior body axis.

Neuromuscular junction • The region where a motor neuron contacts a muscle fiber, creating a synapse.

Neuron (noor' on) [Gr. *neuron*: nerve, sinew] • A cell derived from embryonic ectoderm and characterized by a membrane potential that can change in response to stimuli, generating action potentials. Action potentials are generated along an extension of the cell (the axon), which makes junctions (synapses) with other neurons, muscle cells, or gland cells.

Neurotransmitter • A substance, produced in and released by one neuron, that diffuses across a synapse and excites or inhibits the postsynaptic neuron.

Neurula (nure' you la) [Gr. *neuron*: nerve] • Embryonic stage during formation of the dorsal nerve cord by two ectodermal ridges.

Neutral allele • An allele that does not alter the functioning of the proteins for which it codes.

Neutral theory • A view of molecular evolution that postulates that most mutations do not affect the amino acid being coded for, and that such mutations accumulate in a population at rates driven by genetic drift and mutation rates.

Neutron (new' tron) [E.: neutral] • One of the three most fundamental particles of matter, with mass approximately 1 amu and no electrical charge.

Nicotinamide adenine dinucleotide • (See NAD.)

Nicotinamide adenine dinucleotide phosphate • (See NADP.)

Nitrification • The oxidation of ammonia to nitrite and nitrate ions, performed by certain soil bacteria.

Nitrogenase • In nitrogen-fixing organisms, an enzyme complex that mediates the stepwise reduction of atmospheric N_2 to ammonia.

Nitrogen fixation • Conversion of nitrogen gas to ammonia, which makes nitrogen available to living things. Carried out by certain prokaryotes, some of them free-living and others living within plant roots.

Node [L. *nodus*: knob, knot] • In plants, a (sometimes enlarged) point on a stem where a leaf is or was attached.

Node of Ranvier • A gap in the myelin sheath covering an axons, where the axonal membrane can fire action potentials.

Noncompetitive inhibitor • An inhibitor that binds the enzyme at a site other than the active site. (Contrast with competitive inhibitor.)

Nondisjunction • Failure of sister chromatids to separate in meiosis II or mitosis, or failure of homologous chromosomes to separate in meiosis I. Results in aneuploidy.

Nonpolar molecule • A molecule whose electric charge is evenly balanced from one end of the molecule to the other.

Nonsense (chain-terminating) mutation • Mutations that change a codon for an amino acid to one of the codons (UAG, UAA, or UGA) that signal termination of translation. The resulting gene product is a shortened polypeptide that begins normally at the amino-terminal end and ends at the position of the altered codon. (Contrast with frame-shift mutation, missense mutation, synonymous mutation.)

Nonspecific defenses • Immunologic responses directed against most or all pathogens, generally without reference to the pathogens' antigens. These defenses include the skin, normal flora, lysozyme, the acidic stomach, interferon, and the inflammatory response.

Nonsynonymous mutation • A nucleotide substitution that that changes the amino acid specified (i.e., AGC → AGA, or serine → arginine). (Compare with frame-shift mutation, missense mutation, nonsense mutation.)

Nonsynonymous substitution • The situation when a nonsynonymous mutation becomes widespread in a population. Typically influenced by natural selection. (Contrast with synonymous substitution.)

Nontracheophytes • Those plants lacking well-developed vascular tissue; the liverworts, hornworts, and mosses. (Contrast with tracheophytes.)

Normal flora • The bacteria and fungi that live on animal body surfaces without causing disease.

Norepinephrine • A neurotransmitter found in the central nervous system and also at the postganglionic nerve endings of the sympathetic nervous system. Also called noradrenaline.

Notochord (no' tow kord) [Gr. *notos*: back + *chorde*: string] • A flexible rod of gelatinous material serving as a support in the embryos of all chordates and in the adults of tunicates and lancelets.

Nuclear envelope • The surface, consisting of two layers of membrane, that encloses the nucleus of eukaryotic cells.

Nucleic acid (new klay' ik) [E.: nucleus of a cell] • A long-chain alternating polymer of deoxyribose or ribose and phosphate groups, with nitrogenous bases—adenine, thymine, uracil, guanine, or cytosine (A, T, U, G, or C)—as side chains. DNA and RNA are nucleic acids.

Nucleoid (new' klee oid) • The region that harbors the chromosomes of a prokaryotic cell. Unlike the eukaryotic nucleus, it is not bounded by a membrane.

Nucleolar organizer (new klee' o lar) • A region on a chromosome that is associated with the formation of a new nucleolus following nuclear division. The site of the genes that code for ribosomal RNA.

Nucleolus (new klee' oh lus) [from L. diminutive of *nux*: little kernel or little nut] • A small, generally spherical body found within the nucleus of eukaryotic cells. The site of synthesis of ribosomal RNA.

Nucleoplasm (new' klee o plazm) • The fluid material within the nuclear envelope of a cell, as opposed to the chromosomes, nucleoli, and other particulate constituents.

Nucleosome • A portion of a eukaryotic chromosome, consisting of part of the DNA molecule wrapped around a group of histone molecules, and held together by another type of histone molecule. The chromosome is made up of many nucleosomes.

Nucleotide • The basic chemical unit (monomer) in a nucleic acid. A nucleotide in RNA consists of one of four nitrogenous bases linked to ribose, which in turn is linked to phosphate. In DNA, deoxyribose is present instead of ribose.

Nucleus (new' klee us) [from L. diminutive of *nux*: kernel or nut] • (1) In chemistry, the dense central portion of an atom, made up of protons and neutrons, with a positive charge. Surrounded by a cloud of negatively charged electrons. (2) In cells, the centrally located chamber of eukaryotic cells that is bounded by a double membrane and contains the chromosomes. The information center of the cell.

Null hypothesis • The assertion that an effect proposed by its companion hypothesis does not in fact exist.

Nutrient • A food substance; or, in the case of mineral nutrients, an inorganic element required for completion of the life cycle of an organism.

Oil • A triglyceride that is liquid at room temperature. (Contrast with fat.)

Okazaki fragments • Newly formed DNA strands making up the lagging strand in DNA replication. DNA ligase links the Okazaki fragments to give a continuous strand.

Olfactory • Having to do with the sense of smell.

Oligomer [Gr.: a few units] • A compound molecule of intermediate size, made up of two to a few monomers. (Contrast with monomer, polymer.)

Oligosaccharins • Plant hormones, derived from the plant cell wall, that trigger defenses against pathogens.

Ommatidium [Gr. *omma*: an eye] • One of the units which, collected into groups of up to 20,000, make up the compound eye of arthropods.

Omnivore [L. *omnis*: all, everything + *vorare*: to devour] • An organism that eats both animal and plant material. (Contrast with carnivore, detritivore, herbivore.)

Oncogenic (ong' co jen' ik) [Gr. *onkos*: mass, tumor + *genes*: born] • Causing cancer.

Oocyte (oh' eh site) [Gr. *oon*: egg + *kytos*: cell] • The cell that gives rise to eggs in animals.

Oogenesis (oh' eh jen e sis) [Gr. *oon*: egg + *genesis*: source] • Female gametogenesis, leading to production of the egg.

Oogonium (oh' eh go' nee um) • In some algae and fungi, a cell in which an egg is produced.

Operator • The region of an operon that acts as the binding site for the repressor.

Operon • A genetic unit of transcription, typically consisting of several structural genes that are transcribed together; the operon contains at least two control regions: the promoter and the operator.

Opportunity cost • The sum of the benefits an animal forfeits by not being able to perform some other behavior during the time when it is performing a given behavior.

Opsin (op' sin) [Gr. *opsis*: sight] • The protein protion of the visual pigment rhodopsin. (See rhodopsin.)

Optic chiasm • Stucture on the lower surface of the vertebrate brain where the two optic nerves come together.

Optical isomers • Isomers that differ in the configuration of the four different groups attached to a single carbon atom; so named

because solutions of the two isomers rotate the plane of polarized light in opposite directions. The two isomers are mirror images of one another.

Optimality models • Models developed to determine the structures or behaviors that best solve particular problems faced by organisms.

Order • In taxonomy, the category below the class and above the family; a group of related, similar families.

Organ • A body part, such as the heart, liver, brain, root, or leaf, composed of different tissues integrated to perform a distinct function for the body as a whole.

Organ identity genes • Plant genes that specify the various parts of the flower. See homeotic genes.

Organ of Corti • Structure in the inner ear that transforms mechanical forces produced from pressure waves ("sound waves") into action potentials that are sensed as sound.

Organelles (or' gan els') [L.: little organ] • Organized structures that are found in or on cells. Examples: ribosomes, nuclei, mitochrondria, chloroplasts, cilia, and contractile vacuoles.

Organic • Pertaining to any aspect of living matter, e.g., to its evolution, structure, or chemistry. The term is also applied to any chemical compound that contains carbon.

Organism • Any living creature.

Organizer, embryonic • A region of an embryo which directs the development of nearby regions. In amphibian early gastrulas, the dorsal lip of the blastopore.

Origin of replication • A DNA sequence at which helicase unwinds the DNA double helix and DNA polymerase binds to initiate DNA replication.

Osmoregulation • Regulation of the chemical composition of the body fluids of an organism.

Osmoreceptor • A neuron that converts changes in the osmotic potential of interstial fluids into action potentials.

Osmosis (oz mo' sis) [Gr. *osmos*: to push] • The movement of water through a differentially permeable membrane from one region to another where the water potential is more negative. This is often a region in which the concentration of dissolved molecules or ions is higher, although the effect of dissolved substances may be offset by hydrostatic pressure in cells with semi-rigid walls.

Ossicle (ah' sick ul) [L. *os*: bone] • The calcified construction unit of echinoderm skeletons.

Osteoblasts • Cells that lay down the protein matrix of bone.

Osteoclasts • Cells that dissolve bone.

Otolith (oh' tuh lith) [Gk.*otikos*: ear + *lithos*: stone[• Structures in the vertebrate vestibular apparatus that mechanically stimulate hair cells when the head moves or changes position.

Outgroup • A taxon that separated from another taxon, whose lineage is to be

inferred, before the latter underwent evolutionary radiation.

Oval window • The flexible membrane which, when moved by the bones of the middle ear, produces pressure waves in the inner ear

Ovary (oh' var ee) • Any female organ, in plants or animals, that produces an egg.

Oviduct [L. *ovum*: egg + *ducere*: to lead] • In mammals, the tube serving to transport eggs to the uterus or to outside of the body.

Oviparous (oh vip' uh rus) • Reproduction in which eggs are released by the female and development is external to the mother's body. (Contrast with viviparous.)

Ovulation • The release of an egg from an ovary.

Ovule (oh' vule) [L. *ovulum*: little egg] • In plants, an organ that contains a gametophyte and, within the gametophyte, an egg; when it matures, an ovule becomes a seed.

Ovum (oh' vum) [L.: egg] • The egg, the female sex cell.

Oxidation (ox i day' shun) • Relative loss of electrons in a chemical reaction; either outright removal to form an ion, or the sharing of electrons with substances having a greater affinity for them, such as oxygen. Most oxidation, including biological ones, are associated with the liberation of energy. (Contrast with reduction.)

Oxidative phosphorylation • ATP formation in the mitochondrion, associated with flow of electrons through the respiratory chain.

Oxidizing agent • A substance that can accept electrons from another. The oxidizing agent becomes reduced; its partner becomes oxidized.

P generation • Also called the parental generation. The individuals that mate in a genetic cross. Their immediate offspring are the F_1 generation.

Pacemaker • That part of the heart which undergoes most rapid spontaneous contraction, thus setting the pace for the beat of the entire heart. In mammals, the sinoatrial (SA) node. Also, an artificial device, implanted in the heart, that initiates rhythmic contraction of the organ.

Pacinian corpuscle • A sensory neuron surrounded by sheaths of connective tissue. Found in the deep layers of the skin, where it senses touch and vibration.

Pair rule genes • Segmentation genes that divide the *Drosophila* larva into two segments each.

Paleomagnetism • The record of the changing direction of Earth's magnetic field as stored in lava flows. Used to accurately date extremely ancient events.

Paleontology (pale' ee on tol' oh jee) [Gr. *palaios*: ancient, old + *logos*: discourse] • The scientific study of fossils and all aspects of extinct life.

Pancreas (pan' cree us) • A gland, located near the stomach of vertebrates, that secretes digestive enzymes into the small intestine and releases insulin into the bloodstream.

Pangaea (pan jee' uh) [Gk. *pan*: all, every] • The single land mass formed when all the continents came together in the Permian period. (Contrast with Gondwana.)

Parabronchi • Passages in the lungs of birds through which air flows.

Paradigm • A general framework within which a scientific or philosophical discipline is viewed and within which questions are asked and hypotheses are developed. Scientific revolutions usually involve major paradigm changes. (Contrast with hypothesis, theory.)

Parallel evolution • Evolutionary patterns that exist in more than one lineage. Often the result of underlying developmental processes.

Parapatric speciation [Gr. *para*: beside + *patria*: fatherland] • Development of reproductive isolation when the barrier is not geographic but is a difference in some other physical condition (such as soil nutrient content) that prevents gene flow between the subpopulations. (Contrast with allopatric speciation, sympatric speciation.)

Paraphyletic taxon • A taxon that includes some, but not all, of the descendants of a single ancestor.

Parasite • An organism that attacks and consumes parts of an organism much larger than itself. Parasites sometimes, but not always, kill the host.

Parasitoid • A parasite that is so large relative to its host that only one individual or at most a few individuals can live within a single host.

Parasympathetic nervous system • A portion of the autonomic (involuntary) nervous system. Activity in the parasympathetic nervous system produces effects such as decreased blood pressure and decelerated heart beat. (Contrast with sympathetic nervous system.)

Parathormone • Hormone secreted by the parathyroid glands. Stimulates osteoclast activity and raises blood calcium levels.

Parathyroids • Four glands on the posterior surface of the thyroid that produce and release parathormone.

Parenchyma (pair eng' kyma) [Gr. *para*: beside + *enchyma*: infusion] • A plant tissue composed of relatively unspecialized cells without secondary walls.

Parental investment • Investment in one offspring or group of offspring that reduces the ability of the parent to assist other offspring.

Parsimony • The principle of preferring the simplest among a set of plausible explanations of a phenomenon. Commonly employed in evolutionary and biogeographic studies.

Parthenocarpy • Formation of fruit from a flower without fertilization.

Parthenogenesis (par' then oh jen' e sis) [Gr. *parthenos*: virgin + *genesis*: source] • The production of an organism from an unfertilized egg.

Partial pressure • The portion of the barometric pressure of a mixture of gases that is due to one component of that mixture. For example, the partial pressure of oxygen at sea level is 20.9% of barometric pressure.

Patch clamping • A technique for isolating a tiny patch of membrane to allow the study of ion movement through a particular channel.

Pathogen (path' o jen) [Gr. *pathos*: suffering + *gignomai*: causing] • An organism that causes disease.

Pattern formation • In animal embryonic development, the organization of differentiated tissues into specific structures such as wings.

Pedigree • The pattern of transmission of a genetic trait in a family.

Pelagic zone (puh ladj' ik) [Gr. *pelagos*: the sea] • The open waters of the ocean.

Penetrance • Of a genotype, the proportion of individuals with that genotype who show the expected phenotype.

PEP carboxylase • The enzyme that combines carbon dioxide with PEP to form a 4-carbon dicarboxylic acid at the start of C_4 photosynthesis or of Crassulacean acid metabolism (CAM).

Pepsin [Gr. *pepsis*: digestion] • An enzyme, in gastric juice, that digests protein.

Peptide linkage • The connecting group in a protein chain, –CO–NH–, formed by removal of water during the linking of amino acids, –COOH to –NH$_2$.

Peptidoglycan • The cell wall material of many prokaryotes, consisting of a single enormous molecule that surrounds the entire cell.

Perennial (per ren' ee al) [L. *per*: through + *annus*: a year] • Referring to a plant that lives from year to year. (Contrast with annual, biennial.)

Perfect flower • A flower with both stamens and carpels, therefore hermaphroditic.

Pericycle [Gr. *peri*: around + *kyklos*: ring or circle] • In plant roots, tissue just within the endodermis, but outside of the root vascular tissue. Meristematic activity of pericycle cells produces lateral root primordia.

Periderm • The outer tissue of the secondary plant body, consisting primarily of cork.

Period • (1) A minor category in the geological time scale. (2) The duration of a cyclical event, such as a circadian rhythm.

Peripheral nervous system • Neurons that transmit information to and from the central nervous system and whose cell bodies reside outside the brain or spinal cord.

Peristalsis (pair' i stall' sis) [Gr. *peri*: around + *stellein*: place] • Wavelike muscular contractions proceeding along a tubular organ, propelling the contents along the tube.

Peritoneum • The mesodermal lining of the coelom among coelomate animals.

Permease • A membrane protein that specifically transports a compound or family of compounds across the membrane.

Peroxisome • An organelle that houses reactions in which toxic peroxides are formed. The peroxisome isolates these peroxides from the rest of the cell.

Petal • In an angiosperm flower, a sterile modified leaf, nonphotosynthetic, frequently brightly colored, and often serving to attract pollinating insects.

Petiole (pet' ee ole) [L. *petiolus*: small foot] • The stalk of a leaf.

pH • The negative logarithm of the hydrogen ion concentration; a measure of the acidity of a solution. A solution with pH = 7 is said to be neutral; pH values higher than 7 characterize basic solutions, while acidic solutions have pH values less than 7.

Phage (fayj) • Short for bacteriophage.

Phagocyte • A white blood cell that ingests microorganisms by endocytosis.

Phagocytosis [Gr.: *phagein* to eat; cell-eating] • A form of endocytosis, the uptake of a solid particle by forming a pocket of plasma membrane around the particle and pinching off the pocket to form an intracellular particle bounded by membrane. (Contrast with pinocytosis.)

Pharynx [Gr.: throat] • The part of the gut between the mouth and the esophagus.

Phenotype (fee' no type) [Gr. *phanein*: to show] • The observable properties of an individual as they have developed under the combined influences of the genetic constitution of the individual and the effects of environmental factors. (Contrast with genotype.)

Phenotypic plasticity • The fact that the phenotype of an organism is determined by a complex series of developmental processes that are affected by both its genotype and its environment.

Pheromone (feer' o mone) [Gr. *phero*: carry + *hormon*: excite, arouse] • A chemical substance used in communication between organisms of the same species.

Phloem (flo' um) [Gr. *phloos*: bark] • In vascular plants, the food-conducting tissue. It consists of sieve cells or sieve tubes, fibers, and other specialized cells.

Phosphate group • The functional group –OPO$_3$H$_2$; the transfer of energy from one compound to another is often accomplished by the transfer of a phosphate group.

Phosphodiester linkage • The connection in a nucleic acid strand, formed by linking two nucleotides.

Phospholipids • Cellular materials that contain phosphorus and are soluble in organic solvents. An example is lecithin (phosphatidyl choline). Phospholipids are important constituents of cellular membranes. (See lipids.)

Phosphorylation • The addition of a phosphate group.

Photoautotroph • An organism that obtains energy from light and carbon from carbon dioxide. (Contrast with chemoautotroph, chemoheterotroph, photoheterotroph.)

Photoheterotroph • An organism that obtains energy from light but must obtain its carbon from organic compounds. (Contrast with chemoautotroph, chemoheterotroph, photoautotroph.)

Photon (foe' tohn) [Gr. *photos*: light] • A quantum of visible radiation; a "packet" of light energy.

Photoperiod (foe' tow peer' ee ud) • The duration of a period of light, such as the length of time in a 24-hour cycle in which daylight is present. The regulation of processes such as flowering by the changing length of day (or of night) is known as photoperiodism.

Photoreceptor • (1) A protein (pigment) that triggers a physiological response when it absorbs a photon. (2) A cell that senses and responds to light energy.

Photorespiration • Light-driven uptake of oxygen and release of carbon dioxide, the carbon being derived from the early reactions of photosynthesis.

Photosynthesis (foe tow sin' the sis) [literally, "synthesis out of light"] • Metabolic processes, carried out by green plants, by which visible light is trapped and the energy used to synthesize compounds such as ATP and glucose.

Phototropin • A yellow protein that is the photoreceptor responsible for phototropism.

Phototropism [Gr. *photos*: light + *trope*: a turning] • A directed plant growth response to light.

Phylogenetic tree • Graphic representation of lines of descent among organisms.

Phylogeny (fy loj' e nee) [Gr. *phylon*: tribe, race + *genesis*: source] • The evolutionary history of a particular group of organisms; also, the diagram of the "family tree" that shows genetic linkages between ancestors and descendants.

Phylum (plural: phyla) [Gr. *phylon*: tribe, stock] • In taxonomy, a high-level category just beneath kingdom and above the class; a group of related, similar classes.

Physiology (fiz' ee ol' o jee) [Gr. *physis*: natural form + *logos*: discourse, study] • The scientific study of the functions of living organisms and the individual organs, tissues, and cells of which they are composed.

Phytoalexins • Substances toxic to fungi, produced by plants in response to fungal infection.

Phytochrome (fy' tow krome) [Gr. *phyton*: plant + *chroma*: color] • A plant pigment regulating a large number of developmental and other phenomena in plants; can exist in two different forms, one of which is active and the other is not. Different wavelengths of light can drive it from one form to the other.

Phytoplankton (fy' tow plangk' ton) [Gr. *phyton*: plant + *planktos*: wandering] • The autotrophic portion of the plankton, consisting mostly of algae.

Pigment • A substance that absorbs visible light.

Pilus (pill' us) [Lat. *pilus*: hair] • A surface appendage by which some bacteria adhere to one another during conjugation.

Pinocytosis [Gr.: drinking cell] • A form of endocytosis; the uptake of liquids by engulfing a sample of the external medium into a pocket of the plasma membrane followed by pinching off the pocket to form an intracellular vesicle. (Contrast with phagocytosis and endocytosis.)

Pistil [L. *pistillum*: pestle] • The female structure of an angiosperm flower, within which the ovules are borne. May consist of a single carpel, or of several carpels fused into a single structure. Usually differentiated into ovary, style, and stigma.

Pith • In plants, relatively unspecialized tissue found within a cylinder of vascular tissue.

Pituitary • A small gland attached to the base of the brain in vertebrates. Its hormones control the activities of other glands. Also known as the hypophysis.

Placenta (pla sen' ta) [Gr. *plax*: flat surface] • The organ found in most mammals that provides for the nourishment of the fetus and elimination of the fetal waste products.

Placental (pla sen' tal) • Pertaining to mammals of the subclass Eutheria, a group characterized by the presence of a placenta; contains the majority of living species of mammals.

Plankton [Gr. *planktos*: wandering] • The free-floating organisms of the sea and fresh water that for the most part move passively with the water currents. Consisting mostly of microorganisms and small plants and animals. (Contrast with nekton.)

Plant • A member of the kingdom Plantae. Multicellular, gaining its nutrition by photosynthesis.

Planula (plan' yew la) [L. *planum*: something flat] • The free-swimming, ciliated larva of the cnidarians.

Plaque (plack) [Fr.: a metal plate or coin] • (1) A circular clearing in a turbid layer (lawn) of bacteria growing on the surface of a nutrient agar gel. Produced by successive rounds of infection initiated by a single bacteriophage. (2) An accumulation of prokaryotic organisms on tooth enamel. Acids produced by the metabolism of these microorganisms can cause tooth decay.

Plasma (plaz' muh) [Gr. *plassein*: to mold] • The liquid portion of blood, in which blood cells and other particulates are suspended.

Plasma cell • An antibody-secreting cell that developed from a B cell. The effector cell of the humoral immune system.

Plasma membrane • The membrane that surrounds the cell, regulating the entry and exit of molecules and ions. Every cell has a plasma membrane.

Plasmid • A DNA molecule distinct from the chromosome(s); that is, an extrachromosomal element. May replicate independently of the chromosome.

Plasmodesma (plural: plasmodesmata) [Gr. *plasma*: formed or molded + *desmos*: band] • A cytoplasmic strand connecting two adjacent plant cells.

Plasmolysis (plaz mol' i sis) • Shrinking of the cytoplasm and plasma membrane away from the cell wall, resulting from the osmotic outflow of water. Occurs only in cells with rigid cell walls.

Plastid • Organelle in plants that serves for food manufacture (by photosynthesis) or food storage; bounded by a double membrane.

Platelet • A membrane-bounded body without a nucleus, arising as a fragment of a cell in the bone marrow of mammals. Important to blood-clotting action.

Pleiotropy (plee' a tro pee) [Gr. *pleion*: more] • The determination of more than one character by a single gene.

Pleural membrane [Gk. *pleuras*: rib, side] • The membrane lining the outside of the lungs and the walls of the thoracic cavity. Inflammation of these membranes is a condition known as *pleurisy*.

Podocytes • Cells of Bowman's capsule of the nephron that cover the capillaries of the glomerulus, forming filtration slits.

Poikilotherm (poy' kill o therm) [Gr. *poikilos*: varied + *therme*: heat] • An animal whose body temperature tends to vary with the surrounding environment. (Contrast with homeotherm, heterotherm.)

Point mutation • A mutation that results from a small, localized alteration in the chemical structure of a gene. Such mutations can give rise to wild-type revertants as a result of reverse mutation. In genetic crosses, a point mutation behaves as if it resided at a single point on the genetic map. (Contrast with deletion.)

Polar body • A nonfunctional nucleus produced by meiosis, accompanied by very little cytoplasm. The meiosis which produces the mammalian egg produces in addition three polar bodies.

Polar molecule • A molecule in which the electric charge is not distributed evenly in the covalent bonds.

Polarity • In development, the difference between one end and the other. In chemistry, the property that makes a polar molecule.

Pollen [L.: fine powder, dust] • The fertilizing element of seed plants, containing the male gametophyte and the gamete, at the stage in which it is shed.

Pollination • Process of transferring pollen from the anther to the receptive surface (stigma) of the ovary in plants.

Poly- [Gr. *poly*: many] • A prefix denoting multiple entities.

Polygamy [Gr. *poly*: many + *gamos*: marriage] • A breeding system in which an individual acquires more than one mate. In polyandry, a female mates with more than one male, in polygyny, a male mates with more than one female.

Polygenes • Multiple loci whose alleles increase or decrease a continuously variable phenotypic trait.

Polymer • A large molecule made up of similar or identical subunits called monomers. (Contrast with monomer, oligomer.)

Polymerase chain reaction (PCR) • A technique for the rapid production of millions of copies of a particular stretch of DNA.

Polymerization reactions • Chemical reactions that generate polymers by means of condensation reactions.

Polymorphism (pol' lee mor' fiz um) [Gr. *poly*: many + *morphe*: form, shape] • (1) In genetics, the coexistence in the same population of two distinct hereditary types based on different alleles. (2) In social organisms such as colonial cnidarians and social insects, the coexistence of two or more functionally different castes within the same colony.

Polyp • The sessile, asexual stage in the life cycle of most cnidarians.

Polypeptide • A large molecule made up of many amino acids joined by peptide linkages. Large polypeptides are called proteins.

Polyphyletic group • A group containing taxa, not all of which share the most recent common ancestor.

Polyploid (pol' lee ploid) • A cell or an organism in which the number of complete sets of chromosomes is greater than two.

Polysaccharide • A macromolecule composed of many monosaccharides (simple sugars). Common examples are cellulose and starch.

Polysome • A complex consisting of a threadlike molecule of messenger RNA and several (or many) ribosomes. The ribosomes move along the mRNA, synthesizing polypeptide chains as they proceed.

Polytene (pol' lee teen) [Gr. *poly*: many + *taenia*: ribbon] • An adjective describing giant interphase chromosomes, such as those found in the salivary glands of fly larvae. The characteristic, reproducible pattern of bands and bulges seen on these chromosomes has provided a method for preparing detailed chromosome maps of several organisms.

Pons [L. *pons*: bridge] • Region of the brain stem anterior to the medulla.

Population • Any group of organisms coexisting at the same time and in the same place and capable of interbreeding with one another.

Population density • The number of individuals (or modules) of a population in a unit of area or volume.

Population genetics • The study of genetic variation and its causes within populations.

Population structure • The proportions of individuals in a population belonging to different age classes (age structure). Also, the distribution of the population in space.

Portal vein • A vein connecting two capillary beds, as in the hepatic portal system.

Positive control • The situation in which a regulatory macromolecule is needed to turn transcription of structural genes on. In its absence, transcription will not occur.

Positive cooperativity • Occurs when a molecule can bind several ligands and each one that binds alters the conformation of the molecule so that it can bind the next ligand more easily. The binding of four molecules of O_2 by hemoglobin is an example of positive cooperativity.

Postabsorptive period • When there is no food in the gut and no nutrients are being absorbed.

Postsynaptic cell • The cell whose membranes receive the neurotransmitter released at a synapse.

Predator • An organism that kills and eats other organisms. Predation is usually thought of as involving the consumption of animals by animals, but it can also mean the eating of plants.

Presynaptic excitation/inhibition • Occurs when a neuron modifies activity at a synapse by releasing a neurotransmitter onto the presynaptic nerve terminal.

Prey [L. *praeda*: booty] • An organism consumed as an energy source.

Primary active transport • Form of active transport in which ATP is hydrolyzed, yielding the energy required to transport ions against their concentration gradients. (Contrast with secondary active transport.)

Primary growth • In plants, growth produced by the apical meristems. (Contrast with secondary growth.)

Primary producer • A photosynthetic or chemosynthetic organism that synthesizes complex organic molecules from simple inorganic ones.

Primary succession • Succession that begins in an areas initially devoid of life, such as on recently exposed glacial till or lava flows.

Primary structure • The specific sequence of amino acids in a protein.

Primary wall • Cellulose-rich cell wall layers laid down by a growing plant cell.

Primate (pry' mate) • A member of the order Primates, such as a lemur, monkey, ape, or human.

Primer • A short, single-stranded segment of DNA serving as the necessary starting material for the synthesis of a new DNA strand, which is synthesized from the 3' end of the primer.

Primitive streak • A line running axially along the blastodisc, the site of inward cell migration during formation of the three-layered embryo. Formed in the embryos of birds and fish.

Primordium [L. *primordium*: origin] • The most rudimentary stage of an organ or other part.

Principle of continuity • States that because life probably evolved from nonlife by a continuous, gradual process, all postulated stages in the evolution of life should be derivable from preexisting states. (Compare with signature principle.)

Pro- [L.: first, before, favoring] • A prefix often used in biology to denote a developmental stage that comes first or an evolutionary form that appeared earlier than another. For example, prokaryote, prophase.

Probe • A segment of single stranded nucleic acid used to identify DNA molecules containing the complementary sequence.

Procambium • Primary meristem that produces the vascular tissue.

Progesterone [L. *pro*: favoring + *gestare*: to bear] • A vertebrate female sex hormone that maintains pregnancy.

Prokaryotes (pro kar' ry otes) [L. *pro*: before + Gk. *karyon*: kernel, nucleus] • Organisms whose genetic material is not contained within a nucleus. The bacteria. Considered an earlier stage in the evolution of life than the eukaryotes.

Prometaphase • The phase of nuclear division that begins with the disintegration of the nuclear envelope.

Promoter • The region of an operon that acts as the initial binding site for RNA polymerase.

Proofreading • The correction of an error in DNA replication just after an incorrectly paired base is added to the growing polynucleotide chain.

Prophage (pro' fayj) • The noninfectious units that are linked with the chromosomes of the host bacteria and multiply with them but do not cause dissolution of the cell. Prophage can later enter into the lytic phase to complete the virus life cycle.

Prophase (pro' phase) • The first stage of nuclear division, during which chromosomes condense from diffuse, threadlike material to discrete, compact bodies.

Prostaglandin • Any one of a group of specialized lipids with hormone-like functions. It is not clear that they act at any considerable distance from the site of their production.

Prosthetic group • Any nonprotein portion of an enzyme.

Protease (pro' tee ase) • See proteolytic enzyme.

Protein (pro' teen) [Gr. *protos*: first] • One of the most fundamental building substances of living organisms. A long-chain polymer of amino acids with twenty different common side chains. Occurs with its polymer chain extended in fibrous proteins, or coiled into a compact macromolecule in enzymes and other globular proteins.

Proteolytic enzyme • An enzyme whose main catalytic function is the digestion of a protein or polypeptide chain. The digestive enzymes trypsin, pepsin, and carboxypeptidase are all proteolytic enzymes (proteases).

Protist • Those eukaryotes not included in the kingdoms Animalia, Fungi, or Plantae.

Protobiont • Aggregates of abiotically produced molecules that cannot reproduce but do maintain internal chemical environments that differ from their surroundings.

Protoderm • Primary meristem that gives rise to epidermis.

Proton (pro' ton) [Gr. *protos*: first] • One of the three most fundamental particles of matter, with mass approximately 1 amu and an electrical charge of +1.

Proto-oncogenes • The normal alleles of genes possessing oncogenes (cancer-causing genes) as mutant alleles. Proto-oncogenes encode growth factors and receptor proteins.

Protostome • One of the major lineages of animal evolution. Characterized by spiral, determinate cleavage of the egg, and by schizocoelous development. (Compare with deuterostome.)

Prototroph (pro' tow trofe') [Gr. *protos*: first + *trophein*: to nourish] • The nutritional wild type, or reference form, of an organism. Any deviant form that requires growth nutrients not required by the prototrophic form is said to be a nutritional mutant, or auxotroph.

Protozoa • A group of single-celled organisms classified by some biologists as a single phylum; includes the flagellates, amoebas, and ciliates. This textbook follows most modern classifications in elevating the protozoans to a distinct kingdom (Protista) and each of their major subgroups to the rank of phylum.

Proximal • Near the point of attachment or other reference point. (Contrast with distal.)

Pseudocoelom • A body cavity not surrounded by a peritoneum. Characteristic of nematodes and rotifers.

Pseudogene • A DNA segment that is homologous to a functional gene but contains a nucleotide change that prevents its expression.

Pseudoplasmodium [Gr. *pseudes*: false + *plasma*: mold or form] • In the cellular slime molds such as *Dictyostelium*, an aggregation of single amoeboid cells. Occurs prior to formation of a fruiting structure.

Pseudopod (soo' do pod) [Gr. *pseudes*: false + *podos*: foot] • A temporary, soft extension of the cell body that is used in location, attachment to surfaces, or engulfing particles.

Pulmonary • Pertaining to the lungs.

Punctuated equiilibrium • An evolutionary pattern in which periods of rapid change are separated by longer periods of little or no change.

Pupa (pew' pa) [L.: doll, puppet] • In certain insects (the Holometabola), the encased developmental stage that intervenes between the larva and the adult.

Pupil • The opening in the vertebrate eye through which light passes.

Purine (pure' een) • A type of nitrogenous base. The purines adenine and guanine are found in nucleic acids.

Purkinje fibers • Specialized heart muscle cells that conduct excitation throughout the ventricular muscle.

Pyramid of biomass • Graphical representation of the total body masses at different trophic levels in an ecosystem.

Pyramid of energy • Graphical representation of the total energy contents at different trophic levels in an ecosystem.

Pyrimidine (peer im' a deen) • A type of nitrogenous base. The pyrimidines cytosine, thymine, and uracil are found in nucleic acids.

Pyruvate • A three-carbon acid; the end product of glycolysis and the raw material for the citric acid cycle.

Q_{10} • A value that compares the rate of a biochemical process or reaction over a 10°C range of temperature. A process that is not temperature-sensitive has a Q_{10} of 1. Values of 2 or 3 mean the reaction speeds up as temperature increases.

Quantum (kwon' tum) [L. *quantus*: how great] • An indivisible unit of energy.

Quaternary structure • Of aggregating proteins, the arrangement of polypeptide subunits.

R factor (resistance factor) • A plasmid that contains one or more genes that encode resistance to antibiotics.

Radial symmetry • The condition in which two halves of a body are mirror images of each other regardless of the angle of the cut, providing the cut is made along the center line. Thus, a cylinder cut lengthwise down its center displays this form of symmetry. (Contrast with biradial symmetry.)

Radioisotope • A radioactive isotope of an element. Examples are carbon-14 (^{14}C) and hydrogen-3, or tritium (^{3}H).

Radiometry • The use of the regular, known rates of decay of radioisotopes of elements to determine dates of events in the distant past.

Rain shadow • A region of low precipitation on the leeward side of a mountain range.

Ramet • The repeated morphological units of sessile, modular organisms. (Contrast with genet.)

Random genetic drift • Evolution (change in gene proportions) by chance processes alone.

Rate constant • Of a particular chemical reaction, a constant which, when multiplied by the concentration(s) of reactant(s), gives the rate of the reaction.

Reactant • A chemical substance that enters into a chemical reaction with another substance.

Reaction, chemical • A process in which atoms combine or change bonding partners.

Realized niche • The actual niche occupied by an organism; it differs from the fundamental niche because of the presence of other species.

Receptive field • Of a neuron, the area on the retina from which the activity of that neuron can be influenced.

Receptor potential • The change in the resting potential of a sensory cell when it is stimulated.

Recessive • See dominance.

Reciprocal altruism • The exchange of altruistic acts between two or more individuals. The acts may be separated considerably in time.

Reciprocal crosses • A pair of crosses, in one of which a female of genotype A mates with a male of genotype B and in the other of which a female of genotype B mates with a male of genotype A.

Recognition site (also called a restriction site) • A sequence of nucleotides in DNA to which a restriction enzyme binds and then cuts the DNA.

Recombinant • An individual, meiotic product, or single chromosome in which genetic materials originally present in two individuals end up in the same haploid complement of genes. The reshuffling of genes can be either by independent segregation, or by crossing over between homologous chromosomes. For example, a human may pass on genes from both parents in a single haploid gamete.

Recombinant DNA technology • The application of genetic tools (restriction endonucleases, plasmids, and transformation) to the production of specific proteins by biological "factories" such as bacteria.

Rectum • The terminal portion of the gut, ending at the anus.

Redox reaction • A chemical reaction in which one reactant becomes oxidized and the other becomes reduced.

Reducing agent • A substance that can donate electrons to another substance. The reducing agent becomes oxidized, and its partner becomes reduced.

Reduction (re duk' shun) • Gain of electrons; the reverse of oxidation. Most reductions lead to the storage of chemical energy, which can be released later by an oxidation reaction. Energy storage compounds such as sugars and fats are highly reduced compounds. (Contrast with oxidation.)

Reflex • An automatic action, involving only a few neurons (in vertebrates, often in the spinal cord), in which a motor response swiftly follows a sensory stimulus.

Refractory period • Of a neuron, the time interval after an action potential, during which another action potential cannot be elicited.

Regulative development • A pattern of animal embryonic development in which the fates of the first blastomeres are not absolutely fixed. (Contrast with mosaic development.)

Regulatory gene • A gene that contains the information for making a regulatory macromolecule, often a repressor protein.

Releaser • A sensory stimulus that triggers a fixed action pattern.

Releasing hormone • One of several hypothalamic hormones that stimulates the secretion of anterior pituitary hormone.

REM sleep • A sleep state characterized by dreaming, skeletal muscle relaxation, and rapid eye movements.

Renal [L. *renes*: kidneys] • Relating to the kidneys.

Replication fork • A point at which a DNA molecule is replicating. The fork forms by the unwinding of the parent molecule.

Repressible enzyme • An enzyme whose synthesis can be decreased or prevented by the presence of a particular compound. A repressible opren often controls the syhthesis of such an enzyme.

Repressor • A protein coded by the regulatory gene. The repressor can bind to a specific operator and prevent transcription of the operon.

Reproductive isolating mechanism • Any trait that prevents individuals from two different populations from producing fertile hybrids.

Reproductive isolation • The condition in which a population is not exchanging genes with other populations of the same species.

Resolving power • Of an optical device such as a microscope, the smallest distance between two lines that allows the lines to be seen as separate from one another.

Resource • Something in the environment required by an organism for its maintenance and growth that is consumed in the process of being used.

Resource defense polygamy • A breeding system in which individuals of one sex (usually males) defend resources that are attractive to individuals of the other sex (usually females); individuals holding better resources attract more mates.

Respiration (res pi ra' shun) [L. *spirare*: to breathe] • (1) Cellular respiration; the oxidation of the end products of glycolysis with the storage of much energy in ATP. The oxidant in the respiration of eukaryotes is oxygen gas. Some bacteria can use nitrate or sulfate instead of O_2. (2) Breathing.

Respiratory chain • The terminal reactions of cellular respiration, in which electrons are passed from NAD or FAD, through a series of intermediate carriers, to molecular oxygen, with the concomitant production of ATP.

Resting potential • The membrane potential of a living cell at rest. In cells at rest, the interior is negative to the exterior. (Contrast with action potential, electrotonic potential.)

Restoration ecology • The science and practice of restoring damaged or degraded ecosystems.

Restriction endonuclease • Any one of several enzymes, produced by bacteria, that break foreign DNA molecules at very specific sites. Some produce "sticky ends." Extensively used in recombinant DNA technology.

Restriction map • A partial genetic map of a DNA molecule, showing the points at which particular restriction endonuclease recognition sites reside.

Reticular system • A central region of the vertebrate brain stem that includes complex fiber tracts conveying neural signals between the forebrain and the spinal cord, with collateral fibers to a variety of nuclei that are involved in autonomic functions, including arousal from sleep.

Retina (rett' in uh) [L. *rete*: net] • The light-sensitive layer of cells in the vertebrate or cephalopod eye.

Retinal • The light-absorbing portion of visual pigment molecules. Derived from β-carotene.

Retrovirus • An RNA virus that contains reverse transcriptase. Its RNA serves as a template for cDNA production, and the cDNA is integrated into a chromosome of the mammalian host cell.

Reverse transcriptase • An enzyme that catalyzes the production of DNA (cDNA), using RNA as a template; essential to the reproduction of retroviruses.

RFLP (Restriction fragment length polymorphism) • Coexistence of two or more patterns of restriction fragments (patterns produced by restriction enzymes), as revealed by a probe. The polymorphism reflects a difference in DNA sequence on homologous chromosomes.

Rhizoids (rye' zoids) [Gr. *rhiza*: root] • Hairlike extensions of cells in mosses, liverworts, and a few vascular plants that serve the same function as roots and root hairs in vascular plants. The term is also applied to branched, rootlike extensions of some fungi and algae.

Rhizome (rye' zome) [Gr. *rhizoma*: mass of roots] • A special underground stem (as opposed to root) that runs horizontally beneath the ground.

Rhodopsin • A photopigment used in the visual process of transducing photons of light into changes in the membrane potential of photoreceptor cells.

Ribonucleic acid • See RNA.

Ribosomal RNA (rRNA) • Several species of RNA that are incorporated into the ribosome. Involved in peptide bond formation.

Ribosome • A small organelle that is the site of protein synthesis.

Ribozyme • An RNA molecule with catalytic activity.

Ribulose 1,5-bisphosphate (RuBP) • The compound in chloroplasts which reacts with carbon dioxide in the first reaction of the Calvin-Benson cycle.

Risk cost • The increased chance of being injured or killed as a result of performing a behavior, compared to resting.

RNA (ribonucleic acid) • A nucleic acid using ribose. Various classes of RNA are involved in the transcription and translation of genetic information. RNA serves as the genetic storage material in some viruses.

RNA polymerase • An enzyme that catalyzes the formation of RNA from a DNA template.

RNA splicing • The last stage of RNA processing in eukaryotes, in which the transcripts of introns are excised through the action of small nuclear ribonucleoprotein particles (snRNP).

Rods • Light-sensitive cells (photoreceptors) in the retina. (Contrast with cones.)

Root cap • A thimble-shaped mass of cells, produced by the root apical meristem, that protects the meristem and that is the organ that perceives the gravitational stimulus in root gravitropism.

Root hair • A specialized epidermal cell with a long, thin process that absorbs water and minerals from the soil solution.

rRNA • See ribosomal RNA.

Rubisco (RuBP carboxylase) • Enzyme that combines carbon dioxide with ribulose bisphosphate to produce 3-phosphoglycerate, the first product of C_3 photosynthesis. The most abundant protein on Earth.

Rumen (rew' mun) • The first division of the ruminant stomach. It stores and initiates bacterial fermentation of food. Food is regurgitated from the rumen for further chewing.

Ruminant • An herbivorous, cud-chewing mammal such as a cow, sheep, or deer, having a stomach consisting of four compartments.

S phase • In the cell cycle, the stage of interphase during which DNA is replicated. (Contrast with G_1 phase, G_2 phase.)

Saprobe [Gr. *sapros*: rotten + *bios*: life] • An organism (usually a bacterium or fungus) that obtains its carbon and energy directly from dead organic matter.

Sarcomere (sark' o meer) [Gr. *sark*: flesh + *meros*: a part] • The contractile unit of a skeletal muscle.

Saturated hydrocarbon • A compound consisting only of carbon and hydrogen, with the hydrogen atoms connected by single bonds.

Schizocoelous development • Formation of a coelom during embryological development by a splitting of mesodermal masses.

Schwann cell • A glial cell that wraps around part of the axon of a peripheral neuron, creating a myelin sheath.

Sclereid [Gr. *skleros*: hard] • A type of sclerenchyma cell, commonly found in nutshells, that is not elongated.

Sclerenchyma (skler eng' kyma) [Gr. *skleros*: hard + *kymus*, juice] • A plant tissue composed of cells with heavily thickened cell walls, dead at functional maturity. The principal types of sclerenchyma cells are fibers and sclereids.

Secondary active transport • Form of active transport in which ions or molecules are transported against their concentration gradient using energy obtained by relaxation of a gradient of sodium ion concentration rather than directly from ATP. (Contrast with primary active transport.)

Secondary compound • A compound synthesized by a plant that is not needed for basic cellular metabolism. Typically has an antiherbivore or antiparasite function.

Secondary growth • In plants, growth produced by vascular and cork cambia, contributing to an increase in girth. (Contrast with primary growth.)

Secondary structure • Of a protein, localized regularities of structure, such as the α helix and the β pleated sheet.

Secondary succession • Ecological succession after a disturbance that does not eliminate all the organisms that originally lived on the site.

Secondary wall • Wall layers laid down by a plant cell that has ceased growing; often impregnated with lignin or suberin.

Second law of thermodynamics • States that in any real (irreversible) process, there is a decrease in free energy and an increase in entropy.

Second messenger • A compound, such as cyclic AMP, that is released within a target cell after a hormone or other "first messenger" has bound to a surface receptor on a cell; the second messenger triggers further reactions within the cell.

Secretin (si kreet' in) • A peptide hormone secreted by the upper region of the small intestine when acidic chyme is present. Stimulates the pancreatic duct to secrete bicarbonate ions.

Section • A thin slice, usually for microscopy, as a tangential section or a transverse section.

Seed • A fertilized, ripened ovule of a gymnosperm or angiosperm. Consists of the embryo, nutritive tissue, and a seed coat.

Seed crop • The number of seeds produced by a plant during a particular bout of reproduction.

Seedling • A young plant that has grown from a seed (rather than by grafting or by other means.)

Segmentation genes • In insect larvae, genes that determine the number and polarity of larval segments.

Segment polarity genes • Genes that determine the boundaries and front-to-back organization of the segments in the *Drosophila* larva.

Segregation (genetic) • The separation of alleles, or of homologous chromosomes, from one another during meiosis so that each of the haploid daughter nuclei produced by meiosis contains one or the other member of the pair found in the diploid mother cell, but never both.

Selective permeability • A characteristic of a membrane, allowing certain substances to pass through while other substances are excluded.

Selfish act • A behavioral act that benefits its performer but harms the recipients.

Semelparous organism • An organism that reproduces only once in its lifetime. (Contrast with iteroparous.)

Semen (see' men) [L.: seed] • The thick, whitish liquid produced by the male reproductive organ in mammals, containing the sperm.

Semicircular canals • Part of the vestibular system of mammals.

Semiconservative replication • The common way in which DNA is synthesized. Each of the two partner strands in a double helix acts as a template for a new partner strand. Hence, after replication, each double helix consists of one old and one new strand.

Seminiferous tubules • The tubules within the testes within which sperm production occurs.

Senescence [L. *senescere*: to grow old] • Aging; deteriorative changes with aging; the increased probability of dying with increasing age.

Sensory neuron • A neuron leading from a sensory cell to the central nervous system. (Contrast with motor neuron.)

Sepal (see' pul) • One of the outermost structures of the flower, usually protective in function and enclosing the rest of the flower in the bud stage.

Septum [L.: partition] • A membrane or wall between two cavities.

Sertoli cells • Cells in the seminiferous tubules that nuture the developing sperm.

Serum • That part of the blood plasma that remains after clots have formed and been removed.

Sessile (sess' ul) [L. *sedere*: to sit] • Permanently attached; not moving.

Set point • In a regulatory system, the threshold sensitivity to the feedback stimulus.

Sex chromosome • In organisms with a chromosomal mechanism of sex determination, one of the chromosomes involved in sex determination.

Sex linkage • The pattern of inheritance characteristic of genes located on the sex chromosomes of organisms having a chromosomal mechanism for sex determination.

Sexual selection • Selection by one sex of characteristics in individuals of the opposite sex. Also, the favoring of characteristics in one sex as a result of competition among individuals of that sex for mates.

Shoot • The aerial part of a vascular plant, consisting of the leaves, stem(s), and flowers.

Sieve tube • A column of specialized cells found in the phloem, specialized to conduct organic matter from sources (such as photosynthesizing leaves) to sinks (such as roots). Found principally in flowering plants.

Sieve tube member • A single cell of a sieve tube, containing cytoplasm but relatively few organelles, with highly specialized perforated end walls leading to elements above and below.

Sign stimulus • The single stimulus, or one out of a very few stimuli, by which an animal distinguishes key objects, such as an enemy, or a mate, or a place to nest, etc.

Signal sequence • The sequence of a protein that directs the protein through a particular cellular membrane.

Signal transduction pathway • The series of biochemical steps whereby a stimulus to a cell (such as a hormone or neurotransmitter binding to a receptor) is translated into a response of the cell.

Signature principle • States that because of continuity, prebiotic processes should leave some trace in contemporary biochemistry. (Compare with principle of continuity.)

Silencer • A sequence of eukaryotic DNA that binds proteins that inhibit the transcription of an associated gene.

Silent mutations • Genetic changes that do not lead to a phenotypic change. At the molecular level, these are DNA sequence changes that, because of the redundancy of the genetic code, result in the same amino acids in the resulting protein. See synonymous mutation.

Similarity matrix • A matrix to compare the structures of two molecules constructed by adding the number of their amino acids that are identical or different

Sinoatrial node (sigh' no ay' tree al) • The pacemaker of the mammalian heart.

Sinus (sigh' nus) [L. *sinus*: a bend, hollow] • A cavity in a bone, a tissue space, or an enlargement in a blood vessel.

Skeletal muscle • See striated muscle.

Sliding filament theory • A proposed mechanism of muscle contraction based on formation and breaking of crossbridges between actin and myosin filaments, causing them to slide together.

Small intestine • The portion of the gut between the stomach and the colon, consisting of the duodenum, the jejunum, and the ileum.

Small nuclear ribonucleoprotein particle (snRNP) • A complex of an enzyme and a small nuclear RNA molecule, functioning in RNA splicing.

Smooth muscle • One of three types of muscle tissue. Usually consists of sheets of mononucleated cells innervated by the autonomic nervous system.

Society • A group of individuals belonging to the same species and organized in a cooperative manner; in the broadest sense, includes parents and their offspring.

Sodium–potassium pump • The complex protein in plasma membranes that is responsible for primary active transport; it pumps sodium ions out of the cell and potassium ions into the cell, both against their concentration gradients.

Solute • A substance that is dissolved in a liquid (solvent).

Solute potential • A property of any solution, resulting from its solute contents; it may be zero or have a negative value.

Solution • A liquid (solvent) and its dissolved solutes.

Solvent • A liquid that has dissolved or can dissolve one or more solutes.

Somatic [Gr. *soma*: body] • Pertaining to the body, or body cells (rather than to germ cells).

Somite (so' might) • One of the segments into which an embryo becomes divided longitudinally, leading to the eventual segmentation of the animal as illustrated by the spinal column, ribs, and associated muscles.

Spatial summation • In the production or inhibition of action potentials in a postsynaptic neuron, the interaction of depolarizations and hyperpolarizations produced by several terminal boutons.

Spawning • The direct release of sex cells into the water.

Speciation (spee' shee ay' shun) • The process of splitting one population into two populations that are reproductively isolated from one another.

Species (spee' shees) [L.: kind] • The basic lower unit of classification, consisting of a population or series of populations of closely related and similar organisms. The more narrowly defined "biological species" consists of individuals capable of interbreeding freely with each other but not with members of other species.

Species diversity • A weighted representation of the species of organisms living in a region; large and common species are given greater weight than are small and rare ones. (Contrast with species richness.)

Species richness • The number of species of organisms living in a region. (Contrast with species diversity.)

Specific heat • The amount of energy that must be absorbed by a gram of a substance to raise its temperature by one degree centigrade. By convention, water is assigned a specific heat of one.

Sperm [Gr. *sperma*: seed] • A male reproductive cell.

Spermatocyte (spur mat' oh site) [Gr. *sperma*: seed + *kytos*: cell] • The cell that gives rise to the sperm in animals.

Spermatogenesis (spur mat' oh jen' e sis) [Gr. *sperma*: seed + *genesis*: source] • Male gametogenesis, leading to the production of sperm.

Spermatogonia • Undifferentiated germ cells that give rise to primary spermatocytes and hence to sperm.

Sphincter (sfingk' ter) [Gr. *sphinkter*: that which binds tight] • A ring of muscle that can close an orifice, for example at the anus.

Spindle apparatus • An array of microtubules stretching from pole to pole of a dividing nucleus and playing a role in the movement of chromosomes at nuclear division. Named for its shape.

Spiracle (spy' rih kel) [L. *spirare*: to breathe] • An opening of the treacheal respiratory system of terrestrial arthropods.

Spiteful act • A behavioral act that harms both the actor and the recipient of the act.

Spliceosome • An RNA–protein complex that splices out introns from eukaryotic pre-mRNAs.

Splicing • The removal of introns and connecting of exons in eukaryotic pre-mRNAs.

Spontaneous generation • The idea that life is generated continually from nonliving matter. Usually distinguished from the current idea that life evolved from nonliving matter under primordial conditions at an early stage in the history of earth.

Spontaneous reaction • A chemical reaction which will proceed on its own, without any outside influence. A spontaneous reaction need not be rapid.

Sporangium (spor an' gee um) [Gr. *spora*: seed + *angeion*: vessel or reservoir] • In plants and fungi, any specialized stucture within which one or more spores are formed.

Spore [Gr. *spora*: seed] • Any asexual reproductive cell capable of developing into an adult plant without gametic fusion. Haploid spores develop into gametophytes, diploid spores into sporophytes. In prokaryotes, a resistant cell capable of surviving unfavorable periods.

Sporophyte (spor' o fyte) [Gr. *spora*: seed + *phyton*: plant] • In plants with alternation of generations, the diploid phase that produces the spores. (Contrast with gametophyte.)

Stabilizing selection • Selection against the extreme phenotypes in a population, so that the intermediate types are favored. (Contrast with disruptive selection.)

Stamen (stay' men) [L.: thread] • A male (pollen-producing) unit of a flower, usually composed of an anther, which bears the pollen, and a filament, which is a stalk supporting the anther.

Starch [O.E. *stearc*: stiff] • An α-linked polymer of glucose; used by plants as a means of storing energy and carbon atoms.

Start codon • The mRNA triplet (AUG) that acts as signals for the beginning of translation at the ribosome. (Compare with stop codons. There are a few mnior exceptions to these codons.)

Stasis • Period during which little or no evolutionary change takes place within a lineage or groups of lineages.

Statocyst (stat' oh sist) [Gk. *statos*: stationary + *kystos*: pouch] • An organ of equilibrium in some invertebrates.

Statolith (stat' oh lith) [Gk. *statos*: stationary + *lithos*: stone] • A solid object that responds to gravity or movement and stimulates the mechanoreceptors of a statocyst.

Stele (steel) [Gr. *stele*: pillar] • The central cylinder of vascular tissue in a plant stem.

Stem cell • A cell capable of extensive proliferation, generating more stem cells and a large clone of differentiated progeny cells, as in the formation of red blood cells.

Step cline • A sudden change in one or more traits of a species along a geographical gradient.

Steroid • Any of numerous lipids based on a 17-carbon atom ring system.

Sticky ends • On a piece of two-stranded DNA, short, complementary, one-stranded regions produced by the action of a restriction endonuclease. Sticky ends allow the joining of segments of DNA from different sources.

Stigma [L.: mark, brand] • The part of the pistil at the apex of the style, which is receptive to pollen, and on which pollen germinates.

Stimulus • Something causing a response; something in the environment detected by a receptor.

Stolon • A horizontal stem that forms roots at intervals.

Stoma (plural: stomata) [Gr. *stoma*: mouth, opening] • Small opening in the plant epidermis that permits gas exchange; bounded by a pair of guard cells whose osmotic status regulates the size of the opening.

Stop codons • Triplets (UAG, UGA, UAA) in mRNA that act as signals for the end of translation at the ribosome. (See also start codon. There are a few mnior exceptions to these codons.)

Stratosphere • The part of the atmosphere above the troposphere; extends upward to approximately 50 kilometers above the surface of the earth; contains very little water.

Stratum (plural strata) • A layer or sedimentary rock laid down at a particular time in a past.

Striated muscle • Contractile tissue characterized by multinucleated cells containing highly ordered arrangements of actin and myosin microfilaments. Also known as skeletal muscle.

Stroma • The fluid contents of an organelle, such as a chloroplast.

Stromatolite • A composite, flat-to-domed structure composed of successive mineral layers. Some are known to be produced by the action of bacteria in salt or fresh water, and some ancient ones are considered to be evidence for early life on the earth.

Structural formula • A representation of the positions of atoms and bonds in a molecule.

Structural gene • A gene that encodes the primary structure of a protein.

Style [Gr. *stylos*: pillar or column] • In flowering plants, a column of tissue extending from the tip of the ovary, and bearing the stigma or receptive surface for pollen at its apex.

Sub- [L.: under] • A prefix often used to designate a structure that lies beneath another or is less than another. For example, subcutaneous, subspecies.

Submucosa • (sub mew koe' sah) • The tissue layer just under the epithelial lining of the lumen of the digestive tract. (Contrast with mucosa.)

Substrate (sub' strayte) • (1) The molecule or molecules on which an enzyme exerts catalytic action. (2) The base material on which an organism lives.

Substrate level phosphorylation • ATP formation resulting from direct transfer of a phosphate group to ADP from an intermediate in glycolysis. (Contrast with oxidative phosphorylation.)

Succession • In ecology, the gradual, sequential series of changes in species composition of a community following a disturbance.

Sulcus (plural: sulci) [L. *sulcare*: to plow] • The valleys or creases between the raised portions of the convoluted surface of the brain. (Contrast to gyrus.)

Sulfhydryl group • The —SH group.

Summation • The ability of a neuron to fire action potentials in response to numerous subthreshold postsynaptic potentials arriving simultaneously at differentiated places on the cell, or arriving at the same site in rapid succession.

Surface area-to-volume ratio • For any cell, organism, or geometrical solid, the ratio of surface area to volume; this is an important factor in setting an upper limit on the size a cell or organism can attain.

Surfactant • A substance that decreases the surface tension of a liquid. Lung surfactant, secreted by cells of the alveoli, is mostly phospholipid and decreases the amount of work necessary to inflate the lungs.

Symbiosis (sim' bee oh' sis) [Gr.: to live together] • The living together of two or more species in a prolonged and intimate ecological relationship. (See parasitism, commensalism, mutualism.)

Symmetry • In biology, the property that two halves of an object are mirror images of each other. (See bilateral symmetry and biradial symmetry.)

Sympathetic nervous system • A division of the autonomic (involuntary) nervous system. Its activities include increasing blood pressure and acceleration of the heartbeat. The neurotransmitter at the sympathetic terminals is epinephrine or norepinephrine. (Contrast with parasympathetic nervous system.)

Sympatric speciation (sim pat' rik) [Gr. sym: same + patria: homeland] • The occurrence of genetic reproduction isolation and the subsequent formation of new species without any physical separation of the subpopulation. (Contrast with allopatric speciation, parapatric speciation.)

Symplast • The continuous meshwork of the interiors of living cells in the plant body, resulting from the presence of plasmodesmata. (Contrast with apoplast.)

Symport • A membrane transport process that carries two substances in the same direction across the membrane. (Contrast with antiport.)

Synapse (sin' aps) [Gr. syn: together + haptein: to fasten] • The narrow gap between the terminal bouton of one neutron and the dendrite or cell body of another.

Synapsis (sin ap' sis) • The highly specific parallel alignment (pairing) of homologous chromosomes during the first division of meiosis.

Synaptic vesicle • A membrane-bounded vesicle, containing neurotransmitter, which is produced in and discharged by the presynaptic neuron.

Syngamy (sing' guh mee) [Gr. sun-: together + gamos: marriage] • Union of gametes. Also known as fertilization.

Synonymous mutation • A mutation that substitutes one nucleotide for another but does not change the amino acid specified (i.e., UUA → UUG, both specifying leucine). (Compare with frame-shift mutation, missense mutation, nonsense mutation.)

Synonymous substitution • The situation when a synonymous mutation becomes widespread in a population. Typically not influenced by natural selection, these substitutions can accumulate in a population. (Contrast with nonsynonymous substitution.)

Systematics • The scientific study of the diversity of organisms.

Systemic circulation • The part of the circulatory system serving those parts of the body other than the lungs or gills.

Systemin • The only polypeptide plant hormone; participates in response to tissue damage.

Systole (sis' tuh lee) [Gr.: contraction] • Contraction of a chamber of the heart, driving blood forward in the circulatory system.

T cell • A type of lymphocyte, involved in the cellular immune response. The final stages of its development occur in the thymus gland. (Contrast with B cell; see also cytotoxic T cell, helper T cell, suppressor T cell.)

T cell receptor • A protein on the surface of a T cell that recognizes the antigenic determinant for which the cell is specific.

T tubules • A system of tubules that runs throughout the cytoplasm of muscle fibers, through which action potentials spread.

Target cell • A cell with the appropriate receptors to bind and respond to a particular hormone or other chemical mediator.

Taste bud • A structure in the epithelium of the tongue that includes a cluster of chemoreceptors innervated by sensory neurons.

TATA box • An eight-base-pair sequence, found about 25 base pairs before the starting point for transcription in many eukaryotic promoters, that binds a transcription factor and thus helps initiate transcription.

Taxis (tak' sis) [Gr. taxis: arrange, put in order] • The movement of an organism in a particular direction with reference to a stimulus. A taxis usually involves the employment of one sense and a movement directly toward or away from the stimulus, or else the maintenance of a constant angle to it. Thus a positive phototaxis is movement toward a light source, negative geotaxis is movement upward (away from gravity), and so on.

Taxon • A unit in a taxonomic system.

Taxonomy (taks on' oh me) [Gr. taxis: arrange, classify] • The science of classification of organisms.

Telomeres (tee' lo merz) [Gr. telos: end] • Repeated DNA sequences at the ends of eukaryotic chromosomes.

Telophase (tee' lo phase) [Gr. telos: end] • The final phase of mitosis or meiosis during which chromosomes became diffuse, nuclear envelopes reform, and nucleoli begin to reappear in the daughter nuclei.

Template • In biochemistry, a molecule or surface upon which another molecule is synthesized in complementary fashion, as in the replication of DNA. In the brain, a pattern that responds to a normal input but not to incorrect inputs.

Template strand • In a stretch of double-stranded DNA, the strand that is transcribed.

Temporal summation • In the production or inhibition of action potentials in a postsynaptic neuron, the interaction of depolarizations or hyperpolarizations produced by rapidly repeated stimulation of a single point.

Tendon • A collagen-containing band of tissue that connects a muscle with a bone.

Terrestrial (ter res' tree al) [L. terra: earth] • Pertaining to the land. (Contrast with aquatic, marine.)

Territory • A fixed area from which an animal or group of animals excludes other members of the same species by aggressive behavior or display.

Tertiary structure • In reference to a protein, the relative locations in three-dimensional space of all the atoms in the molecule. The overall shape of a protein. (Contrast with primary, secondary, and quaternary structures.)

Test cross • A cross of a dominant-phenotype individual (which may be either heterozygous or homozygous) with a homozygous-recessive individual.

Testis (tes' tis) (plural: testes) [L.: witness] • The male gonad; that is, the organ that produces the male sex cells.

Testosterone (tes toss' tuhr own) • A male sex steroid hormone.

Tetanus [Gr. tetanos: stretched] • (1) In physiology, a state of sustained, maximal muscular contraction caused by rapidly repeated stimulation. (2) In medicine, an often-fatal disease ("lockjaw") caused by the bacterium Clostridium tetani.

Thalamus • A region of the vertebrate forebrain; involved in integration of sensory input.

Thallus (thal' us) [Gr.: sprout] • Any algal body which is not differentiated into root, stem, and leaf.

Theory • An explanation or hypothesis that is supported by a wide body of evidence. (Contrast with hypothesis, paradigm.)

Thermoneutral zone • The range of temperatures over which an endotherm does not have to expend extra energy to thermoregulate.

Thermoreceptor • A cell or structure that responds to changes in temperature.

Thoracic cavity • The portion of the mammalian body cavity bounded by the ribs, shoulders, and diaphragm. Contains the heart and the lungs.

Thorax • In an insect, the middle region of the body, between the head and abdomen. In mammals, the part of the body between the neck and the diaphragm.

Thrombin • An enzyme that converts fibrinogen to fibrin, thus triggering the formation of blood clots.

Thrombus (throm' bus) [Gk. thrombos: clot] • A blood clot that forms within a blood vessel and remains attached to the wall of the vessel. (Contrast with embolus.)

Thylakoid • A flattened sac within a chloroplast. The membranes of the numerous thylakoids contain all of the chlorophyll in a plant, in addition to the electron carriers of photophosphorylation. Thylakoids stack to form grana.

Thymine • A nitrogen-containing base found in DNA.

Thymus • A ductless, glandular portion of the lymphoid system, involved in development of the immune system of vertebrates.

Thyroid [Gr. *thyreos*: door-shaped] • A two-lobed gland in vertebrates. Produces the hormone thyroxin.

Thyrotropic hormone • A hormone that is produced in the pituitary gland of amphibia such as frogs and transported in the bloodstream to the thyroid gland, inducing the thyroid gland to produce the thyroid hormone that regulates metamorphosis from tadpole to adult frog.

Tight junction • A junction between epithelial cells, in which there is no gap whatever between the adjacent cells. Materials may get through a tight junction only by entering the epithelial cells themselves.

Tissue • A group of similar cells organized into a functional unit and usually integrated with other tissues to form part of an organ such as a heart or leaf.

Tonus • A low level of muscular tension that is maintained even when the body is at rest.

Totipotency • In a cell, the condition of possessing all the genetic information and other capacities necessary to form an entire individual.

Toxigenicity [L. *toxicum*: poison] • The ability of a bacterium to produce chemical substances injurious to the tissues of the host organism.

Trachea (tray' kee ah) [Gr. *trakhoia*: a small tube] • A tube that carries air to the bronchi of the lungs of vertebrates, or to the cells of arthropods.

Tracheid (tray' kee id) • A distinctive conducting and supporting cell found in the xylem of nearly all vascular plants, characterized by tapering ends and walls that are pitted but not perforated.

Tracheophytes [Gr. *trakhoia*: a small tube + *phyton*: plant] • Those plants with xylem and phloem, including psilophytes, club mosses, horsetails, ferns, gymnosperms, and angiosperms. (Contrast with nontracheophytes.)

Trait • One form of a character: Eye color is a character; brown eyes and blue eyes are traits.

Transcription • The synthesis of RNA, using one strand of DNA as the template.

Transcription factors • Proteins that assemble on a eukaryotic chromosome, allowing RNA polymerase II to perform transcription.

Transduction • (1) Transfer of genes from one bacterium to another, with a bacterial virus acting as the carrier of the genes. (2) In sensory cells, the transformation of a stimulus (e.g., light energy, sound pressure waves, chemical or electrical stimulants) into action potentials.

Transfection • Uptake, incorporation, and expression of recombinant DNA.

Transfer cell • A modified parenchyma cell that transports solutes from its cytoplasm into its cell wall, thus moving the solutes from the symplast into the apoplast.

Transfer RNA (tRNA) • A category of relatively small RNA molecules (about 75 nucleotides). Each kind of transfer RNA is able to accept a particular activated amino acid from its specific activating enzyme, after which the amino acid is added to a growing polypeptide chain.

Transformation • Mechanism for transfer of genetic information in bacteria in which pure DNA extracted from bacteria of one genotype is taken in through the cell surface of bacteria of a different genotype and incorporated into the chromosome of the recipient cell.

Transgenic • Containing recombinant DNA incorporated into its genetic material.

Translation • The synthesis of a protein (polypeptide). This occurs on ribosomes, using the information encoded in messenger RNA.

Translocation • (1) In genetics, a rare mutational event that moves a portion of a chromosome to a new location, generally on a nonhomologous chromosome. (2) In vascular plants, movement of solutes in the phloem.

Transpiration [L. *spirare*: to breathe] • The evaporation of water from plant leaves and stem, driven by heat from the sun, and providing the motive force to raise water (plus ions) from the roots.

Transposable element • A segment of DNA that can move to, or give rise to copies at, another locus on the same or a different chromosome.

Triglyceride • A simple lipid in which three fatty acids are combined with one molecule of glycerol.

Triplet • See codon.

Triplet repeat • Occurrence of repeated triplet of bases in a gene, often leading to genetic disease, as does excessive repetition of CGG in the gene responsible for fragile-X syndrome.

Triploblastic • Having three cell layers. (Contrast with diploblastic.)

Trisomic • Containing three, rather than two members of a chromosome pair.

tRNA • See transfer RNA.

Trochophore (troke' o fore) [Gr. *trochos*: wheel + *phoreus*: bearer] • The free-swimming larva of some annelids and mollusks, distinguished by a wheel-like band of cilia around the middle, and indicating an evolutionary relationship between these two groups.

Trophic level • A group of organisms united by obtaining their energy from the same part of the food web of a biological community.

Tropic hormones • Hormones of the anterior pituitary that control the secretion of hormones by other endocrine glands.

Tropism [Gr. *tropos*: to turn] • In plants, growth toward or away from a stimulus such as light (phototropism) or gravity (gravitropism).

Tropomyosin (troe poe my' oh sin) • A protein that, along with actin, constitutes the thin filaments of myofibrils. It controls the interactions of actin and myosin necessary for muscle contraction.

Troposphere • The atmospheric zone reaching upward approximately 17 km in the tropics and subtropics but only to about 10 km at higher latitudes. The zone in which virtually all the water vapor in the atmosphere is located.

Trypsin • A protein-digesting enzyme. Secreted by the pancreas in its inactive form (trypsinogen), it becomes active in the duodenum of the small intestine.

T-tubules • A set of transverse tubes that penetrates skeletal muscle fibers and terminates in the sarcoplasmic reticulum. The T-system transmits impulses to the sacs, which then release Ca^{2+} to initiate muscle contraction.

Tube nucleus • In a pollen tube, the haploid nucleus that does not participate in double fertilization. (Contrast with generative nucleus.)

Tubulin • A protein that polymerizes to form microtubules.

Tumor • A disorganized mass of cells, often growing out of control. Malignant tumors spread to other parts of the body.

Tumor suppressor genes • Genes which, when homozygous mutant, result in cancer. Such genes code for protein products that inhibit cell proliferation.

Twitch • A single unit of muscle contraction.

Tympanic membrane [Gr. *tympanum*: drum] • The eardrum.

Umbilical cord • Tissue made up of embryonic membranes and blood vessels that connects the embryo to the placenta in eutherian mammals.

Understory • The aggregate of smaller plants growing beneath the canopy of dominant plants in a forest.

Unicellular (yoon' e sell' yer ler) [L. *unus*: one + *cella*: chamber] • Consisting of a single cell; as for example a unicellular organism. (Contrast with multicellular.)

Uniport • A membrane transport process that carries a single substance. (Contrast with antiport, symport.)

Unsaturated hydrocarbon • A compound containing only carbon and hydrogen atoms. One or more pairs of carbon atoms are connected by double bonds.

Upwelling • The upward movement of nutrient-rich, cooler water from deeper layers of the ocean.

Urea • A compound serving as the main excreted form of nitrogen by many animals, including mammals.

Ureotelic • Describes an organism in which the final product of the breakdown of nitrogen-containing compounds (primarily proteins) is urea. (Contrast with ammonotelic, uricotelic.)

Ureter (your' uh tur) [Gr. *ouron*: urine] • A long duct leading from the vertebrate kidney to the urinary bladder or the cloaca.

Urethra (you ree' thra) [Gr. *ouron*: urine] • In most mammals, the canal through which urine is discharged from the bladder and which serves as the genital duct in males.

Uric acid • A compound that serves as the main excreted form of nitrogen in some animals, particularly those which must conserve water, such as birds, insects, and reptiles.

Uricotelic • Describes an organism in which the final product of the breakdown of nitrogen-containing compounds (primarily proteins) is uric acid. (Contrast with ammonotelic, ureotelic.)

Urinary bladder • A structure structure that receives urine from the kidneys via the ureter, stores it, and expels it periodically through the urethra.

Urine (you' rin) [Gk. *ouron*: urine] • In vertebrates, the fluid waste product containing the toxic nitrogenous by-products of protein and amino acid metabolism.

Uterus (yoo' ter us) [L.: womb] • The uterus or womb is a specialized portion of the female reproductive tract in certain mammals. It receives the fertilized egg and nurtures the embryo in its early development.

Vaccination • Injection of virus or bacteria or their proteins into the body, to induce immunization. The injected material is usually attenuated (weakened) before injection.

Vacuole (vac' yew ole) [Fr.: small vacuum] • A liquid-filled cavity in a cell, enclosed within a single membrane. Vacuoles play a wide variety of roles in cellular metabolism, some being digestive chambers, some storage chambers, some waste bins, and so forth.

Vagina (vuh jine' uh) [L.: sheath] • In female mammals, the passage leading from the external genital orifice to the uterus; receives the copulatory organ of the male in mating.

van der Waals interaction • A weak attraction between atoms resulting from the interaction of the electrons of one atom with the nucleus of the other atom. This attraction is about one-fourth as strong as a hydrogen bond.

Variable regions • The part of an immunoglobulin molecule or T-cell receptor that includes the antigen-binding site.

Vascular (vas' kew lar) • Pertaining to organs and tissues that conduct fluid, such as blood vessels in animals and phloem and xylem in plants.

Vascular bundle • In vascular plants, a strand of vascular tissue, including conducting cells of xylem and phloem as well as thick-walled fibers.

Vascular ray • In vascular plants, radially oriented sheets of cells produced by the vascular cambium, carrying materials laterally between the wood and the phloem.

Vascular tissue system • The conductive system of the plant, consisting primarily of xylem and phloem. (Contrast with dermal tissue system, ground tissue system.)

Vasopressin • See antidiuretic hormone.

Vector • (1) An agent, such as an insect, that carries a pathogen affecting another species. (2) A plasmid or virus that carries an inserted piece of DNA into a bacterium for cloning purposes in recombinant DNA technology.

Vegetal hemisphere • The lower portion of some animal eggs, zygotes, and embryos, in which the dense nutrient yolk settles. The **vegetal pole** refers to the very bottom of the egg or embryo. (Contrast with animal hemisphere.)

Vegetative • Nonreproductive, or nonflowering, or asexual.

Vein [L. *vena*: channel] • A blood vessel that returns blood to the heart. (Contrast with artery.)

Ventral [L. *venter*: belly, womb] • Toward or pertaining to the belly or lower side. (Contrast with dorsal.)

Ventricle • A muscular heart chamber that pumps blood through the body.

Vernalization [L. *vernalis*: belonging to spring] • Events occurring during a required chilling period, leading eventually to flowering.

Vertebral column • The jointed, dorsal column that is the primary support structure of vertebrates.

Vertebrate • An animal whose nerve cord is enclosed in a backbone of bony segments, called vertebrae. The principal groups of vertebrate animals are the fishes, amphibians, reptiles, birds, and mammals.

Vessel [L. *vasculum*: a small vessel] • In botany, a tube-shaped portion of the xylem consisting of hollow cells (vessel elements) placed end to end and connected by perforations. Together with tracheids, vessel elements conduct water and minerals in the plant.

Vestibular apparatus (ves tib' yew lar) [L. *vestibulum*: an enclosed passage] • Structures associated with the vertebrate ear; these structures sense changes in position or momentum of the head, affecting balance and motor skills.

Vestigial (ves tij' ee al) [L. *vestigium*: footprint, track] • The remains of body structures that are no longer of adaptive value to the organism and therefore are not maintained by selection.

Vicariance (vye care' ee unce) [L. *vicus*: change] • The splitting of the range of a taxon by the imposition of some barrier to dispersal of its members.

Vicariant distribution • A distribution resulting from the disruption of a formerly continuous range by a vicariant event.

Villus (vil' lus) (plural: villi) [L.: shaggy hair] • A hairlike projection from a membrane; for example, from many gut walls.

Virion (veer' e on) • The virus particle, the minimum unit capable of infecting a cell.

Viroid (vye' roid) • An infectious agent consisting of a single-stranded RNA molecule with no protein coat; produces diseases in plants.

Virus [L.: poison, slimy liquid] • Any of a group of ultramicroscopic infectious particles constructed of nucleic acid and protein (and, sometimes, lipid) that can reproduce only in living cells.

Visceral mass • The major internal organs of a mollusk.

Vitamin [L. *vita*: life] • Any one of several structurally unrelated organic compounds that an organism cannot synthesize itself, but nevertheless requires in small quantity for normal growth and metabolism.

Viviparous (vye vip' uh rus) [L. *vivus*: alive] • Reproduction in which fertilization of the egg and development of the embryo occur inside the mother's body. (Contrast with oviparous.)

Waggle dance • The running movement of a working honey bee on the hive, during which the worker traces out a repeated figure eight. The dance contains elements that transmit to other bees the location of the food.

Water potential • In osmosis, the tendency for a system (a cell or solution) to take up water from pure water, through a differentially permeable membrane. Water flows toward the system with a more negative water potential. (Contrast with osmotic potential, turgor pressure.)

Water vascular system • The array of canals and tubelike appendages that serves as the circulatory system, locomotory system, and food-capturing system of many echinoderms; is in direct connection with the surrounding sea water.

Wavelength • The distance between successive peaks of a wave train, such as electromagnetic radiation.

Wild type • Geneticists' term for standard or reference type. Deviants from this standard, even if the deviants are found in the wild, are said to be mutant.

Xanthophyll (zan' tho fill) [Gr. *xanthos*: yellowish-brown + *phyllon*: leaf] • A yellow or orange pigment commonly found as an accessory pigment in photosynthesis, but found elsewhere as well. An oxygen-containing carotenoid.

X-linked (also called sex-linked) • A character that is coded for by a gene on the X chromosome.

Xerophyte (zee' row fyte) [Gr. *xerox*: dry + *phyton*: plant] • A plant adapted to an environment with a limited water supply.

Xylem (zy' lum) [Gr. *xylon*: wood] • In vascular plants, the woody tissue that conducts water and minerals; xylem consists, in various plants, of tracheids, vessel elements, fibers, and other highly specialized cells.

Yeast artificial chromosome • A laboratory-made DNA molecule containing sequences of yeast chromosomes (origin of replication, telomeres, centromere, and selectable markers) so that it can be used as a vector in yeast.

Yolk • The stored food material in animal eggs, usually rich in protein and lipid.

Z-DNA • A form of DNA in which the molecule spirals to the left rather than to the right.

Zooplankton (zoe' o plang ton) [Gr. *zoon*: animal + *planktos*: wandering] • The animal portion of the plankton.

Zoospore (zoe' o spore) [Gr. *zoon*: animal + *spora*: seed] • In algae and fungi, any swimming spore. May be diploid or haploid.

Zygote (zye' gote) [Gr. *zygotos*: yoked] • The cell created by the union of two gametes, in which the gamete nuclei are also fused. The earliest stage of the diploid generation.

Zymogen • An inactive precursor of a digestive enzyme secreted into the lumen of the gut, where a protease cleaves it to form the active enzyme.

Illustration Credits

Authors' Photograph
Christopher Small

Table of Contents Photographs
Golgi: © Biophoto Associates/Science Source/Photo Researchers, Inc.
Fibroblast: © Dennis Kunkel, U. Hawaii.
Plasmid: © A. B. Dowsett/Science Photo Library/Photo Researchers, Inc.
Protein: © James King-Holmes/OCMS/Science Photo Library/Photo Researchers, Inc.
Fossils: © Tom and Therisa Stack/Tom Stack and Assoc.
Starfish: © Darrell Gulin/DRK PHOTO.
Fireweed: © Stephen J. Krasemann/DRK PHOTO.
Blueberries: © Pat O'Hara/DRK PHOTO.
Fetus: © Science Pictures, Ltd. OSF/DRK PHOTO.
Polar bears: © Mike and Lisa Husar/DRK PHOTO.
Giraffes: © BIOS/Peter Arnold, Inc.
Poppies: © Larry Ulrich/DRK PHOTO.
Zebras: © Art Wolfe.

Part-Opener Photographs
Part 1: © K. R. Porter/Science Source/Photo Researchers, Inc.
Part 2: © Conly L. Rieder/BPS.*
Part 3: © Tui de Roy/Minden Pictures.
Part 4: © Dan Dempster/Dembinsky Photo Assoc.
Part 5: © Gerry Ellis/Minden Pictures.
Part 6: © Michael Fogden/DRK PHOTO.
Part 7: © Doug Perrine/DRK PHOTO.

Chapter 1 *Opener*: © Enric Marti/Associated Press. 1.3: © T. Stevens and P. McKinley/Photo Researchers, Inc. 1.4: © Dennis Kunkel, U. Hawaii. 1.5: © Science Pictures Limited/CORBIS. 1.6 *larva*: © Valorie Hodgson/Visuals Unlimited. 1.6 *pupa*: © Dick Poe/Visuals Unlimited. 1.6 *butterfly*: © Bill Beatty/Visuals Unlimited. 1.7: © K. and K. Amman/Planet Earth Pictures. 1.8a: © Staffan Widstrand. 1.8b: © Joe MacDonald/Tom Stack & Assoc. 1.8c: © Steve Kaufman/Peter Arnold, Inc. 1.8d: © Luiz C. Marigo/Peter Arnold, Inc. 1.11: J. S. Bleakney, courtesy of J. S. Boates. 1.12: © T. Leeson/Photo Researchers, Inc. 1.13: © J. S. Boates.

*BPS = Biological Photo Service

Chapter 2 *Opener*: NASA. 2.3 *left*: © SIU/Visuals Unlimited. 2.3 *right*: © SIU/Visuals Unlimited. 2.14: © Art Wolfe. 2.16: © P. Armstrong/Visuals Unlimited.

Chapter 3 *Opener*: © Dennis Kunkel, U. Hawaii. 3.6a,b,c; 3.7: © Dan Richardson. 3.14c *left*: © Biophoto Associates/Photo Researchers, Inc. 3.14c *middle*: © W. F. Schadel/BPS. 3.14c *right*: © CNRI, Science World Enterprises/BPS. 3.15 *upper*: © Robert Brons/BPS. 3.15 *lower*: © Peter J. Bryant/BPS. 3.18: © Dan Richardson.

Chapter 4 *Opener*: © Dennis Kunkel, U. Hawaii. 4.1: After N. Campbell, 1990. *Biology*, 2nd Ed., Benjamin Cummings Publishing Co. 4.3 *upper row*: © David M. Phillips/Visuals Unlimited. 4.3 *middle row, left*: © Conly L. Rieder/BPS. 4.3 *middle row, center*: David Albertini, Tufts U. School of Medicine. 4.3 *middle row, right*: © M. Abbey/Photo Researchers, Inc. 4.3 *bottom row, left*: © D. P. Evenson/BPS. 4.3 *bottom row, center*: © BPS. 4.3 *bottom row, right*: © L. Andrew Staehelin, U. Colorado. 4.4: © J. J. Cardamone Jr. & B. K. Pugashetti/BPS. 4.5: © Stanley C. Holt/BPS. 4.6a: © J. J. Cardamone Jr./BPS. 4.6c: S. Abraham & E. H. Beachey, VA Medical Center, Memphis, TN. 4.7 *centrioles*: © Barry F. King/BPS. 4.7 *mitochondrion*: © K. Porter, D. Fawcett/Visuals Unlimited. 4.7 *rough ER*: © Don Fawcett/Science Source/Photo Researchers, Inc. 4.7 *plasma membrane*: J. David Robertson, Duke U. Medical Center. 4.7 *peroxisome*: © E. H. Newcomb & S. E. Frederick/BPS. 4.7 *nucleolus*: © Richard Rodewald/BPS. 4.7 *golgi apparatus*: © L. Andrew Staehelin, U. Colorado. 4.7 *smooth ER*: © Don Fawcett, D. Friend/Science Source/Photo Researchers, Inc. 4.7 *ribosome*: From Boublik et al., 1990, *The Ribosome*, p. 177. Courtesy of American Society for Microbiology. 4.7 *cell wall*: © David M. Phillips/Visuals Unlimited. 4.7 *chloroplast*: © W. P. Wergin, courtesy of E. H. Newcomb/BPS. 4.8 *upper*: © Richard Rodewald/BPS. 4.8 *lower*: © Don W. Fawcett/Photo Researchers, Inc. 4.9a: © Barry King, U. California, Davis/BPS. 4.9b: © Biophoto Associates/Science Source/Photo Researchers, Inc. 4.10: From Aebi, U., et al., 1986. *Nature* 323:560–564. © Macmillan Publishers Ltd. 4.11: © Don Fawcett/Visuals Unlimited. 4.12: © L. Andrew Staehelin, U. Colorado. 4.13: © K. G. Murti/

Visuals Unlimited. 4.14: © K. Porter, D. Fawcett/Visuals Unlimited. 4.15: © W. P. Wergin, courtesy of E. H. Newcomb/BPS. 4.16a: © John Durham/Science Photo Library/Photo Researchers, Inc. 4.16b: © Ed Reschke/Peter Arnold, Inc. 4.16c: © Chuck Davis/Tony Stone Images. 4.17a: © Richard Shiell/Animals Animals. 4.17a *inset*: © Richard Green/Photo Researchers, Inc. 4.17b: © G. Büttner/Naturbild/OKAPIA/Photo Researchers, Inc. 4.17b *inset*: © R. R. Dute. 4.19: © E. H. Newcomb & S. E. Frederick/BPS. 4.20: © M. C. Ledbetter, Brookhaven National Laboratory. 4.21: Courtesy of Vic Small, Austrian Academy of Sciences, Salzburg. 4.22: © Nancy Kedersha/Immunogen/Science Photo Library/Custom Medical Stock Photo. 4.23: © N. Hirokawa. 4.24: © W. L. Dentler/BPS. 4.25: B. J. Schnapp et al., 1985. *Cell* 40:455. Courtesy of B. J. Schnapp, R. D. Vale, M. P. Sheetz, and T. S. Reese. 4.26: © Barry F. King/BPS. 4.27: © David M. Phillips/Visuals Unlimited. 4.28 *left*: Courtesy of David Sadava. 4.28 *upper right*: © J. Gross, Biozentrum/Science Photo Library/Photo Researchers, Inc. 4.28 *lower right*: © From J. A. Buckwalter and L. Rosenberg, 1983. *Coll. Rel. Res.* 3:489. Courtesy of L. Rosenberg.

Chapter 5 *Opener*: © Reuters Newmedia Inc./CORBIS. 5.2: After L. Stryer, 1981. *Biochemistry*, 2nd Ed., W. H. Freeman. 5.3: © L. Andrew Staehelin, U. Colorado. 5.6a: © D. S. Friend, U. California, San Francisco. 5.6b: © Darcy E. Kelly, U. Washington. 5.6c: Courtesy of C. Peracchia. 5.15: Courtesy of J. Casley-Smith. 5.16: From M. M. Perry, 1979. *J. Cell Sci.* 39:26.

Chapter 6 *Opener*: © Geoff Tompkinson/Science Photo Library/Photo Researchers, Inc. 6.2: © Jonathan Scott/Masterfile. 6.7: © Jeff J. Daly/Visuals Unlimited. 6.15: © The Mona Group. 6.17: © Clive Freeman, The Royal Institution/Photo Researchers, Inc. 6.19: © The Mona Group.

Chapter 7 *Opener*: © Catherine Ursillo/Photo Researchers, Inc. 7.13: © Ephraim Racker/BPS.

Chapter 8 *Opener*: © C. S. Lobban/BPS. 8.1: © C. G. Van Dyke/Visuals Unlimited. 8.15: © Lawrence Berkeley National Laboratory. 8.18, 8.20: © E. H. Newcomb & S. E. Frederick/BPS. 8.21 *left*: © Arthur R. Hill/Visuals Unlimited. 8.21 *right*: © David Matherly/Visuals Unlimited.

Chapter 9 *Opener*: © Nancy Kedersha/Science Photo Library/Photo Researchers, Inc. 9.1*a,c*: © John D. Cunningham, Visuals, Unlimited. 9.1*b*: © David M. Phillips/Visuals Unlimited. 9.2: © Ruth Kavenoff, Designergenes Ltd., P.O. Box 100, Del Mar, CA 90214. 9.3*b*: © John J. Cardamone Jr./BPS. 9.6: © G. F. Bahr/BPS. 9.7 *upper inset*: © A. L. Olins/BPS. 9.7 *lower inset*: © Biophoto Associates/Science Source/Photo Researchers, Inc. 9.8: © Andrew S. Bajer, U. Oregon. 9.9*b*: © Conly L. Rieder/BPS. 9.10*a*: © T. E. Schroeder/BPS. 9.10*b*: © B. A. Palevitz, U. Wisconsin, courtesy of E. H. Newcomb/BPS. 9.11: © Gary T. Cole/BPS. 9.12*a*: © Andrew Syred/Science Photo Library/Photo Researchers, Inc. 9.12*b*: © E. Webber/Visuals Unlimited. 9.12*c*: © Bill Kamin/Visuals Unlimited. 9.13: © Dr. Thomas Ried and Dr. Evelin Schröck, NIH. 9.14: © C. A. Hasenkampf/BPS. 9.15: © Klaus W. Wolf, U. West Indies. 9.19*b*: © Gopal Murti/Photo Researchers, Inc.

Chapter 10 *Opener*: © David H. Wells/CORBIS. 10.2: © R. W. Van Norman/Visuals Unlimited. 10.12: Courtesy the American Netherland Dwarf Rabbit Club. 10.15: © NCI/Photo Researchers, Inc. 10.16: Courtesy of Pioneer Hi-Bred International, Inc. 10.17: After N. Campbell, 1990. *Biology*, 2nd Ed., Benjamin Cummings Publishing Co. 10.26: © Science VU/Visuals Unlimited. *Bay scallops*: © Barbara J. Miller/BPS.

Chapter 11 *Opener*: © From coordinates provided by N. Geacintov, NYU. 11.2: © Lee D. Simon/Photo Researchers, Inc. 11.4: Courtesy of Prof. M. H. F. Wilkins, Dept. of Biophysics, King's College, U. London. 11.6*a*: © A. Barrington Brown/Photo Researchers, Inc. 11.6*b*: © Dan Richardson.

Chapter 12 *Opener*: © David Wrobel/Visuals Unlimited. 12.7: © Dan Richardson. 12.13*b*: Courtesy of J. E. Edström and *EMBO J*. 12.17*a*: © Stanley Flegler/Visuals Unlimited. 12.17*b*: © Stanley Flegler/Visuals Unlimited.

Chapter 13 *Opener*: © Rosenfeld Images LTD/Photo Researchers, Inc. 13.1*a*: © Dennis Kunkel, U. Hawaii. 13.1*b*: © E.O.S./Gelderblom/Photo Researchers, Inc. 13.1*c*: © Dennis Kunkel, U. Hawaii. 13.8: Courtesy of L. Caro and R. Curtiss. 13.21: Based on an illustration by Anthony R. Kerlavage, Institute for Genomic Research. *Science* 269: 449–604 (1995).

Chapter 14 *Opener*: © Andrew Syred/Tony Stone images. 14.8: © Tiemeier et al., 1978. *Cell* 14:237–246. 14.18: Courtesy of Murray L. Barr, U. Western Ontario. 14.19: Courtesy of O. L. Miller, Jr.

Chapter 15 *Opener*: © Victoria Blackie/Tony Stone Images. 15.3 *inset*: © Biophoto Associates/Photo Researchers, Inc. 15.4: © From de Vos et al., 1992. *Science* 255: 306–312. 15.14: © Stephen A. Stricker, courtesy of Molecular Probes, Inc.

Chapter 16 *Opener*: © Yorgos Nikas, Karolinska Institute. 16.4: © Roddy Field, the Roslin Institute. 16.5: Courtesy of T. Wakayama and R. Yanagimachi. 16.9: J. E. Sulston and H. R. Horvitz, 1977. *Dev. Bio.* 56:100. 16.10*b*; 16.12 *left*: Courtesy of J. Bowman. 16.12 *right*: Courtesy of Detlef Weigel. 16.13: Courtesy of W. Driever and C. Nüsslein-Vollhard. 16.20: Courtesy of F. R. Turner, Indiana U.

Chapter 17 *Opener*: Courtesy of Nexia Biotechnologies, Inc. 17.2: © Philippe Plailly/Photo Researchers, Inc. 17.7: Pamela Silver and Jason A. Kahana, courtesy of Chroma Technology. 17.16 *left*: © Custom Medical Stock Photography. 17.16 *right*: Courtesy of Ingo Potrykus, Swiss Federal Institute of Technology. 17.18: © Bettmann/CORBIS.

Chapter 18 *Opener*: Willard Centerwall, from Lyman, F. L. (ed.), 1963. *Phenylketonuria*. Charles C. Thomas, Springfield, IL. 18.5: C. Harrison et al., 1983. *J. Med. Genet.* 20:280. 18.10: Courtesy of Harvey Levy and Cecelia Walraven, New England Newborn Screening Program. 18.13: © P. P. H. Debruyn and Yongock Cho, U. Chicago/BPS.

Chapter 19 *Opener*: © Francis G. Mayer/CORBIS. 19.4: © Dennis Kunkel, U. Hawaii. 19.10: © Dr. Gopal Murti/Science Photo Library/Photo Researchers, Inc. 19.15: A. Liepins, Sloan-Kettering Research Inst. 19.17: Model by Ted Jardetzky, reprinted from Fineschi, B., et al., *Trends in Biochemical Sciences*, 22:377. © 1997, with permission from Elsevier Science.

Chapter 20 *Opener*: © Robert Fried/Tom Stack & Assoc. 20.1: © Richard Coomber/Planet Earth Pictures. 20.5: © François Gohier/The National Audubon Society Collection/Photo Researchers, Inc. 20.6: © W. B. Saunders/BPS. 20.9 *left*: © Ken Lucas/BPS. 20.9 *right*: © Stanley M. Awramik/BPS. 20.10: © Chip Clark. 20.11: © Tom McHugh/Field Museum, Chicago/Photo Researchers, Inc. 20.12: © Chase Studios, Cedarcreek, MO. 20.14: Transparency no. 5800 (3), photo by D. Finnin, painting by Robert J. Barber. Courtesy the Library, American Museum of Natural History.

Chapter 21 *Opener*: © Toshiyuki Yoshino/Nature Production. 21.1: © Science Photo Library/Photo Researchers, Inc. 21.2: Levi, W. 1965. *Encyclopedia of Pigeon Breeds*. T. F. H. Publications, Jersey City, NJ. (*a,b*: photos by R. L. Kienlen, courtesy of Ralston Purina Company; *c,d*: photos by Stauber.). 21.9: © Frank S. Balthis. 21.11: © Lincoln Nutting/The National Audubon Society

Collection/Photo Researchers, Inc. 21.13*a*: © C. Allan Morgan/Peter Arnold, Inc. 21.16: © Based on drawings produced by the Net-Spinner Web Program by Peter Fuchs and Thiemo Krink. 21.17: After D. Futuyma, 1987. *Evolutionary Biology*, 2nd Ed., Sinauer Associates, Inc. 21.20: Courtesy P. Brakefield and S. Carroll, from Brakefield et al., *Nature* 372:458–461. © Macmillan Publishers Ltd. 21.21*a*: © Marilyn Kazmers/Dembinsky Photo Assoc. 21.21*b*: © Randy Morse/Tom Stack and Assoc.

Chapter 22 *Opener*: © Patti Murray/Animals Animals. 22.1*a*: © Gary Meszaros/Dembinsky Photo Assoc. 22.1*b*: © Lior Rubin/Peter Arnold, Inc. 22.7*a*: © Virginia P. Weinland/Photo Researchers, Inc. 22.7*b*: © José Manuel Sánchez de Lorenzo Cáceres. 22.8: © Reed/Williams/Animals Animals. 22.10 *upper, lower*: © Peter J. Bryant/BPS. 22.10 *center*: © Kenneth Y. Kaneshiro, U. Hawaii. 22.13 *left*: © Peter K. Ziminsky/Visuals Unlimited. 22.13 *center*: © Elizabeth N. Orians. 22.13 *right*: © Noble Proctor/The National Audubon Society Collection/Photo Researchers, Inc.

Chapter 23 *Opener*: © Gary Brettnacher/Adventure Photo & Film. 23.3 *left*: © Adam Jones/Dembinsky Photo Assoc. 23.3 *right*: © Brian Parker/Tom Stack & Assoc. 23.10*a*: © Michael Giannechini/Photo Researchers, Inc. 23.10*b*: © Helen Carr/BPS. 23.10*c*: © Skip Moody/Dembinsky Photo Assoc.

Chapter 24 *Opener*: © John Reader/Science Photo Library/Photo Researchers, Inc. 24.2, 24.5: © Richard Alexander, U. Pennsylvania. 24.8: Courtesy of E. B. Lewis.

Chapter 25 *Opener*: © Mehau Kulyk/Science Photo Library/Photo Researchers, Inc. 25.1: © Stanley M. Awramik/BPS. 25.2: © Roger Ressmeyer/CORBIS. 25.5*a*: © Tom & Therisa Stack/Tom Stack & Assoc. 25.5*b*: © Gary Bell/Planet Earth Pictures.

Chapter 26 *Opener*: Photo by Ferran Garcia Pichel, from the cover of *Science* 284 (no. 5413). 26.1: © Kari Lounatmaa/Photo Researchers, Inc. 26.3*a*: © David Phillips/Photo Researchers, Inc. 26.3*b*: © R. Kessel-G. Shih/Visuals Unlimited. 26.3*c*: © Stanley Flegler/Visuals Unlimited. 26.4: © T. J. Beveridge/BPS. 26.5*a*: © J. A. Breznak and H. S. Pankratz/BPS. 26.5*b*: © J. Robert Waaland/BPS. 26.6: © George Musil/Visuals Unlimited. 26.7*a left*: © S. C. Holt/BPS. 26.7*a center*: © David M. Phillips/Visuals Unlimited. 26.7*b left*: © Leon J. LeBeau/BPS. 26.7*b center*: © A. J. J. Cardamone, Jr./BPS. 26.8: © Alfred Pasieka/Photo Researchers, Inc. 26.9: © Wolfgang Baumeister/Science Photo Library/Photo Researchers, Inc. 26.13: © Phil Gates, U. Durham/BPS. 26.14: © S. C. Holt/BPS. 26.15*a*: © Paul W. Johnson/BPS. 26.15*b*: © H. S. Pankratz/BPS. 26.15*c*: © Bill Kamin/Visuals Unlimited. 26.16: © Science VU/Visuals Unlimited. 26.17: © Randall C. Cutlip/BPS. 26.18: © T. J. Beveridge/BPS.

Photo Assoc. 33.22*a*: © Ed Kanze/Dembinsky Photo Assoc. 33.22*b*: © Dave Watts/Tom Stack & Assoc. 33.23*a*: © Art Wolfe. 33.23*b*: © Jany Sauvanet/Photo Researchers, Inc. 33.23*c*: © Hans & Judy Beste/Animals Animals. 33.24*a*: © Rod Planck/Dembinsky Photo Assoc. 33.24*b*: © Joe McDonald/Tom Stack & Assoc. 33.24*c*: © Doug Perrine/Planet Earth Pictures. 33.24*d*: © Erwin & Peggy Bauer/Tom Stack & Assoc. 33.26*a*: © Art Wolfe. 33.26*b*: © Gary Milburn/Tom Stack & Assoc. 33.27*a*: © Steve Kaufman/DRK PHOTO. 33.27*b*: © John Bracegirdle/Masterfile. 33.28*a*: © Art Wolfe. 33.28*b*: © Anup Shah/Dembinsky Photo Assoc. 33.28*c*: © Anup Shah/Dembinsky Photo Assoc. 33.28*d*: © Stan Osolinsky/Dembinsky Photo Assoc. 33.31*a*: © Dembinsky Photo Assoc. 33.31*b*: © Tim Davis/Photo Researchers, Inc. 33.31*c*: © John Downer/Planet Earth Pictures.

Chapter 34 *Opener*: © D. Cavagnaro/Visuals Unlimited. 34.3*a*: © Jan Tove Johansson/Planet Earth Pictures. 34.3*b*: © R. Calentine/Visuals Unlimited. 34.4*a*: © Joyce Photographics/Photo Researchers, Inc. 34.4*b*: © Renee Lynn/Photo Researchers, Inc. 34.4*c*: © C. K. Lorenz/The National Audubon Society Collection/Photo Researchers, Inc. 34.7: © Biophoto Associates/Photo Researchers, Inc. 34.9*a,b*: © Phil Gates, U. Durham/BPS. 34.9*c*: © Biophoto Associates/Photo Researchers, Inc. 34.9*d*: © Jack M. Bostrack/Visuals Unlimited. 34.9*e*: © John D. Cunningham/Visuals Unlimited. 34.9*f*: © J. Robert Waaland/BPS. 34.11*b*, 34.14: © J. Robert Waaland/BPS. 34.16*a*: © Jim Solliday/BPS. 34.16*b*: © Microfield Scientific LTD/Photo Researchers, Inc. 34.16*c*: © Ray F. Evert, U. Wisconsin, Madison. 34.16*d*: © John D. Cunningham/Visuals Unlimited. 34.18*a* left: © Cabisco/Visuals Unlimited. 34.18*a* right: © J. Robert Waaland/BPS. 34.18*b* left: © Cabisco/Visuals Unlimited. 34.18*b* right: © J. Robert Waaland/BPS. 34.20: © J. N. A. Lott/BPS. 34.21: © Jim Solliday/BPS. 34.22: © Phil Gates, U. Durham/BPS. 34.23*b*: © Thomas Eisner, Cornell U. 34.23*c*: © C. G. Van Dyke/Visuals Unlimited.

Chapter 35 *Opener*: © Patti Murray/Animals Animals. 35.5: Brentwood, B., and J. Cronshaw, 1978. *Planta* 140:111–120. 35.6: © Ed Reschke/Peter Arnold, Inc. 35.9*a*: © David M. Phillips/Visuals Unlimited. 35.13: © M. H. Zimmermann.

Chapter 36 *Opener*: © J. H. Robinson/The National Audubon Society Collection/Photo Researchers, Inc. 36.1: © Inga Spence/Tom Stack & Assoc. 36.4: © Kathleen Blanchard/Visuals Unlimited. 36.6: © Hugh Spencer/Photo Researchers, Inc. 36.8: © E. H. Newcomb and S. R. Tandon/BPS. 36.10: © Gilbert S. Grant/Photo Researchers, Inc. 36.11: © Milton Rand/Tom Stack & Assoc.

Chapter 37 *Opener*: © Jeremy Woodhouse/DRK PHOTO. 37.4: © Tom J. Ulrich/Visuals Unlimited. 37.5: © John Eastcott, Yva Momatiuk/DRK PHOTO. 37.6:

© J. N. A. Lott/BPS. 37.8: © J. A. D. Zeevaart, Michigan State U. 37.13: © Ed Reschke/Peter Arnold, Inc. 37.16*a*: © Biophoto Associates/Photo Researchers, Inc. 37.19: © T. A. Wiewandt/DRK PHOTO. 37.22: Dr. Eva Huala, Carnegie Institution of Washington.

Chapter 38 *Opener*: © C. C. Lockwood/Animals Animals. 38.1 *lower*: © J. R. Waaland/BPS. 38.1 *upper*: © Jim Solliday/BPS. 38.2: © Oliver Meckes/Science Source/Photo Researchers, Inc. 38.3: © Stephen Dalton/The National Audubon Society Collection/Photo Researchers, Inc. 38.5: © Bowman, J. (ed.), 1994. *Arabiopsis: An Atlas of Morphology and Development*. Springer-Verlag, New York. Photo by S. Craig & A. Chaudhury. 38.9*a*: © C. P. George/Visuals Unlimited. 38.9*b*: © Tess & David Young/Tom Stack & Assoc. 38.17*a*: © Nigel Cattlin, Holt Studios International/Photo Researchers, Inc. 38.17*b*: © Jerome Wexler/The National Audubon Society Collection/Photo Researchers, Inc.

Chapter 39 *Opener*: Agricultural Research Service, USDA. 39.2: © D. Cavagnaro/Visuals Unlimited. 39.4: © Stan Osolinski/Dembinsky Photo Assoc. 39.7: © Thomas Eisner, Cornell U. 39.8: © Adam Jones/Dembinsky Photo Assoc. 39.9: © J. N. A. Lott/BPS. 39.10, 39.11: © Richard Shiell. 39.12: © Janine Pestel/Visuals Unlimited. 39.13: © Chip Isenhart/Tom Stack & Assoc. 39.14: © J. N. A. Lott/BPS. 39.15: © Robert & Linda Mitchell. 39.16: © Budd Titlow/Visuals Unlimited.

Chapter 40 *Opener*: © S. Asad/Peter Arnold, Inc. 40.3*a,b*: © Biophoto Associates/Science Source/Photo Researchers, Inc. 40.3*c*: © G. W. Willis/BPS. 40.4*a*: © Cabisco/Visuals Unlimited. 40.4*b*: © Biophoto Associates/Science Source/Photo Researchers, Inc. 40.4*c*: © Cabisco/Visuals Unlimited. 40.4*d*: © David M. Phillips/Visuals Unlimited. 40.10*a*: © B. & C. Alexander/Photo Researchers, Inc. 40.10*b*: © Timothy Ransom/BPS. 40.12: © Auscape (Parer-Cook)/Peter Arnold, Inc. 40.16: © G. W. Willis/BPS. 40.17*a*: © Stephen J. Kraseman/DRK PHOTO. 40.17*b*: © Jim Roetzel/Dembinsky Photo Assoc.

Chapter 41 *Opener*: © R. D. Fernald, Stanford U. 41.6*a*: © Associated Press Photo. 41.6*b*: © Bettman/CORBIS. 41.14*a*: Courtesy of Gerhard Heldmaier, Philipps University.

Chapter 42 *Opener*: © Nik Wheeler. 42.1*a*: © Biophoto Associates/Photo Researchers, Inc. 42.1*b*: © Brian Parker/Tom Stack & Assoc. 42.1*c*: © Thomas Eisner, Cornell U. 42.2: © Patricia J. Wynne. 42.3: © David M. Phillips/Science Source/Photo Researchers, Inc. 42.5: © Fred Bavendam/Minden Pictures. 42.6: © David T. Roberts, Nature's Images/The National Audubon Society Collection/Photo Researchers, Inc. 42.7*a*: © Mitsuaki Iwago/Minden Pictures. 42.7*b*: ©

Johnny Johnson/DRK PHOTO. 42.12 *inset*: © P. Bagavandoss/Photo Researchers, Inc. 42.16: © CC Studio/Photo Researchers, Inc.

Chapter 43 *Opener*: © Dave B. Fleetham/Tom Stack & Assoc. 43.5 *inset*: Courtesy of Richard Elinson, U. Toronto. 43.24*a*: © C. Eldeman/Photo Researchers, Inc. 43.24*b*: © Nestle/Photo Researchers, Inc. 43.26: © S. I. U. School of Med./Photo Researchers, Inc.

Chapter 44 *Opener*: © Associated Press Photo. 44.4: © C. Raines/Visuals Unlimited.

Chapter 45 *Opener*: Courtesy of Grace Sours, ATF. 45.4 *left*: © R. A. Steinbrecht. 45.4 *right*: © G. I. Bernard/Animals Animals. 45.6, 45.12: © P. Motta/Photo Researchers, Inc. 45.15*b*: © S. Fisher, U. California, Santa Barbara. 45.19*a*: © Dennis Kunkel, U. Hawaii. 45.22: © Omikron/Science Source/Photo Researchers, Inc. 45.26: © Joe McDonald/Tom Stack & Assoc.

Chapter 46 *Opener*: From Harlow, J. M., 1869. *Recovery from the passage of an iron bar through the head*. Boston: David Clapp & Son. 46.14: David Joel, courtesy of Bio-logic Systems Corp. 46.16: © Wellcome Dept. of Cognitive Neurology/Science Photo Library/Photo Researchers, Inc.

Chapter 47 *Opener*: © AFP/CORBIS. 47.2: © P. Motta/Photo Researchers, Inc. 47.5 *upper*: © CNRI/Photo Researchers, Inc. 47.5 *center*: © G. W. Willis/BPS. 47.5 *lower*: © Michael Abbey/Photo Researchers, Inc. 47.7: © Frank A. Pepe/BPS. 47.12: Courtesy of Jesper L. Andersen. 47.14: © Skip Moody/Dembinsky Photo Assoc. 47.18*a*: © G. Mili. 47.18*b*: © Robert Brons/BPS. 47.22*a*: © Ken Lucas/Visuals Unlimited. 47.22*b*: © Fred McConnaughey/The National Audubon Society Collection/Photo Researchers, Inc.

Chapter 48 *Opener*: © Darrell Gulin/Tony Stone Images. 48.1*a*: © Ed Robinson/Tom Stack & Assoc. 48.1*b*: © Robert Brons/BPS. 48.1*c*: © Tom McHugh/Photo Researchers, Inc. 48.3: © Eric Reynolds/Adventure Photo. 48.5*b*: © Skip Moody/Dembinsky Photo Assoc. 48.5*c*: © Thomas Eisner, Cornell U. 48.9: © Walt Tyler, U. California, Davis. 48.12 *left inset*: © Science Photo Library/Photo Researchers, Inc. 48.12 *right inset*: © P. Motta/Photo Researchers, Inc. 48.15: © Fred Bruemmer/DRK PHOTO.

Chapter 49 *Opener*: © Norbert Wu/DRK PHOTO. 49.9: © Geoff Tompkinson/Photo Researchers, Inc. 49.11: © Dennis Kunkel, U. Hawaii. 49.14*a*: © Chuck Brown/Science Source/Photo Researchers, Inc. 49.14*b*: © Biophoto Associates/Science Source/Photo Researchers, Inc. 49.15: After N. Campbell, 1990. *Biology*, 2nd Ed., Benjamin Cummings Publishing Co. 49.16*a*: © NYU Franklin Research Fund/Phototake. 49.17*b*: © CNRI/Photo Researchers, Inc.

Index